I0042122

SMART MATERIALS DESIGN FOR ELECTROMAGNETIC INTERFERENCE SHIELDING APPLICATIONS

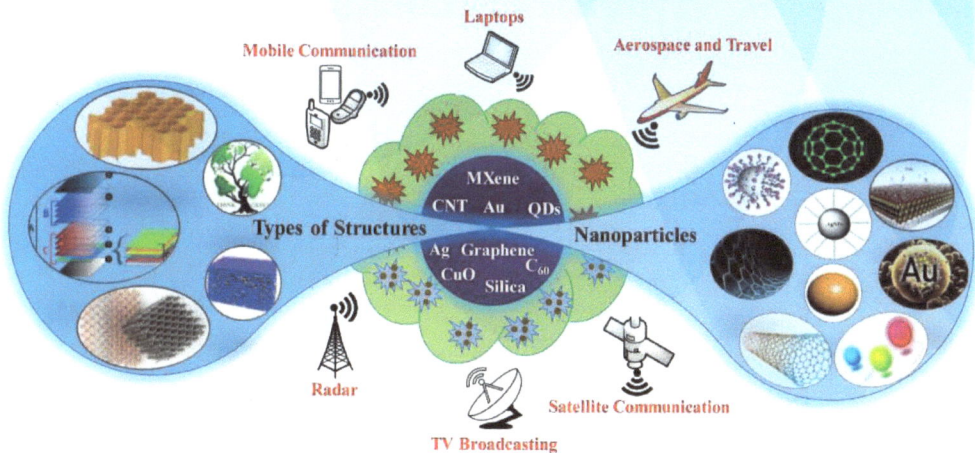

Editors:
Sundeep K. Dhawan
Avanish Pratap Singh
Anil Ohlan
Kuldeep Singh Kakran
Pradeep Sambyal

Bentham Books

Smart Materials Design for Electromagnetic Interference Shielding Applications

Edited by

Sundeep K. Dhawan

Division of Materials Physics and Engineering,
CSIR-National Physical Laboratory
New Delhi-110012
India

Avanish Pratap Singh

Experimental Research Laboratory
Atam Ram Sanatan Dharma College (University of Delhi)
New Delhi – 110021
India

Anil Ohlan

Department of Physics
M.D. University, Rohtak – 124001
India

Kuldeep Singh Kakran

CSIR-Central Electrochemical Research Institute (CECRI) Chennai Unit
CSIR Madras Complex, Taramani, Chennai – 600113
India

&

Pradeep Sambyal

Materials Architecturing Research Center, Korea Institute of Science and Technology,
Hwarangno 14-gil 5, Seongbuk Gu, Seoul 02792
Republic of Korea

Smart Materials Design for Electromagnetic Interference Shielding Applications

Editors: Sundeep K. Dhawan, Avanish Pratap Singh, Anil Ohlan and Kuldeep Singh Kakran

ISBN (Online): 978-981-5036-42-8

ISBN (Print): 978-981-5036-43-5

ISBN (Paperback): 978-981-5036-44-2

© 2022, Bentham Books imprint.

Published by Bentham Science Publishers Pte. Ltd. Singapore. All Rights Reserved.

First published in 2022.

BENTHAM SCIENCE PUBLISHERS LTD.
End User License Agreement (for non-institutional, personal use)

This is an agreement between you and Bentham Science Publishers Ltd. Please read this License Agreement carefully before using the ebook/echapter/ejournal (**"Work"**). Your use of the Work constitutes your agreement to the terms and conditions set forth in this License Agreement. If you do not agree to these terms and conditions then you should not use the Work.

Bentham Science Publishers agrees to grant you a non-exclusive, non-transferable limited license to use the Work subject to and in accordance with the following terms and conditions. This License Agreement is for non-library, personal use only. For a library / institutional / multi user license in respect of the Work, please contact: permission@benthamscience.net.

Usage Rules:

1. All rights reserved: The Work is the subject of copyright and Bentham Science Publishers either owns the Work (and the copyright in it) or is licensed to distribute the Work. You shall not copy, reproduce, modify, remove, delete, augment, add to, publish, transmit, sell, resell, create derivative works from, or in any way exploit the Work or make the Work available for others to do any of the same, in any form or by any means, in whole or in part, in each case without the prior written permission of Bentham Science Publishers, unless stated otherwise in this License Agreement.
2. You may download a copy of the Work on one occasion to one personal computer (including tablet, laptop, desktop, or other such devices). You may make one back-up copy of the Work to avoid losing it.
3. The unauthorised use or distribution of copyrighted or other proprietary content is illegal and could subject you to liability for substantial money damages. You will be liable for any damage resulting from your misuse of the Work or any violation of this License Agreement, including any infringement by you of copyrights or proprietary rights.

Disclaimer:

Bentham Science Publishers does not guarantee that the information in the Work is error-free, or warrant that it will meet your requirements or that access to the Work will be uninterrupted or error-free. The Work is provided "as is" without warranty of any kind, either express or implied or statutory, including, without limitation, implied warranties of merchantability and fitness for a particular purpose. The entire risk as to the results and performance of the Work is assumed by you. No responsibility is assumed by Bentham Science Publishers, its staff, editors and/or authors for any injury and/or damage to persons or property as a matter of products liability, negligence or otherwise, or from any use or operation of any methods, products instruction, advertisements or ideas contained in the Work.

Limitation of Liability:

In no event will Bentham Science Publishers, its staff, editors and/or authors, be liable for any damages, including, without limitation, special, incidental and/or consequential damages and/or damages for lost data and/or profits arising out of (whether directly or indirectly) the use or inability to use the Work. The entire liability of Bentham Science Publishers shall be limited to the amount actually paid by you for the Work.

General:

1. Any dispute or claim arising out of or in connection with this License Agreement or the Work (including non-contractual disputes or claims) will be governed by and construed in accordance with the laws of Singapore. Each party agrees that the courts of the state of Singapore shall have exclusive jurisdiction to settle any dispute or claim arising out of or in connection with this License Agreement or the Work (including non-contractual disputes or claims).
2. Your rights under this License Agreement will automatically terminate without notice and without the

need for a court order if at any point you breach any terms of this License Agreement. In no event will any delay or failure by Bentham Science Publishers in enforcing your compliance with this License Agreement constitute a waiver of any of its rights.

3. You acknowledge that you have read this License Agreement, and agree to be bound by its terms and conditions. To the extent that any other terms and conditions presented on any website of Bentham Science Publishers conflict with, or are inconsistent with, the terms and conditions set out in this License Agreement, you acknowledge that the terms and conditions set out in this License Agreement shall prevail.

Bentham Science Publishers Pte. Ltd.
80 Robinson Road #02-00
Singapore 068898
Singapore
Email: subscriptions@benthamscience.net

CONTENTS

FOREWORD... i

PREFACE.. ii

ACKNOWLEDGEMENTS... v

CHAPTER 1 ELECTROMAGNETIC INTERFERENCE SHIELDING AND ITS
EVALUATION... 1

Avanish Pratap Singh, Monika Mishra, Anil Ohlan and *S.K. Dhawan*

  1.1. INTRODUCTION 2
  1.2. ELECTROMAGNETIC INTERFERENCE SHIELDING 3
  1.3. SHIELDING MECHANISM OF E-FIELD AND H-FIELD 5
  1.4. SHIELDING EFFECTIVENESS 7
    1.4.2 Reflection Loss 9
    1.4.3. Multiple Reflections 9
  1.5. CONDUCTION IN THE MICROWAVE FIELD 10
  1.6. ORIENTATION POLARIZATION 10
    1.6.1. Electronic and Atomic polarization.. 11
    1.6.2. Interfacial or Space Charge Polarization..................................... 11
  **1.7. THEORETICAL CALCULATIONS OF SHIELDING
  EFFECTIVENESS** 11
  1.8. SHIELDING EFFECTIVENESS MEASUREMENT 17
    1.8.1. Shielded Box Method 17
    1.8.2. Open Area or Free Space Method... 18
    1.8.3. Shielded Room Enclosures 19
    1.8.4. Coaxial Transmission Line Method 21
    1.8.5. Vector Network Analyzer.. 21
    1.8.6. Nicholson-Ross-Weir Method 24
    1.8.7. NIST Iterative Technique 26
    1.8.8. Short Circuit Line (SCL) Technique... 26
    1.8.9. Dielectric Measurement Techniques... 27

  1.9. FACTORS AFFECTING THE EMI SHIELDING PERFORMANCE ... 28
    1.9.1. Permittivity and Conductivity.. 29
    1.9.2. Permeability and Magnetization 31
    1.9.3. Attenuation Constant and Eddy Current Losses 31
    1.9.4. Anisotropy, Domain Wall Displacement, Resonance and Snoek's Limit...................... 33
    1.9.5. Morphology 34
    1.9.6. Effect of Thickness... 34
    1.9.7. Impedance Matching 34
    1.9.8. Materials for EMI Shielding 35
      1.9.8.1. Conductors.. 35
      1.9.8.2. Semiconductor 36
      1.9.8.3. Magnetic Materials... 36
      1.9.8.4 Dielectric Composites... 37
      1.9.8.5. Conjugated Polymers.. 37
      1.9.8.6. Dielectric-Conductor Composites.. 38
      1.9.8.7. Magnetic Parameters.. 38
      1.9.8.8. Permeability.. 39
      1.9.8.9. Ferromagnetic Materials 39

   1.9.8.10. Soft Ferrites 40
   1.9.8.11. Hard Ferrites 41
  1.10. COMPOSITES FOR ELECTROMAGNETIC INTERFERENCE SHIELDING 43
  1.11 5th GENERATION (5G) ELECTROMAGNETIC RADIATION AND ITS PREVENTION 47
  CONCLUSION 50
  CONSENT FOR PUBLICATION 51
  CONFLICT OF INTEREST 51
  ACKNOWLEDGEMENTS 51
  REFERENCES 51

CHAPTER 2 LIGHTWEIGHT CARBON COMPOSITE FOAMS FOR ELECTROMAGNETIC INTERFERENCE SHIELDING APPLICATIONS 59
Rajeev Kumar, Tejendra K. Gupta, Neeraj Dwivedi and D.P. Mondal
  2.1. INTRODUCTION 60
  2.2. THEORY OF ELECTROMAGNETIC INTERFERENCE SHIELDING 61
  2.3. METHODS FOR THE FABRICATION OF CARBON COMPOSITE FOAMS 63
   2.3.1. Foaming Method 63
   2.3.2. Pressure Release Method 65
   2.3.3. Template Method 66
  2.4. CARBON FOAMS FOR EMI SHIELDING 71
   2.4.1. Carbon Composite Foams with Carbon Fillers 72
   2.4.2. Carbon Nanotube Reinforced Composite Foams for EMI Shielding 72
   2.4.3. Graphene Reinforced Composite Foams for EMI Shielding 79
   2.4.4. Metals Reinforced Composite Foams for EMI Shielding 85
  CONCLUSION AND FUTURE PERSPECTIVE 95
  CONSENT FOR PUBLICATION 96
  CONFLICT OF INTEREST 96
  ACKNOWLEDGEMENTS 96
  REFERENCES 96

CHAPTER 3 CARBON NANOSTRUCTURES-BASED POLYMER NANOCOMPOSITES FOR EMI SHIELDING APPLICATIONS 109
Tejendra K. Gupta, Rajeev Kumar, Manjeet Singh Goyat and Deepshikha Gupta
  3.1. INTRODUCTION 110
  3.2. EMI PARAMETERS AND SHIELDING MECHANISM 111
  3.3. THERMOPLASTIC AND THERMOSETTING POLYMERS 116
  3.4. COMMON FILLERS USED IN POLYMER NANOCOMPOSITES 117
   3.4.1. Carbon-based Nanostructures (CBNS) 117
   3.4.2. Carbon Nanofiber (CNF) 117
   3.4.3. Carbon Nanotube (CNT) 117
   3.4.4. Graphene 120
   3.4.5. Magnetic Fillers 122
   3.4.6. Dielectric Fillers 122
   3.4.7. Hybrid Nanostructures 122
  3.5. FABRICATION OF THERMOPLASTIC POLYMER NANOCOMPOSITES 123
  3.6. MAIN STRATEGIES FOR THE PROCESSING OF CBNS REINFORCED THERMOPLASTIC PNCS 123
   3.6.1. Solution Processing 124
   3.6.2. Melt Blending 125
   3.6.3. *In-situ* Polymerization 126

3.6.4. Layer-by-layer Approach 127
3.7. MAIN STRATEGIES FOR THE PROCESSING OF CBNS
REINFORCED THERMOSET PNCS 128
    3.7.1. Hand Layup Method 128
    3.7.2. Vacuum Assisted Resin Transfer Moulding 128
    3.7.3. Hot Pressing Method 130
    3.7.4. Bucky Paper Route 130
3.8. EMI SHIELDING PROPERTIES OF CBNS REINFORCED PNCS 131
    3.8.1. Thermoplastic Polymer Nanocomposites-based EMI Shield 132
    3.8.2. Thermoset Polymer Nanocomposites-Based EMI Shield 141
CONCLUSIONS AND FUTURE OUTLOOK 143
CONSENT FOR PUBLICATION 144
CONFLICT OF INTEREST 145
ACKNOWLEDGEMENTS 145
REFERENCES 145

**CHAPTER 4 THERMOPLASTIC POLYURETHANE GRAPHENE NANOCOMPOSITES FOR EMI
SHIELDING** 153
Meenakshi Verma, Veena Choudhary and *S.K. Dhawan*
4.1. INTRODUCTION 154
4.2. EMI SHIELDING EFFECTIVENESS 155
    4.2.1. Reflection Loss 158
    4.2.2. Absorption Loss 159
    4.2.3. Multiple Reflections 160
**4.3. ELECTROMAGNETIC ATTRIBUTES: COMPLEX PERMITTIVITY AND
PERMEABILITY** 160
4.4. MATERIALS FOR EMI SHIELDING 162
    4.4.1. Graphene 163
    4.4.2. Synthesis of Graphene 164
        4.4.2.1 Mechanical Exfoliation 165
        4.4.2.2 Liquid-Phase Exfoliation of Graphite 166
        4.4.2.3. Chemical Method: Exfoliation and Reduction of Graphite Oxide 167
        4.4.2.4. Chemical Vapor Depositions 170
        4.4.2.5. Thermal Decomposition of Silicon Carbide 171
4.5. CARBON NANOTUBES 172
4.6. GRAPHENE-CARBON NANOTUBES HYBRID 172
    4.6.1. Synthesis of Graphene-carbon Nanotubes Hybrid 172
        4.6.1.1. Solution Based Approaches: Simple Sonication and Reduction 172
        4.6.1.2. Chemical Vapor Deposition 173
        4.6.1.3. Preparation by Self-assembly 174
**4.7. FABRICATION/PROCESSING OF CARBON NANOSTRUCTURES-BASED
POLYMER COMPOSITES** 175
    4.7.1. Solution Processing or Solvent Casting 176
    4.7.2. Melt Mixing 176
    4.7.3. *In-situ* Polymerization 177
**4.8. PROPERTIES OF GRAPHENE/GRAPHENE HYBRIDS-BASED POLYMER
NANOCOMPOSITES** 177
    4.8.1. Electrical Conductivity and EMI Shielding Properties CNT Based PU Composites 177
    4.8.2. Graphene Based PU Composites 178

4.9. APPLICATIONS OF GRAPHENE/GRAPHENE HYBRIDS-BASED POLYMER NANOCOMPOSITES 192
SUMMARY, CONCLUSION & FUTURE SCOPE 193
CONSENT FOR PUBLICATION 194
CONFLICT OF INTEREST 195
ACKNOWLEDGEMENTS 195
REFERENCES 195

CHAPTER 5 SYNTHESIS OF POLY (3, 4-ETHYLENE DIOXYTHIOPHENE) CONDUCTING POLYMER COMPOSITES FOR EMI SHIELDING APPLICATIONS 213
M. Farukh, Jasvir Dalal, Anil Ohlan and S. K. Dhawan

5.1. INTRODUCTION 214
    5.1.1. Polythiophene 214
5.2. CONJUGATED POLYMER COMPOSITES FOR EMI SHIELDING 217
    5.2.1. Synthesis: Polymerization of Conjugated Polymers 218
        5.2.1.1. Emulsion Polymerization 219
        5.2.1.2. Synthesis of Conducting PEDOT & PEDOT's Nanocomposites 221
        5.2.1.3. Synthesis of Dodecyl Benzene Sulfonic Acid Doped PEDOT 221
        5.2.1.4. Synthesis of PEDOT Grafted MWCNT Composites 222
5.3. CHARACTERIZATION OF PEDOT AND PEDOT GRAFTED MWCNT COMPOSITES 223
5.4. ELECTROMAGNETIC SHIELDING AND DIELECTRIC STUDIES 228
5.5. PEDOT/MWCNT COMPOSITE REINFORCED POLYURETHANE CONDUCTIVE FILMS: PREPARATION, CHARACTERIZATION AND EMI SHIELDING STUDIES 230
5.6. PREPARATION OF PU CONDUCTIVE SHEETS INCORPORATED WITH PEDOT/OR PEDOT COATED MWCNTS 231
5.7. CHARACTERIZATION OF PEDOT AND PEDOT GRAFTED MWCNT FILLED PU SHEETS 232
5.8. PEDOT/PSS COATED MWCNT BUCKY PAPER: PREPARATION, CHARACTERIZATION AND EMI SHIELDING STUDIES 240
    5.8.1. Preparation of PEDOT/PSS Coated MWCNT Bucky Paper 240
    5.8.2. Synthesis of PEDOT/ RGO Nanocomposites 247
    5.8.3. Dielectric Properties of PEDOT's Composites 248
    5.8.4. PEDOT/Graphene Composites 250
    5.8.5. PEDOT/Graphene/SrF Composites 251
    5.8.6. Shielding Mechanism of PEDOT/graphene Composites 254
    5.8.7. PEDOT/Graphene/SrF Composites 257
CONCLUDING REMARKS 262
CONSENT FOR PUBLICATION 262
CONFLICT OF INTEREST 262
ACKNOWLEDGEMENTS 263
REFERENCES 263

CHAPTER 6 GRAPHENE AND ITS DERIVATIVES BASED NANOCOMPOSITES AS POTENTIAL CANDIDATE TO SWALLOW MICROWAVE POLLUTION 271
Monika Mishra, Avanish Pratap Singh, and S.K. Dhawan

6.1. INTRODUCTION 271
6.2. ELECTROMAGNETIC RADIATION SHIELDING THEORY 276
6.3. FACTORS THAT INFLUENCE THE ELECTROMAGNETIC WAVE ABSORPTION .. 279
    6.3.1. Skin Depth and Quarter Wave Principle 279
    6.3.2. Magnetic Loss and Dielectric Loss Mechanism 279

6.3.3. Filler Type 281
6.4. PREPARATION OF GRAPHENE AND ITS DERIVATIVES 282
6.4.1. Synthesis of Graphene Oxide and its Reduction 282
6.4.2. Graphene-based Nanocomposites -1: Conducting Ferrofluid 283
6.4.2.1. Synthesis of Conducting Ferrofluid 283
6.4.3. Surface Morphology and Microstructural Studies of Composite 285
6.4.4. X-ray Diffraction Analysis 286
6.4.5. Magnetic Properties 287
6.4.6. Raman Studies 288
**6.5. SHIELDING AGAINST ELECTROMAGNETIC INTERFERENCE AND
DIELECTRIC CHARACTERISTICS** 288
**6.6. GRAPHENE BASED NANOCOMPOSITES -2 - (TIN OXIDE NANOPARTICLES
ENGINEERED REDUCED GRAPHENE OXIDE COMPOSITE (SNO2@RGO))** 295
6.6.1. Synthesis of Tin Oxide 295
6.6.2. Synthesis of SnO_2 Decorated RGO Sheets 295
6.6.3. Morphological Analysis 296
6.6.4. Structural Analysis 297
6.6.5. Raman Spectroscopy 298
6.6.6. Shielding Effectiveness Measurement 299
**6.7. GRAPHENE-BASED NANOCOMPOSITES -3 -(G RAPHENE OXIDE/FERROFLUID/
CEMENT COMPOSITES)** 302
CONCLUSION 305
CONSENT FOR PUBLICATION 305
CONFLICT OF INTEREST 305
ACKNOWLEDGEMENTS 306
REFERENCES 306

**CHAPTER 7 UTILIZATION OF FLY ASH COMPOSITES IN ELECTROMAGNETIC SHIELDING
APPLICATIONS** 315
Swati Varshney and S.K. Dhawan

7.1. INTRODUCTION 316
7.2. ELECTROMAGNETIC SHIELDING MECHANISM 318
7.3. EMI SHIELDING MEASUREMENT METHODS 320
7.4. SELECTION OF MATERIAL FOR THE DESIGNING OF MICROWAVE SHIELD 321
7.5. SYNTHESIS OF POLYPYRROLE- Γ-FE2O3-FLY ASH (PFFA) NANOCOMPOSITES . 322
7.5.1. Characterization 323
7.5.1.1. X-ray Diffraction (XRD) Analysis 323
7.5.1.2. Micro Structural Analysis 324
7.5.1.3. Conductivity and Magnetic Measurements 326
7.5.1.4. Dielectric, Permeability and Electromagnetic Interference Shielding Investigations ... 327
7.6. SYNTHESIS OF POLYPYRROLE- Γ-FE2O3-FLY ASH (PFFA) NANOCOMPOSITES 332
7.6.1. Characterization 333
7.6.1.1. Micro Structural Analysis 333
7.6.1.2. X-ray Diffraction (XRD) and Conductivity Analysis 334
7.6.1.3. Thermogravimetric Analysis 336
7.6.1.4. Electromagnetic Interference Shielding and Dielectric Investigations 336
7.7. DESIGNING OF CEMENT COMPOSITE PAINT 339
7.7.1. Characterization 341
7.7.1.1. X-Ray Diffraction (XRD) 341
7.7.1.2. Micro Structural Analysis 342

7.7.1.3. Conductivity and Shore Hardness Test 343
7.7.1.4. Magnetic Measurements 344
7.7.1.5. Electromagnetic Interference Shielding Investigations 344
CONCLUDING REMARKS 346
CONSENT FOR PUBLICATION 347
CONFLICT OF INTEREST 347
ACKNOWLEDGEMENTS 347
REFERENCES 347

CHAPTER 8 FABRICATION AND MICROWAVE SHIELDING PROPERTIES OF FREE-STANDING CONDUCTING POLYMER-CARBON FIBER THIN SHEETS 355
Rakesh Kumar and S K Dhawan
    8.1. INTRODUCTION 356
        8.1.1.Conducting Polymers... 356
        8.1.2. Polyaniline and its Applications 358
    8.2. ELECTROMAGNETIC INTERFERENCE (EMI) SHIELDING 362
        8.2.1. Theory of EMI Shielding and Microwave Absorption 363
            8.2.1.1. Shielding Effectiveness... 363
            8.2.1.2. Absorption Loss (SE$_A$).. 365
            8.2.1.3. Reflection Loss (SE$_R$) 365
            8.2.1.4. Multiple Reflection Correction Factors (SE$_M$) 366
    8.3. I.CPs and THEIR COMPOSITES as SHIELDING MATERIALS 366
    8.4. DESIGN AND DIFFERENT FORMS OF EMI SHIELDS............................ 369
    8.5. BINDERS FOR THE FABRICATION OF POLYANILINE COMPOSITE SHEETS........ 369
    8.6. SCOPE AND OBJECTIVE OF THE CHAPTER 371
    8.7. SYNTHESIS OF POLYANILINE AND POLYANILINE-CARBON FIBER (PA-CF) COMPOSITES 372
        8.7.1. Chemistry of In-situ Chemical Oxidative Emulsion Polymerization 374
        8.7.2. Mechanism of Polymerization 376
    8.8. BLENDING AND MIXING OF PA-CF COMPOSITES WITH NOVOLAC RESIN TO PREPARE POLYANILINE-CARBON FIBER-NOVOLAC (PACN) COMPOSITES............. 377
    8.9. FABRICATION OF FREE-STANDING SELF-SUPPORTED THIN SHEETS OF PACN COMPOSITES 378
        8.9.1. Hydraulic Hot Press & Fabrication Process 378
    8.10. CHARACTERIZATION AND ANALYSIS 380
        8.10.1. UV-vis Spectroscopy 380
        8.10.2. FTIR Spectroscopy... 382
        8.10.3. SEM Analysis 383
        8.10.4. Thermogravimetric Analysis.. 385
        8.10.5. Flexural Strength and Flexural Modulus 387
        8.10.6. Electrical Conductivity (EC).. 389
        8.10.7. EMI Shielding Measurements... 391
    CONCLUSION... 398
    CONSENT FOR PUBLICATION 399
    CONFLICT OF INTEREST 399
    ACKNOWLEDGEMENTS 399
    REFERENCES 399

CHAPTER 9 EMI SHIELDING PROPERTIES OF CONDUCTING POLY (ANILINE-CO-O-TOLUIDINE)-CF-NOVOLAC COMPOSITES 411
Seema Joon and S.K. Dhawan

9.1. INTRODUCTION 411

9.2. SYNTHESIS OF MATERIALS 413

    9.2.1. Synthesis of Poly (Aniline-co-o-Toluidine)-Carbon Fiber (PANIoT-CF) Composite ... 413

    9.2.2. Fabrication of Thin Sheets of Poly (Aniline-co-o-Toluidine)-Carbon Fiber-Novolac (PANIoTCFN) Composites 414

9.3. ANALYSIS OF MATERIALS 414

    9.3.1. XRD Analysis 414

    9.3.2 FTIR Spectral Study 416

    9.3.3. SEM Analysis 420

    9.3.4. Thermogravimetric Analysis... 421

    9.3.5. Flexural Strength... 423

    9.3.6. Conductivity.. 425

9.4. EMI SHIELDING AND DIELECTRIC MEASUREMENTS............................ 428

CONCLUSION 434

CONSENT FOR PUBLICATION 435

CONFLICT OF INTEREST 435

ACKNOWLEDGEMENTS 435

REFERENCES 435

CHAPTER 10 POROUS 2D MXENES FOR EMI SHIELDING 439

    Pradeep Sambyal, Chong Min Koo and *S.K. Dhawan*

10.1. INTRODUCTION 440

10.2 MXENES 441

10.3. SYNTHESIS OF MXENES 442

10.4. STRUCTURAL PROPERTIES OF MXENES 443

10.5. CRYSTAL STRUCTURE 443

10.6. MXENES AS EMI SHIELDING MATERIALS 445

10.7. MXENE POROUS FOAMS FOR EMI SHIELDING 448

10.8. MXENE AEROGELS FOR EMI SHIELDING 453

CONCLUSION.. 459

CONSENT FOR PUBLICATION 460

CONFLICT OF INTEREST 460

ACKNOWLEDGEMENTS 460

REFERENCES 460

CHAPTER 11 NANOSTRUCTURED TWO-DIMENSIONAL (2D) MATERIALS AS POTENTIAL CANDIDATES FOR EMI SHIELDING 465

    Ayushi Saini, Anil Ohlan, S. K. Dhawan and *Kuldeep Singh*

11.1. INTRODUCTION 466

11.2. EMI SHIELDING REQUIREMENT AND MATERIAL APPLICATIONS 467

11.3. EMI TEST METHODS AND STANDARDS 471

    11.3.1. Open Field Test... 471

    11.3.2. Coaxial Transmission Line Test.. 472

    11.3.3. Shielded Box Test.. 472

    11.3.4. Shielded Room Test... 472

11.4. MARKET FOR EMI SHIELDING MATERIALS 474

11.5. GRAPHENE AND ITS PROPERTIES 476

    11.5.1. Synthesis of Graphene by CVD Method... 479

    11.5.2. Epitaxial Growth of Graphene 481

    11.5.3. Chemically Modified Graphene (CMG) 481

11.6 ELECTRICAL PROPERTIES AND PERMITTIVITY OF GRAPHENE COMPOSITES FOR EMI 483
11.7. TWO-DIMENSIONAL MATERIALS BEYOND GRAPHENE 487
    11.7.1. Synthesis of MoS2 and its Composite Application in Electromagnetic Absorption 488
        11.7.1.1. Mechanical Exfoliation 488
        11.7.1.2. Liquid Exfoliation (LE) 488
        11.7.1.3. Chemical vapor deposition (CVD) 489
        11.7.1.4. Hydrothermal or Solvothermal Methods 489
        11.7.1.5. EMI shielding properties of MoS2 490
    11.7.2. 2D Nitrides: Hexagonal Boron Nitride (h-BN) 492
        11.7.2.1 Solid State Reaction 495
        11.7.2.2. Graphene/hexagonal Boron Nitride Nanoparticle Hybrids for Microwave Absorption 496
    11.7.3. 2D Black Phosphorus 497
        11.7.3.1. Preparation of Black Phosphorus 498
    11.7.4. Uses of 2D Black Phosphorus in EMI Shielding 500
    11.7.5. MXene 501
        11.7.5.1. Structure, Synthesis of MXene and its Composite 502
        11.7.5.2. Synthesis of MXene 505
    11.7.6. EMI Shielding Properties of MXene and MXene/polymer Composite 506
    11.7.7. Future Direction of MXene Research 512
CONCLUDING REMARKS 513
CONSENT FOR PUBLICATION 514
CONFLICT OF INTEREST 514
ACKNOWLEDGEMENTS 514
REFERENCES 514
CHAPTER 12 NOVEL RADIATION SHIELDING CONCRETE UTILIZING INDUSTRIAL WASTE FOR GAMMA-RAY SHIELDING 527
Manish Mudgal and Er R.K. Chouhan
12.1. INTRODUCTION 528
12.2. INDUSTRIAL WASTE - RED MUD 530
    12.2.1. Red Mud Generation 531
    12.2.2. Properties and Characterization of Red Mud 534
12.3. CONVENTIONAL SHIELDING AGGREGATES 536
12.4. NOVEL DEVELOPED SYNTHETIC RADIATION SHIELDING AGGREGATES 537
12.5. CHARACTERIZATION STUDIES 538
    12.5.1. Chemical Analysis 538
    12.5.2. XRD Phase Identification 539
    12.5.3. X-ray Diffraction Analysis of Red Mud 540
    12.5.4. X-ray Diffraction Analysis of Developed Synthetic Shielding Aggregates 541
    12.5.5. Scanning Electron Microphotographs 541
    12.5.6. EDXA Analysis 542
    12.5.7. Engineering Properties of Advanced non-toxic Synthetic Radiation Shielding Aggregates 543
12.6. INNOVATIVE RADIATION SHIELDING CONCRETE 545
12.7. GAMMA ATTENUATION CHARACTERISTICS 547
CONCLUSION 550
CONSENT FOR PUBLICATION 550
CONFLICT OF INTEREST 550

ACKNOWLEDGEMENTS 550
REFERENCES 550

SUBJECT INDEX 555

FOREWORD

Present book reflects the importance of EMI shielding in the field of functional microwave materials being relevant to material science and electrical and electronics engineering and ever increasing with the expansion of electronics, mobile communications and satellite communications. It is culmination of the hard work undertaken by the team under leadership of Dr. Dhawan over a long period of time and with great perseverance.

The book "Smart Materials Design for Electromagnetic Interference Shielding Applications", comprises 12 chapters and covers the gamut comprising shielding performance and properties of the light weight carbon composite foams, synthesis and EMI shielding properties of carbon based nanostructures (CBNS) reinforced polymer nanocomposites, microwave absorption properties of graphene and its polymer nanocomposites, microwave absorption properties of graphene and its derivatives based nanocomposites and their composites with polyurethane and metal oxides, conjugated polymer and their composites *viz.* PEDOT and its composites with MWCNT, grapheme and graphene and graphene/strontium ferrite hybrid. Shielding properties of polypyrrole and polypyrrole-fly ash composites; polyaniline and polyaniline-carbon fiber composites and polyaniline-toluidine copolymer composites with carbon fiber are also elaborated in the book. Synthesis, shielding behavior and dielectric attributes of 2D materials such as graphene layered MoS_2, nitrides, black phosphrous and MXenes, gamma ray shielding behavior of industrial waste like red mud in designing geopolymershave been discussed in carefully formatted chapters. The core ideas underlying various designs of microwave shield and their characterization essential to design a nearly perfect shield using different materials, methods and approaches are presented in an interesting manner. This book may serve as an important reference not only for professionals engaged in designing instruments for characterization and commercial microwave materials but for researchers cutting across disciplines interested in microwave materials and measurement methods. Written in a lucid manner with a sense of continuity, the book will surely be able to establish and sustain the interest in the field of Electromagnetic Interference with numerous applications.

D. K. Aswal
Health, Safety and Environment Group
Bhabha Atomic Research Centre
Trombay, Mumbai 400085, India

PREFACE

The development of functional microwave materials is the most active areas in physical science, material science and electrical and electronic engineering and this is parallelly increasing with the expansion of electronics, mobile communication and satellite communication. In recent years, Microwave materials are employed in a variety of applications ranging from communication devices to military satellite services. The design of an ideal microwave shield requires complete understanding of the properties of materials at high frequencies. To dissipate the both electric and magnetic field components of microwave, generally, shield is a composite of magnetic and dielectric materials. Therefore, the characterization of dielectric and magnetic attributes, surface morphology, crystalline size, shape of material *etc.* are very important in microwave electronics. A number of direct and indirect measurement methods and techniques were developed to characterize the microwave shield, in past century. Even numerous textbooks, reviews and articles have been published for understanding the characterization techniques but most of them are either cover characterization techniques or materials. There is a strong need to provide a practical reference text book that support measurement techniques and microwave materials for researchers and engineers. This book presents a detailed discussion on up-to-date measurement methodologies, material synthesis and designing of an effective microwave shield. Basically, this book put all scattered information in the form of reports, journals and advances in this area at one place. This book is likely to be most useful to professional engineers who are engaged in designing characterization related instruments and designing commercial microwave materials in the form of absorbing sheets, fabric, glass *etc.* In spite of this, book also satisfies the need of researchers of other disciplines, biophysicist, and materials scientists, industrial engineer, graduate students, who wish to understand the thoroughly limitations of microwave materials and measurement methods in materials characterization. This book discusses almost all properties of conductors, semiconductors, dielectrics, and magnetic materials. at microwave frequencies. The electromagnetic characterization specially includes permittivity, permeability, mobility, and surface impedance.

The first chapter discusses the fundamentals of microwave shielding and underlying physics. After detailed discussion on electromagnetic properties and characterization techniques, a brief review of the materials and their shielding performance is presented.

Chapter 1 introduces the general properties of various electromagnetic materials and their, Chapter 2 provides a summary of the shielding performance and properties of light weight Carbon Composite Foams. The electrical conductivity in semiconducting range is very important properties for the development of absorption-based microwave shield. Chapter 3 and chapter 8 deal with the measurements of the permittivity and permeability of low-conductivity materials such as Carbon nanostructures-based Polymer nanocomposites and their free-standing sheet like structure for EMI shielding applications. Microwave properties of graphene and its derivatives-based nanocomposites and their composite with thermoplastic polyurethane are discussed in chapter 4 and chapter 6 respectively. Conducting polymer composites and copolymer composites are explored in chapter 5 and chapter 9 respectively. Chapter 7 is concerned with the microwave properties of dielectric flyash composites. In recent years, the research on 2D materials has been active. Chapter 10 and chapter 11 discuss the shielding behavior and dielectric attribute of 2D materials such as graphene and Mxenes. The last chapter of the book discusses the gamma-Ray Shielding properties of Industrial Waste Concrete also known as Synthetic Heavy Density Aggregate materials that are resistant to their properties at high temperatures are often needed in industry and in commercial applications. In general, it is possible to apply the techniques and materials discussed in this book to high-temperature measurements. Derails have been emphasized in the book on the latest development in fourth-generation (4G) as well as fifth-generation (5G) mobile networking technology. 5G mobile communication also produce intense radiations which are harmful for the human health and environment. Similar to the Radar absorbing materials, and defense systems, carbon-based materials such as carbon nanotubes and graphene, 2D materials, metal nanoparticles, porous materials and polymer-based EMI shielding materials have also attracted a lot of attention in the industry for solving the 4G & 5G electromagnetic radiations. In this book, each chapter is designed as a self-contained unit so that readers can easily get detailed knowledge relevant to their research interests. We meshed rivers of literature on microwaves materials and measurements into readable units. Several references have been provided for the convenience of readers who choose to pursue a certain issue in greater depth or revert to the original posts.

In writing this book, we tried to present the core ideas underlying various designs of microwave shield and their characterization so that readers would appreciate the method to design a nearly perfect shield using different materials, methods and approaches. We would like to mention that this textbook is a collection of multiple people's work. For the mentioned materials and designs that may still are under patent, we will neither be responsible nor liable.

There are a lot of people we owe a lot of appreciation for helping us in preparing this book and contributing their valuable chapters. Our most important gratitude goes to Dr. Manish Mudghal, Dr. Monika, Dr. Rajeev, Dr.Tajendra, Dr. Meenkashi, Dr. Rakesh, Dr. Seema and all our EMI shielding team which are the source of inspiration for the finalization of this book.

Sundeep K. Dhawan
CSIR-National Physical Laboratory
Dr. K. S. Krishnan Marg
New Delhi
India

ACKNOWLEDGEMENTS

We thank **Dr. D.K. Aswal**, Distinguished Scientist, Bhabha Atomic Research Centre, Mumbai & Former Director, CSIR-National Physical Laboratory, New Delhi and **Prof. Dr Venu Gopal Achanta,** Director, National Physical Laboratory, New Delhi to allow us to publish the results in the form of book. We also thank Benthan Science Publishers for inviting us to publish our work on EMI shielding in the form of book.

We are highly thankful to **Dr. S.R. Dhakate,** Chief Scientist & Head, Advanced Materials & Devices Metrology Division**,** NPL and all other colleagues and technical staff for their cooperation.

We express our sincere thanks to **Dr.Rajeev Kumar,** Inspire Faculty, CSIR-AMPRI, Bhopal, and **Dr. Tejendra Gupta**, Assistant Professor, Amity University, Noida for excellent work on Carbon family being presented in the book. We thank **Dr. Meenkashi** for her excellent contribution in designing polyurethane graphene composites chapter. Conducting polymer composites for their usage in EMI shielding was nicely contributed by **Dr. Md. Farukh, Dr. Rakesh and Dr Seema Joon** and we express our gratitude for their contribution. We thank **Dr. Monika Mishra** for her contribution on ferrofluid graphene oxide cement composite besides her study on conducting composites which can be used for EMI shielding.

We are thankful to **Dr Manish Mudgal,** Sr. Principal Scientist, AMPRI, Bhopal for presenting his work on gamma ray shielding by utilizing red mud concrete**.** We also express our sincere thanks to **Mr. Brijesh Sharma, Dr. Ridham Dhawan** for their help in the lab work.

Sundeep K. Dhawan
CSIR-National Physical Laboratory
Dr. K. S. Krishnan Marg
New Delhi
India

CHAPTER 1

Electromagnetic Interference Shielding and its Evaluation

Avanish Pratap Singh[1*], Monika Mishra[2], Anil Ohlan[3] and S.K. Dhawan[4]

[1] *Experimental Research Laboratory, Department of Physics, Atma Ram Sanatan Dharma College, Dhaula Kuan, New Delhi – 110021, India*

[2] *Department of Physics, Netaji Subhas University of Technology, Dwarka, Delhi - 110078, India*

[3] *Department of Physics, Maharshi Dayanand University, Rohtak - 124001, Haryana, India*

[4] *Materials Physics & Engineering Division, CSIR-National Physical Laboratory, New Delhi – 110012, India*

Abstract: This chapter introduces the basis of electromagnetic interference shielding theory, techniques for characterizing electromagnetic materials, and the application of studying the electromagnetic properties of materials. The main topics include theoretical shielding effectiveness calculation and its measurement, Shielding Mechanism of E-Field and H-Field, absorption and reflection loss, interfacial and orientation polarization, electronic and atomic polarization, characterization methodology, Nicholson-Ross-Weir technique, short circuit line technique electrical properties, electrical permittivity, and magnetic permeability, and dielectric measurement techniques. In addition, the field method is used to analyze the electromagnetic field. This chapter introduced the concept of microwave shield and described its characteristics to achieve the ideal shield. Next, the basic physics that coordinates the interaction between the material and the electromagnetic field is described in detail. Subsequently, we analyze the general properties of typical electromagnetic materials such as dielectric materials, semiconductors, conductors, magnetic materials, and artificial materials. The last part of this chapter introduces the latest developments in EMI shielding materials in various structural forms, as well as future challenges and guidelines for finding material solutions for next-generation shielding applications. Indeed, advanced materials and process technology are the keys to successful EMI shielding.

Keywords: Composite Shielding Materials, Dielectric Attributes, Dielectric Measurement Techniques, Dielectric Parameters, Electromagnetic shield, EMI

---

*Corresponding author Avanish Pratap Singh:** Experimental Research Laboratory, Department of Physics, Atma Ram Sanatan Dharma College, New Delhi - 10021, India; Tel: 918588819244; E-mail: avanishpratap@gmail.com

Sundeep K. Dhawan, Avanish Pratap Singh, Anil Ohlan, Kuldeep Singh Kakran and Pradeep Sambyal (Ed.)
All rights reserved-© 2022 Bentham Science Publishers

Shielding, Microwave Pollution, Nicolson Ross-Wier methods, Permittivity, Permeability, Skin Depth.

1.1. INTRODUCTION

Electromagnetic interference has become the fourth environmental pollution after noise, water, and air pollution. It is the side-effect of electronic devices and the most serious concern because it is one of the reasons which affect the performance of the electronic systems [1-5]. This EMI is not only detrimental to the functionality of electronic systems but can also adversely impact human health [6, 7]. Some examples of EMI in our day-to-day lives are picture flickering or noise disruptions on televisions, radios, and laptop screens from wireless or cell phone signals [8-11]. In addition, EMI shielding has become more common for microwave-frequency devices such as telecommunications, TV image transmission, microwave processing, weather radar, microwave electronics, radar surveillance systems, and lateral guidance [12-15]. Microwave shielding is, therefore, needed in the form of electronic enclosures (for rooms, computers, aircraft, *etc.*) and radiation source enclosures (for example, telephone receivers) [9, 12, 16]. To protect the workplace, environment, computers and telecommunications equipment, and sensitive circuits, a lightweight EMI shield is necessary [3, 16, 17]. Lightweight carbon composites have gained popularity compared to traditional metal-based EMI shielding materials [18-20]. Many attempts have been made to reduce its impact due to enhanced regulatory controls to regulate the amount of electromagnetic radiation in the atmosphere (owing to health hazards) [11, 21]. Electromagnetic materials have magnetic/dielectric losses, and these losses also depend on frequency impinges on the shield. These three factors are mainly responsible for the reflection, absorption and/or scattering of electromagnetic waves [9, 22, 23]. A brief review is being presented here. Analysis of EMI shielding behavior and dielectric attributes using different fillers like conducting (polymer, graphene, carbon nano tubes, MXenes), magnetic (ferrites), and dielectric (fly ash) materials is also being presented here. Glimpses of developments in fourth-generation (4G) and fifth-generation (5G) mobile networking technology are being presented besides designing specific composites from industrial waste like red mud from the aluminum industry, which can be used in gamma-ray shielding.

1.2. ELECTROMAGNETIC INTERFERENCE SHIELDING

In essence, an electromagnetic interference shield is a fence to block the unwanted electromagnetic (EM) induction caused by alternating voltage/current that attempts to generate corresponding induced signals in the neighboring electronic device, thereby attempting to ruin its performance [24, 25]. The word "shielding" refers to an enclosure that completely encloses or partially encloses an electronic device and prevents EM radiation from an external source from degrading its electronic operation [26, 27]. It can also be used to avoid the external sensitivity of the instrument to internal radiation [9, 28-31]. It can be accomplished in one of two ways, as shown in Fig. (**1.1**). The radiated electromagnetic signals generated by the operating electrical circuits can either be halted or reduced (Fig. **1.1a**) or radiation beyond the product's borders can be avoided (Fig. **1.1b**). So, conceptually, the shield serves as a barrier to electromagnetic field propagation. Furthermore, shielding efficiency is defined as the ratio of field strength without the imposed shield to field strength with the imposed shield at a certain distance from the source [32, 33].

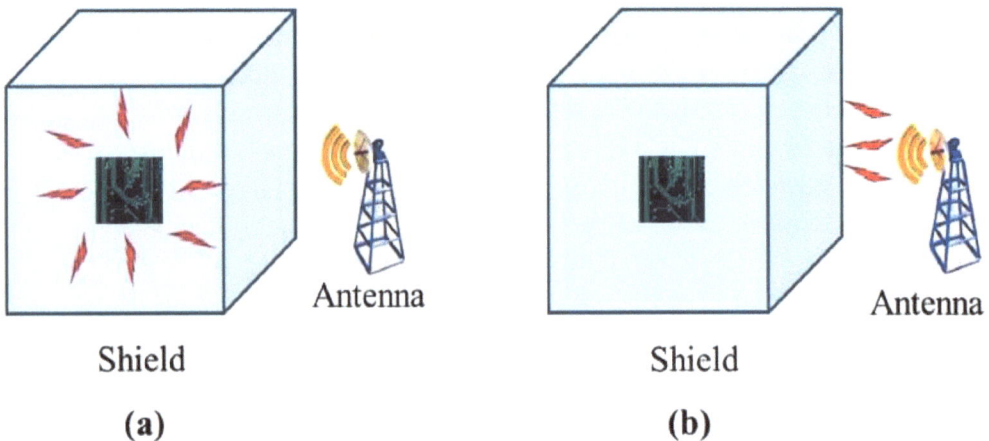

Fig. (1.1). Illustration of the shielded enclosure application (**a**) radiation from the operating electrical circuit and (**b**) the removal of radiation from the external source of radiation.

If an electronic component is positioned within a small, conductive shell positioned in an electric field, it will be protected because the electromagnetic wave's current setup does not work within the shell [34, 35]. This is not because the field was completely absorbed by the shell but because the E-field induced electrical charges along with the shell of various polarities. These charges create an electrical field that within the shell would appear to cancel the original field [34-36]. Shields are made of a highly permeable soft magnetic material, μ>>1, and when it comes to H-fields, suitable thickness weakens the magnetic field in the shielding shell by giving low reluctance. That is because the H-field seems to remain in the magnetic material layer as the magnetic material offers a low-reluctance route; the H-field strength can be reduced by a spherical shell of magnetic material with excellent permeability [34, 37].

Alternatively, a thin, low permeable and conductive shield may also provide efficient shielding at high frequencies for H-fields. This is because the shield has sufficient conductivity; Eddy currents in the shielding screen are caused by an alternating H-field. These eddy currents will create an alternating H-field of the opposite orientation within the body. The impact increases as the frequency rises, resulting in strong shielding efficacy at high frequencies.

Depending on the distance r between the observation point and radiation source, the radiation area of the electromagnetic wave can be divided into three parts for all wavelengths of the electromagnetic wave, as shown in Fig. (**1.2**). The area inside the distance r $> \lambda/2\pi$ is the near-field, and the distance r $> \lambda/2\pi$ is the far-field [38], and the distance r $\cong \lambda/2\pi$ between the two zones is the transition zone. To develop materials for a particular shield application, it is necessary to understand both intrinsic and extrinsic parameters upon which the shielding effect depends and the appropriate theoretical relationships associated with multiple components, reflection, absorption, and multiple internal reflections, *etc.*

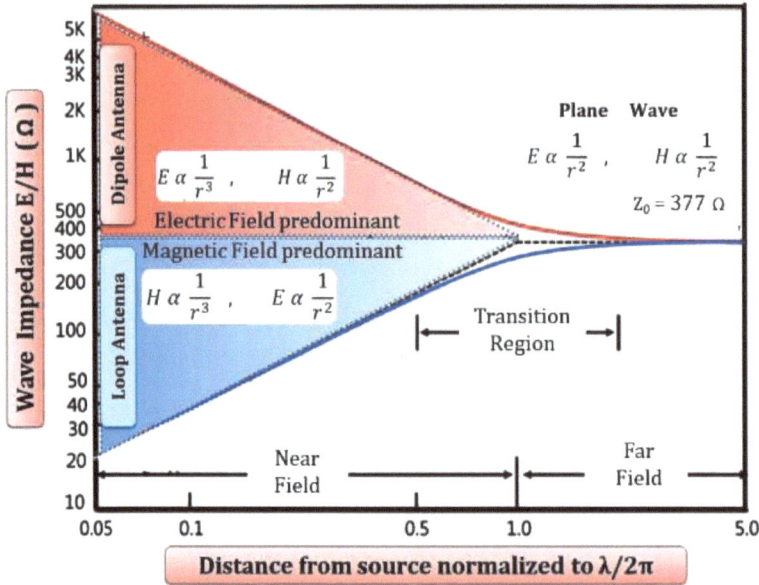

Fig. (1.2). Dependence of the wave impedance relative to the distance between the standard source (Loop antenna or Dipole Antenna) and $\lambda/2\pi$.

1.3. SHIELDING MECHANISM OF E-FIELD AND H-FIELD

Generally, EMI shielding is provided by a conductive barrier or enclosure that can safely either reflect electromagnetic radiations or transmit them to the ground [39]. Michael Faraday was the first who introduced the concept of the zero electric field in a closed conductive box known as the *Faraday cage*, and it forms the basis of E-field shielding technology. When sensitive electronic equipment is kept inside a thin conductive shell, no electromagnetic waves can cross the barrier because the electric field E generates charges of different polarities in the shield, as shown in Fig. (**1.3a**). These generated charges produce an electric field, which tends to cancel out the external electric field [40]. Interestingly, an external electric field penetrates the conductive shield up to a depth; this depth thickness is known as skin depth, and it can be small when the frequency of the electromagnetic wave is high enough [41].

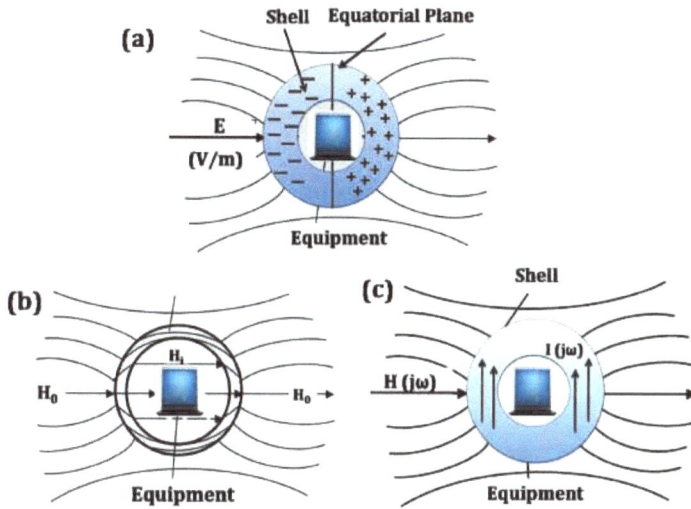

Fig. (1.3). (**a**) Generation of electrical charge on spherical conducting shell containing sensitive electronic equipment in an applied external electric field. (**b**) H-field power (H$_i$ >> H$_0$) within can be diminished by a spherical magnetic shell consisting of strong permeability material while the field stays in the magnetic crust. (**c**) An alternative magnetic field power can be decreased by a thin metallic layer of excellent conductivity as the currents I(jω) generated by field H(jω) can produce a field in the opposite direction of the field applied.

The Faraday effect does not hold true in the case of the H field. Shield made up of high permeable soft magnetic material (μ>>1) with appropriate thickness can be used for H Field shielding, as shown in Fig. (**1.3b**). In other words, a magnetic material's spherical shell is designed with high magnetic permeability. Because magnetic material has a low magneto resistance channel, it will lower the H field strength inside the magnetic cage [39, 40, 42]. A narrow shield having low permeability and weakly conductive material, on the other hand, may successfully protect H-fields at high frequencies. This is because the shielding layer has adequate conductivity; an alternating H-field causes eddy currents in the shielding layer (Fig. **1.3c**). These eddy currents produce alternating H magnetic fields in opposite directions in the shield. The higher the frequency, the better the shielding effect [39, 40]. That is why it is relatively difficult to prevent low-frequency H fields. The H field shielding layer requires the installation of a thick barrier that is usually manufactured using costlier magnetic material. A conductive shield based on the induced current principle is very effective at network frequencies; for example, an aluminum screen is used to prevent law frequency 50 and 60 Hz magnetic fields produced by transformers and other sources [40]. The induced currents generated from impinged *H*-field can only flow when there is no hindrance in their path, while apertures or breaks in the shield act as obstacles and will decrease the shielding effectiveness. Therefore, arrangements of apertures must be made in such a way that it minimizes this effect [39].

1.4. SHIELDING EFFECTIVENESS

The efficacy of the shielding depends on the attenuation of the EMI, the distance of the shield from the source, the frequency, the shield thickness, and the substance of the shield. Shielding efficacy (SE) is usually expressed in decibels (dB) as a function of the magnetic (H), electrical (E), or plane-wave field amplitude (P) logarithm of the incident and exit ratio.

$$SE\,(dB) = 10\,log\{P_T/P_I\} = 20\,log\{E_T/E_I\} = 20\,log\{H_T/H_I\} \qquad \textbf{(1)}$$

where, P_I (H_I or E_I) and P_T (H_T or E_T) are the power (magnetic or electric field) of the incident and transmitted EM waves, respectively. Three methods contribute to the shielding efficiency for all forms of electromagnetic interference. As illustrated in Fig. (**1.4**), impinged radiation is partially reflected from the front surface of the shield, partially absorbed within the shield material, and partially reflected several times from the shield rear surface to the front, aiding the shields' efficacy. The overall shielding effectiveness of a shielding material (SE) is, therefore, equal to the sum of the absorption (SE_A), the reflection (SE_R), and the multiple reflection correction (SE_M) [43-48] and is given by:

$$SE\,(dB) = SE_A + SE_R + SE_M \qquad \textbf{(2)}$$

According to Schelkunoff's theory, SE_M should be overlooked in all practical situations where the shield's thickness is greater than the skin's (δ) depth. The skin depth (δ) for a substance is the distance at which the amplitude of the EM wave decreases to 1/e of its initial amplitude.

1.4.1 Absorption Loss

Absorption loss, SE_A, is a function of the shield's physical properties and is independent of the source field type [25]. As illustrated in Fig. (**1.2**), the amplitude of an electromagnetic wave drops exponentially as it passes through a medium. This decay or absorption loss happens because currents produced in the medium causes ohmic losses and material heating, making it possible to transport E_1 and H_1 as separate signals. $E_1 = E_oe^{-t/\delta}$ and $H_1 = H_oe^{-t/\delta}$. The absorption term SE_A in decibel, therefore, is given by the expression:

$$SE_A(dB) = 20\frac{t}{\delta}\log e = 8.68\frac{t}{\delta} \qquad (3)$$

where permeability (1 for copper) σ is the conductivity relative to copper. The skin depth δ may be stated as:

$$\delta = \sqrt{2/\sigma\omega\mu'} \qquad (4)$$

The skin effect is most noticeable at low frequencies when the fields are more likely to be primarily magnetic and have a lower wave impedance than 377 Ω. A suitable shielding material would have sufficient conductivity and permeability, as well as an adequate thickness, to accomplish the requisite number of skin depths at the lowest worry frequency in terms of absorption loss [11, 25].

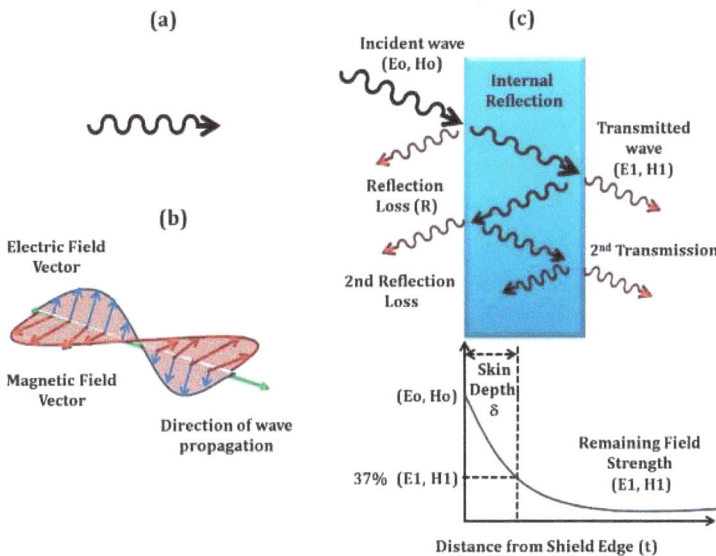

Fig. (1.4) EMI shielding mechanism schematic illustration: **(a)** EM wave, **(b)** magnetic and electric field vector perpendicular to the wave propagation route and **(c)** electromagnetic wave splitting when passing through a shield.

1.4.2 Reflection Loss

The reflection loss is caused by a mismatch between the incident wave and the shield's surface impedance. Consider shielding efficacy for incident electric fields as a separate problem from magnetic, electric, or plane waves to simplify the computation of refection losses. The equations for the three fundamental fields are as follows [25, 44]:

$$R_E = K_1 10 \, log \left(\frac{\sigma}{f^3 r^2 \mu} \right) \tag{5}$$

$$R_H = K_2 10 \, log \left(\frac{f r^2 \sigma}{\mu} \right) \tag{6}$$

$$R_P = K_3 10 \, log \left(\frac{f \mu}{\sigma} \right) \tag{7}$$

where for the magnetic, electric, and plane wave fields, R_H is the reflection losses, expressed in dB; R_E is the relative conductivity to copper and R_P is the relative permeability to free space; f is the frequency in Hz, and r is the distance in meters between the source and the shield.

1.4.3. Multiple Reflections

The factor SE_M can be either positive or negative (in most practical cases, it is always negative) and turn out to be insignificant when the absorption loss $SE_A > 6$ dB [25, 48]. It is usually only relevant when metals/conductors are thin (and at low frequencies, *i.e.*, below approximately 20 kHz). The SE_M of a factor may be written as:

$$SE_M = -20 \, log(1 - e^{-\frac{2t}{\delta}}) \tag{8}$$

1.5. CONDUCTION IN THE MICROWAVE FIELD

A material can have many polarization effects or dielectric mechanisms at the microscopic level that contribute to its inclusive permittivity (Fig. **1.5**). An induced dipole moment related to polarizability occurs when an electric field is applied to a material. There are many modes of polarization that are as follows:

➢Electronic polarization is the displacement of an electronic cloud around an atom.

➢Atomic polarization is the ability of an electric field to change electron distribution and, as a result, the equilibrium position of the atoms in a molecule.

➢Dipolar polarization is caused by the alignment of permanent dipoles.

1.6. ORIENTATION POLARIZATION

Permanent dipole moments are orientated in a random manner in the absence of an electric field, resulting in no polarization. The electric field 'E' will exert torque 'T' on the electric dipole, and the dipole will rotate to align with the electric field, allowing the polarization of orientation. The dielectric losses will be exacerbated by the friction that follows the dipole's direction. At the relaxation frequency that typically happens in the microwave field, the dipole rotation causes a variation in both ε 'and' ε ".

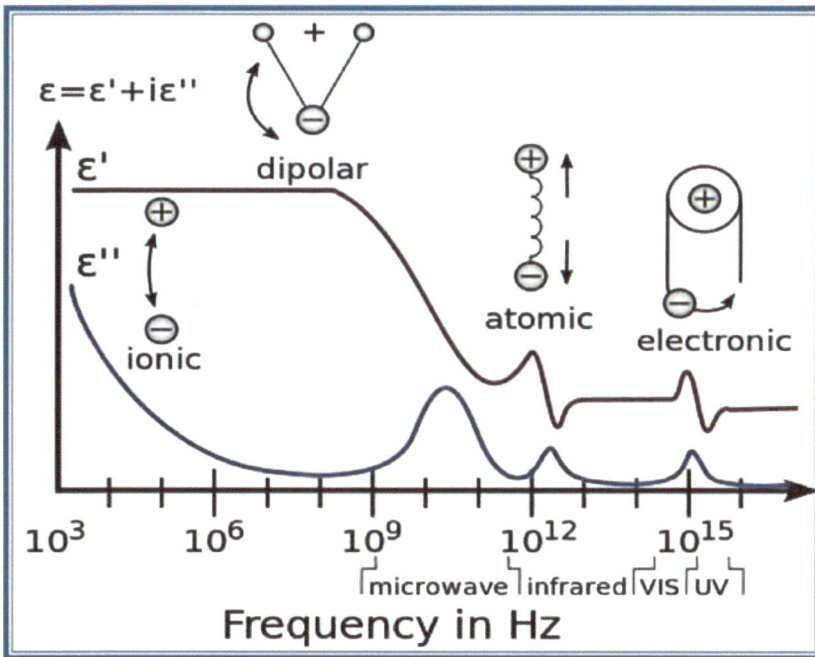

Fig. (1.5). Dielectric mechanisms (Source: http://stability issues.net/2012/02/15/electrolytic-capacitor-esr-ripple-current-and-frequency-2/).

1.6.1. Electronic and Atomic polarization

Electronic polarization happens in neutral atoms when an electric field displaces the nucleus relative to the electrons that surround it. Atomic polarization occurs when adjacent positive and negative ions "stretch" in response to an applied electrical field.

1.6.2. Interfacial or Space Charge Polarization

When the motion of traveling charges is hindered, interfacial or space charge polarization occurs. Inside a material interface, the charges are trapped. Motion can also be impeded when charges on the electrodes may not be freely discharged or replaced. The distortion of the field induced by the accumulation of these charges increases a material's overall capacitance, which tends to increase in ε '.

1.7.   THEORETICAL   CALCULATIONS   OF   SHIELDING EFFECTIVENESS

An electromagnetic shield transmits plane electromagnetic waves in the same way as traditional transmission lines transmit current and voltage. Therefore, the calculation of far-field shielding efficiency (SE) is based on plane wave shielding theory and transmission line theory. It requires the solution of Maxwell's equation, describing the reduction in electromagnetic interference amplitude due to electromagnetic shield. To facilitate this calculation, consider a uniform plane wave characterized by electric and magnetic field strength vectors $\vec{E}$ and $\vec{B}$ that vary within a plane only with x direction, as shown in Fig. (**1.6**). The tangential components of vectors $\vec{E}$ and $\vec{B}$ are continuous across the interface between the shield's interior and its exterior.

To create an expression for the far-field SE, consider a linearly polarized plane wave whose vector fields $\vec{E}$ and $\vec{B}$ are expressed as:

$$\vec{E} = \vec{E}_o e^{i(\omega t \pm kx)} \tag{9}$$

$$\vec{B} = \vec{B}_o e^{i(\omega t \pm kx)} \tag{10}$$

where, $\vec{E}_o$ and $\vec{B}_o$ are the wave's amplitude.

$\vec{E}_o = \vec{E}(0,0)$, x is a location along the wave's travel direction of travel, and t is the time at which the wave is described in a medium of permeability μ and permittivity ε.

$\vec{k} = \frac{2\pi}{\lambda}\,\hat{k}$ is the wave vector in free space and λ is its wavelength, ω is the angular frequency of the wave. The magnetic field strength $\vec{H}$ and the magnetic induction $\vec{B}$ are associated with each other empirically as $\vec{B} = \mu\,\vec{H}$. Maxwell's equations can be utilized to obtain the desired relationship between vectors $\vec{E}$ and $\vec{B}$

$$\nabla \times \vec{E} = -\frac{\partial \vec{B}}{\partial t} \qquad \text{(Faraday law of induction)} \qquad \textbf{(11)}$$

$$\pm ik \times \vec{E} = -i\omega\mu\vec{H} \qquad \textbf{(12)}$$

Assume that the wave is traveling at a speed c; then the angular frequency is related with wave number k by dispersion relation ω=ck on further simplification of Equation 13.

$$\vec{E} = \pm\frac{\omega\mu}{k}\vec{H} \quad = \pm\frac{\mu}{\sqrt{\varepsilon\mu}}\vec{H} \quad = \pm\sqrt{\frac{\mu}{\varepsilon}}\,\vec{H} = \pm Z\vec{H} \qquad \textbf{(13)}$$

The quantity Z is the wave impedance. Now, define a new term q as the wave number of the wave in a conducting medium of conductivity σ, relative permittivity ε_r, and relative permeability μ_r; q is defined as follows:

$$q = \omega\sqrt{\left\{\mu\left(\varepsilon - i\frac{\sigma}{\omega}\right)\right\}} \quad = \frac{\omega}{c}\sqrt{\varepsilon_r\mu_r\left(1 - i\frac{\sigma}{\omega\varepsilon}\right)} \qquad \textbf{(14)}$$

Hence, the complex wave impedance Z for the conducting medium is defined as

$$Z = \frac{\omega\mu}{q} = \frac{\mu}{\sqrt{\left\{\mu\left(\varepsilon - i\frac{\sigma}{\omega}\right)\right\}}} \quad = \sqrt{\frac{\mu}{\varepsilon}\left(1 - i\frac{\sigma}{\omega\varepsilon}\right)} \qquad \textbf{(15)}$$

To evaluate the far-field shielding, consider a plane EM wave propagating in +x direction, which is linearly polarized in the y direction, incident from the –x direction onto a conducting thin metallic sheet of thickness d, orientated in the y-z plane, as shown in Fig. (**1.6**).

In region I, an EM wave with vector fields $\vec{E}$ (Eoe-ikx) and $\vec{H} = \left(\frac{E_0}{Z_0}\right) e^{-ikx}$ is incident on a shield and partially reflected with fields $\vec{E}$ (REoe-ikx) and $\vec{H}$ ($R\frac{E_0}{Z_0} e^{-ikx}$, where R is the reflection amplitude R (complex number). In region III, there is a transmitted wave with fields TEoe-ikx and $T\frac{E_0}{Z_0} e^{-ikx}$. In general, we neglect the magnetic field vector in most of the calculations as it is c times smaller than the electric field vector. The shielding effectiveness of the shield is determined using transmission amplitude T (complex number) as follows:

$$SE(dB) = 10 \log\left[\frac{1}{T^2}\right] \tag{16}$$

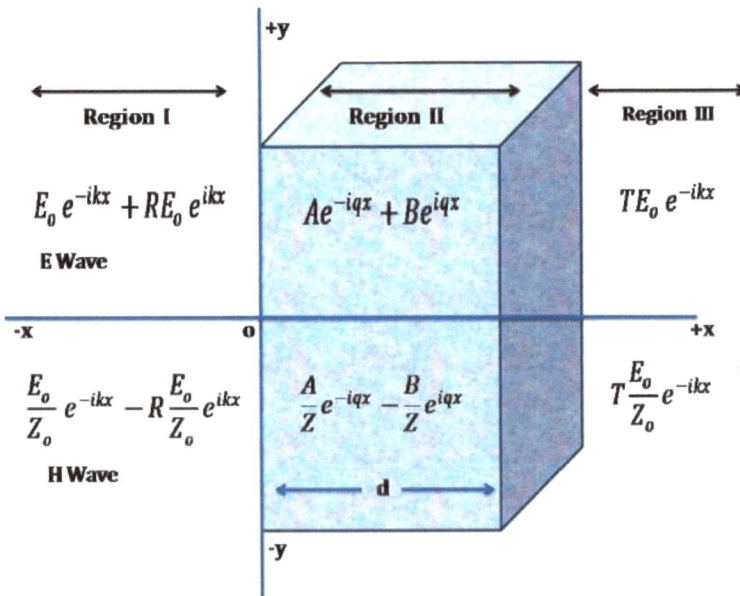

Fig. (1.6). Typical behaviour of E field and H field vector of an EM wave impinged normally on a shield.

Typical electromagnetic boundary conditions are employed to deduce the transmission coefficient, T, *i.e.*, the tangential component of $\vec{E}$, which is always continuous across the boundary. Let the electromagnetic disturbance in region II have a spatial variation of the form

$$Ae^{-iqx} + Be^{iqx} \tag{17}$$

$$\frac{A}{Z}e^{-iqx} - \frac{B}{Z}e^{iqx} \tag{18}$$

Continuity of the tangential component between region I and region II imply the following equation:

$$\vec{E}^{\parallel}{}_{region\,I} = \vec{E}^{\parallel}{}_{region\,II}$$

$$\rightarrow \quad E_o\,e^{-ikx} + RE_o\,e^{ikx} = Ae^{-iqx} + Be^{iqx} \tag{19}$$

$$\vec{H}^{\parallel}{}_{region\,I} = \vec{H}^{\parallel}{}_{region\,II}$$

$$\frac{E_o}{Z_0}\,e^{-ikx} - R\frac{E_o}{Z_0}e^{ikx} = \frac{A}{Z}e^{-iqx} - \frac{B}{Z}e^{iqx} \tag{20}$$

where, $Z_0 = \sqrt{\frac{\mu_0}{\varepsilon_0}} = 377\ \Omega$, which is the impedance of free space.

Now, boundary conditions between region II and region III will give two more equations:

$$\vec{E}^{\parallel}{}_{region\,I} = \vec{E}^{\parallel}{}_{region\,II}$$

$$\rightarrow \quad Ae^{-iqx} + Be^{iqx} = TE_o\,e^{-ikx} \tag{21}$$

$$\vec{H}^{\parallel}{}_{region\,I} = \vec{H}^{\parallel}{}_{region\,II}$$

$$\frac{A}{Z}e^{-iqx} - \frac{B}{Z}e^{iqx} = T\frac{E_o}{Z_0}\,e^{-ikx} \tag{22}$$

On eliminating R from Equations 19 and 20, E_0 in terms of A and B is given:

$$E_0 = \frac{1}{2}\left\{\left(1 + \frac{Z_0}{Z}\right)A + \left(1 - \frac{Z_0}{Z}\right)B\right\} \tag{23}$$

In a similar way, Equations 21 and 22 give A and B in terms of T and E_0:

$$A = \frac{1}{2}\left(1 + \frac{Z}{Z_0}\right)TE_o\, e^{-i(k-q)d} \tag{24}$$

$$B = \frac{1}{2}\left(1 - \frac{Z}{Z_0}\right)TE_o\, e^{-i(k+q)d} \tag{25}$$

On solving Equations 23, 24 and 25 and inserting the value of q, the transmission amplitude T can be written as

$$\frac{1}{|T|^2} = \frac{1}{4}\beta^2\left[\cosh\left(\frac{2d}{\delta}\right) - \cos\left(\frac{2d}{\delta}\right)\right] + \frac{1}{2}\beta\left[\sinh\left(\frac{2d}{\delta}\right) + \sin\left(\frac{2d}{\delta}\right)\right] + \frac{1}{2}\left[\cosh\left(\frac{2d}{\delta}\right) + \cos\left(\frac{2d}{\delta}\right)\right] \tag{26}$$

where δ is the skin depth of the shield, which is a measurement at which the intensity of the incident wave remains 1/e times the original. It decreases with increasing frequency like $\omega^{-1/2}$. For all practical shields, the thickness of the shield should be a bit larger than the skin depth of the shield. At frequencies low enough, d << δ, Equation 26 reduces to

$$\frac{1}{|T|^2} = \left(\frac{\beta d}{\delta}\right)^2 + 2\left(\frac{\beta d}{\delta}\right) + 1 \quad = \left(\frac{\beta d}{\delta} + 1\right)^2 \quad = \left(\frac{1}{2Zd\sigma} + 1\right)^2 \tag{27.1}$$

Therefore, at the low frequency, the shielding effectiveness is independent of frequency and is given by:

$$SE(dB) \cong 20.\log\left(1 + \frac{1}{2}Z_0 d\sigma\right) \tag{28.1}$$

Similarly, at very high frequency, d >> δ, Equation 27 can be reduced to

$$\frac{1}{|T|^2} \cong \frac{1}{8}\beta^2 e^{\frac{2d}{\delta}} \tag{29}$$

Using the above equation, shielding effectiveness comes out to be independent of frequency

$$SE(dB) \cong 10.\log\left(\frac{\sigma}{16\epsilon\omega}\right) + 20.\log_e d \sqrt{\frac{\sigma\mu\omega}{2}} \qquad (30)$$

The first term of the right-hand side represents the reflection of the incident wave, while the second term corresponds to the absorption of the incident wave within the shield. Fig. (**1.7**) shows Formulas 24 and 28 plotted against the frequency range from 1 MHz to 10 GHz.

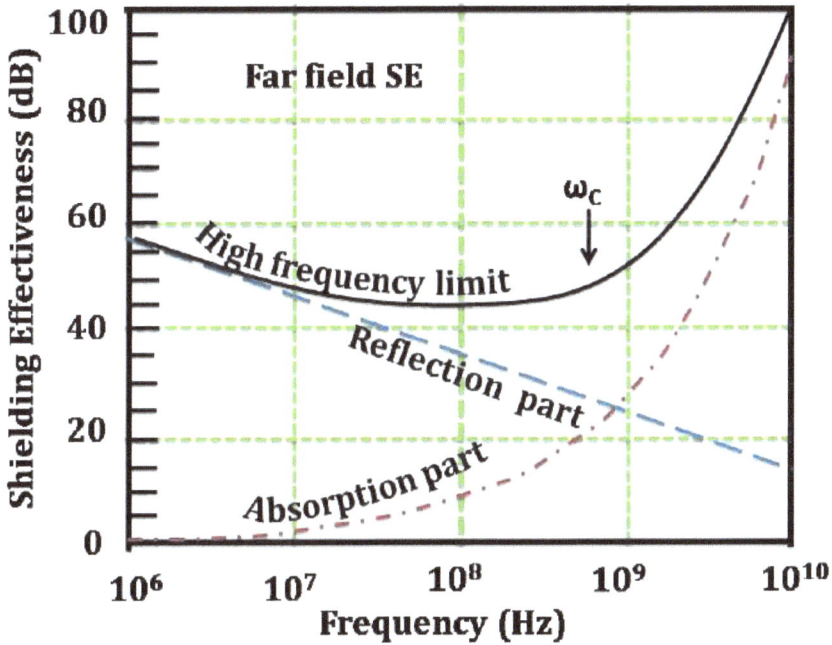

Fig. (1.7). Theoretical predictions of shielding effectiveness with frequency.

For typical calculation, the thickness, d, of the shield was taken as 3 mm, and its electrical conductivity σ was 3 S/cm. Actual far-field SE calculated using Equation 26 was found to be independent of frequency, as shown by the straight solid line. The two dotted lines are shown to the two terms on right side of Equation **28.1**. Curves plotted using Equations **28.1** and 24 converges at the frequency ω_c, where the skin depth δ is equal to the shield thickness d. In a simple way, the analysis of the shielding efficiency of the thin planar barrier can be extended to check the effectiveness of two parallel planar barriers with gaps. According to the specific values of the parameters describing this system (barrier thickness, barrier distance, signal frequency, barrier conductivity, *etc.*), gaps between barriers can cause interference effects that improve or reduce the shielding effect.

1.8. SHIELDING EFFECTIVENESS MEASUREMENT

The methods to measure the shielding effectiveness of various electromagnetic shields like a traditional conductive metallic shield or currently designed multiphase composite shields enriched with high dielectric and magnetic attributes include a variety of standard methods. However, the standardized methods are not always useful as they require well-defined size and shape of the samples that are sometimes impossible to design. It is, therefore, important to design adapted measuring equipment based on standard techniques that can be used to measure complex enclosures or other materials of a complex shape and size. An output value obtained using one measurement method is normally difficult to compare with the value is produced by another measurement method.

For example, a measurement method in which a shielding material sample is placed inside a coaxial transmission line fixture would not necessarily expose the sample to the same impedance of the electromagnetic field as a method in which an antenna radiates an incident plane wave. Transferred impedance data does not offer itself for full-scale enclosure tests due to correlation issues. In general, a shielding efficacy value can be obtained using the following testing methods:

1.Shielded Box Method
2.Open Area or Free Space Method
3.Shielded Room Enclosures
4.Coaxial Transmission Line Method

1.8.1. Shielded Box Method

The shielded box method is often used for measuring the relative shielding effectiveness of similar materials. The test consists of a metal box and an electrically tight seam with a sample port in one wall and receiving antenna in the opposite wall, as illustrated in Fig. (**1.8**). The transmitting antenna is outside the box, and the antenna's signal strength is recorded through open ports and test samples installed over the ports [49]. The disadvantage of this technique is that electrical contact is very poor between the test sample and the shielded box. Another problem is the limited frequency range of around 500 MHz. Results from different research laboratories show a meagre connection.

1.8.2. Open Area or Free Space Method

The open area method or the free space method is used to assess the actual shielding effect of the entire device or electronic assembly. Therefore, this test measures the radiated emissions that are emitted from the product. Shielding effectiveness measurement can be represented in a block diagram, as shown in Fig. (**1.9**). The test does not measure the performance of any specific material, and the result can vary because of differences in the assembly of individual products. The distance between the specimen under test and the antenna should be typically 3 m, 10 m, or 30 m. The measuring distance is important so that the electric field strength is measured in the far-field and not in the near field. At 30 MHz, the wavelength is 10 m. When approaching the near-field or Fresnel area (the area between the near-field and the far-field), the electric field becomes unstable, and the measurement accuracy decreases. Some standards require a particular separation, while others allow two or more different separations. Since the strength of the electromagnetic field changes with distance, the limit value is recalculated with each measurement interval.

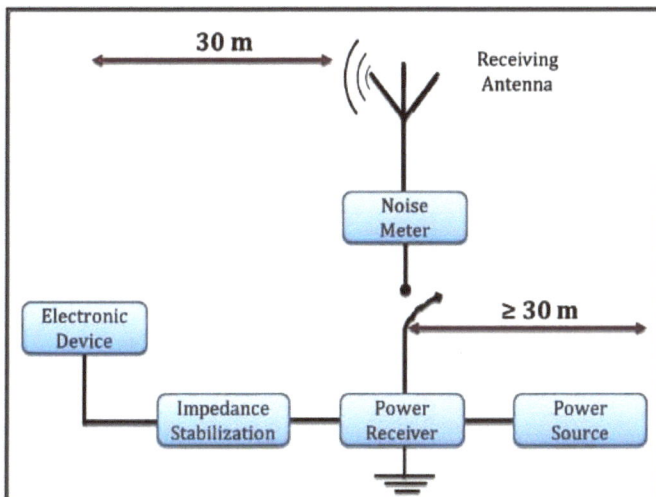

Fig. (1.8). Shielding Effectiveness measurement by shielded box.

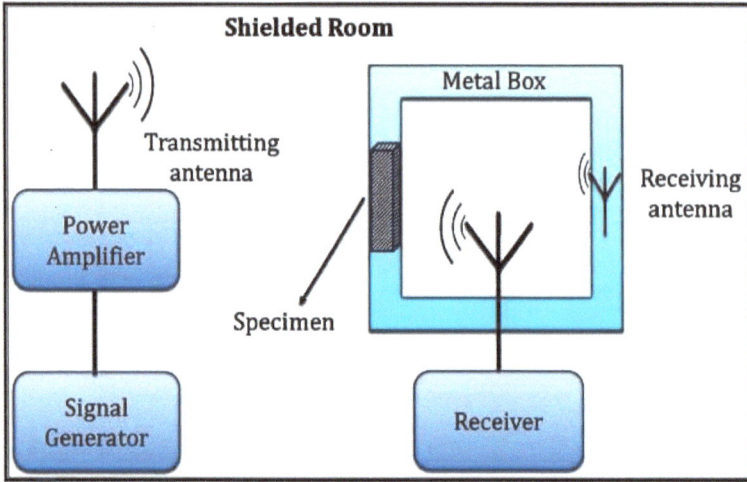

Fig. (1.9). Open area method of shielding effectiveness measurement.

1.8.3. Shielded Room Enclosures

The shielded room method is the most complicated method that was developed to conquer the limitations of the shielded box method. Its principle is the same as the shielded box method. The measuring system, the signal generator, the receiving antenna, the transmitting antenna, and the recorder components are individually isolated in shielded rooms to abolish the possibility of interference. Furthermore, the antennas are placed in room-sized anechoic chambers. The sample size is relatively as large as 2.5 square meters, as shown in Fig. (**1.10**). In this way, results are more reliable and reproducible in the same frequency domain as compared to the shielded box method.

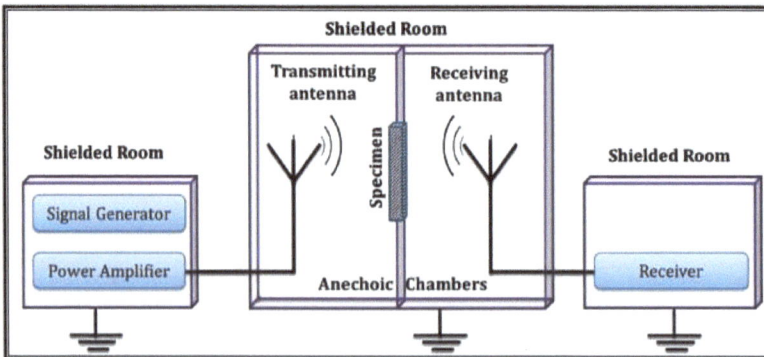

Fig. (1.10). Typical shieling effectiveness measurement set up.

American military standard MIL-STD-285 is one of the standard methods, probably used for all commercial and industrial shieling measurements, as illustrated in Fig. (**1.11**). During its modifications, it is permitted to move the test equipment and antennas around and repeat the test for best results. Following are some of the most used test techniques for shielding efficiency. Upgraded versions of the method MIL-STD-285 have been established to minimize the difficulties with reflections by using EM wave absorbing. The shielded room enclosure is a well-sealed metal box that offers magnetic and electric field shielding effectiveness over a specified range of frequencies. The absorber material is carefully chosen to ensure appropriate attenuation across a wide frequency range; otherwise, reflections or standing waves may occur. New variants, such as MIL-G-83528, are included in the IEEE-STD-299 (1997) standard, which covers a frequency range of a few megahertz to 18 GHz material in the shielded rooms enclosures.

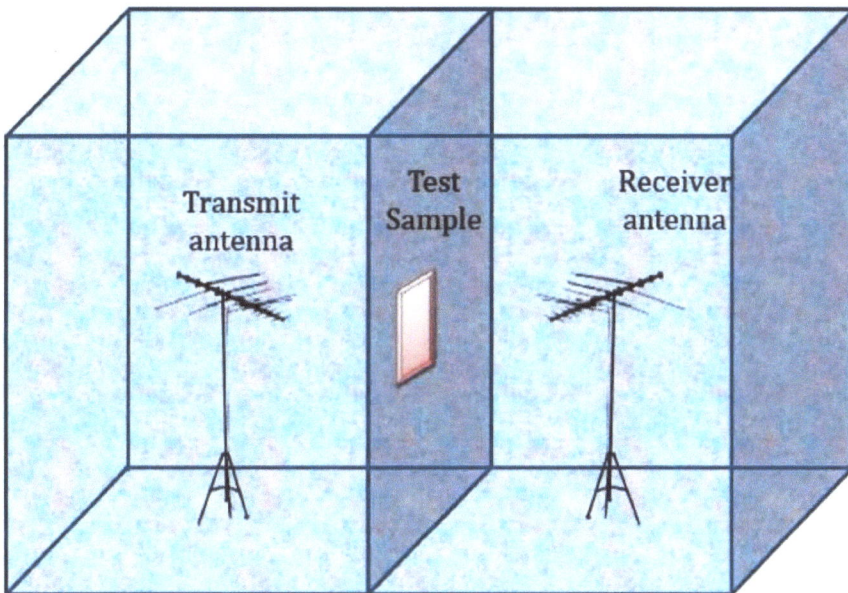

Fig. (1.11). Shieling effectiveness measurement using standard military standard MIL-STD-285.

1.8.4. Coaxial Transmission Line Method

The method of measuring the shielding effectiveness using a coaxial transmission line conquers the limitations of shielded box techniques and is currently the first choice [50]. The main advantage of this method is that the results obtained by different laboratories are approximately the same. In addition, the coaxial transmission line provides reflective, absorbent, and transparent components. The measurement can be made using a spectrum analyzer, which receives the signal of specific frequency generated by a crystal detector, modulated signal generator, and tuning amplifier. Step by step procedure is as follows; the system is initially calibrated to a specific frequency without the sample holder in the line. For this, the variable attenuator is set to maximum, and the signal level is recorded. Then, insert the specimen holder in the line. When the specimen is being tested, the attenuators get reduced until the previous reading is recorded. In this way, the attenuation of the signal achieved is a direct measurement of the shielding effectiveness of the specimen. Various frequency signal is applied to the specimen to obtain a wide spectrum. This method is time consuming and the generation of the spectrum takes several hours. To overcome this, in sweep mode, a tracking generator driven by a spectrum analyzer is used in place of the generator. The spectrum analyzer gives the response as a function of the frequency of the specimen as a single curve on display in a few seconds. The coaxial transmission line technique [50, 51] has been approved as a standard method for evaluating the shielding efficiency of flat specimens by the American Society for Testing and Materials (ASTM D4935-99). Small torus-shaped cells are typical equipment (internal diameter of 50 mm and external diameter of 125 mm). The contact resistance between the circular sample and the holder (thickness 1 mm) must be less than 0.2 cm.

1.8.5. Vector Network Analyzer

The network analyzer is one of the essential tools for analyzing analogue circuits. By computing the amplitudes and phases of transmission and reflection coefficients in an analogue circuit, a network analyzer displays all of the circuit's network properties. Microwave engineers utilize network analyzers to test a wide range of materials, components, circuits, and systems. A computation of the reflection from and/or transmission through a material, coupled with information on its physical dimensions, provides the data needed to define the material's permittivity and permeability. PNA, PNA-L, ENA, and ENA-L vector network analyzers sweep high-frequency stimulus-response data from 300 kHz to 110 GHz or even 325 GHz. A receiver, a signal source, and a display make up a vector network analyzer (Fig.

1.12). The source sends a signal to the substance under test at the required single frequency. The receiver is set to that frequency in order to detect reflected and transmitted signals from the material. The magnitude and phase data are produced by the measured response at that frequency. The source is then switched to the next frequency, and the test is repeated to illustrate the reflection and transmission measurement response as a function of frequency.

At high frequencies, simple components and connecting wires that function effectively at low frequencies react differently. At microwave frequencies, wavelengths become tiny in comparison to the physical dimensions of the devices, resulting in a large phase difference between two closely spaced locations. Transmission line theory must be used to assess the behavior of devices at higher frequencies, replacing low-frequency lumped-circuit element methods. High-frequency issues like radiation loss, dielectric loss, and capacitive coupling complicate microwave circuits.

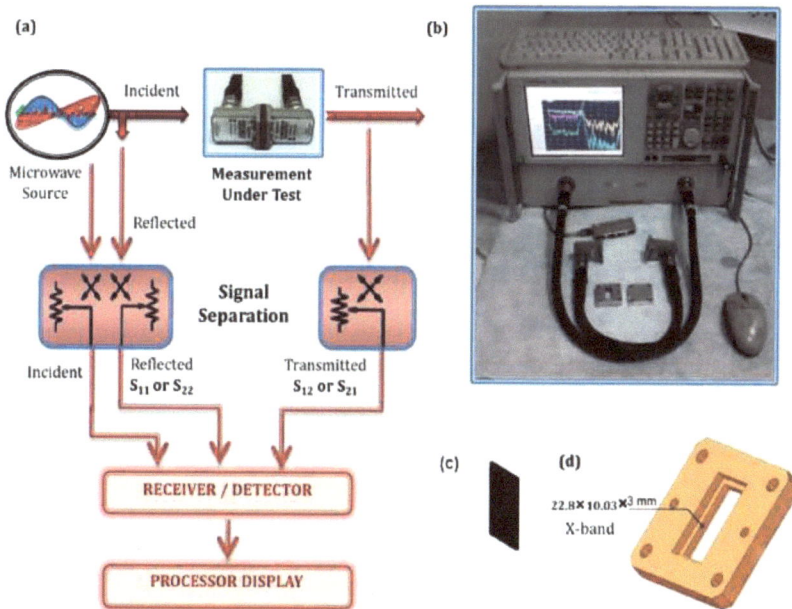

Fig. (1.12). (a) Block diagram of a vector network analyzer, **(b)** Agilent E8362B: vector network analyzer, **(c)** rectangular shaped sample and **(d)** sample holder {Adapted from Singh *et al.* [21], copyright 2015, One Central Press (OCP)}.

In the microwave range, an Agilent E8362B Vector Network Analyzer was used to evaluate electromagnetic shielding, dielectric, and permeability (X-band). Powder samples were crushed into rectangular pellets (2 mm thick) and put into a copper sample container with a volume of $22.86 \times 10.14 \times 6$ mm^3 that was linked between the network analyzer's wave-guide flanges. To account for any loss or power redistribution caused by the sample holder, a full two-port calibration was conducted simultaneously with the sample holder.

There are different approaches for obtaining S-parameter permittivity and permeability. In order to determine dielectric properties, Table **1.1** provides a summary of the conversion techniques using different sets of S-parameters. There are distinct benefits and disadvantages of each conversion technique. The choice of technique depends on many factors, such as the S-parameters measured, the thickness of the sample, the dielectric properties required, the speed of conversion and the accuracy of the results converted. The details of these methods are given below.

Table 1.1. Comparison between the conversion techniques.

Conversion Technique	S-parameters	Dielectric Properties
Nicholson Rose Weir	$(S_{11}, S_{21}, S_{12}, S_{22})$ or a pair $(S_{11}, S_{31},)$	ε_r and μ_r
National Institute of Standards & Technology iterative	$(S_{11}, S_{21}, S_{12}, S_{22})$ or a pair $(S_{11}, S_{31},)$	ε_r and $\mu_r = 1$
New non-iterative	$(S_{11}, S_{21}, S_{12}, S_{22})$ or a pair $(S_{11}, S_{31},)$	ε_r and $\mu_r = 1$
Short Circuit Line	S_{11}	ε_r

1.8.6. Nicholson-Ross-Weir Method

Direct measurement of both permeability and permittivity from the S-parameters is given by the Nicholson-Ross-Weir (NRW) process [52, 53]. The Nicholson-Ross-Weir method calculates the permeability and permittivity directly from the S-parameters. It is the most widely used method for such conversions to be carried out. Reflection coefficient and transmission coefficient calculation involves the measurement of all four (S_{11}, S_{21}, S_{12}, S_{22}) or a pair (S_{11}, S_{21}) of S parameters of the material under test being studied.

For low loss materials, however, the technique diverges at frequencies corresponding to integer multiples of half a wavelength in the sample due to phase uncertainty. It is, therefore, limited to the optimal sample thickness of g/4 and used ideally for short samples. The procedure proposed by the NRW method is deduced from the following equations:

$$S_{11} = \frac{\Gamma\left(1 - T^2\right)}{\left(1 - \Gamma^2 T^2\right)} \text{ and } S_{21} = \frac{T\left(1 - \Gamma_2\right)}{\left(1 - \Gamma_2 T_2\right)}$$

The network analyzer can provide these parameters directly. The reflection coefficient is calculated as follows:

$$\Gamma = X \pm \sqrt{X^2 - 1} \tag{31}$$

where, $\left|\Gamma_1\right| < 1$ is required for finding the correct root and in terms of S-parameter

$$X = \frac{S_{11}^2 - S_{21}^2 + 1}{2S_{11}} \tag{32}$$

The transmission coefficient is calculated as follows:

$$T = \frac{S_{11} + S_{21} - \Gamma}{1 - (S_{11} - S_{21})\Gamma} \tag{33}$$

The permeability is given as:

$$\mu_r = \frac{1 + \Gamma_1}{\Lambda(1 - \Gamma)\sqrt{\dfrac{1}{\lambda_o^2} - \dfrac{1}{\lambda_c^2}}} \tag{34}$$

where λ_o is the free space wavelength and λ_c is the cutoff wavelength and

$$\frac{1}{\Lambda^2} = \left(\frac{\varepsilon_r \mu_r}{\lambda_o^2} - \frac{1}{\lambda_c^2}\right) = -\left[\frac{1}{2\pi L}\ln\left(\frac{1}{T}\right)\right]^2 \tag{35}$$

The permittivity can be defined as

$$\varepsilon_r = \frac{\lambda_o^2}{\mu_r}\left(\frac{1}{\lambda_c^2} - \left[\frac{1}{2\pi L}\ln\left(\frac{1}{T}\right)\right]^2\right) \tag{36}$$

Equations 33 and 34 have an unlimited number of roots since the imaginary part of the term $\ln(1/T)$ is equal to i $(\theta + 2\pi n)$, where n= 0, ± 1, ± 2..., the integer of (L/λ_g). There are two ways for determining n.

The first technique is to analyze group delay, and the second is to estimate from λ_g using the sample's initial values for μ_r^* and ε_r^*. Permittivity may be calculated using these approaches as follows:

$$\varepsilon_r = \mu_r\left(\frac{1 - \Gamma}{1 + \Gamma}\right)^2\left(1 - \frac{\lambda_o^2}{\lambda_c^2}\right) + \frac{\lambda_o^2}{\mu_r\lambda_c^2} \tag{37}$$

where L = length of materialλ_g = wavelength in sample

μ_r = relative permeabilityε_r = relative permittivity

μ_r^* = initial guess permeability e_r^*= initial guess permittivity

γ= propagation constant of material c = velocity of light

f = frequency

1.8.7. NIST Iterative Technique

The calculation is done using the root-finding technique of a Newton-Raphson NIST Iterative technique and is only suitable for permittivity calculation. To measure the reflection and transmission coefficient, it utilizes all four (S_{11}, S_{21}, S_{12}, S_{22}) or a pair (S_{11}, S_{21}) of S-parameters of MUT. When a good initial value of approximation is available, it works well. When the sample thickness is an integer multiple of half a wavelength ($n\lambda/2$), the approach avoids the inaccuracy peaks that occur with the NRW method. It is suitable for analyzing lengthy samples with minimal loss content.

The novel non-iterative technique is quite similar to the NRW approach but has a different formulation and is perfect for determining permittivity for the situation μ_r = 1. To measure the reflection and transmission coefficients, it utilizes all four (S_{11}, S_{21}, S_{12}, S_{22}) S-parameters or only two (S_{11}, S_{21}) S-parameters of MUT. The method has the benefit of being resilient over a wide range of frequencies for any sample length. The method is based on a reduced form of the NRW technique, and no divergence is detected at half-wavelength multiples of the sample frequencies. It does not need an initial permittivity estimate and can conduct the computation quickly.

1.8.8. Short Circuit Line (SCL) Technique

 A one-port calculation on coaxial lines or waveguides is the short circuit line (SCL) technique. Using the same Newton-Raphson numerical method as in the NIST iterative methodology, it performs the measurement and is only suitable for measuring permittivity. It uses only the calculation under test (MUT) parameter S_{11} to determine the coefficient of reflection. In order to achieve an accurate result, the procedure needs a reasonable initial estimate value. For precise measurements, the technique also includes the input of sample length and position.

An effective measurement and conversion technique is needed in order to measure the dielectric properties of the material. The correct methodology for the substance to be tested must be used since particular techniques are relevant to specific materials. The calculation results will not be accepted if the incorrect technique is used.

1.8.9. Dielectric Measurement Techniques

Measuring the complex dielectric properties of materials is important because it can provide materials with electrical or magnetic properties that are useful in many areas. Dielectric property calculation requires measurements of the materials' complex relative permittivity (ε_r) and complex relative permeability (μ_r).

In order to measure the complex permeability and permittivity, several techniques are developed, such as open-ended coaxial probe technique, transmission/reflection line technique, free space technique and resonant technique. The reflection/transmission line technique is commonly used by all of them. Only the fundamental waveguide mode (TEM mode in coaxial line and TE mode in waveguides) is believed to propagate in this technique. A calculation using the reflection/transmission line approach involves putting a sample in a waveguide or coaxial line segment and calculating complex dispersion parameters with a vector network analyzer for the two ports [52-55]. Before making the measurement, calibration must be carried out. The technique includes the reflected (S_{11}) and transmitted signal (S_{21}) calculation. The specific scattering parameters are closely related to the complex permittivity and equation-based permeability of the material. By solving the equations using a program, the conversion of S-parameters to complex dielectric parameters is computed. In certain situations, such as machining, the method involves sample preparation so that the sample fits securely into the waveguide or coaxial line. Transmission line measurement calibrations use various terminations that generate distinct resonant activity in the transmission line. The maximum electric field is needed for good dielectric measurement, which can be accomplished by an open circuit or other capacitive termination, while calibration can be done using either short circuit, open circuit, or matched load termination in coaxial line measurements. Permittivity and permeability of the dielectric substance may be measured using this approach.

The material to be tested (MUT) is placed in a sample holder after the VNA has been calibrated at the connection calibration plane (Fig. **1.13**). To decrease the measurement uncertainty induced by air gaps, the MUT must fit tightly in the sample holder. Two methods may extend the calibration plane to the sample surface. In the first step, the phase factor is manually fed, which is equal to the distance between the calibration plane of the sample surface and the connector. With the characteristics of the VNA, the phase factor can easily be included in the calculation. The VNA will transfer the calibration plane to the surface of the MUT from the connector.

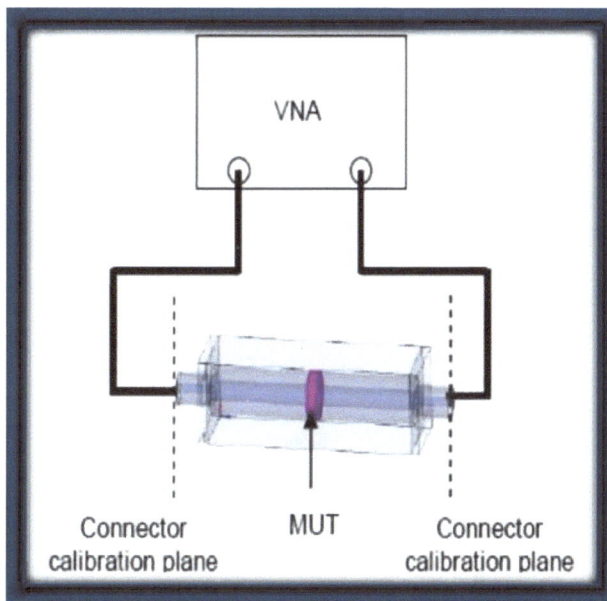

Fig. (1.13). Measurement using transmission, reflection method with a waveguide.

The second phase involves the VNA's de-embedding feature. After the calibration has been completed, the method involves measuring the S-parameter of an empty sample holder. The empty holder's calculated S-parameter is then entered into the network analyzer. The effect of the sample holder on the actual material measurement can be balanced out using the de-embedding feature in the VNA.

1.9. FACTORS AFFECTING THE EMI SHIELDING PERFORMANCE

The EMI performance of the materials depends on various factors such as:

- Permittivity and conductivity

- Permeability and magnetization

- Attenuation constant and Eddy current losses

- Anisotropy, domain wall displacement, resonance, and Snoek's limit

- Morphology

- Thickness

- Impedance matching

1.9.1. Permittivity and Conductivity

The shielding performance is strongly influenced by the permittivity and conductivity of the material. According to the classical electromagnetic theory, shielding effectiveness for the thick shield of thickness 't' having ac conductivity 'σ_s' and magnetic permeability 'μ' can be given as

$$SE\,(dB) = 20\frac{t}{\delta}log\,e + 10log\left(\frac{\sigma_s}{16\omega\varepsilon_0\mu_r}\right) \qquad (38)$$

where δ is skin depth, and ω is the angular frequency. The first term in Equation 36 is the shielding effectiveness due to absorption, and the second term is due to the reflection of the electromagnetic wave.

$$SE_A\,(dB) = 20\frac{t}{\delta}log\,e = 20t\sqrt{\frac{\mu_r\omega\sigma_s}{2}}log\,e \qquad (39)$$

$$SE_R\,(dB) = 10\log\left(\frac{\sigma_s}{16\omega\varepsilon_0\mu_r}\right) \qquad (40)$$

From Equations 37 and 38, it can be seen that the shielding effectiveness due to absorption is directly proportional to the square root of conductivity, whereas shielding effectiveness due to reflation is also directly proportional to the log of conductivity. Therefore, the shielding performance increases with an increase in the conductivity of the material.

The ac conductivity 'σ_s' can be correlated with the permittivity of material as

$$\sigma_s = \omega\varepsilon_0\varepsilon''$$ **(41)**

where, ε'' is the imaginary part of the complex relative permittivity ($\varepsilon_r = \varepsilon' - j\varepsilon''$). Equation 39 indicates that the permittivity is dependent on conductivity. The higher value of permittivity leads to higher conductivity resulting in improved shielding performance. The permittivity can be called an inherent factor for shielding performance. Qualitatively, the real part of complex permittivity (ε') is signifying the storage ability in terms of various polarization such as electronic, orientational, ionic, space charge polarization, *etc.* The imaginary part of permittivity (ε'') measures the dissipation of energy due to conduction phenomena, as shown in Fig. (**1.14**). Thus, if the material possesses higher permittivity, it will absorb more energy from the incident electromagnetic wave, which will be partially utilised in the polarisation and conduction processes, leading to an increase in the shielding performance.

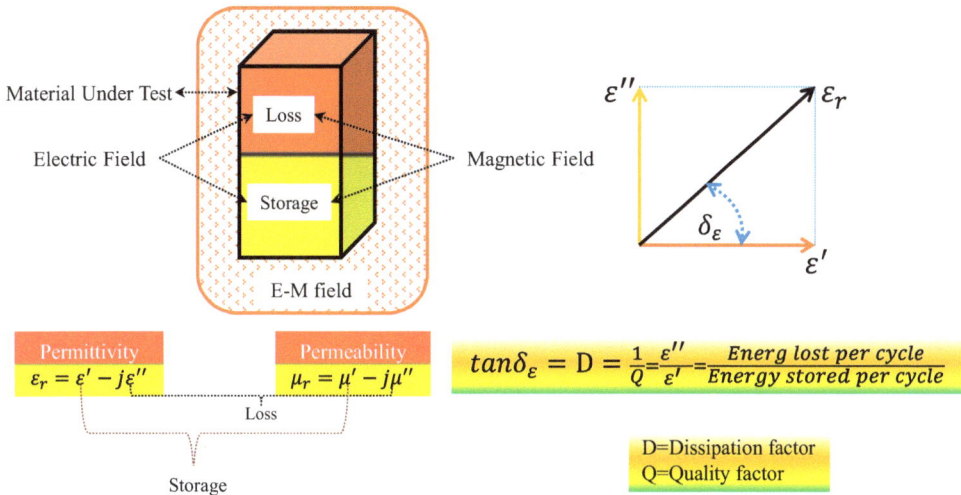

Fig. (1.14). Schematic demonstration of electric and magnetic energy dissipation in the material.

1.9.2. Permeability and Magnetization

As seen from Equations 37 and 38, shielding effectiveness due to absorption is directly proportional to the square root of the permeability, and shielding effectiveness due to reflection is inversely proportional to the log of permeability. Statistically, the square root value of a term is greater than its logarithm, which shows that the EMI shielding performance is mainly dominated by the shielding due to absorption rather than reflection. For example, according to these equations, a material having higher permeability along with the conductivity results in higher absorption of electromagnetic waves, *i.e.*, shielding due to absorption is higher and shielding due to reflection becomes weak according to Equation 38. Therefore, the overall shielding performance significantly increases with the permeability properties of the material.

For qualitative analysis, permeability can be written in terms of the real and imaginary parts as $\mu_r = \mu' - j\mu''$. The real (μ') and imaginary (μ'') parts measure the extent of magnetic energy storage and energy dissipation capacities of the material, as shown in Fig. (**1.14**). Therefore, a material with more permeability utilizes more energy in storage and in dissipation which in turn enhances shielding performance.

Both the terms, permeability and magnetization, are correlated as $M = \mu H$, where M is magnetization, H is the magnetic field strength and μ is the characteristic of the medium/material. The higher value of permeability leads to a higher value of magnetization, *i.e.*, magnetic polarization, on account of the external field. Therefore, magnetization is the parameter that favors the shielding performance through the absorption of electromagnetic waves.

1.9.3. Attenuation Constant and Eddy Current Losses

According to electromagnetic theory, during the propagation of the electromagnetic signal through a non-ideal medium, the amplitude, as well as phase of the signal, is changed, and this measure in terms of propagation constant () can be stated as:

$$\gamma = \alpha + i\beta \tag{42}$$

where α & β are attenuation constant and phase constant, respectively.

The attenuation constant measured change in amplitude of the signal and the phase constant measured the change in phase with the propagation of the signal. Attenuation constant can be expressed as:

$$A_x = A_o e^{-\alpha x} \tag{43}$$

where A_x & A_o are amplitude of signal after a propagation distance x and initial amplitude of the signal, respectively. According to Equation 41, as the electromagnetic wave penetrates the material, the amplitude of the waves decreases exponentially. The amplitude decreases more rapidly with an increase in the value of the attenuation constant. Attenuation constant is the intrinsic property of the material because it is the function of permittivity and permeability of the material, which can be expressed as:

$$\alpha = \frac{\sqrt{2}\pi f}{c}\left\{(\mu''\varepsilon'' - \mu'\varepsilon') + \sqrt{(\mu''\varepsilon'' - \mu'\varepsilon')^2 + (\mu'\varepsilon'' + \mu''\varepsilon')^2}\right\}^{1/2} \tag{44}$$

where c is the velocity of light. The attenuation constant factor contributes to the shielding mechanism due to the absorption part rather than reflection. Thus, the balance values of permittivity and permeability lead to a higher attenuation constant resulting in higher shielding performance that can be achieved from a thin shield. According to Faraday's law, time-varying magnetic flux induces emf, and Ohm's law emf induces the current further. The induced current is known as Eddy current, and the process may be called as Eddy current effect. Such type of induced current does not require a physical medium, resulting in Joule heating that gives rise to a large waste of energy. Energy loss produced due to the Eddy current effect can be examined by using Equation 43:

$$C_o = \mu''/(\mu'^2 f) = {}^2\!/_3\,\pi\mu_0 t^2 \sigma \tag{45}$$

where C_o is a magnetic loss parameter. If the energy loss is contributed by the Eddy current effect, then C_o should remain static with the change in frequency; in reverse, if C_o changes with change in the frequency, then magnetic loss may be recognized with other effects. Thus, the Eddy current effect is advantageous for the shielding performance.

1.9.4. Anisotropy, Domain Wall Displacement, Resonance and Snoek's Limit

As the natural resonance frequency of uniaxial anisotropic materials is proportional to their anisotropy fields, the hard ferrites and their composites have large anisotropy to be used as an absorber in the microwave range. Therefore, higher isotropy may be favored for the shielding effectiveness in the higher frequency range.

A domain wall is an interface between domains of different orientations/ polarization. In the presence of the time-varying magnetic field, the domain got displaced, resulting in a change in the phase of the magnetic moment. The occurrence of phase may be caused due to the magnetic losses in the material. The thickness of the domain wall is dependent on the anisotropy of the material. The width of the domain wall is reduced, and lower anisotropy is required to align the magnetic moments along the crystals' lattice axis.

Resonance is the state of the system, and at this stage, the state is able to absorb energy easily. Thus, the natural frequency of charge carried in the material is comparable to the field where it absorbed more energy from the incoming electromagnetic radiations, resulting in improved shielding performance due to absorption phenomena.

The permeability of the material can be correlated to the anisotropic field as:

$$\mu = \frac{M_S}{H_a} \tag{46}$$

The anisotropy field is HA, while the saturation magnetization is Ms. The permeability remains constant at lower frequencies up to the critical frequency, then begins to decline as the frequency is increased. According to Snoek's law, a higher amount of static permeability results in a lower critical frequency. The frequency of this precession is determined by the coupling of magnetization with the c-axis. The greater the contact, the higher the natural precession frequency. The greater the interaction, the higher is the anisotropic field (HA) or magneto crystalline anisotropy.

1.9.5. Morphology

Morphology is related to the shape, size, and texture of grains in the material. Therefore, morphology is correlated with intrinsic properties (*i.e.*, conductivity, permittivity and permeability) of the material. It is reported that the C-SiC/epoxy composites flake-like texture produces strong interfacial polarization leading to improved electrical as well as electromagnetic shielding performance of the composite [56]. The real part of permittivity of the composite is caused by the interfacial polarization, which is attributed to the heterogeneity of phase that occurred during the incorporation of the filler. Interfacial polarization occurs at the interface of two different phases having different properties, supporting the migration of charges carriers through different phases of composites. If the material possesses porous morphology, it is reported that such composites exhibit higher polarization resulting in higher dielectric properties. Thus, morphology is also an important parameter that affects the shielding performance of material by controlling its intrinsic properties.

1.9.6. Effect of Thickness

The thickness of the sample directly influences the shielding performance due to absorption phenomena, as seen from Equation 37 and the reflection phenomenon is independent of the thickness of the material. For a clear understanding of the concept, the electromagnetic shielding due to absorption phenomena can be correlated with the concept of attenuation constant. From Equation 38, the amplitude of incident electromagnetic wave decreases as x increase, *i.e.*, more propagation distance in the material from its surface, indicating more thickness of the shield.

1.9.7. Impedance Matching

Another important parameter for the shielding performance is the impedance matching characteristics of the materials. It depends on the suitable values of permittivity and permeability of the material approaching the requirement of impedance matching that leads to improving the absorption of EM waves. Impedance matching may be examined from the ratio of the input impedance of the material (Z_{in}) to the impedance of free space (Z_o), *i.e.*, $Z_{in}/Z_o \propto \sqrt{\mu_r/\varepsilon_r}$, depending on the ratio of permittivity and permeability. Thus, if a material possesses either permittivity or permeability, then it is disadvantageous for the absorption of EM waves; conversely, the material strongly reflects the EM waves. The reflection loss (R) can be expressed as:

$$R = \frac{(Z_{in}-Z_o)}{(Z_{in}+Z_o)} \tag{47}$$

From Equation 45, it can be observed that if $Z_{in} = Z_o$, then R approaches zero. Thus, the material that has balanced dielectric and magnetic properties provides improved impedance matching properties that assist the absorption of EM waves.

1.9.8. Materials for EMI Shielding

On the basis of the points mentioned in section 1, the material can be categorized that are advantageous for the EMI shielding application as:

- Conductors

- Semiconductors

- Magnetic materials

- Dielectric composites

- Conjugated polymers

- Dielectric-conductor composites

- Magnetic materials – polymer composites

1.9.8.1. Conductors

Metals are good conductors of electric charge and are widely used for shielding application due to their electrical conductivity as well as thermal conductivity. As per Equations 36 and 37, metal conductors provide effective shielding as absorbers and reflectors due to their higher conductivity. Metal can be used in bulk form or coating form as a shield for the shielding effect from electromagnetic interference. For example, an aluminum sheet is used to build the enclosure for the electronic device for its protection against EMI. Similarly, Mu-metal, an alloy composed of iron (14 %), copper (5 %), chromium (1.5 %) and nickel (79.5 %), are the common metallic material that are used for the EMI shielding. On the other hand, silver, brass, nickel, steel, aluminum, *etc.*, are also used as shielding materials due to their higher conductivity. But their susceptibility for corrosion and high density are the main drawbacks in some applications of shielding.

1.9.8.2. Semiconductor

A semiconductor is a substance with an electrical conductivity that is somewhere between that of a conductor and that of an insulator. The band gap of traditional semiconductors is between 1 and 1.5 eV, but the band gap of wide band gap semiconductors is between 2 and 4 eV. It should be noted that the value of a material's band gap is influenced by particle size as well as the phase of the substance [57]. By changing the ratio of the precursors, He *et al.* [58] produced composites with medium band gap fillers like CuS nanostructures and discovered that the change in nanostructure resulted in the change in EMI shielding performance. The lowest R_L value for the 20 wt. % loading of CuS in paraffin composite was observed to be 76.4 dB at 12.64 GHz for a shield thickness of 3.5 mm. The microwave absorption behavior is attributed to CuS dielectric losses and interfacial polarization due to the composite's heterogeneous structure. Hu *et al.* [34] used solvothermal synthesis to create a flower-like CuS with a band gap of 1.45 eV. High band gap fillers in composites made of $MoO_3/Mo_4O_{11}/MoO_2$ heterogeneous nanobelts (60 wt. %) were produced and integrated into the paraffin matrix by Lyu *et al.* [59], which at a thickness of 4.6 mm has an R_L value of -59.2 dB at 10.8 GHz.

1.9.8.3. Magnetic Materials

Magnetic materials are excellent for shielding applications, especially when the absorption of electromagnetic radiation is required. As per Equation 37, due to the high permeability, the magnetic material performs better as an electromagnetic absorber. Mu-metal, permalloy, stainless steel (No. 430), *etc.*, are example of materials that possess higher permeability. But the synergistic use of magnetic material with carbon material has been found to be an effective shielding material. For example, the magnetic material FeNi along with the carbon nanotubes is a more efficient filler in a polymer matrix instead of the magnetic particle alone or carbon nanotubes alone [60].

1.9.8.4 Dielectric Composites

Dielectric materials showing polarization behavior in response to the electric field are also used as shielding materials due to dielectric properties. The dielectric response is correlated with conductivity, and conductivity is the main element of shielding performance, as indicated by Equation 37. Carbon materials such as graphene, carbon nanotubes, and conducting polymers exhibit higher permeability and are the representative material useful for shielding application. Dielectric composites represent a class of heterogeneous system of multi-constituent materials, in which one material act as host while others act as filler materials to yield the desirable properties. For the dielectric composites, the host and filler materials have a combination of dielectric-dielectric phases. Generally, the dielectric composite structure consists of a multilayered structure due to the incorporation of different fillers resulting in a heterogeneity structure. These types of structures assist the dielectric properties of the materials. It is reported that the PEDOT polymer incorporated with graphene shows improved shielding effectiveness due to absorption because more incorporation of graphene in polymer matrix improves the dielectric properties of the composite, leading to better shielding properties.

1.9.8.5. Conjugated Polymers

Conjugated polymers are another emerging class of materials mainly due to their controllable electrical attributes and dielectric properties. The conjugated polymer-based shielding material is a rapidly growing field of research and acts as an alternative to the metallic shield due to its anticorrosion properties and light weight. They can reduce the problem of shielding due to their higher conductivity and relative permittivity. These parameters can be easily controlled by controlling the concentration of filler in the polymer and dopant concentration. In the past decades, numerous reports on the conjugated polymers composites have been published, showing high shielding efficiency.

1.9.8.6. Dielectric-Conductor Composites

These types of composites are synergistically composed of dielectric materials and conductors. The dielectric material can be chosen from the polymers and other carbon allotropes; also, it can be a metallic oxide. Nickel-coated carbon fiber reinforced polypropylene (dielectric-conductor) composite has been designed and the composite had improved electrical properties and electromagnetic shielding performance due to enhanced dielectric properties. PC/ABS/nickel-coated carbon fiber electrical conducting composites have been synthesized, which has shielding effectiveness of 47 dB at 1.0 GHz, which can be attributed to improved dielectrics properties. The dielectric properties are responsible for various phenomena related to energy loss that results in improved shielding performance of the material.

1.9.8.7. Magnetic Parameters

Fig. (**1.15**) shows a hysteresis loop that may be used to characterize magnetic materials and calculate different characteristics. The saturation polarization, Js, and hence the saturation magnetization, (Ms), may be measured from the first quadrant. The majority of relevant information, on the other hand, may be extracted from the loop's second quadrant, and it is customary to only be present in this quadrant. The remanence, Mr or Jr, is the field generated by the magnet after the magnetizing field has been withdrawn. The inductive coercivity, Hc, is the reverse field necessary to bring the induction to zero, whereas the intrinsic coercivity, jHc, is the reverse field required to bring the magnetization to zero. The maximum energy product, (BH) max, is the maximum value of the product of B and H and is a measure of the maximum amount of beneficial work that the magnet can accomplish. For permanent magnet materials, the Fig. of merit (BH) max is employed.

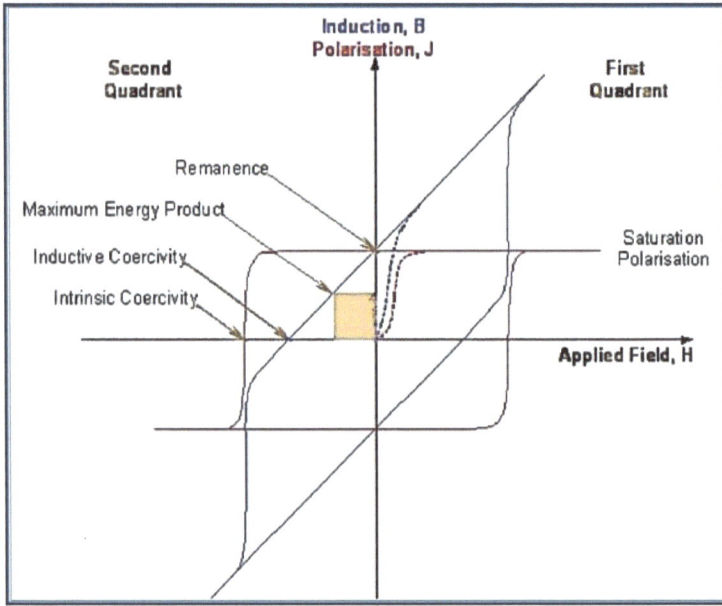

Fig. (1.15). A ferro- or ferri- magnetic material's typical hysteresis loop.

1.9.8.8. Permeability

The ease with which a magnetic flux is established in a component is described by permeability, a material characteristic. It is defined as the ratio of flux density to magnetic force (B/H). The maximum permeability is the point on the B/H curve for the unmagnetized material where the slope is the steepest. A straight line from the origin is typically considered to be tangent to the B/H curve at this point. The relative permeability [$\mu = \mu_{material}/\mu_{air}$] is calculated by dividing the permeability of the substance by the permeability of free space (air).

1.9.8.9. Ferromagnetic Materials

Ferromagnetic materials are commonly utilized in EMC as noise absorbers and microwave absorbers [61-65]. Because of its high value of complicated permeability, gamma-Fe_2O_3 is determined to be the best for producing radar absorption materials among all of them [64, 66]. When it comes to EMC applications, all ferromagnetic materials have particular characteristics that must be understood. Ferrites are magnetic materials made up of oxides that include ferric ions as the primary component. The name is frequently used for materials with the cubic crystal structure of the mineral spinel, but it is also used to describe magnetic oxides in general, regardless of crystal shape. Magnetite, often known as ferrous ferrite, is a naturally occurring ferrite. Spinel ferrite [$MeO.Fe_2O_3$] is the most common soft ceramic magnet, with a typical formula of [$MeO.Fe_2O_3$], where the iron is in the trivalent state Fe^{3+} and Me indicates a divalent metal ion with an ionic radius of around 0.6 to 1Å. Me is one of the divalent ions of the transition elements Mn, Fe, Co, Ni, Cu, and Zn, or Mg and Cd, in simple ferrites.

1.9.8.10. Soft Ferrites

Because of its dielectric and magnetic properties, soft ferrites are commonly utilized in microwave devices such as circulators, isolators, and phase shifters. The magnetic and electrical properties of ferrites may be customized by altering the type and amount of metal ion replacement, which is fascinating. Materials that are easily magnetized and demagnetized are known as soft magnetic materials. The first group is distinguished by the high permeability and flux-multiplying power of magnetically soft materials, which makes them suitable for use in machines and devices. The soft ferrites have a cubic crystal structure and the general formula $MeO.Fe_2O_3$, where Me is a divalent metal like Mg, Mn, or Ni, and Fe_2O_3 is the general formula. To enhance magnetization saturation (Ms), non-magnetic Zn ferrite is frequently added, and all commercial ferrites are mixed ferrites (solid solutions of one ferrite in another). Their Curie points range from about 300 to 600^0C, densities approximately 5 g/cm^3, and Ms from about 100 to 500 emu/cm^3 (100–500 kA/m). They all exhibit <111> easy magnetization directions, little magneto-crystalline anisotropy, and low to moderate magnetostriction [67-69]. Rare earth ions have unpaired 4f electrons due to their orbital structure, which contributes to magnetic anisotropy. Doping rare-earth ions into spinel ferrite can improve their electrical and magnetic properties because 4f-3d couplings between the transition metal and rare earth ions are linked to magneto-crystalline anisotropy in ferrite.

1.9.8.11. Hard Ferrites

Permanent magnets are manufactured from magnetically hard materials, and a high coercivity is necessary since, once magnetized, a permanent magnet must be able to resist the demagnetizing action of stray fields, including its own. The barium and strontium ferrites are the most significant in this category ($BaO.6Fe_2O_3$ and $SrO.6$ Fe_2O_3), which are magnetically hard. These are produced by firing at a temperature above $1000°C$ by taking an appropriate mixture of the oxides [70, 71]. Nano ferrites are used by researchers because they are not only environmentally safe but also have a plentiful natural supply, making them a low-cost material. In the past decade, a number of ferrites like Fe, Mn, Ni, Co ferrites and their multi-component ferrites like Fe_2O_3, Fe_3O_4, CrO_2, $BaFe_{12}O_{19}$, $Mn_{0.5}Zn_{0.5}Fe_2O_4$, MnO_2 have been shown to improve microwave absorption [61, 62, 64, 65, 72, 73]. Fe_3O_4 is determined to be the finest among them for creating radar absorption materials since it has a high permeability value [64, 66].

Singh *et al.* devised a method for fabricating vertically aligned iron oxide infiltrated multiwalled carbon nanotubes (MWCNT forest) between reduced graphene oxide (rGO) nanosheets.

The results showed that using iron oxide nanoparticles in the CNT forest enhances microwave absorption for high-performance EMI shielding applications, opening up new possibilities in this field. When compared to traditional EMI shielding materials, these designed sandwiched network has a higher shielding efficacy. The shielding efficacy of this network of exotic carbons is better than 37dB (> 99.98 percent attenuation) in Ku-band (12.4 GHz to 18 GHz), which is higher than the permissible limit (~30 dB) for techno-commercial application. The M-H curve at room temperature was used to investigate the field dependency of magnetization for a CNT forest containing Fe_3O_4 nanoparticles sandwiched between rGO sheets, as illustrated in Fig. (**1.16a, b, c**). As shown in the inset of Fig. (**1.16d**), the saturation magnetization (Ms) value of Fe_3O_4 synthesized by co-precipitation technique was 50 emu/g at a 5 kOe external field, with a modest coercivity and minimal retentivity and no hysteresis loop, suggesting a superparamagnetic nature. The Ms value has grown from 0.04 emug-1 (pristine CNT) to 4.10 emug-1 when these nano ferrite particles are filled in the gap of the CNT forest. Furthermore, by sandwiching the rGO with sample B, Ms value decreases a little (Fig. **1.16e**). For EMI shielding applications, a suitable magnetization value is necessary to enhance and regulate magnetic losses.

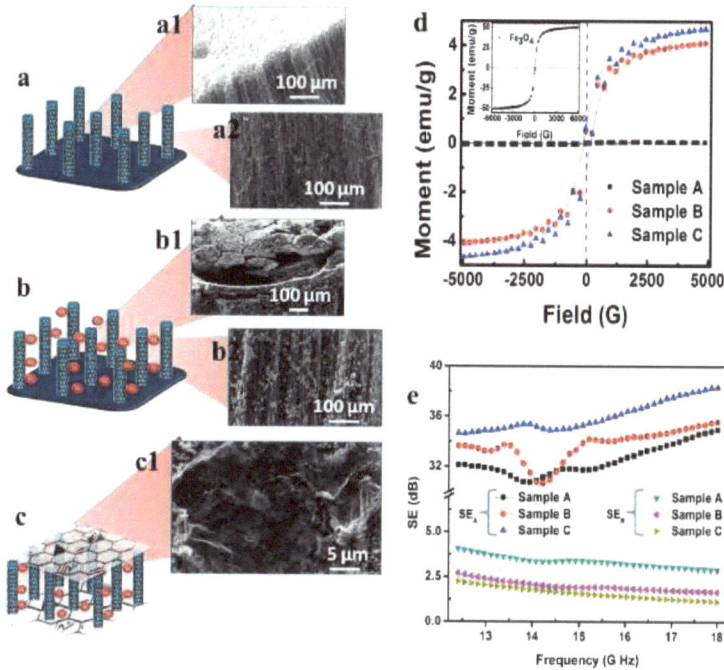

Fig. (1.16). SEM images from the CNT forest (sample A) sandwiched between rGO sheets are shown schematically with their equivalents. (**a**) A CNT forest is depicted schematically. (a1) SEM pictures of the top surface of CNT forest, (a2) SEM image of vertically aligned CNT forest, (**b**) Schematic design of Fe_3O_4 nanoparticle-filled CNT forest, (b1) top view of SEM image of sample B, (b2) side view of Fe_3O_4 nanoparticles adorned with Fe_3O_4 nanoparticles, (**c**) A top-view SEM picture of Fe_3O_4 nanoparticles adorned in CNT forest sandwiched between rGO sheets, (c1) a schematic representation of sample B sandwiched between rGO sheets. (**d**) VSM curves of CNT forest hybrid nanostructure and Fe_3O_4 and the inset shows the VSM plot of Fe_3O_4 nanoparticles (**e**) Frequency dependence of shielding effectiveness (SE_A and SE_R) in the Ku band frequency range (Reproduced from a study [26], with permission from Elsevier).

The average SE_A due to absorption (SE_A) of sample A, sample B, and sample C was determined to be 32.46, 33.69, and 36.01 dB, respectively, indicating that SE_A rises with the presence of Fe_3O_4 nanoparticles and improves further with rGO sandwiching. The SE_R on the other hand, drops from 3.33 to 1.57 dB. As a result, at a critical thickness of 2 mm, the total average SE for samples A, B, and C was 35.79, 35.68, and 37.58 dB, respectively. It is worth noting that magnetic particle infiltration and rGO sandwiching assist the reduction of shielding effectiveness caused by reflection. In comparison to pristine CNT, pure rGO, γ-Fe_2O_3 nanoparticles, CNT-iron oxide composite [8], and rGO/iron oxide composite described earlier, SE is dominated by absorption and exhibits superior microwave absorption characteristics [73].

1.10. COMPOSITES FOR ELECTROMAGNETIC INTERFERENCE SHIELDING

Stealth technology primarily addresses the target's shape and the type of radar-absorbing material (RAM) or microwave absorbing material (MAM). RAM is made up of high-loss energy molecules that collect and dissipate incoming radiation in the form of heat at coordinated frequencies [9, 25]. As a result, MAM or RAM materials play an important role in the field of stealth technology. RAM products are made from materials that can cause dielectric and/or magnetic loss when exposed to electromagnetic radiation [74]. RAM materials are complicated composites containing conductive additives or magnetic fillers that are molded into an insulating binder or matrix. These conductive additions with dielectric loss or magnetic fillers are commonly referred to as absorbers in microwave absorption materials. The absorber is an essential component of microwave absorbing material because it dominates the microwave absorbing characteristics of the material at microwave frequencies. According to the properties of microwave absorbers, microwave absorption materials are classified into electrical and magnetic loss microwave absorbing materials [25]. The absorption loss is, in general, a function of the substance, where the materials' electrical conductivity and magnetic permeability are σ and μ, respectively. Dielectric loss is primarily related to material conductivity, while magnetic loss results not only from electrical conductivity but also from magnetic permeability (μ) [19, 62, 75]. Therefore, relative to that of dielectric loss materials, the magnetic loss materials have a characteristic of high absorption efficiency and wide bandwidth. Carbonyl irons and ferrites are excellent microwave absorption materials for industrial magnetic failure. Their application is limited, however, by their heavy mass. Carbon materials typically act as microwave absorption materials for dielectric failure. The dielectric loss microwave absorbers are advantageous due to their light weight compared to the magnetic loss absorbers; however, their practical application is limited by their poor absorbing performance and narrow bandwidth. Carbon nanotubes, due to their high conductivity (1-10^3 S/cm), lightweight and larger special surfaces belong to promising microwave absorbing materials.

In order to develop new types of microwaves absorbing materials with wide bandwidth, a high absorption coefficient and a thinner coating layer are therefore needed. Thus, in the microwave frequency range (1-18 GHz), the design and preparation of such materials is a hot field in material science. Conducting polymers have also evinced much attention in the potential application as broadband microwave absorbers due to their high conductivity, light weight, processability, durability, and controllable electromagnetic properties by doping nature [64, 76]. In the past few decades, a lot of work has been done on polymer-ferromagnetic composites. The ferrite particles are incorporated in the polymer matrix (polyaniline and polypyrrole) by the *in-situ* or *ex-situ* method, which possesses moderate magnetization and conductivity [25, 77, 78]. In addition, conducting polymers not only reflect but also selectively absorb electromagnetic radiation compared to metals, which has made conducting polymers useful for microwave absorbing materials used for military and civil purposes, such as stealth technology [3, 14, 79]. In fact, because of the reversible electrical properties of conducting polymers by doping/de-doping methods, conducting polymers tend to be one of the few materials capable of microwave absorption, called 'intelligent stealth materials.'

The basic shielding theory found that a multiphase composite containing the optimum concentration of electrically conducting material, dielectric filler, and magnetic material must be an ideal microwave shield. Physical geometry also plays a key role in enhancing the SE alongside this. The composite's mild conductivity (10^{-4} to 10^{-1} S/cm) boosts SE in two respects. First, the front face of the conducting shield reflects the incident EM wave because the interaction of the electric vector with mobile charge carriers on the conducting surface results in ohmic losses (heat). Secondly, polarization due to the migration of charge carriers to form space charges at interfaces or grain boundaries is essential for materials consisting of a high concentration of charge carriers (*i.e.*, with high conductivity). This polarization of the space charge strengthens the polarization effect [80].

The multiphase composite dielectric filler presence raises the dielectric constant of the shield, which is further followed by high dielectric losses. High-dielectric constant ceramic powders such as SiO_2, TiO_2, ZnO, barium titanate, fly ash, strontium titanate, barium strontium titanate, and $PbTiO_3$ were applied to increase the dielectric constant of multiphase composites [75, 81]. In multiphase composite, high permeability materials are intentionally introduced to increase magnetic losses, which are the combined product of natural resonances, eddy current effects, and anisotropy energy present in the composites [72]. The presence of nano ferrite particles in the composite is the key cause of eddy current in the microwave range [25]. Hence, different research groups around the globe tried to build an efficient shield against electromagnetic radiation using a combination of fillers with part conducting fillers (*e.g.*, graphitic materials [82], conducting polymers [83, 84], carbon fiber [22], and CNTs [85, 86]), magnetic part (iron-nickel [87], soft [72] and hard ferrites [19]) and insulating or dielectric part (*e.g.*, TiO_2, $BaTiO_3$, Fly ash) [20, 74].

In this series, graphene/Fe_3O_4/SiO_2/NiO nanosheets combining the sol-gel process and hydrothermal reaction were prepared by Lei Wang *et al.* Around 2-18 GHz, the microwave absorption properties of graphene/Fe_3O_4 and graphene/Fe_3O_4/NiO nanosheets have been investigated. The overall reflection loss for graphene/Fe_3O_4/NiO nanosheet was -51.5 dB at 14.6 GHz for a thickness of 1.8 mm at 14.6 GHz [88]. Another example of multiphase composite is designed by Ren *et al.*, who used graphene/Fe_3O_4/Fe/ZnO quaternary nanocomposite to create a microwave absorbing shield. The results showed that the overall absorption for the quaternary nanocomposite was less than -30 dB for the 2.5-5 mm (5.9-15.2 GHz) thickness range [89]. Yang *et al.* synthesized bowl-like hollow spheres of Fe_3O_4/reduced graphene oxide nanocomposites. These synthesized nanocomposites with a coating layer of thickness of 2 mm displayed a maximum absorption loss of -24 dB at 12.9 GHz and a bandwidth of 4.9 GHz (from 10.8-15.7 GHz frequency) corresponding to a reflection loss of -10 dB at -10 dB [90]. Qin Li *et al.*, by sol-gel auto-combustion method, prepared magnetic composites of barium ferrite coated fly ash cenospheres. With a maximum reflection loss of -15.4 dB at 8.4 GHz with a sample thickness of 2.0 mm, the barium ferrite coated fly ash powder-epoxy composites have excellent microwave absorption properties in the 2-18 GHz frequency range [91]. In order to determine electromagnetic interference (EMI) shielding, Song *et al.* developed flexible graphene/polymer composite films into paraffin-based sand with structures having shielding effectiveness up to 27 dB [11]. The same group synthesized carbon nanosheet/wax composites showing maximum reflection loss up to 60 dB [92].

The EMI shielding of PET-coated single-walled carbon nanotubes synthesized by the spin coating technique with improved EMI shielding in the terahertz field has been tested by Seo *et al.* [93]. Zhang *et al.* prepared poly(methyl methacrylate)-based bulk foams embedded with chemically reduced graphene oxide by blending and then foamed by environmentally benign subcritical CO_2 foaming technique with excellent EMI shielding up to 19 dB [5]. Xu Huang *et al.* successfully synthesized $BaTiO_3$/MWCNTs core/shell heterostructure *via* the solvent-thermal method. The maximum peak value of $BaTiO_3$/MWCNTs heterostructure (10:1) reaches -45 dB at 5.5 GHz when the thickness of the absorber layer is 3.5 mm [81]. Yang *et al.* demonstrated that a small volume of carbon nanotube spread inside a polymer matrix into the void space between carbon nanofibers had a remarkable effect on EMI shielding. The efficacy of shielding has significantly improved to 20.3 dB, which is needed for commercial applications [22]. A novel carbon nanotube-polystyrene foam composite was also synthesized by Yang *et al.* for EMI shielding applications. For 7 percent CNT loading, the EMI shielding offered by the composite was approximately 20 dB. FeNi/RGO nanocomposites have been synthesized successfully by Chen Sun *et al.* by *in-situ* reduction. On the RGO nanosheets, sphere-like FeNi nanoparticles were deposited. FeNi/RGO nanocomposites have outstanding EM absorption properties in the 2-18 GHz range, as shown by the large effective bandwidth of absorption (up to 3.3 GHz, R_L less than 10 dB reflective loss) [94]. Composites of DBSA-doped polyaniline and $BaTiO_3$/$Ni_{0.5}Zn_{0.5}Fe_2O_4$ in epoxy resin matrices were prepared successfully by Das *et al.* A gross reflection loss of -15.78 dB at 10.89 GHz has been shown. Magnetic losses were less important for these composites, and dielectric loss and joule-heating loss added greatly to the failure process [80]. Single-walled carbon nanotube-epoxy composites are made using the *in-situ* process by Li *et al.* For 15 percent SWCNT, the highest EMI shielding was up to 49 dB at 10 MHz and exhibited 15-20 dB in the 500 MHz to 1.5 GHz range [16]. The electric conductivity and EMI shielding properties of the Fe-containing multi-walled carbon nanotube-PMMA composite were tested by Kim *et al.* [95] and got shielding effectiveness of 27 dB for loading of 40 percent MWCNTs. Xiang *et al.* [96] synthesized and analyzed MWCNT/silica composites' microwave shielding characteristics in the Ku band.

The above studies indicate that multi-phase nanocomposite improved microwave shielding can be accomplished either by integrating the desired amount of magnetic and dielectric filler into the semiconducting matrix or by constructing a matrix with a combination of dielectric conducting and magnetic materials. While a lot of studies have been conducted, it remains a difficult problem to look for a shield that can provide full absorption without reflection. Therefore, most documented research groups have reported composites based on a mixture of fillers (such as conducting silicone, graphic fabrics, CNTs and carbon fibre), magnetic material (such as iron nickel, soft ferrites) and insulating or dielectric material (such as TiO_2, fly ash), hexagonal-ferrite/composites of metal/polymer (dielectrics TiO_2, fly ash and ferrites such as Fe_3O_4, γ-Fe_2O_3 of Fe-Ni ferrites), single-wall carbon nanotube–epoxy composites, cement-carbon fiber-graphite composite.

1.11 5th GENERATION (5G) ELECTROMAGNETIC RADIATION AND ITS PREVENTION

The scientific and technological development of wireless communications benefits the community and people in their daily life [97, 98], and electromagnetic interference (EMI) is becoming universal due to extensive use and increasing demand for electronic devices and other powerful equipment's. In the next decade, mobile traffic may increase thousands of times, and billions of connections for communicating devices are expected compared to what we see today [97, 99]. Moreover, the rapid development of essential services, including electronic banking, electronic education, electronic health, and electronic commerce, causes a large growth in data volume and requires low latency and high reliability to support applications [100, 101]. A wide range of data rates has to be supported up to multiple gigabits per second, and tens of megabits per second need to be guaranteed with high availability and reliability [101]. The fourth-generation (4G) embedded version of mobile networking technology is the most recent, and the fifth-generation (5G) is close to being implemented by the end of 2021.

Nowadays, 4G technology has been adopted by almost all the countries over the globe, but with the increase in the number of users and devices, the 4G spectrum has become overloaded. Therefore, 5G technology can accommodate the increasing demand of the users and real-time data calculation. The 5G bands work in the three different band spectrum such as low band spectrum (< 1GHz), which is used for LTE applications with high coverage but limited data speed of 100 Mbps, mid-band spectrum provides fast speed with lower latency as compared to the low band spectrum (but, it fails to penetrate the building, massive MIMO may improve the penetration and cover area under this band) and high band spectrum also called millimeter-wave band provides the maximum performance in all the 5G spectrum bands and provide speed up to 100 Gbps but provides low coverage area. The 5G wireless can work on millimeter waves of 30–100 GHz. The ranges of 5G spectrum, along with their corresponding wavelength, are shown in Fig. (**1.17**).

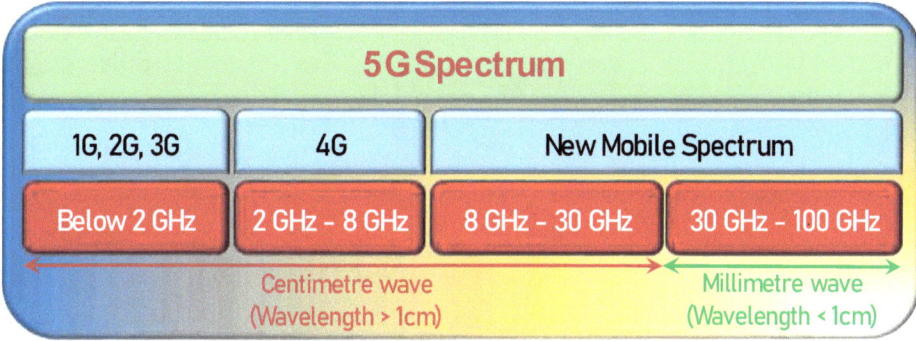

Fig. (1.17). 5G technology with a range of spectrum from 1G to 5G and their wavelength.

Nowadays, many electronic devices are generating interference of external electromagnetic waves. The challenges of public health would be the increasing electromagnetic pollution caused by electronic devices. Therefore, shielding of these radiations is required to protect the society and environment. EMI shielding materials will encircle unprecedented new opportunities and has become the main component of electronic materials.

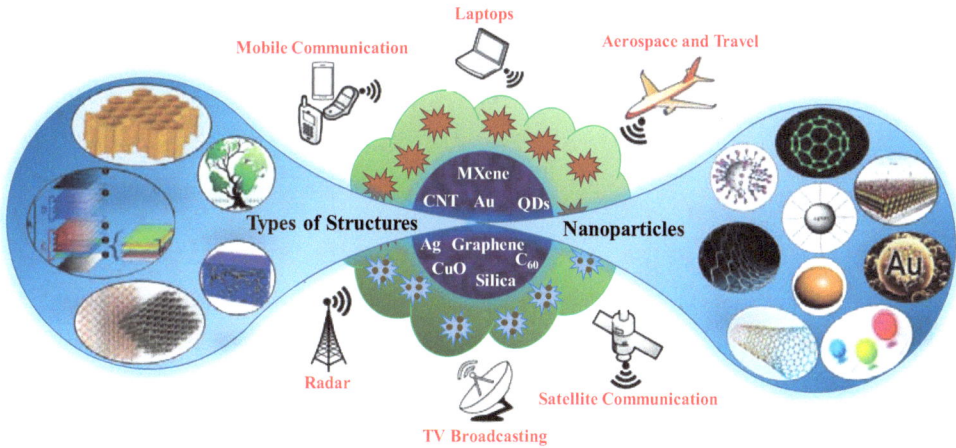

Fig. (1.18). Types of electromagnetic radiations and the types of EMI shielding materials.

As shown in Fig. (**1.18**), carbon nanotube, graphene, 2-dimensional (2D) materials, fullerene, porous materials, layered structures, polymer nanocomposites, metal nanoparticles are the most common materials for EMI shielding. As discussed by the researchers, 5G mobile communication also produces intense radiation, which is harmful to human health and the environment. Similar to the radar absorbing materials and defense systems, carbon-based materials such as carbon nanotubes and graphene, 2D materials, metal nanoparticles, porous materials, and polymer-based EMI shielding materials have also attracted a lot of attention in the industry and academics because of their low density, ease of production, and flexibility. Researchers go over the evolution of polymer-based shielding materials made with polymers as matrices and precursors. The architecture of polymer composites is addressed with features such as homogeneous structure, porous structure, and laminates. Metallic nanowires and nanoparticles carbon-based materials [102-106] have all recently been incorporated into the polymer matrix to yield EMI shielding polymer nanocomposites. While these composites are simple to produce, lightweight, and mechanically flexible, their EMI shielding effectiveness (SE) is necessarily hampered by the loss of electrical conductivity with applied stretch, which is a frequent occurrence in today's stretchy conductors [107]. It has been shown that at 400 percent strain, the EMI SE of a three-dimensional (3D) LM-skeleton-based polymer composite increased to two times, approaching the shielding levels of metallic plates tested under the same circumstances. The enhanced reflection and absorption of EM waves as a result of the stretch-induced augmentation of electrical conductivity and the higher surface-to-volume ratio afforded by the 3D LM design, respectively, results in a rise in EMI SE. 3D LM composite has a record electrical conductivity of $1.1x \ 10^6 \ Sm^{-1}$ at 400 percent strain as well as strong electrical and EMI shielding properties with massive repetitive deformations. In contrast to the relatively narrow frequency area, namely X-band (8.2–12.4 GHz), examined by conventional EMI shielding materials, researchers study the EMI shielding capabilities of the composite throughout an unprecedented frequency range from 2.65–40 GHz, which is important to 5G technology. The EMI SE of a highly stretchy polymer composite embedded with a three-dimensional (3D) liquid-metal (LM) network, which equals the EMI SE of metallic plates over an extraordinarily large frequency range of 2.65–40 GHz, has been reported by Yao *et al.* [108]. The electrical conductivities achieved in the 3D LM composite are among the best in the world for stretchy conductors that have been subjected to substantial mechanical deformations. The material, which has skin-like elastic compliance and hardness, offers a way to address the growing need for soft and human-friendly electronics. A study was done to study the shielding effectiveness of the conducting polymer coated fabrics at the 101 GHz range [109]. The measurements were carried out using a phase log oscillator of a 101 GHz frequency generator, Model No. 956 W4-010-101. S/N 033 using a conical horned antenna, Model No. 458264-1031, S/N 012 and mm wave power receiving setup consisting of a pyramidal horned antenna, Model No. 861 W/387, S/N 483. The shielding effectiveness was measured by noting the power with and without the samples by placing them near the surface of the antenna. The setup for the measurement of shielding effectiveness at 101 GHz is shown in Fig. (**1.19**).

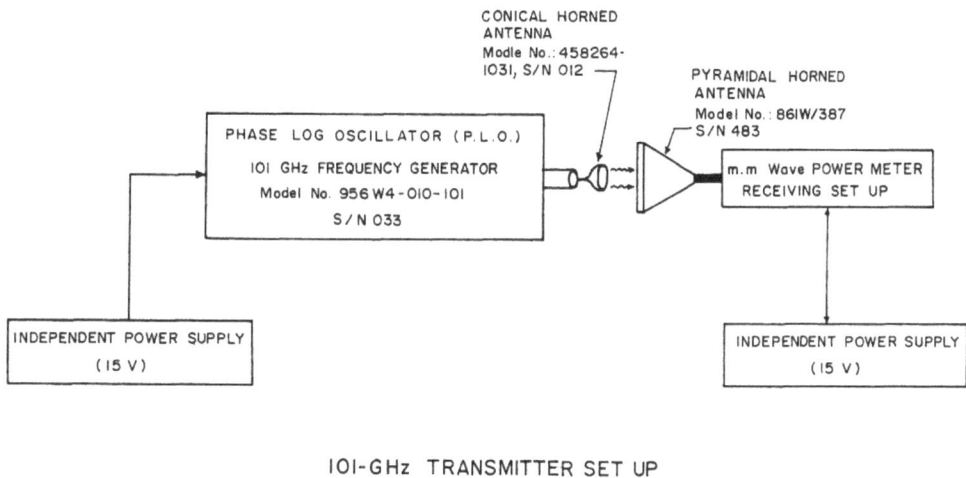

Fig. (1.19). 101 GHz transmitter setup used for the measurement of shielding effectiveness.

EMI shielding studies of conducting fabrics at 101 GHz show shielding effectiveness from 17.77 dB to 48.72 dB, which can be considered an effective solution for controlling electromagnetic pollution in the mm range.

CONCLUSION

Electromagnetic contamination levels have risen considerably in recent years as a result of the development of new electronic devices and communication networks. Every day, new research is published in an attempt to find a technique to reduce electromagnetic interferences (EMI). This chapter covers the fundamentals of electromagnetic interference shielding theory, methodologies for characterizing electromagnetic materials, and strategies for exploring the electromagnetic characteristics of materials. Nicholson-Ross-Weir technique, short circuit line technique, and dielectric measurement techniques are among the main topics covered. This chapter explained the concept of microwave shielding and outlined the requirements that must be met in order to produce an optimal shield. Following that, the fundamental physics that governs the interaction of the material with the electromagnetic field is thoroughly detailed. Following that, we look at the general characteristics of common electromagnetic materials such as dielectrics, semiconductors, conductors, magnetic materials, and artificial materials. Furthermore, due to the rapid advancement of technology, there is a need to search for lighter and more efficient materials. Exploration of novel functional materials capable of successfully blocking or shielding electromagnetic radiation as well as 4G and 5G radiation is a currently active area of study. A lot of research is done in this direction; nanocomposites with suitable filler are one of the best solutions to meet the demand of EMI. These nanocomposites could be shaped into the desired product. The developed nanocomposites can be considered as a potential candidate for the development of products like shielded fabrics, EMI protective caps, maternity belts, shielded structures, microwave absorbing coatings, *etc.*, that can be implemented according to the needs of society.

CONSENT FOR PUBLICATION

Not applicable.

CONFLICT OF INTEREST

The authors declare no conflict of interest, financial or otherwise.

ACKNOWLEDGEMENTS

The authors wish to thank Professor Gyantosh Kumar Jha, Principal Atma Ram Sanatan Dharma College, Delhi, for his keen interest in the work. The authors also thank Dr. Maneesh Kumar and Dr. Kuldeep Kakran for their valuable comments and suggestions.

REFERENCES

[1]   Yang, Y.; Gupta, M.C.; Dudley, K.L.; Lawrence, R.W. Novel carbon nanotube-polystyrene foam composites for electromagnetic interference shielding. *Nano Lett.,* **2005,** *5*(11), 2131-2134.
        http://dx.doi.org/10.1021/nl051375r PMID: 16277439

[2]   Chen, Z.; Xu, C.; Ma, C.; Ren, W.; Cheng, H-M. Lightweight and flexible graphene foam composites for high-performance electromagnetic interference shielding. *Adv. Mater.,* **2013,** *25*(9), 1296-1300.

http://dx.doi.org/10.1002/adma.201204196 PMID: 23300002

[3]   Hemming, L.H. *Architectural Electromagnetic Shielding Handbook: A Design and Specification Guide.,* **2000,**

       http://dx.doi.org/10.1109/9780470544181

[4]   Hanada, E.; Antoku, Y.; Tani, S.; Kimura, M.; Hasegawa, A.; Urano, S.; Ohe, K.; Yamaki, M.; Nose, Y. Electromagnetic interference on medical equipment by low-power mobile telecommunication systems. *Electromagnetic Compatibility, IEEE Transactions,* **2000,** *42*(4), 470-476.

[5]   Zhang, H-B.; Yan, Q.; Zheng, W-G.; He, Z.; Yu, Z-Z. Tough graphene-polymer microcellular foams for electromagnetic interference shielding. *ACS Appl. Mater. Interfaces,* **2011,** *3*(3), 918-924.

       http://dx.doi.org/10.1021/am200021v PMID: 21366239

[6]   Raagulan, K.; Kim, B.M.; Chai, K.Y.J.N. Recent Advancement of Electromagnetic Interference (EMI) Shielding of Two Dimensional (2D) MXene and Graphene Aerogel Composites. *Nanomaterials (Basel),* **2020,** *10*(4), 702.

       http://dx.doi.org/10.3390/nano10040702 PMID: 32276331

[7]   Jang, J.O.; Park, J.W. *Coating material for shielding electromagnetic waves.,* **2002,**

[8]   Che, R.C.; Peng, L.M.; Duan, X.F.; Chen, Q.; Liang, X.L. Microwave Absorption Enhancement and Complex Permittivity and Permeability of Fe Encapsulated within Carbon Nanotubes. *Adv. Mater.,* **2004,** *16*(5), 401-405.

       http://dx.doi.org/10.1002/adma.200306460

[9]   Saini, P.; Arora, M.; Gupta, G.; Gupta, B.K.; Singh, V.N.; Choudhary, V. High permittivity polyaniline-barium titanate nanocomposites with excellent electromagnetic interference shielding response. *Nanoscale,* **2013,** *5*(10), 4330-4336.

       http://dx.doi.org/10.1039/c3nr00634d PMID: 23563991

[10]   Wen, B.; Wang, X.X.; Cao, W.Q.; Shi, H.L.; Lu, M.M.; Wang, G.; Jin, H.B.; Wang, W.Z.; Yuan, J.; Cao, M.S. Reduced graphene oxides: the thinnest and most lightweight materials with highly efficient microwave attenuation performances of the carbon world. *Nanoscale,* **2014,** *6*(11), 5754-5761.

       http://dx.doi.org/10.1039/C3NR06717C PMID: 24681667

[11]   Song, W-L.; Cao, M-S.; Lu, M-M.; Bi, S.; Wang, C-Y.; Liu, J.; Yuan, J.; Fan, L-Z. Flexible graphene/polymer composite films in sandwich structures for effective electromagnetic interference shielding. *Carbon,* **2014,** *66*(0), 67-76.

       http://dx.doi.org/10.1016/j.carbon.2013.08.043

[12]   Dhawan, S.K.; Singh, N.; Rodrigues, D. Electromagnetic shielding behaviour of conducting polyaniline composites. *Sci. Technol. Adv. Mater.,* **2003,** *4*(2), 105-113.

       http://dx.doi.org/10.1016/S1468-6996(02)00053-0

[13]   Duan, Y.; Liu, S.; Guan, H. Investigation of electrical conductivity and electromagnetic shielding effectiveness of polyaniline composite. *Sci. Technol. Adv. Mater.,* **2005,** *6*(5), 513-518.

       http://dx.doi.org/10.1016/j.stam.2005.01.002

[14]   Gupta, T.K.; Singh, B.P.; Singh, V.N.; Teotia, S.; Singh, A.P.; Elizabeth, I.; Dhakate, S.R.; Dhawan, S.K.; Mathur, R.B. MnO$_2$ decorated graphene nanoribbons with superior permittivity and excellent microwave shielding properties. *J. Mater. Chem. A Mater. Energy Sustain.,* **2013,** *2*(12), 4256-4263.

       http://dx.doi.org/10.1039/c3ta14854h

[15]   Yousefi, N.; Sun, X.; Lin, X.; Shen, X.; Jia, J.; Zhang, B.; Tang, B.; Chan, M.; Kim, J.-K. Highly Aligned Graphene/Polymer Nanocomposites with Excellent Dielectric Properties for High Performance Electromagnetic Interference Shielding. *Advanced Materials,* **2014.**

[16]   Li, N.; Huang, Y.; Du, F.; He, X.; Lin, X.; Gao, H.; Ma, Y.; Li, F.; Chen, Y.; Eklund, P.C. Electromagnetic interference (EMI) shielding of single-walled carbon nanotube epoxy composites. *Nano Lett.,* **2006,** *6*(6), 1141-1145.

       http://dx.doi.org/10.1021/nl0602589 PMID: 16771569

[17]   Gupta, T.K.; Singh, B.P.; Mathur, R.B.; Dhakate, S.R. Multi-walled carbon nanotube-graphene-polyaniline multiphase nanocomposite with superior electromagnetic shielding effectiveness. *Nanoscale,* **2014**, *6*(2), 842-851.
http://dx.doi.org/10.1039/C3NR04565J PMID: 24264356

[18]   Ohlan, A.; Singh, K.; Chandra, A.; Dhawan, S. Microwave absorption properties of conducting polymer composite with barium ferrite nanoparticles in 12.4–18. *Appl. Phys. Lett.,* **2008**, *93*(5), 053114-053117.
http://dx.doi.org/10.1063/1.2969400

[19]   Sambyal, P.; Singh, A.P.; Verma, M.; Farukh, M.; Singh, B.P.; Dhawan, S.K. Tailored polyaniline/barium strontium titanate/expanded graphite multiphase composite for efficient radar absorption. *RSC Advances,* **2014**, *4*(24), 12614.
http://dx.doi.org/10.1039/c3ra46479b

[20]   Singh, A.P. S., A. K.; Chandra, A.; Dhawan, S. K., Conduction mechanism in Polyaniline-flyash composite material for shielding against electromagnetic radiation in X-band & Ku band. *AIP Adv.,* **2011**, *1*(2)
http://dx.doi.org/10.1063/1.3608052

[21]   Festa, A.; Panella, M.; Lo Sterzo, E. Roberto; Liparulo, L., Radiofrequency Identification Systems for Healthcare: A Case Study on Electromagnetic Exposures. *J. Clin. Eng.,* **2013**, *38*(3), 125-133.
http://dx.doi.org/10.1097/JCE.0b013e31829a9174

[22]   Yonglai, Y.; Mool, C.G.; Kenneth, L.D. Towards cost-efficient EMI shielding materials using carbon nanostructure-based nanocomposites. *Nanotechnology,* **2007**, *18*(34)345701
http://dx.doi.org/10.1088/0957-4484/18/34/345701

[23]   Singh, K.; Ohlan, A.; Kotnala, R.K.; Bakhshi, A.K.; Dhawan, S.K. Dielectric and magnetic properties of conducting ferromagnetic composite of polyaniline with Fe_2O_3 nanoparticles. *Mater. Chem. Phys.,* **2008**, *112*(2), 651-658.
http://dx.doi.org/10.1016/j.matchemphys.2008.06.026

[24]   Zhu, J.; Wei, S.; Haldolaarachchige, N.; Young, D.P.; Guo, Z. Electromagnetic field shielding polyurethane nanocomposites reinforced with core–shell Fe–silica nanoparticles. *J. Phys. Chem. C,* **2011**, *115*(31), 15304-15310.
http://dx.doi.org/10.1021/jp2052536

[25]   Reddy, B. *Advances in Nanocomposites - Synthesis, Characterization and Industrial Applications.,* **2011**,
http://dx.doi.org/10.5772/604

[26]   Singh, A.P.; Mishra, M.; Hashim, D.P.; Narayanan, T.; Hahm, M.G.; Kumar, P.; Dwivedi, J.; Kedawat, G.; Gupta, A.; Singh, B.P.; Chandra, A.; Vajtai, R.; Dhawan, S.K.; Ajayan, P.M.; Gupta, B.K. Probing the engineered sandwich network of vertically aligned carbon nanotube–reduced graphene oxide composites for high performance electromagnetic interference shielding applications. *Carbon,* **2015**, *85*, 79-88.
http://dx.doi.org/10.1016/j.carbon.2014.12.065

[27]   Saini, P.; Arora, M. Microwave absorption and EMI shielding behavior of nanocomposites based on intrinsically conducting polymers, graphene and carbon nanotubes. **2012**.
http://dx.doi.org/10.5772/48779

[28]   Varshney, S.; Ohlan, A.; Singh, K.; Jain, V.K.; Dutta, V.P.; Dhawan, S.K. Robust Multifunctional Free Standing Polypyrrole Sheet for Electromagnetic Shielding. *Sci. Adv. Mater.,* **2013**, *5*(7), 881-890.
http://dx.doi.org/10.1166/sam.2013.1534

[29]   Gerke, D.D.; Kimmel, B. *Edn Designers Guide to Electromagnetic Compatibility.,* **2002**,

[30]   Ma, M.T.; Kanda, M.; Crawford, M.L.; Larsen, E.B. A review of elecromagnetic compatibility/interference measurement methodologies. *Proc. IEEE,* **1985**, *73*(3), 388-411.
http://dx.doi.org/10.1109/PROC.1985.13164

[31]   Dhia, S.B.; Ramdani, M.; Sicard, E. *Electromagnetic Compatibility of Integrated Circuits: Techniques for low emission and susceptibility.,* **2006**,
http://dx.doi.org/10.1007/b137864

[32]   Das, A.; Kothari, V.; Kothari, A.; Kumar, A.; Tuli, S. Effect of various parameters on electromagnetic shielding effectiveness of textile fabrics. *Indian J. Fibre Text. Res.,* **2009**, *34*(2), 144.

[33]   Singh, A.P.; Mishra, M.; Dhawan, S.J.N. *Conducting multiphase magnetic nanocomposites for microwave shielding application.,* **2015**, , 246-277.

[34]   Bjorklof, D. *Shielding for EMC. Shielding for EMC.,* **1999**,

[35]   Bridges, J.E. *IEEE Trans. EMC,* **1988**, *30*, 289.

[36]   Thomas, D. W.; Denton, A. C.; Konefal, T.; Benson, T.; Christopoulos, C.; Dawson, J. F.; Marvin, A.; Porter, S. J.; Sewell, P. Model of the electromagnetic fields inside a cuboidal enclosure populated with conducting planes or printed circuit boards. *Electromagnetic Compatibility, IEEE Transactions,* **2001**, *43*(2), 161-169.

[37]   Duffin, W.J. *Advanced Electricity and Magnetism.,* **1968**,

[38]   Ahn, S. Magnetic Field Generation.*The On-line Electric Vehicle.,* **2017**, , 81-96.
       http://dx.doi.org/10.1007/978-3-319-51183-2_5

[39]   Tong, X.C. *Advanced materials and design for electromagnetic interference shielding.,* **2016**,
       http://dx.doi.org/10.1201/9781420073591

[40]   Bjorklof, D. *Shielding for EMC.,* **1999**,

[41]   Bridges, J.E. An update on the circuit approach to calculate shielding effectiveness. *IEEE Trans. Electromagn. Compat.,* **1988**, *30*(3), 211-221.
       http://dx.doi.org/10.1109/15.3299

[42]   Duffin, W.J. *Advanced electricity and magnetism for undergraduates.,* **1968**,

[43]   Paul, C.R. *Electromagnetics for Engineers.,* **2004**,

[44]   Ott, H.W. *Noise Reduction Techniques in Electronic Systems.,* (2nd ed.), **1988**,

[45]   Schulz, R. B.; Plantz, V. C.; Brush, D. R. Shielding theory and practice. *IEEE Trans. Electromagn. Compat. EMC,* **1988**, *30*, 187.

[46]   Schelkunoff, S.A. *Electromagnetic Waves.,* **1943**,

[47]   Anoop Kumar, S.; Singh, A.; Saini, P.; Khatoon, F.; Dhawan, S.K. Synthesis, charge transport studies, and microwave shielding behavior of nanocomposites of polyaniline with Ti-doped γ-Fe$_2$O$_3$. *J. Mater. Sci.,* **2012**, *47*(5), 2461-2471.
       http://dx.doi.org/10.1007/s10853-011-6068-5

[48]   Ashokkumar, M.; Narayanan, T.N.; Gupta, B.K.; Leela Mohana Reddy, A.; Avanish, P.; Dhawan, S.K.; Bangaru, C.; Rawat, D.S.; Talapatra, S.; Ajayan, P.M. Conversion of Industrial Bio-Waste into Useful Nanomaterials. *ACS Sustain. Chem.& Eng.,* **2013**, *1*(6), 619-626.
       http://dx.doi.org/10.1021/sc3001564

[49]   Violette, N. *Electromagnetic compatibility handbook.,* **2013**,

[50]   Geetha, S.; Satheesh Kumar, K.; Rao, C.R.; Vijayan, M.; Trivedi, D.J.J.s. EMI shielding: Methods and materials—. *RE:view,* **2009**, *112*(4), 2073-2086.

[51]   Chen, H.; Lee, K.; Lin, J.; Koch, M. J. J. o. M. P. T. Comparison of electromagnetic shielding effectiveness properties of diverse conductive textiles via various measurement techniques. **2007**, *192*, 549-554.
       http://dx.doi.org/10.1016/j.jmatprotec.2007.04.023

[52]   Nicolson, A.M.; Ross, G.F. Measurement of the Intrinsic Properties of Materials by Time-Domain Techniques. *IEEE Trans. Instrum. Meas.,* **1970**, *19*(4), 377-382.
       http://dx.doi.org/10.1109/TIM.1970.4313932

[53]   Weir, W.B. Automatic measurement of complex dielectric constant and permeability at microwave frequencies. *Proc. IEEE,* **1974**, *62*(1), 33-36.
       http://dx.doi.org/10.1109/PROC.1974.9382

[54]   Ohlan, A.; Singh, K.; Chandra, A.; Dhawan, S.K. Microwave absorption properties of conducting polymer composite with barium ferrite nanoparticles in 12.4--18 GHz. *Appl. Phys. Lett.,* **2008**, *93*(5), 053114-053114.
       http://dx.doi.org/10.1063/1.2969400

[55]   Ohlan, A.; Singh, K.; Chandra, A.; Dhawan, S.K. Microwave absorption behavior of core-shell structured poly (3,4-ethylenedioxy thiophene)-barium ferrite nanocomposites. *ACS Appl. Mater. Interfaces,* **2010**, *2*(3), 927-933.
        http://dx.doi.org/10.1021/am900893d PMID: 20356300

[56]   Feng, L.; Huo, P.; Liang, Y.; Xu, T. J. A. M. Photonic metamaterial absorbers: Morphology engineering and interdisciplinary applications. **2020**, *32*(27), 1903787.

[57]   Sushmita, K.; Madras, G.; Bose, S. Polymer Nanocomposites Containing Semiconductors as Advanced Materials for EMI Shielding. *ACS Omega,* **2020**, *5*(10), 4705-4718.
        http://dx.doi.org/10.1021/acsomega.9b03641 PMID: 32201755

[58]   He, S.; Wang, G-S.; Lu, C.; Luo, X.; Wen, B.; Guo, L.; Cao, M-S. *Controllable Fabrication of CuS Hierarchical Nanostructures and Their Optical, Photocatalytic, and Wave Absorption Properties.,* **2013**, *78*(3), 250-258.

[59]   Lyu, L.; Wang, F.; Qiao, J.; Ding, X.; Zhang, X.; Xu, D.; Liu, W.; Liu, J. J. J. o. A. Compounds, Novel synthesis of $MoO_3/Mo_4O_{11}/MoO_2$ heterogeneous nanobelts for wideband electromagnetic wave absorption. **2020**, *817*, 153309.

[60]   Chung, D. D. J. M. C. Physics, Materials for electromagnetic interference shielding. **2020**, 123587.

[61]   Xu, H.; Zhang, H.; Lv, T.; Wei, H.; Song, F. Study on Fe_3O_4/polyaniline electromagnetic composite hollow spheres prepared against sulfonated polystyrene colloid template. *Colloid Polym. Sci.,* **2013**, *291*(7), 1713-1720.
        http://dx.doi.org/10.1007/s00396-013-2906-0

[62]   Ohlan, A.; Singh, K.; Chandra, A.; Dhawan, S.K. Conducting ferromagnetic copolymer of aniline and 3,4-ethylenedioxythiophene containing nanocrystalline barium ferrite particles. *J. Appl. Polym. Sci.,* **2008**, *108*(4), 2218-2225.
        http://dx.doi.org/10.1002/app.27794

[63]   Yang, R-B.; Liang, W-F.; Lin, W-S.; Lin, H-M.; Tsay, C-Y.; Lin, C-K. Microwave absorbing properties of iron nanowire at x-band frequencies. *J. Appl. Phys.,* **2011**, *109*(7)07B527
        http://dx.doi.org/10.1063/1.3561449

[64]   Singh, K.; Ohlan, A.; Saini, P.; Dhawan, S.K. Poly (3,4-ethylenedioxythiophene) γ-Fe_2O_3 polymer composite–super paramagnetic behavior and variable range hopping 1D conduction mechanism–synthesis and characterization. *Polym. Adv. Technol.,* **2008**, *19*(3), 229-236.
        http://dx.doi.org/10.1002/pat.1003

[65]   Belaabed, B.; Wojkiewicz, J.L.; Lamouri, S.; El Kamchi, N.; Lasri, T. Synthesis and characterization of hybrid conducting composites based on polyaniline/magnetite fillers with improved microwave absorption properties. *J. Alloys Compd.,* **2012**, *527*(0), 137-144.
        http://dx.doi.org/10.1016/j.jallcom.2012.02.179

[66]   Sun, X.; He, J.; Li, G.; Tang, J.; Wang, T.; Guo, Y.; Xue, H. Laminated magnetic graphene with enhanced electromagnetic wave absorption properties. *J. Mater. Chem. C Mater. Opt. Electron. Devices,* **2013**, *1*(4), 765-777.
        http://dx.doi.org/10.1039/C2TC00159D

[67]   Smit, J.; Wijn, H.P.J. *Ferrites, New York: Wiley,* **1959**.

[68]   Standley, K.J. *Oxide Magnetic Materials Oxford.,* **1962**,

[69]   Snelling, E.C. *Soft ferrites Properties and Applications.,* **1969**,

[70]   Adelskold, V.; Kemi, A. *Min. Geol. 12A,* **1938**, *29*, 1.

[71]   Went, J.J.; Rathenau, G.W.; Gorter, E.W.; VanOsterhout, G.W. *Philips Techn Rev.,* **1951**, *13*, 194.

[72]   Singh, A.P.; Mishra, M.; Chandra, A.; Dhawan, S.K. Graphene oxide/ferrofluid/cement composites for electromagnetic interference shielding application. *Nanotechnology,* **2011**, *22*(46)465701
        http://dx.doi.org/10.1088/0957-4484/22/46/465701 PMID: 22024967

[73]   Singh, A.P.; Garg, P.; Alam, F.; Singh, K.; Mathur, R.B.; Tandon, R.P.; Chandra, A.; Dhawan, S.K. Phenolic resin-based composite sheets filled with mixtures of reduced graphene oxide, γ-Fe$_2$O$_3$ and carbon fibers for excellent electromagnetic interference shielding in the X-band. *Carbon,* **2012,** *50*(10), 3868-3875.
http://dx.doi.org/10.1016/j.carbon.2012.04.030

[74]   Zhu, Y-F.; Zhang, L.; Natsuki, T.; Fu, Y-Q.; Ni, Q-Q. Facile synthesis of BaTiO3 nanotubes and their microwave absorption properties. *ACS Appl. Mater. Interfaces,* **2012,** *4*(4), 2101-2106.
http://dx.doi.org/10.1021/am300069x PMID: 22409350

[75]   Ohlan, A.; Singh, K.; Chandra, A.; Singh, V.; Dhawan, S. Conjugated polymer nanocomposites: Synthesis, dielectric, and microwave absorption studies. *J. Appl. Phys.,* **2009,** *106*(4), 044305-, 044305-044306.
http://dx.doi.org/10.1063/1.3200958

[76]   Kim, B.H.; Park, D.H.; Joo, J.; Yu, S.G.; Lee, S.H. Synthesis, characteristics, and field emission of doped and de-doped polypyrrole, polyaniline, poly(3,4-ethylenedioxythiophene) nanotubes and nanowires. *Synth. Met.,* **2005,** *150*(3), 279-284.
http://dx.doi.org/10.1016/j.synthmet.2005.02.012

[77]   Sezer, E. *Conducting nanocomposite systems.,* **2008,**
http://dx.doi.org/10.1016/B978-008045052-0.50006-3

[78]   Singh, K.; Ohlan, A.; Saini, P.; Dhawan, S. Poly (3, 4-ethylenedioxythiophene) γ-Fe$_2$O$_3$ polymer composite–super paramagnetic behavior and variable range hopping 1D conduction mechanism–synthesis and characterization. *Polym. Adv. Technol.,* **2008,** *19*(3), 229-236.
http://dx.doi.org/10.1002/pat.1003

[79]   Kanda, K.; Morimoto, M.; Junichi, H.; Takumi, F. *Electromagnetic wave absorbing material.,* **2001,**

[80]   Das, C. K.; Mandal, A. Microwave Absorbing Properties of DBSA-doped Polyaniline/BaTiO3-Ni$_{0.5}$Zn$_{0.5}$Fe$_2$O$_4$ Nanocomposites. **2012,** *1*(1), 43-53.

[81]   Huang, X.; Chen, Z.; Tong, L.; Feng, M.; Pu, Z.; Liu, X. Preparation and microwave absorption properties of BaTiO3@MWCNTs core/shell heterostructure. *Mater. Lett.,* **2013,** *111*(0), 24-27.
http://dx.doi.org/10.1016/j.matlet.2013.08.034

[82]   Mishra, M.; Singh, A.P.; Dhawan, S.K. Expanded graphite-nanoferrite-fly ash composites for shielding of electromagnetic pollution. *J. Alloys Compd.,* **2013,** *557*, 244-251.
http://dx.doi.org/10.1016/j.jallcom.2013.01.004

[83]   Dhawan, S.K.; Singh, K.; Bakhshi, A.K.; Ohlan, A. Conducting polymer embedded with nanoferrite and titanium dioxide nanoparticles for microwave absorption. *Synth. Met.,* **2009,** *159*(21–22), 2259-2262.
http://dx.doi.org/10.1016/j.synthmet.2009.08.031

[84]   Cheng, F.; Tang, W.; Li, C.; Chen, J.; Liu, H.; Shen, P.; Dou, S. Conducting poly(aniline) nanotubes and nanofibers: controlled synthesis and application in lithium/poly(aniline) rechargeable batteries. *Chemistry,* **2006,** *12*(11), 3082-3088.
http://dx.doi.org/10.1002/chem.200500883 PMID: 16429467

[85]   Xiang, C.; Pan, Y.; Liu, X.; Sun, X.; Shi, X.; Guo, J. Microwave attenuation of multiwalled carbon nanotube-fused silica composites. *Appl. Phys. Lett.,* **2005,** *87*(12), 123103-3.
http://dx.doi.org/10.1063/1.2051806

[86]   Yuan, B.; Yu, L.; Sheng, L.; An, K.; Zhao, X. Comparison of electromagnetic interference shielding properties between single-wall carbon nanotube and graphene sheet/polyaniline composites. *J. Phys. D Appl. Phys.,* **2012,** *45*(23), 45.
http://dx.doi.org/10.1088/0022-3727/45/23/235108

[87]   Narayanan, T.; Sunny, V.; Shaijumon, M.; Ajayan, P.; Anantharaman, M. Enhanced microwave absorption in nickel-filled multiwall carbon nanotubes in the S band. *Electrochem. Solid-State Lett.,* **2009,** *12*(4), K21-K24.
http://dx.doi.org/10.1149/1.3065992

[88]   Wang, L.; Huang, Y.; Sun, X.; Huang, H.; Liu, P.; Zong, M.; Wang, Y. Synthesis and microwave absorption enhancement of graphene@Fe_3O_4@SiO_2@NiO nanosheet hierarchical structures. *Nanoscale,* **2014**, *6*(6), 3157-3164.
        http://dx.doi.org/10.1039/C3NR05313J PMID: 24496379

[89]   Ren, Y.-L.; Wu, H.-Y.; Lu, M.-M.; Chen, Y.-J.; Zhu, C.-L.; Gao, P.; Cao, M.-S.; Li, C-Y.; Ouyang, Q-Y. Quaternary nanocomposites consisting of graphene, Fe_3O_4@Fe core@shell, and ZnO nanoparticles: synthesis and excellent electromagnetic absorption properties. *ACS Appl. Mater. Interfaces,* **2012**, *4*(12), 6436-6442.
        http://dx.doi.org/10.1021/am3021697 PMID: 23176086

[90]   Xu, H.-L.; Bi, H.; Yang, R.-B. Enhanced microwave absorption property of bowl-like Fe3O4 hollow spheres/reduced graphene oxide composites. *Journal of Applied Physics,* **2012**, *111*(7), 07A522.

[91]   Li, Q.; Pang, J.; Wang, B.; Tao, D.; Xu, X.; Sun, L.; Zhai, J. Preparation, characterization and microwave absorption properties of barium-ferrite-coated fly-ash cenospheres. *Adv. Powder Technol.,* **2013**, *24*(1), 288-294.
        http://dx.doi.org/10.1016/j.apt.2012.07.004

[92]   Song, W.-L.; Cao, M.-S.; Lu, M.-M.; Liu, J.; Yuan, J.; Fan, L-Z. Improved dielectric properties and highly efficient and broadened bandwidth electromagnetic attenuation of thickness-decreased carbon nanosheet/wax composites. *J. Mater. Chem. C Mater. Opt. Electron. Devices,* **2013**, *1*(9), 1846-1854.
        http://dx.doi.org/10.1039/c2tc00494a

[93]   Seo, M.A.; Yim, J.H.; Ahn, Y.H.; Rotermund, F.; Kim, D.S.; Lee, S.; Lim, H. Terahertz electromagnetic interference shielding using single-walled carbon nanotube flexible films. *Appl. Phys. Lett.,* **2008**, *93*(23)231905
        http://dx.doi.org/10.1063/1.3046126

[94]   Sun, C.; Jiang, W.; Wang, Y.; Sun, D.; Liu, J.; Li, P.; Li, F. Magnetic and electromagnetic absorption properties of FeNi alloy nanoparticles supported by reduced graphene oxide. *physica status solidi (RRL) –. Rapid Research Letters,* **2014**, *8*(2), 141-145.

[95]   Kim, H.M.; Kim, K.; Lee, C.Y.; Joo, J.; Cho, S.J.; Yoon, H.S.; Pejakovic, D.A.; Yoo, J.W.; Epstein, A.J. Electrical conductivity and electromagnetic interference shielding of multiwalled carbon nanotube composites containing Fe catalyst. *Appl. Phys. Lett.,* **2004**, *84*(4), 589-591.
        http://dx.doi.org/10.1063/1.1641167

[96]   Xiang, C.; Pan, Y.; Liu, X.; Shi, X.; Sun, X.; Guo, J. Electrical properties of multiwalled carbon nanotube reinforced fused silica composites. *J. Nanosci. Nanotechnol.,* **2006**, *6*(12), 3835-3841.
        http://dx.doi.org/10.1166/jnn.2006.603 PMID: 17256338

[97]   Hao, H.; Hui, D.; Lau, D. Material advancement in technological development for the 5G wireless communications. *Nanotechnol. Rev.,* **2020**, *9*(1), 683-699.
        http://dx.doi.org/10.1515/ntrev-2020-0054

[98]   Li, Q.C.; Niu, H.; Papathanassiou, A.T.; Wu, G. 5G Network Capacity: Key Elements and Technologies. *IEEE Veh. Technol. Mag.,* **2014**, *9*(1), 71-78.
        http://dx.doi.org/10.1109/MVT.2013.2295070

[99]   Andrews, J.G.; Buzzi, S.; Choi, W.; Hanly, S.V.; Lozano, A.; Soong, A.C.K.; Zhang, J.C. What Will 5G Be? *IEEE J. Sel. Areas Comm.,* **2014**, *32*(6), 1065-1082.
        http://dx.doi.org/10.1109/JSAC.2014.2328098

[100]   Simkó, M.; Mattsson, M-O. 5G Wireless Communication and Health Effects-A Pragmatic Review Based on Available Studies Regarding 6 to 100 GHz. *Int. J. Environ. Res. Public Health,* **2019**, *16*(18), 3406.
        http://dx.doi.org/10.3390/ijerph16183406 PMID: 31540320

[101]   Osseiran, A.; Boccardi, F.; Braun, V.; Kusume, K.; Marsch, P.; Maternia, M.; Queseth, O.; Schellmann, M.; Schotten, H.; Taoka, H.; Tullberg, H.; Uusitalo, M.A.; Timus, B.; Fallgren, M. Scenarios for 5G mobile and wireless communications: the vision of the METIS project. *IEEE Commun. Mag.,* **2014**, *52*(5), 26-35.
        http://dx.doi.org/10.1109/MCOM.2014.6815890

[102]   Abdollahi, A.; Abnavi, A.; Ghasemi, S.; Mohajerzadeh, S.; Sanaee, Z. Flexible free-standing vertically aligned carbon nanotube on activated reduced graphene oxide paper as a high-performance lithium-ion battery anode and supercapacitor. *Electrochim. Acta,* **2019**, *320*134598
http://dx.doi.org/10.1016/j.electacta.2019.134598

[103]   Abouimrane, A.; Compton, O.C.; Amine, K.; Nguyen, S.T. Non-annealed graphene paper as a binder-free anode for lithium-ion batteries. *J. Phys. Chem. C,* **2010**, *114*(29), 12800-12804.
http://dx.doi.org/10.1021/jp103704y

[104]   Aïssa, B.; Laberge, L.L.; Habib, M.A.; Denidni, T.A.; Therriault, D.; El Khakani, M.A. Super-high-frequency shielding properties of excimer-laser-synthesized-single-wall-carbon-nanotubes/polyurethane nanocomposite films. *J. Appl. Phys.,* **2011**, *109*(8)084313
http://dx.doi.org/10.1063/1.3574443

[105]   Andrews, R.; Weisenberger, M. Carbon nanotube polymer composites. *Curr. Opin. Solid State Mater. Sci.,* **2004**, *8*(1), 31-37.
http://dx.doi.org/10.1016/j.cossms.2003.10.006

[106]   Gupta, T.K.; Singh, B.P.; Singh, V.N.; Teotia, S.; Singh, A.P.; Elizabeth, I.; Dhakate, S.R.; Dhawan, S.; Mathur, R. MnO_2 decorated graphene nanoribbons with superior permittivity and excellent microwave shielding properties. *J. Mater. Chem. A Mater. Energy Sustain.,* **2014**, *2*(12), 4256-4263.
http://dx.doi.org/10.1039/c3ta14854h

[107]   Gupta, T.K.; Singh, B.P.; Tripathi, R.K.; Dhakate, S.R.; Singh, V.N.; Panwar, O.S.; Mathur, R.B. Superior nano-mechanical properties of reduced graphene oxide reinforcd polyurethane composites. *RSC Advances,* **2015**, *5*(22), 16921-16930.
http://dx.doi.org/10.1039/C4RA14223C

[108]   Yao, B.; Hong, W.; Chen, T.; Han, Z.; Xu, X.; Hu, R.; Hao, J.; Li, C.; Li, H.; Perini, S.E.; Lanagan, M.T.; Zhang, S.; Wang, Q.; Wang, H. Highly Stretchable Polymer Composite with Strain-Enhanced Electromagnetic Interference Shielding Effectiveness. *Adv. Mater.,* **2020**, *32*(14)e1907499
http://dx.doi.org/10.1002/adma.201907499 PMID: 32080903

[109]   Dhawan, S.K.; Singh, N.; Venkatachalam, S. Shielding Effectiveness of Conducting Polyaniline Coated fabrics at 101 GHz. *Synth. Met.,* **2002**, *123*, 389-393.

CHAPTER 2

Lightweight Carbon Composite Foams for Electromagnetic Interference Shielding Applications

Rajeev Kumar[1*], Tejendra K. Gupta[2], Neeraj Dwivedi[1] and D.P. Mondal[1]

[1] *CSIR-Advanced Materials and Processes Research Institute, Bhopal-462026, India*
[2] *Amity Institute of Applied Sciences, Amity University, Sector-125, Noida 201313, India*

Abstract: In the modern technological era, various electronic devices are being widely used in proportion to the fast-growing demand in society. These devices have created electronic pollutions such as electronic noise, electromagnetic interference (EMI), and radiofrequency interference (RFI), leading to improper functioning of electronic devices. In view of mitigating these detrimental effects, strong EMI shielding materials are required. In recent years, carbon composite foams have attracted worldwide interest due to their outstanding properties such as lightweight, interconnected porosity, high surface area, excellent corrosion resistance, superior electrical and thermal properties that are highly useful for suppressing electromagnetic noises. In this chapter, several carbon foam preparation methods, such as the foaming method, pressure release method, template method, *etc.*, are described. The effect of precursors like pitches, resins, polymers, and biodegradable materials on the microstructure, electrical, and EMI shielding properties of carbon composite foams have been studied. The influences of different fillers such as CNT, graphene, MXene, metals, and magnetic materials on the electrical conductivity and EMI shielding properties of carbon composite foams have been deeply reviewed. The EMI shielding, density, and thickness of carbon composite foam with various loading of fillers are summarized in tables in this chapter, which will provide readers with useful information. In the last section, current challenges and future research directions of this growing field are also discussed.

Keywords: Carbon composite foam, CNTs, Electrical conductivity, EMI shielding, Graphene, Magnetic fillers, Metals, MXene, Pitch, Polymer.

*Corresponding author Rajeev Kumar: CSIR-Advanced Materials & Processes Research Institute, Bhopal 462026, India; Tel: 91 7838352624; E-mail: kumarrajeev4@gmail.com

Sundeep K. Dhawan, Avanish Pratap Singh, Anil Ohlan, Kuldeep Singh Kakran and Pradeep Sambyal (Ed.)
All rights reserved-© 2022 Bentham Science Publishers

2.1. INTRODUCTION

In the modern world, electronics and telecommunication devices are developing rapidly, and these devices are generating more and more electromagnetic (EM) waves. These high-frequency EM waves affect not only the electronic system but also the surrounding environment. They also affect the lives of humans and animals [1-4]. The EMI shielding mechanism depends on the reflection and absorption of the incoming EM wave. For high reflection, the shielding materials must have high electrical conductivity, whereas for achieving high absorption, the shielding materials require magnetic permeability. Conventionally, metals such as aluminum, copper, steel, iron zinc, *etc.*, have been used as an effective EMI shielding material due to their high electrical conductivity [5, 6]. However, heavyweight, easy corrosion and poor flexibility make them an undesirable choice for advanced EMI shielding materials. Further, the high EMI reflection in these metals limits their use in those applications where absorption dominance is required, such as stealth technology or other electronic devices [7]. The main necessities for an EMI shielding material are lightweight, good electrical conductivity, and thermal stability. Recently, carbon composite foams have many advantages over their metal-based counterparts due to their lightweight, open porosity, high surface area, excellent electrical and thermal conductivity, resistance to corrosion, and chemical and thermal stability properties [8-12]. Porosity can play an important role in the composite foam for achieving high EMI shielding, especially absorption, because an open porous structure provides a high chance for multiple internal reflections [13]. However, due to the highly porous structure, carbon foams have low mechanical properties. Also, the EMI shielding properties of polymer-based composite foams depend on electrical conductivity, dielectric and magnetic permeability, and conducting fillers. As we know, most polymers are electrically insulating and cannot prevent EM waves. It is reported that the incorporation of conducting and magnetic fillers in the polymer matrix can improve the electrical conductivity, thermal stability, mechanical strength, and EMI shielding properties [13-18]. A large number of conducting fillers such as graphite, carbon black, carbon nanotubes (CNTs), carbon fibers and graphene, MXene, metals, and magnetic nanoparticles, and different types of industrial waste have been incorporated in composite foam [13, 19-26]. The lightweight composite foams with excellent EMI shielding performance have been reported for next-generation electronic and communication devices [27, 28]. Besides this, the demand for continuous development of portable electronic devices is increasing day by day, not only for lightweight EMI shielding enclosures but also for electrodes in secondary batteries

[29], supercapacitors [8], fuel cells, as sorbents for separation processes and gas storage [30, 31] and as supports for catalytic processes [32, 33]. In this chapter, up-to-date research on the fabrication of lightweight carbon composite foam by different methods and improvement in EMI shielding performance by the incorporation of conducting and magnetic fillers is covered, and the mechanism of EM wave with carbon composites foam is also discussed.

2.2. THEORY OF ELECTROMAGNETIC INTERFERENCE SHIELDING

Electromagnetic interference (EMI) is the disturbance that affects an electrical circuit due to either EM induction or EM radiation generated from an external device. This disturbance may interrupt the smooth functioning of another electronic device. In general, EMI shielding effectiveness (SE) is defined as the logarithmic ratio of the incident and transmitted wave as follows; [34, 35].

$$\text{SE} = 10\log\left(\frac{P_I}{P_O}\right) \tag{1}$$

where P_I is the power of the incoming EM wave, and P_O is the power of the outgoing wave. When this EM wave falls on the surface of shielding material, generally, three mechanisms, *i.e.,* reflection (R) of the wave from the front surface, absorption (A) of the wave inside the shield material, and multiple reflections of the wave at various interfaces of the shield, occur, which are connected as described here;

$$A + R + T = 1 \tag{2}$$

The scattering parameters S_{11} and S_{21} are used to calculate these components using a vector network analyzer (VNA) and can be defined as;

$$R = |S_{11}|^2 \tag{3}$$

$$T = |S_{21}|^2 \tag{4}$$

$$A = 1 - R - T \tag{5}$$

By using equation 5, total shielding effectiveness (SE_T) can be written as:

$$SE_T\,(dB) = SE_R + SE_A + SE_M \tag{6}$$

where SE_R, SE_A, and SE_M are shielding effectiveness due to reflections, absorption, and multiple reflections, respectively. The SE_M can be ignored when the SE_T is > -10 dB. In that case, total shielding (SE_T) depends only on reflection and absorption, as shown below [19].

$$SE_T = SE_R + SE_A \qquad (7)$$

Further, SE_A and SE_R can be described as

$$SE_R = -10 \log \left(\frac{1}{1-R}\right) \qquad (8)$$

$$SE_A = -10 \log \left(\frac{1-R}{T}\right) \qquad (9)$$

Here, SE_R is associated with the impedance matching of two materials. However, SE_A is associated with the dielectric and magnetic permeability of the materials.

Further, SE_R and SE_A can also be described as [34, 36]:

$$SE_R \text{ (dB)} = 10 \log_{10} \left(\frac{\sigma}{16 f \varepsilon_0 \, \mu_r}\right) \qquad (10)$$

$$SE_A \text{ (dB)} = 20 \frac{t}{\delta} \log_{10} e = 8.68 \left(\frac{t}{\delta}\right) = 8.68 \ t \left(\frac{\sigma f \mu_r}{2}\right)^{1/2} \qquad (11)$$

where σ, f, ε, μ, t and δ represent electrical conductivity, angular frequency, permittivity, permeability, thickness, and skin depth of the shielding material, respectively. The skin depth of a material is the penetration depth where the power of the EM wave gets reduced by 37%. Skin depth can be defined as follows [25]:

$$\delta = \sqrt{\frac{2}{f \mu \sigma}} = -8.68 \left(\frac{t}{SE_A}\right) \qquad (12)$$

Several methods are reported for the measurement of shielding effectiveness. The basic principle for these methods is the same. However, the testing setup could differ from each other. The shielding effectiveness of all of these methods is calculated by the measurement of the energy of EM waves that penetrate the material and the energy of the EM wave that is transmitted through the material [37]. The Vector Network Analyzer (VNA) is used for the generation and measurements of EM waves [38]. Generally, four methods are used for the EMI shielding measurement [2], which are (i) open-field method, (ii) shielded box method, (iii) shielded room method, and (iv) coaxial transmission line method. Out of these four different methods, the coaxial transmission line is a well-known method for the EMI shielding measurement due to its repeatability and ability to be used in a wide range of frequencies. The test setup contains VNA, two coaxial cables for receiving and transmitting EM waves, and a sample holder.

2.3. METHODS FOR THE FABRICATION OF CARBON COMPOSITE FOAMS

Carbon composite foams are lightweight, high-performance materials consisting of an interconnected cell wall structure. The carbon foam can be an open cell structure or a closed-cell structure. Different materials such as phenolic resins [39, 40], polyurethanes [41, 42], coals [43, 44], coal-tar pitches [45, 46], petroleum pitches [47, 48], mesophase pitch [49, 50] and various kinds of biomass/raw biomaterials [51, 52] have been used for the fabrication of carbon foam. For the synthesis of carbon foams, different fabrication processes such as blowing agent, temperature, pressure release, and template methods have been reported in the literature [11, 42, 50, 53-55]. The research focused on the development and evaluation of carbon foams has shown that different foaming techniques play an important role in the microstructure and final properties of carbon foam. Herein, in this section, we have addressed three methods for the fabrication of carbon foams.

2.3.1. Foaming Method

In this method, a blowing agent or foaming agent is generally used for the preparation of carbon foam. Foaming agents that produce gas *via* chemical reactions include ammonium bicarbonate, azodicarbonamide, titanium hydride, isocyanates, aluminum nitrate, *etc.* [56]. The mixture of carbon precursor and the foaming agent is heated up to the evaporation temperature of foaming agents. For the decomposition of foaming agents, a lower heating rate (<1°C/ min) is preferred. Afterward, the foam material is heat-treated at various temperatures for oxidation stabilization, carbonization, and graphitization to get the carbon foam. Different foaming agents and space holder materials have been used for the fabrication of carbon composite foams. Prabhakaran *et al.* [57] have synthesized low-density carbon foam by sucrose using aluminum nitrate as a blowing agent and converting it into solid foams followed by dehydration and carbonization.

In another work, Kumar *et al.* [58] used ammonium bicarbonate to prepare graphite foam from coal tar pitch derived semi coke through carbonization and graphitization, as shown in Fig. (**1**). Partially open cells were generated *in situ* during the sintering due to the removal of ammonium bicarbonate, as shown in Fig. (**2a, b**). It can be observed that with graphite foam with high relative density (0.30), the number of pores is relatively less (Fig. **2a**) compared to low relative density (0.22) foam sample (Fig. **2b**).

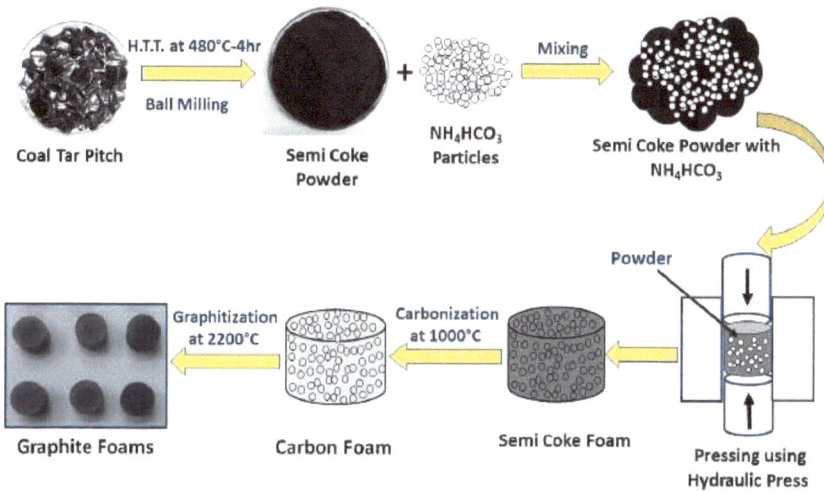

Fig. (1). Schematic illustration for the synthesis of carbon/graphite foam using ammonium bicarbonate. Reproduced with permission [58].

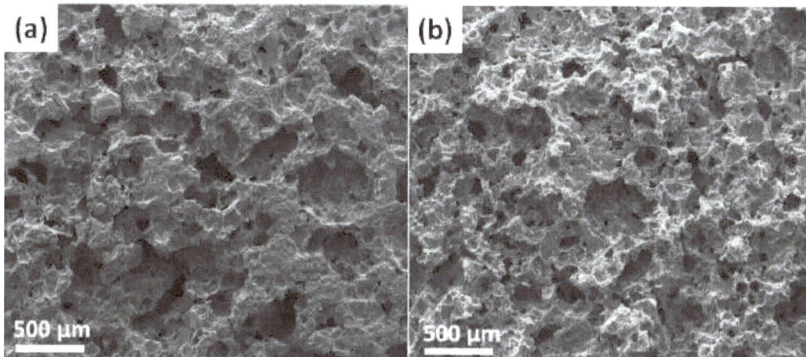

Fig. (2). Microstructure of graphite foam with different relative densities; (**a**) 0.30 and (**b**) 0.22. Reproduced with permission from [58].

2.3.2. Pressure Release Method

The pressure release method is a complex technique for the production of carbon foam because cell size and the cell wall of foam cannot be controlled by this method. The coal-based precursors such as coal tar pitch, petroleum pitch, and mesophase pitch are used for carbon foam [59]. In this method, the carbon precursor is filled in an autoclave and then heated to the melting point of this pitch and held for some time. The volatile gas which is generated from the pitch and inert gas is used to press the foam. The green foam sample taken out from the autoclave is first stabilized in an air atmosphere above the melting point of this pitch. Afterward, the stabilized foam heated up to 1000°C for carbonization and finally graphitized at 2500°C, as displayed in Fig. (**3**) [49, 60].

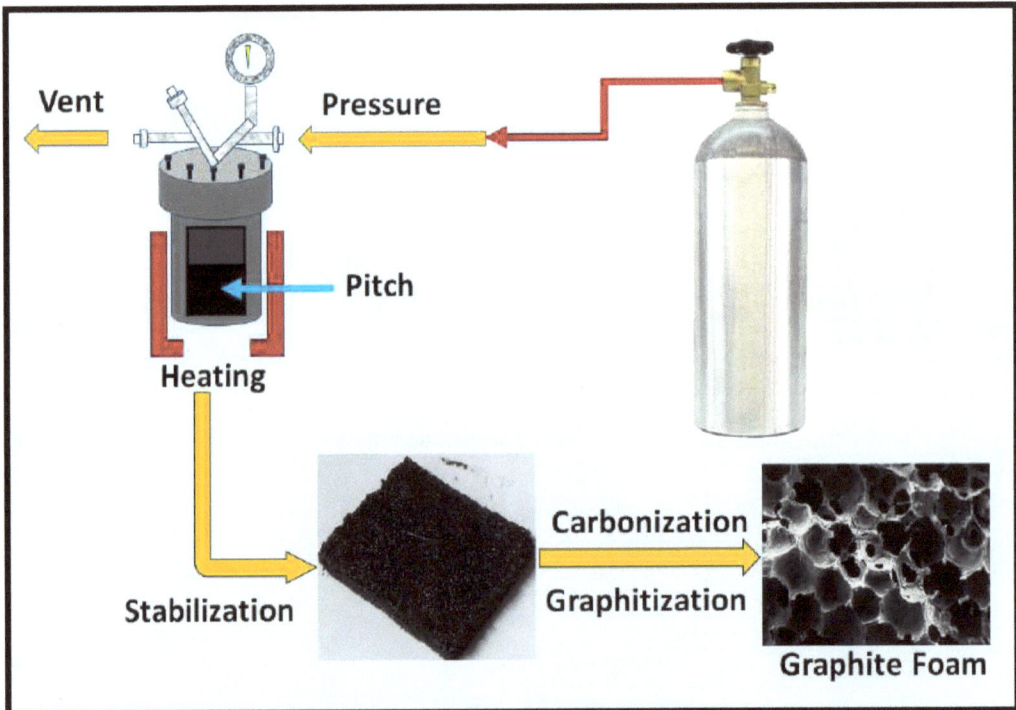

Fig. (3). Schematic drawing for the synthesis of graphite foam using temperature and pressure release method.

Klett *et al.* [49] produced carbon foam from the mesophase pitch (ARA-24) using a high-temperature high-pressure autoclave and foam samples graphitized at

2800°C in an argon atmosphere. Similarly, Chen and his coworkers [44] have demonstrated carbon foam from the various pitch precursors. The green foam samples were carbonized at 1000°C in an inert atmosphere to remove the volatiles and increase the strength. Eksilioglu *et al.* [61] demonstrated that pressure, pressure release time, and sintering temperature can affect the foam properties such as pore morphology, density, and compressive strength. The carbon foams produced at four different pressures (*i.e.,* 3.8, 5.8, 6.8, and 7.8 MPa) are represented. In all these experiments the temperature was kept at 573 K and pressure release time at 5 s. The SEM image revealed a smaller number of pores in carbon foam produced at low pressure (38 bars) compared to foam produced at high pressure. Li *et al.* [62] reported that graphite foams possess low compressive strength because a large number of cracks have been developed during the sintering carbonization and graphitization process. They showed that the compressive strength of graphite foam can be improved by the addition of meso-carbon microbeads (MCMBs) into the mesophase pitch. By adding 55 wt. % of MCMB, the compressive strength of graphite foam increased from 2.8 to 23.7 MPa because cell sizes decreased from 1000 to 100 mm. It is also observed that a smaller number of cracks in graphite foam significantly improved the thermal properties.

2.3.3. Template Method

This is a simple and cost-effective method to fabricate carbon foam in which polyurethane (PU) foam and silica are generally used to synthesize the carbon foam. The template is removed *via* heat treatment by creating a porosity according to the template size [19, 42, 63, 64]. The pitches, resins/polymers are used as a carbon precursor for the impregnation of PU foam. The slurry of carbon precursors (pitch, resin/ polymer) is prepared using water or organic solvent. Then a PU foam is impregnated by the pitch/resin slurry to obtain the impregnated PU foam. The pitch/resin-impregnated foam is then dried to evaporate the water or organic solvent completely. The dried foam is heat-treated at 300°C for oxidation stabilization in the air atmosphere which introduces polymerization of an organic molecule and maintains the shape of foams during high-temperature sintering. The stabilized foam is then carbonized at 1000°C and finally graphitized at 2500°C in an inert atmosphere to get graphite foam as shown in Fig. (**4**).

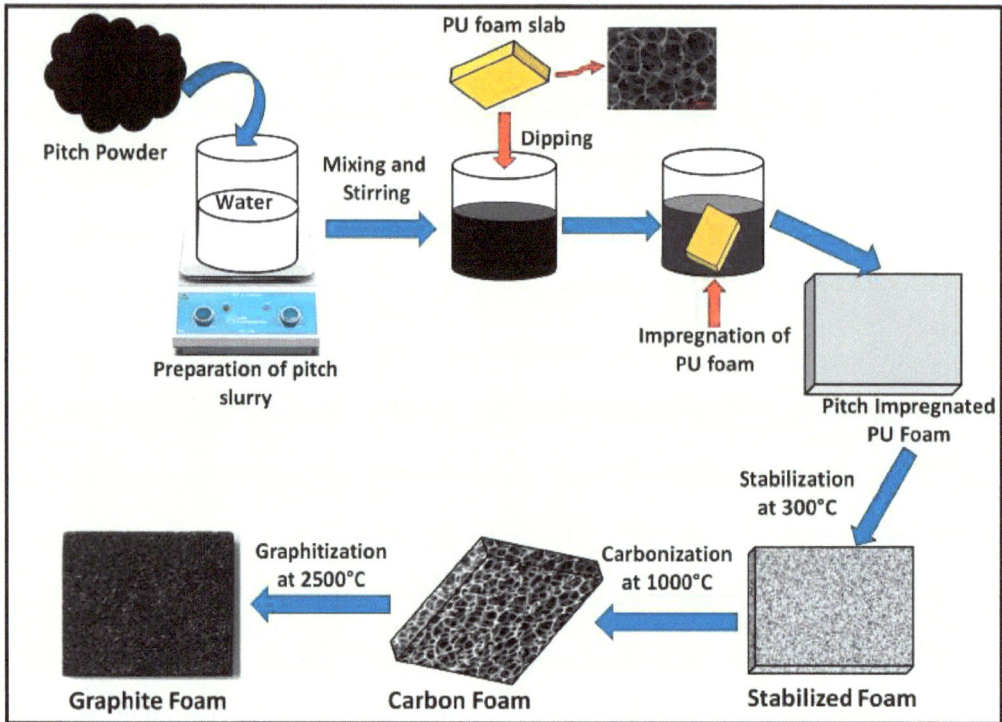

Fig. (4). Schematic illustration for the synthesis of carbon/graphite foam using the PU foam template method.

Taking the advantage of the template method, Inagaki and his co-workers [42] produced graphite foams from polyimide (PI) as the carbon precursor. Initially, PU foam is filled with PI and heat-treated at 200°C for imidization of PU/PI composite foam. The PU/PI foam is heat-treated at 400°C for the removal of PU and then carbonized above 800°C and finally graphitized at 3000°C in an inert atmosphere. Similarly, Sharma *et al.* [13] synthesized carbon foam by dissolving a phenol-formaldehyde resin in acetone and then impregnation of PU foam through resin solution followed by carbonization at 1000°C in an argon atmosphere. The developed carbon foam has an interconnected open cell structure as shown in Fig. (**5**). Further, a pitch-based graphite foam using the PU foam template method has been prepared by Chen *et al.* [55]. In this investigation, a high softening point petroleum pitch is used as a carbon source and PU foam as a template. The pitch was ground into a fine powder using ball milling then mixed with 35% by weight of water to form a homogeneous slurry. The slurry was impregnated into the PU foam slabs. The pitch impregnated foams are then stabilized, carbonized, and finally graphitized at 2800°C in an argon atmosphere. SEM image of the carbon foam shows that cells are open and uniformly distributed.

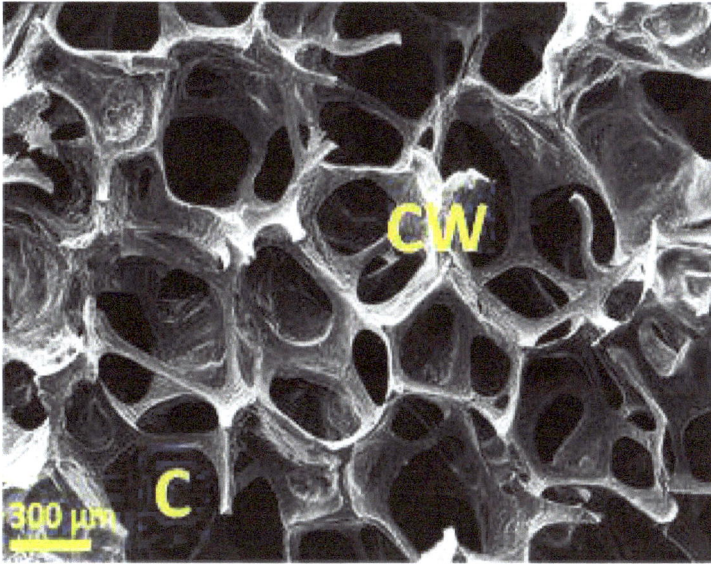

Fig. (5). SEM images of phenolic resin-based carbon foam. Reproduced with permission from reference [13].

Metal foams such as nickel/copper are also used for the fabrication of graphene in the CVD method [65, 66]. In this method, nickel foam is used as a template. The nickel foam is heated to 1000°C in a horizontal tubular furnace under a hydrogen and argon atmosphere to remove the oxide layer. Afterward, CH_4 gas is introduced into the reaction tube at ambient pressure. After that, the sample was cooled down to room temperature to get the graphene on a nickel foam template. To support the graphene foam when the nickel is etched away, a thin layer of PMMA is used to support the graphene structure. Finally, PMMA is dissolved in acetone to obtain free-standing graphene foam.

Various polymer/resin-based carbon foam using the PU foam template has also been recently produced for various applications including thermal insulating, EMI shielding, an electrode for lead-acid batteries, oil spillage, and water purifications

[25, 67, 68]. In addition to that, lightweight porous carbon derived from biomass has recently drawn attention for their sustainable properties, low cost, and high performance. In this direction, Li *et al.* [69] used sugarcane to produce aerogel-like carbon (ALC) by a hydrothermal carbonization process. In this study, the pitch of a sugarcane stalk was cut into suitable dimensions, and put into a Teflon-lined stainless-steel autoclave, and then heat-treated at 180°C to get carbonaceous sugarcane monoliths. After that carbonaceous sugarcane monoliths were freeze-dried for obtaining the aerogel-like sugarcanes (ALSs). To obtain the ALC, the ALS was paralyzed at 800 °C for 1 h in an N_2 atmosphere as shown in Fig. (**6**). The developed ALC was used for EMI shielding applications.

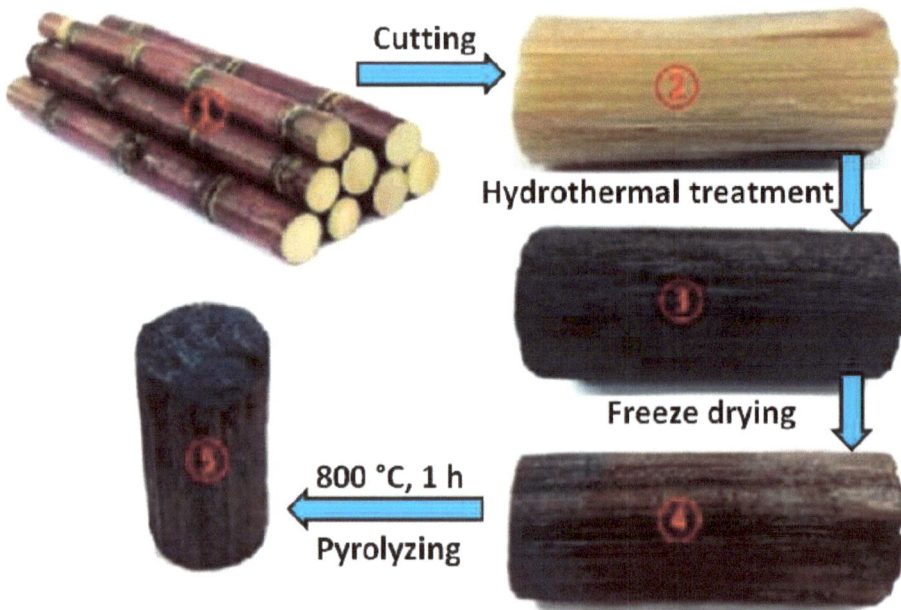

Fig. (6). Schematics representation for the preparation of ALC (**1**) sugarcane stalks, (**2**) sugarcane pith, (**3**) carbonaceous sugarcane, (**4**) aerogel-like sugarcane, (**5**) aerogel-like carbon. Reproduced with permission from reference [69].

Yuan *et al.* [51] investigated lightweight and stiff carbon foam using flour by a simple fermentation and carbonization process as shown in Fig. (**7**). In another work, again Yuan *et al.* [70] presented carbon honeycomb-induced graphene composite foams for EMI shielding. In this study, they used Nomex honeycomb as a template and carbonized it at 1000°C for obtaining the CH structure. The CH structure was then dipped into a graphene oxide solution until all of the cells were filled with graphene oxide. The freeze-drying method was used for graphene composites foam as shown in Fig. (**8**). Similarly, lightweight and efficient microwave absorbing porous carbon/CoFe$_2$O$_4$ composites have been prepared using a loofah sponge [71].

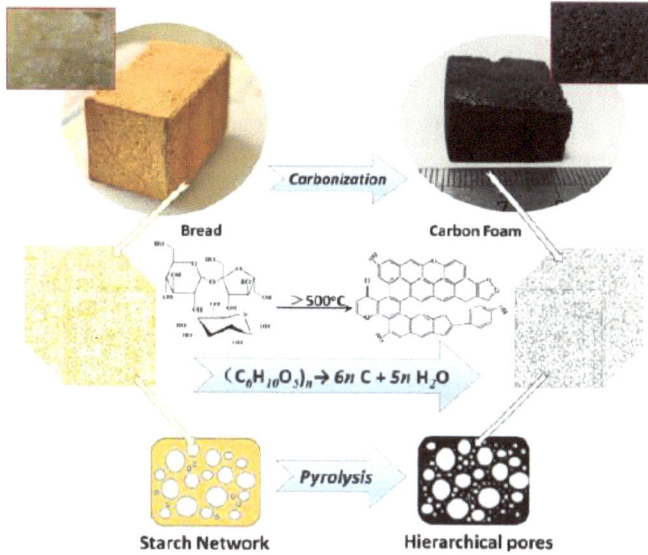

Fig. (7). Schematic Fig. for the synthesis of carbon foam derived from bread. Reproduced with permission from reference [51].

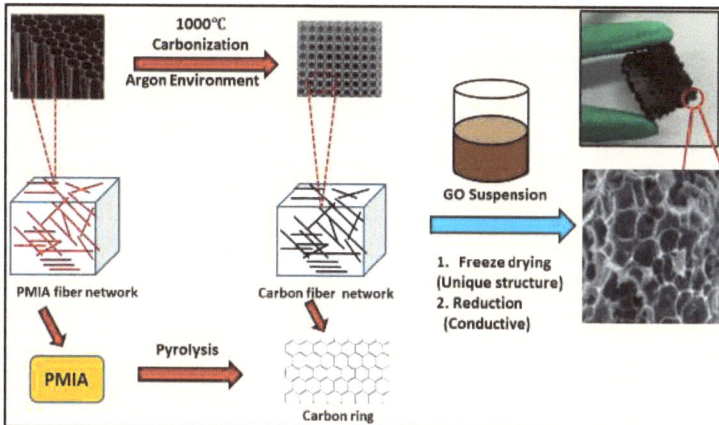

Fig. (8). Schematic illustration for the preparation of carbon honeycomb-induced graphene composite foams.

2.4. CARBON FOAMS FOR EMI SHIELDING

In the past years, conventional shielding materials such as steel, copper, aluminum, zinc, and nickel have been used for achieving the EMI shielding due to high electrical conductivity and dielectric constant. However, metals have high density and high corrosion rate makes them unsuitable for EMI shielding. Alternatively, most of the metals mainly protect from EMI through reflection, preventing their use in dominant absorption such as stealth technology or other electronic devices [72, 73]. Considerable attention has been drawn to lightweight carbon composite foam due to its high surface area, good electrical conductivity, and excellent absorption in the EMI shielding [74-76]. The open porous structure and high surface provide more chances to entrap the EM wave inside the material. The schematic diagram of the EM wave interaction with carbon foams is shown in Fig. (**9**).

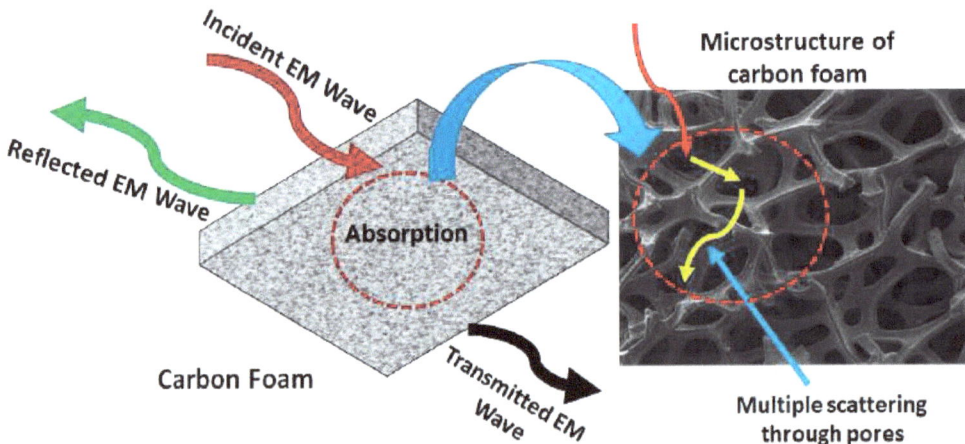

Fig. (9). Electromagnetic wave interaction with carbon foam.

The application of carbon foam as an EMI shielding material is reported by a few authors. Yang *et al.* [77] investigated mesophase-derived carbon foams for microwave absorption properties. They showed that carbon foams are sintered in the temperature range of 400-800°C and described that carbon foams exhibit good microwave absorption (10 dB) which were sintered at 600 and 700°C. Similarly, EMI shielding properties of carbon foam having different pore size has been studied by Fang *et al.* [78]. The resonant cavity perturbation technique is used for calculating the EMI parameters of carbon foam and its corresponding pulverized powder at a frequency of 2.45 GHz. It is observed that with increasing the sintering temperature, the electrical conductivity of carbon foam increased, and their dielectric loss was several times higher than their corresponding crushed powder. Similarly, EMI shielding of carbon foam (GRAFOAM FPA-20 and FRA-10) was measured in the frequency range of 1- 4 GHz [79]. Here, it is important to note that in any shielding material, the SE below 10 dB is said to offer very little or no shielding. However, materials with EMI SE ≥ 20 dB are acceptable for industrial and commercial applications as it attenuates 99 % of impinging EM signals [80]. To make the carbon foam for commercial applications, several attempts have been devoted to improving the EMI shielding properties of carbon composite foams with various conducting and magnetic fillers.

2.4.1. Carbon Composite Foams with Carbon Fillers

The outstanding properties of carbon materials such as lightweight, high surface area, excellent tensile strength, high flexibility, and outstanding electrical and thermal conductivity make them potential candidates for multifunctional applications [81]. The various forms of carbon fillers include graphite, carbon nanotubes (CNTs) carbon fibers, graphene, graphene oxide (GO) have been incorporated in carbon/polymer matrix for achieving excellent EMI shielding performance in composite foam. Keeping in view of the main importance of carbon-based fillers, this section summarizes the current progress of carbon filler reinforced composite foams for EMI shielding applications.

2.4.2. Carbon Nanotube Reinforced Composite Foams for EMI Shielding

After the discovery of carbon nanotubes (CNTs) by Iijima [82, 83], CNTs have great potential due to their lightweight, small diameter, high aspects ratio, exceptional electrical and thermal conductivity, and ultrahigh mechanical strength which make them ideal reinforcement over conventional carbon fillers for high EMI SE properties [84, 85]. A recent investigation reveals that a lower weight percentage of CNTs in the polymer matrix can increase the EMI SE of composite foam. Yang *et al.* [86] demonstrated polystyrene-carbon nanotubes (PS-CNT) composite foams for EMI shielding applications. The polystyrene composite foams prepared using foaming method, were reinforced with the different loading of CNT ranging from 0 to 7 wt. %. The SEM images of PS composite foam containing 5 wt. CNT % are shown in Fig. (**10a, b**). The foam structure was observed throughout the CNT-PS matrix as shown in Fig. (**10a**). The high magnified SEM image of PS-CNT composite foam is shown in Fig. (**10b**). From this image, it was found that CNT is uniformly dispersed in the PS matrix and formed an interconnected conducting network. This conductive network is provided by an electrical conduction path in the PS-CNTs matrix which is responsible for enhancing the electrical conductivity and EMI SE in PS-CNT composite foam. Fig. (**10c**) shows the EMI SE of PS-CNT composite foam at the X band (8.2-12.4 GHz) frequency range. It was obvious that with increasing the CNTs content in composite foam, the EMI SE continuously increased and reached the maximum value of 19.3 dB when the CNTs content was only 7 wt. %. This may be due to the large interfacial area and high aspect ratio of CNTs which further provide a conducting network in composite foam to interact with incident EM waves and lead to high EMI SE in PS-CNT composite foam.

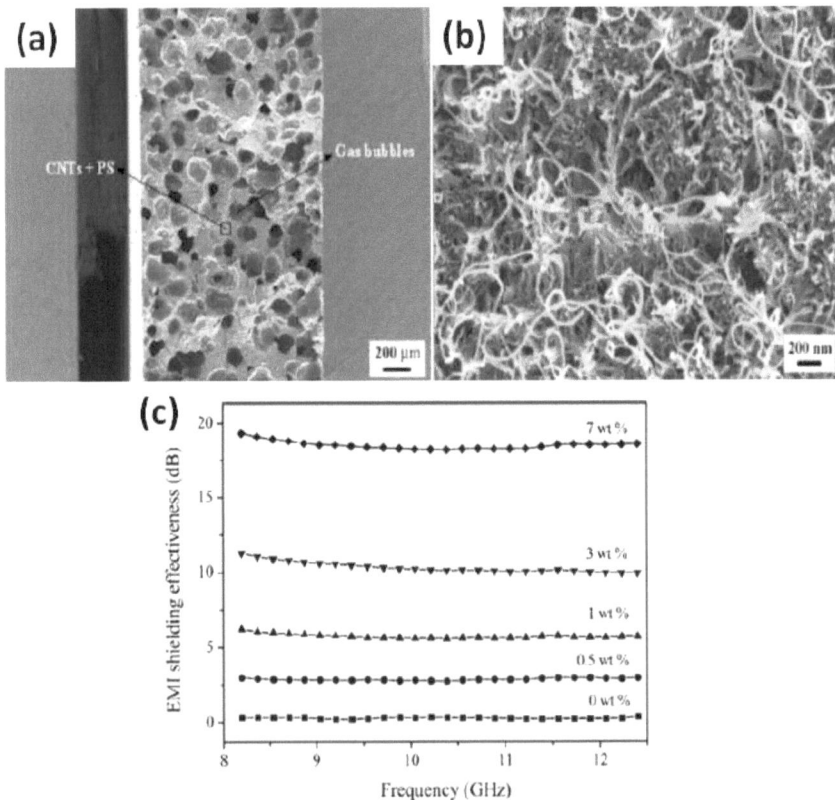

Fig. (10). SEM images of the cross-section of the 5 wt. % CNTs reinforced PS-CNT composite foam (**a**) foam structure, (**b**) CNT networks within the PS matrix can be seen through the high-magnified image (**c**) EMI SE of PS-CNT composite foam at 8.2-12.4 GHz as a function of CNT loadings. Reproduced with permission from reference [86].

As per the EMI shielding mechanism, the total EMI SE is due to reflection, absorption, and multiple reflections. In a particular case, when total SE is >10 dB, shielding effectiveness due to multiple reflections can be neglected. So, to investigate the effect of CNTs on the reflection and absorption of carbon foam, Narasimhan *et al.* [87] reported MWCNT reinforced carbon composite foams. The different contents (0-2.5 wt. %) of MWCNT were dispersed in molten sucrose and foam was developed by foaming agent followed by carbonization at 900°C. The VNA (model; E5071C) is used to measure the EMI SE of foam samples in the X-band frequency range (8.2-GHz) (Fig. **11**). It is observed that the absorption is higher than the reflection in all the carbon foams. When the MWCNTs concentration increases from 0 to 0.25 wt. %, the absorption component increases from 13.8 to 17 dB, and the total EMI SE increases from 20 to 26 dB respectively. Furthermore, the total EMI SE reached a maximum value of 38 dB with high absorption ~28 dB when MWCNTs concentration increases to 2.5 wt. %. The enhanced absorption in carbon foam is mainly due to point defects and dangling bonds in MWCNTs which provide repeated polarization of dipoles [88]. Further, the agglomeration of MWCNTs with sucrose created dielectric loss at the walls of the mesopores which is also responsible for higher EMI SE.

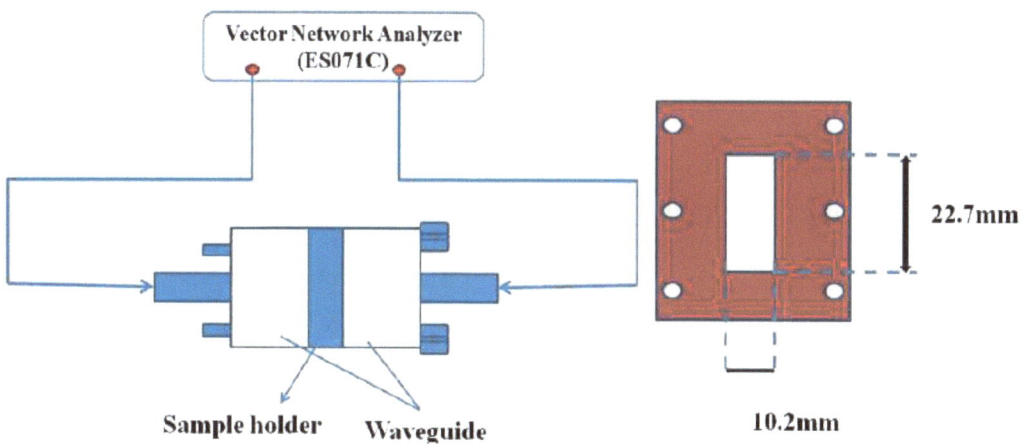

Fig. (11). Schematic setup of VNA for the measurement of EMI shielding of carbon foam.

Further, the effect of CNTs on the EMI SE of pitch-based carbon foam was investigated by Kumar *et al.* [19]. They presented that carbon foams are decorated with MWCNTs *via* two different routes. In the first case, different contents (0.5-2 wt. %) of MWCNTs are incorporated in the coal tar pitch and carbon foams are synthesized using the PU foam template method by heat treatments up to 2500°C in an inert atmosphere. While in the second case, MWCNTs are directly grown over 2500°C heat-treated pristine carbon foams by the chemical vapor deposition technique. The incorporation of MWCNTs in carbon foams exhibited an EMI SE of 72 dB at 8.2 GHz with only 1 wt. % MWCNTs content. In comparison, MWCNTs directly grown on carbon foam display a maximum EMI SE of 85 dB at the same frequency 8.2 GHz with only 0.5 wt. % MWCNTs decoration. The high EMI SE in directly grown MWCNT carbon foam is due to a continuous network of CNTs which helps to increase the electron transport properties in the foam. Further, on the surface of carbon foam, the directly grown MWCNTs quickly react with EM waves which leads to higher EMI shielding.

Li *et al.* [89] have described the EMI SE of epoxy/multi-wall carbon nanotube (EP/MWCNT) composite foams at the Ku band (12-18 GHz). The EMI shielding properties of EP/MWCNTs composite foam compared with solid EP/MWCNTs composite at the same loading of MWCNTs. It was observed that the EMI SE of both EP/MWCNTs solid and foamed composites increased gradually with increasing MWCNTs content. However, EP/MWCNT composite foam has lower EMI SE compared to the solid composite with the same MWCNTs content. The EMI SE of 3 wt. % MWCNT loaded foamed composite is 7.1 dB whereas, in solid composite, 9.5 dB is observed.

A lightweight composite foam can also be synthesized by biodegradable poly (l-lactic) acid (PLLA) with MWCNTs using supercritical carbon dioxide foaming [90]. The morphology of 10 wt. % MWCNTs nanocomposite foam shows the variation in the cell shape, size, and wall thickness compared to neat PLLA foam. It was observed that the EMI SE of composites foams increases significantly with an increase of the MWCNT content. It was found that SE_{total} and SE_A increase with increasing the MWCNT content, while SE_R remains the same (below 1.3 dB). In this case, composite foam containing 10 wt. % MWCNT, the measured value of SE_{total}, SE_A, and SE_R are 24.6, 23.8, and 0.8 dB, respectively. The EMI SE performance of this composite foam is dominated by absorption. There are mainly two reasons for increasing EMI SE in the composite foam. First, the incident EM wave can easily enter into the composite foam due to the open cell structure and gives less reflection. Second, the conductivity of composite foam provides more multiple reflections and thicker cell walls, which could also successfully attenuate the EM wave from one pore to another pore.

Recently, aerogels are highly porous and ultra-lightweight materials consisting of the ultra-high- specific surface area which makes them a prime candidate for lightweight highly efficient EMI materials. As an example, Huang *et al.* synthesized ultra-light and high conducting aerogel of CNTs and cellulose by using the freeze-drying method [91]. The obtained conductive aerogel exhibits an EMI SE of ~20.8 dB with absorption dominating and their corresponding specific EMI SE of ~219 dB cm^3 g^{-1} at a density of 0.095 g cm^{-3}. Similarly, Li *et al.* adopted a freeze-drying method to prepare robust CNT/chitosan (CS) foam for efficient microwave absorption [92]. Different contents of chitosan (CS) *i.e.,* 0, 10, 20, and 40 wt. % were used to prepare CNT/CS foam and referred to as CNT, CNS10, CNS20, and CNS40, respectively. Fig. (**12**) displays the EMI shielding parameters; SE$_T$, SE$_A$, and SE$_R$ of CNT and CNS foam at a frequency of 10.3 GHz. As displayed in Fig. (**12**), it was found that the SE$_A$ first increases and then decreases with a further increase in the CS content, while SE$_R$ is nearly unchanged among all the CS contents. The total EMI SE is mainly due to SE$_A$ (96%) suggesting an absorption dominating mechanism in CNS foams. The superior SE$_A$ is mostly due CNT network in the CS matrix which makes a conducting path for achieving high conductivity and EMI SE in CNS foam. Further, in porous materials, the EMI SE does not only depend on conductivity but also density/thickness, and the shielding (dB) per unit density/thickness is called specific shielding effectiveness (SSE). In some instigations the SSE is calculated by dividing the shielding effectiveness by both density and thickness. Also, a specific EMI SE up to 8556 dB.cm^2.g^{-1} could be achieved in CNS20 foam.

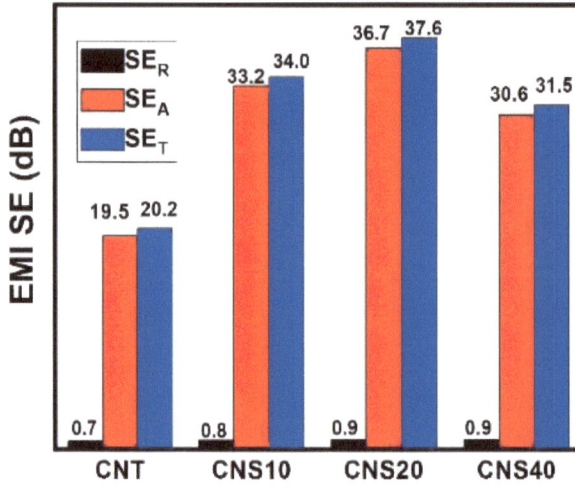

Fig. (12). SE_A, SE_R, and SE_T of pure CNT and CNS foams at the frequency of 10.3 GHz.

Furthermore, the highly porous structure might make the composite foam brittle which greatly influences the overall performance of foam in flexible devices. To solve this problem, Zeng *et al.* [93] demonstrated lightweights, flexible, and porous MWCNT/polymer composite for EMI shielding applications. The porous composite was synthesized from the blend of MWCNT and water-borne polyurethane (WPU) by the freeze-drying method. The MWCNT-WPU composite foam possesses the EMI shielding of 46.7 and 21.7 dB for the density of 32.3 and 9.0 mg cm^{-3}, respectively in the X –band region. Due to very low density and thickness, the composite foam exhibited the highest SSE (10^4 dB cm^2 g^{-1}) reported ever. The EMI SE of various CNT-based composite foams in the X-band frequency range (8.2-12.4 GHz) is shown in Table **1**. From this table, it is observed that the EMI SE not only depends on the fillers but also depends on the properties such as density and thickness of composite foams.

Table 1. EMI SE of carbon foam and CNTs based composite foams in X band (8.2-12.4 GHz) frequency. (PS: Polystyrene; PLLA: Poly (l-lactic acid); WPU: Waterborne polyurethane; PVDF: Poly vinylidene fluoride).

Materials	Density g/cm^3	Thickness (mm)	EMI SE (dB)	References
Resin derived carbon foam	0.34	2.75	22.3	[94]
Pitch derived carbon foam	0.52	2.75	39.0	[76]
Bread derived carbon foam	0.27	5.0	17.2	[51]
Sugarcane derived carbon foam	0.11	10	51	[69]
PS-MWCNT composite foam	0.56	-	19.3	[86]
Sucrose-MWCNTs Composite foam	0.22	5	38	[87]
MWCNTs incorporated carbon foam	0.57	2.75	72	[19]
MWCNTs decorated carbon foam	0.54	2.75	85	[19]
PLLA-MWCNT composite foam	0.30	2.5	77	[90]
Cellulose-CNT aerogel	0.095	2.5	20.8	[91]
CNTs-chitosan foam	0.017	2.5	37.6	[92]
MWCNT-WPU porous composite	0.032	2.3	46.7	[93]
MWCNT-WPU porous composite	0.126	2.4	50	[95]
PVDF-MWCNTs 3D network	0.79	2.0	56.7	[96]
CNT sponge	0.01	1.8	54.8	[97]

2.4.3. Graphene Reinforced Composite Foams for EMI Shielding

Graphene is the thinnest material and is considered the most promising 2D material due to its excellent charge carrier mobility, high thermal conductivity, large specific surface area, and excellent mechanical properties [98, 99]. Graphene or reduced graphene oxide (rGO) has been successfully reinforced with a different type of polymer/carbon matrix for improving the EMI shielding of composite foams [20, 100-105]. The EMI shielding performance of the graphene-based composite foams is summarized in Table **2**.

Table 2: EMI SE of graphene-based composite foams in X band (8.2-12.4 GHz) frequency. (PEI: Polyetherimide; PMMA: Polymethyl methacrylate; rGO: Reduce graphene oxide; PDMS: Polydimethylsiloxane; PVDF: Polyvinylidene fluoride; PI: Polyimide; PU: Polyurethane; ABS: Acrylonitrile butadiene styrene; PEDOT: PSS: Poly (3,4-ethylene dioxythiophene): poly (styrene sulfonate); CH: Carbon honeycomb).

Materials	Density g/cm³	Thickness (mm)	EMI SE (dB)	References
PEI–graphene composite foam	0.30	2.3	11	[106]
Graphene-PMMA composites foam	0.79	2.4	19	[107]
Carbon-rGO composite foam	0.36	2.0	44.6	[20]
Graphene-PDMS composite foam	0.06	1.0	20	[65]
PU-rGO composite foam	0.092	2.5	20	[108]
Graphene-PVDF composite foam	-	-	28	[103]
PI-rGO composite foam	0.28	0.8	21	[109]

(Table 2) cont.....

Carbon foam	-	0.073	51	[110]
Sucrose-graphene composites foam	0.28	5.0	38.6	[105]
Graphene-ABS composite foam	0.012	1.6	42.4	[111]
rGO-aerogel	0.0055	2.5	27.6	[112]
Graphene aerogel	0.0108	4.0	35	[113]
Graphene foam/PEDOT: PSS	0.0182	1.5	90	[104]
Graphene foam	0.06	0.3	25.2	[114]
Graphene foam	0.0339	2.0	39	[115]
Graphene-CNT-PDMS cake composite	-	1.0	67.3	[116]
Graphene-CNT-PDMS composite foam	0.88	1.6	75	[117]
CH-rGO composite foam	0.020	5.0	41.0	[70]
Lignin derived carbon-rGO foam	0.065	2.0	70.5	[118]

Ling *et al.* [106] used the water vapor-induced phase separation process to fabricate PEI/graphene composite foams. The EMI shielding properties of PEI/graphene nanocomposite foams were also discussed and compared with PEI/graphene solid nanocomposite. It is obvious that the EMI SE of both foam and solid composite increases with increasing the graphene content. In comparison, the EMI SE of the solid composite reached up to ~20 dB at 10 wt. % graphene loading, however, the EMI SE value of foam composite was about 11 dB at the same graphene loading. It is also observed that when the graphene loading increases from 3-10 wt. % in solid composite, the SE_A increases from 7.27−19.66 dB, suggesting about 76.2−90.8% of EM energy is absorbed. Whereas in foam composite, the contribution of SE_A to SE_{total} increased, where about 90.6– 98.9 % EM energy is absorbed due to the cellular structure of foam composite. Further, the incident EM wave is attenuated by spherical microscale air bubbles due to reflecting and scattering between the cell wall and nanofillers, and the EM wave is difficult to escape from the foam before being absorbed and transferred to heat. In another investigation, Zhang *et al.* [107] synthesized polymethyl methacrylate (PMMA)-graphene microcellular composite foams using CO_2 as a foaming agent. The composite foam exhibited improved mechanical and EMI SE properties as compared to bulk composite. The microcellular foam exhibits an electrical conductivity of 3.11 S/m at a very low content of graphene sheets (1.8 vol %). The EMI SE of microcellular foam with different loading of graphene sheets is shown in Fig. (**13a**). Neat PMMA foam is displayed with very low EMI SE. However, with increasing the graphene content in PMMA foam, the EMI SE increases significantly. The microcellular foam with 1.8 vol % graphene content shows an EMI SE of 13-19 dB over the frequency range of 8 to 12 GHz. The EMI SE parameters such as SE_{total}, SE_R, and SE_A in PMMA-graphene foam are calculated at the frequency of 9 GHz and plotted in Fig. (**13b**). It is found that both SE_{total} and SE_A increase with increasing the graphene content whereas SE_R is negligible over all the graphene contents.

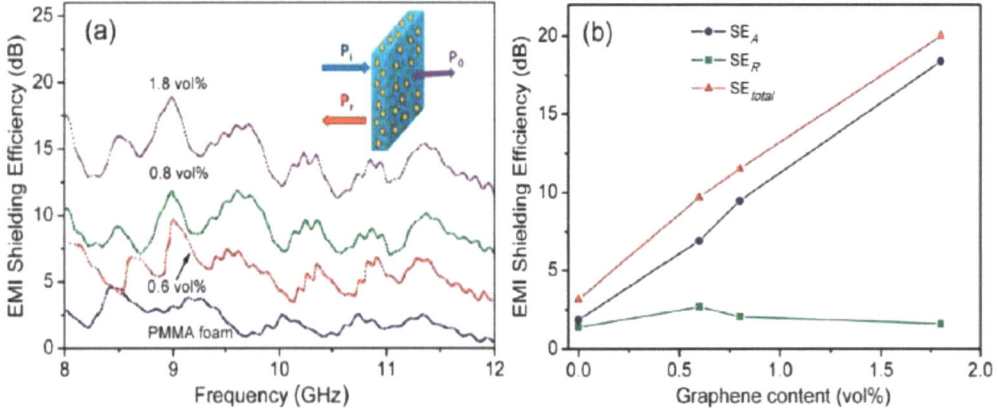

Fig. (13). (a) EMI SE of PMMA-graphene composite foams in X band frequency with different contents of graphene sheets and **(b)** the comparison of SE_{total}, SE_A and SE_R at 9 GHz. Reproduced with permission from reference [107].

Further, graphene-based composite foams also have been investigated for EMI shielding applications. Agrawal *et al.* [20] prepared carbon-rGO composite foams from phenolic resin and rGO using the PU foam template method. Two different processes were adopted to prepare the carbon-rGO composite foam. In one process, 0-4 wt. % of rGO was mixed in phenolic resin and foam was synthesized using the PU foam impregnation method followed by carbonization. In the second process, the as-received carbon foam was decorated with graphene oxide (GO) using dip coating and heat-treated at 1000°C in an inert atmosphere to obtain rGO decorated carbon foam. The electrical conductivity of carbon-rGO composite foams is shown in Fig. (**14a**). The rGO incorporated carbon foam displayed electrical conductivity of 52.4 Scm^{-1} at only 4.0 wt. % rGO loading. Whereas, the electrical conductivity of rGO decorated carbon foam exhibited 58.5 S cm^{-1} with graphene decoration of only 1 wt. %. The variation of absorption, reflection, and the total EMI SE of carbon-rGO composite foam is shown in Fig. (**14b-d**). From these Figures, it is observed that with increasing the rGO content, the SE_T and SE_A increased significantly in both the process of incorporation and decoration. The SE_T, SE_A, and SE_R for pristine carbon foam are 23.2, 12.2, and 12.0 dB respectively at 8.2 GHz. In rGO (4 wt. %) incorporated carbon foam, the value of SE_T and SE_A increased to 44.6 and 37.2 dB respectively at 8.2 GHz. However, in rGO (1 wt. %) decorated carbon foam, the value of SE_T and SE_A are reported to be 50.7 and 44.6 dB respectively at 8.2 GHz. The decoration of rGO on the foam surface makes a heterojunction structure which helps to form a conducting network for faster electron transport leads to high EMI SE in rGO decorated carbon foam. The interconnected open pore structure is also responsible for enhancing microwave absorption. Further, the high impedance mismatch between air, carbon, and graphene sheets helps to transfer the EM energy in the form of heat leading to the enhancement in absorption [119].

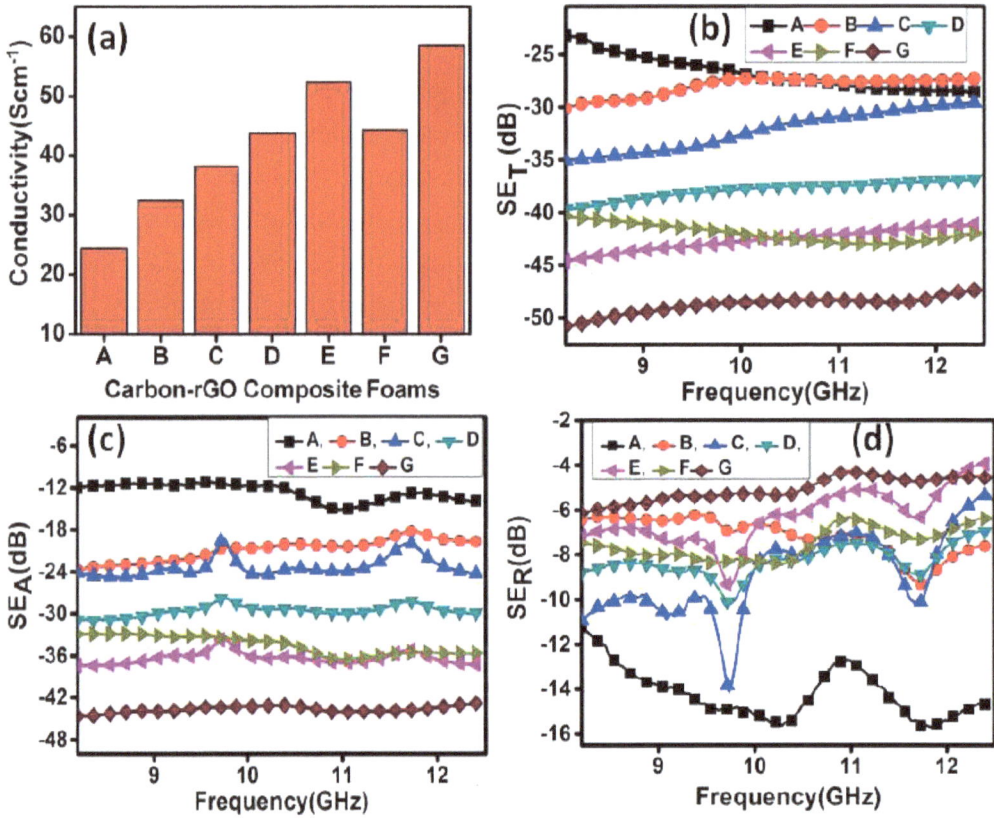

Fig. (14). (a) Electrical conductivity **(b)** Total **(c)** Absorption and **(d)** Reflection of carbon-rGO composite foams in X band frequency range. (A = 0 wt.% rGO incorporated carbon foam; B = 0.5 wt.% rGO incorporated carbon foam; C = 1 wt. % rGO incorporated carbon foam; D = 2 wt.% rGO incorporated carbon foam; E = 4 wt.% rGO incorporated carbon foam; F = 0.5 wt.% rGO decorated carbon foam and G = 1 wt.% rGO decorated carbon foam). Reproduced with permission from reference [20].

Chen and his co-workers [65] used the CVD method to prepare flexible and very thin graphene foam on nickel foam first, and then a thin layer of polydimethylsiloxane (PDMS) is coated on the graphene foam. Afterward, a freestanding graphene/PDMS composite foam is obtained by etching the nickel foam in HCl as shown in Fig. (**15a**). The electrical conductivity of the graphene/ PDMS composite foam is tuned by varying the graphene content *via* changing the flow rate of methane in CVD. The electrical conductivity of composite foam is obtained in the range of ~ 0.6 to ~ 2 S/cm for the graphene layers of ~ 3 to ~ 8 respectively. The digital photograph of flexible graphene/PDMS composite foam is shown in Fig. (**15b**) and its SEM image are shown in Fig. (**15c**) which confirms that composite foam is highly porous and similar to the structure of nickel foam. The obtained EMI SE of very low loading (< 0.8 wt.%) graphene/PDMS composite foam is reached to 20 dB (Fig. **15d**) and specific EMI SE 500 dB cm^3/g in X band frequency range. To check the effect of flexibility on the electrical and EMI SE, the composite foam was bent and tested for electrical conductivity and EMI SE. It was observed that after repeatedly bending (radius ~2.5 mm, 1000 times), the composite foam shows a very minute change in electrical conductivity and EMI SE. Due to high flexibility and excellent EMI SE, the composite foam can be used in aerospace, aircraft, and next-generation flexible portable electronic devices [120].

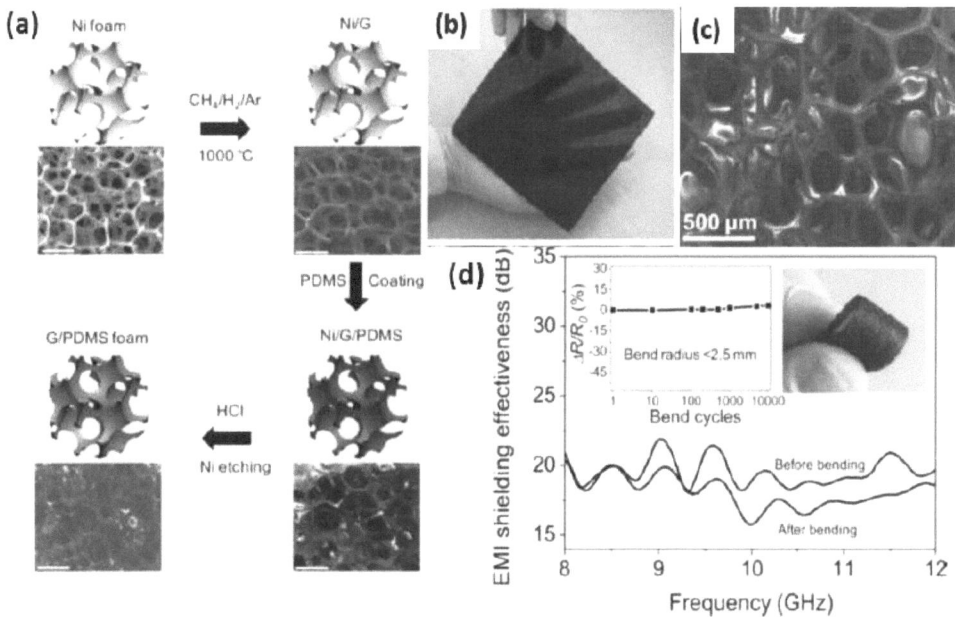

Fig. (15). (a) Schematic representation for the fabricating graphene/PDMS composites foam **(b)** digital photographs of a flexible composite foam **(c)** SEM images of a composite foam **(d)** EMI SE of composite foam before and after repeatedly bending. The change in resistance in composite foam under repeated bending is shown in the left inset. The digital photograph of bent composite foam is shown in the right inset Reproduced with permission from reference [65].

Furthermore, to prepare the flexible graphene polymer composite foam, Gavgani *et al.* [108] used *in-situ* polymerization of polyurethane (PU) in presence of reduced ultra-large graphene oxide (rUL-GO). The electrical, thermal, and mechanical properties of PU/rUL-GO composite foams were studied in detail. Composite foam exhibits an increasing order of conductivity with increasing the loading rUL-GO contents. This may be due to the homogenous dispersion of rUL-GO in the PU matrix. The electrical conductivity of 9.34 S/m is achieved when the rUL-GO content is 2 wt. %, suggesting that these composite foams can be potential candidates for the electronic industry. Results revealed that this flexible PU/rUL-GO composite foam exhibits an EMI SE of ~23 dB in the X band range with only 1 wt. % of rUL-GO content.

2.4.4. Metals Reinforced Composite Foams for EMI Shielding

Metals such as silver, nickel, copper, iron, zinc, *etc.* have been used to develop highly efficient EMI shielding materials due to their ultra-high electrical conductivity [121-123]. In addition to the small size of metal, nanoparticles can be easily dispersed in any matrix. As reported earlier the EMI SE mechanism depends on the reflection and absorption. For reflection shielding materials must be conducted, which can interact with the incoming EM wave. Therefore, metal particles could be the best choice for enhancing the EMI shielding of composite foam. To achieve EMI shielding in composite foam different metals and metal nanoparticles have been incorporated in precursor materials. In some studies, transition metals such as Fe, Ni, and Co have been used for catalytic graphitization of carbon materials which leads to improving electrical conductivity [124, 125]. Recently, some researchers have tried to design carbon-metal composite foams for EMI shielding applications and results are summarized in Table **3**.

Table 3. EMI SE of carbon-metal composite foams in X band (8.2-12.4 GHz) frequency. (PP: Polypropylene; SS: Stainless steel; PVA: Polyvinyl alcohol; PDMS: Polydimethylsiloxane).

Materials	Density g/cm³	Thickness (mm)	EMI SE (dB)	References
Carbon foam with ferrocene	0.61	2.75	81	[76]
Carbon foam nickel nanoparticles	0.56	2.75	61	[126]
Cu-Ni alloy-CNT foam	0.23	1.5	54.6	[22]
Carbon foam with Ni-Zn ferrite	1.72	3.0	42	[127]
Carbon foam with silver particles	0.58	4.0	33	[128]
PP-SS fiber composite foam	-	3.1	48	[129]
Silver plated melamine foam	-	2.0	95.8	[130]
Epoxy-Ni coated carbon fiber CNT foam	-	2.0	40.8	[131]
MXene-graphene hybrid foam	0.0037	0.3	50.7	[132]
MXene-PVA composite foam	0.10	5.0	56	[133]
MXene/rGO hybrid aerogels	0.032	2.0	56.4	[134]
PDMS coated MXene foam	0.019	2.0	70.5	[135]

Kumar *et al.* [76] used iron nanoparticles derived from ferrocene for catalytic graphitization of coal tar pitch-based carbon foam which can able to improve the EMI shielding of carbon foam. In this investigation, different proportions (0, 2, 5, and 10 wt. %) of ferrocene were mixed with coal tar pitch and carbon foam was prepared by PU template method followed by carbonization and graphitization at

2500°C. It is reported that during carbonization, the pitch-based carbon is reacted with iron at 700-900°C and formed iron carbide. Further increasing the temperature above 1400°C (during graphitization), this iron carbide subsequently split into a-Fe and carbon. Above 2000 °C, the iron content removes from carbon and improves the graphitization up to 75 % with only 10 wt. % ferrocene which leads to increases in the electrical conductivity. The carbon foam exhibited an electrical conductivity of 110 Scm^{-1} only at 10 wt. % ferrocene contents. The EMI SE of carbon foam was measured in X-band frequency (8.2-12.4 GHz). The measured EMI SE was found to be increased with increases in ferrocene content. Initially as received carbon foam showed EMI SE of 39 dB and it further increased to 81 dB by using ferrocene only at 10 wt. %. Here it was found that the performance of carbon foam in all the samples is dominated by reflection due to high conductivity. However, the carbon foam having 10 wt. % of ferrocene catalyst shows relatively higher SE$_A$ due to high nano-size porosity which is created by the evaporation of iron nanoparticles above 2000°C.

As mentioned earlier, the EMI SE of ferrocene loaded carbon foam is mostly dominated by reflection. It is well known that the reflection and absorption in the material are mainly due to conductivity and magnetic permeability respectively. Therefore, in another study, Kumar *et al.* [126] used nickel nanoparticles in coat tar pitch for improving the magnetic permeability of carbon foam which significantly increases the microwave absorption. Through this experiment, it is found that carbon foam without nickel nanoparticles exhibited an EMI SE of 25 dB with SE$_A$ of 16.8 at 8.2 GHz. Meanwhile, only the addition of 1 wt. % nickel nanoparticles can enhance the total EMI to 61 dB with a very high SE$_A$ of 48.5 dB at the same frequency. This improvement in absorption is mainly due to the increase in dielectric and magnetic permeability of nickel nanoparticles. Similarly, Liu *et al.* [127] investigated the EMI shielding in Ni-Zn ferrite incorporated carbon foam. The various amounts (0-15) wt. % of Ni-Zn ferrite is incorporated in the coal tar pitch and the foam was synthesized using pressure release method followed by carbonization at 650°C. The carbon foam with 15 wt. Ni-Zn ferrite exhibits a SE$_T$ of 42 dB with a SE$_A$ of 29 dB and SE$_R$ of 13 dB in the X band frequency range.

Later on, Farhan *et al.* [128] fabricated carbon foam decorated with silver particles and *in situ* grown nanowires to achieve EMI shielding. These foams were synthesized by the powder moulding process; here coal tar pitch was used as a densification agent along with silver foils, novolac resin as binder, and PU as pore former. In this process, the conversion of silver particles into spherical particles takes place under the heat treatment at 1200°C. The EMI SE of as-received carbon foam exhibited 17.83 dB, when the carbon foam was decorated with 1.5 wt. % silver nanoparticles, the EMI SE values increase to 33 dB. This enhancement arises due to the improvement of electrical conductivities. As EM radiations enter the silver nanoparticle decorated foam, these silver particles provide fast electronic channels which results in better dissipation of microwave radiations. Ameli *et al.* [129] has developed a facile approach to prepare stainless steel fibers reinforced polypropylene composite foam. The maximum EMI SE was measured to be 48 dB at 1.1 wt. % SS fiber loading which was much higher than that of the solid counterpart of this composite.

Recently, MXene is a novel two-dimensional material and has received considerable attention for EMI shielding because of its excellent electrical conductivity, high aspect ratio, and good mechanical properties [136-138]. Very few research works have been conducted on MXene based composite foam for EMI shielding to date. The first time, Shahzad *et al.* [139] used MXene for EMI shielding application. The MXene ($Ti_3C_2T_x$) film exhibited the EMI SE of 92 dB at a very low thickness of about 45μm in the X band which is the highest SE as compared to other synthetic materials of similar thickness. The high EMI SE of the $Ti_3C_2T_x$ film is mainly due to their outstanding electrical conductivity 4600 Scm^{-1} which could effectively attenuate the entered electromagnetic waves inside the material. Further, MXene based foams also have been demonstrated for EMI shielding application. Hu *et al.* [133] have reported (Ti_2CT_x) MXene/ polyvinyl alcohol (PVA) composite foam for EMI shielding. The (Ti_2CT_x) MXene/PVA composite foams were prepared by the freezing-dry method. The measured SE_T, SE_A, and SE_R of (Ti_2CT_x) MXene/PVA composite foams are 28, 26, and 2 dB respectively in the X band frequency range. Due to the lower reflection (< 2 dB), this composite foam shows absorption dominating the EMI shielding mechanism. Further, the porous structure provides multiple internal reflections and dipole and interfacial polarization led to excellent absorption in composite foam. Further, Zho *et al.* [134] reported ($Ti_3C_2T_x$) MXene/rGO hybrid aerogels for EMI shielding. The hybrid aerogel possesses a high EMI SE of 56.4 dB in the X-band at a 0.74 vol % of $Ti_3C_2T_x$ content.

In another work, Wu *et al.* [135] prepared compressible PDMS coated MXene foam using a freeze-drying method. The different mass ratios (100, 95.24, 86.95, 74.07, and 62.5) of $Ti_3C_2T_x$ were used for the synthesis of foam and referred to as MS100, MS95, MS87, MS74, and MS63. The process for the synthesis of PDMS coated

MXene foam is displayed in Fig. (**16**). The EMI SE of PDMS coated MXene was investigated in X band frequency as a function of MXene contents. It was seen that EMI SE increased with increasing the MXene content and reached the maximum value of 70 dB suggesting that 99.9999% incident EM wave can be blocked [139]. The SE_{total} and SE_A increase with increasing the MXene content leading to absorption dominating mechanism for this composite foam.

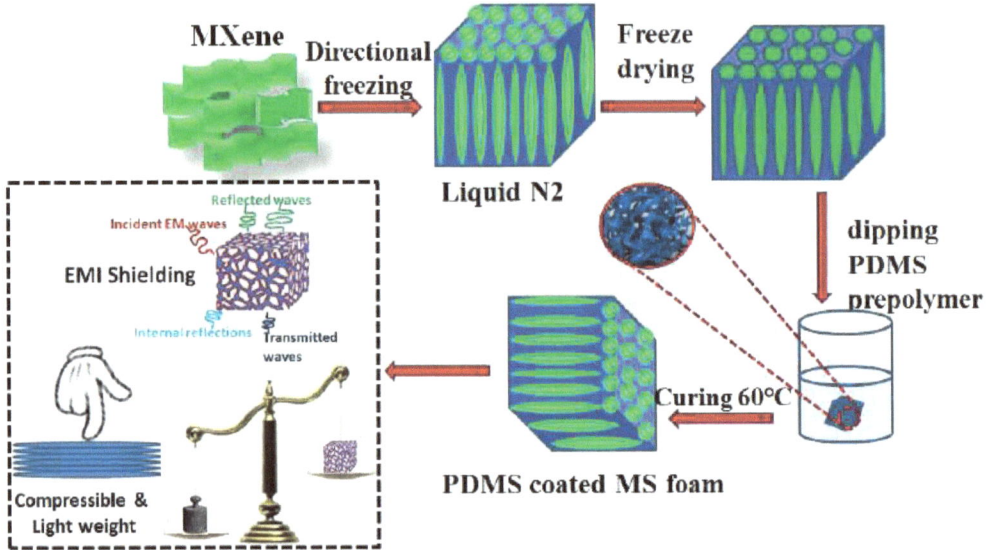

Fig. (16). Schematic figure for the development of PDMS-coated MS foam.

2.4.5. Conducting and Magnetic Fillers Reinforced Composite Foams for EMI Shielding

The synergistic effect of magnetic loss and dielectric loss greatly contributes to the improvement of EMI SE [140, 141], it is necessary to review the most recent advances on carbon composite foam containing dielectric and magnetic fillers, thought to enhance the microwave absorption of composite foams. Because these dielectric and magnetic particles such as ZnO, SiO_2, MnO_2 Fe_2O_3, and Fe_3O_4 help in impedance matching to balance dielectric permittivity and magnetic permeability which is most essential for microwave absorption [142]. However, the eddy-current losses induced by the EM waves may reduce the permeability of magnetic materials. Therefore, the eddy-current phenomenon can be suppressed by using a small amount of conducting fillers which further leads to an increase in the interaction of EM wave with the material. To meet the desired requirements, the composite foams have been modified with conducting magnetic fillers. With the appropriate combination of these two kinds of fillers, a better impedance match at the material's surface could be achieved which will improve the overall microwave absorption.

Shen *et al.* [21] used a water vapor-induced phase separation process to fabricate polyetherimide (PEI)/graphene@Fe_3O_4 composite foam for EMI shielding with strong microwave absorption. The EMI shielding performance of PEI/graphene@Fe_3O_4 composite foams was carried out in the X band frequency range as shown in Fig. (**17a**). It was seen that EMI SE of the composite foam increased on increasing graphene@ Fe_3O_4 content. Initially, EMI SE was found to be 3.5−5.8 dB over the frequency range of 8−12 GHz for the composite foam containing 1.0 wt. % graphene@ Fe_3O_4. The maximum EMI SE obtained was in the range of 14.3−18.2 dB for the composite foam of 10 wt. % graphene@ Fe_3O_4 in the same frequency range. To further explain the EMI shielding mechanism in terms of reflection and absorption, the SE_R and SE_A are calculated at 9.6 GHz and displayed in Fig. (**17b**). It was also observed that both SE_{total} and SE_A were significantly improved with the increase of graphene@Fe_3O_4 content whereas the contribution of SE_R was negligible for all the composite foams. The introduction of Fe_3O_4 nanoparticles tends to decrease the electrical conductivity of the composite foams, but it results in getting proper impedance matching thereby leading to enhanced EM absorption. Moreover, the large conductivity difference between graphene and Fe_3O_4 nanoparticles enhances the interfacial polarization which further contributes to higher microwave absorption [143]. Apart from conductivity, the dielectric and magnetic loss and defects on the graphene are also helpful to increase the microwave absorption.

Similarly, Zhang *et al.* [144] used the carbon dioxide foaming process to prepare PMMA/Fe_3O_4@MWCNTs composite foams for EMI shielding applications. Through the experiment, it is found that PMMA composite foam exhibited an EMI SE of 16 dB in the X-band frequency range at 10 wt. % Fe_3O_4@MWCNTs content. It was also observed that the contribution of absorption to overall shielding was 95.6% of the total EMI SE of PMMA/Fe_3O_4@MWCNTs composite foam. The improved absorption in the composite foam is mainly due to the synergic effect between Fe_3O_4 and MWCNTs and foam structure which provide repeated scattering to attenuate incident EM waves in the form of heat [107]. The EM wave interaction with PMMA/Fe_3O_4@MWCNTs composite foam is schematically present in Fig. (**18**).

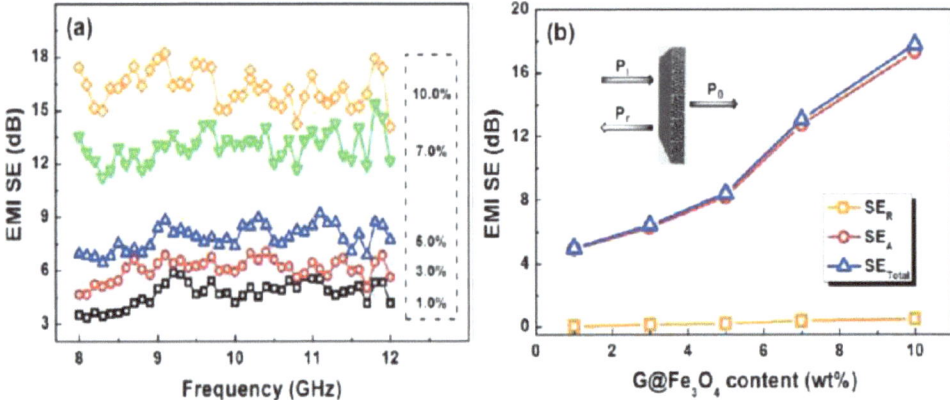

Fig. (17). (a) EMI SE of PEI/G@Fe_3O_4 composite foams *versus* graphene@Fe_3O_4 content in X band frequency range and (**b**) contribution of SE_{total}, SE_R, and SE_A of PEI/G@Fe_3O_4 composite foams at 9.6 GHz. Reproduced with permission from reference [21].

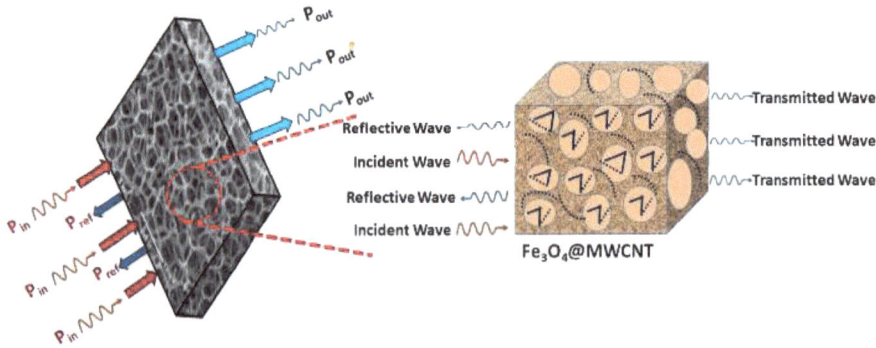

Fig. (18). Schematic diagram of the EM wave interaction with PMMA/Fe_3O_4@MWCNTs composite foam.

In another report, Kumar *et al.* [94] used Fe_3O_4 and Fe_3O_4-ZnO nanoparticles for improving microwave absorption properties in carbon foam. Fe_3O_4 and Fe_3O_4-ZnO nanoparticles were decorated on carbon foam using a dip-coating method. The carbon foam coated with Fe_3O_4-ZnO shows a good EMI SE in a wide frequency range and the values of SE_T and SE_A are 48.5 and 41.5 dB respectively at 8.2 GHz.

This improved microwave absorption is due to impedance matching of magnetic (Fe_3O_4) and dielectric (ZnO) particles and multiple reflections of the EM wave from an interconnected porous structure. Another magnetic filler MnO_2 was used in carbon foam with improved microwave absorption properties for conventional EMI shielding applications [145]. The shielding results showed absorption as the dominant mechanism, the maximum measured SE_A for the composite foam was 40.8 dB whereas SE_R of only 4.2 dB in the X band range at only 4 wt. % MnO_2.

In recent times, global attention towards environmental protection has accelerated the rate of recycling industrial waste and utilizing it in novel products. Industrials waste such as fly ash [146], blast furnace slag [147], rice husk ash [148], waste glass [149] and red mud [150, 151] are a major nuisance for societies everywhere. In this regard, Kumar *et al.* [25] used fly ash derived cenosphere to fabricate carbon composite foam for EMI shielding and thermal properties. Cenospheres are the lightweight and hollow microsphere in which silica and alumina are the major constituents. Through this experiment, it was observed that the EMI SE of carbon foam increases with the increase of cenosphere content. The maximum EMI SE was observed to be 48.6 dB (with SE_A of 42.9 dB) at 8.2 GHz for the composite foam containing 30 wt. % cenosphere. The improved absorption can be attributed to the porosity in foam as well as the cenosphere, the EM wave deeply penetrates inside the composite foam, and attenuation is achieved by the repeated reflection.

In another work, Kumar *et al.* [26] used red mud for the fabrication of carbon-red mud hybrid foam for EMI shielding applications. The different content (0-20 wt. %) red mud was used in hybrid foam and named as CF, CF-RM5, CF-RM10, CF-RM15, and CF-RM20 for 0, 5, 10, 15, and 20 wt. % red mud contents respectively. The EMI SE of hybrid foam was measured in the X band and shown in Fig. (**19a-c**). The EMI SE of hybrid foam without red mud was found to be 22.6 dB due to reflection (SE_R 9.9 dB) and absorption (SE_A 12.7). It was observed that EMI SE increases with the increases in red mud content. The maximum EMI was obtained 51.4 dB with excellent absorption 45.5 dB and very minute reflection 5.9 dB at 8.2 GHz for the hybrid foam containing 20 wt. % red mud. The improvement in absorption is attributed to interfacial polarization, eddy current loss, and magnetic loss in the hybrid foam. The major constituents in the red mud are Fe_2O_3, Al_2O_3, SiO_2, these metal oxides have high dielectric and magnetic loss which further provide the impedance matching and contribute to higher absorption. Besides, the absorption of hybrid foam can further correlate with skin depth. The skin depth of the carbon-red mud hybrid was measured in the X band frequency range and shown in Fig. (**19d**). It was observed that the skin depth of hybrid foam decreases with an increase in red mud content. The low skin depth of 0.38 mm was observed at 8.2 GHz in hybrid foam containing 20 wt. % red mud which ultimately gives rise to high absorption.

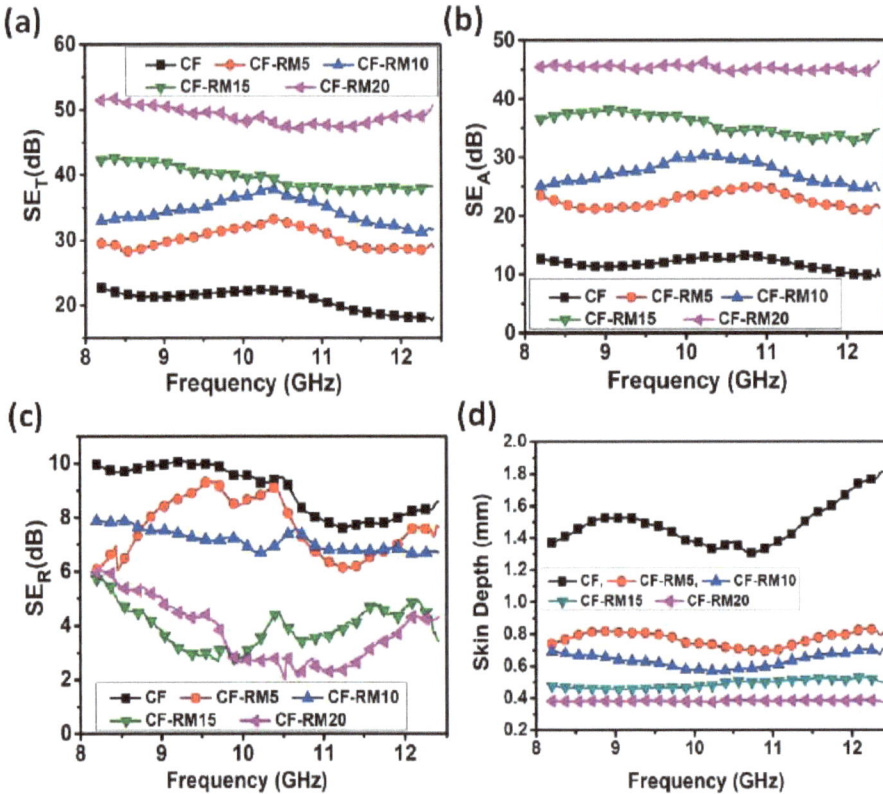

Fig. (19). EMI shielding; (**a**) Total (**b**) Absorption (**c**) Reflection (**d**) skin depth of carbon red mud hybrid foams in X frequency range. Reproduced with permission from reference [26].

Recently, Fang *et al.* [24] reported a novel approach for high-performance EMI shielding using the *in-situ* hollow-Fe_3O_4 sphere grown on graphene foam networks *via* a facile solvothermal method. Then the GF/h-Fe_3O_4 samples with different densities were infiltrated with liquid PDMS at ambient temperature finally, the mixture was vacuumed and cured at 80°C for 4 hours to get the corresponding composites. It was noted that when the loading of GF/h-Fe_3O_4 was up to 12 wt. %, the composite foam exhibits a high EMI SE value of 70.37 dB which is ~22 dB higher compared to GF/PDMS composites. Nevertheless, to further enhance the EMI shielding of composite foam, a hybrid foam containing conductive and magnetic fillers along with metal carbide (MXene) is highly essential. Nguyen *et al.* [152] have developed graphene PDMS composite foam decorated with Fe_3O_4 nanoparticles and MXene ($Ti_3C_2T_X$) nanosheets for EMI and sensing shielding applications. The maximum composite foam exhibits a maximum EMI SE of 80 dB with excellent absorption in X-band (8.2-12.4 GHz). Table **4** summarizes the EMI SE of carbon composite foams with conducting and magnetic fillers in the X band (8.2-12.4 GHz) frequency.

Table 4. EMI SE of carbon composite foams with conducting and magnetic fillers in X band (8.2-12.4 GHz) frequency. (PEI: Polyetherimide; PMMA: Polymethylmethacrylate; rGO: Reduce graphene oxide; EP: Epoxy; PDMS: Polydimethylsiloxane).

Materials	Density g/cm^3	Thickness (mm)	EMI SE (dB)	References
PEI/graphene@Fe_3O_4 composite foam	0.4	2.5	18.2	[21]
Fe_3O_4 coated carbon foam	0.40	2.75	45.7	[94]
Fe_3O_4-ZnO coated carbon foam	0.40	2.75	48.5	[94]
Carbon foam with MnO_2	0.30	2.5	45	[145]
PMMA/Fe_3O_4-MWCNT composite foam	0.20	2.5	16	[145]
Carbon-cenosphere composite foam	0.32	2.0	48.6	[25]
Rubber/MWCNT/Fe_3O_4 composite foam	0.32	2.0	27.5	[23]
Carbon-red mud hybrid foam	0.46	2.0	51.4	[26]

(Table 4) cont.....

Graphene foam/hollow-Fe$_3$O$_4$/PDMS	-	2.0	70.4	[24]
Graphene foam PDMS/MXene@Fe$_3$O$_4$	-	1.0	80	[152]
Fe$_3$O$_4$-CNT/rGO/ EP	-	-	36	[153]

CONCLUSION AND FUTURE PERSPECTIVE

In this chapter, we reviewed the current progress in the development of carbon composite foams for EMI shielding applications. The various methods and strategies to develop lightweight and flexible carbon composite foams for EMI shielding applications have been reviewed. Also, carbon composite foam is modified with various fillers such as CNTs, graphene, metal and magnetic, and MXene and the role of these fillers on the electrical and EMI shielding properties has been discussed in this chapter. We have also discussed the concise EMI shielding mechanism of carbon composite foam. More specifically lightweight graphene and MXene based composite foam have been found studied for high-performance EMI shielding applications. The role of porosity in the composite foams is found to be a benefit in the absorption of EM waves. The EMI shielding, density, and thickness of carbon composite foams with different loading of fillers have been summarized.

Instead of that, lightweight, thinner, and flexible EMI shielding materials need to be developed and explored for next-generation flexible electronic devices. So, the newly developed 2D conducting materials will be key fillers for enhancing the EMI shielding performance of the composite foams. The following research direction could be explored in the future.

(i) The CNTs and graphene materials have already been synthesized which could be further used for the development of efficient EMI shielding materials.

(ii) Recently, MXene and MXene based composite foam has shown outstanding EMI shielding performance because MXene has both conducting and magnetic properties which make it an excellent filler for future EMI shielding applications.

(iii) Industrial waste materials are continuously increasing and polluting the environment. So, there is a lot of scope to explore these wastes in composite foams for EMI shielding applications.

(iv) Instead of carbon, metal, and conducting fillers, newly developed materials like borophene could be useful for the development of EMI shielding materials.

CONSENT FOR PUBLICATION

Not applicable.

CONFLICT OF INTEREST

The authors declare no conflict of interest, financial or otherwise.

ACKNOWLEDGEMENTS

The authors would like to sincerely thank the Director, CSIR-AMPRI, Bhopal for his support. The authors would like to acknowledge the Department of Science and Technology (DST), India for financial support through the Inspire Program (IFA15-MS-51).

REFERENCES

[1]   Chung, D. Carbon materials for structural self-sensing, electromagnetic shielding and thermal interfacing. *Carbon,* **2012**, *50*(9), 3342-3353.
        http://dx.doi.org/10.1016/j.carbon.2012.01.031

[2]   Geetha, S.; Satheesh Kumar, K.; Rao, C.R.; Vijayan, M.; Trivedi, D. EMI shielding: Methods and materials—A review. *J. Appl. Polym. Sci.,* **2009**, *112*(4), 2073-2086.
        http://dx.doi.org/10.1002/app.29812

[3]   Carlberg, M.; Koppel, T.; Ahonen, M.; Hardell, L. Case-control study on occupational exposure to extremely low-frequency electromagnetic fields and the association with meningioma. *BioMed Research International,* **2018**, *2018,* 5912394
        http://dx.doi.org/10.1002/app.29812

[4]   Markham, D. Shielding: quantifying the shielding requirements for portable electronic design and providing new solutions by using a combination of materials and design. *Mater. Des.,* **1999**, *21*(1), 45-50.
        http://dx.doi.org/10.1016/S0261-3069(99)00049-7

[5]   Chung, D.D.L. Materials for electromagnetic interference shielding. *Mater. Chem. Phys.,* **2020**, *255*123587

http://dx.doi.org/10.1016/j.matchemphys.2020.123587

[6]   Wenderoth, K.; Petermann, J.; Kruse, K.D.; ter Haseborg, J.L.; Krieger, W. Synergism on electromagnetic inductance (EMI)-shielding in metal-and ferroelectric-particle filled polymers. *Polym. Compos.,* **1989**, *10*(1), 52-56.

http://dx.doi.org/10.1002/pc.750100108

[7]   Wang, Y.; Jing, X. Intrinsically conducting polymers for electromagnetic interference shielding. *Polym. Adv. Technol.,* **2005**, *16*(4), 344-351.

http://dx.doi.org/10.1002/pat.589

[8]   Vix-Guterl, C.; Frackowiak, E.; Jurewicz, K.; Friebe, M.; Parmentier, J.; Béguin, F. Electrochemical energy storage in ordered porous carbon materials. *Carbon,* **2005**, *43*(6), 1293-1302.

http://dx.doi.org/10.1016/j.carbon.2004.12.028

[9]   Dash, R.; Chmiola, J.; Yushin, G.; Gogotsi, Y.; Laudisio, G.; Singer, J.; Fischer, J.; Kucheyev, S. Titanium carbide derived nanoporous carbon for energy-related applications. *Carbon,* **2006**, *44*(12), 2489-2497.

http://dx.doi.org/10.1016/j.carbon.2006.04.035

[10]   Dhakate, S.R.; Subhedar, K.M.; Singh, B.P. Polymer nanocomposite foam filled with carbon nanomaterials as an efficient electromagnetic interference shielding material. *RSC Advances,* **2015**, *5*(54), 43036-43057.

http://dx.doi.org/10.1039/C5RA03409D

[11]   Gallego, N.C.; Klett, J.W. Carbon foams for thermal management. *Carbon,* **2003**, *41*(7), 1461-1466.

http://dx.doi.org/10.1016/S0008-6223(03)00091-5

[12]   Jiang, Q.; Liao, X.; Yang, J.; Wang, G.; Chen, J.; Tian, C.; Li, G. A two-step process for the preparation of thermoplastic polyurethane/graphene aerogel composite foams with multi-stage networks for electromagnetic shielding. *Composites Communications,* **2020**, *21*, 100416.

[13]   Sharma, A.; Kumar, R.; Patle, V.K.; Dhawan, R.; Abhash, A.; Dwivedi, N.; Mondal, D.P.; Srivastava, A.K. Phenol formaldehyde resin derived carbon-MCMB composite foams for electromagnetic interference shielding and thermal management applications. *Composites Communications,* **2020**, *22*100433

http://dx.doi.org/10.1016/j.coco.2020.100433

[14]   Das, N.C.; Maiti, S. Electromagnetic interference shielding of carbon nanotube/ethylene vinyl acetate composites. *J. Mater. Sci.,* **2008**, *43*(6), 1920-1925.

http://dx.doi.org/10.1007/s10853-008-2458-8

[15]   Al-Saleh, M.H.; Saadeh, W.H.; Sundararaj, U. EMI shielding effectiveness of carbon based nanostructured polymeric materials: a comparative study. *Carbon,* **2013**, *60*, 146-156.

http://dx.doi.org/10.1016/j.carbon.2013.04.008

[16]   Zhang, H.; Jia, Z.; Feng, A.; Zhou, Z.; Zhang, C.; Wang, K.; Liu, N.; Wu, G. Enhanced microwave absorption performance of sulfur-doped hollow carbon microspheres with mesoporous shell as a broadband absorber. *Composites Communications,* **2020**, *19*, 42-50.

http://dx.doi.org/10.1016/j.coco.2020.02.010

[17]   Zhang, T.; Xiao, B.; Zhou, P.; Xia, L.; Wen, G.; Zhang, H. Porous-carbon-nanotube decorated carbon nanofibers with effective microwave absorption properties. *Nanotechnology,* **2017**, *28*(35)355708

http://dx.doi.org/10.1088/1361-6528/aa7ae9 PMID: 28636565

[18]   Saini, P.; Choudhary, V.; Sood, K.; Dhawan, S. Electromagnetic interference shielding behavior of polyaniline/graphite composites prepared by in situ emulsion pathway. *J. Appl. Polym. Sci.,* **2009**, *113*(5), 3146-3155.

http://dx.doi.org/10.1002/app.30183

[19]   Kumar, R.; Dhakate, S.R.; Gupta, T.; Saini, P.; Singh, B.P.; Mathur, R.B. Effective improvement of the properties of light weight carbon foam by decoration with multi-wall carbon nanotubes. *J. Mater. Chem. A Mater. Energy Sustain.,* **2013**, *1*(18), 5727-5735.

http://dx.doi.org/10.1039/c3ta10604g

[20]   Agrawal, P.R.; Kumar, R.; Teotia, S.; Kumari, S.; Mondal, D.P.; Dhakate, S.R. Lightweight, high electrical and thermal conducting carbon-rGO composites foam for superior electromagnetic interference shielding. *Compos., Part B Eng.,* **2019**, *160*, 131-139.

http://dx.doi.org/10.1016/j.compositesb.2018.10.033

[21]   Shen, B.; Zhai, W.; Tao, M.; Ling, J.; Zheng, W. Lightweight, multifunctional polyetherimide/graphene@Fe_3O_4 composite foams for shielding of electromagnetic pollution. *ACS Appl. Mater. Interfaces,* **2013**, *5*(21), 11383-11391.

http://dx.doi.org/10.1021/am4036527 PMID: 24134429

[22]   Ji, K.; Zhao, H.; Zhang, J.; Chen, J.; Dai, Z. Fabrication and electromagnetic interference shielding performance of open-cell foam of a Cu–Ni alloy integrated with CNTs. *Appl. Surf. Sci.,* **2014**, *311*, 351-356.

http://dx.doi.org/10.1016/j.apsusc.2014.05.067

[23]   Yang, J.; Liao, X.; Li, J.; He, G.; Zhang, Y.; Tang, W.; Wang, G.; Li, G. Light-weight and flexible silicone rubber/MWCNTs/Fe_3O_4 nanocomposite foams for efficient electromagnetic interference shielding and microwave absorption. *Compos. Sci. Technol.,* **2019**, *181*107670

http://dx.doi.org/10.1016/j.compscitech.2019.05.027

[24]   Fang, H.; Guo, H.; Hu, Y.; Ren, Y.; Hsu, P-C.; Bai, S-L. In-situ grown hollow Fe3O4 onto graphene foam nanocomposites with high EMI shielding effectiveness and thermal conductivity. *Compos. Sci. Technol.,* **2020**, *188*107975

http://dx.doi.org/10.1016/j.compscitech.2019.107975

[25]   Kumar, R.; Mondal, D.P.; Chaudhary, A.; Shafeeq, M.; Kumari, S. Excellent EMI shielding performance and thermal insulating properties in lightweight, multifunctional carbon-cenosphere composite foams. *Compos., Part A Appl. Sci. Manuf.,* **2018**, *112*, 475-484.

http://dx.doi.org/10.1016/j.compositesa.2018.07.003

[26]   Kumar, R.; Sharma, A.; Pandey, A.; Chaudhary, A.; Dwivedi, N.; Shafeeq M, M.; Mondal, D.P.; Srivastava, A.K. Lightweight carbon-red mud hybrid foam toward fire-resistant and efficient shield against electromagnetic interference. *Sci. Rep.,* **2020**, *10*(1), 9913.

http://dx.doi.org/10.1038/s41598-020-66929-3 PMID: 32555266

[27]   Xie, W.; Cheng, H-F.; Chu, Z-Y. CHEN, Z.-H.; ZHOU, Y.-J., Radar absorbing properties of light radar absorbing materials based on hollow-porous carbon fibers. *J. Inorg. Mater.,* **2009**, 2.

[28]   Chung, D.D.L. Materials for electromagnetic interference shielding. *J. Mater. Eng. Perform.,* **2000**, *9*(3), 350-354.

http://dx.doi.org/10.1361/105994900770346042

[29]   Ji, X.; Lee, K.T.; Nazar, L.F. A highly ordered nanostructured carbon-sulphur cathode for lithium-sulphur batteries. *Nat. Mater.,* **2009**, *8*(6), 500-506.

http://dx.doi.org/10.1038/nmat2460 PMID: 19448613

[30] Otowa, T.; Nojima, Y.; Miyazaki, T. Development of KOH activated high surface area carbon and its application to drinking water purification. *Carbon,* **1997**, *35*(9), 1315-1319.

http://dx.doi.org/10.1016/S0008-6223(97)00076-6

[31] Gomez-Serrano, V.; Macias-Garcia, A.; Espinosa-Mansilla, A.; Valenzuela-Calahorro, C. Adsorption of mercury, cadmium and lead from aqueous solution on heat-treated and sulphurized activated carbon. *Water Res.,* **1998**, *32*(1), 1-4.

http://dx.doi.org/10.1016/S0043-1354(97)00203-0

[32] Joo, S.H.; Choi, S.J.; Oh, I.; Kwak, J.; Liu, Z.; Terasaki, O.; Ryoo, R. Ordered nanoporous arrays of carbon supporting high dispersions of platinum nanoparticles. *Nature,* **2001**, *412*(6843), 169-172.

http://dx.doi.org/10.1038/35084046 PMID: 11449269

[33] Park, G-G.; Yang, T-H.; Yoon, Y-G.; Lee, W-Y.; Kim, C-S. Pore size effect of the DMFC catalyst supported on porous materials. *Int. J. Hydrogen Energy,* **2003**, *28*(6), 645-650.

http://dx.doi.org/10.1016/S0360-3199(02)00140-4

[34] Saini, P.; Choudhary, V.; Singh, B.; Mathur, R.; Dhawan, S. Enhanced microwave absorption behavior of polyaniline-CNT/polystyrene blend in 12.4–18.0 GHz range. *Synth. Met.,* **2011**, *161*(15), 1522-1526.

http://dx.doi.org/10.1016/j.synthmet.2011.04.033

[35] Singh, A.P.; Mishra, M.; Sambyal, P.; Gupta, B.K.; Singh, B.P.; Chandra, A.; Dhawan, S. Encapsulation of γ-Fe_2O_3 decorated reduced graphene oxide in polyaniline core–shell tubes as an exceptional tracker for electromagnetic environmental pollution. *J. Mater. Chem. A Mater. Energy Sustain.,* **2014**, *2*(10), 3581-3593.

http://dx.doi.org/10.1039/C3TA14212D

[36] Gupta, T.K.; Singh, B.P.; Mathur, R.B.; Dhakate, S.R. Multi-walled carbon nanotube-graphene-polyaniline multiphase nanocomposite with superior electromagnetic shielding effectiveness. *Nanoscale,* **2014**, *6*(2), 842-851.

http://dx.doi.org/10.1039/C3NR04565J PMID: 24264356

[37] Tong, X.C. *Advanced materials and design for electromagnetic interference shielding.,* **2016**,

http://dx.doi.org/10.1201/9781420073591

[38] Tseng, C-H.; Chu, T-H. An effective usage of vector network analyzer for microwave imaging. *IEEE Trans. Microw. Theory Tech.,* **2005**, *53*(9), 2884-2891.

http://dx.doi.org/10.1109/TMTT.2005.854251

[39] Ford, W.D., Method of making cellular refractory thermal insulating material. US Patent, 3,121,050. **1964**.

[40] Li, Q.; Chen, L.; Ding, J.; Zhang, J.; Li, X.; Zheng, K.; Zhang, X.; Tian, X. Open-cell phenolic carbon foam and electromagnetic interference shielding properties. *Carbon,* **2016**, *104*, 90-105.

http://dx.doi.org/10.1016/j.carbon.2016.03.055

[41] Googin, J.M.; Napier, J.M.; Scrivner, M.E., Method for manufacturing foam carbon products. US Patent 3345440, **1967**

[42] Inagaki, M.; Morishita, T.; Kuno, A.; Kito, T.; Hirano, M.; Suwa, T.; Kusakawa, K. Carbon foams prepared from polyimide using urethane foam template. *Carbon,* **2004**, *42*(3), 497-502.

http://dx.doi.org/10.1016/j.carbon.2003.12.080

[43]   Stiller, A.H.; Stansberry, P.G.; Zondlo, J.W. Method of making a carbon foam material and resultant product. US Patent 5888469, *1999*.

[44]   Chen, C.; Kennel, E.B.; Stiller, A.H.; Stansberry, P.G.; Zondlo, J.W. Carbon foam derived from various precursors. *Carbon,* **2006**, *44*(8), 1535-1543.
       http://dx.doi.org/10.1016/j.carbon.2005.12.021

[45]   Klett, J.W, Process for making carbon foam. *US Patent 6033506, 2000.*

[46]   Wang, Y.; He, Z.; Zhan, L.; Liu, X. Coal tar pitch based carbon foam for thermal insulating material. *Mater. Lett.,* **2016**, *169*, 95-98.
       http://dx.doi.org/10.1016/j.matlet.2016.01.081

[47]   Klett, J.W.; Burchell, T.D.; Choudhury, A., Pitch-based carbon foam and composites and use thereof. US Patent 7070755, **2006**.

[48]   Rogers, D.K. Petroleum pitch-based carbon foam. US Patent 6833012, **2004**.

[49]   Klett, J.W., Pitch-based carbon foam and composites. US Patent 6387343, **2002**.

[50]   Wang, M.; Wang, C-Y.; Li, T-Q.; Hu, Z-J. Preparation of mesophase-pitch-based carbon foams at low pressures. *Carbon,* **2008**, *46*(1), 84-91.
       http://dx.doi.org/10.1016/j.carbon.2007.10.038

[51]   Yuan, Y.; Ding, Y.; Wang, C.; Xu, F.; Lin, Z.; Qin, Y.; Li, Y.; Yang, M.; He, X.; Peng, Q.; Li, Y. Multifunctional Stiff Carbon Foam Derived from Bread. *ACS Appl. Mater. Interfaces,* **2016**, *8*(26), 16852-16861.
       http://dx.doi.org/10.1021/acsami.6b03985 PMID: 27295106

[52]   Yang, F.; Sun, L.; Zhang, W.; Zhang, Y. One-pot synthesis of porous carbon foam derived from corn straw: atrazine adsorption equilibrium and kinetics. *Environ. Sci. Nano,* **2017**, *4*(3), 625-635.
       http://dx.doi.org/10.1039/C6EN00574H

[53]   Klett, J.; Hardy, R.; Romine, E.; Walls, C.; Burchell, T. High-thermal-conductivity, mesophase-pitch-derived carbon foams: effect of precursor on structure and properties. *Carbon,* **2000**, *38*(7), 953-973.
       http://dx.doi.org/10.1016/S0008-6223(99)00190-6

[54]   Li, J.; Wang, C.; Zhang, C.; Zhan, L.; Qiao, W.; Liang, X-y.; Ling, L-C. Effect of pre-oxidation on microcracks in graphite foams. *Carbon,* **2011**, *1*(49), 354.
       http://dx.doi.org/10.1016/j.carbon.2010.08.021

[55]   Chen, Y.; Chen, B-Z.; Shi, X-C.; Xu, H.; Hu, Y-J.; Yuan, Y.; Shen, N-B. Preparation of pitch-based carbon foam using polyurethane foam template. *Carbon,* **2007**, *45*(10), 2132-2134.
       http://dx.doi.org/10.1016/j.carbon.2007.06.004

[56]   Nagel, B.; Pusz, S.; Trzebicka, B. Review: tailoring the properties of macroporous carbon foams. *J. Mater. Sci.,* **2014**, *49*(1), 1-17.
       http://dx.doi.org/10.1007/s10853-013-7678-x

[57]   Narasimman, R.; Prabhakaran, K. Preparation of low density carbon foams by foaming molten sucrose using an aluminium nitrate blowing agent. *Carbon,* **2012**, *50*(5), 1999-2009.
       http://dx.doi.org/10.1016/j.carbon.2011.12.058

[58]   Kumar, R.; Jain, H.; Chaudhary, A.; Kumari, S.; Mondal, D.; Srivastava, A. Thermal conductivity and fire-retardant response in graphite foam made from coal tar pitch derived semi coke. *Compos., Part B Eng.,* **2019**, *172*, 121-130.
       http://dx.doi.org/10.1016/j.compositesb.2019.05.036

[59]   Klett, J.; McMillan, A.; Gallego, N.; Walls, C. The role of structure on the thermal properties of graphitic foams. *J. Mater. Sci.,* **2004**, *39*(11), 3659-3676.

http://dx.doi.org/10.1023/B:JMSC.0000030719.80262.f8

[60]   Li, S.; Song, Y.; Song, Y.; Shi, J.; Liu, L.; Wei, X.; Guo, Q. Carbon foams with high compressive strength derived from mixtures of mesocarbon microbeads and mesophase pitch. *Carbon,* **2007**, *45*(10), 2092-2097.

http://dx.doi.org/10.1016/j.carbon.2007.05.014

[61]   Eksilioglu, A.; Gencay, N.; Yardim, M.; Ekinci, E. Mesophase AR pitch derived carbon foam: Effect of temperature, pressure and pressure release time. *J. Mater. Sci.,* **2006**, *41*(10), 2743-2748.

http://dx.doi.org/10.1007/s10853-006-7079-5

[62]   Li, S.; Guo, Q.; Song, Y.; Liu, Z.; Shi, J.; Liu, L.; Yan, X. Carbon foams with high compressive strength derived from mesophase pitch treated by toluene extraction. *Carbon,* **2007**, *45*(14), 2843-2845.

http://dx.doi.org/10.1016/j.carbon.2007.09.035

[63]   Yadav, A.; Kumar, R.; Bhatia, G.; Verma, G. Development of mesophase pitch derived high thermal conductivity graphite foam using a template method. *Carbon,* **2011**, *49*(11), 3622-3630.

http://dx.doi.org/10.1016/j.carbon.2011.04.065

[64]   Lee, J.; Han, S.; Hyeon, T. Synthesis of new nanoporous carbon materials using nanostructured silica materials as templates. *J. Mater. Chem.,* **2004**, *14*(4), 478-486.

http://dx.doi.org/10.1039/b311541k

[65]   Chen, Z.; Xu, C.; Ma, C.; Ren, W.; Cheng, H.M. Lightweight and flexible graphene foam composites for high-performance electromagnetic interference shielding. *Adv. Mater.,* **2013**, *25*(9), 1296-1300.

http://dx.doi.org/10.1002/adma.201204196 PMID: 23300002

[66]   Yavari, F.; Chen, Z.; Thomas, A.V.; Ren, W.; Cheng, H-M.; Koratkar, N. High sensitivity gas detection using a macroscopic three-dimensional graphene foam network. *Sci. Rep.,* **2011**, *1*(1), 166.

http://dx.doi.org/10.1038/srep00166 PMID: 22355681

[67]   Kumar, R.; Kumari, S.; Mathur, R.B.; Dhakate, S.R. Nanostructuring effect of multi-walled carbon nanotubes on electrochemical properties of carbon foam as constructive electrode for lead acid battery. *Appl. Nanosci.,* **2015**, *5*(1), 53-61.

http://dx.doi.org/10.1007/s13204-014-0291-8

[68]   Rani Agrawal, P.; Singh, N.; Kumari, S.; Dhakate, S.R. The removal of pentavalent arsenic by graphite intercalation compound functionalized carbon foam from contaminated water. *J. Hazard. Mater.,* **2019**, *377*, 274-283.

http://dx.doi.org/10.1016/j.jhazmat.2019.05.097 PMID: 31173976

[69]   Li, Y-Q.; Samad, Y.A.; Polychronopoulou, K.; Liao, K. Lightweight and highly conductive aerogel-like carbon from sugarcane with superior mechanical and EMI shielding properties. *ACS Sustain. Chem.& Eng.,* **2015**, *3*(7), 1419-1427.

http://dx.doi.org/10.1021/acssuschemeng.5b00340

[70]   Yuan, Y.; Liu, L.; Yang, M.; Zhang, T.; Xu, F.; Lin, Z.; Ding, Y.; Wang, C.; Li, J.; Yin, W.; Peng, Q.; He, X.; Li, Y. Lightweight, thermally insulating and stiff carbon honeycomb-induced graphene composite foams with a horizontal laminated structure for electromagnetic interference shielding. *Carbon,* **2017**, *123*, 223-232.

http://dx.doi.org/10.1016/j.carbon.2017.07.060

[71]   Liu, L.; Yang, S.; Hu, H.; Zhang, T.; Yuan, Y.; Li, Y.; He, X. Lightweight and efficient microwave-absorbing materials based on loofah-sponge-derived hierarchically porous carbons. *ACS Sustain. Chem.& Eng.,* **2018**, *7*(1), 1228-1238.

http://dx.doi.org/10.1021/acssuschemeng.8b04907

[72]   Chen, X.; Liu, L.; Pan, F.; Mao, J.; Xu, X.; Yan, T. Microstructure, electromagnetic shielding effectiveness and mechanical properties of Mg–Zn–Cu–Zr alloys. *Mater. Sci. Eng. B,* **2015**, *197*(Suppl. C), 67-74.

http://dx.doi.org/10.1016/j.mseb.2015.03.012

[73]   Dou, Z.; Wu, G.; Huang, X.; Sun, D.; Jiang, L. Electromagnetic shielding effectiveness of aluminum alloy–fly ash composites. *Compos., Part A Appl. Sci. Manuf.,* **2007**, *38*(1), 186-191.

http://dx.doi.org/10.1016/j.compositesa.2006.01.015

[74]   Singh, A.K.; Shishkin, A.; Koppel, T.; Gupta, N. A review of porous lightweight composite materials for electromagnetic interference shielding. *Compos., Part B Eng.,* **2018**, *149*, 188-197.

http://dx.doi.org/10.1016/j.compositesb.2018.05.027

[75]   Ren, F.; Song, D.; Li, Z.; Jia, L.; Zhao, Y.; Yan, D.; Ren, P. Synergistic effect of graphene nanosheets and carbonyl iron–nickel alloy hybrid filler on electromagnetic interference shielding and thermal conductivity of cyanate ester composites. *J. Mater. Chem. C Mater. Opt. Electron. Devices,* **2018**, *6*(6), 1476-1486.

http://dx.doi.org/10.1039/C7TC05213H

[76]   Kumar, R.; Dhakate, S.R.; Saini, P.; Mathur, R.B. Improved electromagnetic interference shielding effectiveness of light weight carbon foam by ferrocene accumulation. *RSC Advances,* **2013**, *3*(13), 4145-4151.

http://dx.doi.org/10.1039/c3ra00121k

[77]   Yang, J.; Shen, Z.; Hao, Z. Microwave characteristics of sandwich composites with mesophase pitch carbon foams as core. *Carbon,* **2004**, *42*(8-9), 1882-1885.

http://dx.doi.org/10.1016/j.carbon.2004.03.017

[78]   Fang, Z.; Li, C.; Sun, J.; Zhang, H.; Zhang, J. The electromagnetic characteristics of carbon foams. *Carbon,* **2007**, *45*(15), 2873-2879.

http://dx.doi.org/10.1016/j.carbon.2007.10.013

[79]   Moglie, F.; Micheli, D.; Laurenzi, S.; Marchetti, M.; Mariani Primiani, V. Electromagnetic shielding performance of carbon foams. *Carbon,* **2012**, *50*(5), 1972-1980.

http://dx.doi.org/10.1016/j.carbon.2011.12.053

[80]   Jose, G.; Padeep, P. Electromagnetic shielding effectiveness and mechanical characteristics of polypropylene based CFRP. *International Journal on Theoretical and Applied Research in Mechanical Engineering., (IJTARME)*, **2014**, 3, 2319-3182.

[81]   Sun, X.; Sun, H.; Li, H.; Peng, H. Developing polymer composite materials: carbon nanotubes or graphene? *Adv. Mater.,* **2013**, *25*(37), 5153-5176.

http://dx.doi.org/10.1002/adma.201301926 PMID: 23813859

[82]   Iijima, S. Helical microtubules of graphitic carbon. *Nature,* **1991**, *354*(6348), 56-58.

http://dx.doi.org/10.1038/354056a0

[83]   Iijima, S.; Ichihashi, T. Single-shell carbon nanotubes of 1-nm diameter. *Nature,* **1993**, *363*(6430), 603-605.

[84]   Salvetat, J-P.; Bonard, J-M.; Thomson, N.; Kulik, A.; Forro, L.; Benoit, W.; Zuppiroli, L. Mechanical properties of carbon nanotubes. *Appl. Phys., A Mater. Sci. Process.,* **1999**, *69*(3), 255-260.
     http://dx.doi.org/10.1007/s003390050999

[85]   Zeng, S.; Li, X.; Li, M.; Zheng, J.; Shiju, E.; Yang, W.; Zhao, B.; Guo, X.; Zhang, R. Flexible PVDF/CNTs/Ni@ CNTs composite films possessing excellent electromagnetic interference shielding and mechanical properties under heat treatment. *Carbon,* **2019**, *155*, 34-43.
     http://dx.doi.org/10.1016/j.carbon.2019.08.024

[86]   Yang, Y.; Gupta, M.C.; Dudley, K.L.; Lawrence, R.W. Novel carbon nanotube-polystyrene foam composites for electromagnetic interference shielding. *Nano Lett.,* **2005**, *5*(11), 2131-2134.
     http://dx.doi.org/10.1021/nl051375r PMID: 16277439

[87]   Narasimman, R.; Vijayan, S.; Dijith, K.; Surendran, K.; Prabhakaran, K. Carbon composite foams with improved strength and electromagnetic absorption from sucrose and multi-walled carbon nanotube. *Mater. Chem. Phys.,* **2016**, *181*, 538-548.
     http://dx.doi.org/10.1016/j.matchemphys.2016.06.091

[88]   Sun, X-G.; Gao, M.; Li, C.; Wu, Y.; Yellampalli, S. *Microwave absorption characteristics of carbon nanotubes. Carbon Nanotubes-Synthesis, Characterization.,* **2011**, , 265-278.

[89]   Li, J.; Zhang, G.; Ma, Z.; Fan, X.; Fan, X.; Qin, J.; Shi, X. Morphologies and electromagnetic interference shielding performances of microcellular epoxy/multi-wall carbon nanotube nanocomposite foams. *Compos. Sci. Technol.,* **2016**, *129*, 70-78.
     http://dx.doi.org/10.1016/j.compscitech.2016.04.003

[90]   Kuang, T.; Chang, L.; Chen, F.; Sheng, Y.; Fu, D.; Peng, X. Facile preparation of lightweight high-strength biodegradable polymer/multi-walled carbon nanotubes nanocomposite foams for electromagnetic interference shielding. *Carbon,* **2016**, *105*, 305-313.
     http://dx.doi.org/10.1016/j.carbon.2016.04.052

[91]   Huang, H-D.; Liu, C-Y.; Zhou, D.; Jiang, X.; Zhong, G-J.; Yan, D-X.; Li, Z-M. Cellulose composite aerogel for highly efficient electromagnetic interference shielding. *J. Mater. Chem. A Mater. Energy Sustain.,* **2015**, *3*(9), 4983-4991.
     http://dx.doi.org/10.1039/C4TA05998K

[92]   Li, M-Z.; Jia, L-C.; Zhang, X-P.; Yan, D-X.; Zhang, Q-C.; Li, Z-M. Robust carbon nanotube foam for efficient electromagnetic interference shielding and microwave absorption. *J. Colloid Interface Sci.,* **2018**, *530*, 113-119.
     http://dx.doi.org/10.1016/j.jcis.2018.06.052 PMID: 29960904

[93]   Zeng, Z.; Jin, H.; Chen, M.; Li, W.; Zhou, L.; Xue, X.; Zhang, Z. Microstructure design of lightweight, flexible, and high electromagnetic shielding porous multiwalled carbon nanotube/polymer composites. *Small,* **2017**, *13*(34)1701388
     http://dx.doi.org/10.1002/smll.201701388 PMID: 28696564

[94]   Kumar, R.; Singh, A.P.; Chand, M.; Pant, R.P.; Kotnala, R.K.; Dhawan, S.K.; Mathur, R.B.; Dhakate, S.R. Improved microwave absorption in lightweight resin-based carbon foam by decorating with magnetic and dielectric nanoparticles. *RSC Advances,* **2014**, *4*(45), 23476-23484.
     http://dx.doi.org/10.1039/C4RA01731E

[95]   Zeng, Z.; Jin, H.; Chen, M.; Li, W.; Zhou, L.; Zhang, Z. Lightweight and anisotropic porous MWCNT/WPU composites for ultrahigh performance electromagnetic interference shielding. *Adv. Funct. Mater.,* **2016**, *26*(2), 303-310.

http://dx.doi.org/10.1002/adfm.201503579

[96]   Wang, H.; Zheng, K.; Zhang, X.; Ding, X.; Zhang, Z.; Bao, C.; Guo, L.; Chen, L.; Tian, X. 3D network porous polymeric composites with outstanding electromagnetic interference shielding. *Compos. Sci. Technol.,* **2016**, *125*, 22-29.

http://dx.doi.org/10.1016/j.compscitech.2016.01.007

[97]   Lu, D.; Mo, Z.; Liang, B.; Yang, L.; He, Z.; Zhu, H.; Tang, Z.; Gui, X. Flexible, lightweight carbon nanotube sponges and composites for high-performance electromagnetic interference shielding. *Carbon,* **2018**, *133*, 457-463.

http://dx.doi.org/10.1016/j.carbon.2018.03.061

[98]   Geim, A.K. Novoselov, K.S.*Nanoscience and technology: a collection of reviews from nature journals.,* **2010**, , 11-19.

[99]   Neto, A.C.; Guinea, F.; Peres, N.M.; Novoselov, K.S.; Geim, A.K. The electronic properties of graphene. *Rev. Mod. Phys.,* **2009**, *81*(1), 109-162.

http://dx.doi.org/10.1103/RevModPhys.81.109

[100]   Chen, J.; Yao, B.; Li, C.; Shi, G. An improved Hummers method for eco-friendly synthesis of graphene oxide. *Carbon,* **2013**, *64*, 225-229.

http://dx.doi.org/10.1016/j.carbon.2013.07.055

[101]   Yan, D-X.; Ren, P-G.; Pang, H.; Fu, Q.; Yang, M-B.; Li, Z-M. Efficient electromagnetic interference shielding of lightweight graphene/polystyrene composite. *J. Mater. Chem.,* **2012**, *22*(36), 18772-18774.

http://dx.doi.org/10.1039/c2jm32692b

[102]   Liang, J.; Wang, Y.; Huang, Y.; Ma, Y.; Liu, Z.; Cai, J.; Zhang, C.; Gao, H.; Chen, Y. Electromagnetic interference shielding of graphene/epoxy composites. *Carbon,* **2009**, *47*(3), 922-925.

http://dx.doi.org/10.1016/j.carbon.2008.12.038

[103]   Eswaraiah, V.; Sankaranarayanan, V.; Ramaprabhu, S. Functionalized graphene–PVDF foam composites for EMI shielding. *Macromol. Mater. Eng.,* **2011**, *296*(10), 894-898.

http://dx.doi.org/10.1002/mame.201100035

[104]   Wu, Y.; Wang, Z.; Liu, X.; Shen, X.; Zheng, Q.; Xue, Q.; Kim, J-K. Ultralight Graphene Foam/Conductive Polymer Composites for Exceptional Electromagnetic Interference Shielding. *ACS Appl. Mater. Interfaces,* **2017**, *9*(10), 9059-9069.

http://dx.doi.org/10.1021/acsami.7b01017 PMID: 28224798

[105]   Narasimman, R.; Vijayan, S.; Prabhakaran, K. Graphene-reinforced carbon composite foams with improved strength and EMI shielding from sucrose and graphene oxide. *J. Mater. Sci.,* **2015**, *50*(24), 8018-8028.

http://dx.doi.org/10.1007/s10853-015-9368-3

[106]   Ling, J.; Zhai, W.; Feng, W.; Shen, B.; Zhang, J.; Zheng, W. Facile preparation of lightweight microcellular polyetherimide/graphene composite foams for electromagnetic interference shielding. *ACS Appl. Mater. Interfaces,* **2013**, *5*(7), 2677-2684.

http://dx.doi.org/10.1021/am303289m PMID: 23465462

[107]   Zhang, H-B.; Yan, Q.; Zheng, W-G.; He, Z.; Yu, Z-Z. Tough graphene-polymer microcellular foams for electromagnetic interference shielding. *ACS Appl. Mater. Interfaces,* **2011**, *3*(3), 918-924.

http://dx.doi.org/10.1021/am200021v PMID: 21366239

[108] Gavgani, J.N.; Adelnia, H.; Zaarei, D.; Gudarzi, M.M. Lightweight flexible polyurethane/reduced ultralarge graphene oxide composite foams for electromagnetic interference shielding. *RSC Advances,* **2016**, *6*(33), 27517-27527.

http://dx.doi.org/10.1039/C5RA25374H

[109] Li, Y.; Pei, X.; Shen, B.; Zhai, W.; Zhang, L.; Zheng, W. Polyimide/graphene composite foam sheets with ultrahigh thermostability for electromagnetic interference shielding. *RSC Advances,* **2015**, *5*(31), 24342-24351.

http://dx.doi.org/10.1039/C4RA16421K

[110] Li, Y.; Shen, B.; Pei, X.; Zhang, Y.; Yi, D.; Zhai, W.; Zhang, L.; Wei, X.; Zheng, W. Ultrathin carbon foams for effective electromagnetic interference shielding. *Carbon,* **2016**, *100*, 375-385.

http://dx.doi.org/10.1016/j.carbon.2016.01.030

[111] Wang, L.; Wu, Y.; Wang, Y.; Li, H.; Jiang, N.; Niu, K. Laterally compressed graphene foam/acrylonitrile butadiene styrene composites for electromagnetic interference shielding. *Compos., Part A Appl. Sci. Manuf.,* **2020**, *133*105887

http://dx.doi.org/10.1016/j.compositesa.2020.105887

[112] Bi, S.; Zhang, L.; Mu, C.; Liu, M.; Hu, X. Electromagnetic interference shielding properties and mechanisms of chemically reduced graphene aerogels. *Appl. Surf. Sci.,* **2017**, *412*, 529-536.

http://dx.doi.org/10.1016/j.apsusc.2017.03.293

[113] Chen, Y.; Zhang, H-B.; Wang, M.; Qian, X.; Dasari, A.; Yu, Z-Z. Phenolic resin-enhanced three-dimensional graphene aerogels and their epoxy nanocomposites with high mechanical and electromagnetic interference shielding performances. *Compos. Sci. Technol.,* **2017**, *152*, 254-262.

http://dx.doi.org/10.1016/j.compscitech.2017.09.022

[114] Shen, B.; Li, Y.; Yi, D.; Zhai, W.; Wei, X.; Zheng, W. Microcellular graphene foam for improved broadband electromagnetic interference shielding. *Carbon,* **2016**, *102*, 154-160.

http://dx.doi.org/10.1016/j.carbon.2016.02.040

[115] Li, Y.; Zhang, H-B.; Zhang, L.; Shen, B.; Zhai, W.; Yu, Z-Z.; Zheng, W. One-pot sintering strategy for efficient fabrication of high-performance and multifunctional graphene foams. *ACS Appl. Mater. Interfaces,* **2017**, *9*(15), 13323-13330.

http://dx.doi.org/10.1021/acsami.7b02408 PMID: 28350156

[116] Zhu, S.; Xing, C.; Wu, F.; Zuo, X.; Zhang, Y.; Yu, C.; Chen, M.; Li, W.; Li, Q.; Liu, L. Cake-like flexible carbon nanotubes/graphene composite prepared via a facile method for high-performance electromagnetic interference shielding. *Carbon,* **2019**, *145*, 259-265.

http://dx.doi.org/10.1016/j.carbon.2019.01.030

[117] Sun, X.; Liu, X.; Shen, X.; Wu, Y.; Wang, Z.; Kim, J-K. Graphene foam/carbon nanotube/poly (dimethyl siloxane) composites for exceptional microwave shielding. *Compos., Part A Appl. Sci. Manuf.,* **2016**, *85*, 199-206.

http://dx.doi.org/10.1016/j.compositesa.2016.03.009

[118] Zeng, Z.; Zhang, Y.; Ma, X.Y.D.; Shahabadi, S.I.S.; Che, B.; Wang, P.; Lu, X. Biomass-based honeycomb-like architectures for preparation of robust carbon foams with high electromagnetic interference shielding performance. *Carbon,* **2018**, *140*, 227-236.

http://dx.doi.org/10.1016/j.carbon.2018.08.061

[119] Chaudhary, A.; Kumar, R.; Dhakate, S.R.; Kumari, S. Scalable development of a multi-phase thermal management system with superior EMI shielding properties. *Compos., Part B Eng.,* **2019**, *158*, 206-217.

http://dx.doi.org/10.1016/j.compositesb.2018.09.048

[120]   Rogers, J.A.; Someya, T.; Huang, Y. Materials and mechanics for stretchable electronics. *Science,* **2010**, *327*(5973), 1603-1607.

[121]   Sankaran, S.; Deshmukh, K.; Ahamed, M.B.; Pasha, S.K. Recent advances in electromagnetic interference shielding properties of metal and carbon filler reinforced flexible polymer composites: a review. *Compos., Part A Appl. Sci. Manuf.,* **2018**, *114*, 49-71.
        http://dx.doi.org/10.1016/j.compositesa.2018.08.006

[122]   Wanasinghe, D.; Aslani, F. A review on recent advancement of electromagnetic interference shielding novel metallic materials and processes. *Compos., Part B Eng.,* **2019**, *176*107207
        http://dx.doi.org/10.1016/j.compositesb.2019.107207

[123]   Roh, J-S.; Chi, Y-S.; Kang, T.J.; Nam, S. Electromagnetic shielding effectiveness of multifunctional metal composite fabrics. *Text. Res. J.,* **2008**, *78*(9), 825-835.
        http://dx.doi.org/10.1177/0040517507089748

[124]   Maldonado-Hódar, F.; Moreno-Castilla, C.; Rivera-Utrilla, J.; Hanzawa, Y.; Yamada, Y. Catalytic graphitization of carbon aerogels by transition metals. *Langmuir,* **2000**, *16*(9), 4367-4373.
        http://dx.doi.org/10.1021/la991080r

[125]   Ōya, A.; Ōtani, S. Catalytic graphitization of carbons by various metals. *Carbon,* **1979**, *17*(2), 131-137.
        http://dx.doi.org/10.1016/0008-6223(79)90020-4

[126]   Kumar, R.; Kumari, S.; Dhakate, S.R. Nickel nanoparticles embedded in carbon foam for improving electromagnetic shielding effectiveness. *Appl. Nanosci.,* **2014**, *5*(5), 553-561.
        http://dx.doi.org/10.1007/s13204-014-0349-7

[127]   Liu, H.; Wu, J.; Zhuang, Q.; Dang, A.; Li, T.; Zhao, T. Preparation and the electromagnetic interference shielding in the X-band of carbon foams with Ni-Zn ferrite additive. *J. Eur. Ceram. Soc.,* **2016**, *36*(16), 3939-3946.
        http://dx.doi.org/10.1016/j.jeurceramsoc.2016.06.017

[128]   Farhan, S.; Wang, R.; Li, K. Carbon foam decorated with silver particles and in situ grown nanowires for effective electromagnetic interference shielding. *J. Mater. Sci.,* **2016**, *51*(17), 7991-8004.
        http://dx.doi.org/10.1007/s10853-016-0068-4

[129]   Ameli, A.; Nofar, M.; Wang, S.; Park, C.B. Lightweight polypropylene/stainless-steel fiber composite foams with low percolation for efficient electromagnetic interference shielding. *ACS Appl. Mater. Interfaces,* **2014**, *6*(14), 11091-11100.
        http://dx.doi.org/10.1021/am500445g PMID: 24964159

[130]   Xu, Y.; Li, Y.; Hua, W.; Zhang, A.; Bao, J. Light-weight silver plating foam and carbon nanotube hybridized epoxy composite foams with exceptional conductivity and electromagnetic shielding property. *ACS Appl. Mater. Interfaces,* **2016**, *8*(36), 24131-24142.
        http://dx.doi.org/10.1021/acsami.6b08325 PMID: 27553528

[131]   Duan, H.; Zhu, H.; Yang, J.; Gao, J.; Yang, Y.; Xu, L.; Zhao, G.; Liu, Y. Effect of carbon nanofiller dimension on synergistic EMI shielding network of epoxy/metal conductive foams. *Compos., Part A Appl. Sci. Manuf.,* **2019**, *118*, 41-48.
        http://dx.doi.org/10.1016/j.compositesa.2018.12.016

[132] Fan, Z.; Wang, D.; Yuan, Y.; Wang, Y.; Cheng, Z.; Liu, Y.; Xie, Z. A lightweight and conductive MXene/graphene hybrid foam for superior electromagnetic interference shielding. *Chem. Eng. J.,* **2020**, *381*122696

http://dx.doi.org/10.1016/j.cej.2019.122696

[133] Xu, H.; Yin, X.; Li, X.; Li, M.; Liang, S.; Zhang, L.; Cheng, L. Lightweight Ti_2CT_x MXene/Poly(vinyl alcohol) Composite Foams for Electromagnetic Wave Shielding with Absorption-Dominated Feature. *ACS Appl. Mater. Interfaces,* **2019**, *11*(10), 10198-10207.

http://dx.doi.org/10.1021/acsami.8b21671 PMID: 30689343

[134] Zhao, S.; Zhang, H-B.; Luo, J-Q.; Wang, Q-W.; Xu, B.; Hong, S.; Yu, Z-Z. Highly electrically conductive three-dimensional $Ti_3C_2T_x$ MXene/reduced graphene oxide hybrid aerogels with excellent electromagnetic interference shielding performances. *ACS Nano,* **2018**, *12*(11), 11193-11202.

http://dx.doi.org/10.1021/acsnano.8b05739 PMID: 30339357

[135] Wu, X.; Han, B.; Zhang, H-B.; Xie, X.; Tu, T.; Zhang, Y.; Dai, Y.; Yang, R.; Yu, Z-Z. Compressible, durable and conductive polydimethylsiloxane-coated MXene foams for high-performance electromagnetic interference shielding. *Chem. Eng. J.,* **2020**, *381*122622

http://dx.doi.org/10.1016/j.cej.2019.122622

[136] Zhan, X.; Si, C.; Zhou, J.; Sun, Z. MXene and MXene-based composites: synthesis, properties and environment-related applications. *Nanoscale Horiz.,* **2020**, *5*(2), 235-258.

http://dx.doi.org/10.1039/C9NH00571D

[137] Yun, T.; Kim, H.; Iqbal, A.; Cho, Y.S.; Lee, G.S.; Kim, M-K.; Kim, S.J.; Kim, D.; Gogotsi, Y.; Kim, S.O.; Koo, C.M. Electromagnetic Shielding of Monolayer MXene Assemblies. *Adv. Mater.,* **2020**, *32*(9)e1906769

http://dx.doi.org/10.1002/adma.201906769 PMID: 31971302

[138] Zhang, X.; Zhang, Z.; Zhou, Z. MXene-based materials for electrochemical energy storage. *Journal of Energy Chemistry,* **2018**, *27*(1), 73-85.

[139] Shahzad, F.; Alhabeb, M.; Hatter, C.B.; Anasori, B.; Man Hong, S.; Koo, C.M.; Gogotsi, Y. Electromagnetic interference shielding with 2D transition metal carbides (MXenes). *Science,* **2016**, *353*(6304), 1137-1140.

http://dx.doi.org/10.1126/science.aag2421 PMID: 27609888

[140] Ren, Y-L.; Wu, H-Y.; Lu, M-M.; Chen, Y-J.; Zhu, C-L.; Gao, P.; Cao, M-S.; Li, C-Y.; Ouyang, Q-Y. Quaternary nanocomposites consisting of graphene, $Fe_3O_4@Fe$ core@shell, and ZnO nanoparticles: synthesis and excellent electromagnetic absorption properties. *ACS Appl. Mater. Interfaces,* **2012**, *4*(12), 6436-6442.

http://dx.doi.org/10.1021/am3021697 PMID: 23176086

[141] Sun, D.; Zou, Q.; Qian, G.; Sun, C.; Jiang, W.; Li, F. Controlled synthesis of porous Fe_3O_4-decorated graphene with extraordinary electromagnetic wave absorption properties. *Acta Mater.,* **2013**, *61*(15), 5829-5834.

http://dx.doi.org/10.1016/j.actamat.2013.06.030

[142] Singh, R.; Kulkarni, S.G. Nanocomposites based on transition metal oxides in polyvinyl alcohol for EMI shielding application. *Polym. Bull.,* **2014**, *71*(2), 497-513.

http://dx.doi.org/10.1007/s00289-013-1073-2

[143] Chaudhary, A.; Kumar, R.; Teotia, S.; Dhawan, S.K.; Dhakate, S.R.; Kumari, S. Integration of MCMBs/MWCNTs with Fe_3O_4 in a flexible and light weight composite paper for promising EMI shielding applications. *J. Mater. Chem. C Mater. Opt. Electron. Devices,* **2017**, *5*(2), 322-332.
http://dx.doi.org/10.1039/C6TC03241A

[144] Zhang, H.; Zhang, G.; Li, J.; Fan, X.; Jing, Z.; Li, J.; Shi, X. Lightweight, multifunctional microcellular PMMA/Fe_3O_4@MWCNTs nanocomposite foams with efficient electromagnetic interference shielding. *Compos., Part A Appl. Sci. Manuf.,* **2017**, *100*, 128-138.
http://dx.doi.org/10.1016/j.compositesa.2017.05.009

[145] Agarwal, P.R.; Kumar, R.; Kumari, S.; Dhakate, S.R. Three-dimensional and highly ordered porous carbon-MnO_2 composite foam for excellent electromagnetic interference shielding efficiency. *RSC Advances,* **2016**, *6*(103), 100713-100722.
http://dx.doi.org/10.1039/C6RA23127F

[146] Zacco, A.; Borgese, L.; Gianoncelli, A.; Struis, R.P.W.J.; Depero, L.E.; Bontempi, E. Review of fly ash inertisation treatments and recycling. *Environ. Chem. Lett.,* **2014**, *12*(1), 153-175.
http://dx.doi.org/10.1007/s10311-014-0454-6

[147] Ding, L.; Ning, W.; Wang, Q.; Shi, D.; Luo, L. Preparation and characterization of glass–ceramic foams from blast furnace slag and waste glass. *Mater. Lett.,* **2015**, *141*, 327-329.
http://dx.doi.org/10.1016/j.matlet.2014.11.122

[148] Pode, R. Potential applications of rice husk ash waste from rice husk biomass power plant. *Renew. Sustain. Energy Rev.,* **2016**, *53*, 1468-1485.
http://dx.doi.org/10.1016/j.rser.2015.09.051

[149] Phonphuak, N.; Kanyakam, S.; Chindaprasirt, P. Utilization of waste glass to enhance physical–mechanical properties of fired clay brick. *J. Clean. Prod.,* **2016**, *112*, 3057-3062.
http://dx.doi.org/10.1016/j.jclepro.2015.10.084

[150] Samal, S.; Ray, A.K.; Bandopadhyay, A. Proposal for resources, utilization and processes of red mud in India — A review. *Int. J. Miner. Process.,* **2013**, *118*, 43-55.
http://dx.doi.org/10.1016/j.minpro.2012.11.001

[151] Liu, W.; Yang, J.; Xiao, B. Application of Bayer red mud for iron recovery and building material production from alumosilicate residues. *J. Hazard. Mater.,* **2009**, *161*(1), 474-478.
http://dx.doi.org/10.1016/j.jhazmat.2008.03.122 PMID: 18457916

[152] Nguyen, V-T.; Min, B.K.; Yi, Y.; Kim, S.J.; Choi, C-G. MXene (Ti_3C_2TX)/graphene/PDMS composites for multifunctional broadband electromagnetic interference shielding skins. *Chem. Eng. J.,* **2020**, *393,* 124608.
http://dx.doi.org/10.1016/j.cej.2020.124608

[153] Liang, C.; Song, P.; Ma, A.; Shi, X.; Gu, H.; Wang, L.; Qiu, H.; Kong, J.; Gu, J. Highly oriented three-dimensional structures of Fe_3O_4 decorated CNTs/reduced graphene oxide foam/epoxy nanocomposites against electromagnetic pollution. *Compos. Sci. Technol.,* **2019**, *181*107683
http://dx.doi.org/10.1016/j.compscitech.2019.107683

CHAPTER 3

Carbon Nanostructures-based Polymer Nanocomposites for EMI Shielding Applications

Tejendra K. Gupta[1*] **Rajeev Kumar**[2], **Manjeet Singh Goyat**[3] **and Deepshikha Gupta**[1]

[1] *Amity Institute of Applied Sciences, Amity University, Sector-125, Noida 201313, India*

[2] *CSIR-Advanced Materials and Processes Research Institute, Bhopal-462026, India*

[3] *Department of Applied Science, University of Petroleum & Energy Studies, Dehradun 248007, Uttarakhand, India*

Abstract: We have seen a rapid surge in the growth and subsequent drive-in scaling down electronic interfaces with intelligent electronic devices. Any electronic gadget that transmits, distributes, or uses electrical energy produces electromagnetic interference (EMI), which has harmful effects on device performance, human health, and the surrounding environment. This increase in unrestricted EM pollution can also affect human well-being and the surrounding environment if proper shielding is not provided. Therefore, there is an increasing demand for EMI shielding materials due to the rapid increase in EM radiation sources. EMI shielding materials must have the capability to absorb and reflect EM radiation at very high frequencies and act as a shield against the penetration of radiation through them. The polymer matrices are generally electrically insulating; therefore, they cannot provide shielding against EM radiations. Thus, the use of electrically conducting fillers enables the path in polymer composites to shield the EM radiations. This chapter covers the up-to-date research activities targeting EMI shielding based on thermoplastic, and thermoset polymer nanocomposites (PNCs) reinforced with carbon-based nanostructures (CBNS). The first section of this chapter gives a brief overview of the fundamentals of EMI shielding, theoretical aspects of shielding, and different strategies for controlling EM radiations. Other synthesis methods are discussed in the next section, which deals with the preparation of PNCs. Comprehensive justification of potential materials for controlling EMI is also described with nanocomposites based on thermoplastic and thermoset polymer matrices incorporated within CBNS, magnetic, dielectric, and hybrid materials. The synergistic effects of the hybrid fillers may render tunable electrical conductivity and electrical percolation phenomenon in nanocomposites.

***Corresponding author Tejendra K. Gupta:** Amity Institute of Applied Sciences, Amity University, Sector-125, Noida 201313, India; Tel: 91 9999425496; E-mail: tejendra.amu@gmail.com

Sundeep K. Dhawan, Avanish Pratap Singh, Anil Ohlan, Kuldeep Singh Kakran and Pradeep Sambyal (Eds.)
All rights reserved-© 2022 Bentham Science Publishers

Keywords: Carbon-based nanostructures, thermoplastic polymer nanocomposites, thermoset polymer nanocomposites, EMI shielding.

3.1. INTRODUCTION

In the present era, there is an increase in the utilization of wireless and smart electronic devices and equipment in the high-frequency range, which leads to inducing electromagnetic (EM) radiation [1]. Any electronic gadget and device that uses or transmits electrical energy can create electromagnetic interference (EMI); these devices and gadgets have harmful effects on the performance of the devices and the nearby environment. EM radiation generated from any source can also disturb several electrical equipment and circuits, mobile phones, satellite communication systems, televisions, *etc.* The turbulence from these channels affects time, energy, and life. Miscarriages, leukemia, brain issues, anxiety, breast cancer, and a variety of other medical issues are related to continuous exposure to electromagnetic pulses [2, 3]. This EMI problem has become a global concern in terms of control and regulation of EM pollution. Therefore, EMI shielding materials are compulsory nowadays for marketable and defense purposes to protect today's society from EM pollution [4].

Polymer nanocomposites (PNCs) filled with conductive nanostructures have been widely investigated in academia and industry, such as strain sensing [5-11], supercapacitor [12], and EMI shielding [13-16] due to their exceptional characteristics [4, 15-22]. PNCs are also highly beneficial for industrial applications as they have the capability to tailor their properties with fillers. PNCs offer numerous promises with improved mechanical, barrier, flame-retardant, electrical, and magnetic properties by the addition of a lower number of fillers. Recently, researchers have been conducting research to improve the structural and functional properties of PNCs [23]. It is clearly seen that PNCs can offer new functionality in the field of bulk polymeric parts and devices, but the drawbacks are still under investigation, such as difficulties in the dispersion of nanofiller, higher time of processing, and optimization of an effective process.

This chapter describes and covers up-to-date research on the role of carbon-based nanostructures (CBNS) on the EMI shielding properties of thermoplastic and thermoset polymer nanocomposites. According to us, these PNCs can enter the market as a huge EM absorbing material. Based on the EMI shielding overview, the theory of EMI shielding and the mechanism of EMI shielding are also given.

We have categorized our study into various CBNS, such as carbon nanotubes (CNTs), carbon nanofibers (CNFs), graphene, and their hybrids. Also, the addition of other magnetic and dielectric fillers in combination with CBNS is explored in this chapter as these are gaining interest nowadays due to the demand for EM absorbing materials. The different preparation methods of PNCs and the effect of morphology, surface functionalization, and loading of CBNS on EMI shielding performances of polymer matrices are also discussed.

3.2. EMI PARAMETERS AND SHIELDING MECHANISM

The study is carried out to study the EM waves within the microwave region, mainly in the ultra-high (UHF) and super-high (SHF) frequency ranges. Fig. (**1**) shows the frequency ranges of the electromagnetic spectrum. Electromagnetic studies are an essential part of electrical engineering and are still useful today [24]. The traditional electromagnetic radiation emitting devices such as radios and television combined with the signals of current wireless devices create a more congested electromagnetic environment [25]. As the processing speed is increasing day by day, the frequency of use and functioning of such devices are also increasing, thereby making the electromagnetic environment highly saturated; thus, there is an increased demand for shielding [26].

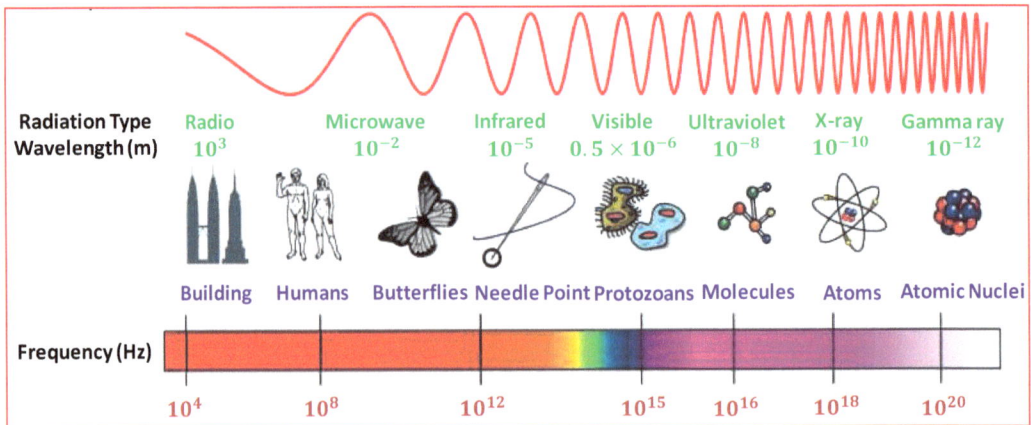

Fig. (1). Electromagnetic spectrum diagram of the ranges of frequencies and wavelengths. https://www.chromatherapylight.com/uploads/7/4/2/9/74293369/electromagnetic_spectrum_wiki.j pg).

EM radiation spans the frequency spectrum with changing properties and applications by combining electric and magnetic energy. These EM waves of various wavelengths and frequencies make up the electromagnetic spectrum, as shown in Fig. (**1**). EMI can be propagated by both conduction and radiation. It can be said to occur when electromagnetic waves affect the electronic device performance and the health of the human being. To achieve effective EMI shielding, a high EMI shielding capability of 99 % absorption (-20 dB) is required for human safety. Therefore, we need to design the EMI shielding enclosures in such a way as to protect the human and the environment from unwanted EM radiations. Among many sources, electronic circuits are the primary source of EMI. High-speed switching in micro-circuitry may generate EMI in the form of short EM pulses. Therefore, familiar sources of EMI include electronic devices, such as radio and calculators, computers, televisions, and communication systems. Such powerful tools can also be the EMI generators if incorporated in electric machinery systems and devices. These unwanted EM signals can also be produced by fluorescent lighting, radar, power lines, and lighting.

The protector from EM radiation can be made up of conductive material that suppresses the radiated EM energy. Shielding effectiveness (SE) calculates the ability of a specific material to attenuate in a specific configuration. It is the logarithm of the incident power (P_I) divided by the transmitted power (P_T). Shielding efficiency (SE_T) is the combination of the three shielding components for disturbing the transmission of EM radiation, such as reflection (SE_R), absorption (SE_A), and multiple internal reflections (SE_M), as shown in Fig. (**2**). Thus, the total EMI shielding (SE_T) of the shield material can be expressed as

$$SE_T = 10 log_{10} \frac{P_I}{P_T} = 20 log_{10} \frac{E_I}{E_T} = 20 log_{10} \frac{H_I}{H_T} \tag{1}$$

where P_I (E_I or H_I) and P_T (E_T or H_T) are the electric and the intensity magnetic fields which are incident and transmitted in the form of electromagnetic waves, respectively, through a shielding material. The reflection power of EM waves is determined by the reflection from the surface of free charge carriers and the multiple reflections of the neighbouring conducting channels. Alternatively, absorption power in a shielding material can be determined by several loss mechanisms related to electrical polarization and magnetization processes [23].

SE$_T$ can be expressed in decibels (dB), and it is the summation of the reflection (SE$_R$), absorption (SE$_A$), and multiple reflections (SE$_M$), *i.e.*,

$$SE_T = SE_A + SE_R + SE_M \qquad (2)$$

However, the increased value of the absorption component of the reflection from the internal surface suppresses the multiple types of reflections (SE$_M$) [27], *i.e.*, when SE$_A$ is ≥ 10 dB; then the SE$_M$ value will be very small, and it can be neglected [23]. Therefore, SE$_T$ can be written as

$$SE_T \approx SE_A + SE_R \qquad (3)$$

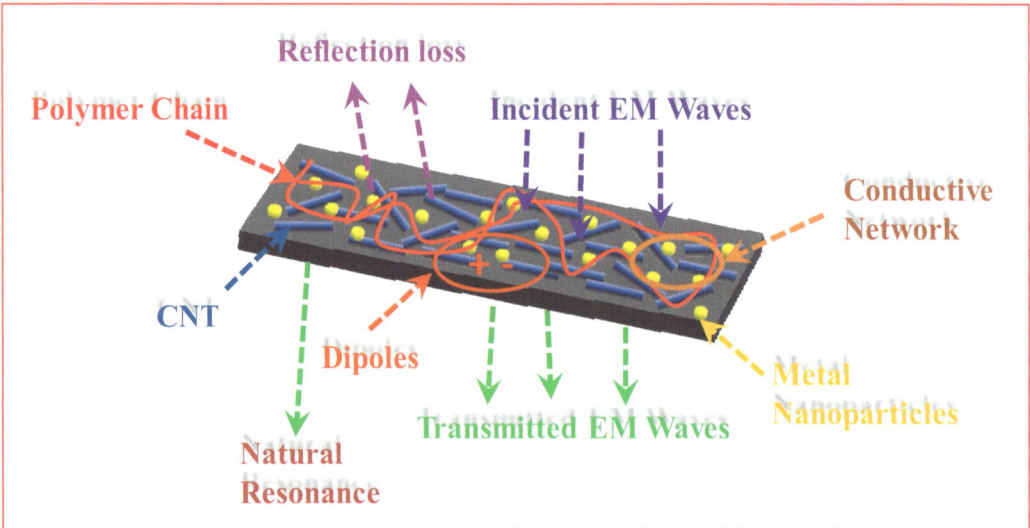

Fig. (2). EMI shielding mechanism in a conductive sheet.

The total EMI shielding (SE$_T$) from the vector network analyzer can be measured using the following relations through scattering parameters, which represent the reflective, absorbing, and transmitted powers [23]. Therefore,

$$SE_T = 10 log_{10} \left(\frac{1}{|S_{12}|^2} \right) = 10 log_{10} \left(\frac{1}{|S_{21}|^2} \right) \qquad (4)$$

where S_{12} and S_{21} are the reverse and forward coefficients of transmission, respectively. With the help of these S_{12} and S_{21} parameters, reflection and absorption parameters of EMI shielding can be calculated as follows [23]:

$$SE_R = 10log_{10}\left(\frac{1}{1-|S_{11}|^2}\right) \tag{5}$$

$$SE_A = 10log_{10}\left(\frac{1-|S_{11}|^2}{|S_{12}|^2}\right) \tag{6}$$

where S_{11} is the forward reflection coefficient. Therefore, it is concluded from EMI theory that sufficient electrical conductivity of a material is required to block the EM radiation.

The shielding contribution due to the reflection component depends on the conductive surface of the material and can be measured as follows [16]:

$$SE_R(dB) = 108 + log_{10}\left(\frac{\sigma_T}{f\mu}\right) = -10log_{10}\left(\frac{\sigma_T}{16\omega\varepsilon_0\mu}\right) \tag{7}$$

where 'σ_T' represents the total electrical conductivity, and 'μ' represents the permeability of the shielding material. Therefore, from Equation 7, it can be said that EMI shielding due to reflection (SE_R) is proportional to the materials' ratio of electrical conductivity and permeability.

However, the absorption part of EMI shielding (SE_A) can fall-off exponentially across the shielding thickness (t), which can be measured as:

$$SE_A (dB) = -8.68 \{t/\delta\} = -8.68\, \alpha t \tag{8}$$

where 'α' is the coefficient of attenuation, which defines the intensity of an EM wave that is reduced by passing over a specific material.

The terms SE$_R$ and SE$_A$ are the reflection and absorption components, respectively, which contribute to the shielding and are expressed mathematically as:

$$\alpha = \frac{4\pi n}{\lambda_0} \tag{9}$$

where n is the refractive index and λ_0 is the wavelength in vacuum of the material.

The total electrical conductivity (σ_T) can also be expressed as $\sigma_T = 2\pi f \varepsilon_0 \varepsilon''$ and $\varepsilon_r = \varepsilon' - j\varepsilon''$, where, $\boldsymbol{\varepsilon'}$ and $\boldsymbol{\varepsilon''}$ describe the real and the imaginary part of the permittivity, respectively and $\boldsymbol{\varepsilon_0}$ is the permittivity in vacuum. It clearly seems that the mechanism of EMI shielding is related to the relative part of permittivity and permeability. Therefore, determination of these relative parts of permittivity and permeability is necessary to explain the EMI shielding mechanism.

The real part of permittivity (ε') measure the stored energy in materials from an external electric field. The imaginary part of permittivity (ε'') is called the loss factor, which measures how the permittivity loss takes place in a material when it comes under the influence of an external electric field. At frequencies below relaxation, the alternating electric field becomes slow, and at this slow alternating electric field, the dipoles can keep pace with the variations in the field. The polarization is able to develop completely and the loss factor (ε'') is directly proportional to the frequency. With an increase in the frequency, the relative permittivity (ε_r) increases continuously but the storage factor (ε') starts to decrease due to the phase lag between the dipole alignment and the electric field. However, both the factors, such as ε' and ε'' drop off at above the relaxation frequency because the electric field becomes very fast and it influences the dipole rotation, and due to this, the orientation polarization disappears. The loss tangent or dissipation factor, $tan\delta$, is a useful dimensionless parameter and is a measure of the loss of the electric energy to the energy stored in a periodic field, *i.e.*, $tan\delta = \frac{\varepsilon''}{\varepsilon'}$.

3.3. THERMOPLASTIC AND THERMOSETTING POLYMERS

Electrically conducting materials were used for EMI shielding applications. Among them, metals have been generally used as materials with good shielding because of high electrical conductivity, and the free electrons present in metals reflect the incident waves and provide shielding from EM radiations. However, metals have several disadvantages like difficult processability, being heavy weight, corrosion, and stretchability issues, which limit the use of metals for EMI shielding. Nowadays, researchers are working on developing smart materials with high electrical conductivity and high saturated magnetization, which help in achieving high-performance EM wave absorbing materials. Therefore, the state-of-the-art and economically reliable research for the growth of EMI shielding materials leads to the preparation of conducting PNCs. These conductive PNCs are light in weight, lower in cost, easily processable, environmentally friendly, and corrosion resistant; therefore, these extraordinary properties make them the ideal choice for EMI shielding applications.

For EMI shielding, thermoplastic polymers offer reliability and value in applications where electromagnetic compatibility is required. Thermoplastic polymers can provide designer products with desired size and shape, flexibility and they can also provide important benefits over unfilled resins, coatings, and metals. Several thermoplastic polymers such as thermoplastic polyurethane (TPU), acrylonitrile butadiene styrene (ABS), polycarbonate (PC), polystyrene (PS), polyaniline (PANI), polypropylene (PP), low-density polyethylene (LDPE), ultrahig-molecular weight polyethylene (UHMWPE), poly (methyl-methacrylate) (PMMA), polyimide (PI), etc., reinforced with CBNS have been studied for EMI shielding applications.

Thermosetting polymers are promising and lightweight materials with superior performance. These are low molecular weight monomers that are converted into 3D cross-linked structures. This cross-linking results in the formation of infusible and insoluble structures and is generally a result of chemical reactions taking place within the structure by the supply of heat. Once the curing reaction is over, it is very difficult to melt a thermoset by applying heat because of its rigid structure. But, if the number of cross-links is less, it may be possible to soften them at elevated temperatures. Thermosets are widely used as matrices to develop high-performance composites. The important roles played by the matrices are (a) complete wetting of fillers, (b) transfer of load from the matrix to fillers, and (c) protecting the filler from the external environment. Epoxy, phenol formaldehyde, and polyester resins reinforced with CBNS have been used for EMI shielding applications.

3.4. COMMON FILLERS USED IN POLYMER NANOCOMPOSITES

3.4.1. Carbon-based Nanostructures (CBNS)

Carbon-based materials as fillers to improve the thermal, mechanical, and electrical properties of polymer matrices are not new in the research areas. Carbon black has previously been used widely as a reinforcing material in polymer matrices to enhance mechanical, thermal, and electrical performances. Moreover, it has extensively been utilized in automotive industries for making race cars tires that decrease the damage due to heat generation [28]. Carbon fiber is a widely used material, which is generally used to make stiffer and conductive PNCs that are used in aerospace and automobile applications [29, 30]. However, carbon-based nanostructures (CBNS) such as carbon nanofibers (CNF), carbon nanotubes (CNTs), graphene, and their derivatives have an important role in the nanoscience community because of their extraordinary properties. These CBNS have been used in several areas such as composite materials for biomedical applications, automobile industries, and different types of sensors and energy storage. CNTs and graphene are continuously gaining attention and have been applied in many fields, and also several potential applications continue to grow.

3.4.2. Carbon Nanofiber (CNF)

Carbon nanofibers (CNF) are 1-dimensional derivatives of carbon, in which layers of graphene are stacked cylindrically with several orientations [23]. They can be prepared by catalytic chemical vapour deposition method in the presence of metal catalysts. CNFs are light in weight, have a high surface area, unique structural properties and high charge transport property, which makes CNF suitable material for EMI shielding.

3.4.3. Carbon Nanotube (CNT)

CNTs have attracted significant attention in the area of material science due to their extraordinary electrical [31-36], mechanical [37-40], and thermal [41] properties. CNTs were used before in several sports applications like tennis racquets, golf clubs, baseball bats, and have been marketed. Nowadays, the production of pure

CNTs in large amounts is straightforward; therefore, CNT can become one of the significant reinforcements for the synthesis of PNCs.

CNT forms when the wrapping of a graphene sheet takes place along its length with various chiral structures. Tube chirality can describe the atomic structure of CNTs, or helicity, defined by the chiral vector and the chiral angle shown in Fig. (**3a**). Chiral vector is equal to $C_h = na_1 + ma_2$, where n and m are the number of unit vectors in two directions in the honeycomb crystal lattice of graphene [42].

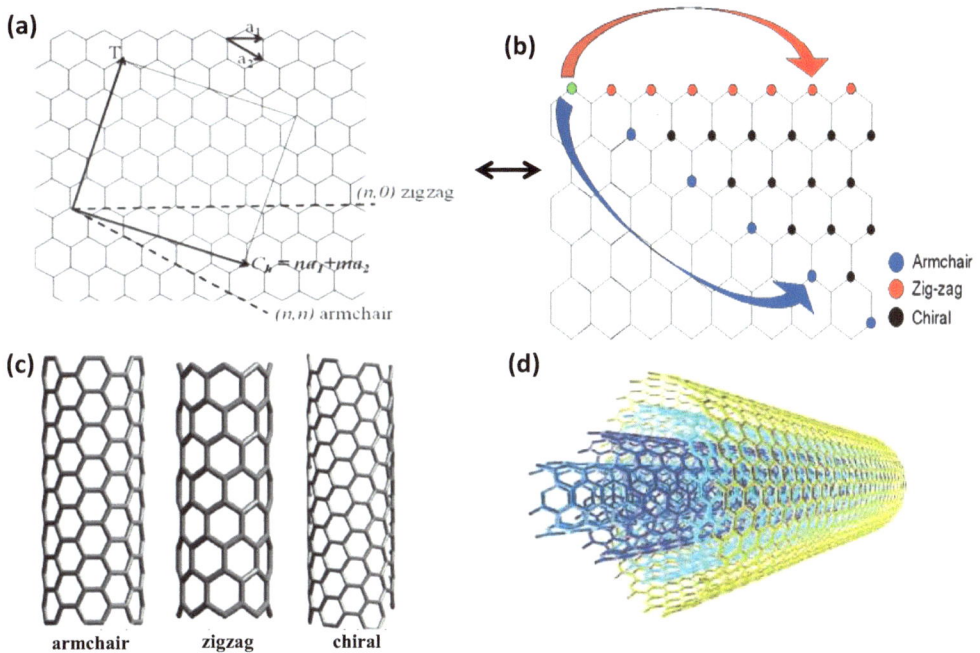

Fig. (**3**). (**a**) Different schemes of rolling of graphene sheet to form CNT, (**b**) formation of armchair, zigzag and chiral CNT, (**c**) structure of armchair, zigzag and chiral CNT and (**d**) coaxial cylinders of SWCNTs [43, 44].

CNTs are called zigzag if, m = 0, armchair if, n = m, otherwise, they are chiral as shown in Fig. (**3b**) and the structure of these three types of CNTs is shown in Fig. (**3c**). These different structures of CNTs impact the properties such as conductance, lattice structure and density. The differences in the conducting properties of CNTs are due to the different band structures and hence the different band gap [45]. Single walled carbon nanotubes (SWCNTs) form by rolling a single layer graphene sheet into a seamless cylinder. Similarly, a multi walled carbon nanotubes (MWCNTs) is assembled by the coaxial cylinders of SWCNTs (see Fig. **3d**), one within another [46]. The tube-to-tube separation is almost the same as the separation between the layers of natural graphite. There are van der Waals forces of attraction between these concentric nanotubes and the length-to-diameter ratio in these CNTs is near around 1000. That is why, they are known as nearly one-dimensional, and their properties vary with structure.

CNTs are the stiffest material till date with extremely high Young's modulus of 1.0 TPa [47, 48]. The expected elongation to failure of CNT is 20 % – 30 %, and tensile strength is almost 100 GPa in the axial direction, which is higher than high strength steel (Young's modulus ~ 200 GPa and tensile strength ~ 1–2 GPa) [48, 49]. The extremely high strength, stiffness and low density suggest that CNTs could be the better filler for composite materials [48].

In real-world applications, the use of CNTs is possible only if the large quantity and high purity CNTs can be obtained using efficient and inexpensive growth methods. Substantial research has been conducted in this field, and several ways have been studied to synthesize CNTs. Now, several methods of synthesis exist for the production of CNTs in which three methods such as arc discharge [50], laser ablation [51], and catalytic chemical vapour deposition (CCVD) [52] are important methods.

Among all of them, the CCVD method is a simple, efficient and inexpensive method for the synthesis of high quality CNTs. In this process, catalytic decomposition of hydrocarbons into vapour takes place at high temperatures. Therefore, this method is also recognized as thermal CVD or catalytic CVD and the schematic diagram is shown in Fig. (**4**). This is a straight forward method for producing CNTs from the pyrolysis of organometallic compounds in a quartz reactor furnace. Many researchers have already used diverse hydrocarbons, types of catalyst and a combination of inert gases for the growth of CNTs [13, 48, 53].

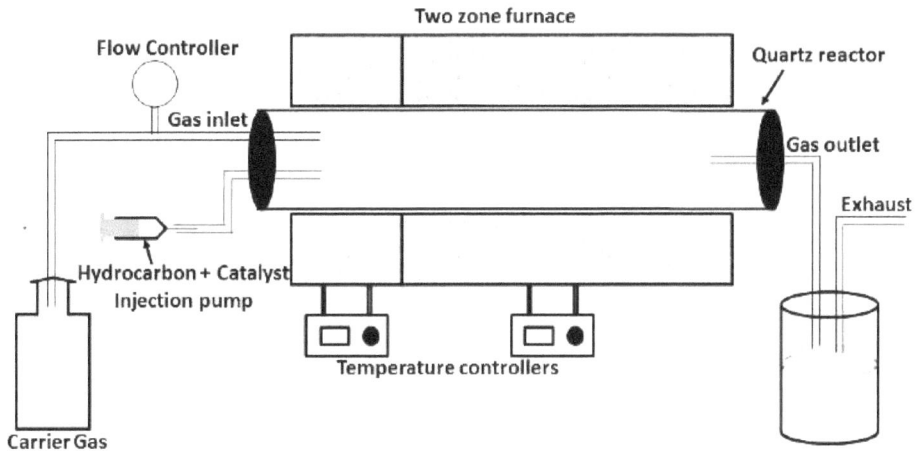

Fig. (4). Representation of Chemical vapour deposition (CVD) Experimental set-up.

3.4.4. Graphene

Graphene can be considered the generation of all the CBNS due to the sheet-like structure, variety of size, morphology and extraordinary properties that are similar to CNTs [54, 55]. Recently, graphene and its derivatives have been widely used as 2D CBNS in several applications such as strain sensing, EMI shielding, solar cells, energy efficient devices, lithium-ion batteries, biomedical applications and supercapacitors [4, 15, 18, 21, 56-58]. Graphene is a 2D CBNS in which sp^2 hybridized carbon atoms are arranged in a hexagonal packed lattice structure with many exceptional like high carrier mobility of $\approx 10,000$ $cm^2V^{-1}S^{-1}$ at room temperature [59], large specific surface area of 2630 m^2g^{-1} [60], good optical transparency of $\approx 97.7\%$, high Young's modulus of ≈ 1 TPa [61] and high thermal conductivity of 3000–5000 $Wm^{-1}K^{-1}$ [55]. There are several methods to synthesize graphene, such as catalytic chemical vapour deposition, microwave exfoliation, mechanical and thermal exfoliation, and other chemical routes. However, other chemical route like Hummer's method is one of the best nominated route for large scale synthesis of graphene in the form of graphene oxide (GO), which can be reduced to graphene (RG) or reduced graphene oxide (RGO) *via* thermal as well as microwave exfoliation [18, 21, 57, 58]. Schematic diagram of improved Hummer's method for the synthesis of GO [62] is shown in Fig. (**5**)**,** in which oxidation reaction of natural graphite flakes is done in the presence of a mixture of conc. H_2SO_4/H_3PO_4. This method is straightforward, and no toxic gases are evolved during the preparation. The addition of $KMnO_4$ could be the reason for the explosion so that it can be avoided by the slow stirring and followed by the slow addition of $KMnO_4$ [63]. Prior to washing, and filtration of the material with HCl and water, it was ultrasonicated for eight hours in ethanol to exfoliate graphite oxide into graphene oxide (GO) sheets and finally dried in an oven at 120°C. To exfoliate GO to RGO, we can use the thermal-mediated method at 1000°C, in which stacked sheets are exfoliated *via* the extrusion of CO_2, which is generated by heating. During the exfoliation process, some topological defects are left behind throughout the plane of the RGO sheet, which might be helpful for better bonding with the polymer matrix [64]. These unique properties of graphene, along with flexibility and modifiable structural properties can be useful for the fabrication of PNCs.

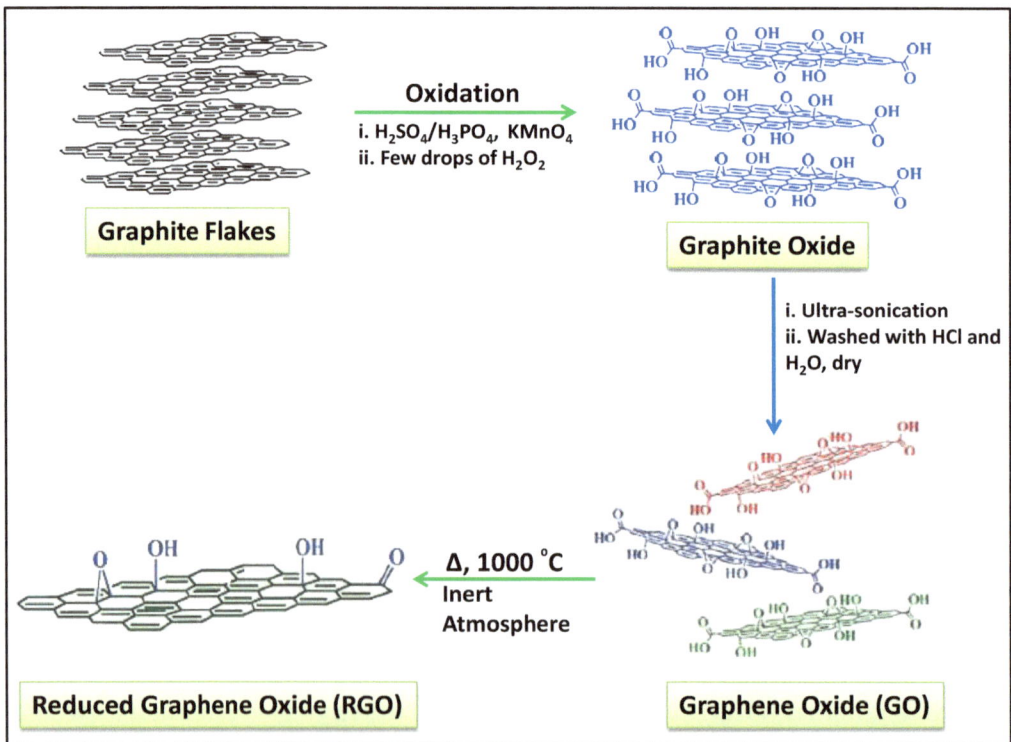

Fig. (5). Synthesis of reduced graphene oxide by chemical route [18, 62, 65].

3.4.5. Magnetic Fillers

The absorption part of EMI shielding mechanism is directly proportional to the magnetic permeability of the materials. Therefore, high magnetic permeability materials such as carbonyl iron, Ni, Co, or Fe metals, $\gamma\text{-}Fe_2O_3$, Fe_3O_4 and metal based ferrofluids are used.

3.4.6. Dielectric Fillers

For the absorption of EM radiation, material should have electrical or magnetic dipoles and also the finite electrical conductivity. In these cases, high dielectric constant materials such as TiO_2, ZnO, $BaTiO_3$, and SiO_2 could be used.

3.4.7. Hybrid Nanostructures

Although CNTs and graphene have outstanding electrical, mechanical and thermal properties as discussed in earlier sections, they have several limitations and challenges. First challenge is the uniform dispersion of CNTs and graphene in polymer matrices which may be due to agglomeration because there is strong van der Waals force of attraction between CNTs. Another challenge is the restacking behavior of graphene. Therefore, to integrate the properties of CNTs and graphene in polymer matrix to prepare high-performance PNCs, the hybrid structures of CNT and graphene could be the perfect choice for the fabrication of PNCs with desired properties because the addition of graphene and CNTs can improve the properties of PNCs significantly, which may be due to the synergistic effect of hybrid fillers [57, 66-68].

Several researchers have worked on the synergetic effect of CNT and graphene and they found that the graphene layer could act as a spacer between CNT and reduce the entanglement of CNT, and the CNT makes a bridge between individual graphene layers. Schematic representation of 3-dimensional graphene-CNT hybrids prepared from CNT and graphene is shown in Fig. (**6**). This combination of hybrid materials may be beneficial to increase the dispersion of CNTs in PNCs and to form the effective conductive networks, which may lead to better electrical properties of PNCs.

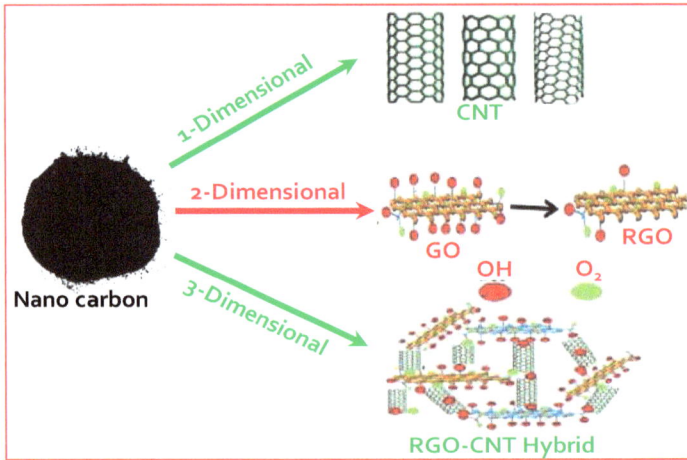

Fig. (6). Schematic representation of 3-dimensional Graphene-CNT hybrids prepared from 1-dimensional CNT and 2-dimensional graphene [57].

3.5. FABRICATION OF THERMOPLASTIC POLYMER NANOCOMPOSITES

Thermoplastic PNCs are highly beneficial for industrial applications because the addition of fillers can tailor their properties. Therefore, due to these exceptional properties, thermoplastic PNCs filled with CNF, CNT, graphene and their hybrids have gained considerable interest in EMI shielding applications.

3.6. MAIN STRATEGIES FOR THE PROCESSING OF CBNS REINFORCED THERMOPLASTIC PNCS

In recent years, great developments have been made to reduce the problems with fabricating of PNCs without losing the properties of nano fillers [69], *i.e.*, to achieve an appropriate dispersion and interfacial interaction of nanosized fillers in nanocomposites. Presently, no universal technique is available that can be used in these conditions. Most of the CBNS reinforced polymer nanocomposites have been fabricated using four common traditional approaches such as solution processing, melt blending, layer-by-layer approach and *in-situ* polymerization [70-72]. However, during the synthesis of PNCs, several additional methods have also been tried to reduce the difficulties of homogeneous dispersion, breaking of bundles of CBNS and interfacial adhesion of CBNS with polymer matrices. All these processing methods for the fabrication of CBNS reinforced PNCs are discussed below in detail.

3.6.1. Solution Processing

Solution processing technique has been usually used as an effective method for the synthesis of CBNS reinforced PNCs on a lab scale and this technique is also useful to fabricate the PNCs without losing the properties of nano-sized fillers like CBNS. In this protocol, initially, polymer granules were dissolved in a suitable solvent and CBNS dispersed in the solvent separately. Dispersed CBNS is then mixed with a polymer solution and then evaporates the solvent with or without vacuum to form a nanocomposite film. In this protocol, vigorous ultrasonication and stirring followed by high-energy homogenization were used to disperse the CBNS into the polymer solution shown in Fig. (**7**). One of the advantages of the solution mixing strategy is the possibility of de-bundling of CBNS and achieving good quality dispersion in the polymer matrix [73]. However, one disadvantage of the solution processing techniques is that this approach is not good for polymers that are insoluble. The use of a large number of solvents contaminated the environment and, therefore, restricts us to use this technique on a large-scale production of CBNS/PNCs. Solvents with low boiling point are generally preferred because of their quick evaporation properties. High boiling point solvents do not easily get removed from PNCs and they also get trapped in the solidifying PNCs.

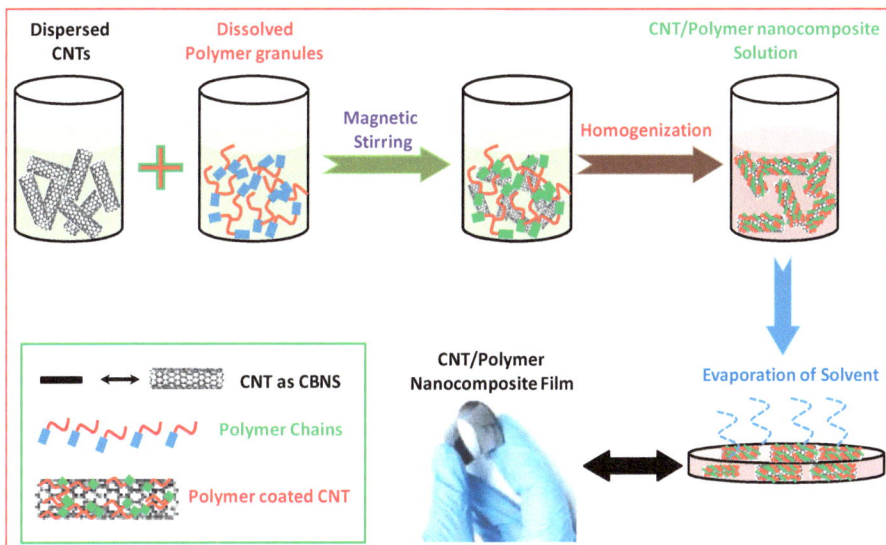

Fig. (7). Schematic diagram of solution processing method for CBNS/polymer nanocomposites.

3.6.2. Melt Blending

The melt processing method is a very good method for the synthesis of PNCs at the industrial level. This is also well suited for polymers, including insoluble ones. In this method, the blending of polymers with CBNS material takes place when we apply the intense shear forces at higher temperatures using a twin-screw extruder and injection moulding [74-76], which is clearly shown by schematic diagram in Fig. (**8a-c**) [77]. This intense shear force de-agglomerates CBNS partially and increases the dispersion in the polymer matrix. Initially, screws and channels slits of the system were cleaned *via* passing some amount of polymer formulation through the system. Then, the micro-compounder was set to "cycle" mode, in such a way that all material was directed into the slit channel as shown in Fig. (**8d**).

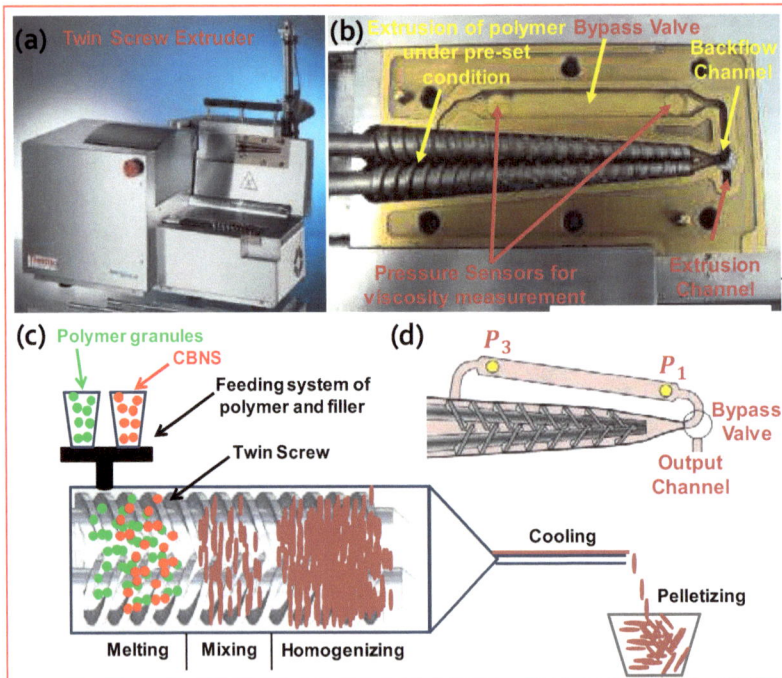

Fig. (8). (a) Extruder system with twin-screw, **(b)** Micro-compounder showing different channels, valves and pressure sensors for viscosity measurement, **(b)** schematic representation of melt mixing process of CBNS with polymer matrix in twin-screw extruder system and **(d)** backflow channel with rheological slit capillary [48].

The melt blending of polymers with CBNS can be done in a continuous or batch process with the help of shear mixing using a twin screw extruder. High shear mixing can be used to prepare the master batches of PNCs. Extruder contains a single or two screws, but an extruder with twin screws is more effective as compared to a single screw because a twin screw extruder increases the uniformity in dispersion. We load the extrusion system with the polymer granules during the process, and inside the extruder system, the polymer granules melt and move forward.

In the hot zone of the system, shearing and melting of the polymer granules between the screws occur. The CBNS as reinforcement are also loaded into the extruder to mix them with the melted polymer in the melting zone, and these CBNS particles combine properly in the polymer matrix due to the shearing force and send them to the homogenization zone [48]. In the end, the mixed CBNS/polymer is transferred to the die mould and then cooled to room temperature and then chopped into granules for further use. The high temperature and shear force affect the polymer assembly and can also break the CBNS particles. Therefore, we must optimize the shear force and temperature properly to get the appropriate dispersion and intrinsic properties of CBNS and polymer matrices. This method is an easy, environmentally friendly, and viable industrial technique with a fast response for the large-scale production of PNCs. In this method, there is no requirement for solvent during the preparation of PNCs. However, this technique is not perfect to separate out of the CBNS bundles. Melt blending is also not appropriate for the materials in which the degradation with temperature takes place [73].

3.6.3. *In-situ* Polymerization

In-situ polymerization is an actual route to enhance the quality of dispersion and adhesion of CBNS with polymer assembly. This method is a feasible choice for the synthesis of insoluble PNCs. In this process, we mix the dispersed CBNS nanofillers in a monomer solution during the formation of the polymer. Then polymerization takes place under certain conditions and in the presence of the initiator [72]. As compared to melt blending and solution mixing and evaporation methods, this method offers excellent dispersion between polymer assembly and CBNS. However, due to the excess requirement of the solvent, this method is less preferred for the processing of PNCs. Fig. (**9**) shows the step-by-step process for the in-situ polymerization of CBNS/PNCs. PNCs can be prepared using high loading of nanofillers and good miscibility with almost all polymers in this method.

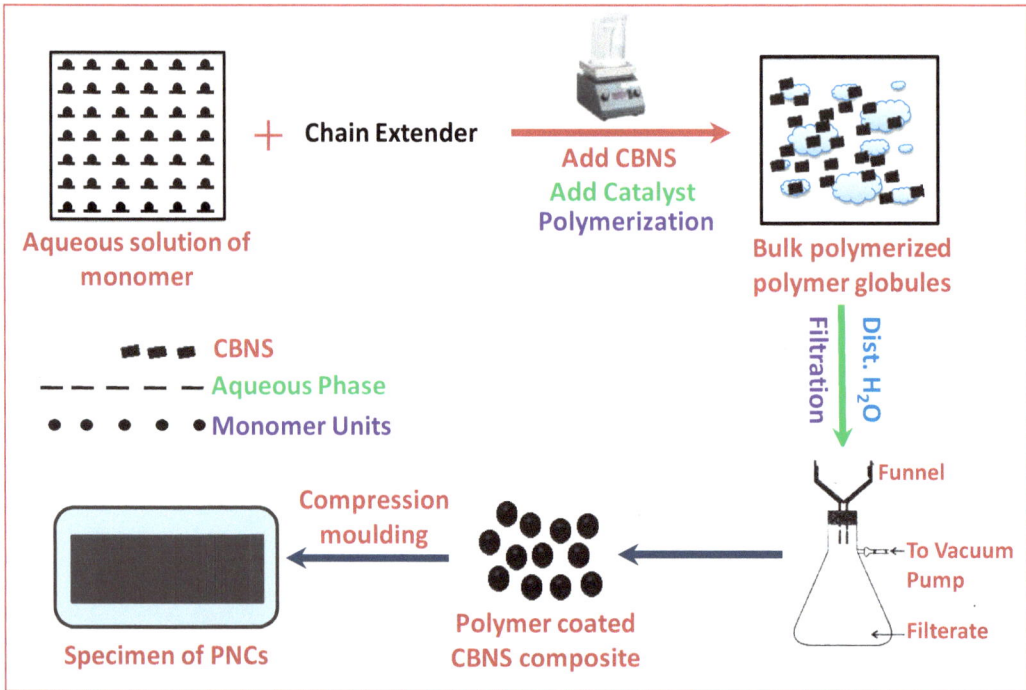

Fig. (9). Representation of the process of *in-situ* polymerization of CBNS/PNCs.

3.6.4. Layer-by-layer Approach

We can prepare the layered composite films on the desired substrate *via* a layer-by-layer (LBL) approach in which the substrate is dipped alternatively into the CBNS dispersion and polymer solution, respectively [78, 79]. A similar method, in which we dip the substrate into the polyelectrolyte solution, which is positively charged and then clean and finally dipped again into the negatively charged polyelectrolyte solution to grow the LBL. A similar method shown in Fig. (**10**) is also used to make the composite films with multiple layers on the substrate's surface by incorporating inorganic nanoparticles into the organic polymers. In this method, first, we make a polymer solution, and then we dip the substrate into it so that a thin layer of polymer can deposit on the surface of the substrate. After this, the polymer-coated substrate is further dipped in CBNS dispersion. To increase the thickness of the polymer composite on the surface in the form of LBL, we can repeat the similar process again and again. This method can help control the CBNS ratio with high nanotube loading [79].

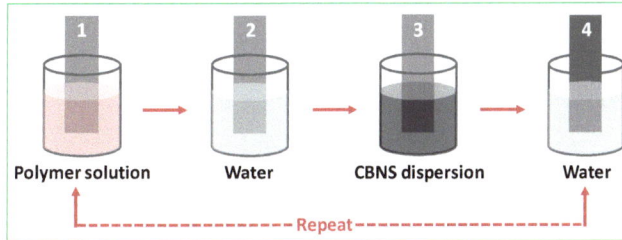

Fig. (10). Schematic representation of LBL film deposition on substrate in which step 1 and step 3 represent the adsorption of positively and negatively charged polyelectrolyte and step 2 and step 4 are washing steps [47].

3.7. MAIN STRATEGIES FOR THE PROCESSING OF CBNS REINFORCED THERMOSET PNCS

Some of the more commonly used methods for fabricating thermoset PNC laminates are hand layup method, vacuum assisted resin transfer moulding, hot pressing and bucky paper routes. These methods are discussed below.

3.7.1. Hand Layup Method

Hand layup is one of the most common and inexpensive fabricating CBNS/fiber/polymer laminates methods. It is an open moulding method in which fibers are placed in the mould in knitted, stitched, woven, or bonded fabrics. The resin reinforced with CBNS is then impregnated using roller or brushes, as shown in Fig. (**11**). The impregnation forces the resin inside the fiber fabric and the laminates of PNC are fabricated *via* this method and cured under the standard atmospheric conditions. The materials such as epoxy, polyester, vinyl ester, and phenolic resin, and any fiber materials are used to prepare the laminates. This process can produce both large and small scale items, including storage tanks, boats, bathing tubs, and car bumpers [80].

3.7.2. Vacuum Assisted Resin Transfer Moulding

Vacuum assisted resin transfer moulding (VARTM) [81] or vacuum assisted resin infusion moulding (VARIM) is a closed mould composite fabrication process. VARTM/VARIM is mainly an extended version of the wet layup process in which we apply the pressure to the laminates to improve their consolidation. It can be done when the laminates are packed in a poly-bag or plastic film.

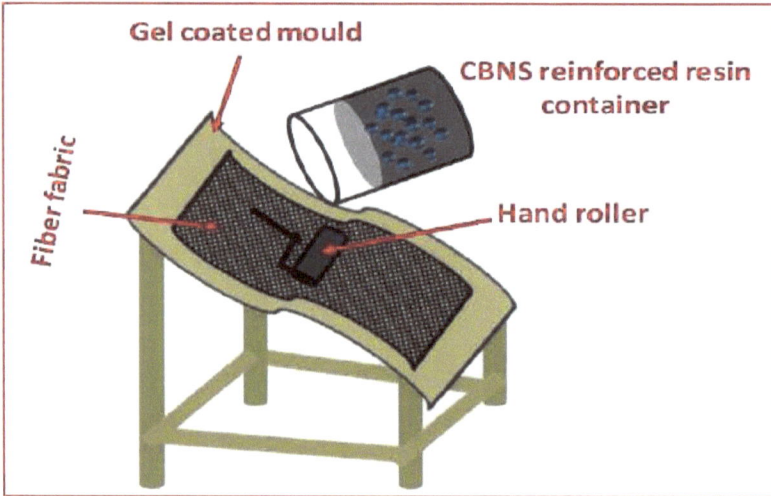

Fig. (11). Wet or hand lay-up process for fabrication of CBNS reinforced PNCs.

A vacuum pump extracts the air under the bag to apply the pressure on the composite laminates to consolidate it, as shown in Fig. (**12**). VARTM is different from the conventional RTM process, where a vacuum-pump system is used to remove the air from the mould and pull the modified resin through the fabric. The VARTM process includes the use of a vacuum to ease the CBNS modified resin flow into a stacked fabric placed on a mould tool covered by a vacuum bag.

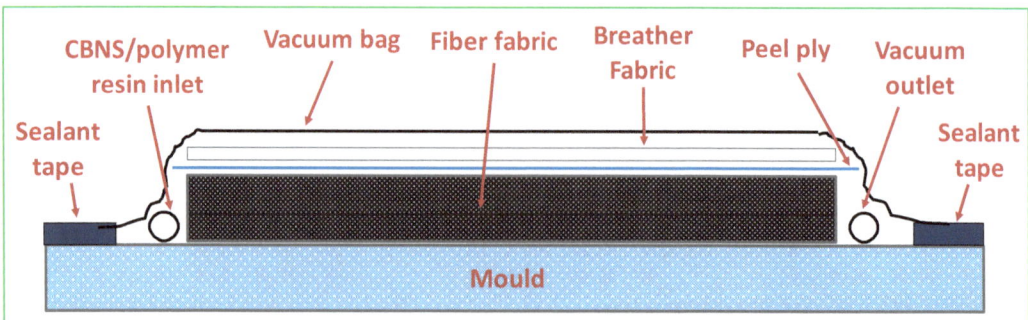

Fig. (12). VARTM method for the fabrication of CBNS modified PNCs.

The CBNS modified resin is passed into the mould under the pressure difference which is created by the vacuum. The CBNS modified resin-infused between the fibers and tows of the fabric until the mould is filled, at which stage the infusion process stops, and then we cure the part at elevated temperature. Generally, epoxy and phenolic resins are used in this process. There are some problems in polyesters and vinyl esters because a large amount of styrene is extracted from the resin by the vacuum pump.

3.7.3. Hot Pressing Method

Hot pressing method is a technique where the reinforcing fiber fabric was impregnated with CBNS modified resin using a simple layup process to get prepregs. These prepregs were then put into a vacuum drying oven to remove entrap air and solvents if used. Then these prepregs were stacked on a metallic plate or mould and then hot pressed with a hot pressing machine [82]. This fabrication method to prepare hybrid PNC laminates is shown in Fig. (**13**).

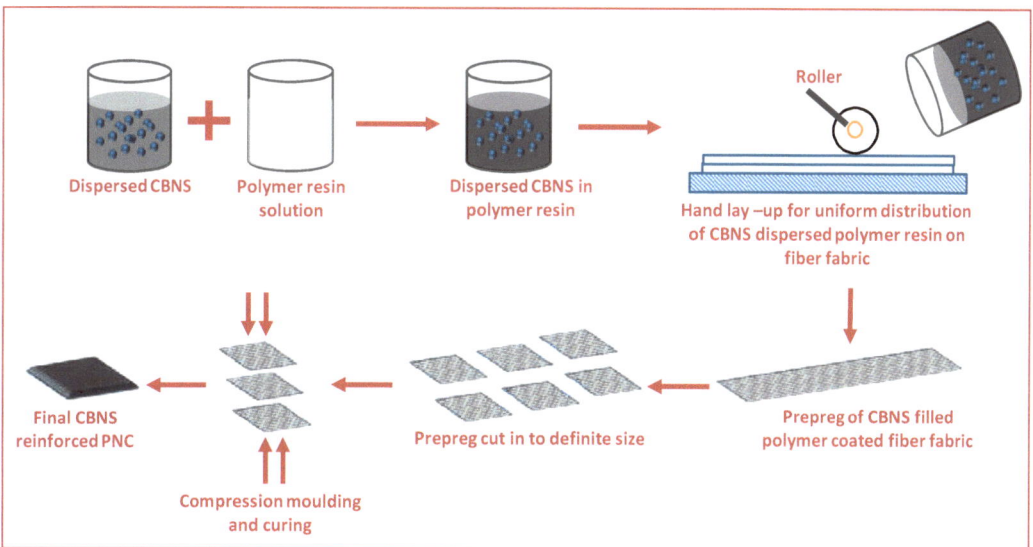

Fig. (13). Fabrication process of CBNS reinforced PNC laminates using hot press method.

3.7.4. Bucky Paper Route

Bucky paper is a porous and thin sheet made by filtration of dispersed CBNS. Polymer is generally infused into the Bucky paper to form the PNCs with high weight percent of CBNS. These CBNS reinforced porous sheets can also enhance the electrical/mechanical properties of PNCs. These Bucky paper sheets can also be sandwiched between the layers of fiber reinforced laminates to enhance desired properties. This process was reported by Teotia *et al.* [83], where they prepared CNT Bucky paper of phenolic resin PNCs, as shown in Fig. (**14**). In this process, dispersion of CNTs takes place in an appropriate solvent and then mixed with dissolved phenolic resin by means of high-speed homogenization.

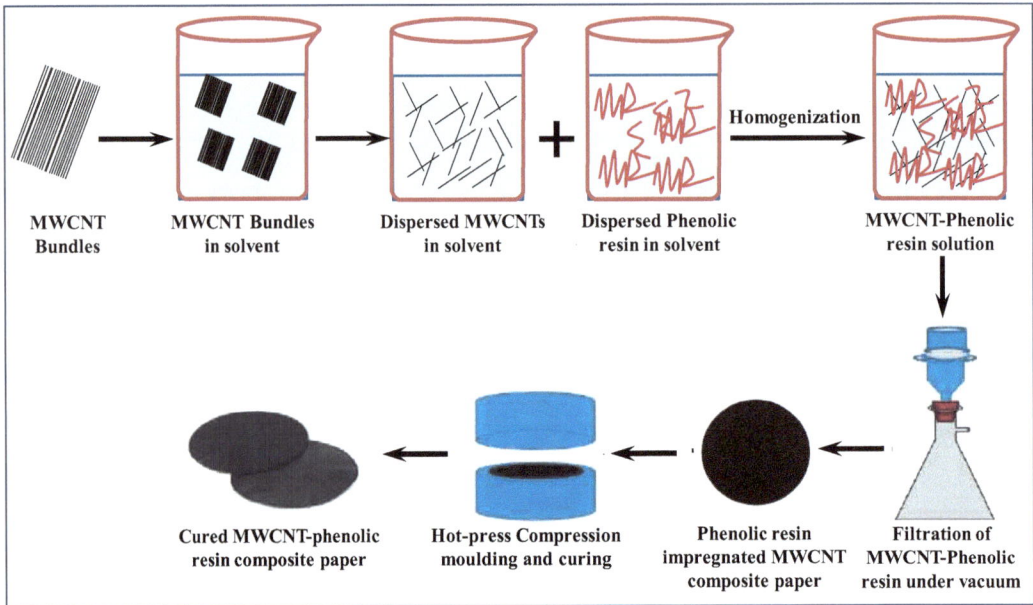

Fig. (14). Schematic diagram of Bucky paper route for the fabrication of CNT–phenolic resin composites [83].

A vacuum filtration system is designed specially to filter the mixture of dispersed CNTs and phenolic resin. A film of CNTs impregnated phenolic resin was obtained, and then dried into a vacuum oven to form the prepreg of CNT-phenolic resin. These prepregs were finally stacked on a metallic plate or mould and then hot pressed using a hydraulic press machine.

3.8. EMI SHIELDING PROPERTIES OF CBNS REINFORCED PNCS

Success in the field of PNCs based materials for EMI shielding has led to the replacement of metal-based EMI shielding materials due to light weight, flexibility, corrosion resistance, and durability. Many studies have described the use of CBNS reinforced thermoplastic and thermoset PNCs for EMI shielding applications. Carbon based materials are generally used as electrical fillers, but can also be used as dielectric materials. Therefore, according to EMI shielding theory, CBNS can enhance the total EMI shielding properties and can also reduce the reflection coefficient of the microwave radiation.

EMI shielding involves blocking of EM radiation using conducting, dielectric and magnetic materials. Depending upon the processing method and type of CBNS used, a comprehensive study of different polymer-based PNCs research on EMI shielding has been presented here. Here, the role of thermoplastic and thermoset PNCs filled with CBNS is discussed individually in reference to the EMI shielding properties.

3.8.1. Thermoplastic Polymer Nanocomposites-based EMI Shield

A variety of thermoplastic polymeric materials have been employed for EMI shielding. Thermoplastic polymeric materials have always been the candidates of choice as matrix for the fabrication of EMI shielding materials because of their properties such as light weight, easily processable, stretchability and can work in long strain range.

Carbon fibers have been used as a reinforcement for PNCs for a long time due to their excellent conductivity and exceptional mechanical properties. With the development of nanotechnology, carbon nanofibers (CNF) with diameter between 50-100 nm have been used to fabricate polymer composites for EMI shielding applications. Lee at al.[84] prepared CNF and carbon black (CB) reinforced poly Vinylidene Fluoride (PVDF) coating of 25-50 μm thickness *via* solution mixing method and they reported the total EMI shielding effectiveness of 14.5 dB at 50 μm thickness in 0-1.6 GHz range.

In a study by Mondal *et al.* [85], nanocomposites of chlorinated polyethylene (CPE) reinforced with CNF and CB were prepared *via* a solution mixing process and EMI shielding properties have been evaluated. They have correlated the surface morphology of conductive filler in the enhancement of the final properties of materials. The surface morphology of functionalized CNF is taken by field-emission scanning electron microscopy (FESEM), as shown in Fig. (**15a**). It is clearly seen from the FESEM micrographs that CNF are entangled with each other and Fig. (**15b**) shows that CPE powder is completely dissolved in tetrahydrofuran (THF) and addition of functionalized CNF in CPE suspension makes stable composite suspension.

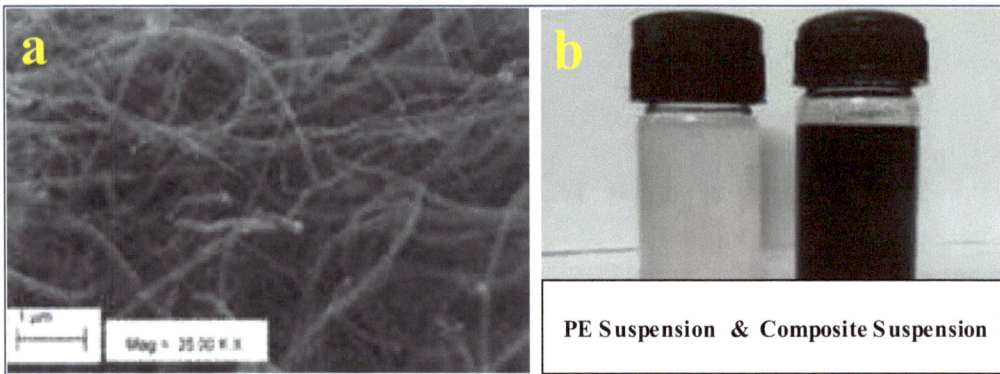

Fig. (15). (**a**) FE-SEM micrograph of functionalized CNF and (**b**) stable suspensions of PE and CNF-PE composite in THF. Reproduced with permission [15, 85].

In Fig. (**16a**), EMI shielding properties of CNF/CB/CPE nanocomposites in 8.2-12.4 GHz range are represented with the varying CNF/CB content. It is clearly found that the total EMI shielding (SE_T) value of pure CPE matrix was lower than -2dB because matrix is completely insulating and it is nearly transparent to the incident EM radiations, while the SE_T value of CNF/CB/CPE nanocomposites was found to be -20.2 dB (> 99.9% attenuation) on 7.0 wt. % of CNF/CB loading and -33 dB (attenuation > 99.99%) at 15 wt. % of CNF/CB loading at a thickness of 1 mm [85].

This improvement in the SET value is due to the increase in the conducting channels in the PNCs with the incorporation of CNF/CB content in the CPE matrix. As we know that the total shielding efficiency is the sum of SE_A and SE_R, and the schematic of the penetration of EM waves is shown in Fig. (**16b**). This schematic diagram shows that some EM waves are absorbed by the CNF/CB/CPE sheet, some are transmitted, and some are reflected from the surface of the PNC sheet.

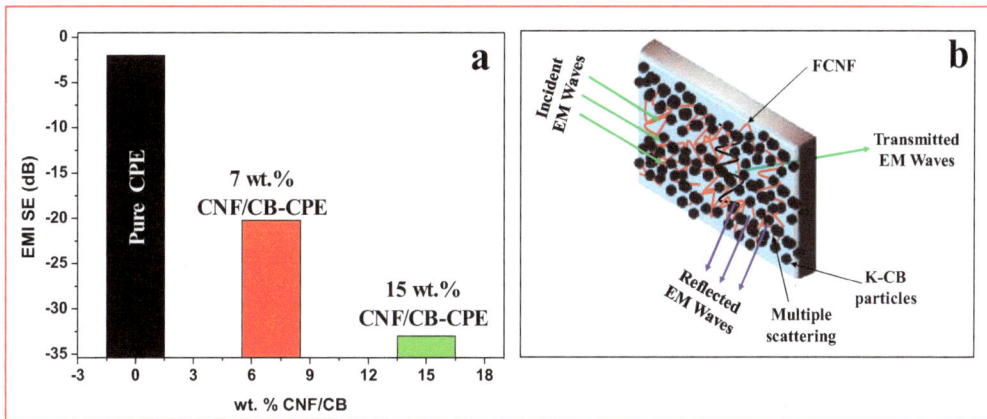

Fig. (16). (a) EMI shielding of the CNF/CB reinforced CPE nanocomposites in the frequency range of 8.2-12.4 GHz and **(b)** Schematic mechanism of penetration of EM waves [85].

Al-Saleh *et al.* [86] have prepared acrylonitrile-butadiene-styrene (ABS) nanocomposite filled with CNF using a solution mixing process. They have found that the maximum total EMI shielding value in 8.2-12.4 GHz range is - 35 dB at 15 wt. % loading of CNF.

It is known that the high aspect ratio and hollow core of CNF are produced when hydrocarbons are decomposed on a metal catalyst. The typical diameter and length of CNFs are 50–200 nm and up to 100 μm, respectively. Electrical resistivity of the order of 10^{-4} Ω.cm was found for the graphitized CNF. CNT is another type of CBNS, and the CNT are smaller in diameter and have high aspect ratio as compared to CNF [86]. Therefore, the electrical percolation threshold of CNT reinforced PNCs is lesser than that of CNF reinforced PNCs. In the same study of Al-Saleh *et al.* [86], they have prepared CNT/ABS nanocomposites vis solvent mixing method and they have measured the EMI shielding properties of these nanocomposites. They found the total EMI shielding values of 50 dB at 15 wt. % loading of MWCNTs, which was considerably better than that of CNF/ABS nanocomposites. In the end, we found that the amount of MWCNT required to formulate nanocomposites with EMI SE of 20 dB (which is suitable for commercial applications) is only about 2 wt.%. But, to attain the same level of EMI shielding, 10 wt. % CNF is required to formulate ABS nanocomposites.

Several studies have also been conducted on thermoplastic polyurethane (TPU) because TPU is generally used in household appliances, scratch-less surface coating on aerospace and automotive vehicles, and electronic panels. Zhang *et al.* [87] described the EMI shielding properties of MWCNT/TPU nanocomposites in different frequency bands such as K-band, V-band, and Q-band, respectively. MWCNT filled TPU nanocomposites were prepared *via* solution mixing and evaporation technique with varying wt. % of MWCNTs, and they found that at 6.7 wt. % of MWCNTs loading in TPU matrix, the electrical conductivity has reached a value of 0.1 Scm^{-1} [87].

At this higher value of electrical conductivity, we measure the total EMI shielding in different bands such as K-band (18-26.5 GHz), Q-band (33-50 GHz), and V-band (50-75 GHz). The results showed that the total EMI shielding effectiveness exceeded over -30 dB in the entire frequency band. The sample thickness of 3.0 mm and the maximum value reached up to -65 dB at the same loading in the V-band frequency range. Several types of research show that the reduction of signal strength by - 30 dB to - 40 dB fulfills 50 % to 95 % of the requirements in the automotive applications and computer industries [88].

Liu *et al.* [17] prepared SWCNTs reinforced PU nanocomposites *via* solution mixing method with varying wt. % of SWCNTs, and they found that at 20 wt. % loading of SWCNTs in PU nanocomposites, total EMI shielding effectiveness reached a value of -17 dB, and at this loading of SWCNTs, the mechanism shifts towards the absorption dominating mechanism. The lower value of total EMI SE is due to the lower value of electrical conductivity (2.2×10^{-4} Scm^{-1}). The main reasons for this low electrical conductivity may be the short length and purity of SWCNTs.

In another study, Liu *et al.* [89] measured the total EMI shielding of - 22 dB at 5.0 wt. % loading of SWCNTs in PU nanocomposites. The SWCNT-PU nanocomposites were fabricated by the *in-situ* polymerization method instead of solvent mixing. In this process, SWCNTs were uniformly coated on the polymer matrix and provided the conducting network. These conducting channels have participated in the improvement in the total EMI shielding values of the resulting nanocomposites. The electrical conductivity of the composites was at 5.0 wt. % SWCNT loading. These experimental results were matched perfectly with the theoretical values and showed their unlimited potential for radar wave absorption.

Gupta *et al.* [14] has also prepared TPU nanocomposites using long length MWCNTs (length ~ 200 μm) *via* solution mixing and evaporation technique and they have reported the high electrical conductivity of 7.9 Scm^{-1} at 10 wt. % loading of MWCNTs in TPU matrix (Fig. **17a**). This electrical conductivity increased the total EMI shielding of the resulting nanocomposites in the X-band frequency range. The total EMI shielding value of - 41.6 dB was achieved at 10wt. % loading of MWCNTs at 10 GHz frequency as shown in Fig. (**17b**). In this work, the length of MWCNTs has played an essential role in enhancing electrical conductivity because tube-to-tube connectivity was perfect and provided an excellent conductive path, and increased the number of charge transport in nanocomposites. This absorption dominated by the high value of EMI shielding, could be the potential materials for the futuristic EMI shielding.

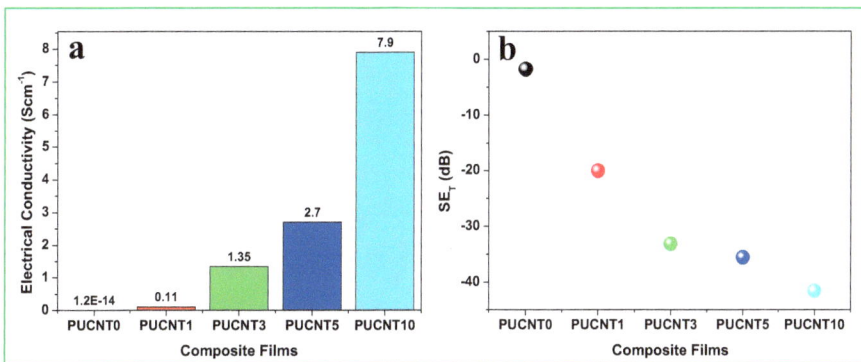

Fig. (17). Change in electrical conductivity with respect to MWCNTs loading in TPU nanocomposites **(a)** and total EMI shielding of MWCNT/TPU nanocomposites in X-band frequency range **(b)** [14].

In another study by Gupta *et al.* [13], they initially functionalize the MWCNTs *via* ultra-sonication and refluxing method using nitric acid followed by filtration and washing, as shown in Fig. (**18a-b**).

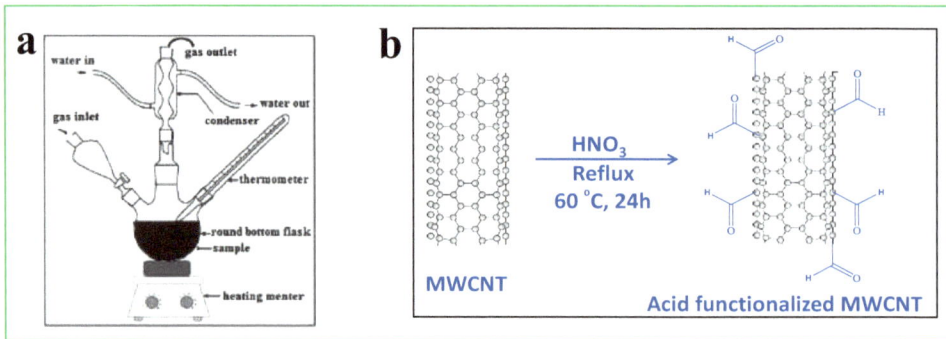

Fig. (18). (a) Functionalization set-up used for the oxidation of MWCNTs and **(b)** Schematic representation of acid functionalization of MWCNTs.

Acid functionalized MWCNTs filled TPU nanocomposites enhanced the EMI shielding properties along with the nanomechanical properties together due to strong interfacial bonding of functionalized MWCNTs with PU matrix as shown in transmission electron micrographs (TEM) in Fig. (**19a-b**). The electrical conductivity of acid functionalized MWCNTs filled TPU nanocomposites was found to be lower compared to that of pristine MWCNTs filled TPU nanocomposites, which is shown in Fig. (**20a**). This reduction in the electrical conductivity may be due to the defects generated on the surface of MWCNTs after functionalization. However, the functionalization of MWCNTs creates a strong interface between MWCNT and TPU assembly and therefore, it enhances the nanomechanical performance of TPU nanocomposites.

Fig. (**20b**) shows that the total EMI shielding value of - 29 dB was achieved at 10 wt. % loading of surface-modified MWCNTs in TPU matrix. According to the EMI theory, - 20 dB is sufficient for commercial applications. Hence, the value achieved in this study was enough to block 99.9 % of EM radiation. These TPU nanocomposites can be used as a good EMI shield along with the hard and scratch proof coatings on automotive vehicles and aerospace industries.

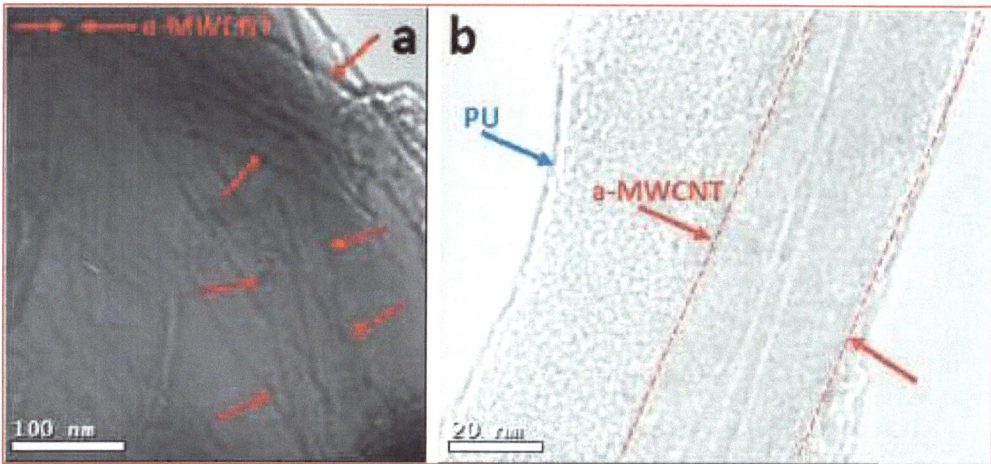

Fig. (19). TEM micrograph of **(a)** acid functionalized MWCNT embedded in the PU matrix for 10 wt. % loading of MWCNTs and **(b)** a single functionalized MWCNT coated with TPU matrix. Reproduced with permission [13].

In another research article, Gupta *et al.* [56] have reported the MWCNT/graphene reinforced hybrid polyaniline (PANI) nanocomposites. Initially, graphite was ball milled with PANI to convert graphite into multilayer graphene, and then nanocomposites were fabricated using the solution mixing method. These multilayer graphene act as a bridge between MWCNTs to protect them from aggregation because graphene between CNTs reduces the Van der Waal force of attraction and increases the dispersion into the polymer matrix. The maximum value of total EMI SE of - 98 dB was achieved for 10 wt. % loaded MWCNTs/PANI nanocomposites. The extremely high and absorption dominated value of EMI shielding could be due to the combined effect of the increase in space charge polarization and decrease in the carrier mobility.

Transmission electron microscopic study (TEM) in Fig. **(21a)** shows the presence of in-situ generated multilayer graphene in PANI matrix and Fig. **(21b)** shows that graphene-PANI is coated on MWCNTs in the nanocomposites [56].

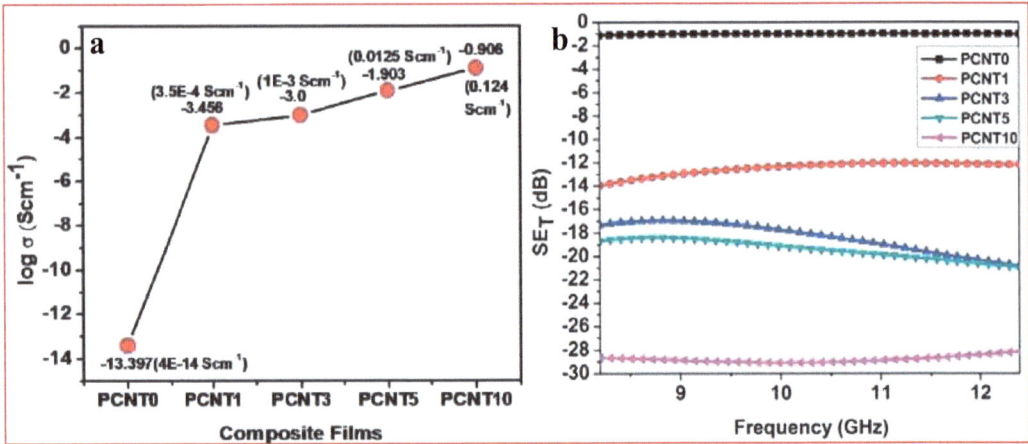

Fig. (20). Change in electrical conductivity with respect to functionalized MWCNTs loading in TPU nanocomposites **(a)** and total EMI shielding of functionalized MWCNT/TPU nanocomposites in X-band frequency range **(b)**. Reproduced with permission [13].

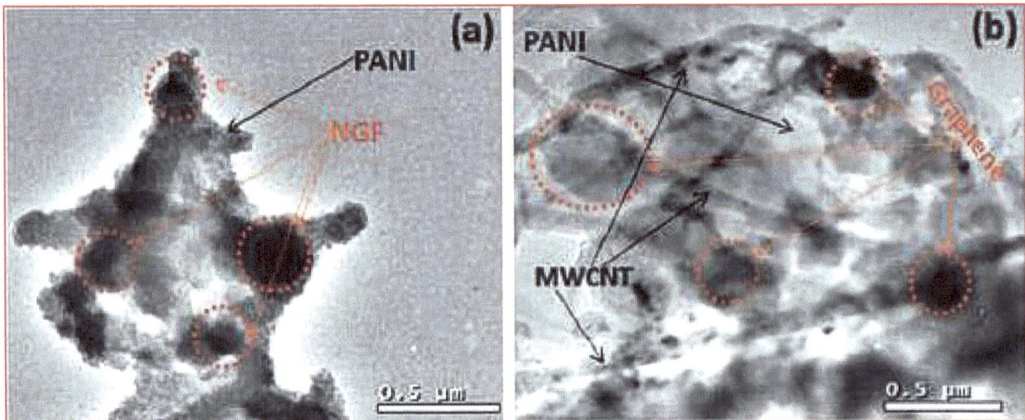

Fig. (21). (a) PANI coated graphene nanocomposite and (b) MWCNT/PANI-graphene hybrid nanocomposite. Reproduced with permission [56].

In other studies on TPU nanocomposites, Hoang *et al.* [90] used planetary ball milling process to disperse MWCNTs homogeneously in TPU matrix and achieved - 25 dB of total EMI shielding at 25 wt. % of MWCNTs loading when the thickness was only 100 μm and the electrical conductivity at this loading was 0.5 Scm⁻¹. This high electrical conductivity may be due to the uniform dispersion of MWCNTs in the TPU matrix.

Bhattacharya *et al.* [91] have reported that the nanocomposites of 15:15 wt. % ratio of TiO_2 coated MWCNT/Fe_3O_4 incorporated TPU were prepared *via* solution blending technique and investigated the microwave absorption properties. They showed that the incorporation of TiO_2 coated MWCNTs and Fe_3O_4 enhances the real part of permittivity and permeability of the composites. These dielectric TiO_2 and magnetic Fe_2O_3 participate in the enhancement of the absorption part of the EMI shielding. The maximum reflection loss of this hybrid nanocomposite at 10.98 GHz was found to be -42.53 dB and this superior improvement in the absorption was due to presence of both dielectric and magnetic filler. These materials could also be useful as Radar Absorbing Materials (RAM). Bhattacharya *et al.* [92], in another report, have shown the role of both graphene and MWCNT in RADAR absorption material, but graphene is found to be more superior to MWCNT. They have done a detailed study on microwave absorption and explained the permittivity and permeability nature of the material. From these above studies, it was found that the dielectric loss is an essential part to contribute to EMI shielding as compared to the magnetic part.

Chauhan *et al.* [93] prepared poly (ether-ketone) (PEK) composites with varying wt. % of MWCNTs *via* melt compounding using a twin-screw extruder. The EMI SE of these nanocomposites was measured in the 26.5-40 GHz range and found that the total EMI SE was -38 dB at 5.0 wt. % of MWCNTs loading in which -34 dB was the contribution due to absorption. The solution blending approach was used to fabricate graphene/TPU nanocomposites, and total EMI SE was found to be -21 dB in the X-band frequency range at 5.5 vol. % loading of graphene and thickness of the sample was 3.0 mm [94].

Yadav *et al.* [95] presented the EMI shielding properties of $NiFe_2O_4$ nanoparticles/reduced graphene oxide (RGO)/polypropylene (PP) nanocomposites. They found the EMI SE value of 29.4 dB at a thickness of 2.0 mm. This enhancement in EMI shielding properties of nanocomposites may be due to increased RGO content and high conductivity, eddy current loss, dipole and interfacial polarization, and natural resonance. Mechanical mixing and melt treatment were done for the fabrication of these nanocomposites.

Mishra *et al.* [96] prepared a conducting composite of ferro fluid and reduced graphene *via* co-precipitation technique and found a high microwave shielding value of 41 dB in X-band frequency range. These prepared conducting composites of ferro fluid/reduced graphene could be a promising candidate for the next generation microwave absorbing materials for fabrication of EM radiation protected polymer nanocomposites. This enhancement in EMI SE was due to the combined effect of magnetic losses such as natural resonance and eddy currents which may be due to ferro fluid and dielectric losses such as natural resonance, electron polarization related relaxation, dipole relaxation, interfacial polarization, residual defects in reduced graphene and high electrical conductivity which may be due to reduced graphene oxide [96].

In a study by Singh *et al.* [97], polyaniline/fly ash composite were prepared *via* chemical oxidative polymerization route and the microwave absorption properties of these nanocomposites in X-band and Ku-band frequency range were measured. Fly ash acts as dielectric material, therefore, the maximum EMI SE of 32 dB was observed, which clearly depends upon the dielectric nature of fly ash.

3.8.2. Thermoset Polymer Nanocomposites-Based EMI Shield

The EMI shielding properties of thermoset polymers are also explored because these polymers are also used for defense, automobile and communication industries. Therefore, thermoset polymers can also be used as EM radiation shielding materials if these polymers are reinforced with some conductive fillers. Several studies have been done on thermoset PNCs reinforced with CBNS for EMI shielding applications and a few of them are discussed here.

Nanni *et al.* [98] have prepared CNF reinforced epoxy nanocomposites *via* solution mixing and vacuum filtration technique with different wt. % of CNF loading. The EMI shielding measurements were done in X-band frequency range and they found that the EMI SE at 4 wt. % loading of CNF was found to be -18.3 dB at a thickness of 1mm.

In a study of Xialo *et al.* [99], CNF of diameter 50-150 nm were prepared *via* wet spinning of PAN/PMMA polymer blends and nanocomposites with epoxy were fabricated using solution mixing and casting method and EMI shielding performance in X-band frequency range were evaluated. They found that the reflection loss increased with an increase in the content of CNF and finally the maximum reflection loss of -34 dB at 8.0 wt. % of CNF loading with a thickness at 2.1 mm as shown in Fig. (**22a**), as shown in Fig. (**22b**) Reflection loss with variation in thickness of 8.0 wt. % CNF loaded epoxy samples were also measured and it was found that the reflection loss was maximum (-34 dB) at a thickness of 2.1 mm between the thickness of 1.9 mm and 2.5 mm where reflection loss was -27.5 dB and -15.8 dB, respectively.

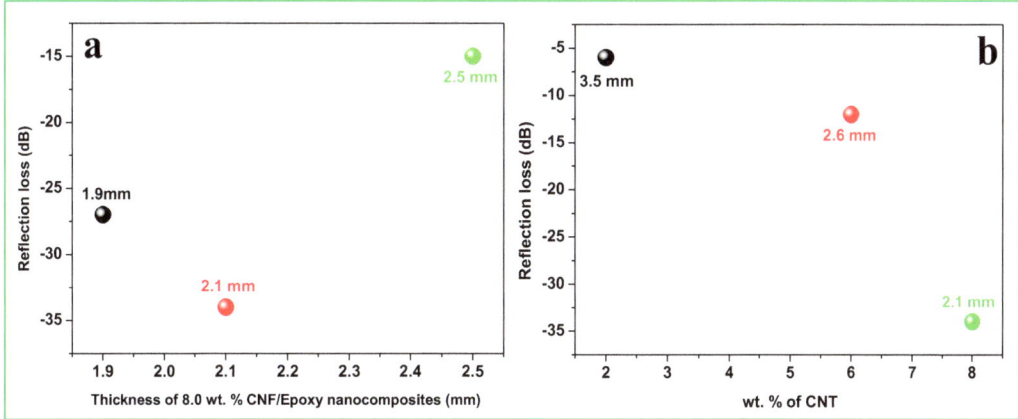

Fig. (22). Reflection loss for CNF/Epoxy nanocomposites with variation in wt. % of CNF **(a)** and with variation of thickness **(b)** in X-band. Reproduced with permission [99].

Teotia *et al.* [83], prepared MWCNT/phenolic composite paper *via* novel bucky paper approach and obtained a very high electrical conductivity of 78 Scm^{-1} at a thickness of 140 µm. At this high electrical conductivity value, the total EMI SE value of 32.4 dB in the frequency range of 12.4-18 GHz (Ku-band) was found. Li *et al.* worked on thermoset polymer nanocomposites and they prepared SWCNT/epoxy nanocomposites that measured the EMI shielding and electrical conductivity by incorporating the different aspect ratios and wall defects SWNTs [100]. The maximum value of EMI SE for SWNT-epoxy composites was found at 15 wt. % loading of SWNTs-long in epoxy matrix. The EMI SE was found to be -49 dB at 10 MHz and -15 to -20 dB in 500 MHz to 1.5 GHz range.

Liang *et al.* [101] prepared graphene reinforced epoxy nanocomposites *via* solution mixing method and X-band EMI SE of 21 dB was found at 15 wt. % loading of graphene.

Singh *et al.* [102] prepared high MWCNT loaded epoxy nanocomposites *via* prepreg and compression moulded nanocomposite laminates. The total EMI SE measured in X-band frequency range was -19 dB for a thickness of 0.35 mm and - 60 dB for a thickness of 1.75 mm.

Singh *et al.* [103] in their other study have prepared MWCNT/epoxy nanocomposites and EMI SE were measured in Ku-band frequency range. The effect of length of MWCNTs on EMI shielding properties of epoxy nanocomposites were discussed in this study. They have prepared these nanocomposites *via* industrial viable fast dispersion of MWCNTs in epoxy matrix *via* high-speed homogenization which is a modified approach of solution mixing technique. Two different lengths of MWCNTs such as 350 μm for long length and 1.5 μm for short length were taken in this study.

The percolation threshold for long MWCNTs was 0.02 wt. % and for short MWCNTs, it was 0.11. The absorption dominated EMI SE of -16 dB for long MWCNTs and -11.5 dB for short MWCNTs was observed with 0.5 wt. % of MWCNT loading in Ku-band frequency range [103].

In this article of Seng *et al.* [104], the MWCNT/polyester nanocomposites were prepared *via* solution mixing method and dielectric and EMI shielding properties of these nanocomposites were presented and compared with the simulated data. The electrical conductivity, dielectric properties and tangent loss were increased with increase in MWCNT content. The results show that the increase in electrical conductivity of nanocomposites is responsible for enhancement of EMI SE. The average values of EMI SE of these MWCNTs/polyester nanocomposites at 1 wt. % of MWCNTs addition was 3 dB and at 20 wt. % MWCNTs addition was 35.2 dB. The experimental and simulated results of the EMI SE in the entire frequency range were compared with variation of MWCNTs content in polyester matrix at 3 mm thickness.

CONCLUSIONS AND FUTURE OUTLOOK

The extraordinary properties of carbon-based nanostructures have opened up new ways for developing multifunctional high-performance nanocomposites. EMI shielding is the process of obstructing EM waves into a selected area *via* using EMI shields. Polymer shields are one such material that can shield EM waves and can be used as hybrid materials for advanced applications. Researches on CBNS reinforced PNCs propose that CBNS have great value in modifying the properties of PNCs. Several factors such as types and purity of filler, dispersion in polymer matrices, aspect ratio, surface modification, and interfacial adhesion between filler and polymer matrix could make the CBNS reinforced PNCs superior as compared to other materials. The frequency-dependent EMI shielding properties of PNCs and adjustment in the morphological structure and processing parameters are discussed, along with the mechanism of EMI shielding. Authors have reported good EMI shielding properties of thermoplastic and thermoset PNCs incorporated with CBNS and hybrid fillers. The effect of dielectric and magnetic fillers on the absorption part of EMI SE has also been explored and discussed in this chapter. To make efficient EMI shielding PNCs, the materials must exhibit good electrical conductivity and dielectric constant. These requirements can only be reached when CBNS are dispersed uniformly into the polymer matrix. The solution mixing method is the more appropriate method to disperse CBNS and other fillers in the polymer matrix than melt mixing.

Though considerable research with diverse approaches has done in the area of CBNS reinforced PNCs for EMI shielding applications, there is still a significant demand and a lot of work is required in this field to achieve excellent EMI shielding with desirable properties such as ultrathin and lightweight structure, environmental stability, and wide absorption bandwidth coverage. Hence, in the present scenario of ever-increasing demand for EMI shielding materials, it is highly praiseworthy if thermoplastic and thermoset polymers-based nanocomposites are fabricated *via* a modified approach. They satisfy the entire spectrum of technological as well as commercial specifications. Other limitations are difficulty in processing and generation of defects during synthesis, and surface functionalization of fillers may affect the physical properties of the nanocomposites. Poor thermal stability of PNCs hinders their applications in high-temperature processes. It was also not clearly explained how temperature affects the EMI shielding properties of materials PNCs. Another future challenge is that the researcher's EMI shielding values are good enough for commercial purposes. Still, with stealth technology and RADAR absorbing materials, there is a need for new advanced materials having excellent microwave absorbing capabilities.

CONSENT FOR PUBLICATION

Not applicable.

CONFLICT OF INTEREST

The authors declare no conflict of interest, financial or otherwise.

ACKNOWLEDGEMENTS

T. K. Gupta is grateful to the Amity Institute of Applied Sciences, Amity University, Noida, India, for encouraging during this COVID-19 pandemic. This chapter was written during COVID-19 pandemic; therefore, the support of the family members and kids is highly appreciable.

REFERENCES

[1]   Al-Saleh, M.H.; Sundararaj, U. Electromagnetic interference shielding mechanisms of CNT/polymer composites. *Carbon,* **2009**, *47*(7), 1738-1746.
        http://dx.doi.org/10.1016/j.carbon.2009.02.030

[2]   Li, Y.; Chen, C.; Zhang, S.; Ni, Y.; Huang, J. Electrical conductivity and electromagnetic interference shielding characteristics of multiwalled carbon nanotube filled polyacrylate composite films. *Appl. Surf. Sci.,* **2008**, *254*(18), 5766-5771.
        http://dx.doi.org/10.1016/j.apsusc.2008.03.077

[3]   Lin, C-T.; Swanson, B.; Kolody, M.; Sizemore, C.; Bahns, J. Nanograin magnetoresistive manganite coatings for EMI shielding against directed energy pulses. *Prog. Org. Coat.,* **2003**, *47*(3-4), 190-197.
        http://dx.doi.org/10.1016/S0300-9440(03)00138-3

[4]   Gupta, T.K.; Singh, B.P.; Singh, V.N.; Teotia, S.; Singh, A.P.; Elizabeth, I.; Dhakate, S.R.; Dhawan, S.K.; Mathur, R.B. MnO2 decorated graphene nanoribbons with superior permittivity and excellent microwave shielding properties. *J. Mater. Chem. A Mater. Energy Sustain.,* **2014**, *2*(12), 4256-4263.
        http://dx.doi.org/10.1039/c3ta14854h

[5]   Wu, X.; Han, Y.; Zhang, X.; Lu, C. Highly Sensitive, Stretchable, and Wash-Durable Strain Sensor Based on Ultrathin Conductive Layer@Polyurethane Yarn for Tiny Motion Monitoring. *ACS Appl. Mater. Interfaces,* **2016**, *8*(15), 9936-9945.
        http://dx.doi.org/10.1021/acsami.6b01174 PMID: 27029616

[6]   Zhao, J.; Dai, K.; Liu, C.; Zheng, G.; Wang, B.; Liu, C.; Chen, J.; Shen, C. A comparison between strain sensing behaviors of carbon black/polypropylene and carbon nanotubes/polypropylene electrically conductive composites. *Compos., Part A Appl. Sci. Manuf.,* **2013**, *48*, 129-136.
        http://dx.doi.org/10.1016/j.compositesa.2013.01.004

[7]   Liu, H.; Huang, W.; Gao, J.; Dai, K.; Zheng, G.; Liu, C.; Shen, C.; Yan, X.; Guo, J.; Guo, Z. Piezoresistive behavior of porous carbon nanotube-thermoplastic polyurethane conductive nanocomposites with ultrahigh compressibility. *Appl. Phys. Lett.,* **2016**, *108*(1), 011904.
        http://dx.doi.org/10.1063/1.4939265

[8]   Lin, L.; Liu, S.; Fu, S.; Zhang, S.; Deng, H.; Fu, Q. Fabrication of highly stretchable conductors *via* morphological control of carbon nanotube network. *Small,* **2013**, *9*(21), 3620-3629.
        http://dx.doi.org/10.1002/smll.201202306 PMID: 23630114

[9] Eswaraiah, V.; Balasubramaniam, K.; Ramaprabhu, S. Functionalized graphene reinforced thermoplastic nanocomposites as strain sensors in structural health monitoring. *J. Mater. Chem.,* **2011**, *21*(34), 12626-12628.
http://dx.doi.org/10.1039/c1jm12302e

[10] Ding, D.; Wei, H.; Zhu, J.; He, Q.; Yan, X.; Wei, S.; Guo, Z. Strain sensitive polyurethane nanocomposites reinforced with multiwalled carbon nanotubes. *Energy and Environment Focus,* **2014**, *3*(1), 85-93.
http://dx.doi.org/10.1166/eef.2014.1093

[11] Liu, H.; Li, Y.; Dai, K.; Zheng, G.; Liu, C.; Shen, C.; Yan, X.; Guo, J.; Guo, Z. Electrically conductive thermoplastic elastomer nanocomposites at ultralow graphene loading levels for strain sensor applications. *J. Mater. Chem. C Mater. Opt. Electron. Devices,* **2016**, *4*(1), 157-166.
http://dx.doi.org/10.1039/C5TC02751A

[12] Li, D.; Liu, Y.; Lin, B.; Lai, C.; Sun, Y.; Yang, H.; Zhang, X. Synthesis of ternary graphene/molybdenum oxide/poly (p-phenylenediamine) nanocomposites for symmetric supercapacitors. *RSC Advances,* **2015**, *5*(119), 98278-98287.
http://dx.doi.org/10.1039/C5RA18979A

[13] Gupta, T.K.; Singh, B.P.; Dhakate, S.R.; Singh, V.N.; Mathur, R.B. Improved nanoindentation and microwave shielding properties of modified MWCNT reinforced polyurethane composites. *J. Mater. Chem. A Mater. Energy Sustain.,* **2013**, *1*(32), 9138-9149.
http://dx.doi.org/10.1039/c3ta11611e

[14] Gupta, T.K.; Singh, B.P.; Teotia, S.; Katyal, V.; Dhakate, S.R.; Mathur, R.B. Designing of multiwalled carbon nanotubes reinforced polyurethane composites as electromagnetic interference shielding materials. *J. Polym. Res.,* **2013**, *20*(6), 1-7.
http://dx.doi.org/10.1007/s10965-013-0169-6

[15] Gupta, T.K.; Singh, B.P.; Mathur, R.B.; Dhakate, S.R. Multi-walled carbon nanotube-graphene-polyaniline multiphase nanocomposite with superior electromagnetic shielding effectiveness. *Nanoscale,* **2014**, *6*(2), 842-851.
http://dx.doi.org/10.1039/C3NR04565J PMID: 24264356

[16] Singh, B.P.; Prabha, P.; Saini, P.; Gupta, T.; Garg, P.; Kumar, G.; Pande, I.; Pande, S.; Seth, R.K.; Dhawan, S.K.; Mathur, R.B. Saini, T. Gupta, P. Garg, G. Kumar, I. Pande, S. Pande, R. K. Seth, S. K. Dhawan, and R. B. Mathur, "Designing of multiwalled carbon nanotubes reinforced low density polyethylene nanocomposites for suppression of electromagnetic radiation. *J. Nanopart. Res.,* **2011**, *13*(12), 7065-7074.
http://dx.doi.org/10.1007/s11051-011-0619-1

[17] Liu, Z.; Bai, G.; Huang, Y.; Ma, Y.; Du, F.; Li, F.; Guo, T.; Chen, Y. Reflection and absorption contributions to the electromagnetic interference shielding of single-walled carbon nanotube/polyurethane composites. *Carbon,* **2007**, *45*(4), 821-827.
http://dx.doi.org/10.1016/j.carbon.2006.11.020

[18] Gupta, T.K.; Singh, B.P.; Tripathi, R.K.; Dhakate, S.R.; Singh, V.N.; Panwar, O.S.; Mathur, R.B. Superior nano-mechanical properties of reduced graphene oxide reinforced polyurethane composites. *RSC Advances,* **2015**, *5*(22), 16921-16930.
http://dx.doi.org/10.1039/C4RA14223C

[19] Sharma, S.; Rawal, J.; Dhakate, S.R.; Singh, B.P. Synergistic bridging effects of graphene oxide and carbon nanotube on mechanical properties of aramid fiber reinforced polycarbonate composite tape. *Compos. Sci. Technol.,* **2020**, *199*, 108370.
http://dx.doi.org/10.1016/j.compscitech.2020.108370

[20] Kumar, S.; Gupta, T.K.; Varadarajan, K.M. Strong, stretchable and ultrasensitive MWCNT/TPU nanocomposites for piezoresistive strain sensing. *Compos., Part B Eng.,* **2019**, *177*, 107285.
http://dx.doi.org/10.1016/j.compositesb.2019.107285

[21]   Gupta, T.K.; Choosri, M.; Varadarajan, K.M.; Kumar, S. Self-sensing and mechanical performance of CNT/GNP/UHMWPE biocompatible nanocomposites. *J. Mater. Sci.,* **2018**, *53*(11), 7939-7952.
http://dx.doi.org/10.1007/s10853-018-2072-3

[22]   Arif, M.F.; Kumar, S.; Gupta, T.K.; Varadarajan, K.M. Strong linear-piezoresistive-response of carbon nanostructures reinforced hyperelastic polymer nanocomposites. *Compos., Part A Appl. Sci. Manuf.,* **2018**, *113*, 141-149.
http://dx.doi.org/10.1016/j.compositesa.2018.07.021

[23]   Bhattacharjee, Y.; Biswas, S.; Bose, S. Chapter 5 - Thermoplastic polymer composites for EMI shielding applications, **2020**.

[24]   Schwarz, S. *Electromagnetics for engineers*; Oxford University Press, USA: USA, **1990**.

[25]   V. P., Kodali *Engineering electromagnetic compatibility: principles measurements technologies and computer models; 2nd ed.",,* **2001**,

[26]   Chung, D.D.L. Materials for electromagnetic interference shielding. *J. Mater. Eng. Perform.,* **2000**, *9*(3), 350-354.
http://dx.doi.org/10.1361/105994900770346042

[27]   Hochi, K.; Uesaka, K. *Rubber composition and tire having tread comprising thereof*; Patents, G., Ed, **2006**.

[28]   Das, T.K.; Ghosh, P.; Das, N.C. Preparation, development, outcomes, and application versatility of carbon fiber-based polymer composites: a review. *Adv. Compos. Hybrid Mater.,* **2019**, *2*(2), 1-20.
http://dx.doi.org/10.1007/s42114-018-0072-z

[29]   Rajak, D.K.; Pagar, D.D.; Menezes, P.L.; Linul, E. Fiber-reinforced polymer composites: Manufacturing, properties, and applications. *Polymers (Basel),* **2019**, *11*(10), 1667.
http://dx.doi.org/10.3390/polym11101667 PMID: 31614875

[30]   Peng, L-M.; Zhang, Z.; Wang, S. Carbon nanotube electronics: recent advances. *Mater. Today,* **2014**, *17*(9), 433-442.
http://dx.doi.org/10.1016/j.mattod.2014.07.008

[31]   Nilsson, J.; Neto, A.H.; Guinea, F.; Peres, N.M. Electronic properties of graphene multilayers. *Phys. Rev. Lett.,* **2006**, *97*(26), 266801.
http://dx.doi.org/10.1103/PhysRevLett.97.266801 PMID: 17280447

[32]   Li, X.; Cheng, Y.; Zhao, L.; Zhang, Q.; Wang, M-S. Structural and electrical properties tailoring of carbon nanotubes *via* a reversible defect handling technique. *Carbon,* **2018**, *133*, 186-192.
http://dx.doi.org/10.1016/j.carbon.2018.03.029

[33]   Pokharel, P.; Xiao, D.; Erogbogbo, F.; Keles, O.; Lee, D.S. A hierarchical approach for creating electrically conductive network structure in polyurethane nanocomposites using a hybrid of graphene nanoplatelets, carbon black and multi-walled carbon nanotubes. *Compos., Part B Eng.,* **2019**, *161*, 169-182.
http://dx.doi.org/10.1016/j.compositesb.2018.10.057

[34]   Han, S.; Meng, Q.; Araby, S.; Liu, T.; Demiral, M. Mechanical and electrical properties of graphene and carbon nanotube reinforced epoxy adhesives: experimental and numerical analysis. *Compos., Part A Appl. Sci. Manuf.,* **2019**, *120*, 116-126.
http://dx.doi.org/10.1016/j.compositesa.2019.02.027

[35]   Jyoti, J.; Kumar, A.; Dhakate, S.; Singh, B.P. Dielectric and impedance properties of three dimension graphene oxide-carbon nanotube acrylonitrile butadiene styrene hybrid composites. *Polym. Test.,* **2018**, *68*, 456-466.
http://dx.doi.org/10.1016/j.polymertesting.2018.04.003

[36]   Ruoff, R.S.; Lorents, D.C. *Mechanical and thermal properties of carbon nanotubes*; , **1995**.
http://dx.doi.org/10.1016/0008-6223(95)00021-5

[37] Han, Z.; Fina, A. Thermal conductivity of carbon nanotubes and their polymer nanocomposites: A review. *Prog. Polym. Sci.,* **2011**, *36*(7), 914-944.
http://dx.doi.org/10.1016/j.progpolymsci.2010.11.004

[38] Cha, J.; Kim, J.; Ryu, S.; Hong, S.H. Comparison to mechanical properties of epoxy nanocomposites reinforced by functionalized carbon nanotubes and graphene nanoplatelets. *Compos., Part B Eng.,* **2019**, *162*, 283-288.
http://dx.doi.org/10.1016/j.compositesb.2018.11.011

[39] Milowska, K.Z.; Burda, M.; Wolanicka, L.; Bristowe, P.D.; Koziol, K.K.K. Carbon nanotube functionalization as a route to enhancing the electrical and mechanical properties of Cu-CNT composites. *Nanoscale,* **2018**, *11*(1), 145-157.
http://dx.doi.org/10.1039/C8NR07521B PMID: 30525144

[40] Zare, Y.; Rhee, K.Y. Following the morphological and thermal properties of PLA/PEO blends containing carbon nanotubes (CNTs) during hydrolytic degradation. *Compos., Part B Eng.,* **2019**, *175*, 107132.
http://dx.doi.org/10.1016/j.compositesb.2019.107132

[41] Dai, H. Carbon nanotubes: synthesis, integration, and properties. *Acc. Chem. Res.,* **2002**, *35*(12), 1035-1044.
http://dx.doi.org/10.1021/ar0101640 PMID: 12484791

[42] Rafiee, R.; Pourazizi, R. Evaluating the influence of defects on the young's modulus of carbon nanotubes using stochastic modeling. *Mater. Res.,* **2014**, *17*(3), 758-766.
http://dx.doi.org/10.1590/S1516-14392014005000071

[43] Kanoun, O.; Müller, C.; Benchirouf, A.; Sanli, A.; Dinh, T.N.; Al-Hamry, A.; Bu, L.; Gerlach, C.; Bouhamed, A. Flexible carbon nanotube films for high performance strain sensors. *Sensors (Basel),* **2014**, *14*(6), 10042-10071.
http://dx.doi.org/10.3390/s140610042 PMID: 24915183

[44] Mittal, V. *Polymer Nanotube Nanocomposites Synthesis, Properties, and Applications,* 1st ed; Scrivener Publishing LLC, **2010**.
http://dx.doi.org/10.1002/9780470905647

[45] Rao, C.N.R.; Voggu, R.; Govindaraj, A. Selective generation of single-walled carbon nanotubes with metallic, semiconducting and other unique electronic properties. *Nanoscale,* **2009**, *1*(1), 96-105.
http://dx.doi.org/10.1039/b9nr00104b PMID: 20644865

[46] Yu, M-F.; Files, B.S.; Arepalli, S.; Ruoff, R.S. Tensile loading of ropes of single wall carbon nanotubes and their mechanical properties. *Phys. Rev. Lett.,* **2000**, *84*(24), 5552-5555.
http://dx.doi.org/10.1103/PhysRevLett.84.5552 PMID: 10990992

[47] Gupta, T.K.; Kumar, S. Fabrication of carbon nanotube/polymer nanocomposites, **2018**.
http://dx.doi.org/10.1016/B978-0-323-48221-9.00004-2

[48] Yakobson, B.I.; Brabec, C.J.; Berhnolc, J. Nanomechanics of carbon tubes: Instabilities beyond linear response. *Phys. Rev. Lett.,* **1996**, *76*(14), 2511-2514.
http://dx.doi.org/10.1103/PhysRevLett.76.2511 PMID: 10060718

[49] Journet, C.; Maser, W.; Bernier, P.; Loiseau, A.; de La Chapelle, M.L.; Lefrant, S.; Deniard, P.; Lee, R.; Fischer, J. *Large-scale production of single-walled carbon nanotubes by the electric-arc technique,* **1997**.
http://dx.doi.org/10.1038/41972

[50] Rinzler, A.; Liu, J.; Dai, H.; Nikolaev, P.; Huffman, C.; Rodriguez-Macias, F.; Boul, P.; Lu, A.H.; Heymann, D.; Colbert, D. *Large-scale purification of single-wall carbon nanotubes: process, product, and characterization*; , **1998**.

[51] Ren, Z.F.; Huang, Z.P.; Xu, J.W.; Wang, J.H.; Bush, P.; Siegal, M.P.; Provencio, P.N. Synthesis of large arrays of well-aligned carbon nanotubes on glass. *Science,* **1998**, *282*(5391), 1105-1107.

http://dx.doi.org/10.1126/science.282.5391.1105 PMID: 9804545

[52]   Garg, P.; Singh, B.P.; Kumar, G.; Gupta, T.; Pandey, I.; Seth, R.K.; Tandon, R.P.; Mathur, R.B. Effect of dispersion conditions on the mechanical properties of multi-walled carbon nanotubes based epoxy resin composites. *J. Polym. Res.,* **2011,** *18*(6), 1397-1407.
http://dx.doi.org/10.1007/s10965-010-9544-8

[53]   Muschi, M.; Serre, C. Progress and challenges of graphene oxide/metal-organic composites. *Coord. Chem. Rev.,* **2019,** *387,* 262-272.
http://dx.doi.org/10.1016/j.ccr.2019.02.017

[54]   Balandin, A.A.; Ghosh, S.; Bao, W.; Calizo, I.; Teweldebrhan, D.; Miao, F.; Lau, C.N. Superior thermal conductivity of single-layer graphene. *Nano Lett.,* **2008,** *8*(3), 902-907.
http://dx.doi.org/10.1021/nl0731872 PMID: 18284217

[55]   Gupta, T.K.; Singh, B.P.; Mathur, R.B.; Dhakate, S.R. Multi-walled carbon nanotube-graphene-polyaniline multiphase nanocomposite with superior electromagnetic shielding effectiveness. *Nanoscale,* **2014,** *6*(2), 842-851.
http://dx.doi.org/10.1039/C3NR04565J PMID: 24264356

[56]   Jyoti, J.; Singh, B.P.; Chockalingam, S.; Joshi, A.G.; Gupta, T.K.; Dhakate, S. Synergetic effect of graphene oxide-carbon nanotube on nanomechanical properties of acrylonitrile butadiene styrene nanocomposites. *Mater. Res. Express,* **2018,** *5*(4), 045608.
http://dx.doi.org/10.1088/2053-1591/aabd2e

[57]   Umrao, S.; Gupta, T.K.; Kumar, S.; Singh, V.K.; Sultania, M.K.; Jung, J.H.; Oh, I-K.; Srivastava, A. Microwave-Assisted Synthesis of Boron and Nitrogen co-doped Reduced Graphene Oxide for the Protection of Electromagnetic Radiation in Ku-Band. *ACS Appl. Mater. Interfaces,* **2015,** *7*(35), 19831-19842.
http://dx.doi.org/10.1021/acsami.5b05890 PMID: 26287816

[58]   Novoselov, K.S.; Geim, A.K.; Morozov, S.V.; Jiang, D.; Zhang, Y.; Dubonos, S.V.; Grigorieva, I.V.; Firsov, A.A. Electric field effect in atomically thin carbon films. *Science,* **2004,** *306*(5696), 666-669.
http://dx.doi.org/10.1126/science.1102896 PMID: 15499015

[59]   Stoller, M.D.; Park, S.; Zhu, Y.; An, J.; Ruoff, R.S. Graphene-based ultracapacitors. *Nano Lett.,* **2008,** *8*(10), 3498-3502.
http://dx.doi.org/10.1021/nl802558y PMID: 18788793

[60]   Lee, C.; Wei, X.; Kysar, J.W.; Hone, J. Measurement of the elastic properties and intrinsic strength of monolayer graphene. *Science,* **2008,** *321*(5887), 385-388.
http://dx.doi.org/10.1126/science.1157996 PMID: 18635798

[61]   Marcano, D.C.; Kosynkin, D.V.; Berlin, J.M.; Sinitskii, A.; Sun, Z.; Slesarev, A.; Alemany, L.B.; Lu, W.; Tour, J.M. Improved synthesis of graphene oxide. *ACS Nano,* **2010,** *4*(8), 4806-4814.
http://dx.doi.org/10.1021/nn1006368 PMID: 20731455

[62]   Dreyer, D.R.; Park, S.; Bielawski, C.W.; Ruoff, R.S. The chemistry of graphene oxide. *Chem. Soc. Rev.,* **2010,** *39*(1), 228-240.
http://dx.doi.org/10.1039/B917103G PMID: 20023850

[63]   Nakajima, T.; Mabuchi, A.; Hagiwara, R. A new structure model of graphite oxide. *Carbon,* **1988,** *26*(3), 357-361.
http://dx.doi.org/10.1016/0008-6223(88)90227-8

[64]   Garg, B.; Bisht, T.; Ling, Y-C. Graphene-based nanomaterials as heterogeneous acid catalysts: a comprehensive perspective. *Molecules,* **2014,** *19*(9), 14582-14614.
http://dx.doi.org/10.3390/molecules190914582 PMID: 25225721

[65]   Yang, S-Y.; Lin, W-N.; Huang, Y-L.; Tien, H-W.; Wang, J-Y.; Ma, C-C.M.; Li, S-M.; Wang, Y-S. Synergetic effects of graphene platelets and carbon nanotubes on the mechanical and thermal properties of epoxy composites. *Carbon,* **2011,** *49*(3), 793-803.

http://dx.doi.org/10.1016/j.carbon.2010.10.014

[66] Li, J.; Wong, P-S.; Kim, J-K. Hybrid nanocomposites containing carbon nanotubes and graphite nanoplatelets. *Mater. Sci. Eng. A,* **2008**, *483*, 660-663.
http://dx.doi.org/10.1016/j.msea.2006.08.145

[67] Vinayan, B.; Nagar, R.; Raman, V.; Rajalakshmi, N.; Dhathathreyan, K.; Ramaprabhu, S. Synthesis of graphene-multiwalled carbon nanotubes hybrid nanostructure by strengthened electrostatic interaction and its lithium ion battery application. *J. Mater. Chem.,* **2012**, *22*(19), 9949-9956.
http://dx.doi.org/10.1039/c2jm16294f

[68] Jyoti, J.; Singh, B.P.; Chockalingam, S.; Joshi, A.G.; Gupta, T.K.; Dhakate, S.R. Synergetic effect of graphene oxide-carbon nanotube on nanomechanical properties of acrylonitrile butadiene styrene nanocomposites. *Mater. Res. Express,* **2018**, *5*(4), 045608.
http://dx.doi.org/10.1088/2053-1591/aabd2e

[69] Seyhan, A.T.; Tanoğlu, M.; Schulte, K. Tensile mechanical behavior and fracture toughness of MWCNT and DWCNT modified vinyl-ester/polyester hybrid nanocomposites produced by 3-roll milling. *Mater. Sci. Eng. A,* **2009**, *523*(1-2), 85-92.
http://dx.doi.org/10.1016/j.msea.2009.05.035

[70] Breuer, O.; Sundararaj, U. Big returns from small fibers: a review of polymer/carbon nanotube composites. *Polym. Compos.,* **2004**, *25*(6), 630-645.
http://dx.doi.org/10.1002/pc.20058

[71] Moniruzzaman, M.; Winey, K.I. Polymer nanocomposites containing carbon nanotubes. *Macromolecules,* **2006**, *39*(16), 5194-5205.
http://dx.doi.org/10.1021/ma060733p

[72] Coleman, J.N.; Khan, U.; Blau, W.J.; Gun'ko, Y.K. Small but strong: a review of the mechanical properties of carbon nanotube–polymer composites. *Carbon,* **2006**, *44*(9), 1624-1652.
http://dx.doi.org/10.1016/j.carbon.2006.02.038

[73] Kim, H.; Miura, Y.; Macosko, C.W. Graphene/polyurethane nanocomposites for improved gas barrier and electrical conductivity. *Chem. Mater.,* **2010**, *22*(11), 3441-3450.
http://dx.doi.org/10.1021/cm100477v

[74] Lee, J-H.; Kim, S.K.; Kim, N.H. Effects of the addition of multi-walled carbon nanotubes on the positive temperature coefficient characteristics of carbon-black-filled high-density polyethylene nanocomposites. *Scr. Mater.,* **2006**, *55*(12), 1119-1122.
http://dx.doi.org/10.1016/j.scriptamat.2006.08.051

[75] Kalaitzidou, K.; Fukushima, H.; Drzal, L.T. A new compounding method for exfoliated graphite–polypropylene nanocomposites with enhanced flexural properties and lower percolation threshold. *Compos. Sci. Technol.,* **2007**, *67*(10), 2045-2051.
http://dx.doi.org/10.1016/j.compscitech.2006.11.014

[76] Wang, W-P.; Pan, C-Y. Preparation and characterization of polystyrene/graphite composite prepared by cationic grafting polymerization. *Polymer (Guildf.),* **2004**, *45*(12), 3987-3995.
http://dx.doi.org/10.1016/j.polymer.2004.04.023

[77] Cooper, C.A.; Ravich, D.; Lips, D.; Mayer, J.; Wagner, H.D. Distribution and alignment of carbon nanotubes and nanofibrils in a polymer matrix. *Compos. Sci. Technol.,* **2002**, *62*(7-8), 1105-1112.
http://dx.doi.org/10.1016/S0266-3538(02)00056-8

[78] Srivastava, S.; Kotov, N.A. Composite Layer-by-Layer (LBL) assembly with inorganic nanoparticles and nanowires. *Acc. Chem. Res.,* **2008**, *41*(12), 1831-1841.
http://dx.doi.org/10.1021/ar8001377 PMID: 19053241

[79] Byrne, M.T.; Gun'ko, Y.K. Recent advances in research on carbon nanotube-polymer composites. *Adv. Mater.,* **2010**, *22*(15), 1672-1688.
http://dx.doi.org/10.1002/adma.200901545 PMID: 20496401

[80] Elkington, M.; Bloom, D.; Ward, C.; Chatzimichali, A.; Potter, K. Hand layup: understanding the manual process. *Advanced Manufacturing: Polymer & Composites Science*, **2015**, *1*, 138-151.

[81] Sánchez, M.; Campo, M.; Jiménez-Suárez, A.; Ureña, A. Effect of the carbon nanotube functionalization on flexural properties of multiscale carbon fiber/epoxy composites manufactured by VARIM. *Compos., Part B Eng.*, **2013**, *45*(1), 1613-1619.
      http://dx.doi.org/10.1016/j.compositesb.2012.09.063

[82] Han, X.; Zhao, Y.; Sun, J.; Li, Y.; Zhang, J.; Hao, Y. Effect of graphene oxide addition on the interlaminar shear property of carbon fiber-reinforced epoxy composites. *N. Carbon Mater.*, **2017**, *32*(1), 48-55.
      http://dx.doi.org/10.1016/S1872-5805(17)60107-0

[83] Teotia, S.; Singh, B.P.; Elizabeth, I.; Singh, V.N.; Ravikumar, R.; Singh, A.P.; Gopukumar, S.; Dhawan, S.K.; Srivastava, A.; Mathur, R.B. Multifunctional, robust, light-weight, free-standing MWCNT/phenolic composite paper as anodes for lithium ion batteries and EMI shielding material. *RSC Advances*, **2014**, *4*(63), 33168-33174.
      http://dx.doi.org/10.1039/C4RA04183F

[84] Lee, B.O.; Woo, W.J.; Song, H.S.; Park, H-S.; Hahm, H-S.; Wu, J-P.; Kim, M-S. EMI Shielding Properties of Carbon Nanofiber Filled Poly Vinylidene Fluoride Coating Materials. *J. Ind. Eng. Chem.*, **2001**, *7*, 305-309.

[85] Mondal, S.; Ravindren, R.; Bhawal, P.; Shin, B.; Ganguly, S.; Nah, C.; Das, N.C. Combination effect of carbon nanofiber and ketjen carbon black hybrid nanofillers on mechanical, electrical, and electromagnetic interference shielding properties of chlorinated polyethylene nanocomposites. *Compos., Part B Eng.*, **2020**, *197*, 108071.
      http://dx.doi.org/10.1016/j.compositesb.2020.108071

[86] Al-Saleh, M.H.; Saadeh, W.H.; Sundararaj, U. EMI shielding effectiveness of carbon based nanostructured polymeric materials: A comparative study. *Carbon*, **2013**, *60*, 146-156.
      http://dx.doi.org/10.1016/j.carbon.2013.04.008

[87] Zhang, C-S.; Ni, Q-Q.; Fu, S-Y.; Kurashiki, K. Electromagnetic interference shielding effect of nanocomposites with carbon nanotube and shape memory polymer. *Compos. Sci. Technol.*, **2007**, *67*(14), 2973-2980.
      http://dx.doi.org/10.1016/j.compscitech.2007.05.011

[88] Simon, R.M. *Emi Shielding Through Conductive Plastics*; , **1981**.
      http://dx.doi.org/10.1080/03602558108067695

[89] Liu, Z.; Bai, G.; Huang, Y.; Li, F.; Ma, Y.; Guo, T.; He, X.; Lin, X.; Gao, H.; Chen, Y. Microwave Absorption of Single-Walled Carbon Nanotubes/Soluble Cross-Linked Polyurethane Composites. *J. Phys. Chem. C*, **2007**, *111*(37), 13696-13700.
      http://dx.doi.org/10.1021/jp0731396

[90] Hoang, A.S. Electrical conductivity and electromagnetic interference shielding characteristics of multiwalled carbon nanotube filled polyurethane composite films. *Advances in Natural Sciences: Nanoscience and Nanotechnology*, **2011**, *2*(2), 025007.
      http://dx.doi.org/10.1088/2043-6262/2/2/025007

[91] Bhattacharya, P.; Sahoo, S.; Das, C. *Microwave absorption behaviour of MWCNT based nanocomposites in X-band region*; , **2013**.
      http://dx.doi.org/10.3144/expresspolymlett.2013.20

[92] Bhattacharya, P.; Das, C.K.; Kalra, S.S. Graphene and MWCNT: potential candidate for microwave absorbing materials. *Journal of Materials Science Research*, **2012**, *1*, 126.

[93] Chauhan, S.S.; Verma, M.; Verma, P.; Singh, V.P.; Choudhary, V. Multiwalled carbon nanotubes reinforced poly (ether-ketone) nanocomposites: Assessment of rheological, mechanical, and electromagnetic shielding properties. *Polym. Adv. Technol.*, **2018**, *29*(1), 347-354.

http://dx.doi.org/10.1002/pat.4120

[94]   Verma, M.; Verma, P.; Dhawan, S.K.; Choudhary, V. Tailored graphene based polyurethane composites for efficient electrostatic dissipation and electromagnetic interference shielding applications. *RSC Advances,* **2015**, *5*(118), 97349-97358.
http://dx.doi.org/10.1039/C5RA17276D

[95]   Yadav, R.S.; Kuřitka, I.; Vilčáková, J.; Machovský, M.; Škoda, D.; Urbánek, P.; Masař, M.; Gořalik, M.; Urbánek, M.; Kalina, L.; Havlica, J. Polypropylene nanocomposite filled with spinel ferrite NiFe2O4 nanoparticles and in-situ thermally-reduced graphene oxide for electromagnetic interference shielding application. *Nanomaterials (Basel),* **2019**, *9*(4), 621.
http://dx.doi.org/10.3390/nano9040621 PMID: 30995813

[96]   Mishra, M.; Singh, A.P.; Singh, B.P.; Singh, V.N.; Dhawan, S.K. Conducting ferrofluid: a high-performance microwave shielding material. *J. Mater. Chem. A Mater. Energy Sustain.,* **2014**, *2*(32), 13159-13168.
http://dx.doi.org/10.1039/C4TA01681E

[97]   Singh, A.P.; S, A.K.; Chandra, A.; Dhawan, S.K. A. K. S., A. Chandra, and S. K. Dhawan, "Conduction mechanism in Polyaniline-flyash composite material for shielding against electromagnetic radiation in X-band & Ku band. *AIP Adv.,* **2011**, *1*(2), 022147.
http://dx.doi.org/10.1063/1.3608052

[98]   Nanni, F.; Travaglia, P.; Valentini, M. Effect of carbon nanofibres dispersion on the microwave absorbing properties of CNF/epoxy composites. *Compos. Sci. Technol.,* **2009**, *69*(3-4), 485-490.
http://dx.doi.org/10.1016/j.compscitech.2008.11.026

[99]   Lv, X.; Yang, S.; Jin, J.; Zhang, L.; Li, G.; Jiang, J. Preparation and Electromagnetic Properties of Carbon Nanofiber/Epoxy Composites. *J. Macromol. Sci. Part B Phys.,* **2010**, *49*(2), 355-365.
http://dx.doi.org/10.1080/00222340903355750

[100]   Li, N.; Huang, Y.; Du, F.; He, X.; Lin, X.; Gao, H.; Ma, Y.; Li, F.; Chen, Y.; Eklund, P.C. Electromagnetic interference (EMI) shielding of single-walled carbon nanotube epoxy composites. *Nano Lett.,* **2006**, *6*(6), 1141-1145.
http://dx.doi.org/10.1021/nl0602589 PMID: 16771569

[101]   Liang, J.; Wang, Y.; Huang, Y.; Ma, Y.; Liu, Z.; Cai, J.; Zhang, C.; Gao, H.; Chen, Y. Electromagnetic interference shielding of graphene/epoxy composites. *Carbon,* **2009**, *47*(3), 922-925.
http://dx.doi.org/10.1016/j.carbon.2008.12.038

[102]   Singh, B.P.; Prasanta, V.; Choudhary, V.; Saini, P.; Pande, S.; Singh, V.N.; Mathur, R.B. Choudhary, P. Saini, S. Pande, V. N. Singh, and R. B. Mathur, "Enhanced microwave shielding and mechanical properties of high loading MWCNT–epoxy composites. *J. Nanopart. Res.,* **2013**, *15*(4), 1554.
http://dx.doi.org/10.1007/s11051-013-1554-0

[103]   Singh, B.P.; Saini, K.; Choudhary, V.; Teotia, S.; Pande, S.; Saini, P.; Mathur, R.B. Effect of length of carbon nanotubes on electromagnetic interference shielding and mechanical properties of their reinforced epoxy composites. *J. Nanopart. Res.,* **2013**, *16*(1), 2161.
http://dx.doi.org/10.1007/s11051-013-2161-9

[104]   Seng, L.Y.; Wee, F.H.; Rahim, H.A.; Malek, F.; You, K.Y.; Liyana, Z.; Jamlos, M.A.; Ezanuddin, A.A.M. EMI shielding based on MWCNTs/polyester composites. *Appl. Phys., A Mater. Sci. Process.,* **2018**, *124*(2), 140.
http://dx.doi.org/10.1007/s00339-018-1564-y

CHAPTER 4

Thermoplastic Polyurethane Graphene Nanocomposites for EMI Shielding

Meenakshi Verma[1]*, Veena Choudhary[2] and S.K. Dhawan[3]

[1]Department of Chemistry, Kalindi College, University of Delhi, New Delhi, India

[2] Department of Materials Science and Engineering, Indian Institute of Technology Delhi, Hauz Khas, New Delhi 110016, India

[3]Division of Materials Physics and Engineering, CSIR-National Physical Laboratory, New Delhi-110012, India

Abstract: Interference and chaos among the various electromagnetic signals are becoming the primary challenge of the current era that relies on wireless communication. Electromagnetic pollution is the overabundance of electromagnetic radiation emitted by electronic devices, like cell phones, cordless phones, Wi-Fi routers, or Bluetooth-enabled equipment, and our relationship with these devices has become more and more intimate. The potential effects of electromagnetic pollution, both in terms of its interaction with electronic devices as well as biological species, are serious concerns for the research community. EMI shielding reduces electromagnetic interference among the electronic components. Therefore, protection from such harmful radiations must be acquired by either blocking or shielding these unavoidable severe electromagnetic radiations. Metals have been typically used as the material of choice for shielding applications, but heavy weight, corrosion susceptibility, and cumbersome processing methods make them unsuitable for both researchers and users. Alternatively, polymer nanocomposites have gained tremendous attention as electromagnetic interference (EMI) shielding materials owing to their facile synthesis, ease of processing, and low cost. Different thermoplastic and thermoset polymer matrices have been explored for the development of lightweight composite material for EMI shielding applications. Among the thermoplastic polymers, thermoplastic polyurethanes (TPU) have attracted a great deal of recognition due to their combination of properties, such as flexibility, stretchability, transparency, good wear and weather resistance, better abrasion and chemical resistance, and better mechanical properties. Although graphene and carbon nanotubes have been explored as conducting fillers in polyurethane matrix for the development of EMI shields, no reports are available

---

*Corresponding author Meenkashi Verma: Department of Chemistry, Kalindi College, University of Delhi, New Delhi, India; Tel: 91 9910796869; E-mail: meenakshi.iitd@gmail.com

Sundeep K. Dhawan, Avanish Pratap Singh, Anil Ohlan, Kuldeep Singh Kakran and Pradeep Sambyal (Ed.)
All rights reserved-© 2022 Bentham Science Publishers

using a combination of these fillers along with magnetic nanoparticles in thermoplastic polyurethane matrix.

Keywords: Electromagnetic interference shielding, Graphene, Polymer nanocomposites, Polyurethane.

4.1. INTRODUCTION

Electromagnetic (EM) radiation is a form of energy propagated through free space or a material medium in the form of electromagnetic waves, such as radio waves, visible light, and gamma rays. Electromagnetic pollution from EM emitting radiations has become a worldwide concern because it is harmful to human health as well as has the potential to cause disturbances in electronic devices. Electromagnetic radiations are everywhere, hence unavoidable. Whether natural or manmade, these emitted fields can interfere with the operation of other nearby electronic equipment. This situation is known as electromagnetic interference (EMI). Man-made EMI originates from power lines, electric motors, fluorescent bulbs, *etc.*, while natural sources include lightning and electrostatic discharges. Driven by the proliferation of electronics and instrumentation in commercial and industrial applications, EMI problems have impacted almost all electrical and electronic systems from daily life to military activity and space exploration. These disturbances may lead to the loss of valuable time, energy, resources, money, or even precious human life. Therefore, the shielding mechanism must be provided to understand the spurious electromagnetic noises or pollution [1].

EMI shielding means using a shield (a shaped conducting material) to partially or completely envelop an electronic circuit, *i.e.*, an EMI emitter or susceptor. Therefore, it limits the amount of EMI radiation from the external environment that can penetrate the circuit, and conversely, it influences how much EMI energy generated by the circuit can escape into the external environment [2]. A variety of materials have been used for shielding with a wide range of electrical conductivity, magnetic permeability, and geometries. It is well known that, because of good mechanical properties and electrical conductivity, metals are the most common EM shielding materials and are extensively used in EM shielding applications [3, 4]. For example, an inner core conductor is generally surrounded by a wire mesh using the shielded cable. The shielding prevents the signals transmitted by the inner conductor and stops them from leaking or being disturbed. Another example is the door with a built-in metal screen of the microwave oven. This screen isolates the oven's metal housing, stops the crossing of EM wave, and allows the visible light to pass through. However, weight considerations decrease the viability of metal shields in portable electronics. It is disadvantageous to use internal metal shrouds lap-top computer cases [5]. The demand for low-cost and low-weight shielding materials has shifted the focus to plastics. Most polymer resins are electrically insulating, therefore, typically incapable of providing EM shielding. Through the addition of conductive fillers, such as conductive metal fibers or carbon fibers, the electrical conductivity of these resins is increased, and acceptable shielding ability is obtained [6-8]. An electrically conductive composite can be used for computer cases and cell phone housings without the need for an extra metallic shield. These devices retain the light weight desired by consumers and meet the Federal Communications Commission guidelines [5].

4.2. EMI SHIELDING EFFECTIVENESS

EMI shielding can be specified in terms of reduction in the magnetic (and electric) field or plane-wave strength caused by shielding. The effectiveness of a shield and its resulting EMI attenuation is based on the frequency, the distance of the shield from the source, thickness of the shield, and shield material. Shielding effectiveness is the ratio of impinging energy (electromagnetic waves) to residual energy. All electromagnetic waves consist of two essential components, a magnetic field (H) and an electric field (E). These two fields are perpendicular to each other, and the direction of wave propagation is at right angles to the plane containing the two components. The relative magnitude depends upon the waveform and its source. The ratio of E to H is called wave impedance. The intrinsic impedance of free space is 377Ω [6].

Shielding effectiveness (SE) is the ratio of the field before and after attenuation of electric and magnetic field and can be expressed as:

$$SE\ (dB) = 20\ log_{10}\left(\frac{E_T}{E_I}\right) = 20\ log_{10}\left(\frac{H_T}{H_I}\right) \qquad (1.1)$$

where E and H are electric and magnetic fields, respectively, and the subscripts T and I refer to the transmitted and incident waves, respectively.

Further electromagnetic radiation consists of coupled electric and magnetic fields. The electric field produces forces on the charge carriers (*i.e.*, electrons) within the conductor or polymeric composites. As soon as an electric field is applied to the surface of an EMI shielding material, it induces a current that causes displacement of charge inside the shielding material that cancels the applied field inside, at which point the current stops. The EMI shielding mechanism is broadly characterized as reflection, absorption, multiple internal reflection, and transmission.

Reflection is a primary mechanism of shielding; the presence of mobile charge carriers (electrons or holes) is responsible for shielding effectiveness due to reflection, which interacts with the electromagnetic fields in the radiation. Although conductivity is a major factor for reflection, and high conductivity is not required for this purpose, the volume resistivity of 1 Ω cm is sufficient for reflection loss.

Absorption is the secondary mechanism of EMI shielding; electric and/or magnetic dipoles which interact with the electromagnetic fields are responsible for this phenomenon. The electric dipoles may be provided by dielectric materials or materials having high dielectric constant values. The magnetic dipoles may be provided by magnetic materials or other materials having a high value of magnetic permeability [9, 10].

Typically, the reflection loss is a function of the ratio σ_r/μ_r while the absorption loss is a function of the product of $\sigma_r.\mu_r$, where σ_r is the electrical conductivity relative to copper and μ_r is the relative magnetic permeability. Silver, copper, gold, and aluminum are admirable for reflection due to their high conductivity. Super paramagnetic and dielectric materials are for outstanding absorption loss as they have high magnetic permeability. Both reflection and absorption losses depend on frequency as the reflection loss decreases and absorption increases with increasing frequency.

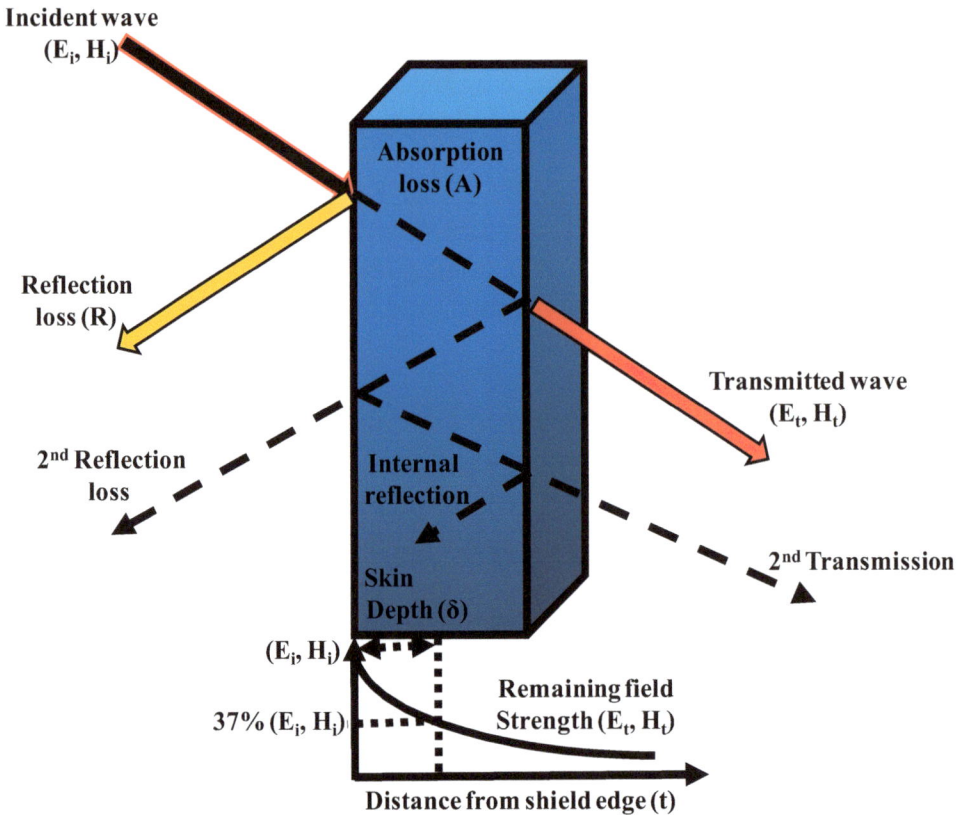

Fig. (1). Graphical representation of EMI shielding.

Other than reflection and absorption, a mechanism, *i.e.*, shielding due to multiple reflections, refers to the reflection phenomena at various surfaces or interfaces in the shield. This mechanism requires the presence of a large surface area or interface area in the shield. The example of a shield with a large surface area is a porous or foam material, while the example of a shield with a large interface area is a composite material containing filler having a large surface area. However, loss due to multiple reflections can be neglected when the distance between reflecting surfaces or interfaces is large enough as compared to the skin depth. All the losses (reflection, absorption, and multiple internal reflections) are commonly expressed in decibel (dB) [11].

The attenuation of an electromagnetic wave occurs by three mechanisms, as shown in Equations 1.2,

$$SE\ (dB) = (SE_R + SE_A + SE_M) = 10\ log_{10}\left(\frac{P_T}{P_I}\right) \qquad (1.2)$$

where P_I and P_T are the power of incident and transmitted EM waves, respectively. In actual practice, three different phenomena named reflection (SE_R), absorption (SE_A), and multiple reflections (SE_M) contribute towards SE_T, as shown in Fig. (**1**).

4.2.1. Reflection Loss

The reflection loss (SE_R) is defined as the mismatch between the incident waves and the surface impedance of the shield. The complex phenomena of reflection losses can be simplified by considering shielding effectiveness for incident electric fields as a separate problem from that of electric, magnetic, or plane waves. The equations for three principal fields are given as:

$$R_E = K_1 10\ log\left(\frac{\sigma_r}{f^3 r^2 \mu_r}\right) \qquad (1.3)$$

$$R_H = K_2 10\ log\left(\frac{fr^2 \sigma_r}{\mu_r}\right) \qquad (1.4)$$

$$R_P = K_3 10\ log\left(\frac{\sigma_r}{f \mu_r}\right) \qquad (1.5)$$

The magnitude of reflection loss for three principal fields can be given by the following generalized expression:

$$SE_R = C + 10 log\left(\frac{\sigma_r}{\mu_r}\frac{1}{f^n r^m}\right) \qquad (1.6)$$

where R_E, R_H, and R_P are reflection losses for the electric, magnetic, and plane wave fields, respectively, (in dB), σ_r is the relative conductivity referred to copper, f is the frequency (in Hz), μ_r is the relative permeability referred to free space while r is the distance from the source to the shielding material. The values of constants C, n, and m are listed in Table **4.1** for calculating reflection losses (in dB) for plane waves, electric fields, and magnetic fields, respectively.

Table 1. Values for C, m, and n for predicting SE$_R$ for the three-principal field plane.

Type of field	C	n	M
Electric field	322	3	2
Magnetic field	14.6	-1	-2
Plane wave field	168	1	0

4.2.2. Absorption Loss

Absorption loss (SE$_A$) is a function of the physical characteristics of the shield and is independent of the type of source field. Therefore, the absorption term SE$_A$ is the same for all three waves. The amplitude of an electromagnetic wave decreases exponentially when it passes through a medium (as shown in Fig. **1**). This decay or absorption loss occurs because currents induced in the medium produce ohmic losses and heating of the material, where E$_1$ and H$_1$ can be expressed as:

$$E_1 = E_0 e^{-t/\delta} \tag{1.7a}$$

$$H_1 = H_0 e^{-t/\delta} \tag{1.7b}$$

The distance required by the wave to be attenuated to 1/e or 37 % is defined as the skin depth. Therefore, the absorption term SE$_A$ in decibel is given by the expression:

$$SE_A = -20\left(\frac{t}{\delta}\right) log_{10} e = -8.68\left(\frac{t}{\delta}\right) = -8.68\sqrt{\sigma f \mu} \tag{1.8}$$

where t is the thickness of the shield in mm; f is the frequency in MHz; μ is relative permeability (1 for copper); σ is conductivity relative to copper. The skin depth δ can be expressed as:

$$\delta = \frac{1}{\sqrt{\pi \mu_r \sigma_r f}} \tag{1.9}$$

The absorption loss of one skin depth in a shield is approximately 9 dB. Skin effect is especially important at low frequencies, where the fields experienced are more likely to be predominantly magnetic with lower wave impedance than 377 Ω. From the absorption loss point of view, a good material for a shield will have high conductivity and high permeability along with a sufficient thickness to achieve the required number of skin depths at the lowest frequency of concern.

4.2.3. Multiple Reflections

If thin material is used as a shield, the reflected wave from the second boundary is re-reflected off the first boundary, and then it returns to the second boundary to be reflected again, as shown in Fig. (**1**). This can be neglected in the case of a thick shield because the absorption loss is high. As the wave reaches the second boundary for the second time, it is of negligible amplitude because by then, it has passed through the thickness of the shield three times. The factor SE_M can be expressed as

$$SE_M = 20 log_{10}\left(1 - e^{-2t/\delta}\right) \tag{1.10}$$

Therefore, when SE_A is greater than 10 dB, SE_M can be safely neglected. In practical calculation, SE_M can also be neglected for electric fields and plane waves.

4.3. ELECTROMAGNETIC ATTRIBUTES: COMPLEX PERMITTIVITY AND PERMEABILITY

The electromagnetic attributes of a shield, *i.e.*, complex permittivity $[\varepsilon^* = (\varepsilon'-j\varepsilon'')]$ and permeability $[\mu^* = (\mu'-j\mu'')]$ can be retrieved from the experimental S-parameters $[S_{11}$ (or S_{22}) and S_{21} (or S_{12})] using suitable algorithms and models as given in Table **2**.

Each of the given conversion techniques has different advantages and limitations. The selection of technique depends on several factors such as the measured S-parameters, sample length, desired output properties, speed of conversion, and accuracies in the converted results. Among the above-mentioned procedures, Nicholson-Ross-Weir (NRW) technique is a widely used regressive/iterative analysis as it provides direct calculation of both the permittivity (ε^*) and permeability (μ^*) from the input S-parameters. Parameter ε' (real permittivity) represents the charge storage (or dielectric constant), whereas ε'' (imaginary permittivity) is a measure of dielectric dissipation or losses. Similarly, μ' and μ'' represent magnetic storage and losses, respectively.

Nicholson-Ross-Weir (NRW) technique provides a direct calculation of both the permittivity and permeability from the S-parameters [12]. It is the most commonly used technique for performing such conversions. Measurement of reflection coefficient and transmission coefficient requires all four (S_{11}, S_{21}, S_{12}, S_{22}) or a pair (S_{11}, S_{21}) of S-parameters of the material under test (MUT).

Table 2. Conversion techniques, S-parameters, and output attributes.

Conversion Technique	Input S-parameters	Output Attributes
Nicolson-Ross-Weir (NRW)	S_{11}, S_{21}, S_{12} and S_{22} or/and S_{11} S_{21} (or S_{22} and S_{12})	μ_r and ε_r
NIST iterative	S_{11}, S_{21}, $S_{12,}$ and S_{22} or pair S_{11} S_{21} (or S_{22} and S_{12})	μ_r and $\varepsilon_r=1$
New non-iterative	S_{11}, S_{21}, S_{12} and S_{22} or/and S_{11} S_{21} (or S_{22} and S_{12})	μ_r and $\varepsilon_r=1$
Short circuit line (SCL)	S_{11}, S_{22}	ε_r

NIST Iterative Technique: performs the calculation using Newton-Raphson's root finding technique and is suitable for permittivity calculation only. It utilizes all four (S_{11}, S_{21}, S_{12}, S_{22}) or a pair (S_{11}, S_{21}) of S parameters of MUT to calculate the reflection and transmission coefficient. It works well if a good initial guess is available. The technique bypasses the inaccuracy peaks that exist in the NRW technique when the sample thickness is an integer multiple of one-half of the wavelength. It is suitable for long samples and characterizing low-loss materials.

New Non-Iterative Technique: It is quite similar to the NRW technique but with a different formulation, and it is suitable for permittivity calculation for the case permeability $\mu_r=1$. It utilizes all four (S_{11}, S_{21}, S_{12}, S_{22}) or just two (S_{11}, S_{21}) S-parameters of MUT to calculate the reflection and transmission coefficients. The technique has the advantage of being stable over a whole range of frequencies for an arbitrary sample length. The technique is based on a simplified version of the NRW technique, and no divergence is observed at frequencies corresponding to multiples of one-half wavelength in the sample. It does not need an initial estimation of permittivity and can perform the calculation very fast.

Short Circuit Line (SCL) Technique: is a one-port measurement on coaxial lines or waveguides. It performs the calculation using the same Newton-Raphson's numerical approach as in the NIST iterative technique and is suitable for permittivity calculation only. It utilizes only the S_{11} parameter of MUT to calculate the reflection coefficient. The technique requires a good initial guess in order to obtain an accurate result. The technique also requires the input of sample length and position for accurate measurements. In order to measure the dielectric properties of the material, the appropriate measurement and conversion technique is required. It is necessary to use the right technique for the material to be measured because specific technique is applicable to specific material. If the wrong technique is used, the measurement results will not be satisfactory.

The incident and transmitted waves in a two port Vector network analyzer (VNA) can be mathematically represented by complex scattering parameters (or S-parameters) *i.e.*, S_{11} (or S_{22}) and S_{12} (or S_{21}) respectively which in-turn can be conveniently correlated with reflectance (R) and transmittance (T) *i.e.*, $T = |E_T/E_I|^2 = |S_{12}|^2 = |S_{21}|^2$, $R = |E_R/E_I|^2 = |S_{11}|^2 = |S_{22}|^2$, giving absorbance (A) as: $A = (1-R-T)$. When SE_A is greater than 10 dB, SE_M becomes negligible ($\sim$ -1.0 dB) and can be safely neglected so that SE_T can be expressed as: $SE_T = SE_R + SE_A$. In addition, the intensity of EM wave inside the shield after primary reflection is based on quantity (1-R), which can be used for normalization of absorbance (A) to yield effective absorbance as: $A_{eff} = 1-R-T)/(1-R)$]. Therefore, experimental reflection and absorption losses can be expressed as:

$$SE_R = 10log(1 - R) \qquad (1.11)$$

$$SE_A = 10log\left(\frac{T}{1-R}\right) \qquad (1.12)$$

Therefore, from the knowledge of reflected and transmitted signals, VNA can easily compute reflection and absorption loss components of total shielding.

4.4. MATERIALS FOR EMI SHIELDING

Metals have been typically used as the material of choice for shielding applications. However, the applicability of metals for designing of highly portable devices has been reduced due to their weight limitations, corrosion susceptibility, flexibility issues and cumbersome fabrication methods. Further metals behave as good EM wave reflector but bad absorber due to their shallow skin depth [13, 14]. These weaknesses hindered them for the practical applications in the EM signal shielding field. Therefore, it is beneficial to develop new materials capable of providing EMI shielding but can be easy to manufacture, economical, light weight, flexible and portable.

Through the addition of conductive fillers to electrically insulating polymer resins, electrically conductive composites are formed which can be used for providing light weight shielding materials. Compared to conventional metal-based EMI shielding materials, carbon-based conductive polymer composites are attractive due to its light weight, resistance to corrosion, flexibility and processing advantages [15-17]. The utility of different types of carbon fillers for shielding applications has been thoroughly studied [16, 18-21] Amongst the various carbon fillers (*e.g.*, graphite, carbon black or carbon fibers), carbon black is commonly used as conducting filler in polymer composites [22-25]. In using carbon black as filler, a major disadvantage is the requirement of high amount of carbon black (up to 30 to 40 %) to achieve desired conductivity, which leads to deterioration of the mechanical properties of polymer [26].

Two very important carbon nanofillers as carbon nanotubes (CNTs) and graphene have been found very suitable candidate for such type of applications because of the conductive composite can be formed at low CNTs and graphene loading, due to low percolation thresholds [27-30]. The small diameter, high aspect ratio, high conductivity and mechanical strength of CNTs and graphene make them an excellent option for creating conductive composites for high-performance EMI shielding materials [31-34].

4.4.1. Graphene

Graphene, *i.e.*, an isolated planar sheet of carbon hexagons consisting of sp^2 hybridized C–C bond with a π-electron cloud, has attracted intense attention from scientists and engineers from different backgrounds [35]. It is an allotrope of carbon and the first known example of a truly two-dimensional (2D) crystal as shown in Fig. (**2**). Thin flakes consisting of a few layers of carbon atoms, including monolayer graphene, could be very important because of their interesting physical and structural characteristics [36-39]. Also, promising potential applications in technological fields have been reported, such as components in microelectronic devices [40-42], transparent conductive films [43-45], gas sensors [46-49], gas storage [50, 51], heat dissipation [52, 53], energy storage [54, 55], solar cells [56], and reinforcement for polymer films [57].

Single layer graphene is found to be the strongest material with Young's modulus of 1 TPa and ultimate strength of 130 GPa [35]. It has a thermal conductivity of 5000 W/ (mK), which corresponds to the upper bound of the highest values reported for single walled carbon nanotubes (SWCNT) bundles.

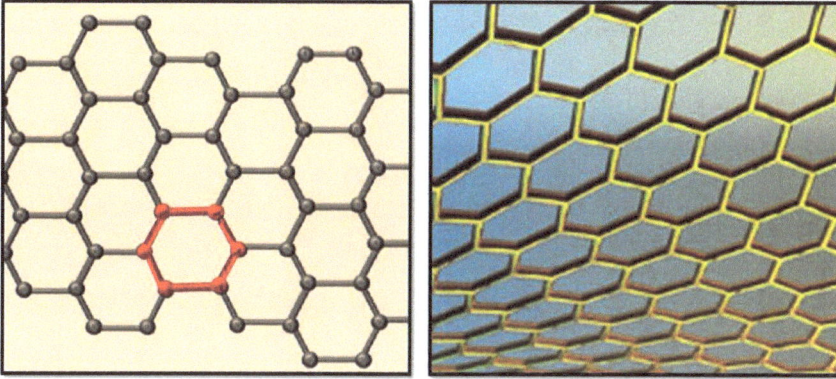

Fig. (2). Structure of graphene.

Moreover, single-layer graphene has very high electrical conductivity, up to 6000 Scm^{-1}, and unlike CNT, chirality is not a factor in its electrical conductivity [58]. Graphene is almost transparent, it absorbs only 2.3% of the light intensity, independent of the wavelength in the optical domain [59]. Thus, suspended graphene does not have any color.

4.4.2. Synthesis of Graphene

Several techniques for production of graphene are now known to us. Each method has its own benefits and related drawbacks. The deciding factor is whether one wishes to synthesize defect free graphene (purity) or graphene with defects (containing oxygen species onto the surface). The drawback of graphene materials with defects is the loss of some of the interesting properties of graphene. On the other hand, defects could provide numerous application opportunities. Therefore, not only the quantity of graphene but the type of applications dictates graphene's preparative methods. Several methods have been developed over the years for the synthesis of graphene which include mechanical exfoliation [60], chemical exfoliation [61], chemical vapor deposition [62] and thermal decomposition of silicon carbide [63]. One of the economical yet effective routes to produce bulk quantities of graphene is chemical reduction of exfoliated graphite oxide [64-67]. Basically two approaches are followed to obtain graphene: top down approach which includes mechanical exfoliation or chemical exfoliation to obtain graphene from graphite by disrupting the Van der Waals interaction between the graphitic layers [68, 69] and bottom up approach by chemical vapor deposition method or by epitaxial growth using SiC substrates [68].

4.4.2.1 Mechanical Exfoliation

The first graphene was produced by the mechanical exfoliation also known as micromechanical cleavage (or peeling) of thin layers of thick graphite crystals [60, 70]. It has been done with the aim of synthesizing a two-dimensional crystal, and thus various properties of the thin sheets obtained have been determined [71]. The cleavage was carried out as follows: a graphite crystal, fixed on a glass plate by using double-sided adhesive tape, was cleaved by using another piece of adhesive tape shown in Fig. (**3a-c**).

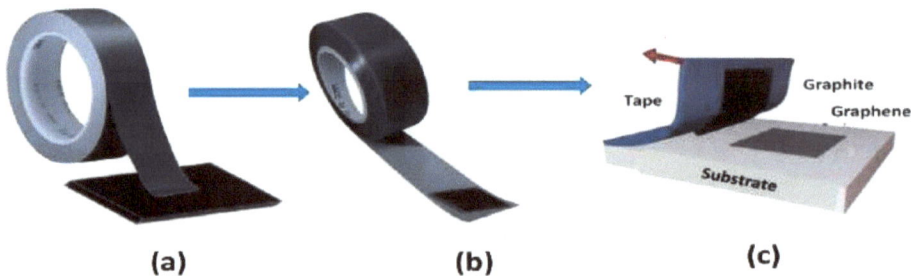

(a)   (b)   (c)

Fig. (3). Micromechanical exfoliation of 2D crystals. (**a**) Adhesive tape is pressed against a 2D crystal so that the top few layers are attached to the tape (**b**). (**c**) The tape with crystals of layered material is pressed against a surface of choice. Upon peeling off, the bottom layer is left on the substrate [71].

This cleavage process was repeated until the sheet became transparent, and finally the thin sheet was recovered by washing in an organic solvent to remove the adhesive tape on the substrate. By repeatedly peeling highly oriented pyrolytic graphite flakes with a thickness between 3 nm and 100 mm, some containing a part of a single layer, were obtained [39, 60, 70, 71]. Careful selection of the initial graphite material and the use of freshly cleaved and cleaned surfaces of graphite and SiO_2 make the isolation and detection of graphene crystallites and monolayer successful [70]. Mechanical exfoliation provides pristine graphene with excellent electrical properties. Hence this technique is best suited for fundamental research studies. This technique, however, is not suitable for bulk synthesis of graphene since size, thickness and location of graphene crystallites cannot be controlled in this technique and the technique is not high-yielding [39, 70].

4.4.2.2 Liquid-Phase Exfoliation of Graphite

Liquid-phase exfoliation method is an important technique and one of the most feasible approaches for industrial production of graphene due to its versatility, potential scalability and low cost [61]. This approach is inspired by the theoretical and experimental studies on the solvent dispersion of carbon nanotubes which showed that nanotubes could be effectively dispersed in solvents whose surface energy matched that of the nanotubes [72, 73]. In this method, graphite is exfoliated into single graphene layers and individual sheets are stabilized in solution by means of ultrasonication in organic solvents. Ultrasonication is an effective tool for exfoliation of graphite. Graphite powder is dispersed in a suitable solvent and the solution is ultrasonicated to obtain colloidal graphene suspensions. The powder fragments into nanosheets, which are stabilized against aggregation by the solvent [72]. The dispersion and exfoliation is possible because energy required to exfoliate graphene *i.e.*, to overcome the interacting forces between the graphene layers is balanced by the solvent–graphene interaction for solvents whose surface energies match that of graphene [66, 72, 74]. The graphitic basal structure is broken and small graphite fragments intercalated by solvent molecules are produced [74]. Highly polar organic solvents such as dimethylformamide (DMF) [75], N-methyl-2-pyrrolidone (NMP) [76], pyridine [77], acetonitrile [78] *etc.* have been successfully used for this method. Exfoliation by ultrasonication is generally followed by centrifugation to increase the percentage of the monolayers of graphene. Surfactants can also be employed to stabilize exfoliated graphene in water [72]. Liquid-phase exfoliation yields graphene essentially free of defects and oxides. Liquid-phase exfoliation methods can be used to deposit graphene on different substrates and can be employed to produce graphene-based composites or films, which are key components for many applications, such as thin-film transistors, conductive transparent electrodes for indium tin oxide replacement in light-emitting diodes or photovoltaics [61, 72].

4.4.2.3. Chemical Method: Exfoliation and Reduction of Graphite Oxide

The most common approach to graphite exfoliation is the use of strong oxidizing agents to yield graphite oxide. Although the exact structure of graphite oxide is difficult to determine, it is clear that for graphite oxide, the previously contiguous aromatic lattice of graphene is interrupted by epoxides, hydroxyls, ketone, carbonyls, and carboxylic groups [79]. The disruption of lattice is reflected in an increase in interlayer spacing from 0.335 nm for graphite to more than 0.625 nm for graphite oxide [80]. Brodie first demonstrated the synthesis of graphite oxide by adding a portion of potassium chlorate to a slurry of graphite in fuming nitric acid [81]. Staudenmaier improved on this protocol by using concentrated sulfuric acid as well as fuming nitric acid and adding the chlorate in multiple aliquots over the course of the reaction [82]. This small change in the procedure made the production of highly oxidized graphite oxide in a single reaction vessel significantly more practical.

Hummers reported the method most commonly used today: the graphite is oxidized by treatment with $KMnO_4$ and $NaNO_3$ in concentrated H_2SO_4 [83]. Recently, Marcano *et al.* proposed an improved method for the preparation of graphite oxide [69]. They found that excluding $NaNO_3$, increasing the amount of $KMnO_4$, and performing the reaction in a 9:1 mixture of H_2SO_4/H_3PO_4 improves the efficiency of the oxidation process as shown in Fig. (**5B**). This improved method provides a greater amount of hydrophilic oxidized graphene material as compared to Hummers' method or Hummers' method with additional $KMnO_4$. Moreover, even though graphite oxide produced by this method is more oxidized than that prepared by Hummers' method, when both are reduced in the same chamber with hydrazine, chemically converted graphene (CCG) produced from this new method is equivalent in its electrical conductivity. In contrast to Hummers' method, the improved method does not generate toxic gas and the temperature is easily controlled. This improved synthesis of graphite oxide may be important for large-scale production of graphite oxide as well as the construction of devices composed of the subsequent CCG.

Graphite oxide is a layered structure of graphene oxide (GO) sheets that are strongly hydrophilic such that intercalation of water molecules between the layers readily occurs and it is easily exfoliated in aqueous media by mechanical stirring or more effectively by ultrasonication to obtain exfoliated oxidized single graphene of GO [64-66, 84, 85]. The exfoliation is accomplished due to the strong interactions between water and the oxygen functionalities introduced into the basal plane during oxidation [74]. Sonication results in near-complete exfoliation of graphite oxide to GO to produce stable aqueous colloidal suspensions [64, 66]. These GO layers are precursors for the production of graphene by the removal of the oxygen groups *i.e.,* by reduction of GO as shown in Fig. (**4**). Graphite oxide can also be dispersed directly in several polar solvents such as ethylene glycol, DMF, NMP and tetrahydrofuran (THF) and fully exfoliated into individual, single-layer GO sheets by sonication to obtain stable GO dispersions [66, 86].

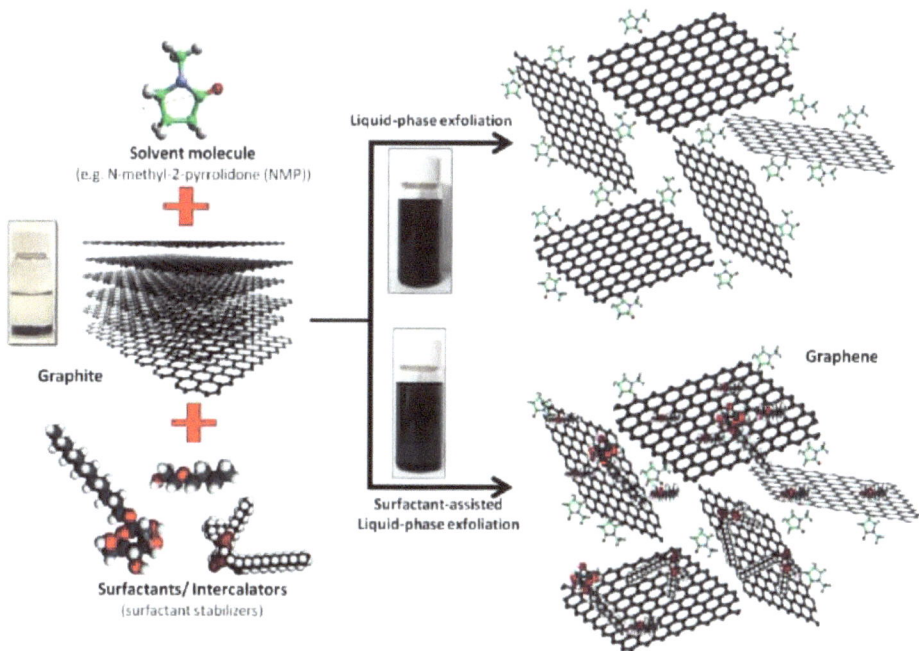

Fig. (4). Schematic representation of liquid-phase exfoliation in the absence (top-right) and presence of surfactant (bottom right).

The reduction of obtained GO to graphene is usually conducted by either thermal or chemical approaches with yields > 50% [64]. The main aim of reduction of GO is to produce electrically conducting graphene-like materials similar to the pristine graphene.

Reduction of GO or the deoxygenating process of GO generally involves low temperature chemical reduction at < 100°C using suitable reducing agents or thermal annealing at temperatures >1000°C [65-67, 74]. A variety of reducing agents have been studied for chemical reduction of colloidal dispersions of GO in water but complete reduction of GO into graphene has not been achieved [66]. Strong alkaline agents such as hydrazine hydrate [86] and sodium borohydride [87] are effective reducing agents as shown in Fig. **5**. One of the disadvantages of using chemical methods of reduction, by $NaBH_4$ and hydrazine in particular, is the introduction of heteroatomic impurities (Fig. **5 (C)**) [64]. Electrochemical reduction of GO (Fig. **5 (D)**) seems to be a better option for getting electrochemically reduced graphene oxide (ER-GO) and is a viable option for getting graphene which can find high usage applications.

After removal of the oxygen groups, reduced graphene oxide (RGO) can be further graphitized by annealing at elevated temperatures. In this process defects that remain after reduction are rearranged and the aromatic character of the monolayers increases.

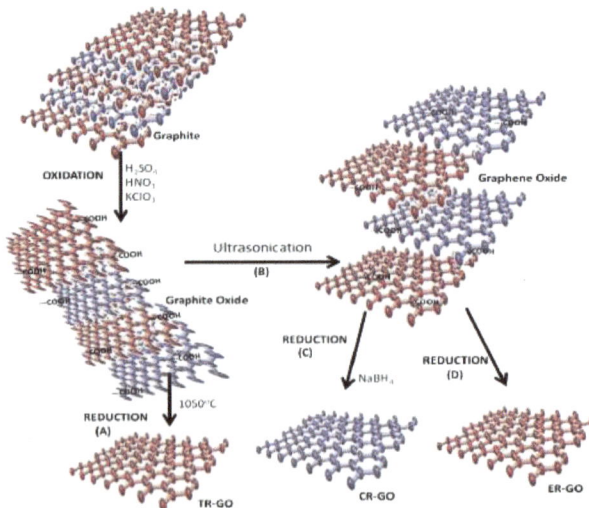

Fig. (5). Schematic representation of the step-wise synthesis of single/few layers graphene from graphite: Oxidation of graphite in the first step is followed by chemical exfoliation (B) to graphene oxide. The graphene oxide is then subjected to reduction by oxidative treatment is followed by thermal treatment (A), chemical treatment (C) or electrochemical) means (D.

Thermal reduction is achieved through rapid heating of dry graphite oxide under inert atmosphere and high temperature up to 1050 °C for 30 s which leads to reduction and exfoliation of graphite oxide producing thermally reduced graphene oxide (TR-GO) sheets (Fig. **5 (A)**) [64, 85]. Exfoliation takes place when the pressure generated by the gas (CO_2) evolved due to the decomposition of the carboxyl sites of graphite oxide exceeds Van der Waals forces holding the GO sheets together [85]. About 30 % weight loss is associated with the decomposition of oxygen groups and evaporation of water [88]. The exfoliation leads to volume expansion of 100-300 times producing very low-bulk density TR-GO sheets. Because of the structural defects caused by the loss of CO_2, these sheets are highly wrinkled. The advantage of thermal reduction methods is the ability to produce chemically modified graphene sheets without the need for dispersion in a solvent. TR-GO has C/O ratio of about 10/1 compared to 2/1 for graphite oxide. Despite the wrinkled sheet defective structure of graphene sheets and some residual functional sites after the reduction step, the graphene produced is electrically conducting [88].

The reduction of GO (chemically or thermally) usually leads to incomplete restoration of the sp^2 hybridized carbon bonds and the presence of residual oxygen functional groups result in poor electrical conductivity. Experiments have shown the existence of a significant amount of oxygen in the RGO, indicating that RGO is not the same as pristine graphene [66]. However, GO provides potential for the production of chemically modified graphene on bulk scale through a variety of chemical modifications due to the presence of reactive oxygen functional groups which provide sites for useful chemical functionalization reactions [64].

4.4.2.4. Chemical Vapor Deposition

The preparation of graphene or thin flakes of graphite through the chemical vapor deposition (CVD) of hydrocarbon gases onto the crystal surface of different transition metal (Ni, Cu *etc.*) is a promising, economical and straight forward synthesis method [62, 89]. Methane is commonly used as hydrocarbon source catalytically decomposed on the substrate (usually copper) at high temperature (~1000 °C) and low pressure leading to its deposition on a substrate in a furnace in the presence of hydrogen gas. Copper substrates are generally preferred over other catalytic surfaces like Fe and Ni because of the low solubility of carbon in copper and also copper substrates provide nucleating sites for adsorbed and diffusing carbon species which grow to a continuous graphene sheet as shown in Fig. (**6**).

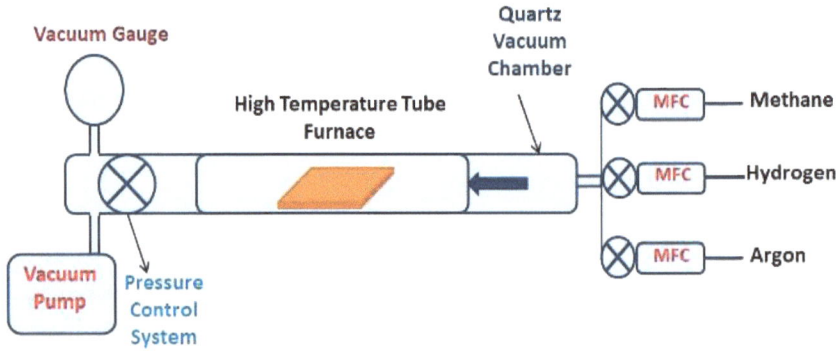

Fig. (6). Schematic of a common setup for chemical vapor deposition of graphene.

Graphene grown on transition metals must be transferred onto insulating substrates for device fabrication and electronic characterization [62]. This is done by first spin-coating a thin polymeric layer, such as poly (methyl methacrylate) on top of the as-grown graphene. The polymeric layer provides a supportive framework for graphene before the transfer. Then the underneath Cu substrate is etched away by iron chloride ($FeCl_3$) solution. The floating membrane is scooped and placed on a desired substrate. After drying, the polymeric film is dissolved using acetone or chloroform as solvent.

4.4.2.5. Thermal Decomposition of Silicon Carbide

Epitaxial growth of monolayer or few layer graphene on silicon carbide (SiC) by thermal decomposition are attractive for large-scale production and suitable for existing electronic device technology and show promising route towards epitaxial graphene based electronics [63, 90]. This method involves the conversion of SiC substrate to graphene *via* sublimation of silicon atoms on the surface at high temperatures of about 1000-1600°C in ultra-high vacuum condition and graphitization of the remaining carbon atoms [63, 74, 91]. When SiC substrates are annealed at high temperatures, Si atoms selectively desorb from the surface and the C atoms left behind naturally form graphene. Riedl *et al.* [63] have reviewed the controlled growth of epitaxial graphene layers on SiC and the manipulation of their electronic structure. Berger *et al.* [91] studied the transport and structural properties of graphene layers grown epitaxially on hexagonal SiC and found that the electronic properties of epitaxial graphene are found to be closely related to pristine graphene, revealing high mobility charge carriers [70].

4.5. CARBON NANOTUBES

Carbon nanotubes (CNTs), a fascinating material with outstanding properties has inspired the research community for wide range of potential applications in many areas. CNTs are endowed with exceptionally high material properties, very close to their theoretical limits, such as electrical and thermal conductivity, strength, stiffness, toughness and low density. Various synthesis methods now exist to produce CNTs. The three main production methods used for synthesis of CNTs are arc discharge, laser ablation and chemical vapor deposition (CVD). Various researchers have already reported the properties and detailed synthesis techniques of CNT [92].

4.6. GRAPHENE-CARBON NANOTUBES HYBRID

In order to combine the merits of the 2D graphene and 1D CNTs, many attempts have recently been made to obtain graphene–CNT hybrid materials.

4.6.1. Synthesis of Graphene-carbon Nanotubes Hybrid

4.6.1.1. Solution Based Approaches: Simple Sonication and Reduction

For the preparation of graphene-CNT hybrid, two simple methods were often used

(1) Preparation of GO-CNT hybrid by ultrasonication of GO and CNT mixture

(2) Subsequent reduction of GO-CNT hybrid to graphene-CNT hybrid [93-97]

The reduction of incorporated GO in GO-CNT hybrid can also be performed by various methods including chemical [97], thermal, electrochemical [98], photochemical and hydrothermal methods [96]. Zhang *et al.* prepared graphene-CNT hybrid composite by ultrasonication of GO and MWNCT mixture and followed by thermal reduction of the resulting GO-MWCNT composite [95]. Yen and coworkers reported a two-step solution-based method at room temperature for the preparation of graphene–MWCNT hybrid material comprising graphene and acid-treated MWCNT [99]. Briefly, the preparation methods involve: (1) synthesis of graphene oxide from graphite by Staudenmaier's method and consequent thermal reduction to graphene at 1050 °C; (2) Mixing of graphene with acid treated MWCNTs and ultrasonication to get the final hybrid.

A non-covalent π-π stacking interaction operating between graphene sheets and CNTs were revealed, which help to avoid the aggregation of individual graphene sheets. Sui and co-workers reported a green method for the fabrication of CNT–graphene hybrid aerogels by supercritical CO_2 drying of the hybrid hydrogel precursors attained by heating the mixtures of GO and CNTs with the aid of Vitamin C [100]. The graphene-CNT aerogels encompass light weight, high conductivity, large BET surface area and large volume with hierarchically porous structure. Liu *et al.* prepared the hybrid nanofiller system consisting Cu^{2+} coordinated graphene-MWCNT network by solution mixing [101]. The graphene sheets were separated and bridged by nanotubes networks *via* coordination of Cu^{2+} ions and the resulting Cu^{2+}- coordinated graphene-MWCNT network can be easily introduced to a range of polymer matrices by simple solution mixing.

It was revealed that electrochemical reduction of GO to graphene was appreciably improved after the incorporation of CNT in comparison with the electrochemical reduction of pristine GO [98, 102]. During the electrochemical reduction by cyclic voltammogram, the onset potential of the cathodic peak appeared for the GO-CNT hybrid (-0.3 V) is much lower than that of pristine GO (-0.7 V). The plausible reason for the significant improvement in the electrochemical reduction of GO in the hybrid was due to the incorporated CNT which bridges the graphene sheets and acts as a conducting wire between graphene sheets.

4.6.1.2. Chemical Vapor Deposition

Numerous efforts were made to prepare graphene-CNT or RGO-CNT by CVD method [103-106]. CVD method of preparation often provide uniformly grown CNTs onto the surface of the graphene sheets *via* strong interactions, which ultimately avoids restacking of graphene sheets and provide high stability to the hybrid material. Chen and co-workers reported *in-situ* growth of multilayered graphene-CNT composite by CVD process [105]. The chemical vapor reduction and deposition reactions were carried out at 500 °C for varied times (2, 5 and 1 h) to obtain various lengths of CNTs on GO sheets. The best performance achieved in the case of graphene-CNT with shortest length grown at 2 min. Therefore, the key factor to prepare a graphene-CNT hybrid of better electrochemical process is tuning the time duration of the CVD process which makes possible to grow CNTs of different lengths.

As well, 3D graphene-CNT sandwich structures consisting CNT pillars grown in between the graphene layers has been prepared by CVD method 106 as: (1) GO and CNTs were prepared by modified Hummers' method and thermal reduction respectively and mixed together in the solution form with the mass ratio of 1:10, (2) subjected to ultrasonication, filtrated and desiccated, (3) CVD process in horizontal quartz tubular reactor at 750 °C for 1 h in Ar atmosphere with a flow rate of 300 sccm. The authors used the obtained graphene-CNT hybrid for the supercapacitors applications and achieved good performance with the maximum specific capacitance value of 385 F g^{-1}. Lee *et al.* reported plasma-enhanced CVD approach for the preparation of graphene-CNT [107]. The process includes:(1) thermal reduction of GO onto silicon wafer (2) deposition of nano-patterned Fe catalyst onto the GO film using self-assembled block-copolymer templates (3) growth of vertical CNTs by plasma-enhanced CVD growth at 600°C. During the CVD process, simultaneously the underlying GO sheets were thermally reduced.

4.6.1.3. Preparation by Self-assembly

Efforts were also made to prepare graphene-CNT hybrid film by self-assembly processes involving simple steps [108, 109]. Huang *et al.* [108] prepared graphene-CNT hybrid by self-assembly on a Ti substrate by simple casting method, the process involving: (1) preparation of GO and MWCNTs by modified Hummers' method and CVD process respectively, (2) mixing of GO and MWCNT and following ultrasonication to acquire GO-CNT hybrid, (3) self-assembly on a Ti sheet to get the hybrid as shown in Fig. (7).

Li and co-workers proposed vacuum-assisted self-assembly to prepare RGO-MWCNT hybrid sandwich from a dispersion of GO and MWCNT followed by thermal reduction at 200 °C [109]. Thus, graphene-CNT hybrid nanomaterials are special kind of hybrid materials with superior performance than the pristine CNT or graphene materials and new ideas in the preparation and processing of graphene-CNT hybrid materials are yet to be explored.

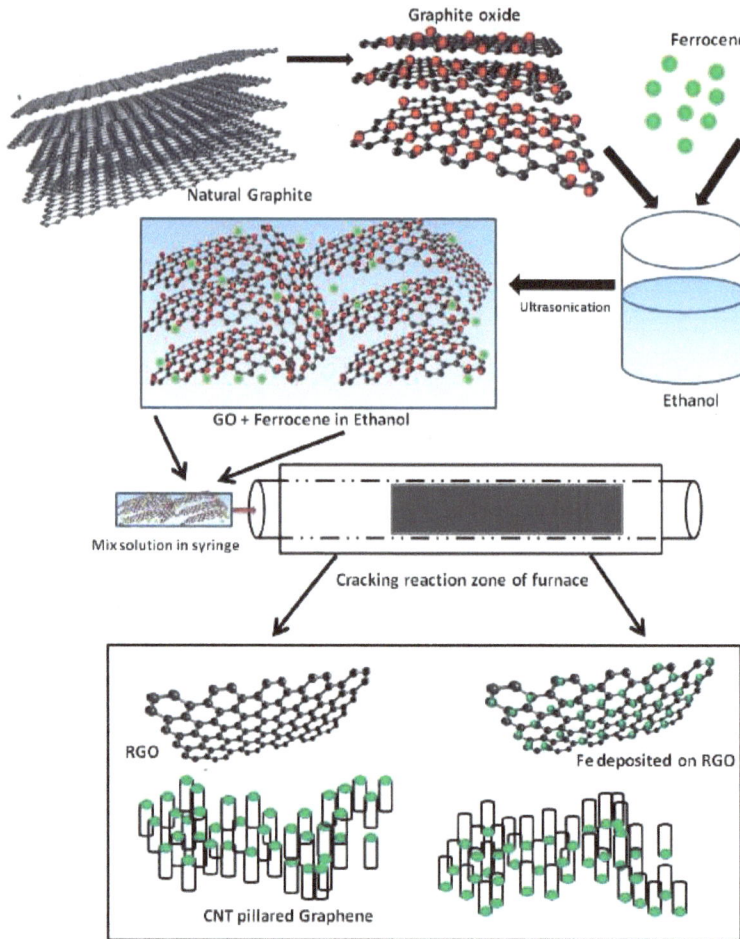

Fig. (7). Mechanism of pillaring of CNTs on GO.

4.7. FABRICATION/PROCESSING OF CARBON NANOSTRUCTURES-BASED POLYMER COMPOSITES

In general, three methods are adopted to prepare carbon nanostructures-based polymer composites:

➢ Solution processing or solvent casting

➢ Melt mixing

➢ In-situ polymerization

4.7.1. Solution Processing or Solvent Casting

Solvent casting is a widely used technique for the preparation of polymer nanocomposites using carbon fillers such as graphene and CNTs *etc*. In solution blending, filler dispersion in the polymers is carried out in a suitable solvent medium. Rigorous mixing of filler with polymer in a solvent significantly enhances de-aggregation and dispersion. Mainly three steps are involved in this mixing technique: (1) dispersion of filler in a suitable solvent (2) mixing with the polymer (at room temperature or elevated temperature) and (3) recovery of the nanocomposite by precipitating or casting a film. For this technique suitable solvent is required in which polymer can dissolve and filler can be well dispersed. To enhance the dispersion of filler, magnetic stirring, shear mixing, reflux or most commonly, ultrasonication is used. Sonication can be provided in two forms, mild sonication in a bath or high-power sonication. The degree of dispersion of graphene nanosheets by solution mixing generally depends on the level of exfoliation. Suitable surfactants may also be used to facilitate the dispersion of fillers. Polycarbonate, poly (methyl methacrylate), polyvinylidene fluoride, low density polyethylene, silicone rubber, polyurethane, ethylene vinyl acetate, polystyrene, and polyimides are the examples of matrices used in the preparation of graphene nanocomposites by this technique. However, the main limitations of this approach are the large amounts of solvent and the high temperatures required to dissolve the polymer during the process, in addition to concerns regarding environmental issues.

4.7.2. Melt Mixing

Relative to solution mixing, melt mixing is often considered more economical (because no solvent is used) and is more compatible with many current industrial practices. In melt mixing, a polymer melt and filler (in a dried powder form) are mixed under high shear conditions.; however, studies suggest that, to date, such methods do not provide the same level of dispersion of the filler as solvent mixing or *in situ* polymerization methods [110]. Several studies report melt mixing using TRGO [111] and graphene nanoplatelets (GNPs) [112-114] as filler, where these materials could be fed directly into an extruder and dispersed into a polymer matrix without the use of any solvents or surfactants. The very low bulk density of RGO makes handling of the dry powders difficult and poses a processing challenge (such as for feeding into processing equipment such as a melt extruder) [88] and in one study a solution mixing process was used to disperse the RGO in the polymer prior to compounding in order to circumvent this issue [115]. In a different approach to 'premix' the polymer and filler prior to mixing, GNPs were sonicated in a non-solvent, such that polymer particles were uniformly coated with GNPs prior to melt mixing, which was reported to lower the electrical percolation threshold of a GNP/polypropylene composite [116]. Notably, for composites incorporating GO platelets as filler, melt processing and molding operations may cause substantial reduction of the platelets due to their thermal instability [117].

4.7.3. *In-situ* Polymerization

In-situ polymerization methods for production of polymer composites generally involve mixing of filler in neat monomer, or a solution of monomer, followed by polymerization in the presence of the dispersed filler. These efforts are often followed with precipitation/extraction or solution casting to generate samples for testing. Many reports using *in-situ* polymerization methods have produced composites with covalent linkages between the matrix and filler. Unlike what has been reported for solution mixing methods, a high level of dispersion of graphene-based filler has been achieved *via in-situ* polymerization without a prior exfoliation step. However, *in-situ* polymerization has also been used to produce non-covalent composites of a variety of polymers, such as polyethylene [118], poly(methyl methacrylate) [119] and polypyrrole [120].

4.8. PROPERTIES OF GRAPHENE/GRAPHENE HYBRIDS-BASED POLYMER NANOCOMPOSITES

Recently, interest in carbon-based nanomaterials has revolutionized the field of polymer composites due to their ability to impart conductivity to the insulating host polymer matrix, along with providing excellent thermal and structural properties. Graphene is potential nanofiller that can dramatically improve the properties of polymer-based composites at a very low filler content. In this section, we present the mechanical, thermal, electrical and EMI shielding properties of graphene/graphene hybrid-based polymer composites.

4.8.1. Electrical Conductivity and EMI Shielding Properties CNT Based PU Composites

Ji *et al.* fabricated a flexible and flame retarding CNT based thermoplastic polyurethane composites *via* co-extrusion technology which exhibited maximum shielding effectiveness of -38.5 dB with less than 4 wt. % CNT. Jiang *et al.* utilized the technique of supercritical CO_2 foaming to fabricate lightweight and flexible thermoplastic polyurethane/reduced graphene oxide composite foams for electromagnetic interference shielding achieving shielding effectiveness of 21.8 dB with only 3.17 vol. % RGO loading owing to the multistage cellular structure with good conductive network. An inexpensive light weight radiation shielding material composed of conductive black (Ketjen-600JD) loaded polyurethane composite foam showing 65.6 dB at only 2 wt. % carbon black has been reported by Ghosh *et al.* The influence of carbon nanotubes length on the EMI shielding properties of PU composite foam has been investigated by Sang *et al.* These composite foams exhibited a density of 0.30 g/cm^3 and excellent specific EMI shielding efficiency of 102.7 dB/(g/cm^3) over the X-band. Gupta *et al.* reported a maximum EMI shielding effectiveness value of -41.6 dB at 10 wt. % loading of MWCNT, indicating the usefulness of this material for EMI shielding in the X-band [123]. In another study, Gupta *et al.* prepared acid modified multiwalled carbon nanotubes (a-MWCNT) reinforced polyurethane (PU) composite films using a solvent casting technique which showed a value of ~29 dB for the 10 wt. % MWCNT loaded sample having a thickness of 1.5 mm [124].

4.8.2. Graphene Based PU Composites

Due to its unique features such as remarkable structural flexibility, high electrical conductivity, thermal stability, and excellent mechanical properties, graphene has acquired tremendous consideration as a prospective conductive filler [121]. Graphene and graphene-like nanomaterials are extremely intriguing candidates for a wide range of possible applications in a variety of technological fields such as polymer nanocomposites [122], supercapacitors [123-125], nanoelectronics [126], energy storage devices [127-129], batteries [130-134] and sensors [135]. Due to the above mentioned excellent properties combined with its very low cost, graphene filled polymer composites are potential alternatives for electromagnetic shielding, antistatic, corrosion resistant coating and other applications that demand mechanical and functional attributes such as stiffness and barrier properties [136]. To obtain high quality graphene sheets, approaches such as micromechanical cleavage [60, 137, 138] and ultrasonication assisted exfoliations of graphite [139, 140], epitaxial growth [141-143], chemical vapor deposition [144, 145] and solution based chemical or thermal reduction of graphene oxide [64, 65, 86, 146, 147] have been widely explored. Among these, the preparation of graphene oxide (GO) followed by thermal or chemical reduction has been considered as one of the most economical and efficient ways for bulk scale production of graphene from natural graphite. Graphene's excellent electrical conductivity and large surface area make it a potential material for electromagnetic shields that absorb incident electromagnetic waves. Due to the fast expansion of the electronic industry, graphene and graphene-based composites have garnered great attention in recent years in the effort to create effective microwave absorbers and electromagnetic shields [10, 148-150]. Because of its high dielectric loss and low density, graphene is extremely attractive as an electromagnetic wave absorber at high frequencies in the gigahertz region [151].

Thermoplastic polyurethanes (TPU) have received a lot of attention among thermoplastic polymers because of its combination of characteristic properties such as flexibility, stretchability, transparency, good abrasion and chemical resistance, good wear and weather resistance, and good mechanical properties [152]. Polyurethanes are commonly utilized in flexible displays [153], smart clothing [154], electronic textiles, durable elastomeric wheels and tires (such as roller coaster, escalator and skateboard wheels) [155], high performance adhesives, hard plastic parts (*e.g.*, for electronic instruments) and surface coatings and sealants [156]. The application range of TPU may be further expanded by incorporating carbon-based fillers into the material, which not only improves the mechanical properties but also develops conductivity in non-conducting matrix for additional applications such as electrostatic dissipation and EMI shielding materials [157-159].

Recently, much effort has been taken towards achieving uniform dispersion of graphene sheets in polyurethane matrix to realize superior mechanical, physical, thermal and electrical characteristics. Several approaches, such as melt blending, solution compounding and *in situ* polymerization have been applied to prepare graphene/polyurethane nanocomposites. Various properties of these nanocomposites such as mechanical, thermal, electrical, *etc.* have been widely investigated. Kim *et al.* [110] investigated the effects of different processes and dispersion techniques of exfoliated graphite on the gas barrier and electrical properties of TPU/graphene nanocomposites. Nguyen *et al.* [160] fabricated TPU/functionalized graphene sheet (FGS) nanocomposites using *in-situ* intercalative polymerization followed by a casting film process. They concluded that FGS has a strong affinity for TPU and it was an effective and convenient novel material for the modification of TPU. Quan *et al.* [136] prepared graphite nanoparticles filled polyurethane nanocomposites *via* a solution blending technique. It was further observed that graphite nanoparticles can act as intumescent flame retardant and significantly reduced the heat release rate, thus improving the flame retardancy of TPU matrix. Gupta *et al.* [161] investigated the mechanical properties of reduced graphene oxide (RGO) reinforced polyurethane composites and a significant increase in the hardness and elastic modulus of these composites was observed suggesting their usefulness in structural applications such as hard and scratch-less coatings. Gao *et al.* [162] fabricated graphite nanoplatelet anchored polyurethane nanofiber composite using ultrasonication induced uniform decoration technique and achieved enhanced thermal stability and hardness along with improved electrical conductivity. Yousefi *et al.* [163] used water based colloidal dispersions of GO and polyurethane latex to produce graphene reinforced polymer matrix nanocomposites with a high degree of orientation (self-alignment) of graphene sheets in the polymer matrix. Li *et al.* [164] reported titanate functionalized graphene reinforced water borne polyurethane coatings for superior anticorrosion properties. Bian *et al.* [159] prepared microwave exfoliated graphite oxide (MEGO) reinforced thermoplastic polyurethane nanocomposite *via* melt blending and established structure property relationships. They concluded that MEGO is an effective and convenient new material that can be used for the modification of TPU in terms of thermal stability, mechanical properties, electrical conductivity *etc.* and could also be used in place of other nanosized conductive fillers, such as carbon nanotubes, which cost more. Liao *et al.* [165] employed co-solvent blending method to synthesize aqueous reduced graphene/thermoplastic polyurethane nanocomposites and reported enhanced elastic modulus and low percolation of these nanocomposites as compared to the ones prepared by conventional solvent blending.

Graphene dramatically enhances the properties of polymer-based composites at a very low loading and its most intriguing attribute is the extremely high surface conductivity leading to the formation of numerous electrically conductive polymer composites. When incorporated as fillers with insulating polymer matrix, graphene greatly enhances the electrical conductivity of the host polymer matrix. The electrical conductivity can be tuned by several orders of magnitude on addition of very small amount of graphene, preserving other performance properties of polymers such as thermal and mechanical properties. The filled composite materials exhibit a non- linear increase of the electrical conductivity as a function of the filler concentration. The fillers are able to form a conducting network leading to a sudden rise of the electrical conductivity of the composite at a certain loading fraction, known as percolation threshold [166].

Various factors influence the electrical conductivity and the percolation threshold of the composites such as processing methods, concentration of filler, aggregation of filler, the presence of functional groups and aspect ratio of graphene sheets, distribution in the matrix, wrinkles and folds *etc.* [166]. Moreover, it is reported that TRGO has higher electrical conductivity than chemically reduced graphene oxide (CRGO) due to the absence of oxygenated functional groups. Kim *et al.* [110] has studied the effect of thermal and chemical reduction of GO on electrical properties of graphene/PU composites. The lower percolation threshold of < 0.5 vol.% was reported for TRGO while > 2.7 vol.% percolation threshold was reported for graphite. However, CRGO and GO did not show decrease in surface resistance due to loss of electrical conductivity after graphite oxidation. Nanocomposites based on FGS and WPU prepared by *in-situ* method exhibited an increase of 10^5 fold compared to pristine WPU due to the homogeneous dispersion of FGS in the insulating WPU matrix [167]. An abrupt change in the electrical conductivity caused by the formation of a conducting channel throughout the WPU polymer matrix was observed and the percolation threshold was obtained at a FGS loading of only 2 wt. %. Interestingly, recent work by Gao and coworkers on the preparation of GNP decorated PU nanofiber mat showed that the GNPs are designed to be located on the fiber surface and exhibited a good electrical property [162]. It is observed that electro spun nanofibers became conductive after GNP decoration and the electrical conductivity of the mat reaches up to 3.8×10^{-1} S/m as compared to the pure PU nanofiber mat (conductivity $\sim 10^{-12}$ S/m). According to Nguyen *et al.* [160], FGS can be finely dispersed in TPU matrix simply by solution mixing, without any chemical surface modification of FGS. An electrical conductivity of 10^{-4} S/cm was achieved in the nanocomposite containing only 2 parts of FGS per 100 parts of TPU. Ding *et al.* [168] reported the hydrothermal method for the conversion of GO to graphene in water and prepared WPU nanocomposites with poly (vinyl pyrrolidone) stabilized graphene solution. They observed that the electrical conductivity of WPU can be greatly improved with the incorporation of graphene, and when the filling amount of graphene is 4.0 wt. %, the conductivity of the composite reaches 8.30×10^{-4} S/cm. Hence, the use of graphene as a conductive filler to polymers open up new avenues in the development of conducting nanocomposites that can be widely applied in anti-static materials, electromagnetic interference (EMI) shielding, chemical sensor, bipolar plates for fuel cells, radio-frequency interference shielding for electronic devices and electrostatic dissipation and conductive coatings.

Polymer nanocomposites based on conductive nanofillers are promising advanced materials for protection from electromagnetic interference (EMI) in cell phones, laptops, aircraft, electronics, military and medical devices because of their low density, design flexibility, ease of processing and high conductivity at low filler loading. Apart from imparting electrical conductivity to the insulating host polymer matrix, conducting fillers also improve the thermal conductivity desired for quick dissipation of heat generated due to the interaction of incident microwave radiation with shield material. The EMI shielding capabilities of polymeric materials filled with high aspect ratio conductive nanofillers, such as SWCNTs [169, 170], MWCNTs [157, 171-173], carbon nanofibers [174-176] and graphene or RGO [149, 177] have been recently investigated. Very few studies have been reported on graphene based polyurethane composites for EMI shielding in X and Ku band. Nanni *et al.* reported an average value of -20 dB of shielding effectiveness in the X band for TPU with 20 wt. % exfoliated graphite and a sample thickness of 4 mm [178]. Yang *et al.* investigated the EMI shielding effectiveness of polydopamine coated graphene nanosheets embedded in polyurethane matrix and obtained -17.6 dB at 1.2 GHz in the frequency range 30MHz-1.8GHz with 4.75 vol. percent loading [179]. Hsiao *et al.* adopted the L-b-L assembly approach to fabricate electro spun water based polyurethane with sulfonate groups/HI reduced graphene to achieve an EMI shielding efficiency of -34dB in the X-band for a 1 mm thick sample [180]. In another course of investigation, Hsiao and his co-workers synthesized water based polyurethane composites utilizing non- covalently modified exfoliated graphene nanosheets and observed a maximum shielding effectiveness of - 32dB at 7.7 wt. % filler content in the frequency range of 8.2-12.4 GHz [181]. Composites containing covalently modified graphene nanosheets dispersed in a water-based polyurethane matrix with grafted sulfonate functional groups demonstrated EMI shielding effectiveness of-38 dB over the frequency of 8.2 to 12.4 GHz [182]. From the above studies on EMI shielding performance of graphene filled polyurethane composites, it was concluded that either very high graphene content or graphene extensively modified by using complicated reactions is required for achieving high shielding effectiveness. Thus, above two factors serve to restrict utilization and commercialization of such composites. So, the present work aims to develop thermoplastic polyurethane nanocomposites for EMI shielding in the 8.2-18 GHz region by using thermally reduced graphene oxide prepared by simple and scalable oxidation and reduction of natural graphite.

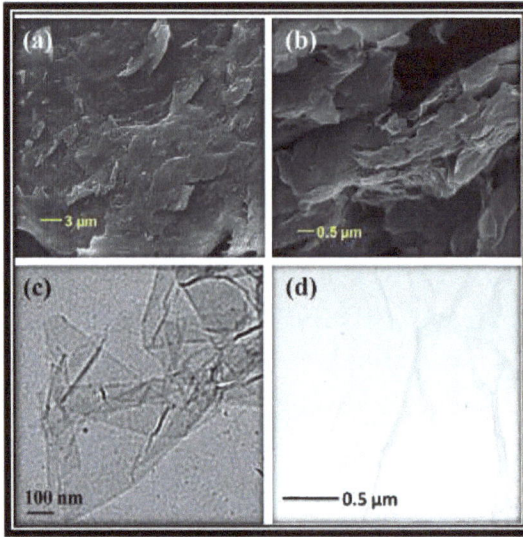

Fig. (8). SEM and TEM micrographs of GO (**a** and **c**) and RGO (**b** and **d**) (Reproduced from Ref. [123] with permission from The Royal Society of Chemistry).

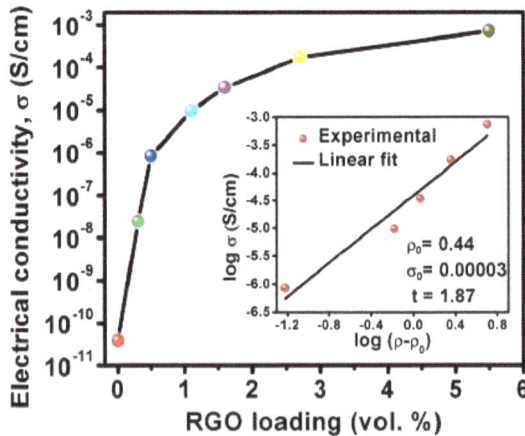

Fig. (9). Variation in electrical conductivity with RGO loading in the TPU matrix. The inset shows the log(s) *vs.* log (r -r₀) plot. (Reproduced from Ref. [123] with permission from The Royal Society of Chemistry).

In this context, Verma *et al.* fabricated reduced thermally reduced graphene oxide based thermoplastic polyurethane nanocomposites using solvent casting method and evaluated their potential as capable EMI shields. RGO appeared as a large, ultrathin and transparent silk curtain wave like structure with a large number of wrinkles on the surface as seen from TEM image. SEM image of GO and after reduction to RGO (Fig. **8a** and **8b**) shows the randomly aggregated, crumped texture of few layered graphene sheets associated with each other, whereas TEM images of GO & RGO (Fig. **8c** and **d**) clearly shows transparent film microstructure and ultrathin silk curtain wave like structure with a large number of wrinkles on the surface. The electrical conductivity of nanocomposite films was measured and a value in the order of 7.3×10^{-4} S cm^{-1} was observed for 5.5 vol. % RGO/PU composite as compared to neat TPU (3.9×10^{-11} S cm^{-1}) as presented in Fig. (**9**). It was observed that electrical conductivity of TPU increased significantly on increasing the graphene content. Even the percolation threshold was evaluated by plotting electrical conductivity as a function of RGO loading and evaluating data fitting using scaling law. On plotting, log σ *vs.* log (ρ – ρ$_o$) as shown in the inset of Fig. (**9**), a straight line is observed which shows excellent fit with the data. An EMI shielding effectiveness of ~-21 dB in the X-band for 3 mm thickness was achieved at 5.5 vol. % graphene loading as shown in Fig. (**10a**). EMI shielding behaviour of composites was analyzed by plotting total shielding effectiveness (SET), shielding effectiveness due to absorption (SEA) and shielding effectiveness due to reflection (SER) as a function of RGO loading as shown in Fig. (**10b**). It was also observed that the designed PUG composite 5.5 shows an absorption efficiency 0f 98 % (Fig. **10c**) indicating that the electromagnetic energy incident on the composite shield attenuates and dissipates the EM energy in the form of heat energy. To elaborate more on the shielding capability of the PUG composites, electromagnetic attributes, such as complex permittivity of the composites were also discussed. The results reveal that when the RGO loading level increases, the real part of permittivity increases from 3.3 to 12.6 and the imaginary part of permittivity increases from 0.1 to 8.6 for the PUG composites as shown in Fig. (**10d** and **e**). It is proposed that the increase in the dielectric properties is a direct consequence of the increased electrical conductivity and electrical polarization in the PUG composites because the relative complex permittivity, a measure of a material's polarizability, induces dipolar and electric polarization during EM wave activation. The tan δ value of 0.65 – 0.7 was observed for the PUG 10 composite as shown in Fig. (**10f**). To elaborate more on the shielding capability of the PUG composites, electromagnetic attributes, such as the complex permittivity of the composites were also discussed. The results reveal that when the RGO loading level increases, the real part of permittivity increases from 3.3 to 12.6 and the imaginary part of permittivity increases from 0.1 to 8.6 for the PUG composites. It is proposed that the increase in dielectric properties is a direct consequence of the increased electrical conductivity and electric polarization in PUG composites because the relative complex permittivity, a measure of a material's polarizability, induces dipolar and electric polarization during EM wave activation. The tan d values of PUG composites increased with increasing RGO content and a tan d value of 0.65–0.7 was observed for the PUG 10 composite.

In order to effectively harness the potential of graphene as superior functional filler, individual single layer graphene sheet from their aggregates need to be assembled within the polymer matrix so that the properties of composites could be tailored as per the desired application. Many attempts have recently been made to employ carbon nanotubes (CNT) as spacers between graphene sheets, which not only increases the interlayer spacing but also bridges the defects for electron transfer, resulting in a graphene-CNT hybrid material (GCNT) that combines the synergistic effect of one dimensional CNT and two dimensional graphene [106].

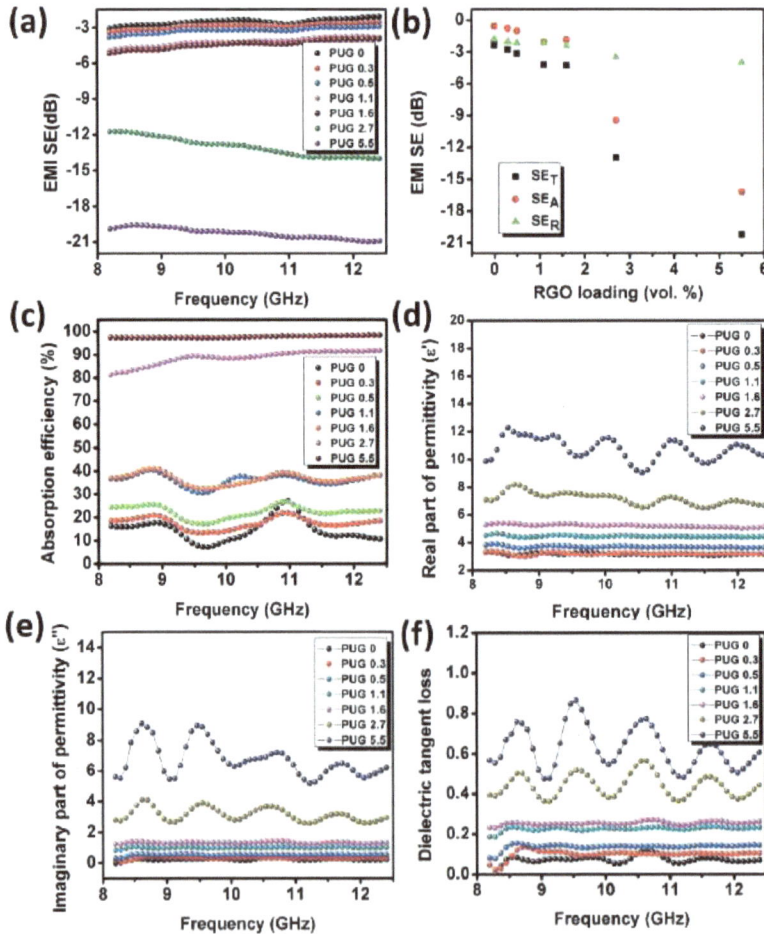

Fig. (10). (a) Variation in the EMI shielding effectiveness with frequency for PUG nanocomposites, **(b)** variation in SE$_T$, SE$_A$ and SE$_R$ with RGO loading at 10.5 GHz, **(c)** variation in absorption efficiency, and the frequency dependence of **(d)** real and **(e)** imaginary parts of the permittivity and **(f)** dielectric loss tangent for PUG nanocomposites. (Reproduced from Ref. [123] with permission from The Royal Society of Chemistry).

Recently, GCNT hybrid nanostructures have been investigated as anode materials for lithium-ion batteries [183], hydrogen storage materials [184], supercapacitors [106, 185] and reinforcements in polymer composites [186-192]. Zhang *et al.* [193] reported a novel, water dispersible, three-dimensional GCNT hybrid prepared by direct reduction of GO sheets in the presence of acid-treated CNTs. They reported high performance poly (vinyl alcohol)/GCNT nanocomposites with enhanced structural properties and thermal degradation temperature, implying a potential flame retardant property of GCNT hybrid. Lee *et al.* [194] explored the utilization of electrostatic force-driven randomly stacked three-dimensional hybrid electrocatalysts of multiwalled carbon nanotubes (MWCNT) hybridized with reduced graphene oxide (RGO) to boost oxygen utility during oxygen reduction processes. Kim *et al.* [195] explored the role of GCNT hybrid filler prepared by thermal chemical vapor deposition approach on the dielectric performance of cyanoethyl pullulan polymer, obtaining a dielectric constant of 32 with a dielectric loss of 0.051 at 100 Hz for a 0.062 wt. % loaded GCNT sample. Vinayan *et al.* [196] reported the fabrication of a hybrid nanomaterial comprised of poly(diallyl dimethyl ammonium chloride) modified solar exfoliated graphene and negatively surface charged carbon nanotubes for application as an anode in lithium ion batteries. Kamalia *et al.* [197] reported a novel strategy for large-scale production of 3D networks of CNT pillared graphene nanostructures (GCNT) utilizing chemical vapor deposition technique for bio-functional optical properties, suggesting that GCNT may be used in optoelectronics, biomedical, and ultrafast optical sensing. Ding *et al.* [20] investigated the electromagnetic wave absorbing characteristics (frequency range 8.2-12.4 GHz) of a novel absorber based on poly(vinyl pyrrolidone)@GCNT synthesized by ultrasonication infiltration, achieving a maximum reflection loss of -26.5 dB at 11.29 GHz.

In the present work, we illustrate the effect of hybridization of graphene nanoplatelets and CNTs on the enhancement of attenuation of electromagnetic radiations for polyurethane based composite materials. The mechanism involving the formation of GCNT hybrid has been proposed. Composites of GCNT (at varying loading) and thermoplastic polyurethane were prepared and explored for their potential as an effective and light weight EMI shielding material in the frequency range of 12.4 -18GHz (Ku band). The mechanism of shielding was thoroughly assessed by evaluating the contribution of reflection and absorption to the total shielding. In order to demonstrate the hybridization of GO sheets with functionalized carbon nanotubes, scanning electron microscopy (SEM) and transmission electron microscopy (TEM), high- resolution transmission electron microscopy (HRTEM) of the graphene oxide sheets, functionalized carbon nanotubes and GO-CNT hybrid sheets was carried out as shown in Fig. (**11a** to **i**). It can be seen from the SEM micrograph of GO (Fig. **11a**) that it has flaky structure overlapped with each other whereas carbon nanotubes micrograph has long, fibrous and stringy nature as shown in Fig. (**11b**). However, SEM micrograph of graphene-carbon nanotube hybrid (Fig. **11c**) reveals the formation of hybrid nanostructure with tubular networks of CNTs adsorbed on the graphene sheets. The TEM and HRTEM images of GO, CNTs and GCNT hybrid (Fig. **11d** to **i**) reveals that both the laminated structure of GO and tubular structure of CNTs are retained in the GCNT hybrid and in the hybrid, thin wrinkled sheet structure of graphene nanoplatelets are being covered with randomly arranged CNTs. Fig. (**12a**) shows the shielding effectiveness of PUGCNT nanocomposites with varying loadings of GCNT hybrid fillers in the frequency range 12.4 to 18 GHz. The results revealed that EMI shielding behavior of polyurethane sheet shows a shielding effectiveness of -3 dB. However, loading of GCNT hybrid in the PU matrix shows a dramatic increase in EMI SE and a value of SE of the order of -47 dB is obtained for the PUGCNT 10 nanocomposite. This increase in shielding attenuation is due to the formation of conducting network of hybrid composites in the PU matrix. Although total shielding effectiveness is an important parameter used to quantify the shielding efficiency of a shield, it does not provide information on contributions of shielding effectiveness due to absorption (SEA) and shielding effectiveness due to reflection (SER). Fig. (**12b**) shows the shielding effectiveness of the PUGCNT composite due to absorption, reflection and total SE as a function of GCNT loading.

It was observed that SEA increases more rapidly compared to SER which shows that more mobile charge carriers are available in the nanocomposite which can form conductive network that can attenuate the penetrating electromagnetic wave. Moreover, this behavior suggests that the shielding effectiveness due to absorption (SE$_A$) is the main contributor for EMI shielding in the PUGCNT nanocomposite. It was also observed that EM waves are absorbed and transformed into thermal energy by the continuous GCNT network. The designed PUGCNT 10 nanocomposite exhibited an absorption efficiency of more than 99.9 % as shown in Fig. (**12c**) which indicates that most of the electromagnetic energy incident on the shield attenuates and dissipates in the form of heat energy. The electromagnetic attributes (complex permittivity) were also evaluated to further understand the reasons behind the observed increase in shielding effectiveness. Fig. (**12d-f**) shows the real part of permittivity (ε'), imaginary part of permittivity (ε'') and tangent loss (tan δ) as a function of frequency for the PUGCNT nanocomposites at a thickness of 3 mm.

The results show that the real and imaginary part of permittivity of the nanocomposite with 10 wt. % loading of GCNT hybrid in PU matrix lies in the range of 35.1-41.4 and 14-21.1 respectively, while PU had a dielectric constant and dielectric loss of 3.2 and 0.5 respectively. The increase in real part of permittivity, imaginary part of permittivity with increasing GCNT loading may be attributed to the increase in electrical conductivity and space charge polarization of the nanocomposite. The ratio of imaginary to real part is dissipation factor which is represented by tan δ. Since the tan δ value indicates the absorptive property of the material *i.e.*, the ability of the material to convert applied energy into heat, materials with high tan δ values are used as microwave absorbing material. Fig. (**12f**) shows the plots of tan δ *vs* frequency of the PUGCNT nanocomposite as a function of GCNT loading. The tan δ value of 0.35 – 0.59 was observed for the PUGCNT 10 composite.

Fig. (**13**) depicts the electrical conductivity of PUGCNT nanocomposites plotted as a function of GCNT amount. The conductivity of cast polyurethane is 3.9×10^{-11} S/cm. However, the conductivity increases exponentially at low loading of GCNT followed by slow increase at higher loadings as shown in Fig. (**13**). The electrical conductivity of PU increased by 5 times at a loading of 0.5 % of GCNT. On increasing the loading of GCNT beyond 2 wt. % results in rather saturated conductivity. The maximum electrical conductivity of PUGCNT 10 nanocomposite was found to be 9.5×10^{-2} S/cm. The electromagnetic attributes (complex permittivity) were also evaluated to further understand the reasons behind the observed increase in shielding effectiveness.

Table **3** provides a comparison of EMI shielding performance of recently published results on CNT or graphene based polyurethane composites. It is clearly evident from the table that the SET value of - 47 dB of PUGCNT composites in the present work is the highest among the reported values of SET for polyurethane-based composites. The EMI shielding effectiveness for PURGOCNT (prepared by 1:1 physical mixture of RGO and CNT and loading level is 10 wt. %) was lower *i.e.*, - 32 dB as compared to - 47 dB for PUGCNT 10 which can be attributed to the synergistic effects originated as a result of hybridization of graphene nanoplatelets with FCNT. The proposed interactive mechanism of electromagnetic waves GCNT hybrid is shown in Fig. (**14**) through a schematic representation. From these observations, it is realized that one dimensional CNTs and two-dimensional graphene nanoplatelets were united to design a three-dimensional hybrid nanocomposite that led to increase in the EMI shielding behavior of PU nanocomposite.

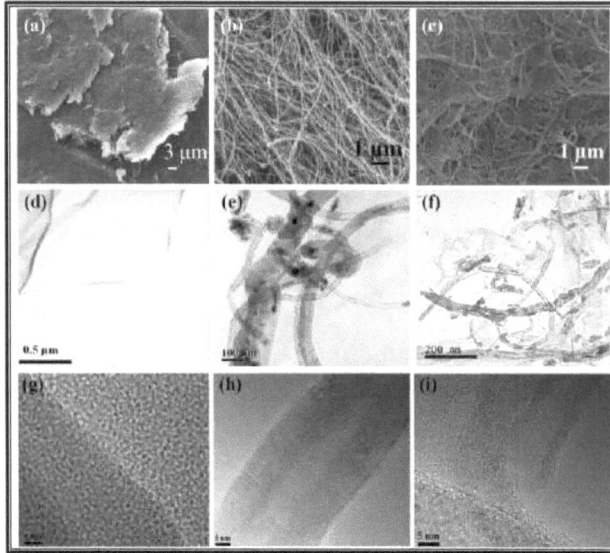

Fig. (11). SEM, TEM and HRTEM micrographs of (**a** and **d**) GO, (**b** and **e**) FCNT and (**c** and **f**) GCNT (Reproduced with the permission of ref. [124]).

Table 3. Electromagnetic attenuation analysis of graphene or carbon nanotubes based polyurethane nanocomposites.

Filler	Frequency (GHz)	Filler Concentration (wt. %)	Shielding Effectiveness (dB)	Reference
SWCNT	8.2-12.4	20	-17	Liu *et al.* [170]
SWCNT	2-18	5	22	Liu *et al.*[198]
MWCNT	8.2-12.4	10	-41.6	Gupta *et al.* [173]

(Table 3) cont.....

MWCNT	8.2-12.4	10	-21.8	Ramoa *et al.* [191]
MWCNT	13-16	9	-35	Jin *et al.* [199]
MWCNT	8.2-12.4	22	-20	Hoang *et al.* [172]
Acid functionalized MWCNT	8.2-12.4	10	-29	Gupta *et al.* [157]
Polydopamine coated graphene	1.2	4.75 vol.%	-17.6	Yang *et al.* [179]
Hydrogen iodide reduced graphene	8.2-12.4	-	34	Hsiao *et al.* [180]
Covalently modified graphene	8.2-12.4	-	38	Hsiao *et al.* [182]
Non covalently modified graphene	8.2-12.4	7.7	32	Hsiao *et al.* [181]
Reduced graphene oxide	8.2-12.4	5.5 vol%	-21	Verma *et al.* [200]
Graphene nanoplatelets/CNT hybrid	12.4-18	10	-47	Verma *et al.* [201]
RGO + CNT	12.4-18	10	-32	Verma *et al.* [201]

High value of SE_T is the consequence of the synergistic effect caused by the formation of an extended conjugation network with the FCNTs bridging the gaps between the graphene nanoplatelets and inhibiting the face-to-face aggregation of graphene nanoplatelets. Moreover, the lessening of the stacking effect and aggregation of graphene nanoplatelets improves the polymer contact area and interfacial interactions resulting in a synergistic enhancement in nanocomposite properties.

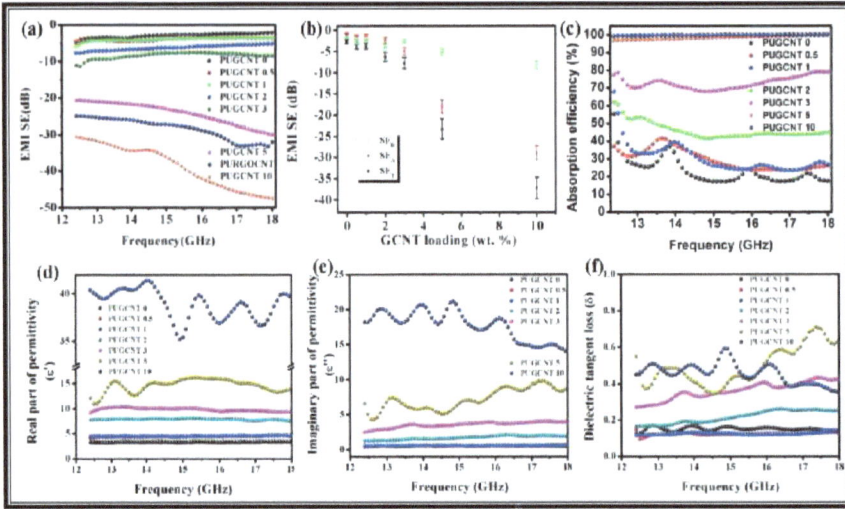

Fig. (12). (**a**) Variation in EMI shielding effectiveness with frequency for PUGCNT nanocomposites, (**b**) Variation in SE$_T$, SE$_A$ and SE$_R$ with GCNT loading at 15.2 GHz, (**c**) variation in absorption efficiency, frequency dependence of (**d**) dielectric constant and (**e**) dielectric loss and dielectric tangent loss for PUGCNT nanocomposites. (Reproduced with the permission of ref. [124]).

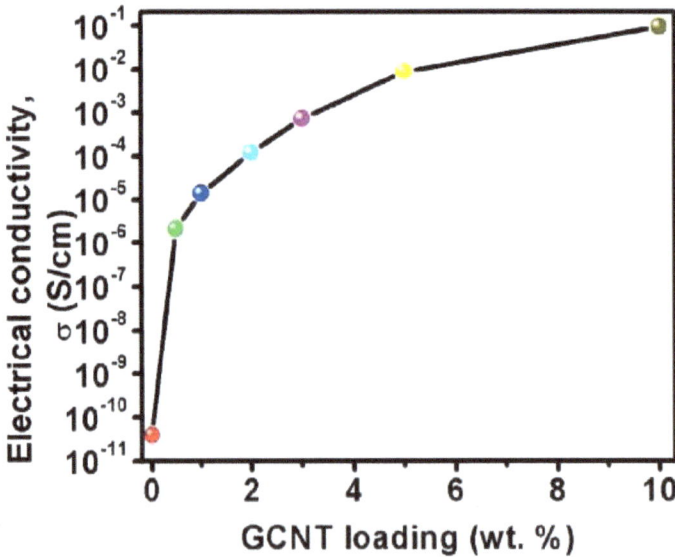

Fig. (13). Effect of GCNT content on the electrical conductivity of TPU. (Reproduced with the permission of ref. [124]).

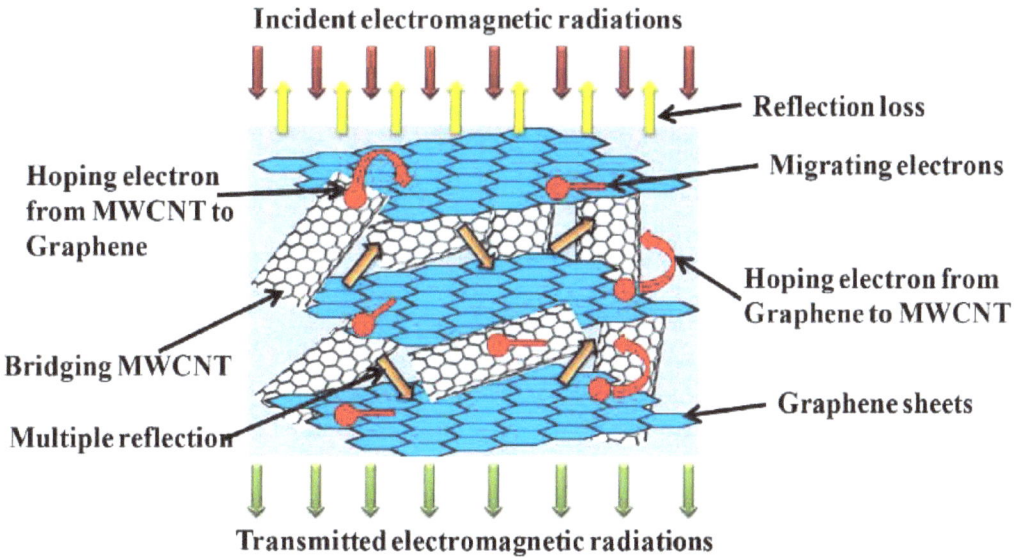

Fig. (14). Schematic representation of the proposed EMI shielding mechanism in PUGCNT nanocomposites (Reproduced with the permission of ref. [124]).

4.9. APPLICATIONS OF GRAPHENE/GRAPHENE HYBRIDS-BASED POLYMER NANOCOMPOSITES

Though numerous challenges exist in developing a fundamental understanding of graphene/graphene hybrids and their polymer composites, these materials have already been explored for a wide range of applications. With their unique combination of properties, graphene and graphene derivatives have opened up a new age of advanced multifunctional materials. With high conductivity and mechanical properties, graphene/polymer composites have possible uses as antistatic coatings [202-204], EMI shielding materials [148, 205-209], electrostatic discharge material [200, 210], *etc.* The increased thermal conductivity and stability and low coefficient of thermal expansion make graphene/polymer composites promising for applications as thermal interfacial materials, which are commonly used to fill air gaps between two surfaces of electronic components and the base plates of heat sinks, and thus minimize the thermal contact resistance between them and dissipate heat efficiently [211, 212]. The high stability and flame retardancy make them promising as flame retardant materials to decrease the heat release and delay time to ignition or resist the spread of fire [213, 214]. In addition, a small graphene loading in polymer matrices can also modify the dielectric [215], electrochemical [54, 216], and catalytical properties [217, 218] of graphene/polymer composites, which can be potentially used in flexible electronics [219, 220], supercapacitors [221, 222], batteries [223], biomedical devices [224] *etc.* The other commercial applications of graphene polymer composites are: lightweight gasoline tanks, plastic containers, more fuel efficient aircraft and car parts, stronger wind turbines, medical implants and sports equipment [225]. The discovery of graphene as nanofiller has opened a new dimension for the production of light weight, low cost, and high-performance composite materials for a range of applications.

SUMMARY, CONCLUSION & FUTURE SCOPE

Following the development of new electronic systems and communication networks, the levels of electromagnetic contamination have increased dramatically in the recent years. Every day, new studies appear searching for a way to mitigate the electromagnetic interferences (EMI). Moreover, the rapid evolution of technology forces the field to search for lighter and more efficient materials. The exploration of new functional materials that enable to effectively block or shield electromagnetic energy is an active field of research nowadays. Polymers offer several advantages over traditional metals and ceramics used for EMI shielding. They can be easily shaped; it is possible to prepare a variety of configurations and formulations and they are substantially lighter. Although polymers are electromagnetically transparent, different strategies are available to convert them into active electromagnetic shields. Graphite, carbon black and carbon fibers were the first to be combined with polymers for the fabrication of EMI shields. The attention soon shifted to nanocarbons, since with lower-weight fractions, more conductive composites could be obtained. Thereby, nanocarbon fillers are adequate to shield in the GHz range. In this context, carbon nanofibers, carbon nanotubes and graphene, which have higher specific surface area and aspect ratio than their microscale analogues, are promising candidates for the preparation of efficient EMI shielding composites.

The developed PUG nanocomposites exhibited sufficient electrical conductivity which is required to attain desirable EMI shielding values in microwave frequency range so that the composites can meet the commercial application demands. The EMI shielding values of -21 dB in the X-band (8-12.4 GHz) and - 38 dB in the Ku-band [12.4 - 18 GHz] achieved for 10 wt. % loading of RGO indicate that these polyurethane nanocomposites have great potential as an effective and light weight shielding materials for protection from electromagnetic radiations, in making electromagnetic shielding bags for packaging of electronic circuits and variety of applications. The electrical conductivity of PUGCNT composites increased with increasing amount of GCNT and it was higher as compared to PUG composites. The maximum electrical conductivity has been achieved up to 9.5×10^{-2} S/cm for 10 wt.% PUGCNT composite, which is 9 orders of magnitude higher than pure insulating polyurethane. The significant improvement in electrical conductivity is thus responsible to achieve an EMI shielding value of - 35dB in the X-band [8-12.4 GHz] and - 47 dB in Ku-band [12.4- 18 GHz] indicating the usefulness of these composites for techno-commercial application demands. The developed PUGCNT nanocomposites open up a new avenue to design promising futuristic lightweight EMI shielding materials.

These nanocomposites could be used to make a sandwich fabric using a cotton cloth which can be shaped into a desired product The developed nanocomposites can be considered as a potential candidate for the development of products like caps, maternity belts, *etc.* that can be implemented to the needs of the society.

The discovery of graphene as nanofiller has opened new horizons for the production of light weight, low cost and high-performance composite materials for a range of applications. It is obvious that graphene is a promising filler to improve the mechanical, electrical and thermal properties of polymers and such conducting graphene/polymer composites have highlighted their potential for making various sensors, conductive electrodes for solar cells, antistatic coatings, electromagnetic interference shielding, *etc.*

It can be concluded from the present studies that selective incorporation of conducting fillers within the polymer matrix enhanced the microwave shielding properties. However, to make a perfect microwave absorber, further modification of electromagnetic attributes is required so as to eliminate the input impedance mismatch and to realize zero reflection.

To take full advantage of excellent properties of graphene and CNTs, these should be aligned in a particular direction. A systematic study related to alignment of graphene and CNTs for achieving superior mechanical properties should be carried out.

CONSENT FOR PUBLICATION

Not applicable.

CONFLICT OF INTEREST

The authors declare no conflict of interest, financial or otherwise.

ACKNOWLEDGEMENTS

Declared none.

REFERENCE

[1] Tong, X.C. *Advanced Materials and Design for Electromagnetic Interference Shielding*; Taylor & Francis Group, LLC, **2009**.

[2] Bellucci, S.; Micciulla, F. *Brief Introduction to Nanocomposites for Electromagnetic Shielding*; Advanced Nanomaterials for Aerospace Applications, **2014**, p. 227.

[3] Hou, C.; Li, T.; Zhao, T.; Liu, H.; Liu, L.; Zhang, W. Electromagnetic wave absorbing properties of multi-wall carbon nanotube/Fe 3 O 4 hybrid materials. *N. Carbon Mater.*, **2013**, *28*(3), 184-190.
http://dx.doi.org/10.1016/S1872-5805(13)60075-X

[4] Wang, L.; Li, J.; Liu, Y. Preparation of electromagnetic shielding wood-metal composite by electroless nickel plating. *J. For. Res.*, **2006**, *17*(1), 53-56.
http://dx.doi.org/10.1007/s11676-006-0013-5

[5] Janda, N.B. *Development of a predictive shielding effectiveness model for carbon fiber/nylon based composites*; Michigan Technological University, **2004**.

[6] Huang, J.C. EMI shielding plastics: a review. *Adv. Polym. Technol.*, **1995**, *14*(2), 137-150.
http://dx.doi.org/10.1002/adv.1995.060140205

[7] Bigg, D.; Stutz, D. Plastic composites for electromagnetic interference shielding applications. *Polym. Compos.*, **1983**, *4*(1), 40-46.
http://dx.doi.org/10.1002/pc.750040107

[8] Bigg, D. Mechanical properties of particulate filled polymers. *Polym. Compos.*, **1987**, *8*(2), 115-122.
http://dx.doi.org/10.1002/pc.750080208

[9] Singh, A.P.; Mishra, M.; Hashim, D.P.; Narayanan, T.; Hahm, M.G.; Kumar, P.; Dwivedi, J.; Kedawat, G.; Gupta, A.; Singh, B.P.; Chandra, A.; Vajtai, R.; Dhawan, S.K.; Ajayan, P.M.; Gupta, B.K. Probing the engineered sandwich network of vertically aligned carbon nanotube–reduced graphene oxide composites for high performance electromagnetic interference shielding applications. *Carbon*, **2015**, *85*, 79-88.
http://dx.doi.org/10.1016/j.carbon.2014.12.065

[10] Verma, M.; Singh, A.P.; Sambyal, P.; Singh, B.P.; Dhawan, S.K.; Choudhary, V. Barium ferrite decorated reduced graphene oxide nanocomposite for effective electromagnetic interference shielding. *Phys. Chem. Chem. Phys.*, **2015**, *17*(3), 1610-1618.
http://dx.doi.org/10.1039/C4CP04284K PMID: 25437769

[11] Saini, P.; Choudhary, V.; Singh, B.P.; Mathur, R.B.; Dhawan, S.K. Polyaniline–MWCNT nanocomposites for microwave absorption and EMI shielding. *Mater. Chem. Phys.*, **2009**, *113*(2-3), 919-926.

http://dx.doi.org/10.1016/j.matchemphys.2008.08.065

[12]   Dubrovskiy, S.; Gareev, K. Measurement method for detecting magnetic and dielectric properties of composite materials at microwave frequencies **2015**.

http://dx.doi.org/10.1109/EIConRusNW.2015.7102223

[13]   Chung, D. Flexible graphite for gasketing, adsorption, electromagnetic interference shielding, vibration damping, electrochemical applications, and stress sensing. *J. Mater. Eng. Perform.,* **2000**, *9*(2), 161-163.

http://dx.doi.org/10.1361/105994900770346105

[14]   Chung, D. Materials for electromagnetic interference shielding. *J. Mater. Eng. Perform.,* **2000**, *9*(3), 350-354.

http://dx.doi.org/10.1361/105994900770346042

[15]   Singh, B.; Saini, K.; Choudhary, V.; Teotia, S.; Pande, S.; Saini, P.; Mathur, R. Effect of length of carbon nanotubes on electromagnetic interference shielding and mechanical properties of their reinforced epoxy composites. *J. Nanopart. Res.,* **2014**, *16*(1), 1-11.

http://dx.doi.org/10.1007/s11051-013-2161-9

[16]   Zhou, H.; Wang, J.; Zhuang, J.; Liu, Q. Synthesis and electromagnetic interference shielding effectiveness of ordered mesoporous carbon filled poly (methyl methacrylate) composite films. *RSC Advances,* **2013**, *3*(45), 23715-23721.

http://dx.doi.org/10.1039/c3ra44267e

[17]   Pande, S.; Chaudhary, A.; Patel, D.; Singh, B.P.; Mathur, R.B. Mechanical and electrical properties of multiwall carbon nanotube/polycarbonate composites for electrostatic discharge and electromagnetic interference shielding applications. *RSC Advances,* **2014**, *4*(27), 13839-13849.

http://dx.doi.org/10.1039/c3ra47387b

[18]   Che, R.C.; Peng, L.M.; Duan, X.F.; Chen, Q.; Liang, X.L. Microwave Absorption Enhancement and Complex Permittivity and Permeability of Fe Encapsulated within Carbon Nanotubes. *Adv. Mater.,* **2004**, *16*(5), 401-405.

http://dx.doi.org/10.1002/adma.200306460

[19]   Das, N.; Khastgir, D.; Chaki, T.; Chakraborty, A. Electromagnetic interference shielding effectiveness of carbon black and carbon fibre filled EVA and NR based composites. *Compos., Part A Appl. Sci. Manuf.,* **2000**, *31*(10), 1069-1081.

http://dx.doi.org/10.1016/S1359-835X(00)00064-6

[20]   Ding, L.; Zhang, A.; Lu, H.; Zhang, Y.; Zheng, Y. Enhanced microwave absorbing properties of PVP@multi-walled carbon nanotubes/graphene three-dimensional hybrids. *RSC Advances,* **2015**, *5*(102), 83953-83959.

http://dx.doi.org/10.1039/C5RA14494A

[21]   Joshi, A.; Bajaj, A.; Singh, R.; Alegaonkar, P.S.; Balasubramanian, K.; Datar, S. Graphene nanoribbon-PVA composite as EMI shielding material in the X band. *Nanotechnology,* **2013**, *24*(45), 455705.

http://dx.doi.org/10.1088/0957-4484/24/45/455705 PMID: 24140728

[22]   Huang, J.C. Carbon black filled conducting polymers and polymer blends. *Adv. Polym. Technol.,* **2002**, *21*(4), 299-313.

http://dx.doi.org/10.1002/adv.10025

[23]   Mittal, V. *Polymer Nanotubes Nanocomposites: Synthesis, Properties and Applications*; John Wiley & Sons, **2014**.

http://dx.doi.org/10.1002/9781118945964

[24]   Thostenson, E.T.; Li, C.; Chou, T-W. Nanocomposites in context. *Compos. Sci. Technol.,* **2005**, *65*(3-4), 491-516.

http://dx.doi.org/10.1016/j.compscitech.2004.11.003

[25] Luo, X.; Chung, D. Electromagnetic interference shielding using continuous carbon-fiber carbon-matrix and polymer-matrix composites. *Compos., Part B Eng.,* **1999**, *30*(3), 227-231.

http://dx.doi.org/10.1016/S1359-8368(98)00065-1

[26] Colbert, D.T. Single-wall nanotubes: a new option for conductive plastics and engineering polymers. *Plast. Addit. Compd.,* **2003**, *5*, 8-25.

[27] Pang, H.; Chen, T.; Zhang, G.; Zeng, B.; Li, Z-M. An electrically conducting polymer/graphene composite with a very low percolation threshold. *Mater. Lett.,* **2010**, *64*(20), 2226-2229.

http://dx.doi.org/10.1016/j.matlet.2010.07.001

[28] Sandler, J.; Kirk, J.; Kinloch, I.; Shaffer, M.; Windle, A. Ultra-low electrical percolation threshold in carbon-nanotube-epoxy composites. *Polymer (Guildf.),* **2003**, *44*(19), 5893-5899.

http://dx.doi.org/10.1016/S0032-3861(03)00539-1

[29] Zhang, Q.; Rastogi, S.; Chen, D.; Lippits, D.; Lemstra, P.J. Low percolation threshold in single-walled carbon nanotube/high density polyethylene composites prepared by melt processing technique. *Carbon,* **2006**, *44*(4), 778-785.

http://dx.doi.org/10.1016/j.carbon.2005.09.039

[30] Liu, H.; Gao, J.; Huang, W.; Dai, K.; Zheng, G.; Liu, C.; Shen, C.; Yan, X.; Guo, J.; Guo, Z. Electrically conductive strain sensing polyurethane nanocomposites with synergistic carbon nanotubes and graphene bifillers. *Nanoscale,* **2016**, *8*(26), 12977-12989.

http://dx.doi.org/10.1039/C6NR02216B PMID: 27304516

[31] Ajayan, P.M.; Tour, J.M. Materials science: nanotube composites. *Nature,* **2007**, *447*(7148), 1066-1068.

http://dx.doi.org/10.1038/4471066a PMID: 17597753

[32] Yang, Y.; Gupta, M.C.; Dudley, K.L.; Lawrence, R.W. Novel carbon nanotube-polystyrene foam composites for electromagnetic interference shielding. *Nano Lett.,* **2005**, *5*(11), 2131-2134.

http://dx.doi.org/10.1021/nl051375r PMID: 16277439

[33] Huang, Y.; Li, N.; Ma, Y.; Du, F.; Li, F.; He, X.; Lin, X.; Gao, H.; Chen, Y. The influence of single-walled carbon nanotube structure on the electromagnetic interference shielding efficiency of its epoxy composites. *Carbon,* **2007**, *45*(8), 1614-1621.

http://dx.doi.org/10.1016/j.carbon.2007.04.016

[34] Yuan, B.; Yu, L.; Sheng, L.; An, K.; Zhao, X. Comparison of electromagnetic interference shielding properties between single-wall carbon nanotube and graphene sheet/polyaniline composites. *J. Phys. D Appl. Phys.,* **2012**, *45*(23), 235108.

http://dx.doi.org/10.1088/0022-3727/45/23/235108

[35] Inagaki, M.; Kim, Y.A.; Endo, M. Graphene: preparation and structural perfection. *J. Mater. Chem.,* **2011**, *21*(10), 3280-3294.

http://dx.doi.org/10.1039/C0JM02991B

[36] Kobayashi, Y.; Fukui, K-i.; Enoki, T.; Kusakabe, K.; Kaburagi, Y. Observation of zigzag and armchair edges of graphite using scanning tunneling microscopy and spectroscopy. *Phys. Rev. B Condens. Matter Mater. Phys.,* **2005**, *71*(19), 193406.

http://dx.doi.org/10.1103/PhysRevB.71.193406

[37] Ferrari, A.C.; Meyer, J.C.; Scardaci, V.; Casiraghi, C.; Lazzeri, M.; Mauri, F.; Piscanec, S.; Jiang, D.; Novoselov, K.S.; Roth, S.; Geim, A.K. Raman spectrum of graphene and graphene layers. *Phys. Rev. Lett.,* **2006**, *97*(18), 187401.

http://dx.doi.org/10.1103/PhysRevLett.97.187401 PMID: 17155573

[38] Stolyarova, E.; Rim, K.T.; Ryu, S.; Maultzsch, J.; Kim, P.; Brus, L.E.; Heinz, T.F.; Hybertsen, M.S.; Flynn, G.W. High-resolution scanning tunneling microscopy imaging of mesoscopic graphene sheets on an insulating surface. *Proc. Natl. Acad. Sci. USA,* **2007**, *104*(22), 9209-9212.
http://dx.doi.org/10.1073/pnas.0703337104 PMID: 17517635

[39] Allen, M.J.; Tung, V.C.; Kaner, R.B. Honeycomb carbon: a review of graphene. *Chem. Rev.,* **2010**, *110*(1), 132-145.
http://dx.doi.org/10.1021/cr900070d PMID: 19610631

[40] Dujardin, E.; Thio, T.; Lezec, H.; Ebbesen, T.W. Fabrication of mesoscopic devices from graphite microdisks. *Appl. Phys. Lett.,* **2001**, *79*(15), 2474-2476.
http://dx.doi.org/10.1063/1.1407306

[41] Zhang, Y.; Han, H.; Wang, N.; Zhang, P.; Fu, Y.; Murugesan, M.; Edwards, M.; Jeppson, K.; Volz, S.; Liu, J. Improved heat spreading performance of functionalized graphene in microelectronic device application. *Adv. Funct. Mater.,* **2015**, *25*(28), 4430-4435.
http://dx.doi.org/10.1002/adfm.201500990

[42] Dabrowski, J.; Lippert, G.; Lupina, G. Graphene for Silicon Microelectronics: Ab Initio Modeling of Graphene Nucleation and Growth, **2016**.
http://dx.doi.org/10.1007/978-3-319-25340-4_8

[43] Lee, Y.; Ahn, J-H. Graphene-based transparent conductive films. *Nano,* **2013**, *8*(3), 1330001.
http://dx.doi.org/10.1142/S1793292013300016

[44] Kim, S-M.; Joo, P.; Ahn, G.; Cho, I.H.; Kim, D.H.; Song, W.K.; Kim, B-S.; Yoon, M-H. Transparent conducting films based on reduced graphene oxide multilayers for biocompatible neuronal interfaces. *J. Biomed. Nanotechnol.,* **2013**, *9*(3), 403-408.
http://dx.doi.org/10.1166/jbn.2013.1511 PMID: 23620995

[45] Kobayashi, T.; Bando, M.; Kimura, N.; Shimizu, K.; Kadono, K.; Umezu, N.; Miyahara, K.; Hayazaki, S.; Nagai, S.; Mizuguchi, Y.; Murakami, Y.; Hobara, D. Production of a 100-m-long high-quality graphene transparent conductive film by roll-to-roll chemical vapor deposition and transfer process. *Appl. Phys. Lett.,* **2013**, *102*(2), 023112.
http://dx.doi.org/10.1063/1.4776707

[46] Yuan, W.; Shi, G. Graphene-based gas sensors. *J. Mater. Chem. A Mater. Energy Sustain.,* **2013**, *1*(35), 10078-10091.
http://dx.doi.org/10.1039/c3ta11774j

[47] Varghese, S.S.; Lonkar, S.; Singh, K.; Swaminathan, S.; Abdala, A. Recent advances in graphene based gas sensors. *Sens. Actuators B Chem.,* **2015**, *218*, 160-183.
http://dx.doi.org/10.1016/j.snb.2015.04.062

[48] Mishra, S.K.; Tripathi, S.N.; Choudhary, V.; Gupta, B.D. SPR based fibre optic ammonia gas sensor utilizing nanocomposite film of PMMA/reduced graphene oxide prepared by *in situ* polymerization. *Sens. Actuators B Chem.,* **2014**, *199*, 190-200.
http://dx.doi.org/10.1016/j.snb.2014.03.109

[49] Mishra, S.K.; Tripathi, S.N.; Choudhary, V.; Gupta, B.D. Surface Plasmon Resonance-Based Fiber Optic Methane Gas Sensor Utilizing Graphene-Carbon Nanotubes-Poly (Methyl Methacrylate) Hybrid Nanocomposite. *Plasmonics,* **2015**, *10*(5), 1147-1157.
http://dx.doi.org/10.1007/s11468-015-9914-5

[50] Gadipelli, S.; Guo, Z.X. Graphene-based materials: synthesis and gas sorption, storage and separation. *Prog. Mater. Sci.,* **2015**, *69*, 1-60.
http://dx.doi.org/10.1016/j.pmatsci.2014.10.004

[51]   Kumar, R.; Suresh, V.M.; Maji, T.K.; Rao, C.N. Porous graphene frameworks pillared by organic linkers with tunable surface area and gas storage properties. *Chem. Commun. (Camb.),* **2014**, *50*(16), 2015-2017.

http://dx.doi.org/10.1039/c3cc46907g PMID: 24412955

[52]   Han, N.; Cuong, T.V.; Han, M.; Ryu, B.D.; Chandramohan, S.; Park, J.B.; Kang, J.H.; Park, Y-J.; Ko, K.B.; Kim, H.Y.; Kim, H.K.; Ryu, J.H.; Katharria, Y.S.; Choi, C.J.; Hong, C.H. Improved heat dissipation in gallium nitride light-emitting diodes with embedded graphene oxide pattern. *Nat. Commun.,* **2013**, *4*(1), 1452.

http://dx.doi.org/10.1038/ncomms2448 PMID: 23385596

[53]   Aravind, S.J.; Ramaprabhu, S. Graphene–multiwalled carbon nanotube-based nanofluids for improved heat dissipation. *RSC Advances,* **2013**, *3*(13), 4199-4206.

http://dx.doi.org/10.1039/c3ra22653k

[54]   Raccichini, R.; Varzi, A.; Passerini, S.; Scrosati, B. The role of graphene for electrochemical energy storage. *Nat. Mater.,* **2015**, *14*(3), 271-279.

http://dx.doi.org/10.1038/nmat4170 PMID: 25532074

[55]   El-Kady, M.F.; Kaner, R.B. Scalable fabrication of high-power graphene micro-supercapacitors for flexible and on-chip energy storage. *Nat. Commun.,* **2013**, *4*(1), 1475.

http://dx.doi.org/10.1038/ncomms2446 PMID: 23403576

[56]   Yin, Z.; Zhu, J.; He, Q.; Cao, X.; Tan, C.; Chen, H.; Yan, Q.; Zhang, H. *Graphene-Based Materials for Solar Cell Applications*; Advanced Energy Materials, **2014**, Vol. 4, .

[57]   Mittal, G.; Dhand, V.; Rhee, K.Y.; Park, S-J.; Lee, W.R. A review on carbon nanotubes and graphene as fillers in reinforced polymer nanocomposites. *J. Ind. Eng. Chem.,* **2015**, *21*, 11-25.

http://dx.doi.org/10.1016/j.jiec.2014.03.022

[58]   Mukhopadhyay, P.; Gupta, R.K. *Graphite, Graphene, and their polymer nanocomposites*; CRC Press, **2012**.

http://dx.doi.org/10.1201/b13051

[59]   Kumar, N.; Kumbhat, S. *Essentials in Nanoscience and Nanotechnology*; John Wiley & Sons, **2016**.

http://dx.doi.org/10.1002/9781119096122

[60]   Novoselov, K. S.; Geim, A. K.; Morozov, S. V.; Jiang, D.; Zhang, Y.; Dubonos, S. V.; Grigorieva, I. V.; Firsov, A. A. Electric field effect in atomically thin carbon films *Science,* **2004**, *306*, 666-669.

http://dx.doi.org/10.1126/science.1102896

[61]   Ciesielski, A.; Samorì, P. Graphene *via* sonication assisted liquid-phase exfoliation. *Chem. Soc. Rev.,* **2014**, *43*(1), 381-398.

http://dx.doi.org/10.1039/C3CS60217F PMID: 24002478

[62]   Kumar, A.; Lee, C.H. *Synthesis and biomedical applications of graphene: present and future trends*; Advances in Graphene Science, **2013**, pp. 5772-5578.

[63]   Riedl, C.; Coletti, C.; Starke, U. Structural and electronic properties of epitaxial graphene on SiC (0 0 0 1): a review of growth, characterization, transfer doping and hydrogen intercalation. *J. Phys. D Appl. Phys.,* **2010**, *43*(37), 374009.

http://dx.doi.org/10.1088/0022-3727/43/37/374009

[64]   Dreyer, D.R.; Park, S.; Bielawski, C.W.; Ruoff, R.S. The chemistry of graphene oxide. *Chem. Soc. Rev.,* **2010**, *39*(1), 228-240.

http://dx.doi.org/10.1039/B917103G PMID: 20023850

[65]   Pei, S.; Cheng, H-M. The reduction of graphene oxide. *Carbon,* **2012**, *50*(9), 3210-3228.

http://dx.doi.org/10.1016/j.carbon.2011.11.010

[66]   Park, S.; Ruoff, R.S. Chemical methods for the production of graphenes. *Nat. Nanotechnol.,* **2009**, *4*(4), 217-224.
        http://dx.doi.org/10.1038/nnano.2009.58 PMID: 19350030

[67]   Zhu, Y.; Murali, S.; Cai, W.; Li, X.; Suk, J.W.; Potts, J.R.; Ruoff, R.S. Graphene and graphene oxide: synthesis, properties, and applications. *Adv. Mater.,* **2010**, *22*(35), 3906-3924.
        http://dx.doi.org/10.1002/adma.201001068 PMID: 20706983

[68]   Neto, A.C.; Guinea, F.; Peres, N.M.; Novoselov, K.S.; Geim, A.K. The electronic properties of graphene. *Rev. Mod. Phys.,* **2009**, *81*(1), 109-162.
        http://dx.doi.org/10.1103/RevModPhys.81.109

[69]   Marcano, D.C.; Kosynkin, D.V.; Berlin, J.M.; Sinitskii, A.; Sun, Z.; Slesarev, A.; Alemany, L.B.; Lu, W.; Tour, J.M. *Improved Synthesis of Graphene Oxide,* **2010**.
        http://dx.doi.org/10.1021/nn1006368

[70]   Geim, A.K.; Novoselov, K.S. The rise of graphene. *Nat. Mater.,* **2007**, *6*(3), 183-191.
        http://dx.doi.org/10.1038/nmat1849 PMID: 17330084

[71]   Novoselov, K.; Neto, A.C. Two-dimensional crystals-based heterostructures: materials with tailored properties. *Phys. Scr.,* **2012**, *2012*, 014006.
        http://dx.doi.org/10.1088/0031-8949/2012/T146/014006

[72]   Coleman, J.N. Liquid exfoliation of defect-free graphene. *Acc. Chem. Res.,* **2013**, *46*(1), 14-22.
        http://dx.doi.org/10.1021/ar300009f PMID: 22433117

[73]   Coleman, J.N.; Khan, U.; Blau, W.J.; Gun'ko, Y.K. Small but strong: a review of the mechanical properties of carbon nanotube–polymer composites. *Carbon,* **2006**, *44*(9), 1624-1652.
        http://dx.doi.org/10.1016/j.carbon.2006.02.038

[74]   Spyrou, K.; Rudolf, P. *An introduction to graphene.* **2014**.
        http://dx.doi.org/10.1002/9783527672790.ch1

[75]   Blake, P.; Brimicombe, P.D.; Nair, R.R.; Booth, T.J.; Jiang, D.; Schedin, F.; Ponomarenko, L.A.; Morozov, S.V.; Gleeson, H.F.; Hill, E.W.; Geim, A.K.; Novoselov, K.S. Graphene-based liquid crystal device. *Nano Lett.,* **2008**, *8*(6), 1704-1708.
        http://dx.doi.org/10.1021/nl080649i PMID: 18444691

[76]   Hernandez, Y.; Nicolosi, V.; Lotya, M.; Blighe, F.M.; Sun, Z.; De, S.; McGovern, I.T.; Holland, B.; Byrne, M.; Gun'Ko, Y.K.; Boland, J.J.; Niraj, P.; Duesberg, G.; Krishnamurthy, S.; Goodhue, R.; Hutchison, J.; Scardaci, V.; Ferrari, A.C.; Coleman, J.N. High-yield production of graphene by liquid-phase exfoliation of graphite. *Nat. Nanotechnol.,* **2008**, *3*(9), 563-568.
        http://dx.doi.org/10.1038/nnano.2008.215 PMID: 18772919

[77]   Bourlinos, A.B.; Georgakilas, V.; Zboril, R.; Steriotis, T.A.; Stubos, A.K. Liquid-phase exfoliation of graphite towards solubilized graphenes. *Small,* **2009**, *5*(16), 1841-1845.
        http://dx.doi.org/10.1002/smll.200900242 PMID: 19408256

[78]   Qian, W.; Hao, R.; Hou, Y.; Tian, Y.; Shen, C.; Gao, H.; Liang, X. Solvothermal-assisted exfoliation process to produce graphene with high yield and high quality. *Nano Res.,* **2009**, *2*(9), 706-712.
        http://dx.doi.org/10.1007/s12274-009-9074-z

[79]   He, H.; Klinowski, J.; Forster, M.; Lerf, A. A new structural model for graphite oxide. *Chem. Phys. Lett.,* **1998**, *287*(1-2), 53-56.
        http://dx.doi.org/10.1016/S0009-2614(98)00144-4

[80]   Hontoria-Lucas, C.; Lopez-Peinado, A.; López-González, J.D.; Rojas-Cervantes, M.; Martin-Aranda, R. Study of oxygen-containing groups in a series of graphite oxides: physical and chemical characterization. *Carbon,* **1995**, *33*(11), 1585-1592.

http://dx.doi.org/10.1016/0008-6223(95)00120-3

[81]   Brodie, B.C. On the Atomic Weight of Graphite. *Philos. Trans. R. Soc. Lond.,* **1859**, *149*, 249-259.
        http://dx.doi.org/10.1098/rstl.1859.0013

[82]   Staudenmaier, L. Verfahren zur darstellung der graphitsäure. *Ber. Dtsch. Chem. Ges.,* **1898**, *31*(2),
        1481-1487.
        http://dx.doi.org/10.1002/cber.18980310237

[83]   Hummers, W. S.; Offeman, R. E. Preparation of Graphitic Oxide *Journal of the American Chemical
        Society,* **1958**, *80*, 1339.
        http://dx.doi.org/10.1021/ja01539a017

[84]   Stankovich, S.; Dikin, D.A.; Dommett, G.H.; Kohlhaas, K.M.; Zimney, E.J.; Stach, E.A.; Piner, R.D.;
        Nguyen, S.T.; Ruoff, R.S. Graphene-based composite materials. *Nature,* **2006**, *442*(7100), 282-286.
        http://dx.doi.org/10.1038/nature04969 PMID: 16855586

[85]   Kim, H.; Abdala, A.A.; Macosko, C.W. Graphene/polymer nanocomposites. *Macromolecules,* **2010**,
        43(16), 6515-6530.
        http://dx.doi.org/10.1021/ma100572e

[86]   Stankovich, S.; Dikin, D.A.; Piner, R.D.; Kohlhaas, K.A.; Kleinhammes, A.; Jia, Y.; Wu, Y.; Nguyen,
        S.T.; Ruoff, R.S. Synthesis of graphene-based nanosheets *via* chemical reduction of exfoliated graphite
        oxide. *Carbon,* **2007**, *45*(7), 1558-1565.
        http://dx.doi.org/10.1016/j.carbon.2007.02.034

[87]   Shin, H.J.; Kim, K.K.; Benayad, A.; Yoon, S.M.; Park, H.K.; Jung, I.S.; Jin, M.H.; Jeong, H.K.; Kim,
        J.M.; Choi, J.Y.; Lee, Y.H. Efficient reduction of graphite oxide by sodium borohydride and its effect
        on electrical conductance. *Adv. Funct. Mater.,* **2009**, *19*(12), 1987-1992.
        http://dx.doi.org/10.1002/adfm.200900167

[88]   Schniepp, H. C.; Li, J.-L.; McAllister, M. J.; Sai, H.; Herrera-Alonso, M.; Adamson, D. H.;
        Prud'homme, R. K.; Car, R.; Saville, D. A.; Aksay, I. A. Functionalized single graphene sheets derived
        from splitting graphite oxide *The Journal of Physical Chemistry,* **2006**, *110*, 8535-8539.
        http://dx.doi.org/10.1021/jp060936f

[89]   Zhao, L.; Rim, K.T.; Zhou, H.; He, R.; Heinz, T.F.; Pinczuk, A.; Flynn, G.W.; Pasupathy, A.N.
        Influence of copper crystal surface on the CVD growth of large area monolayer graphene. *Solid State
        Commun.,* **2011**, *151*(7), 509-513.
        http://dx.doi.org/10.1016/j.ssc.2011.01.014

[90]   Hibino, H.; Kageshima, H.; Nagase, M. Epitaxial few-layer graphene: towards single crystal growth.
        J. Phys. D Appl. Phys., **2010**, *43*(37), 374005.
        http://dx.doi.org/10.1088/0022-3727/43/37/374005

[91]   Berger, C.; Wu, X.; First, P.N.; Conrad, E.H.; Li, X.; Sprinkle, M.; Hass, J.; Varchon, F.; Magaud, L.;
        Sadowski, M.L. Dirac particles in epitaxial graphene films grown on SiC, **2008**.
        http://dx.doi.org/10.1007/978-3-540-74325-5_12

[92]   Choudhary, V.; Singh, B.; Mathur, R. Carbon nanotubes and their composites.*Syntheses and
        applications of carbon nanotubes and their composites*; InTech, Rijeka ISBN, **2013**, pp. 978-953.
        http://dx.doi.org/10.5772/52897

[93]   Cheng, Q.; Tang, J.; Ma, J.; Zhang, H.; Shinya, N.; Qin, L-C. Graphene and carbon nanotube composite
        electrodes for supercapacitors with ultra-high energy density. *Phys. Chem. Chem. Phys.,* **2011**, *13*(39),
        17615-17624.
        http://dx.doi.org/10.1039/c1cp21910c PMID: 21887427

[94]   Yang, S-Y.; Chang, K-H.; Tien, H-W.; Lee, Y-F.; Li, S-M.; Wang, Y-S.; Wang, J-Y.; Ma, C-C.M.; Hu, C-C. Design and tailoring of a hierarchical graphene-carbon nanotube architecture for supercapacitors. *J. Mater. Chem.,* **2011**, *21*(7), 2374-2380.
http://dx.doi.org/10.1039/C0JM03199B

[95]   Zhang, D.; Yan, T.; Shi, L.; Peng, Z.; Wen, X.; Zhang, J. Enhanced capacitive deionization performance of graphene/carbon nanotube composites. *J. Mater. Chem.,* **2012**, *22*(29), 14696-14704.
http://dx.doi.org/10.1039/c2jm31393f

[96]   Wang, Y.; Wu, Y.; Huang, Y.; Zhang, F.; Yang, X.; Ma, Y.; Chen, Y. Preventing Graphene Sheets from Restacking for High-Capacitance Performance *The Journal of Physical Chemistry C,* **2011**, *115*, 23192-23197.
http://dx.doi.org/10.1021/jp206444e

[97]   Lu, X.; Dou, H.; Gao, B.; Yuan, C.; Yang, S.; Hao, L.; Shen, L.; Zhang, X. A flexible graphene/multiwalled carbon nanotube film as a high performance electrode material for supercapacitors. *Electrochim. Acta,* **2011**, *56*(14), 5115-5121.
http://dx.doi.org/10.1016/j.electacta.2011.03.066

[98]   Qiu, L.; Yang, X.; Gou, X.; Yang, W.; Ma, Z-F.; Wallace, G.G.; Li, D. Dispersing carbon nanotubes with graphene oxide in water and synergistic effects between graphene derivatives. *Chemistry,* **2010**, *16*(35), 10653-10658.
http://dx.doi.org/10.1002/chem.201001771 PMID: 20680948

[99]   Yen, M-Y.; Hsiao, M-C.; Liao, S-H.; Liu, P-I.; Tsai, H-M.; Ma, C-C.M.; Pu, N-W.; Ger, M-D. Preparation of graphene/multi-walled carbon nanotube hybrid and its use as photoanodes of dye-sensitized solar cells. *Carbon,* **2011**, *49*(11), 3597-3606.
http://dx.doi.org/10.1016/j.carbon.2011.04.062

[100]   Sui, Z.; Meng, Q.; Zhang, X.; Ma, R.; Cao, B. Green synthesis of carbon nanotube-graphene hybrid aerogels and their use as versatile agents for water purification. *J. Mater. Chem.,* **2012**, *22*(18), 8767-8771.
http://dx.doi.org/10.1039/c2jm00055e

[101]   Liu, Y-T.; Dang, M.; Xie, X-M.; Wang, Z-F.; Ye, X-Y. Synergistic effect of Cu2+-coordinated carbon nanotube/graphene network on the electrical and mechanical properties of polymer nanocomposites. *J. Mater. Chem.,* **2011**, *21*(46), 18723-18729.
http://dx.doi.org/10.1039/c1jm13727a

[102]   Mani, V.; Devadas, B.; Chen, S-M. Direct electrochemistry of glucose oxidase at electrochemically reduced graphene oxide-multiwalled carbon nanotubes hybrid material modified electrode for glucose biosensor. *Biosens. Bioelectron.,* **2013**, *41*, 309-315.
http://dx.doi.org/10.1016/j.bios.2012.08.045 PMID: 22964382

[103]   Dong, X.; Li, B.; Wei, A.; Cao, X.; Chan-Park, M.B.; Zhang, H.; Li, L-J.; Huang, W.; Chen, P. One-step growth of graphene–carbon nanotube hybrid materials by chemical vapor deposition. *Carbon,* **2011**, *49*(9), 2944-2949.
http://dx.doi.org/10.1016/j.carbon.2011.03.009

[104]   Li, S.; Luo, Y.; Lv, W.; Yu, W.; Wu, S.; Hou, P.; Yang, Q.; Meng, Q.; Liu, C.; Cheng, H-M. Vertically Aligned Carbon Nanotubes Grown on Graphene Paper as Electrodes in Lithium-Ion Batteries and Dye-Sensitized Solar Cells. *Adv. Energy Mater.,* **2011**, *1*(4), 486-490.
http://dx.doi.org/10.1002/aenm.201100001

[105]   Chen, S.; Chen, P.; Wang, Y. Carbon nanotubes grown *in situ* on graphene nanosheets as superior anodes for Li-ion batteries. *Nanoscale,* **2011**, *3*(10), 4323-4329.
http://dx.doi.org/10.1039/c1nr10642b PMID: 21879120

[106]   Fan, Z.; Yan, J.; Zhi, L.; Zhang, Q.; Wei, T.; Feng, J.; Zhang, M.; Qian, W.; Wei, F. A three-dimensional carbon nanotube/graphene sandwich and its application as electrode in supercapacitors. *Adv. Mater.,* **2010**, *22*(33), 3723-3728.
      http://dx.doi.org/10.1002/adma.201001029 PMID: 20652901

[107]   Lee, D.H.; Kim, J.E.; Han, T.H.; Hwang, J.W.; Jeon, S.; Choi, S-Y.; Hong, S.H.; Lee, W.J.; Ruoff, R.S.; Kim, S.O. Versatile carbon hybrid films composed of vertical carbon nanotubes grown on mechanically compliant graphene films. *Adv. Mater.,* **2010**, *22*(11), 1247-1252.
      http://dx.doi.org/10.1002/adma.200903063 PMID: 20437513

[108]   Huang, Z-D.; Zhang, B.; Oh, S-W.; Zheng, Q-B.; Lin, X-Y.; Yousefi, N.; Kim, J-K. Self-assembled reduced graphene oxide/carbon nanotube thin films as electrodes for supercapacitors. *J. Mater. Chem.,* **2012**, *22*(8), 3591-3599.
      http://dx.doi.org/10.1039/c2jm15048d

[109]   Li, Y-F.; Liu, Y-Z.; Yang, Y-G.; Wang, M-Z.; Wen, Y-F. Reduced graphene oxide/MWCNT hybrid sandwiched film by self-assembly for high performance supercapacitor electrodes. *Appl. Phys., A Mater. Sci. Process.,* **2012**, *108*(3), 701-707.
      http://dx.doi.org/10.1007/s00339-012-6953-z

[110]   Kim, H.; Miura, Y.; Macosko, C.W. *Graphene/Polyurethane Nanocomposites for Improved Gas Barrier and Electrical Conductivity,* **2010**.
      http://dx.doi.org/10.1021/cm100477v

[111]   Zhang, H-B.; Zheng, W-G.; Yan, Q.; Yang, Y.; Wang, J-W.; Lu, Z-H.; Ji, G-Y.; Yu, Z-Z. Electrically conductive polyethylene terephthalate/graphene nanocomposites prepared by melt compounding. *Polymer (Guildf.),* **2010**, *51*(5), 1191-1196.
      http://dx.doi.org/10.1016/j.polymer.2010.01.027

[112]   Kim, I.H.; Jeong, Y.G. Polylactide/exfoliated graphite nanocomposites with enhanced thermal stability, mechanical modulus, and electrical conductivity. *J. Polym. Sci., B, Polym. Phys.,* **2010**, *48*(8), 850-858.
      http://dx.doi.org/10.1002/polb.21956

[113]   Kim, S.; Do, I.; Drzal, L.T. Thermal stability and dynamic mechanical behavior of exfoliated graphite nanoplatelets-LLDPE nanocomposites. *Polym. Compos.,* **2010**, *31*(5), 755-761.
      http://dx.doi.org/10.1002/pc.20781

[114]   Kalaitzidou, K.; Fukushima, H.; Drzal, L.T. Mechanical properties and morphological characterization of exfoliated graphite–polypropylene nanocomposites. *Compos., Part A Appl. Sci. Manuf.,* **2007**, *38*(7), 1675-1682.
      http://dx.doi.org/10.1016/j.compositesa.2007.02.003

[115]   Steurer, P.; Wissert, R.; Thomann, R.; Mülhaupt, R. Functionalized graphenes and thermoplastic nanocomposites based upon expanded graphite oxide. *Macromol. Rapid Commun.,* **2009**, *30*(4-5), 316-327.
      http://dx.doi.org/10.1002/marc.200800754 PMID: 21706607

[116]   Kalaitzidou, K.; Fukushima, H.; Drzal, L.T. A new compounding method for exfoliated graphite–polypropylene nanocomposites with enhanced flexural properties and lower percolation threshold. *Compos. Sci. Technol.,* **2007**, *67*(10), 2045-2051.
      http://dx.doi.org/10.1016/j.compscitech.2006.11.014

[117]   Jeong, H-K.; Lee, Y.P.; Jin, M.H.; Kim, E.S.; Bae, J.J.; Lee, Y.H. Thermal stability of graphite oxide. *Chem. Phys. Lett.,* **2009**, *470*(4-6), 255-258.
      http://dx.doi.org/10.1016/j.cplett.2009.01.050

[118]   Fim, F.C.; Guterres, J.M.; Basso, N.R.; Galland, G.B. Polyethylene/graphite nanocomposites obtained by *in situ* polymerization. *J. Polym. Sci. A Polym. Chem.,* **2010**, *48*(3), 692-698.

http://dx.doi.org/10.1002/pola.23822

[119] Jang, J.Y.; Kim, M.S.; Jeong, H.M.; Shin, C.M. Graphite oxide/poly (methyl methacrylate) nanocomposites prepared by a novel method utilizing macroazoinitiator. *Compos. Sci. Technol.,* **2009,** *69*(2), 186-191.
http://dx.doi.org/10.1016/j.compscitech.2008.09.039

[120] Gu, Z.; Li, C.; Wang, G.; Zhang, L.; Li, X.; Wang, W.; Jin, S. Synthesis and characterization of polypyrrole/graphite oxide composite by *in situ* emulsion polymerization. *J. Polym. Sci., B, Polym. Phys.,* **2010,** *48*(12), 1329-1335.
http://dx.doi.org/10.1002/polb.22031

[121] Singh, K.; Ohlan, A.; Pham, V.H.; R, B.; Varshney, S.; Jang, J.; Hur, S.H.; Choi, W.M.; Kumar, M.; Dhawan, S.K.; Kong, B-S.; Chung, J.S. B R, S. Varshney, J. Jang, S. H. Hur, W. M. Choi, M. Kumar, S. K. Dhawan, B.-S. Kong, and J. S. Chung, "Nanostructured graphene/Fe3O4 incorporated polyaniline as a high performance shield against electromagnetic pollution. *Nanoscale,* **2013,** *5*(6), 2411-2420.
http://dx.doi.org/10.1039/c3nr33962a PMID: 23400248

[122] Musico, Y.L.F.; Santos, C.M.; Dalida, M.L.P.; Rodrigues, D.F. Improved removal of lead(ii) from water using a polymer-based graphene oxide nanocomposite. *J. Mater. Chem. A Mater. Energy Sustain.,* **2013,** *1*(11), 3789-3796.
http://dx.doi.org/10.1039/c3ta01616a

[123] Sun, M.; Wang, G.; Yang, C.; Jiang, H.; Li, C. A graphene/carbon nanotube@[small pi]-conjugated polymer nanocomposite for high-performance organic supercapacitor electrodes. *J. Mater. Chem. A Mater. Energy Sustain.,* **2015,** *3*(7), 3880-3890.
http://dx.doi.org/10.1039/C4TA06728B

[124] Wang, H.; Hao, Q.; Yang, X.; Lu, L.; Wang, X. A nanostructured graphene/polyaniline hybrid material for supercapacitors. *Nanoscale,* **2010,** *2*(10), 2164-2170.
http://dx.doi.org/10.1039/c0nr00224k PMID: 20689894

[125] Choi, B.G.; Yang, M.; Hong, W.H.; Choi, J.W.; Huh, Y.S. *3D Macroporous Graphene Frameworks for Supercapacitors with High Energy and Power Densities,* **2012.**
http://dx.doi.org/10.1021/nn3003345

[126] Berger, C.; Song, Z.; Li, T.; Li, X.; Ogbazghi, A. Y.; Feng, R.; Dai, Z.; Marchenkov, A. N.; Conrad, E. H.; First, P. N.; de Heer, W. A. Ultrathin Epitaxial Graphite: 2D Electron Gas Properties and a Route toward Graphene-based Nanoelectronics *The Journal of Physical Chemistry B,* **2004,** *108*, 19912-19916.
http://dx.doi.org/10.1021/jp040650f

[127] Bhattacharya, P.; Dhibar, S.; Hatui, G.; Mandal, A.; Das, T.; Das, C.K. Graphene decorated with hexagonal shaped M-type ferrite and polyaniline wrapper: a potential candidate for electromagnetic wave absorbing and energy storage device applications. *RSC Advances,* **2014,** *4*(33), 17039-17053.
http://dx.doi.org/10.1039/c4ra00448e

[128] Shen, J.; Han, K.; Martin, E.J.; Wu, Y.Y.; Kung, M.C.; Hayner, C.M.; Shull, K.R.; Kung, H.H. Upper-critical solution temperature (UCST) polymer functionalized graphene oxide as thermally responsive ion permeable membrane for energy storage devices. *J. Mater. Chem. A Mater. Energy Sustain.,* **2014,** *2*(43), 18204-18207.
http://dx.doi.org/10.1039/C4TA04852K

[129] Zhang, F.; Zhang, T.; Yang, X.; Zhang, L.; Leng, K.; Huang, Y.; Chen, Y. A high-performance supercapacitor-battery hybrid energy storage device based on graphene-enhanced electrode materials with ultrahigh energy density. *Energy Environ. Sci.,* **2013,** *6*(5), 1623-1632.
http://dx.doi.org/10.1039/c3ee40509e

[130] Diao, G.; Zhu, S.; Chen, M.; Ren, W.; Yang, J.; Qu, S.; Li, Z. Microwave Assisted Synthesis [small alpha]-Fe2O3/Reduced Graphene Oxide as Anode Materials for High Performance Lithium Ion Batteries. *New J. Chem.,* **2015**.

[131] Dong, Y.; Zhang, Z.; Xia, Y.; Chui, Y-S.; Lee, J-M.; Zapien, J.A. Green and facile synthesis of Fe3O4 and graphene nanocomposites with enhanced rate capability and cycling stability for lithium ion batteries. *J. Mater. Chem. A Mater. Energy Sustain.,* **2015**, *3*(31), 16206-16212.

http://dx.doi.org/10.1039/C5TA03690A

[132] Perumal Veeramalai, C.; Li, F.; Xu, H.; Kim, T.W.; Guo, T. One pot hydrothermal synthesis of graphene like MoS2 nanosheets for application in high performance lithium ion batteries. *RSC Advances,* **2015**, *5*(71), 57666-57670.

http://dx.doi.org/10.1039/C5RA07478A

[133] Wang, B.; Wang, G.; Wang, H. Hybrids of Mo2C nanoparticles anchored on graphene sheets as anode materials for high performance lithium-ion batteries. *J. Mater. Chem. A Mater. Energy Sustain.,* **2015**.

http://dx.doi.org/10.1039/C5TA06949A

[134] Zhang, Y.; Jiang, L.; Wang, C. Facile synthesis of SnO2 nanocrystals anchored onto graphene nanosheets as anode materials for lithium-ion batteries. *Phys. Chem. Chem. Phys.,* **2015**, *17*(31), 20061-20065.

http://dx.doi.org/10.1039/C5CP03305E PMID: 26186479

[135] Weaver, C.L.; Li, H.; Luo, X.; Cui, X.T. A graphene oxide/conducting polymer nanocomposite for electrochemical dopamine detection: origin of improved sensitivity and specificity. *J. Mater. Chem. B Mater. Biol. Med.,* **2014**, *2*(32), 5209-5219.

http://dx.doi.org/10.1039/C4TB00789A PMID: 32261663

[136] Quan, H.; Zhang, B.; Zhao, Q.; Yuen, R.K.; Li, R.K. Facile preparation and thermal degradation studies of graphite nanoplatelets (GNPs) filled thermoplastic polyurethane (TPU) nanocomposites. *Compos., Part A Appl. Sci. Manuf.,* **2009**, *40*(9), 1506-1513.

http://dx.doi.org/10.1016/j.compositesa.2009.06.012

[137] Lu, X.; Yu, M.; Huang, H.; Ruoff, R.S. Tailoring graphite with the goal of achieving single sheets. *Nanotechnology,* **1999**, *10*(3), 269-272.

http://dx.doi.org/10.1088/0957-4484/10/3/308

[138] Novoselov, K.S.; Jiang, D.; Schedin, F.; Booth, T.J.; Khotkevich, V.V.; Morozov, S.V.; Geim, A.K. Two-dimensional atomic crystals *Proceedings of the National Academy of Sciences of the United States of America, vol. 102,* pp. 10451-10453.**2005**,

http://dx.doi.org/10.1073/pnas.0502848102

[139] Hernandez, Y.; Nicolosi, V.; Lotya, M.; Blighe, F.M.; Sun, Z.; De, S.; McGovern, I.T.; Holland, B.; Byrne, M.; Gun'Ko, Y.K.; Boland, J.J.; Niraj, P.; Duesberg, G.; Krishnamurthy, S.; Goodhue, R.; Hutchison, J.; Scardaci, V.; Ferrari, A.C.; Coleman, J.N. Y. K. Gun'Ko, J. J. Boland, P. Niraj, G. Duesberg, S. Krishnamurthy, R. Goodhue, J. Hutchison, V. Scardaci, A. C. Ferrari, and J. N. Coleman, "High-yield production of graphene by liquid-phase exfoliation of graphite. *Nat. Nanotechnol.,* **2008**, *3*(9), 563-568.

http://dx.doi.org/10.1038/nnano.2008.215

[140] Lotya, M.; Hernandez, Y.; King, P. J.; Smith, R. J.; Nicolosi, V.; Karlsson, L. S.; Blighe, F. M.; De, S.; Wang, Z.; McGovern, I. T.; Duesberg, G. S.; Coleman, J. N. Liquid Phase Production of Graphene by Exfoliation of Graphite in Surfactant/Water Solutions *Journal of the American Chemical Society,* **2009**, *131*, 3611-3620.

http://dx.doi.org/10.1021/ja807449u

[141]   Berger, C.; Song, Z.; Li, X.; Wu, X.; Brown, N.; Naud, C.; Mayou, D.; Li, T.; Hass, J.; Marchenkov, A. N.; Conrad, E. H.; First, P. N.; de Heer, W. A. Electronic Confinement and Coherence in Patterned Epitaxial Graphene *Science,* **2006**, *312*, 1191-1196.
       http://dx.doi.org/10.1126/science.1125925

[142]   Sutter, P.W.; Flege, J-I.; Sutter, E.A. Epitaxial graphene on ruthenium. *Nat. Mater.,* **2008**, *7*(5), 406-411.
       http://dx.doi.org/10.1038/nmat2166 PMID: 18391956

[143]   Pan, Y.; Zhang, H.; Shi, D.; Sun, J.; Du, S.; Liu, F.; Gao, H. Highly Ordered, Millimeter-Scale, Continuous, Single-Crystalline Graphene Monolayer Formed on Ru (0001). *Adv. Mater.,* **2009**, *21*(27), 2777-2780.
       http://dx.doi.org/10.1002/adma.200800761

[144]   Aizawa, T.; Souda, R.; Otani, S.; Ishizawa, Y.; Oshima, C. Anomalous bond of monolayer graphite on transition-metal carbide surfaces. *Phys. Rev. Lett.,* **1990**, *64*(7), 768-771.
       http://dx.doi.org/10.1103/PhysRevLett.64.768 PMID: 10042073

[145]   Kim, K.S.; Zhao, Y.; Jang, H.; Lee, S.Y.; Kim, J.M.; Kim, K.S.; Ahn, J-H.; Kim, P.; Choi, J-Y.; Hong, B.H. Large-scale pattern growth of graphene films for stretchable transparent electrodes. *Nature,* **2009**, *457*(7230), 706-710.
       http://dx.doi.org/10.1038/nature07719 PMID: 19145232

[146]   Li, D.; Müller, M.B.; Gilje, S.; Kaner, R.B.; Wallace, G.G. Processable aqueous dispersions of graphene nanosheets. *Nat. Nanotechnol.,* **2008**, *3*(2), 101-105.
       http://dx.doi.org/10.1038/nnano.2007.451 PMID: 18654470

[147]   Eda, G.; Fanchini, G.; Chhowalla, M. Large-area ultrathin films of reduced graphene oxide as a transparent and flexible electronic material. *Nat. Nanotechnol.,* **2008**, *3*(5), 270-274.
       http://dx.doi.org/10.1038/nnano.2008.83 PMID: 18654522

[148]   Yousefi, N.; Sun, X.; Lin, X.; Shen, X.; Jia, J.; Zhang, B.; Tang, B.; Chan, M.; Kim, J-K. Highly aligned graphene/polymer nanocomposites with excellent dielectric properties for high-performance electromagnetic interference shielding. *Adv. Mater.,* **2014**, *26*(31), 5480-5487.
       http://dx.doi.org/10.1002/adma.201305293 PMID: 24715671

[149]   Yan, D-X.; Ren, P-G.; Pang, H.; Fu, Q.; Yang, M-B.; Li, Z-M. Efficient electromagnetic interference shielding of lightweight graphene/polystyrene composite. *J. Mater. Chem.,* **2012**, *22*(36), 18772-18774.
       http://dx.doi.org/10.1039/c2jm32692b

[150]   Tung, T.T.; Feller, J-F.; Kim, T.; Kim, H.; Yang, W.S.; Suh, K.S. Electromagnetic properties of Fe3O4-functionalized graphene and its composites with a conducting polymer. *J. Polym. Sci. A Polym. Chem.,* **2012**, *50*(5), 927-935.
       http://dx.doi.org/10.1002/pola.25847

[151]   Sun, X.; He, J.; Li, G.; Tang, J.; Wang, T.; Guo, Y.; Xue, H. Laminated magnetic graphene with enhanced electromagnetic wave absorption properties. *J. Mater. Chem. C Mater. Opt. Electron. Devices,* **2013**, *1*(4), 765-777.
       http://dx.doi.org/10.1039/C2TC00159D

[152]   Husić, S.; Javni, I.; Petrović, Z.S. Thermal and mechanical properties of glass reinforced soy-based polyurethane composites. *Compos. Sci. Technol.,* **2005**, *65*(1), 19-25.
       http://dx.doi.org/10.1016/j.compscitech.2004.05.020

[153]   Ummartyotin, S.; Juntaro, J.; Sain, M.; Manuspiya, H. Development of transparent bacterial cellulose nanocomposite film as substrate for flexible organic light emitting diode (OLED) display. *Ind. Crops Prod.,* **2012**, *35*(1), 92-97.
       http://dx.doi.org/10.1016/j.indcrop.2011.06.025

[154]   Mondal, S.; Hu, J. Temperature stimulating shape memory polyurethane for smart clothing. *Indian J. Fibre Text. Res.,* **2006**, *31*, 66.

[155]   Ifeyinwa, M.C.; Reginald, U. A Survey on the Use of Nigeria Local Palm Oil for the Production of Polyol for Polyurethane Foam Production in Nigeria. *Europe,* •••, *3*, 5.

[156]   Meier-Westhues, U. *Polyurethanes: coatings, adhesives and sealants*; Vincentz Network GmbH & Co KG, **2007**.

[157]   Gupta, T.K.; Singh, B.P.; Dhakate, S.R.; Singh, V.N.; Mathur, R.B. Improved nanoindentation and microwave shielding properties of modified MWCNT reinforced polyurethane composites. *J. Mater. Chem. A Mater. Energy Sustain.,* **2013**, *1*(32), 9138-9149.
        http://dx.doi.org/10.1039/c3ta11611e

[158]   Gupta, T. K.; Singh, B. P.; Teotia, S.; Katyal, V.; Dhakate, S. R.; Mathur, R. B. Designing of multiwalled carbon nanotubes reinforced polyurethane composites as electromagnetic interference shielding materials *Journal of Polymer Research,* **2013**, *20*, 1-7.
        http://dx.doi.org/10.1007/s10965-013-0169-6

[159]   Bian, J.; Lin, H.L.; He, F.X.; Wei, X.W.; Chang, I.T.; Sancaktar, E. Fabrication of microwave exfoliated graphite oxide reinforced thermoplastic polyurethane nanocomposites: Effects of filler on morphology, mechanical, thermal and conductive properties. *Compos., Part A Appl. Sci. Manuf.,* **2013**, *47*, 72-82.
        http://dx.doi.org/10.1016/j.compositesa.2012.12.009

[160]   Nguyen, D.A.; Lee, Y.R.; Raghu, A.V.; Jeong, H.M.; Shin, C.M.; Kim, B.K. Morphological and physical properties of a thermoplastic polyurethane reinforced with functionalized graphene sheet. *Polym. Int.,* **2009**, *58*(4), 412-417.
        http://dx.doi.org/10.1002/pi.2549

[161]   Gupta, T.K.; Singh, B.P.; Tripathi, R.K.; Dhakate, S.R.; Singh, V.N.; Panwar, O.; Mathur, R.B. Superior nano-mechanical properties of reduced graphene oxide reinforced polyurethane composites. *RSC Advances,* **2015**, *5*(22), 16921-16930.
        http://dx.doi.org/10.1039/C4RA14223C

[162]   Gao, J.; Hu, M.; Dong, Y.; Li, R.K. Graphite-nanoplatelet-decorated polymer nanofiber with improved thermal, electrical, and mechanical properties. *ACS Appl. Mater. Interfaces,* **2013**, *5*(16), 7758-7764.
        http://dx.doi.org/10.1021/am401420k PMID: 23910565

[163]   Yousefi, N.; Gudarzi, M.M.; Zheng, Q.; Aboutalebi, S.H.; Sharif, F.; Kim, J-K. Self-alignment and high electrical conductivity of ultralarge graphene oxide–polyurethane nanocomposites. *J. Mater. Chem.,* **2012**, *22*(25), 12709-12717.
        http://dx.doi.org/10.1039/c2jm30590a

[164]   Li, Y.; Yang, Z.; Qiu, H.; Dai, Y.; Zheng, Q.; Li, J.; Yang, J. Self-aligned graphene as anticorrosive barrier in waterborne polyurethane composite coatings. *J. Mater. Chem. A Mater. Energy Sustain.,* **2014**, *2*(34), 14139-14145.
        http://dx.doi.org/10.1039/C4TA02262A

[165]   Liao, K-H.; Park, Y.T.; Abdala, A.; Macosko, C. Aqueous reduced graphene/thermoplastic polyurethane nanocomposites. *Polymer (Guildf.),* **2013**, *54*(17), 4555-4559.
        http://dx.doi.org/10.1016/j.polymer.2013.06.032

[166]   Khanam, P. N.; Ponnamma, D.; AL-Madeed, M. Electrical Properties of Graphene Polymer Nanocomposites. *Graphene-Based Polymer Nanocomposites in Electronics,* **2015**, , 25-47.

[167]   Lee, Y.R.; Raghu, A.V.; Jeong, H.M.; Kim, B.K. Properties of waterborne polyurethane/functionalized graphene sheet nanocomposites prepared by an *in situ* method. *Macromol. Chem. Phys.,* **2009**, *210*(15), 1247-1254.

http://dx.doi.org/10.1002/macp.200900157

[168]   Ding, J.; Fan, Y.; Zhao, C.; Liu, Y.; Yu, C.; Yuan, N. Electrical conductivity of waterborne polyurethane/graphene composites prepared by solution mixing. *J. Compos. Mater.,* **2011**. 0021998311413835.

[169]   Park, S.H.; Thielemann, P.; Asbeck, P.; Bandaru, P.R. Enhanced dielectric constants and shielding effectiveness of, uniformly dispersed, functionalized carbon nanotube composites. *Appl. Phys. Lett.,* **2009**, *94*(24), 243111.

http://dx.doi.org/10.1063/1.3156032

[170]   Liu, Z.; Bai, G.; Huang, Y.; Ma, Y.; Du, F.; Li, F.; Guo, T.; Chen, Y. Reflection and absorption contributions to the electromagnetic interference shielding of single-walled carbon nanotube/polyurethane composites. *Carbon,* **2007**, *45*, 821-827.

http://dx.doi.org/10.1016/j.carbon.2006.11.020

[171]   Kuan, C-F.; Lin, K-C.; Chiang, C-L.; Chen, C-H.; Peng, H-C.; Kuan, H-C. Effect of modification method and processing condition on the properties of multiwall carbon nanotube/acrylonitrile-butadiene-styrene nanocomposite. *Adv. Sci. Lett.,* **2013**, *19*(2), 559-561.

http://dx.doi.org/10.1166/asl.2013.4771

[172]   Hoang, A. S. Electrical conductivity and electromagnetic interference shielding characteristics of multiwalled carbon nanotube filled polyurethane composite films. *Advances in Natural Sciences: Nanoscience and Nanotechnology,* **2011**, *2*, 025007.

http://dx.doi.org/10.1088/2043-6262/2/2/025007

[173]   Gupta, T.; Singh, B.; Teotia, S.; Katyal, V.; Dhakate, S.; Mathur, R. Designing of multiwalled carbon nanotubes reinforced polyurethane composites as electromagnetic interference shielding materials. *J. Polym. Res.,* **2013**, *20*(6), 1-7.

http://dx.doi.org/10.1007/s10965-013-0169-6

[174]   Yang, Y.; Gupta, M.C.; Dudley, K.L.; Lawrence, R.W. A comparative study of EMI shielding properties of carbon nanofiber and multi-walled carbon nanotube filled polymer composites. *J. Nanosci. Nanotechnol.,* **2005**, *5*(6), 927-931.

http://dx.doi.org/10.1166/jnn.2005.115 PMID: 16060155

[175]   Al-Saleh, M.H.; Sundararaj, U. Morphological, electrical and electromagnetic interference shielding characterization of vapor grown carbon nanofiber/polystyrene nanocomposites. *Polym. Int.,* **2013**, *62*(4), 601-607.

http://dx.doi.org/10.1002/pi.4317

[176]   Al-Saleh, M.H.; Sundararaj, U. A review of vapor grown carbon nanofiber/polymer conductive composites. *Carbon,* **2009**, *47*(1), 2-22.

http://dx.doi.org/10.1016/j.carbon.2008.09.039

[177]   Liang, J.; Wang, Y.; Huang, Y.; Ma, Y.; Liu, Z.; Cai, J.; Zhang, C.; Gao, H.; Chen, Y. Electromagnetic interference shielding of graphene/epoxy composites. *Carbon,* **2009**, *47*(3), 922-925.

http://dx.doi.org/10.1016/j.carbon.2008.12.038

[178]   Valentini, M.; Piana, F.; Pionteck, J.; Lamastra, F.R.; Nanni, F. Electromagnetic properties and performance of exfoliated graphite (EG) – Thermoplastic polyurethane (TPU) nanocomposites at microwaves. *Compos. Sci. Technol.,* **2015**, *114*, 26-33.

http://dx.doi.org/10.1016/j.compscitech.2015.03.006

[179]   Yang, L.; Phua, S.L.; Toh, C.L.; Zhang, L.; Ling, H.; Chang, M.; Zhou, D.; Dong, Y.; Lu, X. Polydopamine-coated graphene as multifunctional nanofillers in polyurethane. *RSC Advances,* **2013**, *3*(18), 6377-6385.

http://dx.doi.org/10.1039/c3ra23307c

[180]   Hsiao, S-T.; Ma, C-C.M.; Liao, W-H.; Wang, Y-S.; Li, S-M.; Huang, Y-C.; Yang, R-B.; Liang, W-F. *Lightweight and Flexible Reduced Graphene Oxide/Water-Borne Polyurethane Composites with High Electrical Conductivity and Excellent Electromagnetic Interference Shielding Performance*, **2014**.
        http://dx.doi.org/10.1021/am502412q

[181]   Hsiao, S-T.; Ma, C-C.M.; Tien, H-W.; Liao, W-H.; Wang, Y-S.; Li, S-M.; Huang, Y-C. Using a non-covalent modification to prepare a high electromagnetic interference shielding performance graphene nanosheet/water-borne polyurethane composite. *Carbon,* **2013**, *60*, 57-66.
        http://dx.doi.org/10.1016/j.carbon.2013.03.056

[182]   Hsiao, S-T.; Ma, C-C.M.; Tien, H-W.; Liao, W-H.; Wang, Y-S.; Li, S-M.; Yang, C-Y.; Lin, S-C.; Yang, R-B. *Effect of Covalent Modification of Graphene Nanosheets on the Electrical Property and Electromagnetic Interference Shielding Performance of a Water-Borne Polyurethane Composite*, **2015**.
        http://dx.doi.org/10.1021/am508069v

[183]   Wang, D.; Choi, D.; Li, J.; Yang, Z.; Nie, Z.; Kou, R.; Hu, D.; Wang, C.; Saraf, L.V.; Zhang, J.; Aksay, I.A.; Liu, J. Self-assembled TiO2-graphene hybrid nanostructures for enhanced Li-ion insertion. *ACS Nano,* **2009**, *3*(4), 907-914.
        http://dx.doi.org/10.1021/nn900150y PMID: 19323486

[184]   Ghazinejad, M.; Guo, S.; Paul, R.K.; George, A.S.; Penchev, M.; Ozkan, M.; Ozkan, C.S. Synthesis of graphene-CNT hybrid nanostructures **2011**.
        http://dx.doi.org/10.1557/opl.2011.1346

[185]   Kim, Y-S.; Kumar, K.; Fisher, F.T.; Yang, E-H. Out-of-plane growth of CNTs on graphene for supercapacitor applications. *Nanotechnology,* **2012**, *23*(1), 015301.
        http://dx.doi.org/10.1088/0957-4484/23/1/015301 PMID: 22155846

[186]   Patole, A.S.; Patole, S.P.; Jung, S-Y.; Yoo, J-B.; An, J-H.; Kim, T-H. Self assembled graphene/carbon nanotube/polystyrene hybrid nanocomposite by *in situ* microemulsion polymerization. *Eur. Polym. J.,* **2012**, *48*(2), 252-259.
        http://dx.doi.org/10.1016/j.eurpolymj.2011.11.005

[187]   Chatterjee, S.; Nafezarefi, F.; Tai, N.; Schlagenhauf, L.; Nüesch, F.; Chu, B. Size and synergy effects of nanofiller hybrids including graphene nanoplatelets and carbon nanotubes in mechanical properties of epoxy composites. *Carbon,* **2012**, *50*(15), 5380-5386.
        http://dx.doi.org/10.1016/j.carbon.2012.07.021

[188]   Li, W.; Dichiara, A.; Bai, J. Carbon nanotube–graphene nanoplatelet hybrids as high-performance multifunctional reinforcements in epoxy composites. *Compos. Sci. Technol.,* **2013**, *74*, 221-227.
        http://dx.doi.org/10.1016/j.compscitech.2012.11.015

[189]   Im, H.; Kim, J. Thermal conductivity of a graphene oxide–carbon nanotube hybrid/epoxy composite. *Carbon,* **2012**, *50*(15), 5429-5440.
        http://dx.doi.org/10.1016/j.carbon.2012.07.029

[190]   Pradhan, B.; Srivastava, S.K. Synergistic effect of three-dimensional multi-walled carbon nanotube–graphene nanofiller in enhancing the mechanical and thermal properties of high-performance silicone rubber. *Polym. Int.,* **2014**, *63*(7), 1219-1228.
        http://dx.doi.org/10.1002/pi.4627

[191]   Ramôa, S.D.; Barra, G.M.; Oliveira, R.V.; de Oliveira, M.G.; Cossa, M.; Soares, B.G. Electrical, rheological and electromagnetic interference shielding properties of thermoplastic polyurethane/carbon nanotube composites. *Polym. Int.,* **2013**, *62*(10), 1477-1484.
        http://dx.doi.org/10.1002/pi.4446

[192] Yang, S-Y.; Lin, W-N.; Huang, Y-L.; Tien, H-W.; Wang, J-Y.; Ma, C-C.M.; Li, S-M.; Wang, Y-S. Synergetic effects of graphene platelets and carbon nanotubes on the mechanical and thermal properties of epoxy composites. *Carbon,* **2011**, *49*(3), 793-803.

http://dx.doi.org/10.1016/j.carbon.2010.10.014

[193] Zhang, C.; Huang, S.; Tjiu, W.W.; Fan, W.; Liu, T. Facile preparation of water-dispersible graphene sheets stabilized by acid-treated multi-walled carbon nanotubes and their poly (vinyl alcohol) composites. *J. Mater. Chem.,* **2012**, *22*(6), 2427-2434.

http://dx.doi.org/10.1039/C1JM13921E

[194] Lee, J-S.; Jo, K.; Lee, T.; Yun, T.; Cho, J.; Kim, B-S. Facile synthesis of hybrid graphene and carbon nanotubes as a metal-free electrocatalyst with active dual interfaces for efficient oxygen reduction reaction. *J. Mater. Chem. A Mater. Energy Sustain.,* **2013**, *1*(34), 9603-9607.

http://dx.doi.org/10.1039/c3ta12520c

[195] Kim, J-Y.; Kim, T.; Suk, J.W.; Chou, H.; Jang, J-H.; Lee, J.H.; Kholmanov, I.N.; Akinwande, D.; Ruoff, R.S. Enhanced dielectric performance in polymer composite films with carbon nanotube-reduced graphene oxide hybrid filler. *Small,* **2014**, *10*(16), 3405-3411.

http://dx.doi.org/10.1002/smll.201400363 PMID: 24789173

[196] Vinayan, B.; Nagar, R.; Raman, V.; Rajalakshmi, N.; Dhathathreyan, K.; Ramaprabhu, S. Synthesis of graphene-multiwalled carbon nanotubes hybrid nanostructure by strengthened electrostatic interaction and its lithium ion battery application. *J. Mater. Chem.,* **2012**, *22*(19), 9949-9956.

http://dx.doi.org/10.1039/c2jm16294f

[197] Kamaliya, R.; Singh, B.P.; Gupta, B.K.; Singh, V.N.; Gupta, T.K.; Gupta, R.; Kumar, P.; Mathur, R.B. Large scale production of three dimensional carbon nanotube pillared graphene network for bi-functional optical properties. *Carbon,* **2014**, *78*, 147-155.

http://dx.doi.org/10.1016/j.carbon.2014.06.062

[198] Liu, Z.; Bai, G.; Huang, Y.; Li, F.; Ma, Y.; Guo, T.; He, X.; Lin, X.; Gao, H.; Chen, Y. Microwave Absorption of Single-Walled Carbon Nanotubes/Soluble Cross-Linked Polyurethane Composites *The Journal of Physical Chemistry C, 111*, 13696-13700.**2007**,

http://dx.doi.org/10.1021/jp0731396

[199] Jin, X.; Ni, Q-Q.; Natsuki, T. Composites of multi-walled carbon nanotubes and shape memory polyurethane for electromagnetic interference shielding. *J. Compos. Mater.,* **2011**, *45*(24), 2547-2554.

http://dx.doi.org/10.1177/0021998311401106

[200] Verma, M.; Verma, P.; Dhawan, S.; Choudhary, V. Tailored graphene based polyurethane composites for efficient electrostatic dissipation and electromagnetic interference shielding applications. *RSC Advances,* **2015**, *5*(118), 97349-97358.

http://dx.doi.org/10.1039/C5RA17276D

[201] Verma, M.; Chauhan, S. S.; Dhawan, S. K.; Choudhary, V. Graphene nanoplatelets/carbon nanotubes/polyurethane composites as efficient shield against electromagnetic polluting radiations *Composites Part B: Engineering,* **2017**, *120*

http://dx.doi.org/10.1016/j.compositesb.2017.03.068

[202] Wang, H.; Xie, G.; Fang, M.; Ying, Z.; Tong, Y.; Zeng, Y. Electrical and mechanical properties of antistatic PVC films containing multi-layer graphene. *Compos., Part B Eng.,* **2015**, *79*, 444-450.

http://dx.doi.org/10.1016/j.compositesb.2015.05.011

[203] Wang, Y.; Liu, M.; Liu, Y.; Luo, J.; Lu, X.; Sun, J. A novel mica-titania@ graphene core-shell structured antistatic composite pearlescent pigment. *Dyes Pigments,* **2017**, *136*, 197-204.

http://dx.doi.org/10.1016/j.dyepig.2016.08.035

[204] HU, N.; ZHAO, L.-j.; FAN, J.; HU, X.-l.; LI, Q. Preparation and Properties Characterization of Waterborne Acrylic Resin/Graphene Oxide Antistatic Coating. *China Plastics IndustryChina Plastics Industry,* **2015,** *8,* 034.

[205] Song, W.-L.; Fan, L.-Z.; Cao, M-S.; Lu, M-M.; Wang, C-Y.; Wang, J.; Chen, T-T.; Li, Y.; Hou, Z-L.; Liu, J.; Sun, Y-P. Facile fabrication of ultrathin graphene papers for effective electromagnetic shielding. *J. Mater. Chem. C Mater. Opt. Electron. Devices,* **2014,** *2*(25), 5057-5064.

http://dx.doi.org/10.1039/C4TC00517A

[206] Song, W.-L.; Cao, M-S.; Lu, M-M.; Bi, S.; Wang, C-Y.; Liu, J.; Yuan, J.; Fan, L-Z. Flexible graphene/polymer composite films in sandwich structures for effective electromagnetic interference shielding. *Carbon,* **2014,** *66,* 67-76.

http://dx.doi.org/10.1016/j.carbon.2013.08.043

[207] Yan, D.X.; Pang, H.; Li, B.; Vajtai, R.; Xu, L.; Ren, P.G.; Wang, J.H.; Li, Z.M. Structured Reduced Graphene Oxide/Polymer Composites for Ultra-Efficient Electromagnetic Interference Shielding. *Adv. Funct. Mater.,* **2015,** *25*(4), 559-566.

http://dx.doi.org/10.1002/adfm.201403809

[208] Yan, D-X.; Pang, H.; Xu, L.; Bao, Y.; Ren, P-G.; Lei, J.; Li, Z-M. Electromagnetic interference shielding of segregated polymer composite with an ultralow loading of *in situ* thermally reduced graphene oxide. *Nanotechnology,* **2014,** *25*(14), 145705.

http://dx.doi.org/10.1088/0957-4484/25/14/145705 PMID: 24633439

[209] Gupta, T.K.; Singh, B.P.; Mathur, R.B.; Dhakate, S.R. Multi-walled carbon nanotube-graphene-polyaniline multiphase nanocomposite with superior electromagnetic shielding effectiveness. *Nanoscale,* **2014,** *6*(2), 842-851.

http://dx.doi.org/10.1039/C3NR04565J PMID: 24264356

[210] Verdejo, R.; Bernal, M.M.; Romasanta, L.J.; Lopez-Manchado, M.A. Graphene filled polymer nanocomposites. *J. Mater. Chem.,* **2011,** *21*(10), 3301-3310.

http://dx.doi.org/10.1039/C0JM02708A

[211] Shahil, K.M.; Balandin, A.A. Graphene-multilayer graphene nanocomposites as highly efficient thermal interface materials. *Nano Lett.,* **2012,** *12*(2), 861-867.

http://dx.doi.org/10.1021/nl203906r PMID: 22214526

[212] Park, W.; Guo, Y.; Li, X.; Hu, J.; Liu, L.; Ruan, X.; Chen, Y.P. High-Performance Thermal Interface Material Based on Few-Layer Graphene Composite. *J. Phys. Chem. C,* **2015,** *119*(47), 26753-26759.

http://dx.doi.org/10.1021/acs.jpcc.5b08816

[213] Han, Y.; Wu, Y.; Shen, M.; Huang, X.; Zhu, J.; Zhang, X. Preparation and properties of polystyrene nanocomposites with graphite oxide and graphene as flame retardants. *J. Mater. Sci.,* **2013,** *48*(12), 4214-4222.

http://dx.doi.org/10.1007/s10853-013-7234-8

[214] Gavgani, J.N.; Adelnia, H.; Gudarzi, M.M. Intumescent flame retardant polyurethane/reduced graphene oxide composites with improved mechanical, thermal, and barrier properties. *J. Mater. Sci.,* **2014,** *49*(1), 243-254.

http://dx.doi.org/10.1007/s10853-013-7698-6

[215] Tong, W.; Zhang, Y.; Zhang, Q.; Luan, X.; Duan, Y.; Pan, S.; Lv, F.; An, Q. Achieving significantly enhanced dielectric performance of reduced graphene oxide/polymer composite by covalent modification of graphene oxide surface. *Carbon,* **2015,** *94,* 590-598.

http://dx.doi.org/10.1016/j.carbon.2015.07.005

[216] Xiong, G.; Meng, C.; Reifenberger, R.G.; Irazoqui, P.P.; Fisher, T.S. A Review of Graphene-Based Electrochemical Microsupercapacitors. *Electroanalysis,* **2014,** *26*(1), 30-51.

http://dx.doi.org/10.1002/elan.201300238

[217] Wang, H.; Yang, Y.; Liang, Y.; Zheng, G.; Li, Y.; Cui, Y.; Dai, H. Rechargeable $Li–O_2$ batteries with a covalently coupled $MnCo_2O_4$–graphene hybrid as an oxygen cathode catalyst. *Energy Environ. Sci.,* **2012**, *5*(7), 7931-7935.

http://dx.doi.org/10.1039/c2ee21746e

[218] Huang, C.; Bai, H.; Li, C.; Shi, G. A graphene oxide/hemoglobin composite hydrogel for enzymatic catalysis in organic solvents. *Chem. Commun. (Camb.),* **2011**, *47*(17), 4962-4964.

http://dx.doi.org/10.1039/c1cc10412h PMID: 21431118

[219] Bonaccorso, F.; Sun, Z.; Hasan, T.; Ferrari, A. Graphene photonics and optoelectronics. *Nat. Photonics,* **2010**, *4*(9), 611-622.

http://dx.doi.org/10.1038/nphoton.2010.186

[220] Wang, D-W.; Li, F.; Zhao, J.; Ren, W.; Chen, Z-G.; Tan, J.; Wu, Z-S.; Gentle, I.; Lu, G.Q.; Cheng, H-M. Fabrication of graphene/polyaniline composite paper *via in situ* anodic electropolymerization for high-performance flexible electrode. *ACS Nano,* **2009**, *3*(7), 1745-1752.

http://dx.doi.org/10.1021/nn900297m PMID: 19489559

[221] Wang, Y.; Shi, Z.; Huang, Y.; Ma, Y.; Wang, C.; Chen, M.; Chen, Y. Supercapacitor devices based on graphene materials. *J. Phys. Chem. C,* **2009**, *113*(30), 13103-13107.

http://dx.doi.org/10.1021/jp902214f

[222] Wu, Q.; Xu, Y.; Yao, Z.; Liu, A.; Shi, G. Supercapacitors based on flexible graphene/polyaniline nanofiber composite films. *ACS Nano,* **2010**, *4*(4), 1963-1970.

http://dx.doi.org/10.1021/nn1000035 PMID: 20355733

[223] Song, Z.; Xu, T.; Gordin, M.L.; Jiang, Y-B.; Bae, I-T.; Xiao, Q.; Zhan, H.; Liu, J.; Wang, D. Polymer-graphene nanocomposites as ultrafast-charge and -discharge cathodes for rechargeable lithium batteries. *Nano Lett.,* **2012**, *12*(5), 2205-2211.

http://dx.doi.org/10.1021/nl2039666 PMID: 22449138

[224] Santos, C.M.; Mangadlao, J.; Ahmed, F.; Leon, A.; Advincula, R.C.; Rodrigues, D.F. Graphene nanocomposite for biomedical applications: fabrication, antimicrobial and cytotoxic investigations. *Nanotechnology,* **2012**, *23*(39), 395101.

http://dx.doi.org/10.1088/0957-4484/23/39/395101 PMID: 22962260

[225] Das, T.K.; Prusty, S. Graphene-based polymer composites and their applications. *Polym. Plast. Technol. Eng.,* **2013**, *52*(4), 319-331.

http://dx.doi.org/10.1080/03602559.2012.751410

CHAPTER 5

Synthesis of Poly (3, 4-ethylene dioxythiophene) Conducting Polymer Composites for EMI Shielding Applications

M. Farukh[1*], Jasvir Dalal[2], Anil Ohlan[3] and S. K. Dhawan[4]

[1] *Higher Institute of Plastics Fabrication, Riyadh- 14331, Saudi Arabia*

[2] *Department of Physics, Chaudhary Ranbir Singh University, Jind, India*

[3] *Department of Physics, Maharshi Dayanand University, Rohtak, India*

[4] *Materials Physics & Engineering Division, CSIR-National Physical Laboratory, New Delhi – 110012, India*

Abstract: This chapter gives a brief oversight of the preparation, characterization, and electromagnetic interference shielding studies of conducting polymer and MWCNT composites. The different approaches have been implemented to fabricate composites, which are used as EMI shielding materials. The key emphasis has been given to PEDOT conducting polymer, which is synthesized *via* emulsion polymerization. The topographical and chemical analyses of polymer composite samples were characterized by scanning electron microscopy, transmission electron microscopy, Fourier transform infrared spectroscopy, X-ray diffraction, thermo gravimetric analysis, and Raman spectroscopy. The dielectric and electromagnetic shielding measurements of the polymer composites were conducted by a vector network analyzer. In addition, PEDOT/MWCNT coated polyurethane foam usage as an antistatic material has also been discussed in this chapter.

Keywords: Anti-static, Electromagnetic shielding and dielectric measurements, Emulsion polymerization, Fourier transform infrared spectroscopy, Multi-walled carbon nanotube, Poly (3, 4-ethylene dioxythiophene), Scanning electron microscopy, Thermogravimetric analysis, Transmission electron microscopy, X-ray diffraction.

---

*Corresponding author **M. Farukh:** Higher Institute of Plastics Fabrication, Department of Basic Technology, New Industrial Area, Riyadh- 14331, Saudi Arabia; Tel: +966557106223;
E-mail: mohdfrk4@gmail.com

Sundeep K. Dhawan, Avanish Pratap Singh, Anil Ohlan, Kuldeep Singh Kakran and Pradeep Sambyal (Ed.)
All rights reserved-© 2022 Bentham Science Publishers

5.1. INTRODUCTION

With the growing demand for advanced technologies, researchers are developing novel functional materials with synergistic properties. Conducting polymers (CPs) and conducting polymer composites (CPCs) have attracted huge curiosity over the past few decades in view of their fascinating electrical, magnetic, and optical properties. Besides these unique properties, they possess light weight, flexibility, and processability characteristics that make them distinctive materials. The CPs have immensely improved our society in the field of solar cells, batteries, bio-sensing, gas sensing, field emission, EMI shielding, corrosion protection, anti-static and electro-chromic devices [1].

In comparison to pristine CPs, CPCs are prepared by incorporating nanomaterials into the matrix of CPs, thereby greatly enhancing their properties. Various methodologies have been adopted by researchers to fabricate CPCs. The inclusion of conducting, magnetic, or dielectric nanomaterial fillers in the CP matrix can be attained by ex-*in-situ* physical blending processes or by in-*in-situ* emulsion polymerization technique [2]. Ex-*in-situ* blending processes are usually associated with poor dispersion of nanomaterials into the CP matrix. Such problems can be overcome by utilizing in-*in-situ* emulsion polymerization techniques where the concentration of surfactant and slow addition of oxidant into the reaction mixture controls the dispersion and polymerization process.

Among these innovative materials, also called "synthetic metals," polyaniline (PANI), polypyrrole (PPy) and polythiophene (PTh) are the most studied due to peculiar characteristics such as good conductivity and high environmental stability [3], making them particularly attractive for applications in various fields like photovoltaic devices, batteries, electrodes, sensors, EMI shielding, *etc*. [4–6].

5.1.1. Polythiophene

Tourillon and Garnier were the first to synthesize polythiophene through electro-polymerization and reported electronic conductivity (10 - 100 S/cm), thereby initiating the new era of polythiophene chemistry [7]. The structure of the polymer is illustrated in Fig. (**1**). Polythiophene is an interesting polymer due to ease of synthesis, facile processability, and environmental stability. The presence of benzidine moieties in the polymer chain produces toxic products on degradation [8, 9] but is disadvantageous due to being insoluble in any organic solvents [10]. To overcome these problems, numerous substituents of atoms/molecules have been carried out, resulting in new derivatives with better physical and electronic properties.

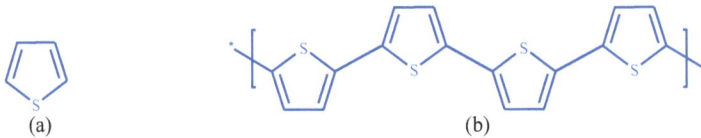

Fig. (1). a) Thiophene and **b**) polythiophene.

Polythiophene derivatives, particularly on poly (3-substituted thiophene/s) and poly (3,4-disubstituted thiophene/s), have been studied and reported [11]. The substituted polythiophene has been extensively studied as compared to unsubstituted thiophene due to its lower oxidation potential. The alkyl substitutions at the 3rd and 4th positions of the thiophene rings are reported to result in a significant increase in conjugation length and electronic conductivity [11,12]. The poly (3,4-ethylene dioxythiophene) and its derivatives are apparently potential materials having lower bandgap due to the availability of two electron-rich oxygen atoms near the thiophene ring [7]. As compared to other substituted thiophenes, these alkylene dioxy thiophenes have outstanding stability and higher conductivity in the doped state [13].

PEDOT was first synthesized by Bayer AG research laboratories in 1980 [10], commercialized under the trade name of BAYTRON P, and was prepared through electrochemical polymerization methods [14,15]. The polymer so produced was insoluble yet possessed interesting properties. The insolubility problem of the polymer was resolved using water-soluble polyelectrolyte and surfactants, for instance, polystyrene sulfonic acid (PSSA) and dodecyl benzene sulfonic acid (DBSA) acting as counter ions with PEDOT [16,17]. The resultant PEDOT/PSS and PEDOT/DBSA possessed good conductivity [18], optical transparency in doped form [19], good film-forming properties [20], high visible light transparency [21], and excellent environmental stability [22]. PEDOT/PSS found various technological applications, including antistatic coating for photographic films [23], hole transport layer in organic photovoltaic and light-emitting diodes [24,25], *etc*. The conducting layer of PEDOT/PSS can be applied to the electro-luminescent and organic field-effect transistors [26]. To overcome the insolubility problem of

PEDOT, Bayer AG polymerized 3,4-ethylene dioxythiophene (EDOT) monomer in an aqueous polystyrene sulfonic acid (PSSA) solution using $Na_2S_2O_8$ as oxidant, resulting in the formation of PEDOT/PSS, as shown in Fig. (**2**) [27].

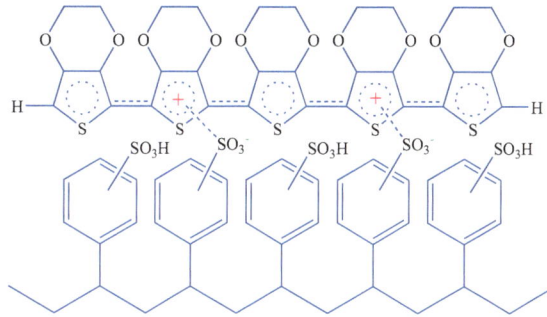

Fig. (2). Chemical structure of PEDOT/PSS.

PEDOT/PSS is an aqueous dispersion of negatively charged saturated polystyrene sulfonate (PSS) and positively charged conjugated PEDOT. PSS is a polymeric surfactant that also acts as a dopant in the polymer chain, which allows it to disperse and stabilize the PEDOT in water and other solvents. Different solution processing techniques such as spray technique, spin coating, screen printing, inkjet printing, *etc.*, can be used to develop a thin film on flexible and stiff surfaces [7].

Multi-walled carbon nanotubes (MWCNTs) are tubular, hollow nanostructures of carbon composed of a single sheet of pure graphite with a diameter of 0.8 to 60 nm and having a length of several micrometers. They have excellent thermal, physical, electrical, and mechanical properties. They are used for the fabrication of CNT-based field-effect transistors (CNFETs) and are potential candidates for the next generation of electronic devices. Digital systems based on CNFET are expected to be able to outperform silicon-based CMOS technologies. The high aspect ratio and exceptional electrical conductivity of CNTs make them a suitable candidate to impart higher conductivity at a lower concentration as compared to conventional filler materials like chopped carbon fiber, carbon black, metal nanoparticles, and SS fiber [28]. These exceptional material properties of CNT reinforced polymer composites make them suitable candidates for EMI shielding applications [29].

The PEDOT/MWCNT composites not only display the advantages of the individual components, such as electrical conductivity, magnetic, dielectric, and charge density, but also the synergistic effects of PEDOT and MWCNTs. The interaction between the PEDOT polymer and MCNTs can be attributed to the covalent or non-covalent bonding. The free electrons of the heteroatom of PEDOT interact with MWCNTs *via* non-covalent, van der Waals, or hydrophobic forces, whereas covalent bonding can be achieved by the surface or chemical functionalization of MWCNTs.

CP grafted MWCNT composites are considered a useful approach to attenuate electromagnetic waves. Xiaoxia B. *et al.* reported highly conducting leaflet-like PEDOT/SWCNT composite, which exhibited reflection loss of 19.9 dB at 11.2 GHz and 40.9 dB at the 14.1 GHz [30]. Parveen S. *et al.* prepared polyaniline MWCNT nanocomposites by in *in-situ* polymerization and showed absorption governed SE value of − 27.5 to − 39.2 dB in the Ku-band [31]. Moreover, composites based on PEDOT grafted CNTs have been reported by many research groups for various applications. Ramona G. *et al.* deposited CNTs/PEDOT on microelectrode by the electro-polymerization of ethylene-dioxythiophene (EDOT) in the presence of suspension of CNTs and polystyrene sulfonate and used these electrodes for stimulation and recording in cardiac and neurophysiological research [32]. An amperometric sensor for the selective detection of hydroquinone in cosmetics using CNTs doped PEDOT composite was developed by Guiyun X. *et al.* [33]. It was observed that, as compared to the bare electrode, the oxidation potential of hydroquinone on the PEDOT/CNT modified carbon paste electrode was much lower, and its charge transfer rate constant was significantly increased from 0.45 to 1.84 s^{-1} for the oxidation of hydroquinone. Jeng-Yu L. *et al.* reported the fabrication of low-cost, flexible, and efficient counter electrodes composed of a PEDOT decorated CNTs/polypropylene composite plate for dye-sensitized solar cells having 6.82% energy conversion efficiency [34].

5.2. CONJUGATED POLYMER COMPOSITES FOR EMI SHIELDING

In recent years, in order to protect electronic equipment linked up with security or communication system, metals were preferred for EMI shielding because of their high conductivity, leading to the use of EMI shielding as a reflector [35]. A metallic shield protects the devices by preventing the emission of e-m radiations emanating from equipment [36]. For this purpose, enclosures are made up of an alloy of nickel, iron, copper, and chromium possessing high permeability. The metal shield provides effective shielding performance, but susceptibility to corrosion limits its application, especially in sea environments due to Rusty Bolt Effect [37]. Galvanic corrosion also occurs in two different metals used for shielding, resulting in resulting in the shield's decreased protection [38].

Stealth technology is one of the important applications of EMI shielding in the defense sector and requires high microwave absorbing capacity material for the shield. The high energy loss is the main requirement of these materials in the microwave frequency range to dissipate energy as heat by absorbing incident radiations [39]. For this purpose, the materials should have magnetization and electrical conductivity for the absorption of e-m radiations [40]. Accordingly, only the materials with the capability of high energy losses during impinging of e-m radiations are selected [41]. The composites synthesized by incorporating conductive and magnetic filler into a conjugated or insulating polymer matrix find application as absorbing materials. According to electromagnetic theory, absorption loss in a material is directly related to the product of magnetic permeability 'μ' and electrical conductivity 'σ'. Hence, a material possessing magnetic loss property has a higher capability of microwave absorption with broadband than dielectric materials [42].

5.2.1. Synthesis: Polymerization of Conjugated Polymers

The chemical oxidative polymerization route is one of the best methods for the fabrication of conjugated polymers. It is an easy process that produces high yield and has been extensively used for commercial production. In this method, a monomer is oxidized using a strong oxidizing agent like ammonium per-oxy disulphate [43], ferric ions [44], tetra butyl ammonium persulphate [45], benzoyl peroxide [46], chloroauric acid [47], hydrogen peroxide [48], *etc.*, resulting in the formation of cations, which further react with the remaining monomers and yield high molecular weight polymers or oligomers. Some properties of the polymer, such as molecular weight, cross-linking, and conductivity, are highly dependent on dopants as well as its concentration. Different surfactants, such as dodecylbenzene sulfonic acid (DBSA), para toluene sulphonic acid (PTSA) and β-naphthalene sulphonic acid, are used for the synthesis of polymer with different morphology [49].

5.2.1.1. Emulsion Polymerization

The chemical reaction where polymerization occurs through micelles formation consisting of monomers and initiator radicals in heterogeneous phase is known as emulsion polymerization. Oil and water emulsion is the most common example of emulsion polymerization in which the oil drops (monomer droplets) are emulsified homogeneously in water. In this process, hydrophobic monomers are involved for emulsification in a heterogeneous aqueous solvent, and the reaction is initiated by a free radical soluble in water. At the end of emulsion polymerization, a blueish viscous fluid is obtained, termed as "polymer dispersion" or "synthetic latex." The term "latex" is " the colloidal dispersion of organic/inorganic polymer particles in an aqueous medium." Latex comprises 40-60% solid polymer particles in the aqueous phase or approximately 10^{15} polymer particles per mL. Usually, the particles are spherical in diameter in the nm range. A particle consists of $1-10^4$ macromolecules, and a single macromolecule is composed of 10^2-10^6 monomer units [50–53]. Mainly, emulsion polymerization comprises three ingredients, *i.e.,* a) monomer, b) surfactant and c) initiator.

A monomer is the smallest unit of polymer structure and the basic element of emulsion polymerization, which could be positive or negative or free radical undergoing polymerization in liquid or gaseous form depending upon the reaction conditions. In the case of highly soluble monomers, particle formation is difficult, and the rate of reaction kinetics gets reduced.

The surfactant is a dispersion medium (for monomer) in which emulsion polymerization proceeds. A surfactant consists of two parts; the first is the tail, and the second is the head. The behavior of the tail and head either is hydrophobic (water repellant) or hydrophilic (water attractive) depending upon the nature of the surfactant, *i.e.,* whether it is anionic, cationic or nonionic. Generally, a surfactant with a low critical micelle concentration (CMC) is preferred, and if the concentration is above the CMC, the polymerization rate increases dramatically. A minimum molar ratio or optimized ratio of surfactant with respect to monomer is used. The most common surfactants are fatty acid, aryl aliphatic, sodium lauryl sulfate, sodium dodecyl benzene sulfonic acid, *etc.*

Initiators are the chemically reducing or oxidizing agents used to produce free radicals, which initiate the emulsion polymerization process [54]. The radical mechanism stage is normally followed by both chain-growth emulsion polymerization and step-growth emulsion polymerization. The free radicals are generated in two ways, *i.e.*, a) thermal decomposition and 2) redox reactions. Commonly used free radical initiators are peroxy disulphate, *i.e.*, $(NH_4)_2S_2O_8$, $Na_2S_2O_8$, $K_2S_2O_8$, *etc.* These initiators are water-soluble and decomposed thermally into two anionic sulfate radicals $(S_2O_8^{-2} \rightarrow SO_4^{-*} + SO_4^{-*})$ that initiate the polymerization. In addition to the water-soluble initiators, some oil-soluble compounds such as azo-bis-isobutyronitrile (AIBN) and benzoyl peroxide may be used to produce free radicals in emulsion polymerization as a thermal initiator. The whole procedure of emulsion polymerization is depicted in Fig. (**3**).

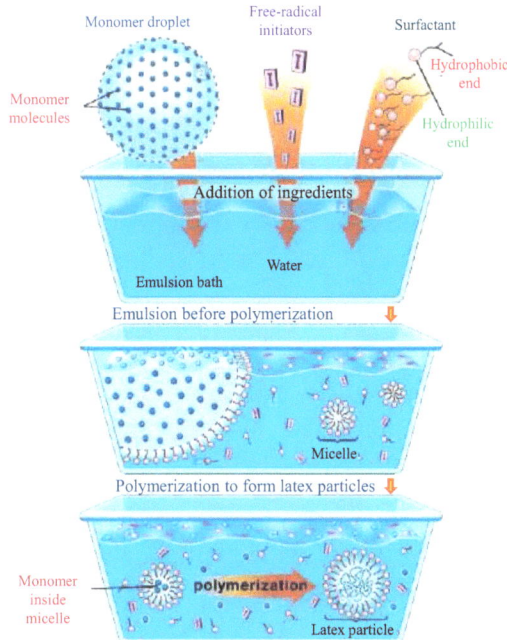

Fig. (3). Emulsion polymerization process [55].

5.2.1.2. Synthesis of Conducting PEDOT & PEDOT's Nanocomposites

The synthesis of PEDOT was carried out through emulsion polymerization techniques (Fig. **4**) [56, 57]. The polymerization of EDOT was allowed to take place in the aqueous phase in which ammonium sulfate was used as a reaction initiator or an oxidizing agent [58]. During polymerization, the pH of the solution was maintained at 1. The molar ratio of monomer, oxidant, and dopant is taken as 1:1:3. The polymerization reaction temperature is maintained at -2 °C to obtain high molecular weight. Using this synthesis technique, the polymer has an effective high yield, and also a bulk quantity of polymer can be obtained.

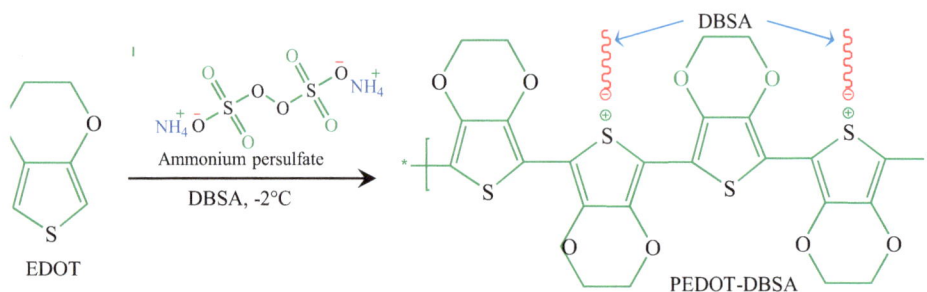

Fig. (4). Polymerization reaction of EDOT to PEDOT.

5.2.1.3. Synthesis of Dodecyl Benzene Sulfonic Acid Doped PEDOT

Synthesis of dodecyl benzene sulfonic acid (DBSA) doped PEDOT was carried out using emulsion polymerization [59]. In emulsion polymerization, 0.1 M of EDOT monomer and 0.3 M of DBSA dopant were homogenized in an aqueous medium for an hour to form EDOT–DBSA micelles. The polymerization was initiated by the drop-wise addition of oxidant, 0.1 M aqueous ammonium persulfate solution, at -2 °C with constant stirring for 8 h. The mechanism of polymerization of EDOT monomer to PEDOT polymer is shown in Fig. (**5**). A bluish-green suspension containing precipitates of PEDOT was obtained thereafter. The resulting suspension was demulsified using isopropyl alcohol due to the formation of stable micro-emulsions of PEDOT, water, and DBSA and then filtered in the Buchner funnel. The filtered precipitate obtained was washed with distilled water and then dried at 60 °C in a vacuum oven.

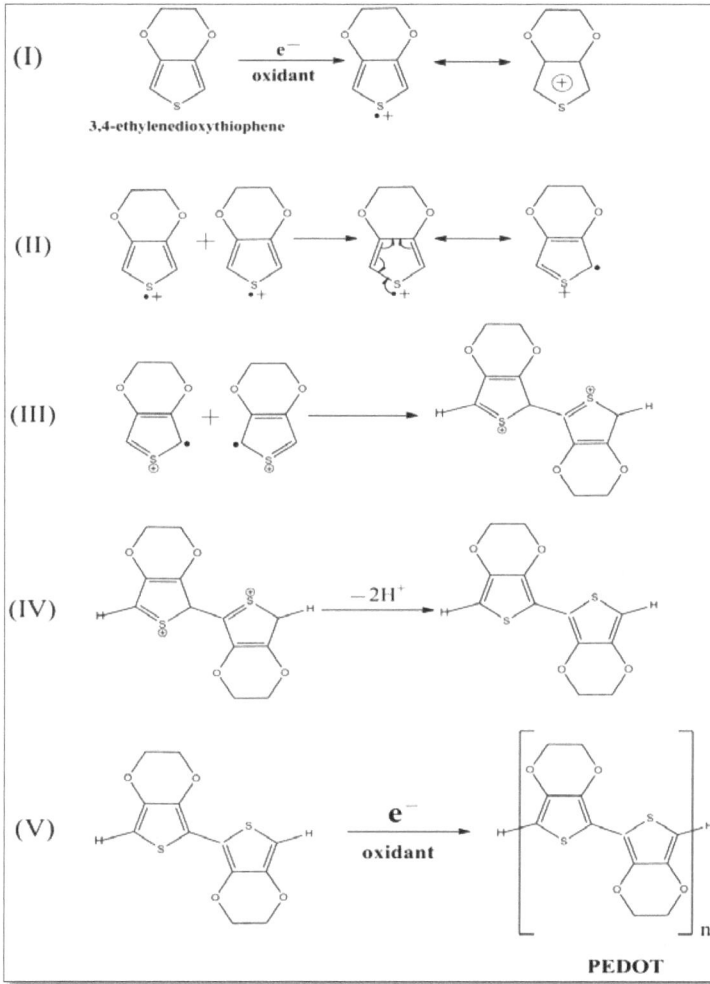

Fig. (5). Mechanism of formation of PEDOT polymer by emulsion polymerization.

5.2.1.4. Synthesis of PEDOT Grafted MWCNT Composites

The preparation of PEDOT grafted MWCNT composite by in-*in-situ* emulsion polymerization is shown in Fig. (**6**) [60]. In this synthesis, 0.3 M DBSA was homogenized in water for an hour to form a stable aqueous emulsion. To this aqueous emulsion, a calculated amount of MWCNTs was added (5.0, 10, and 15 wt. %) and then homogenized at 12000 rpm for 2 h. The wt. % of filler was taken with respect to the EDOT weight. After that, 0.1 M of EDOT was added and stirred for hours. At last, the oxidant APS (0.1 M) was added drop-wise with vigorous stirring while keeping the temperature of the reactor at -2 °C. A blackish precipitate of PEDOT grafted MWCNT composite was obtained after 8 hours of stirring. The precipitate obtained was treated with isopropyl alcohol (de-emulsifier) with vigorous stirring for 2 h. The resulting precipitate was filtered, washed thoroughly with distilled water, and dried at 60 °C in a vacuum oven.

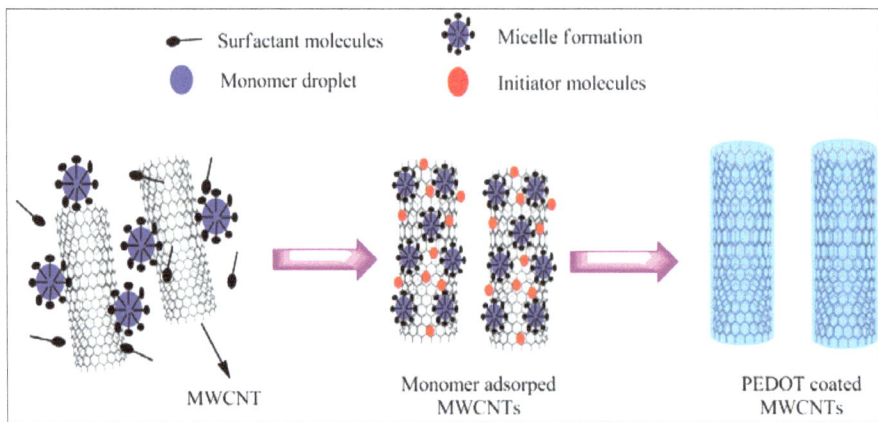

Fig. (6). Preparation of PEDOT grafted MWCNT composite by in-*in-situ* emulsion polymerization.

5.3. CHARACTERIZATION OF PEDOT AND PEDOT GRAFTED MWCNT COMPOSITES

Electron microscopy images of PEDOT and PEDOT grafted MWCNT composites are presented in Fig. (**7**). Fig. (**7a**) shows the SEM micrographs of PEDOT in which a regular granular/particulate morphology of the polymer is observed. Fig. (**7b**) shows an SEM micrograph of PEDOT grafted MWCNT composite, where a sharp change in the morphology of PEDOT is observed, justified by the presence of fibrillar MWCNT in the PEDOT polymer matrix. Fig. (**7c**) is high-resolution transmission electron microscopy (HRTEM) image of MWCNTs in the PEDOT polymer matrix. In Fig. (**7d**), Fe particles as impurities are observed that are responsible for contributing char residues in the thermo gravimetric analysis (TGA) thermograms and enhanced wave absorption behavior of the PEDOT grafted MWCNT composites.

XRD patterns of MWCNTs, PEDOT, and PEDOT grafted MWCNT composites with different loadings (5.0 %, 10 %, and 15 %) are shown in Fig. (**8**). The diffraction peaks observed at 2θ value 26.5° and 42.4° observed for MWCNTs along planes (002) and (100), respectively, attributed to the hexagonal graphitic structure [61]. The peak at 26.5° corresponds to the interlayer spacing 0.34 nm of the CNTs and the peak at 42.4° shows the reflections of the carbon atoms. For pure PEDOT, a broad diffraction peak is observed at 24.5°, which reveals its amorphous nature [59]. For the composites, the characteristic peaks of MWCNT disappear. This could be due to the uniform layer of PEDOT over the surface of MWCNTs.

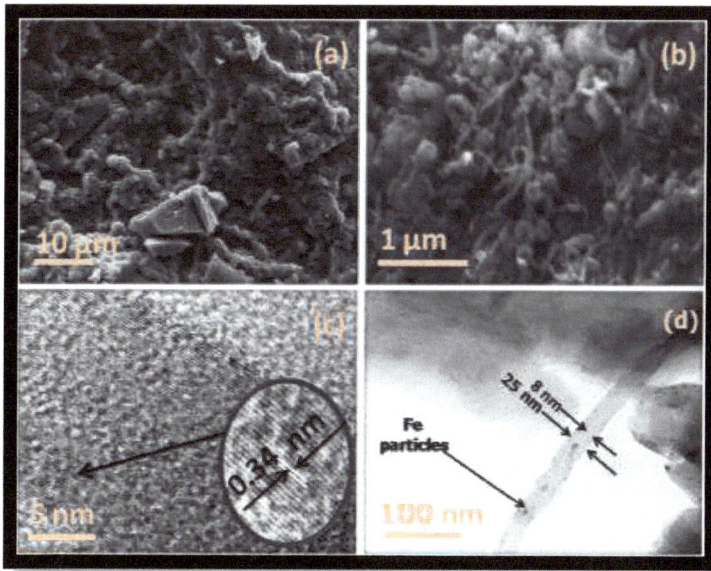

Fig. (7). Electron Micrographs (**a**) SEM of PEDOT, (**b**) SEM of PEDOT grafted MWCNT composite (**c**) HRTEM of PEDOT grafted MWCNT (**d**) TEM micrograph presents MWCNTs diameter of approximately 25 nm and Fe particles. Reprinted from reference [60]. Copyright (2015) Elsevier.

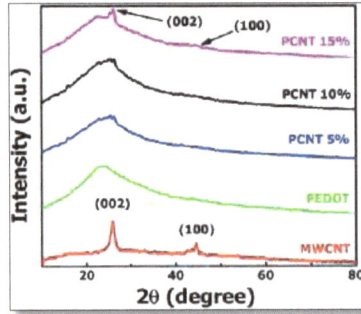

Fig. (8). XRD patterns of MWCNT, PEDOT, and PEDOT/MWCNT composites. Reprinted from reference [60]. Copyright (2015) Elsevier.

Raman spectra of chemically synthesized PEDOT, MWCNT, and PEDOT grafted MWCNT composites with 5%, 10%, and 15% are presented in Fig. (**9**). The important bands related to the PEDOT structure are strongly visible in the spectra. The bands near 1561 and 1518 cm^{-1} can be assigned to $C_\alpha = C_\beta$ anti-symmetric vibrations in the aromatic thiophene ring [62]. The band at 1432 cm^{-1} is due to symmetric $C_\alpha - C_\beta$ stretching deformations. The band at 1362 cm^{-1} is due to $C_\beta - C_\beta$ stretching deformations in the aromatic thiophene ring. However, a number of very low-intensity bands of PEDOT are present but are not clearly visible in the spectra. These peaks are due to $C_\alpha - C_\alpha$ inter-ring stretching at 1259 cm^{-1}, $C - O - C$ deformation at 1100 cm^{-1}, oxyethylene ring deformation at 995 cm^{-1}, $C - S - C$ deformation at 702 cm^{-1}, oxyethylene deformation at 575 cm^{-1}, and SO_2 bending at 433 cm^{-1} [63]. MWCNTs show characteristic bands of graphite at 1580 cm^{-1} (G-band), assigned to the in-plane vibration of the $C - C$ bond, and at 1345 cm^{-1} (D-band), due to the presence of disorder in carbon systems. The MWCNT spectrum also demonstrates a band at 2700 cm^{-1} due to overtone of the D band called the G' band [64]. However, in the Raman spectra of PEDOT grafted MWCNT composites with 5%, 10%, and 15% loadings, a blue shift of peaks is observed. The band at 1432 cm^{-1} is shifted to 1442 cm^{-1}, while the band at 1561 cm^{-1} is shifted to 1574 cm^{-1}. The shifting is ascribed to a strong interaction between polymer and MWCNTs. Moreover, the presence of G' band in the composites shows MWCNTs entrenched PEDOT matrix.

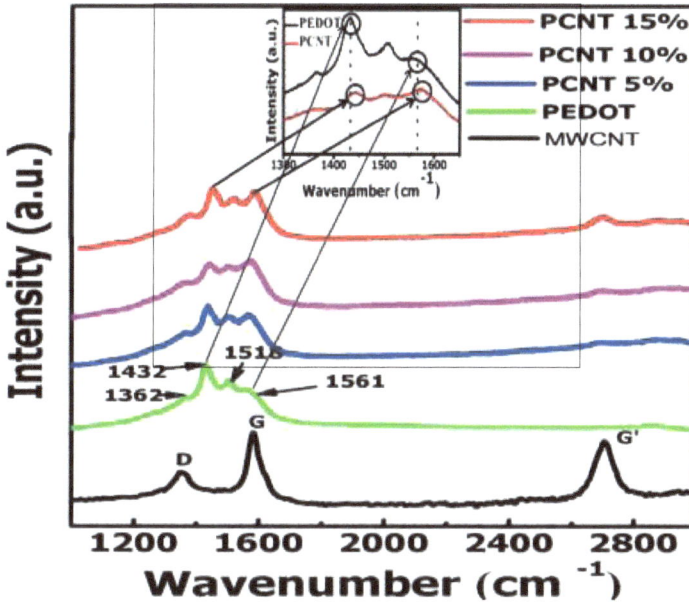

Fig. (9). Raman spectra of MWCNTs, PEDOT, and PEDOT grafted MWCNT composites. Reprinted from reference [60]. Copyright (2015) Elsevier.

Fig. (**10**) shows the FTIR spectra of MWCNT, PEDOT and PEDOT grafted MWCNT composites. In the FTIR spectrum of PEDOT, the observed vibrational bands at around 1312 cm^{-1} and 1514 cm^{-1} are due to C - C or C = C stretching of quinoid structure and ring stretching of thiophene ring, respectively. The peaks at 1187, 1137, and 1049 cm^{-1} arise due to the C−O−C bond stretching in the ethylene dioxy group. Further, the FTIR peaks at 978 and 832 cm^{-1} are due to the deformation modes in the C−S−C thiophene ring. The main characteristic peak of PEDOT (1330 cm^{-1}) is also present in the spectra near 1320 cm^{-1} [65]. The FTIR spectrum of MWCNTs is shown in the inset shows the characteristic peak near 1600 cm^{-1} elucidating the existence of carbon-carbon double bonds [66]. FTIR spectrum of PEDOT grafted MWCNT composite shows all bands of PEDOT [67]. The shifting of peaks is observed in the spectra suggesting some weak forces acting between MWCNTs and PEDOT due to the wrapping or capping of polymer over MWCNTs. The spectra of composites do not show new peaks confirming that there is no chemical reaction between MWCNTs and PEDOT.

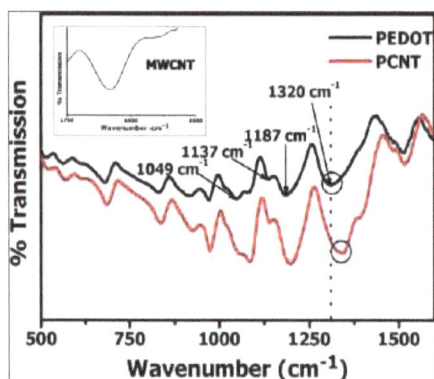

Fig. (10). FTIR spectra of MWCNTs, PEDOT, and PEDOT grafted MWCNT composite. Reprinted from reference [60]. Copyright (2015) Elsevier.

Thermo gravimetric analysis (TGA) of PEDOT and PEDOT grafted MWCNT composites are presented in Fig. (**11**). TGA was performed to evaluate the role of thermally stable fillers in improving the thermal stability of composites. Thermograms of PEDOT and PEDOT grafted MWCNT composites were recorded by heating the samples in temperature programming mode from 25 to 600°C under a nitrogen atmosphere with a heating rate of 15 °C/minute. Pristine MWCNTs are thermally stable and show no degradation over the entire range of temperature (25-600 °C). PEDOT and PEDOT grafted MWCNT composites show a three-step degradation pattern in the recorded temperature range. At 100 °C, for PEDOT and PEDOT grafted MWCNT composites, the first stage of weight loss is due to the loss of absorbed water molecules. At the second stage, PEDOT shows a sudden weight loss in the temperature ~240 °C, which is basically due to the removal of $-SO_3H$ functional group of the dopant molecules from the polymer chain [68]. In the PEDOT/MWCNT composites, the dopant degradation takes place slightly later at~ 260 °C temperature. In the third stage, PEDOT shows a consistent decrease in weight, at~ 380 °C due to the degradation of the polymer backbone. In PEDOT grafted MWCNT composites, the degradation temperature of the polymer chain is shifted to 400 °C. The improvement in the thermal stability of the composites is certainly attributed to the presence of MWCNTs in the polymer matrix. MWCNTs act as a nucleation site to the monomer units (EDOT); their inclusion enhances polymer crystallinity [69]. Apart from this, MWCNTs possess excellent thermal conductivity properties, which, when present in the polymer matrix, restrict the thermal motion of the polymer chain and delay the degradation of the polymer. The residues remain at 600 °C due to the presence of thermally stable inert materials like MWCNTs, iron particle residues of MWCNTs, and carbonized polymer.

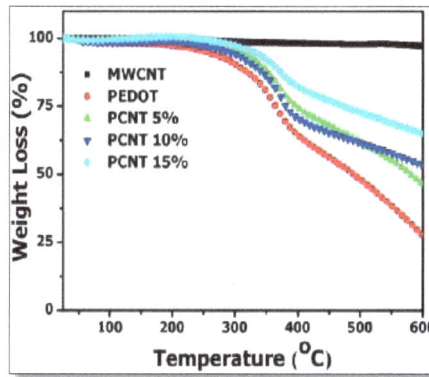

Fig. (11). Thermograms of MWCNTs, PEDOT, and PEDOT grafted MWCNTs composites. Reprinted from reference [60]. Copyright (2015) Elsevier.

5.4. ELECTROMAGNETIC SHIELDING AND DIELECTRIC STUDIES

Total shielding effectiveness (SE), absorption shielding effectiveness (SE$_A$), and reflection shielding effectiveness (SE$_R$) of the PEDOT and PEDOT grafted MWCNT composites in the 12.4-18 GHz frequency range (Ku-band) is shown in Fig. (**12**). From Fig. (**12a**), it is observed that SE of the PEDOT grafted MWCNT composites increases with loadings of MWCNTs due to improved conductivity, permittivity, and permeability properties. The SE$_A$ of the EM wave is enormously enhanced from − 20 to − 58 dB by the addition of MWCNTs in the PEDOT matrix. This is due to the increase in conductivity and capacitive coupling effects. The SE$_R$ of the PEDOT/MWCNT composites is presented in Fig. (**12b**), which shows that SE$_R$ increases slightly from − 4.0 to 8.0 dB with the increase in MWCNTs loading. This is due to an increase in the conductivity of the composites.

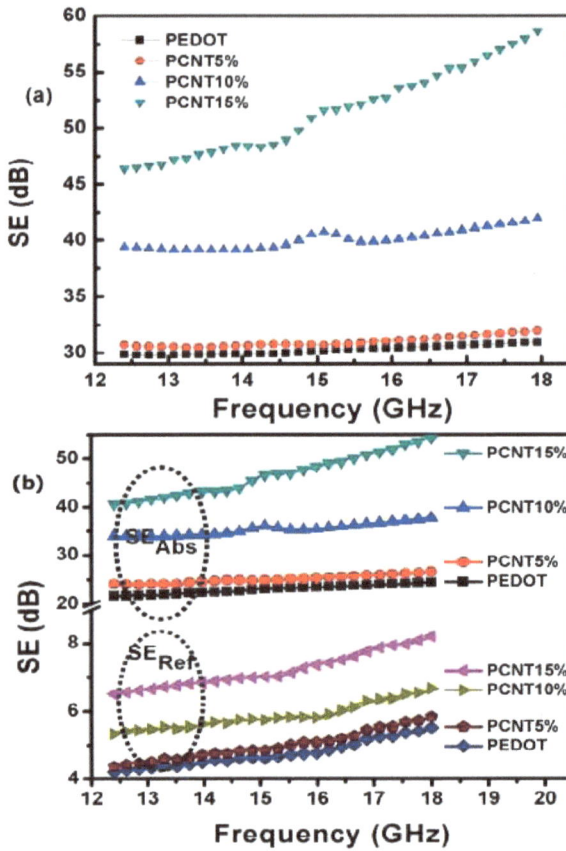

Fig. (12). (a) Total shielding effectiveness, **(b)** Reflection shielding effectiveness and absorption shielding effectiveness of PEDOT, PEDOT grafted MWCNT composites. Reprinted from reference [60]. Copyright (2015) Elsevier.

Complex permittivity and permeability PEDOT/MWCNT composites have been calculated using scattering parameters (S_{11} and S_{21}) based on theoretical calculations given in the Nicholson, Ross, and Weir method [70]. The real part (ε') is due to the polarization that takes place in the material, and the imaginary part (ε'') is related to the dissipation of energy. Fig. **(13)** shows the complex dielectric parameters of PEDOT and PEDOT grafted MWCNT composite with 15% loading; the values of ε' and ε'' decreases with increasing frequency. The higher values of ε' and ε'' arise due to the differences in the relative dielectric constant of MWCNT and the PEDOT. This results in the accumulation of more space charge and strong orientation polarization, which consequently leads to the improved values of wave absorption.

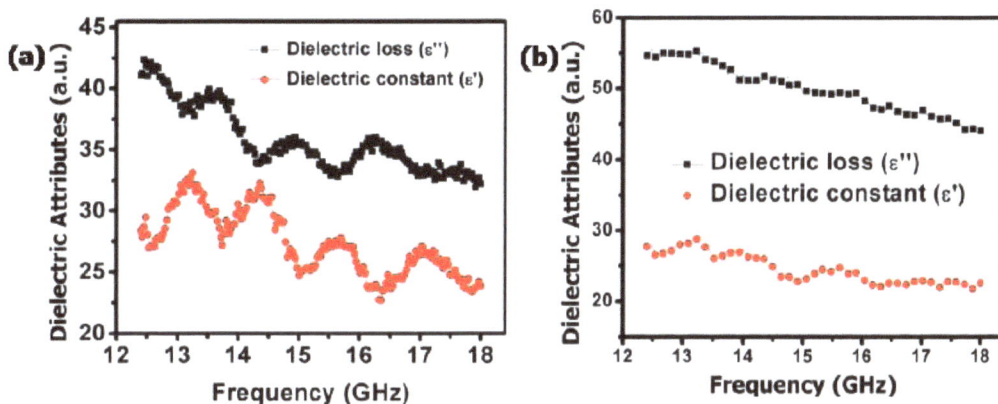

Fig. (13). Complex parameters of (a) PEDOT and (b) PEDOT grafted MWCNT 15% loading composite. Reprinted from reference [60]. Copyright (2015) Elsevier.

5.5. PEDOT/MWCNT COMPOSITE REINFORCED POLYURETHANE CONDUCTIVE FILMS: PREPARATION, CHARACTERIZATION AND EMI SHIELDING STUDIES

Polyurethane (PU) is a very versatile polymer known for its unique properties, such as good elasticity, high impact strength, and elongation [71]. It is used in various forms such as coatings, adhesives, thermoplastic elastomers, and composites [72]. It consists of a hard segment which is composed of an alternating diisocyanate and chain extender molecules (*i.e.*, diol or diamine), and a soft segment which is composed of a linear, long-chain diol. The most important feature of PU's structure is micro-phase separation arising from the thermodynamic incompatibility of hard and soft segments. In the previous section, the synthesis, characterization, and EMI shielding studies of PEDOT grafted MWCNT composites were discussed. The studies demonstrated that the PEDOT grafted MWCNT composites possess very good electrical conductivity and EMI shielding effectiveness values. However, these composites have limited practical applications as they lack mechanical strength, which limits their scope for technological applications. In this section, PEDOT grafted MWCNT composites were utilized as reinforcement filler in the PU polymer matrices to fabricate PU conductive sheets by solution casting technique.

5.6. PREPARATION OF PU CONDUCTIVE SHEETS INCORPORATED WITH PEDOT/OR PEDOT COATED MWCNTS

Polyurethane composite sheets incorporated with PEDOT/or PEDOT coated MWCNTs were prepared by the solution casting technique as shown in Fig. (**14**) [73]. Initially, PU granules were dissolved in dimethyl formamide (DMF) solvent at 40 °C for 8 hours; meanwhile, PEDOT/or PEDOT grafted MWCNTs filler was dispersed in another beaker in DMF. The dispersed PEDOT/or PEDOT coated MWCNTs filler in DMF and PU dissolved in DMF were mixed thoroughly for 2 hours. The mixer was then poured into a Petri dish and then dried in an oven at 50 °C for 12 hours [74]. Several compositions of the PU sheet composites have been formulated, which are given below:

In Table **1,** PUP is PU sheet filled with PEDOT and PUPCNT is PU sheet filled with PEDOT grafted MWCNTs composite.

Table 1. Compounding formulations of PU sheets containing different wt.% loading of PEDOT and PEDOT grafted MWCNT filler in PU matrix. Reprinted from reference [73]. Copyright (2015) The Royal Society of Chemistry.

S. No.	Sample	PU	PEDOT	PEDOT Grafted MWCNT Composite
1.	PU	100%	0	0
2.	PUP1	90%	10%	0
3.	PUP2	80%	20%	0
4.	PUP3	70%	30	0
5.	PUPCNT1	90%	0	10%
6.	PUPCNT2	80%	0	20%
7.	PUPCNT3	70%	0	30%

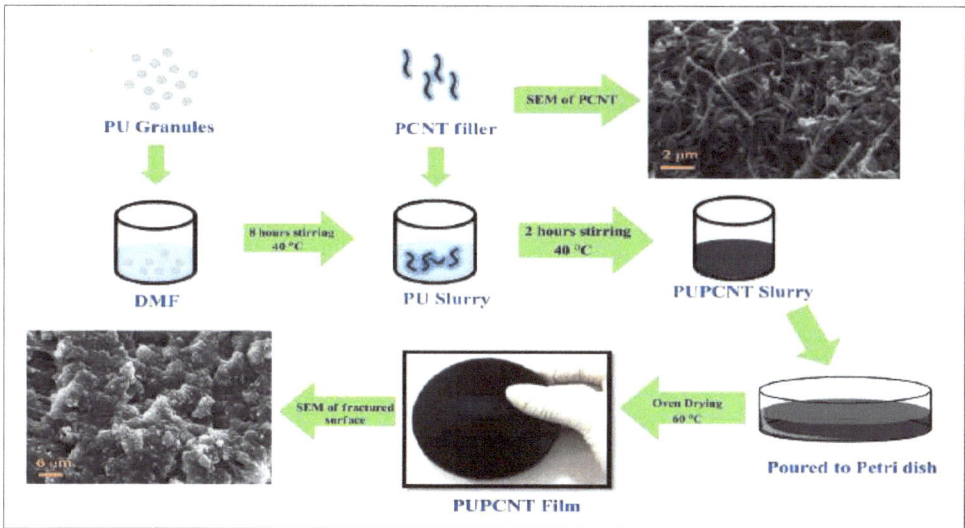

Fig. (14). Preparation of PU sheets reinforced with PEDOT grafted MWCNTs composite. Reprinted from reference [73]. Copyright (2015) The Royal Society of Chemistry.

5.7. CHARACTERIZATION OF PEDOT AND PEDOT GRAFTED MWCNT FILLED PU SHEETS

Fig. (15) shows the SEM images of the PEDOT grafted MWCNTs composite filler at low and high magnifications, respectively. The PCNT presents a homogenous structure and looks like rods of snowflakes. The nano-sized MWCNTs, having high surface area, serve as nucleation sites for the polymerization of EDOT monomers to form a coating over the MWCNTs. The micrographs show that the PEDOT is uniformly coated over the MWCNTs, thereby forming a high aspect ratio filler, which is responsible for the good electrical conductivity of the PUPCNT composites.

Fig. (15). SEM image of PEDOT grafted MWCNT composite incorporated in PU sheets. Reprinted from reference [73]. Copyright (2015) The Royal Society of Chemistry

The fractured cross-sectional surface of the PU conductive sheets was also observed in the SEM. Fig. (**16a**) shows the fractured surface of PUP3, with an indiscriminate distribution of PEDOT filler in the PU matrix. In Fig. (**16b**), the fractured surface of PUPCNT1 with uniformly distributed PCNT filler in the PU matrix can be distinctly observed. The PCNT filler, which appears as white bright lines embedded in the PU matrix, is marked with arrows in Fig. (**16b**). The white bright lines increased with the increase of PCNT loading in the PU matrix (see Fig. **16c**) and (d), respectively. It is interesting to note that PCNT filler exhibits good dispersion except for one or two clusters in the PUPCNT2 and PUPCNT3 composite due to excessive loading.

TEM images were further used to examine the core-shell structure and thickness of PEDOT coating over MWCNTs. Fig. (**17a**) shows the TEM image of the PCNT filler, where individual PCNTs can be clearly observed. Fig. (**17b**) TEM image further confirms that MWCNTs are packed underneath the PEDOT layer. The diameter of PCNT filler and embedded MWCNT is about ~150 and ~46 nm, respectively, revealing a thick coating of the PEDOT layer (~ 45 nm). The presence of Fe particles can also be observed inside the MWCNTs, which are remnants of ferrocene catalysts used during the preparation of MWCNTs. Fig. (**17c**) shows the

PCNT filler incorporated PU matrix, where individually distributed PCNT filler can be clearly observed. Fig. (**17d**) shows the HRTEM image of PCNT filler with an inset of the image showing the spacing of 0.34 nm between the concentric carbon nanotubes in the PEDOT matrix.

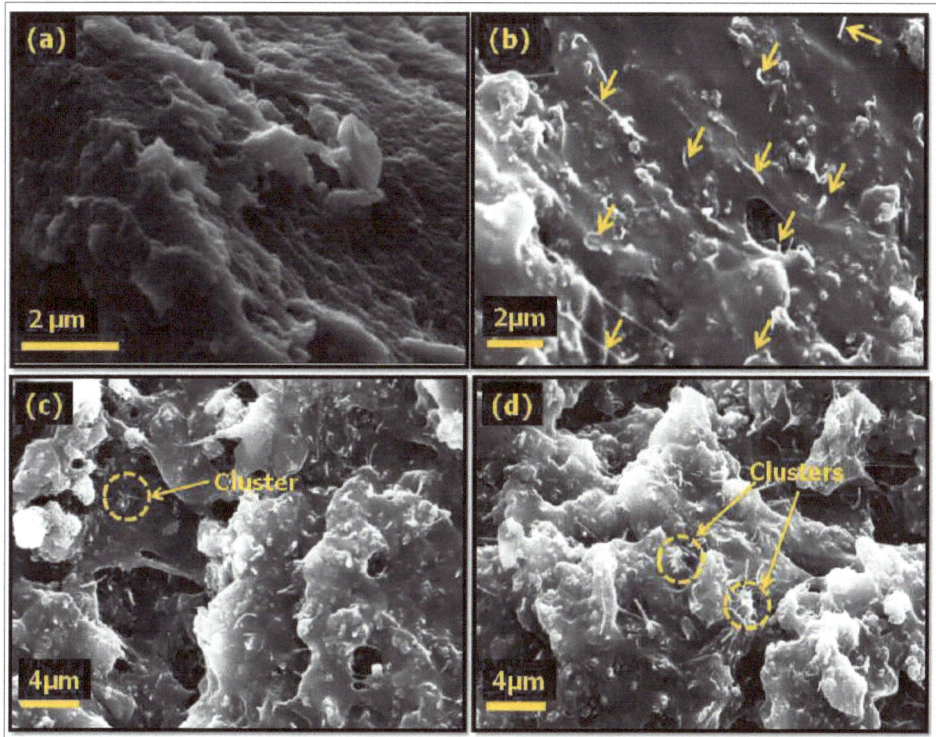

Fig. (16). SEM images of the fractured surface of PU filled with PEDOT and PCNT filler (**a**) PUP3 (**b**) PUPCNT1 (**c**) PUPCNT2 and (**d**) PUPCNT3. Reprinted from reference [73]. Copyright (2015) The Royal Society of Chemistry.

Fig. (**18**) shows the XRD patterns of MWCNT, PU, PEDOT, PUP, and PUPCNT sheet composites. The XRD pattern of MWCNT shows two peaks at a 2θ value of 26.1° and 43.2° for (002) and (100) planes of carbon atoms. These peaks imply the interlayer spacing of 0.34 and 0.21 nm, respectively. PU shows two peaks at 21.1° and 23.4° because of the crystals of polycaprolactone (PCL) corresponding to (110) and (200) planes [75]. A broad diffraction peak at 24.5° is observed for PEDOT due to its amorphous nature. Since PEDOT does not show any significant peak, the

PUP presents a much broader peak than PEDOT and PU. In the PUPCNT composites, the peaks of PCL and MWCNTs are present at their respective 2θ values, which further confirm the co-existence of PCNT in the PU matrix.

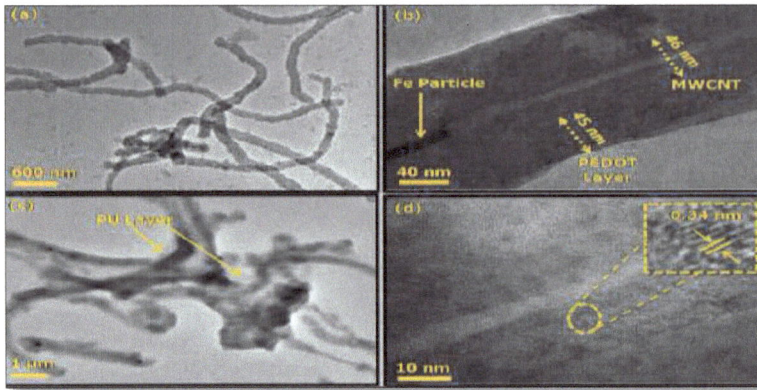

Fig. (17). TEM images (**a**) low magnification image of PCNT filler (**b**) high magnification image of PCNT (**c**) PUPCNT1, and (**d**) HRTEM of PCNT. Reprinted from reference [73]. Copyright (2015) The Royal Society of Chemistry.

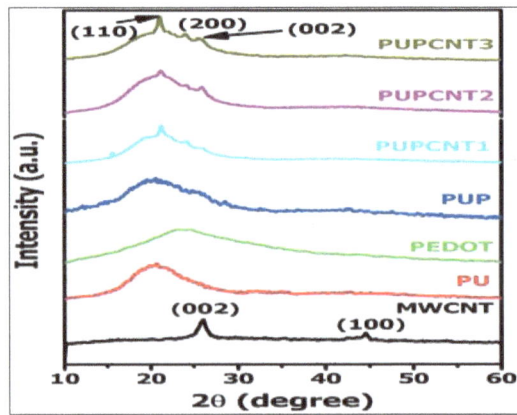

Fig. (18). XRD patterns of pristine MWCNT, PU, PEDOT, PUP, and PUPCNT sheet composites. Reprinted from reference [73]. Copyright (2015) The Royal Society of Chemistry.

FTIR spectra of PU, PUP3, and PUPCNT3 recorded using the ATR accessory are shown in Fig. (**19**). All three samples have characteristic peaks of PU. The characteristic peaks observed near 3320 and 1722 cm^{-1} arise due to the stretching vibrations of $-NH$ and $-C = O$ groups in PU [76]. Other peaks observed in the PU spectra are assigned in the table given in Fig. (**6**). FTIR spectrum of PEDOT shows bands at 678, 829, 917, and 967 cm^{-1} which are due to the deformation modes of C − S − C in the thiophene ring; the peaks at 1052, 1084, 1134, and 1187 cm^{-1} arises due to bending vibration of C − O − C of ethylene group; the peaks at 1312 and 1512 cm^{-1} corresponds to C − C or C = C stretching of quinoid structure and ring stretching of thiophene ring, respectively. In spite of some minor differences, PU, PUP3, and PUPCNT3 exhibit similar spectra due to the presence of significant amounts of PU in the matrix.

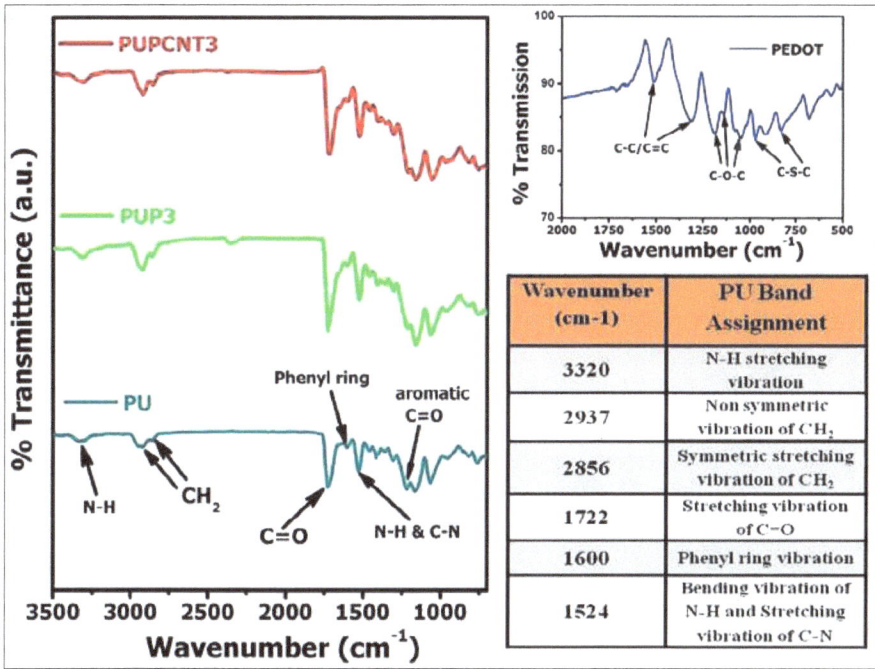

Fig. (19). FTIR spectra of PU sheet composites. Reprinted from reference [73]. Copyright (2015) The Royal Society of Chemistry.

Fig. (**20**) shows the effect of filler loading on the tensile strength of PUP and PUPCNT composites. It is well known that pure conducting polymers are mechanically very weak. The blending of PEDOT filler in the PU matrix deteriorates the strength of PUP composites consistently. This can be attributed to the weak interaction between filler and matrix. It is striking to note that the addition of 10 % PCNT filler significantly increases the tensile strength from 20 MPa to 26 MPa. This enhancement can be attributed to the homogeneous dispersion of PCNT and effective load transfer from the matrix to MWCNTs embedded in PEDOT. However, with a further increase in PCNT loading to 20 and 30 wt. %, the tensile strength decreases gradually because of the plasticization effect on the mechanical properties, caused by the thick coating of the PEDOT layer, which has a very poor mechanical strength. Hence, tensile strength decreases with increasing loading of PCNT filler since the amount of PEDOT contributing to PCNT is greater than MWCNTs.

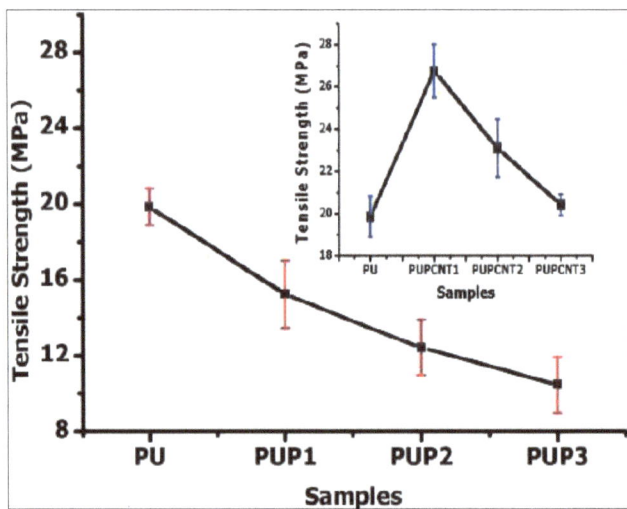

Fig. (20). Tensile strength studies of PU sheet composites. Reprinted from reference [73]. Copyright (2015) The Royal Society of Chemistry.

The shielding effectiveness value for the PU, PUP, and PUPCNT composites in the frequency range of 12.4-18 GHz is shown in Fig. **(21)**. It is clearly visible from Fig. **(21a)** that the PU film sample provides negligible attenuation to EM waves, whereas the solution mixing of PEDOT polymer in the PU matrix provides 3.98 dB, 5.75 dB, and 7.41 dB of attenuation at mid-frequency of 15.2 GHz, for the composites; PUP1, PUP2, and PUP3, respectively. On the other hand, solution mixing of PCNT filler in the PU matrix results in linear enhancement of the SE_T of the PUPCNT composites. The maximum shielding effectiveness obtained with PUPCNT3 was 44.25 dB at 15.2 GHz, which is due to the formation of conducting networks throughout the electrically insulating PU matrix by the addition of PCNT filler having a high aspect ratio. The shielding effectiveness of electrically conductive polymer composites strongly depends upon conductivity, dispersion, aspect ratio, and loading of conductive filler in the polymer matrix. To investigate the shielding mechanism, the total shielding effectiveness is further resolved into reflection loss (SE_R) and absorption loss (SE_A) components, as shown in Fig. (**21b** and **c**).

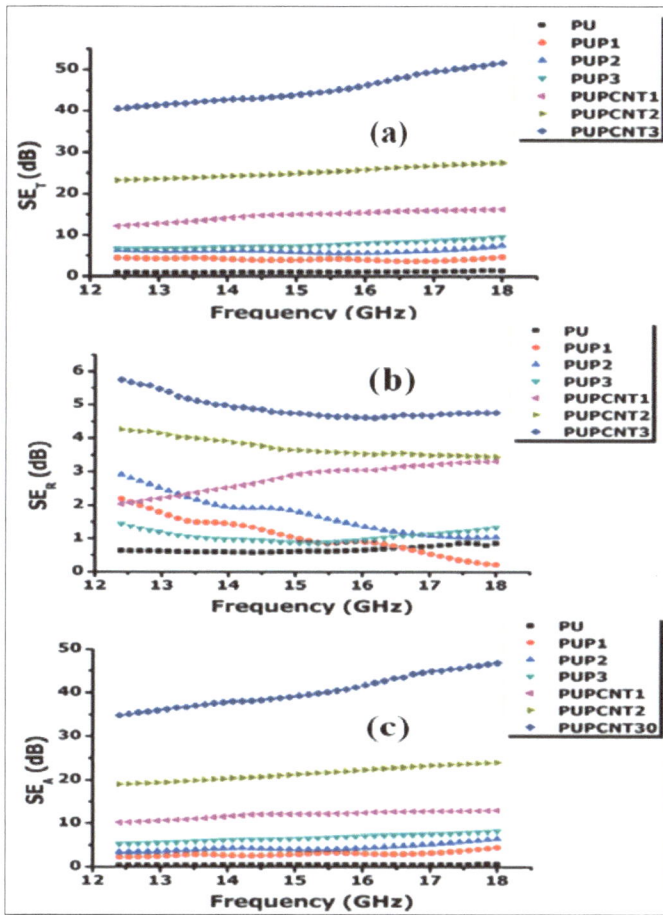

Fig. (21). (**a**) Frequency dependence of total shielding effectiveness (SE_T), (**b**) losses due to reflection (SE_R), and (**c**) absorption (SE_A) of different samples of PU composites. Reprinted from reference [73]. Copyright (2015) The Royal Society of Chemistry.

From the experimental measurements, it is observed that the shielding effectiveness due to absorption and reflection (SE_A and SE_R) increases with increasing loading of the fillers (PEDOT and PCNT). This can be attributed to the increase in the conductivity of the composites. It is interesting to note that the contribution of SE_A for the attenuation of EM waves is greater than the SE_R due to an increase in conductivity, dielectric losses as well as magnetic permeability (entrapped Fe catalyst particles in the MWCNTs). Fig. (**21b**) shows an anomalous behaviour of SE_R of the samples in the applied frequency band. The SE_R of PU is almost constant throughout the frequency range due to its insulating nature, whereas the SE_R of PUP3 and PUPCNT1 increases with frequency. It is because the samples with the loading of 30 % of PEDOT and 10 % of PCNT in PU matrix, respectively, were able to reach the percolation threshold (formation of continuous electrically conducting network in the matrix) with uniform conductivity. The decrease in SE_R of PUP1 and PUP2 with an increase in frequency can be attributed to the inability of PEDOT filler with 10 % and 20% loading in PU matrix respectively to form the electrically conducting path, whereas, in the case of PUPCNT2 and PUPCNT3, the decrease in SE_R is because of cluster formation due to excessive loading of PCNT filler in the PU matrix which causes heterogeneity at the nano-level (non-uniform conductive network) thereby resulting in decreased SE_R values. The total shielding effectiveness of PUPCNT is greater than PUP composites, and it increases systematically with increasing loading of PCNT.

PUP composites failed to reach the minimum EMI shielding (20 dB) required for most commercial applications in EMI measurements. Antistatic measurements of PUP composites were conducted by JCI 155v5 charge decay test unit. Fig. (**22**) shows the static decay time of PUP1, PUP2, and PUP3 samples measured by applying a positive voltage of 5000 V. From the measurements, it is observed that PUP1, PUP2, and PUP3 composites show a decay time of less than 0.2 seconds for the initial peak voltage to reach 1/e (about 37%), which meets the required time limit less than half a second in the 1/e criterion. The 10 % criterion was also measured (initial peak voltage to 10 % within 2 seconds), and it is observed that all samples show an excellent 10 % criterion decay time of less than 0.8 seconds. Hence, antistatic results indicate that loading of PEDOT in PU increases the conductivity of composites required for static charge dissipation.

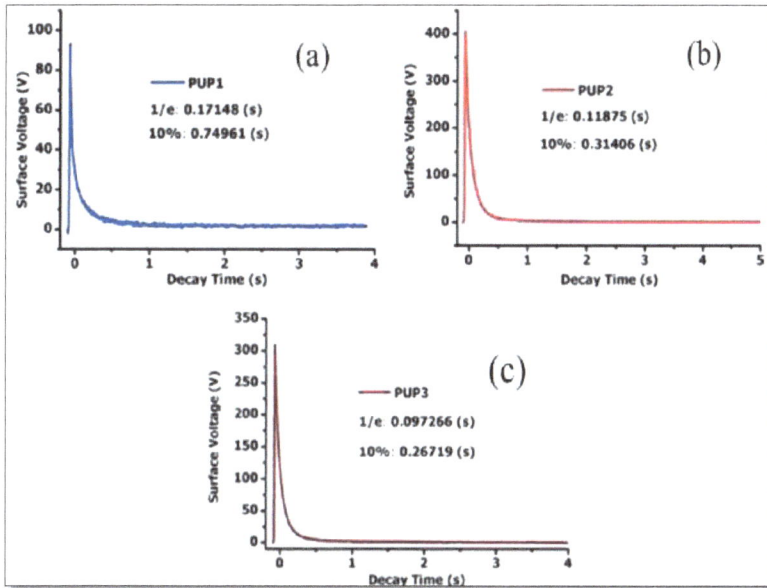

Fig. (22). Static charge decay time of (**a**) PUP1, (**b**) PUP2, and (**c**) PUP3 samples at charging voltage of 5.0 kV. Reprinted from reference [73]. Copyright (2015) The Royal Society of Chemistry.

5.8. PEDOT/PSS COATED MWCNT BUCKY PAPER: PREPARATION, CHARACTERIZATION AND EMI SHIELDING STUDIES

5.8.1. Preparation of PEDOT/PSS Coated MWCNT Bucky Paper

MWCNTs are used as such without any purification for the preparation of bucky paper [77]. Initially, 0.25 g of MWCNTs is dispersed in acetone solvent by using a homogenizer for 15 minutes. After that, 5, 10, 20 ml of PEDOT/PSS was added to the dispersion of the MWCNTs and again homogenized for half an hour. So that PEDOT/PSS is coated over the MWCNTs. The PEDOT contains aromatic thiophene rings in the polymeric chain, which allows strong π-π stacking with the hybridized carbon on the surface of the MWCNTs. Moreover, PSS is a surfactant that helps MWCNTs to disperse, and it's long-chain surrounds the surface of MWCNTs to inhibit stacking [78]. The suspension of PEDOT/PSS and MWCNTs is then filtered using vacuum infiltration technique. Filter paper Whatman no. 44 is used for filtration. Finally, after filtration, the PEDOT/PSS coated MWCNT bucky paper is allowed to dry in a vacuum oven at 60°C. A schematic representation of the preparation of PEDOT/PSS coated MWCNT bucky paper is shown in Fig. (**23**).

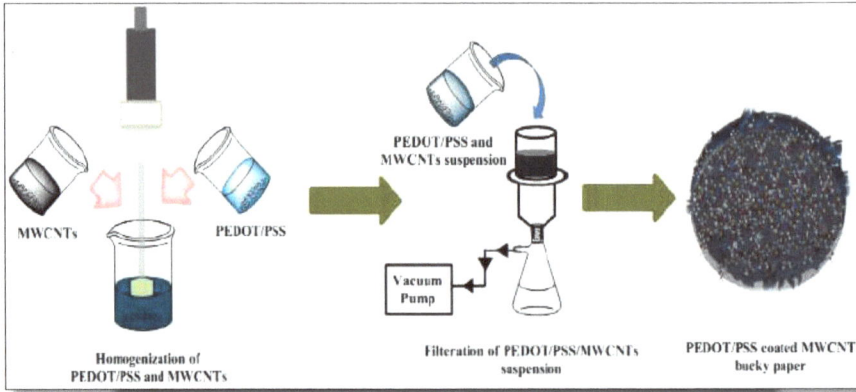

Fig. (23). Preparation of PEDOT: PSS coated MWCNT bucky paper by vacuum infiltration technique.

SEM with energy dispersive X-ray analysis (EDX) enables both high magnification characterization of surfaces and elemental composition analysis. Fig. (**24a**) shows the digital image of the PEDOT/PSS coated MWCNTs bucky paper. From the naked eyes, all bucky paper looks like smooth, uniform, thin black sheets. Fig. (**24b**) is the EDX pattern; the presence of 18% of sulfur atoms confirms the presence of thiophene and polystyrene sulfonate moieties at the surface of MWCNTs bucky paper.

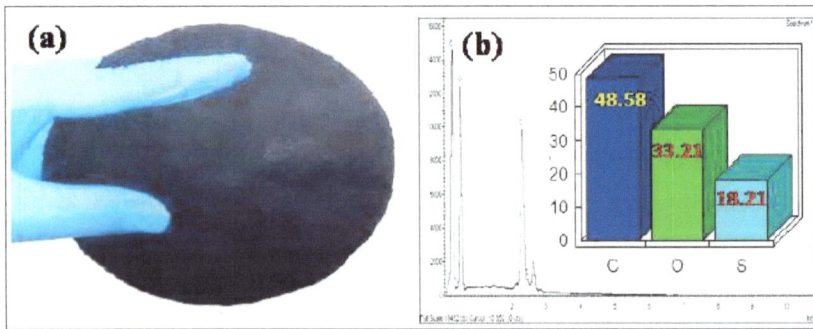

Fig. (24). (a) Digital photograph of the PEDOT: PSS coated MWCNT bucky paper **(b)** EDX analysis of PEDOT: PSS coated MWCNT bucky paper.

Fig. (**25a**) shows the low magnification SEM image of the bucky paper, which reveals a slightly uneven rough surface of the paper. Fig. (**25b**) shows the high magnification image of the bucky paper, where individual MWCNTs are clearly visible. The MWCNTs form a random, extremely interconnected macroporous structure. It is also observed that most of the MWCNTs are unbundled, while bundles of MWCNTs can also be seen. The diameter of the MWCNTs is ~ 25-30 nm, which is in agreement with the MWCNTs synthesized by the CVD technique. The macropores do not have any particular shape due to the random entrenched arrangement of MWCNTs. The inset shows the morphology of the individual PEDOT/PSS coated MWCNT analyzed by using TEM, which clearly shows the nano-tubular structure of the crystalline MWCNTs and the PEDOT/PSS thin layer coated as a sheath at the surface of MWCNTs.

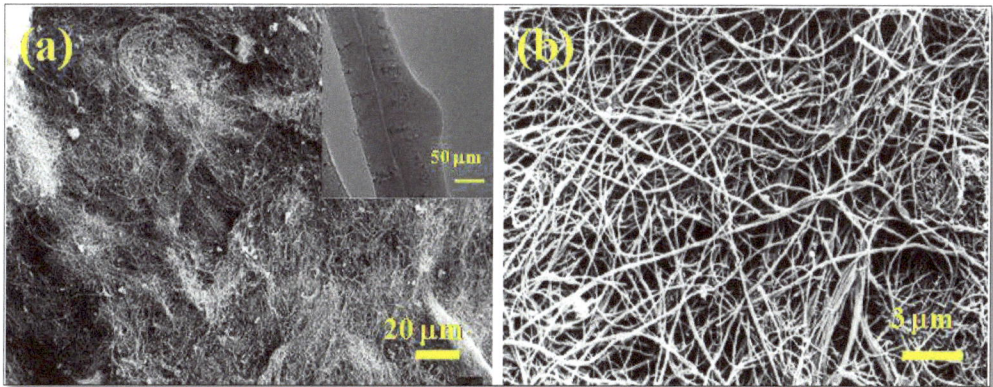

Fig. (25). (**a**) Low magnification SEM image of PEDOT/CNT2 bucky paper (**b**) High magnification SEM image of PEDOT/CNT2 bucky paper.

Fig. (**26**) demonstrates the XRD patterns of PEDOT/PSS coated bucky papers. The samples show characteristic peaks of MWCNTs observed at 2θ value $26.2°$ and $42.5°$ for the corresponding planes (002) and (100) respectively, attributed to the hexagonal graphitic structure [79]. The peak of PEDOT/PSS is not appeared in the PEDOT/CNT1 and PEDOT/CNT2 bucky papers because of the very low concentrations of PEDOT/PSS, whereas PEDOT/CNT3 shows a broad peak at $12°$ due to the amorphous PEDOT/PSS.

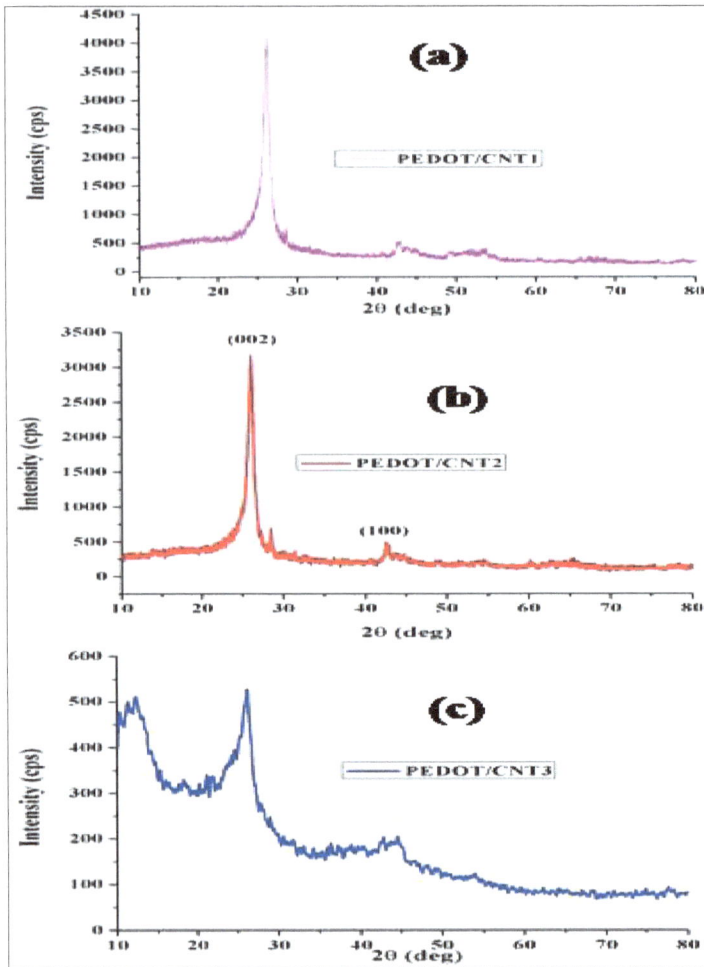

Fig. (26). XRD patterns of PEDOT: PSS coated Bucky papers.

Fig. (**27a** and **b**) shows Raman spectrums of PEDOT/CNT1 and PEDOT/CNT2, which show no peaks of PEDOT/PSS but peaks associated with MWCNTs are clearly visible. They show peaks at ~1310 (D-band) and 1580 cm^{-1} (G-band) due to the disorder in the carbon system and in-plane vibration of the C – C bond because of graphite, respectively. The band around 2700 cm^{-1} was also appearing in these samples, which is an overtone of the D-band known as the G′ band. However, in the PEDOT/CNT3 sample (Fig. **27c**), bands related to the PEDOT structure are strongly visible in the spectrum. The bands at 1570 and 1520 cm^{-1} are due to the $C_\alpha = C_\beta$ antisymmetric vibrations in the thiophene ring. The strong band at 1419 cm^{-1} is because of symmetric $C_\alpha - C_\beta$ stretching deformations. Several other peaks are also present in the spectrum at 990, 695, 574, and 436 cm^{-1} due to the deformation of oxyethylene ring, C–S–C deformation, oxyethylene deformation, and SO$_2$ bending, respectively. The presence of these peaks of PEDOT confirms the presence of PEDOT sheath over the surface of MWCNTs.

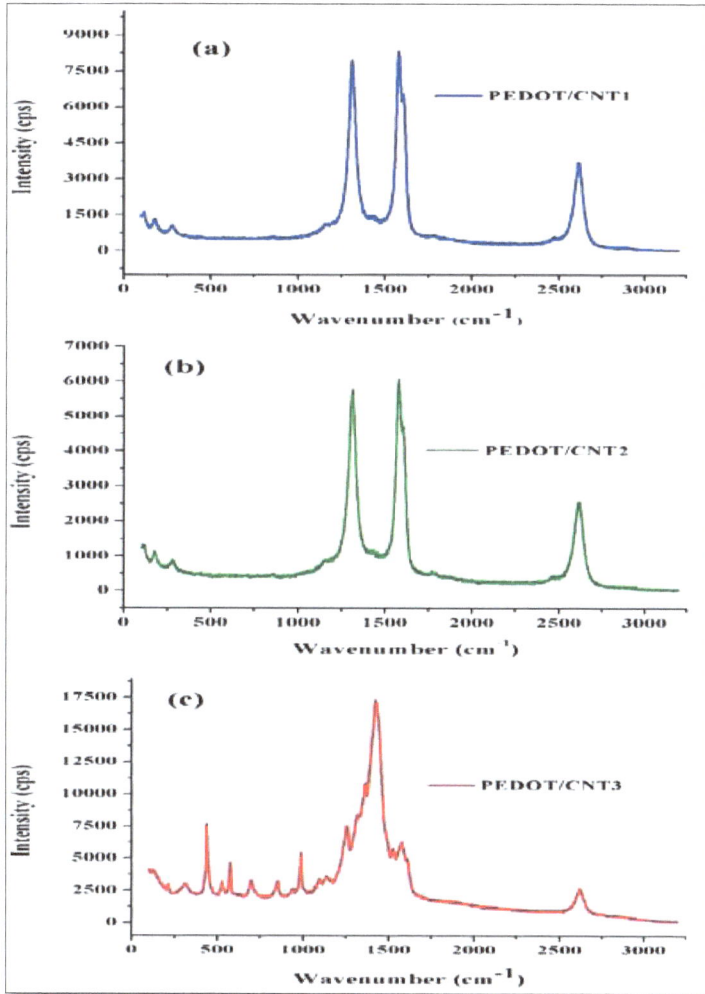

Fig. (27). Raman spectrum of PEDOT: PSS coated MWCNT Bucky paper.

TGA of the PEDOT/PSS coated MWCNT bucky papers were carried out in a nitrogen atmosphere from 30-800°C in order to determine the concentration of PEDOT/PSS in bucky papers (Fig. **28**). MWCNTs are thermally very stable and cannot degrade in an inert atmosphere up to 800°C. All the bucky papers show three major weight losses due to the degradation of PEDOT/PSS. The first step is at about 100°C, which indicates the loss of adsorbed water molecules. The second loss takes place at about 180°C due to the loss of dopant molecules from the polymer chain. Finally, the third major loss is near 310 °C, attributed to the degradation of the polymeric backbone. From the thermogram, it is clearly seen that the addition of PEDOT/PSS gradually decreases the stability of the PEDOT/PSS coated MWCNT bucky papers.

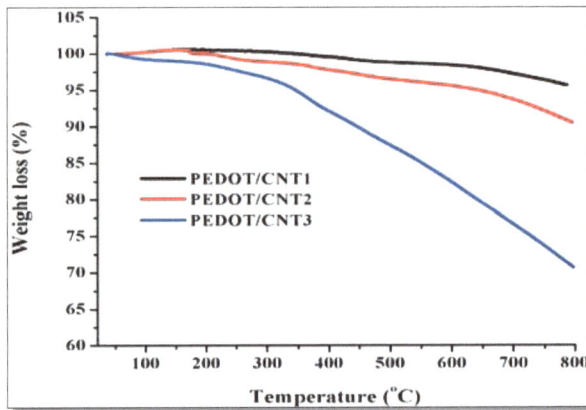

Fig. (28). Thermogravimetric analysis of the PEDOT: PSS coated MWCNT bucky paper to determine PEDOT: PSS concentration.

From (**29a**), it is observed that the total shielding effectiveness of the bucky papers increases with increases in the concentration of PEDOT/PSS. The maximum shielding effectiveness of 55 dB was obtained with PEDOT/CNT3 at 12.4 GHz frequency, whereas PEDOT/CNT2 and PEDOT/CNT1 showed 52 and 31 dB of shielding effectiveness, respectively as shown in Table **2**. To understand the shielding mechanism, the total shielding effectiveness is further resolved into SE_R and SE_A components, as shown in Fig. (**29b** and **c**), respectively. It is clearly observed from the figure that the contributions to the shielding effectiveness values mainly come from the absorption rather than the reflection mechanism. It is interesting to note that by increasing the concentration of PEDOT/PSS in the bucky paper, the SE_A increases linearly. On increasing PEDOT/PSS loading from 5.2 - 26 wt.% in the bucky paper, the SE_A increases from 62 % to 77 %, respectively. This is due to the fact that the addition of PEDOT/PSS adds more polarons and bipolarons in the bucky paper, which enhances the SE_A significantly.

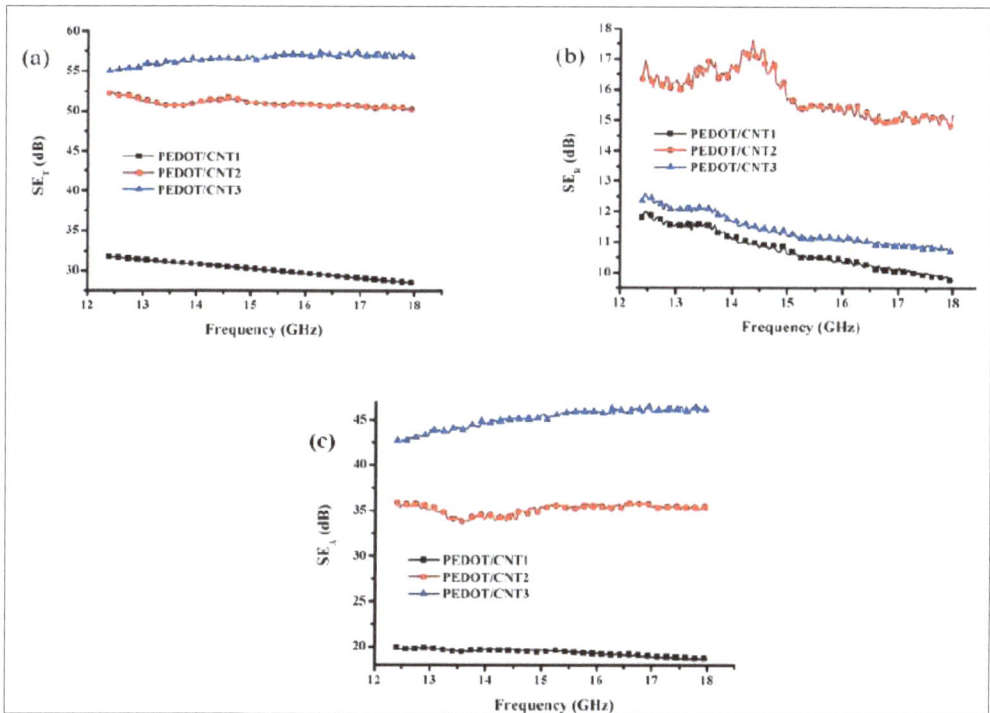

Fig. (29). (**a**) Variation of total shielding effectiveness of PEDOT: PSS coated MWCNT bucky paper with frequency, (**b**) shielding due to reflection & (**c**) shielding due to absorption.

Table 2. Percentage of PEDOT/PSS loading in MWCNT bucky paper with a contribution of shielding effectiveness values at 12.4 GHz frequency.

S. No.	Sample	PEDOT/PSS	MWCNTs	SE_T	SE_A	SE_R
1.	PEDOT/CNT1	0.013 g	0.25	31.73	19.94	11.79
2.	PEDOT/CNT2	0.026 g	0.25	52.17	35.79	16.38
3.	PEDOT/CNT3	0.065 g	0.25	55.11	42.66	12.45

5.8.2. Synthesis of PEDOT/ RGO Nanocomposites

The PEDOT nanocomposites with reduced graphene oxide (RGO) have been synthesized *via* in-*in-situ* emulsion polymerization using DBSA, which works as a dopant as well as a surfactant as described in Fig. (**30**) [80,81]. An aqueous solution of DBSA, EDOT, and filler was prepared using an IKA T25 ultra homogenizer. Afterward, the emulsion was transferred to a reactor attached to a chiller maintained at temperature -2 °C for polymerization. The polymerization of nanocomposite was initiated by adding APS (0.1 M) drop-wise into the emulsion with continuous stirring, and the reaction continued for 12 hours resulting in the dark bluish emulsion. The emulsion was de-emulsified by adding isopropyl alcohol with slow stirring. Afterward, the solution was filtered washed with distilled water, and the precipitates were dried in an oven at 65 °C, resulting in the formation of PEDOT composites.

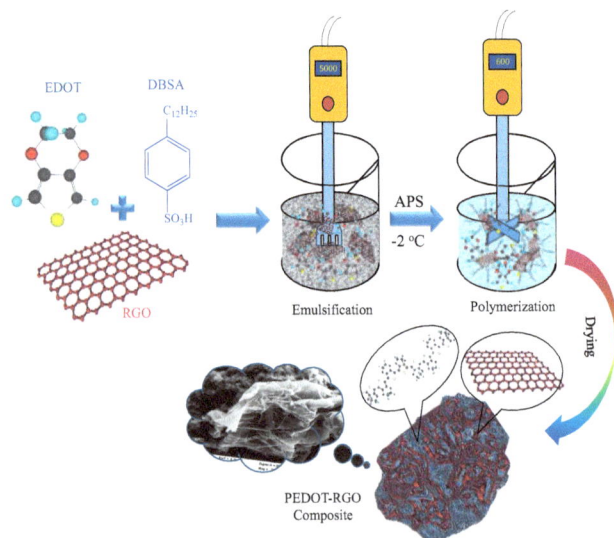

Fig. (30). Schematic representation of the polymerization of PEDOT/Graphene nanocomposites. Reproduced from reference [80]. Copyright (2016) Elsevier.

The graphene was incorporated in the PEDOT matrix as conductive filler in different wt. ratio to investigate the effect of graphene concentration on the shielding parameters of PEDOT/Graphene nanocomposites. Further, to improve the shielding properties, the PEDOT is simultaneously incorporated with Graphene &

dielectric filler and then with graphene & magnetic filler. The flowchart of the overall process of PEDOT's nanocomposites synthesis is illustrated in Fig. (**31**).

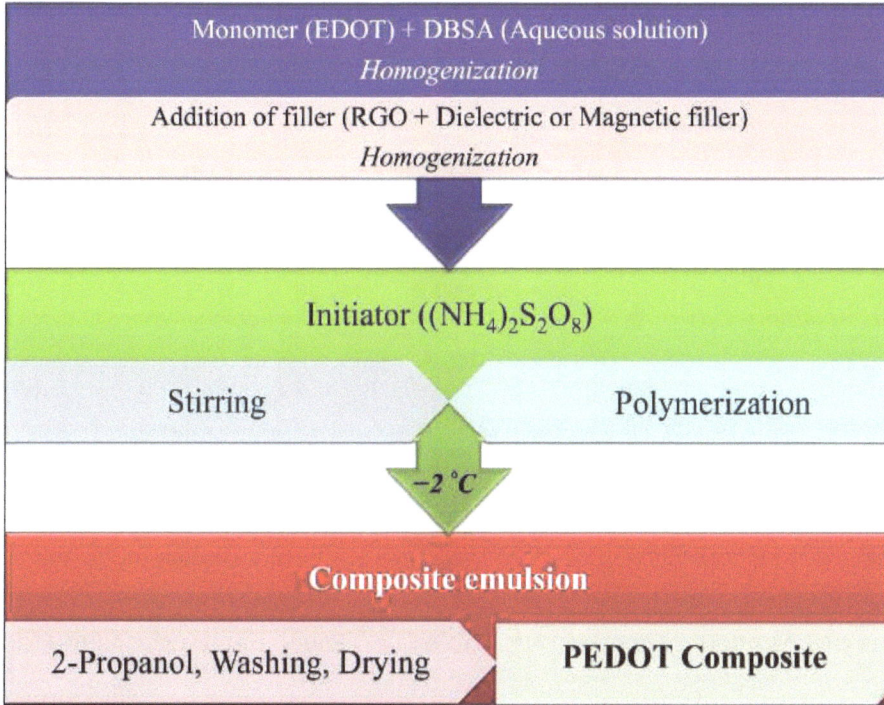

Fig. (31). Illustration of synthesis process for PEDOT's nanocomposites.

5.8.3. Dielectric Properties of PEDOT's Composites

The electromagnetic shielding properties of PEDOT's composites depend upon the magnetic and dielectric losses assessed from the dispersive complex permeability ($\mu_r = \mu' - j\mu''$) and permittivity ($\varepsilon_r = \varepsilon' - j\varepsilon''$) respectively. These parameters (μ_r & ε_r) are estimated from the S_{11} & S_{21} (scattering parameters) following Nicholson-Ross-Weir algorithm [82, 83]. The real and imaginary elements of the dispersive complex permittivity and permeability measure the energy storage and energy dissipation ability of the materials, respectively, as depicted in Fig. (**32**). [81].

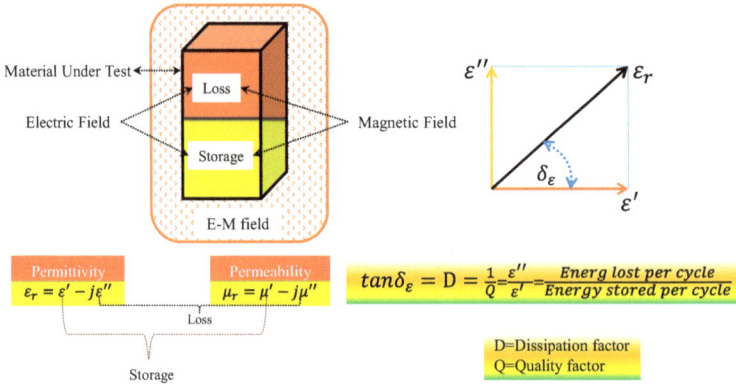

Fig. (32). Schematic demonstration of electric and magnetic energy dissipation in the material.

The dielectric parameters are subject to polarizations such as interfacial polarization, ionic polarization, space charge polarization, dipolar polarization, *etc.*, and conductivity of the specimen that depends on the frequency [84]. In the case of higher frequency regime (microwave frequency band), ε' strongly depends upon space-charge and interfacial polarizations [85]. In these samples, heterogeneity in the material is one main factor in controlling the space-charge and interfacial polarization. As per the hypothesis of Maxwell-Wanger, the interfacial polarization is instigated by the charges accumulated at the interface of two distinct materials due to differences in their conductivity and dielectric properties [86]. The addition of filler material in the PEDOT polymer creates a large number of heterogeneous interfaces in the polymer system that leads to more attenuation of electromagnetic waves, and the attenuation at such interfaces is proportional to the square of electromagnetic wave intensity [87]. It can be attributed to enhanced interfacial polarization, which is increased by the accumulation of charges at the interface of polymer and filler materials due to their distinct conductivity and dielectric constant.

The conduction mechanism of conjugated polymers involved two types of charge carriers; polaron and bipolarons. The charge carriers can move along polymeric chains having higher mobility than the bound charges, *i.e.*, dipoles. Fundamentally, the dipoles are countable for the strong dipolar orientation polarization in the materials. The polarization is correlated with the real part of the permittivity.

When electromagnetic radiation passes through the shield, two types of current are induced, *i.e.*, conduction and displacement current, due to two different types of charge carriers [40]. For the conduction current, the moveable charge carriers are responsible and directly contribute to the imaginary part of the permittivity ($\varepsilon'' = \sigma_s/\omega\varepsilon_o$). Whereas, the displacement current is attributed by bounded charge carriers that are responsible for different polarizations and contribute to the real part of permittivity. Mainly two polarizations, orientational polarization and space charge polarization, are the major contributors to the conjugated polymer.

5.8.4. PEDOT/Graphene Composites

Fig. (**33**) depicts the plot of ε' & ε'' *versus* frequency. Values of the real part of permittivity vary from 22.1 to 16.5 with the increase in frequency for pure PEDOT samples. The higher value of ε' is observed for the PG4 composite sample that possesses a higher content of RGO; it may be due to the enhancement in both the space charge and interfacial polarization. Because the addition of RGO in the polymer leads to heterogeneity (in terms of conducting grains bound by the polymer) in the material, which in turn increases space-charge and interfacial polarization. A fall in the value of ε' & ε" with the increase in frequency is observed; it may be due to the inability of the dipoles to synchronize with the frequency of the field at the higher values. It results in a decrease in orientational polarization that leads to a lower value of permittivity.

As mentioned above ε'' is a function of the conductivity of the material that increases with an increase in the amount of filler material (RGO) in the polymer system. For a more understanding of the losses, the overall dielectric loss factor, *i.e.*, tangent loss ($\varepsilon''/\varepsilon'$) is calculated and plotted *versus* frequency as depicted in the inset of Fig. (**33**). The value of the dielectric constant for the samples (PEDOT and PG composites) is more than 0.9, which reveals the lossy character of the samples. The values of dielectric loss for the PG1 and PG2 samples are observed lower than PEDOT may be due to the incorporation of a very small concentration of RGO in these samples. This happening can be explained by correlating it with conductivity. It is possible that a lower amount of RGO is not sufficient to increase the conductivity as well as ε'' by a considerable value. However, the increase in

ε' is attributed to enhanced heterogeneity, thereby increase in space-charge and interfacial polarization. Moreover, a higher value of dielectric loss of the samples upsurges the microwave absorption abilities.

5.8.5. PEDOT/Graphene/SrF Composites

Further, the addition of SrF filler with graphene in the polymer system gives rise to more heterogeneity in the material. As a result, space-charge polarization increases. Due to the dependence of polarization on the frequency, the values of ε' & ε'' goes down with the increase in frequency because of lessening in space-charge polarization as well as orientation polarization, with the rise in frequency of field as reflected from Fig. (**34a**). At 12.4 GHz, the values of ε' & ε'' changing in between 14.8 – 23.0 & 8.4 – 26.8 respectively for all PSrG composite samples. It is observed that the relative complex permittivity rises as the ratio of SrF filler material increases in the composite, mainly due to the creation of new interfaces inside the composite. Thus, the combined effect of SrF together with graphene and PEDOT results in enhanced absorption of microwaves.

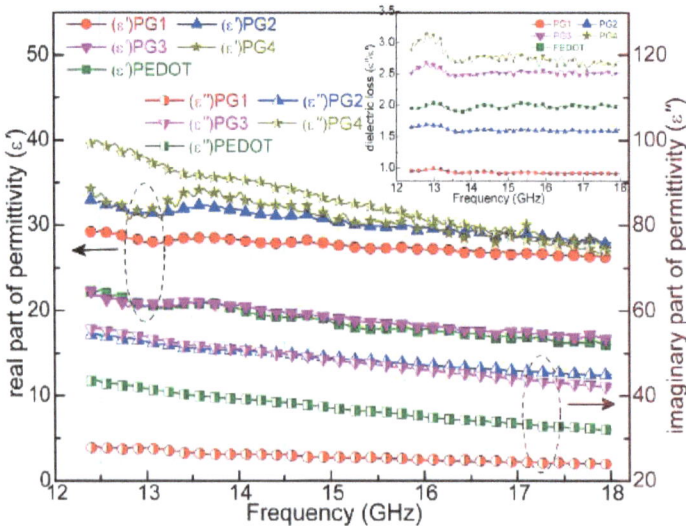

Fig. (33). Plots of ε' & ε'' *versus* frequency, in Ku-band, whereas the inset illustrates the change in dielectric loss ($\tan\delta_\varepsilon = \varepsilon''/\varepsilon'$) with frequency. Reproduced from reference [80], Copyright (2016) Elsevier.

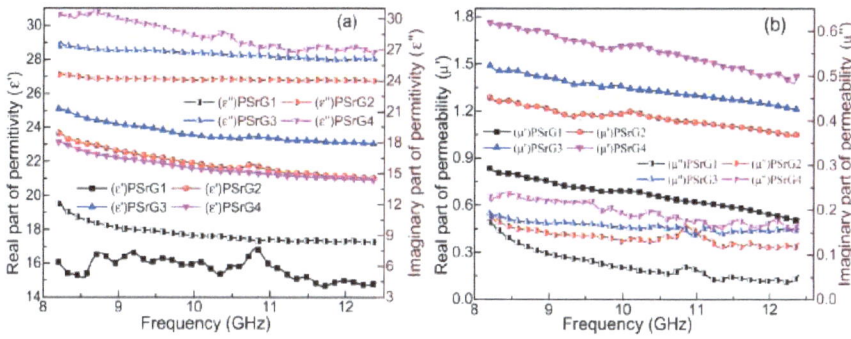

Fig. (34). Plots of (**a**) ε' & ε'' and (**b**) μ' & μ'' *versus* frequency of all PSrG composites samples. Reproduced from reference [88], Copyright (2019) John Wiley and Sons.

As seen from Fig. (**34b**), the variation in μ' & μ'' with respect to frequency, shows decreasing trends in this frequency band, although its values increase with the concentration of ferrite content in the polymer. The higher value of permeability is achieved with respect to the higher ratio of ferrite content in the polymer, *i.e.*, for the PSrG4 sample due to higher magnetic losses. The magnetic losses collectively arise from domain-wall resonance, eddy current effect, magnetic hysteresis loop, and domain wall resonance [84]; among them eddy current effect, domain-wall resonance, and hysteresis loss may occur in the microwave frequencies. The permeability of the strontium ferrite can be written as [89]:

$$\frac{(\mu-1)}{4\pi} = \frac{2M_S}{3H_A} \tag{1}$$

whereas M_S and H_A are saturation magnetization and anisotropy, respectively. According to Snoek's law, the permeability remains unchanged in a certain frequency regime, then lessens for further frequency increase [90]. The lower critical value of frequency occurs with respect to the higher static value of permeability; at this frequency of this precession, magnetization is coupled with the c-axis. The strong coupling refers to a higher precession frequency. The magnitude of this interaction correlated with anisotropic field (H_A) or magneto-crystalline anisotropy which is higher for strontium ferrite. But, the μ'' attributes maximum loss at resonance frequency. At higher frequency, the magnetic dipoles try to synchronize with the frequency of the field, but the induced magnetic field (M) legs behind the magnetization field (M) resulting in phase difference between them that assists magnetic loss in the material. The magneto-crystalline anisotropy is varying considerably with particle size, caused by surface defect and microstructure defect [90,91]. The smaller particle owing higher surface area, *i.e.*, more space is able to be used for interaction with the electromagnetic wave; as a result, large numbers of electromagnetic waves are absorbed. The saturation magnetization and hysteresis loss increased considerably with the rise in the concentration of SrF particles. Therefore, the sample with higher content of SrF is exhibited higher magnetic loss, consequently higher absorption of microwaves.

As mentioned earlier, the dissipation of EM waves is correlated with the complex dielectric and magnetic parameters of the materials. The overall loss, *i.e.*, tangent loss factor, is estimated for both dielectric loss as well as magnetic loss and plotted *versus* frequency in Fig. (**35**). A higher dielectric loss of 1.29 comes in observation for the PSrG4 sample confirming its lossy character. As per the above discussion, the magnetic loss is also observed to increase with the ratio of SrF particles because of the induced losses due to the eddy current effect and hysteresis effect. Fig. (**36**) depicts the happening of various phenomena during the interaction of electromagnetic waves with the material. Both types of losses (dielectric and magnetic loss) in the Ku-band of frequency leads to improved impedance matching properties of the material, produced an enhanced result for the microwave absorption, arising from the synergic effect of SrF & graphene (acts as cores in the core-shell morphology of the composite) PEDOT acts as a shell.

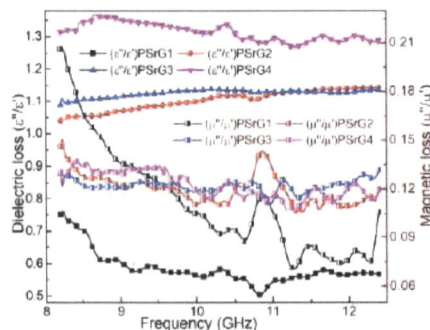

Fig. (35). Plots of dielectric loss (tan $\delta_\varepsilon = \varepsilon''/\varepsilon'$) and magnetic loss (tan $\delta_\mu = \mu''/\mu'$) *versus* frequency. Reproduced from reference [88], Copyright (2019) John Wiley and Sons.

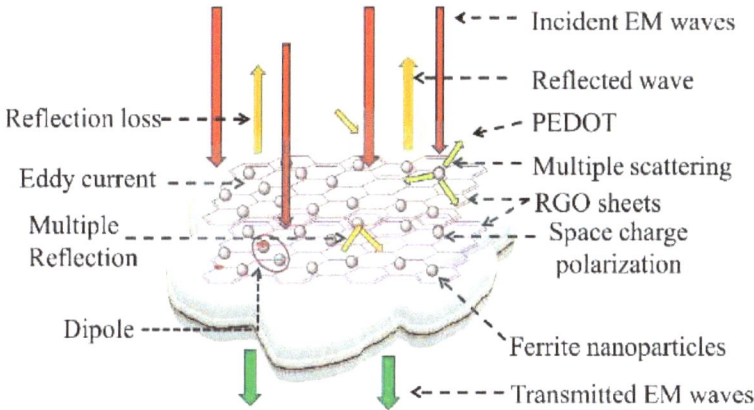

Fig. (36). Pictorial demonstration of various phenomena occurred during the interaction of the electromagnetic wave with the PSrG composite samples that contribute to the attenuation of microwaves. Reproduced from reference [88], Copyright (2019) John Wiley and Sons.

5.8.6. Shielding Mechanism of PEDOT/graphene Composites

To understand the shielding mechanism of PG composites, SE_A and SE_R are estimated, as depicted in Fig. (**37**). For shielding and dielectric measurements, the thickness of samples is taken as ~2.5 mm. The shielding properties of the composites are dominated by absorption rather than reflection mechanism due to the electrical conducting character of the samples. The higher value of SE_A sample is found to be 34.9 dB at 15 GHz for PG4, while it is 34.7, 31.4, 34, and 28.4 dB for PG3, PG2, PG1, and PEDOT samples, respectively.

The enhanced shielding response of the PEDOT's composites is due to the high aspect ratio of graphene and impedance matching behavior of the material. In all the samples, the resonance peak is absent, but as the variation in the plots of PG1 and PG2 is seen, it seems that the resonance peak may occur at higher frequencies out of this frequency regime. It is also observed that the variation in SE_A is very small, suggesting that the material can be useful as broadband microwave absorbers. The values of SE_R are 8, 5.8, 5.2, 4.2, and 5.2 dB for PG4, PG3, PG2, PG1, and PEDOT, respectively observed at 15 GHz; these values are much lower than the values of SE_A for the respective sample. It indicates that the shielding properties of the composites are dominated by absorption phenomena rather than reflection. Also, the shielding due to absorption is increased as the ratio of RGO increases in the composite; however, the variation is small caused by the lower ratio of RGO incorporated in the composites.

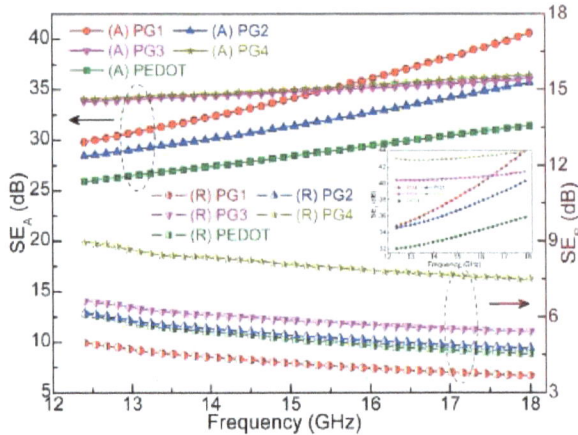

Fig. (37). Plots of shielding effectiveness due to absorption and reflection of PEDOT and PG composites *versus* frequency. The inset shows plots of total shielding effectiveness. Reproduced from reference [80], Copyright (2016) Elsevier.

The trend of total shielding effectiveness for all samples depicted in the inset of Fig. (**37**). is analogous to the trend, as observed for the SE_A, because shielding properties are dominated by the absorption phenomena. Moreover, thickness is another important parameter in shielding. The effect of thickness on SE is examined, and results are shown in Fig. (**38a**). The electromagnetic wave that enters inside the shield attenuates exponentially as $e^{-\alpha z}$, whereas, α is the attenuation constant, and 'z' is the penetration depth measured from the surface of the shield. Therefore, the attenuation of the electromagnetic wave is higher for thick shields. As seen from the Figure, shielding effectiveness due to absorption increases from ~20 dB to ~35 dB with the change in thickness of sample from 1.5 mm to 2.5 mm respectively and decreases after that. For an electrically conducting sample, shielding effectiveness due to reflection and absorption is given by the following expressions [92, 93].

$$SE_A \ (dB) = \ 20 \ \frac{d}{\delta} \log e = 20d\sqrt{\frac{\mu_r \omega \sigma_s}{2}} \log e \tag{2}$$

$$SE_R \ (dB) = 10 \log \left(\frac{\sigma_s}{16\omega\varepsilon_0\mu_r}\right) \tag{3}$$

whereas d is the thickness of the sample, $\delta = 1/\beta = \sqrt{2/\omega\mu\sigma_s}$ is the skin depth, $\sigma_s = \omega\varepsilon_0\varepsilon''$ is the microwave conductivity dependent upon frequency, and $\omega = 2\pi\nu$ is the angular frequency.

According to Equations 2 & 3, both SE_R & SE_A are dependent upon the conductivity. Further, conductivity depends upon the ε'' (dielectric loss) of the material. The absorption loss is directly proportional to the square root of the product of conductivity and relative permeability and to the nature of the interaction of electromagnetic waves with electric & magnetic dipoles in the material. The reflection loss is directly proportional to the log of the ratio of conductivity to relative permeability and depends upon the behavior of mobile charge carriers toward the electromagnetic waves [94]. From the above discussion, it is obvious that the reflection loss dominating in conducting materials with lower permeability and in the conducting materials with higher permeability absorption loss plays an important role in the shielding properties.

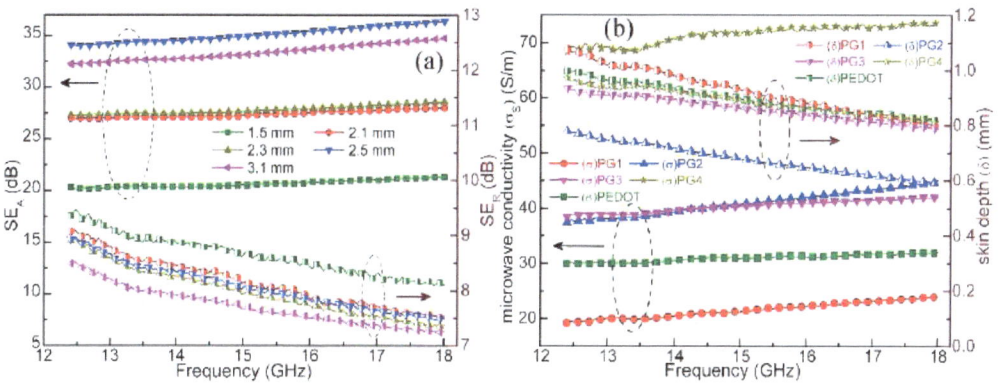

Fig. (38). (a) Plots of SE_A & SE_R with the frequency for different thicknesses of PG4 sample, (b) Plots of microwave conductivity and skin depth with the frequency. Reproduced from reference [80], Copyright (2016) Elsevier.

In these polymer composites, polaron, bipolarons, and electric dipoles are the main charge carriers that contribute to the conductivity; thus, these composites are lossy in nature. The skin depth and conductivity of PEDOT and its composites calculated from the dielectric data plotted with the frequency in Fig. (**38b**), the variation in both parameters satisfy the nature of Equations 2 & 3. As seen from the figure, the skin depth is observed to decrease with frequency, whereas conductivity is found to increase with an increase in frequency. These variations in the skin depth ($\delta = \sqrt{2/\omega\mu\sigma_s}$) and microwave conductivity ($\sigma_s = \omega\varepsilon_0\varepsilon''$) with the frequency is analogues with these theoretical relations. The higher value of conductivity is evaluated for the sample that possesses a higher ratio of RGO in PEDOT polymer, *i.e.*, for PG4 sample, as consequently, PG4 attributed higher dielectric loss and higher shielding effectiveness due to absorption $SE_A \propto (\sigma_s)^{1/2}$ (Fig. **39**).

5.8.7. PEDOT/Graphene/SrF Composites

From Fig. (**40a**), the maximum value of SE_A of 39.22 dB is observed for the PSrG4 sample, and the respective SE_R is 3.07 dB at 12.4 GHz. From the results, it is concluded that shielding effectiveness is mainly governed by the absorption due to the dielectric and magnetic loss in the samples. According to equations 2 & 3, SE_A & SE_R both depend upon the conductivity. As a result, it indirectly depends upon the dielectric loss of the materials because dielectric parameters are functions of the conductivity [95]. Therefore, SE_A increases with the increase in conductivity and permeability of the sample. It is interesting to discuss here that the graphene itself can act as a microwave absorber due to its large aspect ratio and conductivity, but absorption of microwave may decrease because of its higher electrical conductivity.

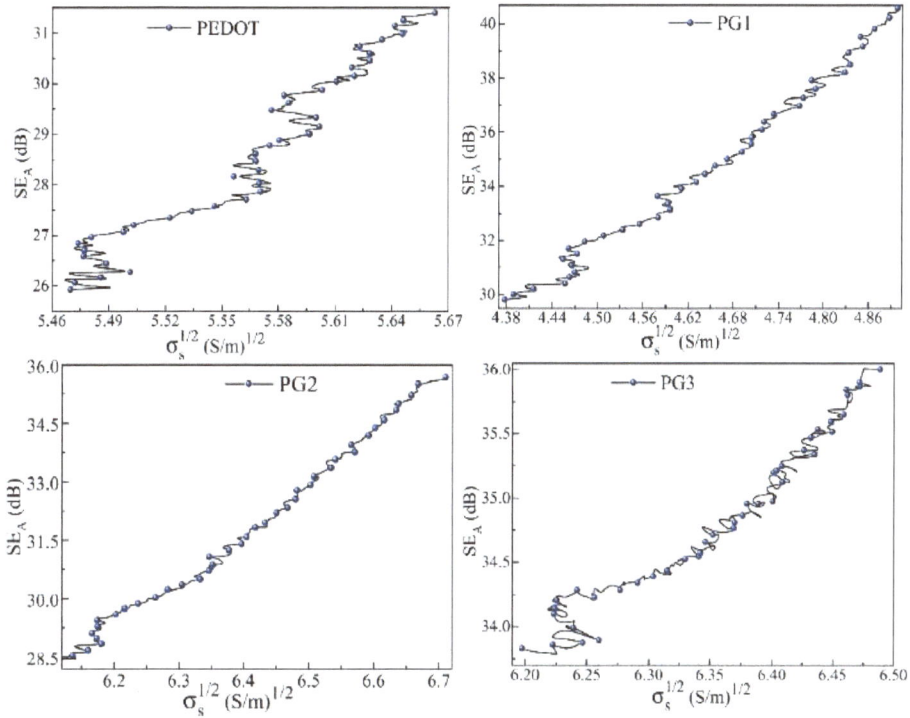

Fig. (39). Variation in SE_A ($SE_A \propto \sqrt{\sigma_s}$) of PEDOT and composites with microwave conductivity.

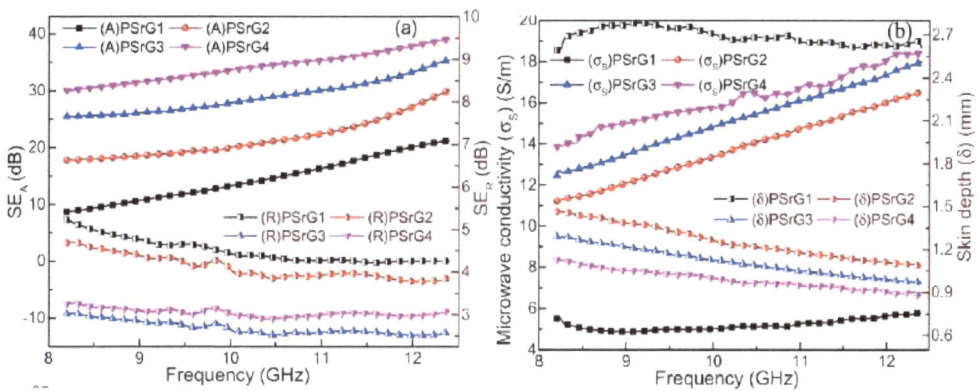

Fig. (40). (**a**) Dependence of SE_A and SE_R of PSrG nanocomposites on frequency, (**b**) Variation of microwave conductivity ($\sigma_S = \omega\varepsilon_o\varepsilon''$) and skin depth ($\delta = 1/\sqrt{\omega\mu\sigma_S}$) with frequency. Reproduced from reference [88], Copyright (2019) John Wiley and Sons.

To examine the variation in microwave shielding parameters due to microwave conductivity, the conductivity and skin depth are plotted with frequency, as depicted in Fig. (**40b**). As observed from Fig. (**40**), the conductivity increases with an increase in the concentration of strontium particles in the polymer matrix may be due to the increase in the number of electrons hopping at the Sr^{2+} ions site [96,97], simultaneously, enhanced conductivity gives rise to dielectric loss. According to equation $\varepsilon'' = \sigma s/\omega \varepsilon_o$, the dielectric loss in RGO is linked to the electron polarization process due to the higher electric mobility of the charge carrier. The higher carrier mobility and conductivity give rise to the skin effect in the material. As a result the electromagnetic waves get reflected from the surface. Therefore, the addition of strontium along with RGO in the PEDOT matrix significantly reduces the skin depth and increases the impedance matching properties of the material that in turn improves the absorption of microwaves.

Fig. (**41a**) reflects the overall effect of strontium ferrite concentration on the shielding performance of the material. From the figure, it can be seen that an increase in the concentration of ferrite particle in the polymer matrix increase the shielding effectiveness due to absorption, whereas shielding due to reflection remains almost invariant with a ratio of the ferrite particles with respect to the monomer of polymer in the composite. The data listed in Table **3** suggest that a higher concentration of strontium ferrite leads to a higher shielding performance due to the absorption of microwave waves.

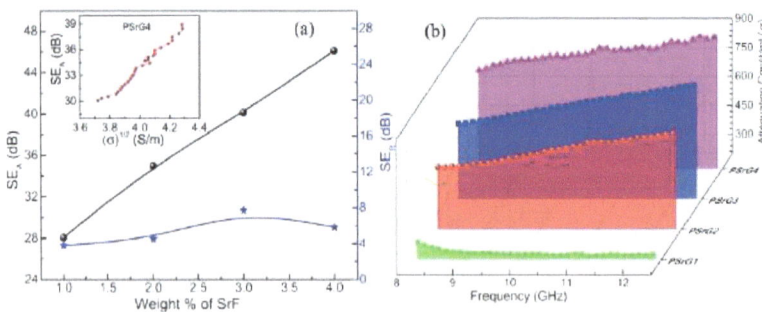

Fig. (**41**). (**a**) Dependence of shielding parameters on the concentration of filler material, and the inset shows the plot between SE_A and microwave conductivity for PSrG4 sample. (**b**) Variation in attenuation constant with the concentration of filler. Reproduced from reference [88], Copyright (2019) John Wiley and Sons.

From the inset of Fig. (**41a**), the linear variation can be seen between the shielding effectiveness due to absorption and microwave conductivity assessed from the dielectric data for the PSrG4 sample which is analogues with the theoretical equation 2. The plot between skin depth and frequency depicted in Fig. (**40b**) confirms the major contribution of absorption phenomena in total shielding effectiveness. The skin depth decreases from 0.24 to 0.19 mm as the frequency increases. This value is very small in comparison to the thickness of the sample, *i.e.*, 2.5 mm. This lessening in skin depth is useful to increase the absorption of microwaves. Moreover, the results of electromagnetic waves absorption are measured corresponding to different concentrations of strontium ferrite particle can be defended by calculating the value of attenuation constant (α), following the equation [84]:

$$\alpha = \frac{\sqrt{2}\pi f}{c}\left\{(\mu''\varepsilon'' - \mu'\varepsilon') + \sqrt{(\mu''\varepsilon'' - \mu'\varepsilon')^2 + (\mu'\varepsilon'' + \mu''\varepsilon')^2}\right\}^{1/2} \quad \textbf{(4)}$$

where c is the velocity of light and f is the frequency. As seen from Fig. (**41b**), among PSrG composites, PSrG4 samples exhibit a higher value of attenuation constant resulting in higher absorption of incident electromagnetic waves. It is observed that α increases as the concentration of ferrite particles increases in the composite. The thickness of the sample plays an important role in the shielding result; it can be confirmed for Fig. (**42**), which shows that SE_A increases from 33.6 to 59.9 dB as the thickness of the PSrG4 sample changes from 2.5 to 4.66 mm at 10 GHz. Thus, the attenuation constant also increases with thickness. The results indicate that a major ratio of incident electromagnetic radiations is absorbed by the PSrG composite attributed to improved impedance matching properties. The energy absorbed by the composite sample may be partially used as field energy storage, and the residual part of the energy is dissipated as heat [87, 96].

Table 3. SE$_A$, SE$_R$, permittivity, permeability, dielectric & magnetic loss, and conductivity at 10 GHz frequency. Reproduced from reference [58] Copyright (2019) John Wiley and Sons.

Sample	ε_r		tanδ_ε	μ_r		tanδ_μ	σ_s (S/m)	SE$_A$ (dB)	SE$_R$ (dB)
	ε'	ε''		μ'	μ''				
PSrG1	16	9.08	0.56	0.68	0.07	0.102	0.79	13.24	4.46
PSrG2	1.92	24.10	1.09	1.18	0.13	0.109	2.14	20.09	4.05
PSrG3	23.55	26.67	1.13	1.34	0.16	0.119	2.36	28.00	2.62
PSrG4	21.50	28.40	1.32	1.60	0.19	0.119	2.50	33.60	3.01

It is observed that the enhanced electromagnetic shielding properties of the PEDOT's composites may be attributed to i) the existence of the number of hetero-interfaces due to incorporation of fillers (RGO & SrF) in the polymer matrix resulting in an increase in polarization and correlated relaxation phenomena, and ii) the improved impedance properties of the composites which majorly contribute in absorption phenomena of electromagnetic shielding. In addition, as the ratio of filler increases in the composite, the dielectric-magneto losses are significantly increased, resulting in higher electromagnetic shielding effectiveness performance of the composite sample.

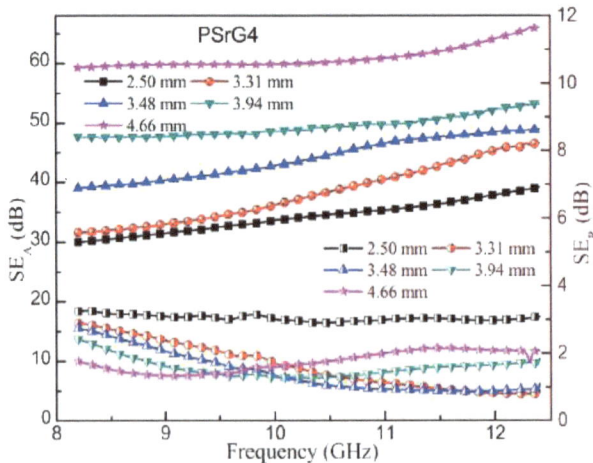

Fig. (42). Comparison of the SE$_A$ and SE$_R$ value of PSrG4 nanocomposites calculated with different thicknesses of sample in X-band. Reproduced from reference [88], Copyright (2019) John Wiley and Sons.

CONCLUDING REMARKS

The PEDOT's nanocomposites have been synthesized *via* in-*in-situ* emulsion polymerization and investigated to analyze the shielding performance of composites against EM radiations. Advanced PEDOT grafted MWCNT composites were synthesized by an in-*in-situ* emulsion polymerization route. The PEDOT grafted MCNT composite was utilized as an effective filler material for conventional polymer for EMI shielding applications in the Ku-Band. The characterization of composites was confirmed by various techniques, which confirm PEDOT grafted MWCNTs were distributed in the PU matrix. The shielding results of the composites suggest that the PEDOT grafted MWCNT filler could also be used in some other polymer matrix. The high-performance has been achieved through the incorporation of conducting & dielectric (RGO) and magnetic filler ($SrFe_{12}O_{19}$). The conducting network comprising the RGO and PEDOT matrix, which interconnects the filler nanoparticles, has been developed during the polymerization process. The dielectric and magnetic losses in nanocomposites result in a total SE of 42.29 dB and 39.22 dB due to absorption at 12.4 GHz showing that the major part of attenuation of EM radiation comes from the absorption. The high absorption ability of composites is ascribed to the synergetic effects of PEDOT, RGO, and SrF due to increased conduction loss, polarization, and magnetic loss. As a result, it can be accomplished that the assimilation of dielectric and magnetic filler in the PEDOT matrix results in improved microwave absorbing material and may find applications in next-generation EM shielding and stealth technology.

CONSENT FOR PUBLICATION

Not applicable.

CONFLICT OF INTEREST

The authors declare no conflict of interest, financial or otherwise.

ACKNOWLEDGEMENTS

The authors are grateful to the Director CSIR-National Physical Laboratory, New Delhi, for his kind support and encouragement. The authors also thank the technicians of the laboratory for their help in offering the resources in running the program.

REFERENCES

[1]   Das, T.K.; Prusty, S. Review on Conducting Polymers and Their Applications. *Polym. Plast. Technol. Eng.,* **2012**, *51*(14), 1487-1500.
        http://dx.doi.org/10.1080/03602559.2012.710697

[2]   Safadi, B.; Andrews, R.; Grulke, E.A. Multiwalled Carbon Nanotube Polymer Composites: Synthesis and Characterization of Thin Films. *J. Appl. Polym. Sci.,* **2002**, *84*(14), 2660-2669.
        http://dx.doi.org/10.1002/app.10436

[3]   Reynolds, J.R.; Skotheim, T.A.; Elsenbaumer, R.L. *Handbook of Conducting Polymers*; Marcel Dekker, **1998**.

[4]   Sun, B.; Marx, E.; Greenham, N.C. Photovoltaic Devices Using Blends of Branched CdSe Nanoparticles and Conjugated Polymers. *Nano Lett.,* **2003**, *3*(7), 961-963.
        http://dx.doi.org/10.1021/nl0342895

[5]   Karami, H.; Mousavi, M.F.; Shamsipur, M. A New Design for Dry Polyaniline Rechargeable Batteries. *J. Power Sources,* **2003**, *117*(1–2), 255-259.
        http://dx.doi.org/10.1016/S0378-7753(03)00168-X

[6]   Duong, N.H.; Nguyen, T.T.; Nguyen, D.T.; Le, H.T. Effect of TiO2 on the gas sensing features of TiO2/PANi nanocomposites. *Sensors (Basel),* **2011**, *11*(2), 1924-1931.
        http://dx.doi.org/10.3390/s110201924 PMID: 22319389

[7]   Elschner, A.; Kirchmeyer, S.; Lovenich, W.; Merker, U. *PEDOT: Principles and Applications of an Intrinsically Conductive Polymer*; CRC Press, **2010**.
        http://dx.doi.org/10.1201/b10318

[8]   Kinlen, P.J.; Liu, J.; Ding, Y.; Graham, C.R.; Remsen, E.E. Emulsion Polymerization Process for Organically Soluble and Electrically Conducting Polyaniline. *Macromolecules,* **1998**, *31*(6), 1735-1744.
        http://dx.doi.org/10.1021/ma971430l

[9]   Bremer, L.G.B.; Verbong, M.; Webers, M.A.M.; Van Doorn, M. Preparation of Core-Shell Dispersions with a Low Tg Polymer Core and a Polyaniline Shell. *Synth. Met.,* **1997**, *84*(1–3), 355-356.
        http://dx.doi.org/10.1016/S0379-6779(97)80779-5

[10]   Jonas, F.; Schrader, L. Conductive Modifications of Polymers with Polypyrroles and Polythiophenes. *Synth. Met.,* **1991**, *41*(3), 831-836.
        http://dx.doi.org/10.1016/0379-6779(91)91506-6

[11]   Roncali, J. Conjugated Poly(Thiophenes): Synthesis, Functionalization, and Applications. *Chem. Rev.,* **1992**, *92*(4), 711-738.
        http://dx.doi.org/10.1021/cr00012a009

[12]   Mastragostino, M.; Arbizzani, C.; Bongini, A, Barbarella G, Zambianchi M. Polymer based electrochromic devices – I. Poly (3-methylthiophene). *Electroch. Acta.,* **1993**, *38*(1), 135-140.
https://doi.org/10.1016/0013-4686 (93)80020-Z

[13]   Monk, P.; Mortimer, R.; Rosseinsky, D. *Electrochromism and Electrochromic Devices (Google EBook)*; Cambridge University Press, **2007**, 2007.

[14]   Tamburri, E.; Orlanducci, S.; Toschi, F.; Terranova, M.L.; Passeri, D. Growth Mechanisms, Morphology, and Electroactivity of PEDOT Layers Produced by Electrochemical Routes in Aqueous Medium. *Synth. Met.,* **2009**, *159*(5), 406-414.
http://dx.doi.org/10.1016/j.synthmet.2008.10.014

[15]   Lefebvre, M.; Qi, Z.; Rana, D.; Pickup, P.G. Chemical Synthesis, Characterization, and Electrochemical Studies of Poly (3, 4-Ethylenedioxythiophene)/Poly (Styrene-4-Sulfonate) Composites. *Chem. Mater.,* **1999**, *11*(2), 262-268.
http://dx.doi.org/10.1021/cm9804618

[16]   Heywang, G.; Materials, F. J.-A. undefined. Poly (Alkylenedioxythiophene) s—New, Very Stable Conducting Polymers. *Wiley Online Libr,* **1992**.

[17]   De, A.; Sen, P.; Poddar, A.; Das, A. Synthesis, Characterization, Electrical Transport and Magnetic Properties of PEDOT–DBSA–Fe 3 O 4 Conducting Nanocomposite. *Synth. Met.,* **2009**, *159*(11), 1002-1007.
http://dx.doi.org/10.1016/j.synthmet.2008.12.030

[18]   Kim, Y.H.; Sachse, C.; Machala, M.L.; May, C.; Müller-Meskamp, L.; Leo, K. Highly Conductive PEDOT: PSS Electrode with Optimized Solvent and Thermal Post-treatment for ITO-free Organic Solar Cells. *Adv. Funct. Mater.,* **2011**, *21*(6), 1076-1081.
http://dx.doi.org/10.1002/adfm.201002290

[19]   Sotzing, G.A.; Lee, K. Poly (Thieno [3, 4-b] Thiophene): A p-and n-Dopable Polythiophene Exhibiting High Optical Transparency in the Semiconducting State. *Macromolecules,* **2002**, *35*(19), 7281-7286.
http://dx.doi.org/10.1021/ma020367j

[20]   Jönsson, S.K.M.; Birgerson, J.; Crispin, X.; Greczynski, G.; Osikowicz, W.; Van Der Gon, A.W.D.; Salaneck, W.R.; Fahlman, M. The Effects of Solvents on the Morphology and Sheet Resistance in Poly (3, 4-Ethylenedioxythiophene)-Polystyrenesulfonic Acid (PEDOT–PSS) Films. *Synth. Met.,* **2003**, *139*(1), 1-10.
http://dx.doi.org/10.1016/S0379-6779(02)01259-6

[21]   Zhang, W.; Zhao, B.; He, Z.; Zhao, X.; Wang, H.; Yang, S.; Wu, H.; Cao, Y. High-Efficiency ITO-Free Polymer Solar Cells Using Highly Conductive PEDOT: PSS/Surfactant Bilayer Transparent Anodes. *Energy Environ. Sci.,* **2013**, *6*(6), 1956-1964.
http://dx.doi.org/10.1039/c3ee41077c

[22]   Nardes, A.M.; Kemerink, M.; De Kok, M.M.; Vinken, E.; Maturova, K.; Janssen, R.A.J. Conductivity, Work Function, and Environmental Stability of PEDOT: PSS Thin Films Treated with Sorbitol. *Org. Electron.,* **2008**, *9*(5), 727-734.
http://dx.doi.org/10.1016/j.orgel.2008.05.006

[23]   Kirchmeyer, S.; Reuter, K. Scientific Importance, Properties and Growing Applications of Poly (3, 4-Ethylenedioxythiophene). *J. Mater. Chem.,* **2005**, *15*(21), 2077-2088.
http://dx.doi.org/10.1039/b417803n

[24]   De Jong, M.P.; Van Ijzendoorn, L.J.; De Voigt, M.J.A. Stability of the Interface between Indium-Tin-Oxide and Poly (3, 4-Ethylenedioxythiophene)/Poly (Styrenesulfonate) in Polymer Light-Emitting Diodes. *Appl. Phys. Lett.,* **2000**, *77*(14), 2255-2257.
       http://dx.doi.org/10.1063/1.1315344

[25]   Wakizaka, D.; Fushimi, T.; Ohkita, H.; Ito, S. Hole Transport in Conducting Ultrathin Films of PEDOT/PSS Prepared by Layer-by-Layer Deposition Technique. *Polymer (Guildf.),* **2004**, *45*(25), 8561-8565.
       http://dx.doi.org/10.1016/j.polymer.2004.10.007

[26]   Kar, P. *Doping in Conjugated Polymers*; John Wiley & Sons, **2013**.
       http://dx.doi.org/10.1002/9781118816639

[27]   York, W.N. *Handbook of Oligo- and Polythiophenes*; WILEY-VCH, **1999**.

[28]   Ibrahim, K.S. Carbon Nanotubes-Properties and Applications: A Review. *Carbon Lett.,* **2013**, *14*(3), 131-144.
       http://dx.doi.org/10.5714/CL.2013.14.3.131

[29]   Kim, H.M.; Kim, K.; Lee, C.Y.; Joo, J.; Cho, S.J.; Yoon, H.S.; Pejaković, D.; Yoo, J.W.; Epstein, J. Electrical Conductivity and Electromagnetic Interference Shielding of Multiwalled Carbon Nanotube Composites Containing Fe Catalyst. *Appl. Phys. Lett.,* **2004**, *84*(4), 589-591.
       http://dx.doi.org/10.1063/1.1641167

[30]   Bai, X.; Hu, X.; Zhou, S.; Li, L.; Rohwerder, M. Controllable Synthesis of Leaflet-like Poly (3, 4-Ethylenedioxythiophene)/Single-Walled Carbon Nanotube Composites with Microwave Absorbing Property. *Compos. Sci. Technol.,* **2015**, *110*, 166-175.
       http://dx.doi.org/10.1016/j.compscitech.2015.02.010

[31]   Saini, P.; Choudhary, V.; Singh, B.P.; Mathur, R.B.; Dhawan, S.K. Polyaniline–MWCNT Nanocomposites for Microwave Absorption and EMI Shielding. *Mater. Chem. Phys.,* **2009**, *113*(2–3), 919-926.
       http://dx.doi.org/10.1016/j.matchemphys.2008.08.065

[32]   Gerwig, R.; Cesare, P.; Kraushaar, U.; Stett, A.; Stelzle, M. Advanced Cardiac and Neuronal Recording Using PEDOT-CNT MEA. **2012**.

[33]   Xu, G.; Li, B.; Luo, X. Carbon Nanotube Doped Poly(3,4-Ethylenedioxythiophene) for the Electrocatalytic Oxidation and Detection of Hydroquinone. *Sens. Actuators B Chem.,* **2013**, *176*, 69-74.
       http://dx.doi.org/10.1016/j.snb.2012.09.001

[34]   Chang, C-Y.; Anuratha, K.S.; Lin, Y-H.; Xiao, Y.; Hasin, P.; Lin, J-Y. Potential-Reversal Electrodeposited MoS2 Thin Film as an Efficient Electrocatalytic Material for Bifacial Dye-Sensitized Solar Cells. *Sol. Energy,* **2020**, *206*, 163-170.
       http://dx.doi.org/10.1016/j.solener.2020.06.001

[35]   Geetha, S.; Satheesh Kumar, K.K.; Rao, C.R.K.; Vijayan, M.; Trivedi, D.C. EMI Shielding: Methods and Materials—A Review. *J. Appl. Polym. Sci.,* **2009**, *112*(4), 2073-2086.
       http://dx.doi.org/10.1002/app.29812

[36]   Bigg, D.M.; Seymour, E.J.B.R.B. *Conductive Polymers*; Plenum: New York, **1981**, Vol. 15, .

[37]   Bailey, J.; Carlson, A.L.; Chandler, G.; Derzon, M.S.; Dukart, R.J.; Hammel, B.A.; Johnson, D.J.; Lockner, T.R.; Maenchen, J.; McGuire, E.J.; Mehlhorn, T.A.; Nelson, W.E.; Ruggles, L.E.; Stygar, W.A.; Wenger, D.F. Observation of Kα. X-Ray Satellites from a Target Heated by an Intense Ion Beam. *Laser Part. Beams,* **1990**, *8*(4), 555-562.

http://dx.doi.org/10.1017/S0263034600008983

[38]   Niranjanappa, A.C.; Biliya, R.; Trivedi, D.C. Seminar on State of the Art in EMI—EMC and Future Trends. India, **1996**.

[39]   Knott, E.F. *Radar Cross Section Measurements*; Springer Science & Business Media, **2012**.

[40]   Dhawan, S.K.; Ohlan, A.; Singh, K. Designing of Nano Composites of Conducting Polymers for EMI Shielding.*Advances in Nanocomposites - Synthesis, Characterization and Industrial Applications*; Reddy, B.S.R., Ed.; INTECH, **2011**.
http://dx.doi.org/10.5772/14752

[41]   Ohlan, A.; Singh, K.; Chandra, A.; Singh, V.N.; Dhawan, S.K. Conjugated Polymer Nanocomposites: Synthesis, Dielectric, and Microwave Absorption Studies. *J. Appl. Phys.,* **2009**, *106*(4), 044305.
http://dx.doi.org/10.1063/1.3200958

[42]   Ohlan, A.; Singh, K.; Chandra, A.; Dhawan, S.K. Microwave Absorption Properties of Conducting Polymer Composite with Barium Ferrite Nanoparticles in 12.4–18 GHz. *Appl. Phys. Lett.,* **2008**, *93*(5), 053114.
http://dx.doi.org/10.1063/1.2969400

[43]   Ohlan, A.; Singh, K.; Dhawan, S.K. Shielding and Dielectric Properties of Sulfonic Acid-Doped π-Conjugated Polymer in 8.2-12.4 GHz Frequency Range. *J. Appl. Polym. Sci.,* **2010**, *115*(1), 498-503.
http://dx.doi.org/10.1002/app.30806

[44]   Varshney, S.; Ohlan, A.; Singh, K.; Jain, V.K.; Dutta, V.P.; Dhawan, S.K. Robust Multifunctional Free Standing Polypyrrole Sheet for Electromagnetic Shielding. *Sci. Adv. Mater.,* **2013**, *5*(7), 881-890.
http://dx.doi.org/10.1166/sam.2013.1534

[45]   Kogan, I.; Fokeeva, L.; Shunina, I.; Estrin, Y.; Kasumova, L.; Kaplunov, M.; Davidova, G.; Knerelman, E. An Oxidizing Agent for Aniline Polymerization. *Synth. Met.,* **1999**, *100*, 303.

[46]   Palaniappan, S. Benzoyl Peroxide Oxidation Route to Polyaniline Salts—Part I. *Polym. Adv. Technol.,* **2004**, *15*(3), 111-117.
http://dx.doi.org/10.1002/pat.424

[47]   Wang, Y.; Liu, Z.; Han, B.; Sun, Z.; Huang, Y.; Yang, G. Facile synthesis of polyaniline nanofibers using chloroaurate acid as the oxidant. *Langmuir,* **2005**, *21*(3), 833-836.
http://dx.doi.org/10.1021/la047442z PMID: 15667157

[48]   Kim, B-K.; Kim, Y.H.; Won, K.; Chang, H.; Choi, Y.; Kong, K.; Rhyu, B.W.; Kim, J-J.; Lee, J-O. Electrical Properties of Polyaniline Nanofibre Synthesized with Biocatalyst. *Nanotechnology,* **2005**, *16*(8), 1177-1181.
http://dx.doi.org/10.1088/0957-4484/16/8/033

[49]   Freund, M.S.; Deore, B.A. *Self-Doped Conducting Polymers*; John Wiley & Sons, **2007**.
http://dx.doi.org/10.1002/9780470061725

[50]   Urban, D.; Takamura, K. *Polymer Dispersions and Their Industrial Applications*; Wiley Online Library, **2002**.
http://dx.doi.org/10.1002/3527600582

[51]   Eliseeva, V.I.; Ivanchev, S.S.; Kuchanov, S.I.; Lebedev, A.V. *Emulsion Polymerization and Its Applications in Industry*; Springer Science & Business Media, **2012**.

[52]   Piirma, I.; Chang, M. Emulsion Polymerization of Styrene: Nucleation Studies with Nonionic Emulsifier. *J. Polym. Sci. A Polym. Chem.,* **1982**, *20*(2), 489-498.
       http://dx.doi.org/10.1002/pol.1982.170200222

[53]   Lovell, P.A.; El-Aasser, M.S. *Emulsion Polymerization and Emulsion Polymers*; Wiley, **1997**.

[54]   Singh, K.; Ohlan, A.; Kotnala, R.K.; Bakhshi, A.K.; Dhawan, S.K. Dielectric and Magnetic Properties of Conducting Ferromagnetic Composite of Polyaniline with γ-Fe2O3 Nanoparticles. *Mater. Chem. Phys.,* **2008**, *112*(2), 651-658.
       http://dx.doi.org/10.1016/j.matchemphys.2008.06.026

[55]   The Editors of Encyclopaedia Britannica. Polymerization.

[56]   Ohlan, A.; Singh, K.; Chandra, A.; Dhawan, S.K. Conducting Ferromagnetic Copolymer of Aniline and 3,4- Ethylenedioxythiophene Containing Nanocrystalline Barium Ferrite Particles. *J. Appl. Polym. Sci.,* **2008**, *108*(4), 2218-2225.
       http://dx.doi.org/10.1002/app.27794

[57]   Singh, K.; Ohlan, A.; Saini, P.; Dhawan, S.K. Poly (3, 4-Ethylenedioxythiophene) γ-Fe2O3 Polymer Composite--Super Paramagnetic Behavior and Variable Range Hopping 1D Conduction Mechanism--Synthesis and Characterization. *Polym. Adv. Technol.,* **2008**, *19*(3), 229-236.
       http://dx.doi.org/10.1002/pat.1003

[58]   Elschner, A.; Kirchmeyer, S.; Lovenich, W.; Merker, U.; Reuter, K. *PEDOT: Principles and Applications of an Intrinsically Conductive Polymer*; CRC Press, **2010**.
       http://dx.doi.org/10.1201/b10318

[59]   Choi, J.W.; Han, M.G.; Kim, S.Y.; Oh, S.G.; Im, S.S. Im, S. S. Poly(3,4-Ethylenedioxythiophene) Nanoparticles Prepared in Aqueous DBSA Solutions. *Synth. Met.,* **2004**, *141*(3), 293-299.
       http://dx.doi.org/10.1016/S0379-6779(03)00419-3

[60]   Farukh, M.; Singh, A.P.; Dhawan, S.K. Enhanced Electromagnetic Shielding Behavior of Multi-Walled Carbon Nanotube Entrenched Poly (3, 4-Ethylenedioxythiophene) Nanocomposites. *Compos. Sci. Technol.,* **2015**, *114*, 94-102.
       http://dx.doi.org/10.1016/j.compscitech.2015.04.004

[61]   Deng, J.; Ding, X.; Zhang, W.; Peng, Y.; Wang, J.; Long, X.; Li, P.; Chan, A.S.C. Carbon Nanotube–Polyaniline Hybrid Materials. *Eur. Polym. J.,* **2002**, *38*(12), 2497-2501.
       http://dx.doi.org/10.1016/S0014-3057(02)00165-9

[62]   Boussoualem, M.; King, R.C.Y.; Brun, J-F.; Duponchel, B.; Ismaili, M.; Roussel, F. Electro-Optic and Dielectric Properties of Optical Switching Devices Based on Liquid Crystal Dispersions and Driven by Conducting Polymer [Poly (3, 4-Ethylene Dioxythiophene): Polystyrene Sulfonate (PEDOT: PSS)]-Coated Electrodes. *J. Appl. Phys.,* **2010**, *108*(11), 113526.
       http://dx.doi.org/10.1063/1.3518041

[63]   Farah, A.A.; Rutledge, S.A.; Schaarschmidt, A.; Lai, R.; Freedman, J.P.; Helmy, A.S. Conductivity Enhancement of Poly (3, 4-Ethylenedioxythiophene)-Poly (Styrenesulfonate) Films Post-Spincasting. *J. Appl. Phys.,* **2012**, *112*(11), 113709.
       http://dx.doi.org/10.1063/1.4768265

[64]   Costa, S.; Borowiak-Palen, E.; Kruszynska, M.; Bachmatiuk, A.; Kalenczuk, R.J. Characterization of Carbon Nanotubes by Raman Spectroscopy. *Mater. Sci.,* **2008**, *26*(2), 433-441.

[65]   Mousavi, Z.; Alaviuhkola, T.; Bobacka, J.; Latonen, R-M.; Pursiainen, J.; Ivaska, A. Electrochemical Characterization of Poly (3, 4-Ethylenedioxythiophene)(PEDOT) Doped with Sulfonated Thiophenes. *Electrochim. Acta*, **2008**, *53*(11), 3755-3762.
http://dx.doi.org/10.1016/j.electacta.2007.09.010

[66]   Maiti, S.; Shrivastava, N. K.; Suin, S.; Khatua, B. B. Polystyrene / MWCNT / Graphite Nanoplate Nanocomposites : Efficient Electromagnetic Interference Shielding Material through Graphite Nanoplate − MWCNT − Graphite Nanoplate Networking. **2013**. No. ii

[67]   Wang, Y.Y.; Cai, K.F.; Shen, S.; Yao, X. In-Situ Fabrication and Enhanced Thermoelectric Properties of Carbon Nanotubes Filled Poly (3, 4-Ethylenedioxythiophene) Composites. *Synth. Met.*, **2015**, *209*, 480-483.
http://dx.doi.org/10.1016/j.synthmet.2015.08.034

[68]   Singh, A.P.; Mishra, M.; Sambyal, P.; Gupta, B.K.; Singh, B.P.; Chandra, A.; Dhawan, S.K. Encapsulation of γ-Fe 2 O 3 Decorated Reduced Graphene Oxide in Polyaniline Core–Shell Tubes as an Exceptional Tracker for Electromagnetic Environmental Pollution. *J. Mater. Chem. A Mater. Energy Sustain.*, **2014**, *2*(10), 3581-3593.
http://dx.doi.org/10.1039/C3TA14212D

[69]   Cadek, M.; Coleman, J.N.; Barron, V.; Hedicke, K.; Blau, W.J. Morphological and Mechanical Properties of Carbon-Nanotube-Reinforced Semicrystalline and Amorphous Polymer Composites. *Appl. Phys. Lett.*, **2002**, *81*(27), 5123-5125.
http://dx.doi.org/10.1063/1.1533118

[70]   Soleimani, H.; Abbas, Z.; Yahya, N.; Soleimani, H.; Ghotbi, M.Y. Determination of Complex Permittivity and Permeability of Lanthanum Iron Garnet Filled PVDF-Polymer Composite Using Rectangular Waveguide and Nicholson–Ross–Weir (NRW) Method at X-Band Frequencies. *Measurement*, **2012**, *45*(6), 1621-1625.
http://dx.doi.org/10.1016/j.measurement.2012.02.014

[71]   Akindoyo, J.O.; Beg, M.; Ghazali, S.; Islam, M.R.; Jeyaratnam, N.; Yuvaraj, A.R. Polyurethane Types, Synthesis and Applications–a Review. *RSC Advances*, **2016**, *6*(115), 114453-114482.
http://dx.doi.org/10.1039/C6RA14525F

[72]   Oertel, G.; Abele, L. *Polyurethane Handbook: Chemistry, Raw Materials, Processing, Application*; Properties, **1994**.

[73]   Farukh, M.; Dhawan, R.; Singh, B.P.; Dhawan, S.K. Sandwich Composites of Polyurethane Reinforced with Poly (3, 4-Ethylene Dioxythiophene)-Coated Multiwalled Carbon Nanotubes with Exceptional Electromagnetic Interference Shielding Properties. *RSC Advances*, **2015**, *5*(92), 75229-75238.
http://dx.doi.org/10.1039/C5RA14105B

[74]   Gupta, T.K.; Singh, B.P.; Dhakate, S.R.; Singh, V.N.; Mathur, R.B. Improved Nanoindentation and Microwave Shielding Properties of Modified MWCNT Reinforced Polyurethane Composites. *J. Mater. Chem. A Mater. Energy Sustain.*, **2013**, *1*(32), 9138-9149.
http://dx.doi.org/10.1039/c3ta11611e

[75]   McClory, C.; McNally, T.; Brennan, G.P.; Erskine, J. Thermosetting Polyurethane Multiwalled Carbon Nanotube Composites. *J. Appl. Polym. Sci.*, **2007**, *105*(3), 1003-1011.
http://dx.doi.org/10.1002/app.26144

[76]   Trovati, G.; Sanches, E.A.; Neto, S.C.; Mascarenhas, Y.P.; Chierice, G.O. Characterization of Polyurethane Resins by FTIR, TGA, and XRD. *J. Appl. Polym. Sci.*, **2010**, *115*(1), 263-268.

http://dx.doi.org/10.1002/app.31096

[77] Pandit, B.; Dhakate, S.R.; Singh, B.P.; Sankapal, B.R. Free-Standing Flexible MWCNTs Bucky Paper: Extremely Stable and Energy Efficient Supercapacitive Electrode. *Electrochim. Acta,* **2017**, *249*, 395-403.
http://dx.doi.org/10.1016/j.electacta.2017.08.013

[78] Zhou, J.; Lubineau, G. Improving electrical conductivity in polycarbonate nanocomposites using highly conductive PEDOT/PSS coated MWCNTs. *ACS Appl. Mater. Interfaces,* **2013**, *5*(13), 6189-6200.
http://dx.doi.org/10.1021/am4011622 PMID: 23758203

[79] Chang, C.; Phillips, E.M.; Liang, R.; Tozer, S.W.; Wang, B.; Zhang, C.; Chiu, H. Alignment and Properties of Carbon Nanotube Buckypaper/Liquid Crystalline Polymer Composites. *J. Appl. Polym. Sci.,* **2013**, *128*(3), 1360-1368.

[80] Dalal, J.; Gupta, A.; Lather, S.; Singh, K.; Dhawan, S.K.; Ohlan, A. Poly (3, 4-Ethylene Dioxythiophene) Laminated Reduced Graphene Oxide Composites for Effective Electromagnetic Interference Shielding. *J. Alloys Compd.,* **2016**, *682*(May), 52-60.
http://dx.doi.org/10.1016/j.jallcom.2016.04.276

[81] Dalal, J.; Lather, S.; Gupta, A.; Dahiya, S.; Maan, A.S.; Singh, K.; Dhawan, S.K.; Ohlan, A. EMI Shielding Properties of Laminated Graphene and PbTiO3 Reinforced Poly (3, 4-Ethylenedioxythiophene) Nanocomposites. *Compos. Sci. Technol.,* **2018**, *165*, 222-230.
http://dx.doi.org/10.1016/j.compscitech.2018.07.016

[82] Nicolson, A.M.; Ross, G.F. Measurement of the Intrinsic Properties of Materials by Time-Domain Techniques. *IEEE Trans. Instrum. Meas.,* **1970**, *19*(4), 377-382.
http://dx.doi.org/10.1109/TIM.1970.4313932

[83] Weir, W.B. Automatic Measurement of Complex Dielectric Constant and Permeability at Microwave Frequencies. *Proc. IEEE,* **1974**, *62*(1), 33-36.
http://dx.doi.org/10.1109/PROC.1974.9382

[84] Ding, D.; Wang, Y.; Li, X.; Qiang, R.; Xu, P.; Chu, W.; Han, X.; Du, Y. Rational Design of Core-Shell Co@ C Microspheres for High-Performance Microwave Absorption. *Carbon,* **2017**, *111*, 722-732.
http://dx.doi.org/10.1016/j.carbon.2016.10.059

[85] Chen, Y.J.; Gao, P.; Zhu, C.L.; Wang, R.X.; Wang, L.J.; Cao, M.S.; Fang, X.Y. Synthesis, Magnetic and Electromagnetic Wave Absorption Properties of Porous Fe 3 O 4/Fe/SiO 2 Core/Shell Nanorods. *J. Appl. Phys.,* **2009**, *106*(5), 54303.
http://dx.doi.org/10.1063/1.3204958

[86] Maxwell, J.C. *A Treatise on Electricity and Magnetism*; Clarendon press, **1881**, Vol. 1, .

[87] Qin, F.; Brosseau, C. A Review and Analysis of Microwave Absorption in Polymer Composites Filled with Carbonaceous Particles. *J. Appl. Phys.,* **2012**, *111*(6), 4.
http://dx.doi.org/10.1063/1.3688435

[88] Dalal, J.; Lather, S.; Gupta, A.; Tripathi, R.; Maan, A.S.; Singh, K.; Ohlan, A. Reduced Graphene Oxide Functionalized Strontium Ferrite in Poly(3,4-Ethylenedioxythiophene) Conducting Network: A High-Performance EMI Shielding Material. *Adv. Mater. Technol.,* **2019**, *4*(7), 1-11.
http://dx.doi.org/10.1002/admt.201900023

[89]   Malyshev, A.V.; Petrova, A.B.; Surzhikov, A.P.; Sokolovskiy, A.N. Effect of Sintering Regimes on the Microstructure and Magnetic Properties of LiTiZn Ferrite Ceramics. *Ceram. Int.,* **2018**, *45*(2), 2719-2724.
http://dx.doi.org/10.1016/j.ceramint.2018.09.114

[90]   Ling, W.; Chen, G.; Lei, P.; Yao, Y.; Li, L.; Huang, Y.; Wei, H.; Du, J. Low-Firing Behavior, Microstructure, and Electromagnetic Properties of a Ferroelectric-Ferromagnetic Composite Material with Multiple Doping. *J. Alloys Compd.,* **2018**, *750*, 479-489.
http://dx.doi.org/10.1016/j.jallcom.2018.04.025

[91]   Poperechny, I.S.; Raikher, Y.L. Ferromagnetic Resonance in Core-Shell Nanoparticles with Multitype Exchange Anisotropy. *Phys. Rev. B,* **2018**, *98*(1), 14434.
http://dx.doi.org/10.1103/PhysRevB.98.014434

[92]   Colaneri, N.F.; Schacklette, L.W. EMI Shielding Measurements of Conductive Polymer Blends. *IEEE Trans. Instrum. Meas.,* **1992**, *41*(2), 291-297.
http://dx.doi.org/10.1109/19.137363

[93]   Singh, K.; Ohlan, A.; Bakhshi, A.K.; Dhawan, S.K. Synthesis of Conducting Ferromagnetic Nanocomposite with Improved Microwave Absorption Properties. *Mater. Chem. Phys.,* **2010**, *119*(1–2), 201-207.
http://dx.doi.org/10.1016/j.matchemphys.2009.08.060

[94]   Wan, M. others. In: *Conducting Polymers with Micro or Nanometer Structure*; Springer, **2008**.

[95]   Colin, X. *Advanced Materials for Electromagnetic Interference Shielding*; CRC press, **2009**.

[96]   Biswas, S.; Arief, I.; Panja, S.S.; Bose, S. Absorption-Dominated Electromagnetic Wave Suppressor Derived from Ferrite-Doped Cross-Linked Graphene Framework and Conducting Carbon. *ACS Appl. Mater. Interfaces,* **2017**, *9*(3), 3030-3039.
http://dx.doi.org/10.1021/acsami.6b14853 PMID: 28036170

[97]   Biswas, S.; Kar, G.P.; Bose, S. Engineering nanostructured polymer blends with controlled nanoparticle location for excellent microwave absorption: a compartmentalized approach. *Nanoscale,* **2015**, *7*(26), 11334-11351.
http://dx.doi.org/10.1039/C5NR01785H PMID: 26067647

CHAPTER 6

Graphene and its Derivatives Based Nanocomposites as Potential Candidate to Swallow Microwave Pollution

Monika Mishra[1*], Avanish Pratap Singh[2], and S.K. Dhawan[3]

[1] Department of Physics, Netaji Subhas University of Technology, Dwarka, Delhi, 110078, India

[2] Experimental Research Laboratory,Department of Physics, Atma Ram Sanatan Dharma College, DhaulaKuan, New Delhi – 110021, India

[3] Materials Physics & Engineering Division, CSIR-National Physical Laboratory, New Delhi – 110012, India

Abstract: Graphene is composed of a single atomic layer of carbon with excellent mechanical, electrical, and optical properties. It can be widely used in the areas of physics, chemistry, energy, information, equipment manufacturing, and electromagnetic interference shielding. This chapter described the concepts, structures, properties, manufacturing methods, and applications of graphene composites in shielding against electromagnetic interference. The continuous 3D conductive array of graphene composites can effectively improve the electronic and ionic transmission of the material. Therefore, adding graphene oxide to the composite material will significantly improve the performance of the material for better conduction and EMI shielding. This chapter summarizes the latest developments in the electromagnetic shielding performance of graphene composites and highlights their prospects.

Keywords: Dielectric Attributes, Ferrofluid, Graphene, Microwave Pollution, Reduced Graphene Oxide, Skin Depth.

6.1. INTRODUCTION

The rapid expansion of electronic gadgets employing wi-fi and blue-tooth technology, such as tablets, laptops, cell phones, smart television, *etc.*, as well as augmented growth of transient power resources, offered electromagnetic pollution to modern society [1-5]. This undesired electromagnetic non-ionizing radiation may

---
*Corresponding author Monika Mishra:** Department of Physics, Netaji Subhas University of Technology, Dwarka, Delhi, 110078, India; Tel: 91 8588850486; E-mail: monika.mishra@nsut.ac.in

Sundeep K. Dhawan, Avanish Pratap Singh, Anil Ohlan, Kuldeep Singh Kakran and Pradeep Sambyal (Ed.)
All rights reserved-© 2022 Bentham Science Publishers

be carcinogenic to biological species and has always been a burning topic of debate. Extreme mobile phone use has been linked to the development of brain tumors.

Radiation frequencies have been linked to symptoms, such as decreased sperm count, headaches, tiredness, irritability, memory loss, and skin irritations in investigations [6]. Acute RF signal exposure has been shown to alter glucose metabolism in the brain [7]. The thermo-physical impact of radiation absorption was validated using Nuclear Magnetic Resonance (NMR) imaging of bovine brain tissue [8]. Migratory birds' magnetic compass orientation is disrupted by man-made electromagnetic noise [9]. Cell phone usage was categorized as "possibly carcinogenic to humans" by the International Agency for Research on Cancer (IARC), a division of the World Health Organization (WHO), in 2011 [10, 11]. Therefore, the World Health Organization (WHO) has expressed worry about non-ionizing radiations by citing, "Everyone is exposed to a complex mix of weak electric and magnetic fields, both at home and at work, from generation and transmission of electricity, domestic appliances and industrial equipment, to telecommunications and broadcasting'' [12, 13]. To minimize radiation pollution for creating a safer and healthier environment, a quest for efficient electromagnetic interference (EMI) shielding materials has been triggered. It is an emerging demand to develop some new materials, which have better absorption characteristics in the microwave frequency range. Therefore, selective frequency radiation shields play an important role in the proper operation of electronic devices in all industrial and strategic sectors.

Recently, electromagnetic interference (EMI) shielding in the microwave range has become the most important concern for researchers. Therefore, they develop lightweight, portable, flexible, corrosion-resistant, cost-effective, and easy to process EMI shielding materials to protect sensitive circuits and defend the workspace and environment from unwanted radiation coming from electronic equipment [14-18]. Several important factors play a key role in controlling the overall performance of the shielding materials, such as electrical conductivity, magnetization, and dielectric attributes including relative permittivity as well as the permeability of the composites, *etc.* The presence of mobile charge carriers, electric and magnetic dipoles is a prerequisite of the electromagnetic shields. In all such shields, the shielding effectiveness is fundamentally ruled by reflection and absorption mechanisms [19]. The electromagnetic radiation interacts with mobile charge carriers of materials and gets reflected. This can be efficiently provided by graphene. The absorption of radiation takes place through interaction with electric and magnetic dipoles present in the material as delineated in Fig. (**1**). The absorption loss depends on the value of σ_r/μ_r, *i.e.*, the absorption loss is maximum when $\mu_r = \sigma_r$. Thus, the shielding mechanism of nanocomposites will be studied by quantifying the contribution of absorption and reflection loss to the total EMI shielding effectiveness (SE) along with the correlation among the conductivity, tan δ, absorption loss, reflection loss, magnetic properties, and composite morphology.

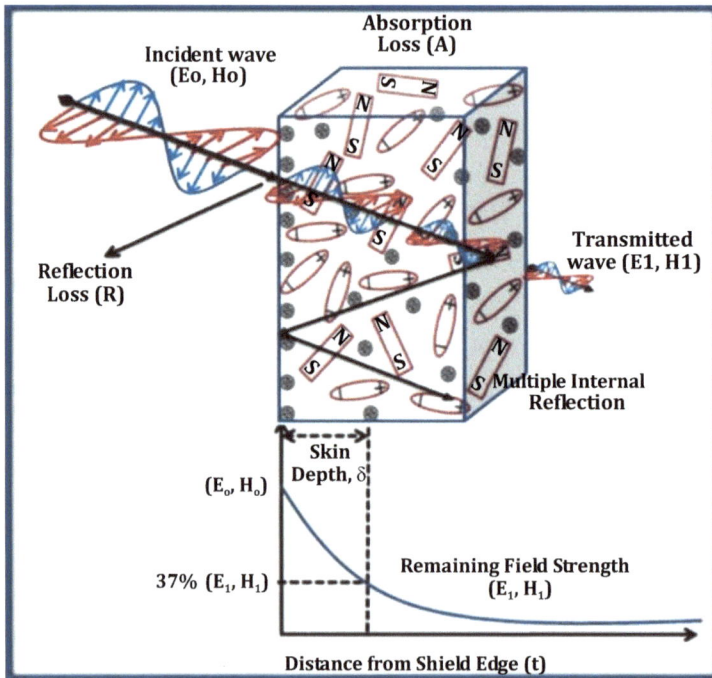

Fig. (1). Schematic representation of EMI shielding through a pool of mobile charges, electric and magnetic dipoles.

Occasionally, multiple reflections can be a secondary mechanism in the case of materials with large surface/interface areas. Therefore, a doctrine for finding good shielding material is to make a pool of mobile charges, electric and magnetic dipoles. Combining materials with mobility (*e.g.*, graphene, CNTs, carbon fiber, metals, conducting polymers, *etc.*), dielectric materials (*e.g.*, SiO_2, TiO_2, barium titanate, barium strontium titanate, polymer and various oxides) and magnetic (Fe_3O_4, Fe_2O_3, $BaFe_{12}O_{19}$, ferrofluid, *etc.*) in proper ratio have been a good strategy [20-27]. Therefore, mixed morphology composites with two, three and many phases have been studied in the last few decades to solve the said electromagnetic radiation pollution problem [28]. In this direction, graphene and its derivatives (Graphite [29], Graphite oxide, exploited graphite [30], graphene oxide [31, 32], reduced graphene oxide, graphene [33], Carbon nanotubes [34, 35] *etc.*) are attracting significant attention of the researchers to meet EMI problem because these structures are associated with high electrical & thermal conductivity, corrosion resistive, lightweight, and easy processability [29, 36-41]. The lightweight materials are especially enviable in the aircraft, spacecraft, and automobile industry to fulfill the requirement of strength and energy savings.

The potential of graphene, along with magnetic and dielectric filler, as microwave shields has been demonstrated by various scientific groups. Graphene-based nanocomposites are architectured using dielectric and magnetic nanoparticles. Such composites emerged as a new class of exciting materials from both perspectives of science and technology [42-44]. The multifunctional composites can combine the advantages of both graphene and nanoparticles. Therefore, a series of efforts using partially reduced graphene oxide with different matrices have been investigated for microwave shielding, as shown in Table **1**.

Table 1. Electromagnetic shielding properties of graphene derivatives composites.

Sample Details	Frequency (GHz)	Shielding Performance (dB)	Reference
Functionalized graphene polyvinylidene fluoride composites	8-12	20	Eswaraiah *et al.* [45]
15 wt. % RGO–hematite composite in wax	6.1	35.8	Chen *et al.* [46]
Graphene/epoxy composites	8.2–12.4	20	Liang *et al.* [15]

(Table 1) cont.....

Fe_3O_4-graphene hybrid	17.2	30.1	Li *et al*. [47]
Graphene foam composites	8-12	30	Chen *et al*. [48]
Graphene–Fe_3O_4 composite	11.5	23	Hu *et al*. [49]
MnO_2 decorated graphene nanoribbons	12.4-18	57	Gupta *et al*. [50]
Polyaniline-gold/graphene oxide composite	2-12	90–120	Basavaraja *et al*. [51]
Carbon nanotube–reduced graphene oxide composites	12.4-18	37	Singh *et al*. [52]
Multilayer graphene oxide aerogel film/polymer composites	8.2–12.4	53	Han *et al*. [53]
Fe/CNTs/epoxy composite	11	25	Che *et al*. [54]
Graphene@Fe_3O_4 composite in polyetherimide	8-12	41.5	Shen *et al*. [55]
Ultrathin Flexible Graphene Film	8-12	20	Shen *et al*. [31]
Barium ferrite-reduced graphene oxide	12.4-18	32	Verma *et al*. [20]
Graphene/Fe_3O_4@Fe/ZnO	7.3	30	Ren *et al*. [56]
Multilayer graphene/polymer composite film	8.2–12.4	27	Song *et al*. [57]
Fe_3O_4 hollow spheres/graphene oxide	12.9	24	Xu *et al*. [58]

(Table 1) cont.....

Graphene oxide/ferrofluid/cement composite	8.2–12.4	46	Singh *et al.* [59]
Graphene foam/CNT/poly (dimethyl siloxane) composites	8.2–12.4	75	Sun *et al.* [51]
Reduced graphene oxide and polystyrene composite	8.2	45.1	Yan *et al.* [60]
Graphene nanosheet/water-borne polyurethane composite	12	32	Hsiao *et al.* [61]

The role of graphene anchored with ferrites nanocomposites (conducting ferrofluid) [62] and tin oxide nanoparticles architectured reduced graphene oxide composite (SnO$_2$@RGO) [63] for microwave shielding is thoroughly discussed in this chapter. The conducting ferrofluid may combine the advantages of both conducting reduced graphene oxide and magnetic nanoparticles, while SnO$_2$@RGO has good conductivity and dielectric loss. More importantly, these functionalized nanocomposites exhibit new or enhanced properties, which warrant them as promising materials with applications in wide areas. In addition, this chapter also establishes recent developments and challenges in shielding research and creates ground for developing futuristic radar absorbing materials.

6.2. ELECTROMAGNETIC RADIATION SHIELDING THEORY

This section establishes the mathematical formalism for calculating EMI shielding and related EM absorption mechanisms using scattering parameters (S_{11}, S_{22}, S_{12} and S_{21}) obtained from two-port measurements from a vector network analyzer, as shown in Fig. (**2**). These scattering parameters were used to calculate the absorption coefficient (A), reflection coefficient (R), transmittance coefficient (T) and absorption efficiency of the shield as, $R = |S_{11}|^2 = |S_{22}|^2$, $T = |S_{21}|^2 = |S_{12}|^2$ and A=1-R-T. The effective absorbance (A$_{eff}$) can be described as $A_{eff} = (1-R-T)/(1-R)$ with respect to the power of the effectively incident EM wave inside the shielding material.

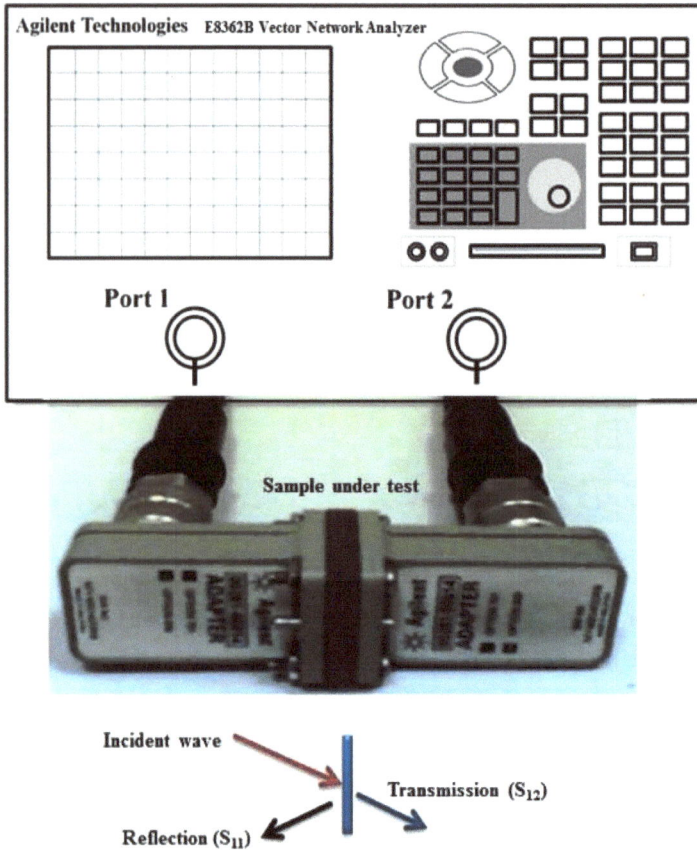

Fig. (2). Schematic of EMI shielding measurement setup.

The electromagnetic attenuation offered by absorbing materials may depend on the three mechanisms, which are a reflection of the incoming wave, absorption of the wave as it passes through the material's thickness, and multiple reflections of the waves at various interfaces, as shown in Fig. (**1**). Therefore, the total EMI SE of any material is the sum of the contributions of the absorption (SE_A), reflection (SE_R), and multiple reflections (SE_M) of the EM energy [64-73]:

$$SE\ (dB) = 10\log(P_T/P_I) = SE_R + SE_A + SE_M$$

where, P_I and P_T are the power of incident and transmitted EM waves, respectively. According to the Schelkunoff's theory, SE_M can be ignored in all practical applications, where the shield is thicker than the skin depth (δ). Furthermore, SE_R and SE_A can be defined as [73-76]:

$$SE_R = -10\log(1-R)$$

$$SE_A = -10\log(1-A_{eff}) = -10\log(T/1-R)$$

According to the EM theory, for electrically thick samples (t > δ), frequency (ω) dependence of far-field losses can be expressed in terms of total conductivity (σ_T), $\sigma_T = \sigma_{dc} + \sigma_{ac}$, real permeability (μ'), skin depth (δ) and thickness (t) of the shield SE_R and SE_A can be modified as [77]:

$$SE_R(dB) = 10\log\{(1-R)/16\omega\varepsilon_0\mu'\}$$

$$SE_A(dB) = 20(t/\delta)\log e = 8.68t\sqrt{\sigma\omega\mu/2} = 8.68(t/\delta)$$

SE_A becomes more dominant as compared to the SE_R in the microwave range. This may be caused by the shallow skin depth and high conductivity (σ_{ac}) values at such high frequencies [77, 78].

In terms of dielectric losses, the SE_A can also be written as [74]:

$$SE_A(dB) = 20d\sqrt{\frac{\mu\omega^2\varepsilon_o\varepsilon''}{2}}.\log e$$

In terms of tan delta, the above equation can be written as:

$$SE_A(dB) = 20d\sqrt{\frac{\mu\omega^2\varepsilon_o\varepsilon'\tan delta}{2}}.\log e$$

i.e., SE_A is directly proportional to tan delta.

6.3. FACTORS THAT INFLUENCE THE ELECTROMAGNETIC WAVE ABSORPTION

6.3.1. Skin Depth and Quarter Wave Principle

The absorption of an electromagnetic wave depends on the attenuation ability of the shield. The skin depth (δ) can be understood by a distance up to which the intensity of the EM wave decreases to $1/e$ of its original strength. A material has zero skin depth when it is an ideal electric conductor. While copper (metals) shows skin depth of the order of 0.65 μm at microwave frequencies (10 GHz). Thus, the lower the microwave electrical conductivity, the higher is the ability of the electromagnetic field to propagate across the material. Skin depth (δ) depends on the frequency of the incident wave. Moreover, the frequency, electrical conductivity ($\sigma_{ac} = \omega\varepsilon_0\varepsilon''$), magnetic permeability and skin depth can be related with the following equation:

$$\delta = \sqrt{2/\omega\mu_o\mu_r\sigma}_{ac}$$

This is related to angular frequency, relative permeability where μ_o is the permeability of free space ($4\pi \times 10^7$ H/m) and μ_r is the relative permeability (a dimensionless parameter) of the material. As the electromagnetic wave propagates within the material, it can be expressed using the equation of propagating wavelength in the material, λ, as follows:

$$\lambda = \frac{\lambda_o}{\sqrt{|\varepsilon||\mu|}}\lambda_o$$

λ is the wavelength in free space and ε and μ are the modulus of ε and μ, respectively. The maximum loss occurs with a quarter wavelength (0.25λ) thickness of the material. However, results show a variation between 0.25 λ and 0.3 λ which depends on the magnitude of ε and μ [79]. Micheli *et al.* showed the peak shift up to 0.5λ for the magnetic component addition where $|\mu|>|\varepsilon|$ [80].

6.3.2. Magnetic Loss and Dielectric Loss Mechanism

The electrical parameter of interest for shielding and absorption of EM waves is the complex permittivity (ε^*) and permeability (μ^*) of the samples. Or, equivalently, the dielectric constant (ε') and the conductivity ($\sigma_{ac}=\omega\varepsilon_0\varepsilon''$) of the nanocomposite. So, the complex permittivity and permeability of the shield are calculated using S-parameters obtained from the vector network analyzer by the coaxial line method. The observed shielding effectiveness could be explained in terms of dielectric losses. The complex permittivity and permeability reflect the amount of polarization (storage ability of the electric and magnetic energy) and loss (dissipated electric and magnetic energy) to the EM wave [73, 76, 81]. According to the EM theory, dielectric losses are the result of complex phenomena like natural resonance, dipole relaxation, electronic polarization and its relaxation and certainly the unique structure of the shield. When the frequency of the applied field is increased, the electrons present in the shield are unable to reorient fast enough to respond to the applied electric field and improve the dielectric constant. Nanoparticles having high dielectric and magnetic properties are anchored on the surface of RGO sheets, act as the polarized center, and improve polarization which results in enhanced microwave shielding. The high electrical conductivity of graphene composites also enhances the shielding properties. In the effective anisotropy energy, the parallel (RGO sheets are in a plane) and random alignment of RGO sheets are of particular importance. According to physics principles, the polarization intensity is directly proportional to the displacement of positive and negative charges during the activation by an EM wave. When EM waves incident perpendicular to the RGO plane, the effective anisotropy energy is higher as polarization intensity is higher in the plane direction. Therefore, contribution to the total EMI SE is higher when EM wave incident perpendicular to RGO plane, as compared to when EM wave incident parallel to RGO plane [82]. Nanoparticles act as tiny dipoles which get polarized in the presence of an EM field and result in better microwave shielding. Anisotropy energy of the small size materials [83], especially in the nanoscale, would be higher due to the surface anisotropic field due to the small size effect [84]. The higher anisotropy energy also contributes to the enhancement of microwave shielding.

Interfacial polarization occurs in heterogeneous media due to the accumulation of charges at the interfaces, formation of dipoles. Interfaces among nanoparticles and RGO sheets further enhance the dielectric losses. According to the reported results in the literature, the defects in RGO may construct localized states near the Fermi level to increase the radiation attenuation [73, 85]. The microstructure investigations show that the RGO contains clustered defects and residual bonds arising from the oxidation process. Furthermore, the existence of residual defects/groups in RGO sheet [39] and multiple reflections within the shield enhances the microwave shielding ability of the composites. There are more oxygenic functional groups on the RGO, which might increase the EM wave attenuation performance. Meanwhile, owing to the thinnest or thickness and high electron mobility, RGO has higher hopping conductivity, which induces strong polarization and loss conductance towards the EM wave.

The dielectric loss of the composites could be explained by the Debye relaxation laws [73, 86].

$$\varepsilon'' = \frac{(\varepsilon_s - \varepsilon_\infty)\omega\tau}{1 + \omega^2\tau^2} + \frac{\sigma}{\omega\varepsilon_0}$$

The given equation describes the relaxation response of an ideal dipolar system. The low and high-frequency limits of ε'' (ω) are ε_s and ε_∞ respectively, and the relaxation time τ, indicating that the dipolar dynamics are present. The last component compensates for dc charge transport losses, where σ_{dc} is the composite's dc conductivity. Since the greater polarization created by numerous surface functional groups, dielectric loss rises owing to the return of electric conductivity after chemical reduction and thinner thickness. In contrast, graphene layers provide more conductive channels for electron transport in composites, which greatly contributes to dielectric loss. Moreover, there are more conductive channels inside graphene composites, the effective conductivity rises as the graphene weight % in the composite increases.

6.3.3. Filler Type

The selection of filler to be incorporated in the graphene matrix is a great challenge. The volume ratio of filler in graphene composites also plays an important role in getting better and higher absorption. Generally, dielectric and magnetic fillers are employed to obtain maximum absorption of electromagnetic energy [30, 87]. Soft ferrites are considered to be the best magnetic material for consuming microwave radiation due to their excellent magnetic and dielectric properties. To effectively suppress EMI, ferrite materials are incorporated into graphene and polymer matrices as well as a complex of both [44, 53, 88]. Therefore, research has been carried out to synthesize ferrite-graphene composites in order to examine the effects of ferrite nanoparticles and their volume fractions on the microwave absorbing properties [20, 30]. Ferrite particles act as the absorption center, which favors energy attenuation and heat loss, while graphene offers excellent thermal conductivity to transfer heat energy. Thus, radar absorbing material includes the combination of materials so that the SE should be greater than 30 dB and 80 dB for commercial and strategic applications, respectively.

6.4. PREPARATION OF GRAPHENE AND ITS DERIVATIVES

There are many methods for the preparation of high-quality graphene in large quantities. These methods may be categorized in two ways, the first one is the top-down approach which includes mechanical exfoliation [89], wherein graphene oxide is oxidized/exfoliated chemically and then reduced [90], while the second one, is bottom-up approach includes methods like epitaxial growth on appropriate substrates [91], chemical vapor deposition [92], and arc discharging methods [93]. Every method has its advantages and limitations. To produce graphene on a large scale, the chemical reduction of graphene oxide (GO) is one of the best methods.

6.4.1. Synthesis of Graphene Oxide and its Reduction

Brodie succeeded in oxidizing graphite using nitric acid fumes and potassium chlorate under controlled cooling conditions in 1859 [94]. Staudenmaier refined this technique in 1898 by combining concentrated sulfuric acid with nitric acid in an optimum ratio and adding the chlorate in several aliquots during the reaction [95]. Hummers invented a technique for graphite oxidation utilizing $KMnO_4$ and $NaNO_3$ in concentrated H_2SO_4 in 1958 for the mass synthesis of graphene, which is still widely used today [96]. Marcano *et al.* enhanced the Hummers method, dubbed "modified Hummer's method." The sole oxidant is $KMnO_4$, with the acidic medium consisting of a concentrated H_2SO_4 and H_3PO_4 (9:1) combination [97]. The improved Hummers method has two primary benefits over the previous ones: first, it has a much higher oxidizing efficiency, and second, it prevents the production of hazardous gases like NO_2 and N_2O_4. The graphene oxide produced by this approach is more oxidized and has a more uniform structure than that produced by Hummer's method. In an open system, graphite can also be oxidized by benzoyl peroxide (BPO) at 110°C for 10 minutes [98].

6.4.2. Graphene-based Nanocomposites -1: Conducting Ferrofluid

6.4.2.1. Synthesis of Conducting Ferrofluid

RGO was initially produced utilizing the Hummers method for the synthesis of conducting ferrofluid. In a nutshell, concentrated H_2SO_4 is mixed with graphite powder (5 g) and $NaNO_3$ (5 g) (230 ml). The $KMnO_4$ (40 g) is gradually added while stirring and cooling the mixture until it reaches a temperature of no more than 20°C. After 2 hours of stirring at 35°C, deionized water (200 ml) is added. The addition of a considerable volume of deionized water (300 ml) and a 30 percent H_2O_2 solution (30 ml) to the reaction causes intense effervescence and a temperature increase to 100°C, after which the color of the suspension changes to bright yellow. To remove metal ions, the suspension is washed with a 1:10 HCl solution (100 ml). The paste is dried at 60 °C. The powder is distributed in distilled water, then stirred and ultrasonicated for three hours. After that, graphene oxide is reduced using hydrazine hydrate to form RGO.

To make different conducting ferrofluid composites, obtained RGO was combined in 1.0 M $FeCl_3$ solution in 1 wt. %, 5 wt. %, 10 wt. %, and 15 wt. % ratios, and designated CFF01, CFF05, CFF10, and CFF15 respectively. To homogenize the entire solution, the mixture was ultrasonicated for 1 hour. It was finally magnetically stirred. In 20 minutes, a 1M ammonia solution was stirred vigorously until the pH value of the solution reached 9-10. After 2 hours, an external magnet was placed beneath the beaker to separate the water from the Fe_3O_4 nanoparticles sandwiched between graphene layers. To avoid the presence of contaminants in the final product, the precipitate was washed twice with distilled water as a solvent to get ammonium and chloride-free particles. The contents of the beaker were centrifuged at 4000 rpm for 15 minutes to separate the supernatant liquid. After that, the supernatant liquid was decanted and centrifuged until all that was left was a thick black precipitate. In the liquid medium, the conducting ferrofluid remained suspended (with adding surfactant 25% TMAH in water). When the solution was placed near a strong magnet, it exhibited spikes, and it was used for future research.

Fig. (**3a**) illustrates a schematic picture of conducting ferrofluid composites synthesis. Fig. (**3b** and **3c**) depict the conducting ferrofluid and the spikes in the conducting ferrofluid when an external bar magnet is present, respectively.

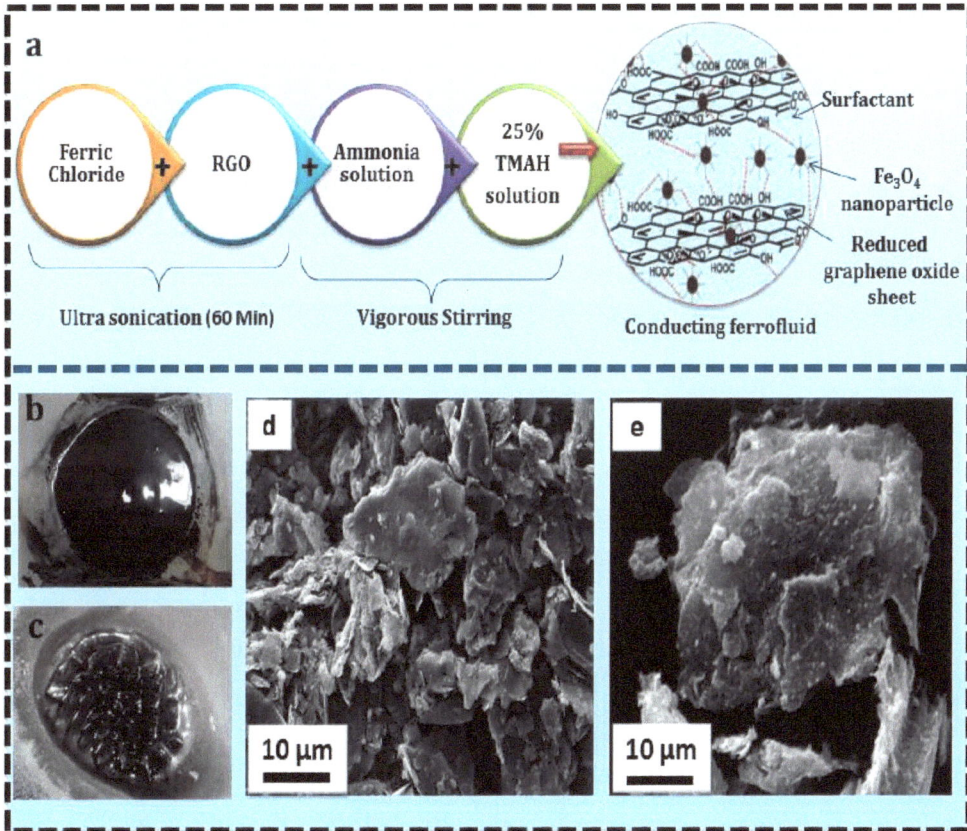

Fig. (3). Schematic representation of the synthesis of conducting ferrofluid (**a**) High-resolution optical graph of (**b**) & **c**) showing conducting ferrofluid (CFF10) with and without the presence of an external magnet, respectively, (**d**) shows SEM micrograph of RGO sheets and (**e**) shows SEM micrograph of conducting ferrofluid (CFF10). Reproduced from Ref. [62] with permission from the Royal Society of Chemistry.

6.4.3. Surface Morphology and Microstructural Studies of Composite

The surface morphology of RGO and conducting ferrofluid was studied using a scanning electron microscope. Fig. (**3d**) shows a two-dimensional planar surface of RGO, while Fig. (**3e**) shows a scanning electron microscope picture of conducting ferrofluid (CFF10), confirming the presence of Fe_3O_4 nanoparticles on the surface of RGO. Agglomerated graphene-based nanosheets are produced when the aqueous graphene oxide solution is reduced. The RGO was created by randomly aggregating folded and wrinkled sheets into a disordered solid., as seen in Fig. (**3d**). Fe_3O_4 is formed on graphene sheets with homogeneous distribution in bunches when *in situ* synthesis of Fe_3O_4 is carried out by chemical co-precipitation, as shown in Fig. (**3e**). These Fe_3O_4 nanoparticles are not single but clustered on the coiled and thin wrinkled sheets, resulting in a homogenous distribution on the RGO surface. Furthermore, the surface morphology verifies the homogenous distribution of RGO and Fe_3O_4 particles with significant surface contacts. Characteristic peaks in the XRD and Raman spectra of the CFF10 composite give additional evidence for the presence and interactions between RGO and Fe_3O_4 nanoparticles.

Fig. (**4a**) shows a TEM image of RGO nanosheets that are less than 3 nm thick in one dimension. The TEM picture of produced magnetic ferrofluid nanoparticles is shown in Fig. (**4b**). It was possible to make large-scale Fe_3O_4 nanoparticles with a homogeneous size of 5–20 nm. Almost every particle is spherical and evenly distributed. The 2D surface of RGO was adorned with a high quantity of Fe_3O_4 nanoparticles, as shown in Fig. (**4c**), and both the contour of RGO and Fe_3O_4 nanoparticles can be seen clearly. In comparison to the agglomerated morphology of virgin Fe_3O_4 nanoparticles, it was also established that Fe_3O_4 nanoparticles grew on the RGO sheets and were spread across the surface. Fe_3O_4 nanoparticles can deposit in an ordered, dense, and even way on both sides of these sheets. Fe_3O_4 nanoparticles were also used to cover the whole surface of the RGO sheets. Furthermore, Fe_3O_4 nanoparticles are positively connected to RGO sheets: even sonication was used during the preparation of TEM specimens, showing that excellent Fe_3O_4 nanoparticle adhesion to RGO sheets was achieved. The HRTEM picture of the CFF10 composite inset in Fig. (**4c**) shows lattice fringes with an interplanar spacing of 0.252 nm, which can be attributed to the cubic spinel crystal Fe_3O_4's (311) planes.

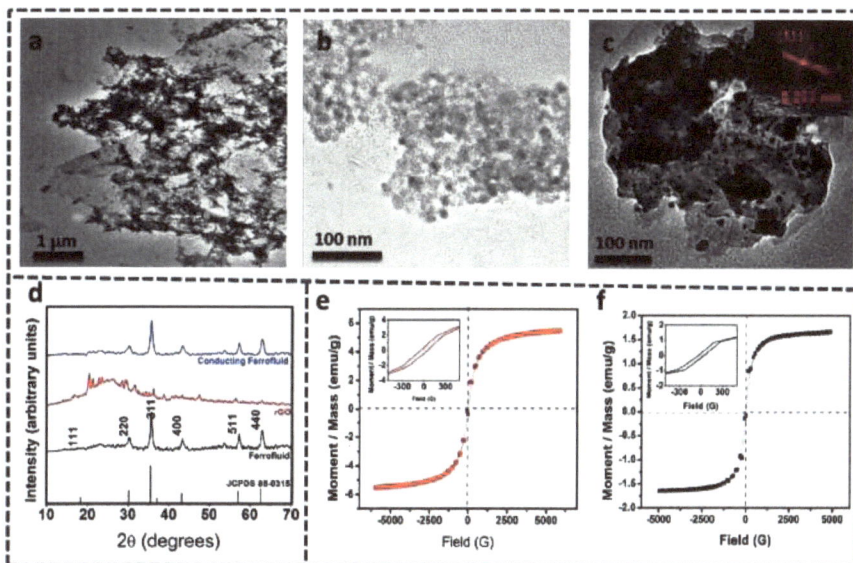

Fig. (4). TEM images of RGO **(a)**, ferrofluid **(b)** conducting ferrofluid (CFF10) **(c),** while the inset shows the fringes in an HRTEM image. XRD patterns of RGO, ferrofluid, and conducting ferrofluid (CFF10) **(d)**, VSM curves of ferrofluid **(e)** conducting ferrofluid (CFF10). Reproduced from Ref. [62] with permission from the Royal Society of Chemistry.

6.4.4. X-ray Diffraction Analysis

X-ray diffraction was used to study the phase composition and crystalline structure of the produced samples. The graphitic structure (002) of short-range order in stacked graphene sheets might be attributed to the diffraction peak of as-synthesized RGO at $2\theta \cong 25°$, as illustrated in Fig. (**4d**). There is no graphene oxide signal, indicating that GO has been effectively reduced to RGO. Fe_3O_4 main peaks were found at 30.45° (d = 2.94), 35.67° (d = 2.51), 43.34° (d = 2.08), 57.34° (d = 1.60), and 62.96° (d = 1.47 °A), matching to the (2 2 0), (3 1 1), (4 0 0), (5 1 1), and (4 4 0) reflections of the pure cubic spinel crystal structure of Fe_3O_4 [99]. All of the Fe_3O_4 peaks that have been seen have been matched to the typical XRD pattern (JCPDS no. 88-0315). The existence of Fe_3O_4 nanoparticles in the RGO sheets is confirmed by the ferrofluid peaks detected in CFF10. Scherrer's formula [30] was used to calculate the crystallite size (D) of ferrofluid nanoparticles and conduct ferrofluid composite (CFF10). Ferrofluid nanoparticles have an average size of 15.2 nm for pure ferrofluid and 12.9 nm for CFF10 composite.

6.4.5. Magnetic Properties

At ambient temperature, the magnetic characteristics of the as-prepared ferrofluid and conducting ferrofluid were investigated using a vibrating sample magnetometer (VSM). The M–H curve has been used to describe the magnetic characteristics of ferrofluid and CFF10. In both situations, very low coercivity is seen with little retentivity, indicating that there is no hysteresis loop, indicating that Fe_3O_4 nanoparticles are super paramagnetic. Figs. (**4 e & f**) demonstrate that the saturation magnetization value (Ms) and remnant magnetization value (Mr) for ferrofluid were higher than those for conducting ferrofluid composite CFF10, which can be attributed to the nanoscale size of Fe_3O_4 particles and the presence of RGO sheets. The magnetization hysteresis loop has an S-shaped pattern, which is typical of super paramagnetic materials. At a 6 kOe external field, the saturation magnetization (Ms) values of the aqueous ferrofluid and CFF10 composite were determined to be 5.499 emug^{-1} and 1.66 emug^{-1}, respectively. Because of the surfactant coating on individual ferrofluid nanoparticles, the super paramagnetic behavior in conducting ferrofluid composites might be described by assuming that all ferrofluid nanoparticles behave in an atomized form. Furthermore, agglomeration is prevented by the difference in magnetic properties of particles caused by a weak structure, which is favored by the combined action of interfacial forces (Vander Waals or acid–base), magnetic attractions between particles immersed in the liquid, and steric repulsion produced between particles. As a result, all nanoparticles exhibit the same behavior as single–domain nanoparticles with sup [99, 100], which are common in soft ferromagnetic materials.

6.4.6. Raman Studies

Raman spectroscopy is a powerful tool for determining how two components interact or bond [33]. Raman spectroscopy was used using a 514.5 nm wavelength laser in the spectral region of 100-2000 cm^{-1} to reveal the graphitic structure of RGO and the interaction between Fe_3O_4 and RGO. Fe_3O_4, RGO, and conducting ferrofluid (CFF10) Raman spectra are shown in Fig. (**5**). The Raman spectrum of pure Fe_3O_4 (ferrofluid) exhibits all of Fe_3O_4's distinctive bands in the low-frequency area, including Eg mode (213, 274, 384, 477 cm^{-1}), and A1g mode (583 cm^{-1}), proving Fe$_3$O$_4$'s presence. The D band (disorder-induced band) and the G band (tangential mode of graphitic structure) bands in the Raman spectra of RGO confirm the graphitic nature of RGO. The peak location of RGO (changing from 1360 to 1348 cm^{-1} in D band and 1604 to 1593 cm^{-1} in G band) and Fe$_3$O$_4$ in the Raman spectrum of CFF10 is slight to the left. Interaction between these components is evidenced by a slight shifting in the bands. Furthermore, due to the presence of Fe$_3$O$_4$ nanoparticles in the hybrid nanostructure material, the low-frequency mode of Fe$_3$O$_4$ has been observed in CFF10.

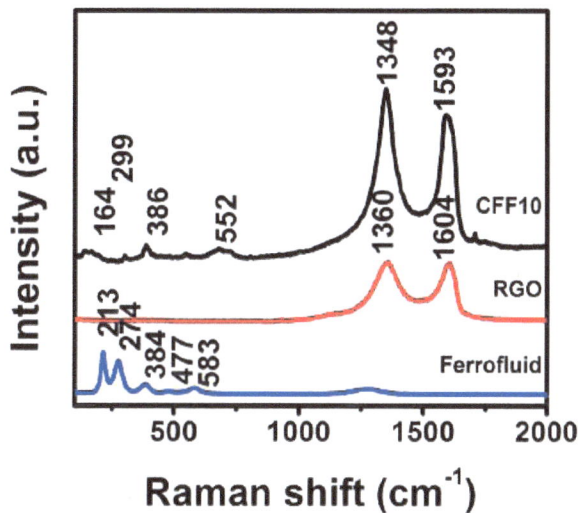

Fig. (5). Raman spectra of Ferrofluid, RGO and conducting ferrofluid (CFF10). Reproduced from Ref. [62] with permission from the Royal Society of Chemistry.

6.5. SHIELDING AGAINST ELECTROMAGNETIC INTERFERENCE AND DIELECTRIC CHARACTERISTICS

With shifting frequency and RGO weight ratio, the plotted curves aim to highlight the conducting ferrofluid's excellent EM wave attenuation characteristics. The computed T, absorption efficiency, R, and A values of the CFF composites with a thickness of 3 mm in the frequency range of 8–12 GHz are shown in Fig. (**6a, b, c,** and **d**). The transmittance coefficient (T) of CFF composites decreases as the RGO weight ratio in the composite increases. As demonstrated in Fig. (**6a**), the transmittance coefficient for all samples is less than 0.2. In other words, because the A and R coefficients are higher, more EM waves are swallowed by the CFF composites, resulting in a larger fall in T values. The A values of the CFF composites were in the range of 0.15~0.47, according to the plots in Fig. (**6d**).

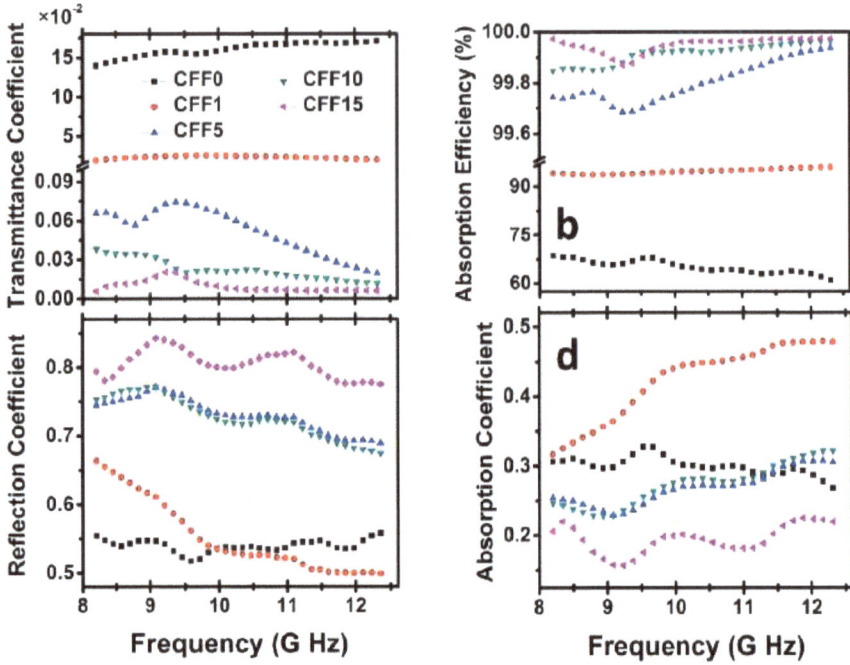

Fig. (6). The as synthesized conducting ferrofluid's transmission coefficient (**a**), absorption efficiency (**b**), reflection coefficient (**c**), and absorption coefficient (**d**). Reproduced from Ref. [62] with permission from the Royal Society of Chemistry.

The maximum SE_{Ref} value of the CFF composites is roughly 7 dB, and SE_{Ref} values increase with increasing RGO weight ratio. Furthermore, the CFF composites' SE_{Abs} values are larger than their SE_{Ref} values, indicating that the CFF composites' primary absorption characteristic towards EM waves is higher.

Fig. (**7**) depicts the as-obtained SE in terms of EMI. The EMI SE of CFF composites is affected by the frequency and weight percent of RGO, as seen in Fig. (**7a**) SE plots. With increasing RGO weight ratio, the SE of the CFF composites rises at 10.2 GHz, the SE values of CFF composites with 0, 1, 5, 10, and 15 wt. percent

RGO are 7.87, 15.96, 32.21, 36.73, and 41.20 dB. When compared to the other graphene @ magnetic nanoparticle as reported by Hu *et al.* [49] (SE_{max} is ca. 23 dB at 11.5 GHz), Xu *et al.* [58] (SE_{max} 24dB at 12.9GHz), Li *et al.* [47] (SE_{max} is 30.1 dB at 17.2 GHz) and Chen *et al.* [46] (SE_{max} 35.8dB at 6.5 GHz) conducting ferrofluid demonstrate excellent shielding properties. In general, the EMI shielding properties of composites are determined by the composition, shape, size, and microstructure of the fillers [73, 81], as well as reflection from the material's surface, absorption of EM energy, and propagation routes of the EM wave. The CFF composites have strong R and A coefficients, indicating that the CFF's potential EMI shielding capacity is mostly based on reflection and absorption. The SE in dB can also be determined using the equation [70], $SE = SE_A + SE_R$, in order to disclose the SE of reflection (SE_{Ref}) and absorption (SE_{Abs}) in detail.

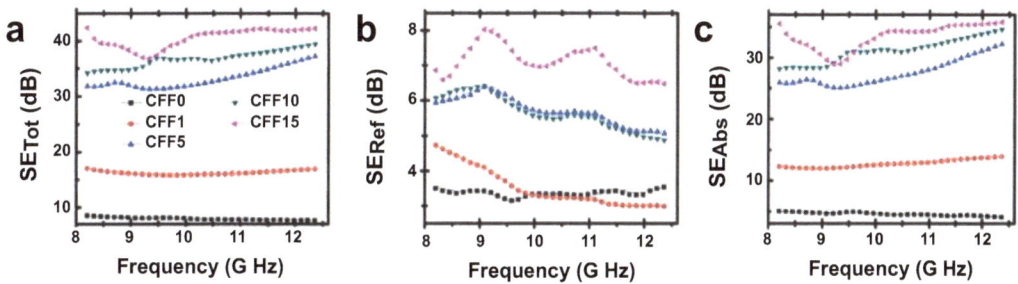

Fig. (7). Frequency dependent EMI SE_{Tot} (**a**), EMI SE_{Ref} (**b**) and EMI SE_{Abs} (**c**). Reproduced from Ref. [62] with permission from the Royal Society of Chemistry.

Fig. (**7 b** and **c**) show the SE_{Ref} and SE_{Abs} as obtained. The SE_{Abs} values significantly rise with increasing RGO weight ratio over the whole frequency range, as shown in the sets of Fig. (**7c**). At 10.2 GHz, the composites with 0, 1, 5, 10, and 15 wt. percent RGO have SE_{Abs} values of 4.49, 12.70, 26.49, 31.29, and 34.23 dB, respectively. The SE_{Ref} values of CFF composites reach a maximum of roughly 7 dB, and SE_{Ref} values rise with increasing RGO weight ratio. Furthermore, the CFF composites' SE_{Abs} values are larger than their SE_{Ref} values, indicating that the CFF composites' primary absorption characteristic towards EM waves is higher.

The permittivity (real part' and imaginary part") and loss tangent (tan $\delta\varepsilon = \varepsilon''/\varepsilon'$) of CFF composites with varied weight ratios were measured in the 8.2-12.4 GHz frequency range, as shown in Fig. (**8a**, **b**, and **c**). In Fig. (**8a**), the ε' values of the CFF composites are generally higher for corresponding higher weight % ratio. The ε' values of the CFF composites are in the range of 12.62-11.88, 24.81-23.43, 57.48-64.48, 71.56-67.97 and 79.04-86.02, when loaded with 0, 1, 5, 10, and 15 wt. % RGO, respectively. Since ε' is an expression of the polarization of a material, which consists of the interface polarization and orientation polarization under the EM field. Furthermore, due to the residual bonds and clustered defects introduced *via* the chemical conversion process, the electrons are not evenly distributed, leading to the orientation polarization, which further enhances the ε'. The ε'' of the CFF composites at each weight ratio is depicted in Fig. (**8b**). Furthermore, when the weight ratio rises, the ε'' of the CFF composites rises as well. With the load 0, 1, 5, 10, and 15 wt.% RGOs, the values of the ε'' vary from 5.84-5.26, 18.84-18.90, 41.71-56.10, 41.18 - 39.87, and 49.66 - 52.74 respectively. When loaded with 0, 1, 5, 10, and 15 wt. % RGOs, the tan $\delta\varepsilon$ values of the CFF composites are ≤ 1, as shown in Fig. (**8c**), which is due to the greater conductivity and polarization of the RGOs, as shown in Fig. (**8**). In the X-band frequency range, Fig. (**8d**, **e**, and **f**) illustrate permeability (real part' and imaginary part") and magnetic loss tangent (tan $\delta\mu = \mu''/\mu'$) of CFF composites with varied weight ratios of RGO. The μ' values of the CFF composites are in the range of 0.84-1.01, 1.12-1.11, 0.78-0.95, 1.09-0.98 and 1.35-1.30 when loaded with 0, 1, 5, 10, and 15 weight % RGO, respectively. The values of μ' are close to one for all the samples. The small variation in permeability value for different samples is similar to the earlier reported results [46]. The μ'' have values in the range of 0.29-0.05, 0.19-0.05, 0.67-0.36, 0.41-0.33 and 0.58-0.65, corresponding to the loading of 0 wt. %, 1 wt. %, 5wt. %, 10 wt. %, and 15 wt. % of RGO, respectively. The tan δ_μ values of the CFF composites is ≤ 1 when loaded with 0, 1, 5, 10, and 15 weight % of RGO, respectively, as shown in Fig. (**8f**). The magnetic loss (tan δ_M) is a result of eddy current effects, natural resonances and anisotropy energy present in the composites. The presence of nano ferrite particles in the composite is the major source of eddy current in the microwave spectrum. The tiny size of Fe_3O_4 on the RGO sheet is responsible for the natural resonances in the X-band. Because of the small size effect, the anisotropy energy of small size materials [83], especially at the nanoscale, would be greater owing to the anisotropic surface field [84]. The increased anisotropy energy also aids in the improvement of microwave absorption.

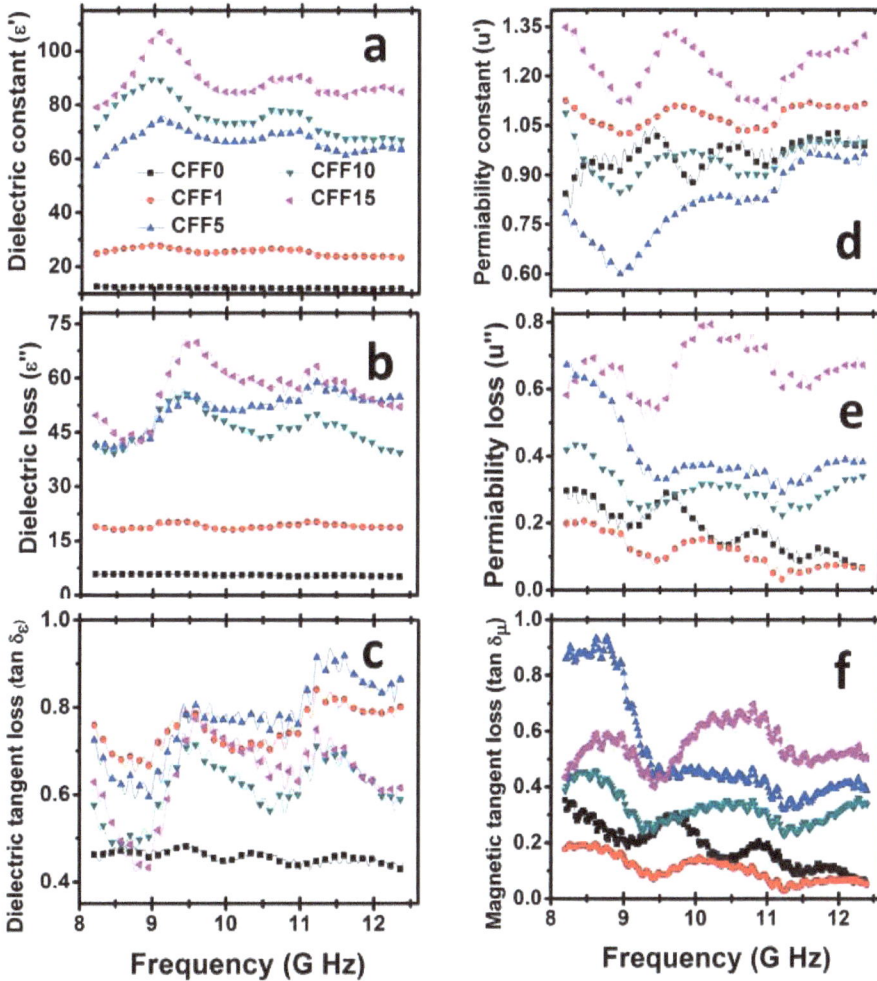

Fig. (8). (a) The complex permittivity's real parts and **(b)** imaginary parts, **(c)** dielectric loss tangents, **(d)** the complex permeability's real parts and **(e)** imaginary parts, **(f)** the corresponding magnetic loss tangents of conducting ferrofluids composites. Reproduced from Ref. [62] with permission from the Royal Society of Chemistry.

The microstructure of the fillers is linked to their dielectric characteristics and microwave attenuation capability [73, 75]. The improved microwave attenuation capabilities of the CFF composites are due to three primary causes, which are schematically illustrated in Fig. (**9**). To begin with, as previously said, the RGO's very thin and high polarity may provide more possibilities to induce polarization *via* EM waves, which enhances both the efficiency and the quality of the polarization of ε' and ε" [81]. Second, the thinner RGO sheets enhance the number of conductive pathways inside the composites, resulting in high-efficiency microwave attenuation due to microwave conversion to heat [73]. Fig. (**8 a-c**) show that the CFF composites have improved dielectric properties, indicating that polarization has a significant impact on dielectric properties, and that the conductivity of the composites plays a significant role in the dielectric properties and associated microwave attenuation performances.

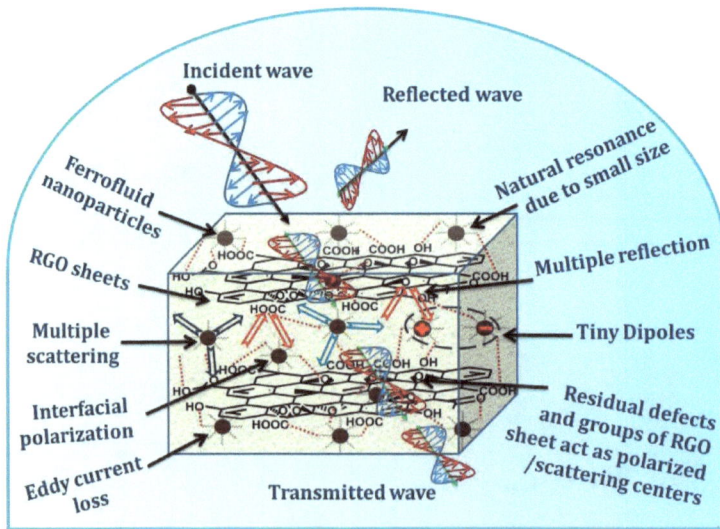

Fig. (9). Microwave absorbing mechanism of conducting ferrofluid composites. Reproduced from Ref. [62] with permission from the Royal Society of Chemistry.

And finally, the RGO sheets in the composites are exceedingly thin, flexible, and corrugated, increasing the EM wave propagation pathways inside the composites. With multiple internal reflection modes, the EM wave dispersed by corrugated graphene layers has considerably improved attenuation capabilities. Furthermore, when the mass ratio of the RGO grows, so does the EMI SE of the composites. The directional motion of the charge carriers in the RGO network generated oscillatory current when the microwave propagated in the composites, using a lot of EM wave energy [73]. As a result, the augmentation of the SE is due to greater scattering combined with higher conductivity. In addition, sufficient states produced by defects and residual bonds on RGO sheets improve microwave absorption. Conductivity is the most important factor in EM wave attenuation when the RGO has a high mass ratio. The RGO's conductivity is helpful for energy attenuation, and boosting conductivity would improve the composites' energy conversion efficiency. As a result, the findings in this paper point to the essential concepts for developing high-performance EMI shielding materials.

The σ_{dc} values of the CFF15 composite were significantly greater than those of other composites, as shown in Fig. (**10a**). While the RGO's greater volume contributes significantly to achieving higher conductivity and shielding characteristics at the same time. The SE_{Abs} and SE_{Ref} have been shown to be linked to the thickness of the shield in earlier research. Using complex permittivity and complex permeability, the optimal thickness for conducting ferrofluid composites was determined. The ideal thickness for an efficient shield is equal to the skin depth, which may be determined using the formula $\delta = \sqrt{2/\omega^2 \varepsilon_o \varepsilon'' \mu'}$. The optimal thickness required for conducting ferrofluid composites is less than 2 mm, as shown in Fig. (**10b**).

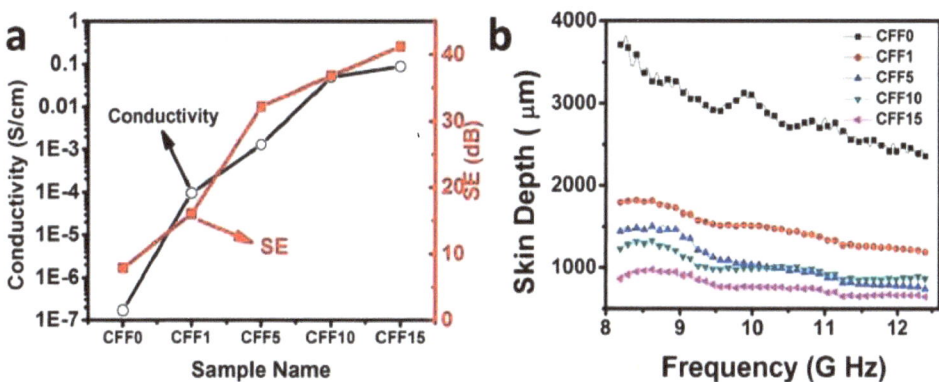

Fig. (10). Variation of dc electrical conductivity (dc) and total EMI SE in conducting ferrofluid composites (**a**), dc electrical conductivity (dc) and total EMI SE in conducting ferrofluid composites (**b**). Reproduced from Ref. [62] with permission from the Royal Society of Chemistry.

6.6. GRAPHENE BASED NANOCOMPOSITES-2-(TIN OXIDE NANOPARTICLES ENGINEERED REDUCED GRAPHENE OXIDE COMPOSITE (SNO2@RGO))

6.6.1. Synthesis of Tin Oxide

To make an $SnCl_2$–HCl solution, $SnCl_2.2H_2O$ (112.81 g) was dissolved in HCl (37 wt. %, 10 ml) and diluted with pure water (500 ml). To get a homogeneous solution with a pH of around 10, a small quantity of ammonia (25%) was added to the $SnCl_2.HCl$ solution at a regulated pace (0.01-0.1 ml/min) with vigorous stirring. The resultant solution was then agitated for 6 hours at 90°C. Centrifugation was used to separate the precipitate, which was then rinsed with distilled water to eliminate any excess chloride ions. The resultant product was dried for 24 hours in a vacuum oven at 110°C, then calcined (at 400°C for 5 hours in N_2 atmosphere) to improve the crystallite of SnO_2 [101].

6.6.2. Synthesis of SnO2 Decorated RGO Sheets

To make SnO_2 adorned RGO, scientists used a chemical reduction of graphene oxide (GO) in the presence of $SnCl_2$. GO had already been synthesized. In our earlier paper, we covered the precise technique of GO synthesis [30]. In a nutshell, graphite powder (5 g) and $NaNO_3$ (5 g) are combined together in concentrated H_2SO_4 (230 ml). The $KMnO_4$ (40 g) is progressively added with stirring and cooling, with the temperature of the mixture not exceeding 20 °C. The mixture is then agitated for 2 hours at 35°C before being added to 200 mL of deionized water. A significant amount of deionized water (300 ml) and a 30 % H_2O_2 solution (30 ml) are added to the reaction, generating rapid effervescence and a temperature increase to 100°C, after which the color of the suspension changes to bright yellow. To eliminate metal ions, the suspension is rinsed with HCl solution. The paste is dried at 60 °C. To make GO solution, 2 gram of GO powder is dispersed in 500 ml distilled water, then stirred and ultrasonicated for 1 hour. The following is a typical SnO_2 adorned RGO preparation procedure: $SnCl_2$ (14 gram) $2H_2O$ was added to a 500 ml HCl solution (37 wt% HCl, 100 ml/l). The mixture was then sonicated for 1 hour with the GO solution. At 90°C, the resultant mixture was agitated for 6 hours. After numerous washes with water and centrifugation, the SnO_2@RGO was recovered and dried in a vacuum oven at 100°C. Now, calcination (400°C for 5 hours in N_2) is being used to enhance the crystallinity of SnO_2 [101]. The following is an example of a potential reaction mechanism:

$$SnCl_2.2H_2O + GO + H_2O + 2HCl \rightarrow SnO_2@RGO.$$

6.6.3. Morphological Analysis

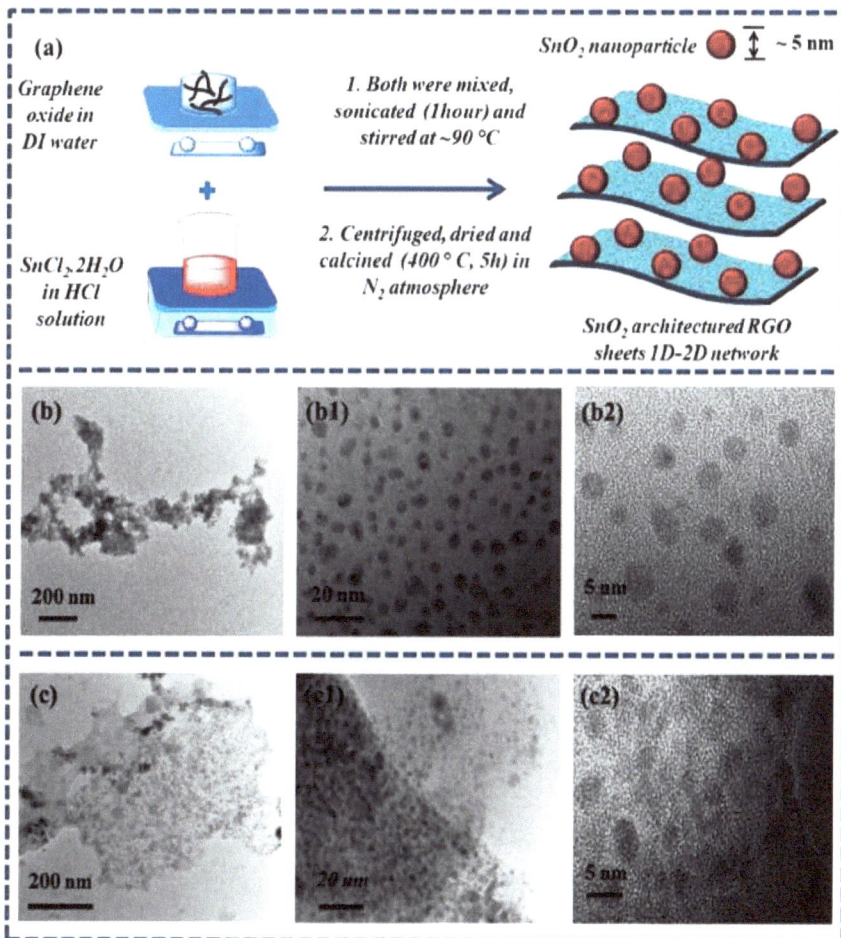

Fig. (11). (**a**) Graphic representation of architecting SnO_2 nanoparticles on the of RGO surface sheets, (**b**) At different magnifications TEM image of SnO_2 nanoparticles and (**c**) TEM image of as-synthesized $SnO_2@RGO$ composite Reproduced from Ref. [63] with permission from the Royal Society of Chemistry.

Fig. (**11a**) shows a schematic illustration of SnO_2 nanoparticles being used to decorate RGO. SnO_2 nanoparticles and SnO_2 coated RGO transmission electron microscopy (TEM) pictures are shown in Fig. (**11b and c**). SnO_2 nanoparticles at various magnifications are seen in Fig. (**11b**). All particles are 3-5 nm in diameter, according to higher magnification pictures (Fig. **11b1 & b2**). The shape of both RGO and SnO_2 nanoparticles can be observed in Fig. (**11 c**), indicating that the RGO sheet was evenly adorned with a considerable number of SnO_2 nanoparticles. Higher magnification pictures (Fig. **11 c2 & c3**) indicated that the SnO_2 nanoparticles had formed on the surface of the RGO sheet and were evenly dispersed across it. On both sides of these sheets, SnO_2 nanoparticles may be seen in abundance. Most significantly, the RGO sheets have no blank regions that aren't covered in SnO_2 nanoparticles. Because sonication was used during the fabrication of the TEM samples, these SnO_2 nanoparticles are tightly connected to RGO sheets, suggesting good bonding between RGO and SnO_2 nanoparticles. The inclusion of dielectric nanoparticles on a conducting surface helps to improve electromagnetic wave shielding.

6.6.4. Structural Analysis

XRD was used to characterize the crystal structures of GO, RGO, and RGO-SnO_2, and the findings are shown in Fig. (**12a**). The (002) reflection of stacked GO sheets is seen in the powder X-ray diffraction pattern of GO, which has a diffraction peak of about 10.2°. The interlayer spacing for GO is 0.86 nm, which is higher than the interlayer spacing for graphite, which is 0.34 nm. It's because oxygen-containing groups were introduced to the GO sheets. When GO is chemically reduced to RGO, the XRD peak broadens and shifts to about 24.9°, corresponding to an interlayer spacing of around 0.36 nm. This indicates that residual oxygenated groups are present on the RGO sheets. There are no diffraction peaks of GO for SnO_2@RGO, showing that the stannous ions have reduced GO to RGO. The diffraction patterns and relative intensities of produced SnO_2 matched those of normal SnO_2 (JCPDS 41-1445), proving that the nanoparticles were SnO2 and indicating that the diffraction peaks of crystalline SnO_2 nanoparticles are easily identifiable. It may correspond to the tetragonal SnO_2 phase (JCPDS 41-1445). The composite XRD peaks at $2\theta = 26.5$, 33.9, and 51.6 may be indexed to the SnO_2 diffraction planes

(110), (101), and (211), respectively. The Scherrer equation was used to compute the mean particle size (D) of the SnO_2 nanoparticles, which was found to be 3–5 nm in the SnO_2@RGO composite.

6.6.5. Raman Spectroscopy

Raman spectroscopy is a strong technique for determining how well two components interact or bind together [102]. Raman spectroscopy in the spectrum region of 100-3300 cm^{-1} was used to investigate the graphitic structure of GO, RGO, and the interaction between RGO and SnO_2 nanoparticles. The Raman spectra of GO, RGO, and SnO_2@RGO composite are shown in Fig. (**12b**). The D band (disorder-induced band), the G band (the tangential mode of graphitic structure), and the G' (or 2D) band are three significant distinctive peaks in the Raman spectra of GO and RGO.

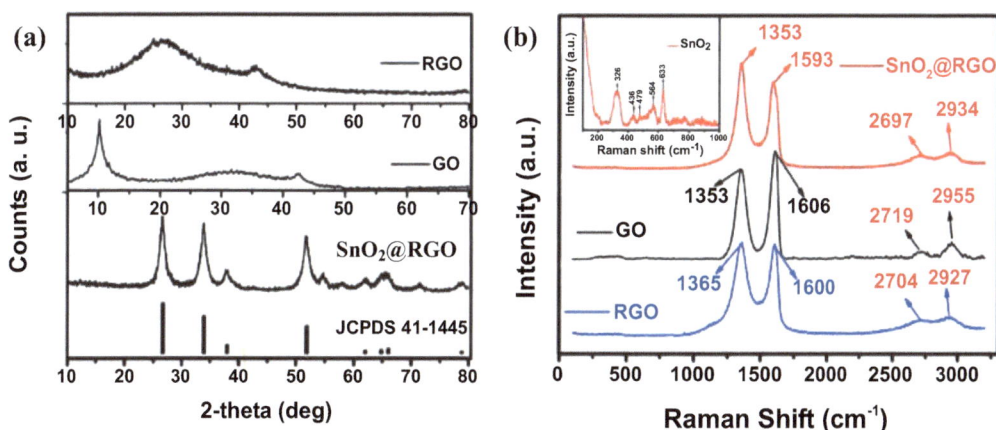

Fig. (12). (**a**) SnO_2@RGO, GO, and RGO X-ray diffraction patterns (**b**) Raman spectra of RGO, GO, and SnO_2 coated RGO sheets; the Raman spectrum of SnO_2 nanoparticles is shown in the inset image. Reproduced from Ref. [63] with permission from the Royal Society of Chemistry.

The G and D bands are caused by the doubly degenerate zone center E2g mode (1580-1600 cm^{-1}) and the breathing modes of six atom rings that emerge at 1350 cm^{-1} owing to graphite imperfections, respectively [103]. The D, G, and 2D peak position values in RGO's Raman spectra corroborate the creation of RGO. The Raman spectrum of pure SnO_2 nanoparticles displays all of SnO_2's distinctive low-frequency bands, including 436, 479 (Eg), 564 (S1), and 633 (A1g) [104]. Because of the very strong graphitic peaks found in RGO sheets, all of these peaks are suppressed in the SnO$_2$@RGO composite. Furthermore, the red shift in RGO peaks from 1365 to 1353 cm^{-1} in D band and 1600 to 1593 cm^{-1} in G band clearly shows the interaction between RGO and SnO$_2$ nanoparticles. Interaction between these components is observed by slight shifting in the bands. In addition, there is a minor shift in the G' band. SnO$_2$@RGO has a higher ID/IG value (1.22) than RGO (1.02). It also suggests that SnO$_2$ interacts with the RGO sheet during the synthesis of SnO$_2$@RGO because SnO$_2$ interacts with the available defect sites of RGO sheet during the synthesis of SnO$_2$ nanoparticles on the RGO surface or on the further occurrence of defect sites during the architecting of SnO$_2$ nanoparticles on RGO surface.

6.6.6. Shielding Effectiveness Measurement

The SE varies with frequency in the 8.2-12.4 GHz range, as seen in Fig. (**13 a-d**). Fig. (**13a**) shows that at a critical thickness of 3 mm, the SE for SnO$_2$@RGO composite (45.8 dB) is substantially greater than RGO (32.5 dB) and SnO$_2$ (14.2 dB). It was hypothesized that adding dielectric filler to RGO improves its SE. In comparison to SnO$_2$ and RGO separately, SnO$_2$@RGO is a better alternative. We also looked at the influence of different composite thicknesses on EMI shielding performance. In the X-band sample holder, rectangular pallets of various thicknesses (1.0 mm, 2.0 mm, 3.0 mm, and 4.0 mm, respectively) were inserted, and SE was measured in the frequency range of 8.2-12.4 GHz. Fig. (**13b**)_ shows a plot of microwave SE *vs* frequency for SnO$_2$@RGO with four different thicknesses. For a sample with a thickness of 1 mm, the SE$_T$ value is 29 dB, and for a sample with a thickness of 4.0 mm, it is 62 dB.

Fig. (**13c**) shows the predicted SE$_A$ and SE$_R$ from experimental scattering parameters (S_{11}, S_{22}, S_{12}, S_{21}).   Over the whole frequency range, the values of SE$_A$ for SnO$_2$@RGO composite rise significantly with increasing thickness. The SE$_A$ values of the composite with thicknesses of 1, 2, 3, and 4 mm are 22.58-23.15, 29.19-35.90, 37.80-42.09, and 45.39-80.14, respectively. SE$_R$, on the other hand, has relatively low numbers when compared to SE$_A$. Furthermore, with increasing thickness, the change in SE$_R$ is relatively modest. As a result, the mechanism of shielding for SnO$_2$@RGO composite is primarily the absorption of EM waves due to the highly conductive composite, and the absorption is proportional to the composite material thickness. Fig. (**13d**) illustrates the total changes in % attenuation as a function of thickness, with changes in reflection loss being minor in comparison to absorption. In the microwave range, the SE_A becomes more dominating than the SE_R, according to shielding theory. The thin skin depth and high conductivity (σ_{ac}) values at such high frequencies may be to blame [77, 78]. SnO$_2$@RGO has high electrical conductivity (13.74 S/cm), although it is significantly lower than that of graphene, which has been reported theoretically. This is due to the fact that charge carrier inter transport in SnO$_2$@RGO is a complicated process including electron tunneling and hopping, as opposed to a single graphene layer [73, 74, 82, 105-107]. Furthermore, the conductivity of the SnO$_2$@RGO composite is satisfactory because the insulating SnO$_2$ particles prevent electron transport in the SnO$_2$@RGO composite. The large multi-peaks in Fig. (**13c**) curves indicate the presence of natural resonance, which is induced by the increased surface anisotropy of the tiny SnO$_2$ particles.

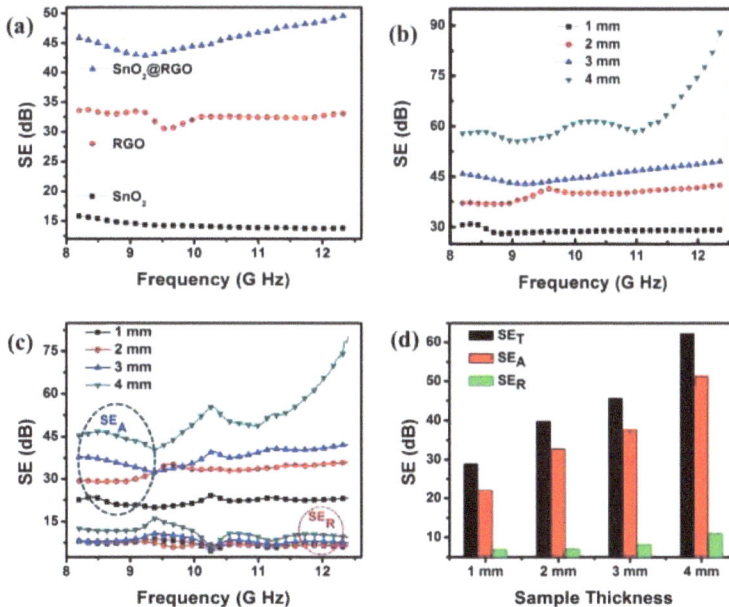

Fig. (13). (**a**) EMI shielding efficacy SE_T of SnO$_2$, RGO, and SnO$_2$@RGO composites at a critical thickness of 3 mm, (**b**) SET behavior of SnO$_2$@RGO composites at various thicknesses (**c**) SE_A and SE_R frequency behavior for various thicknesses of SnO$_2$@RGO composite, and (**d**) SE_T, SE_A, and SE_R for various thicknesses of SnO$_2$@RGO composite. Reproduced from Ref. [63] with permission from the Royal Society of Chemistry.

In comparison to pure RGO, pristine SnO_2 nanoparticles, γ-Fe_2O_3 nanoparticles, and RGO/iron oxide composite, MnO_2 decorated graphene nanoribbon described before, the inclusion of SnO_2 nanoparticles exhibits improved microwave shielding capabilities [108, 109].

The microwave shielding characteristics and dielectric loss mechanism of the SnO_2@RGO composite were investigated using dielectric attributes. Fig. (**14**) illustrates the permittivity (ε'), permittivity loss (ε''), and tangent loss (tan $\delta = \varepsilon''/\varepsilon'$) for SnO_2@RGO composite at a thickness of 3 mm as a function of frequency. High permittivity levels are a measure of polarization intensity or a material's capacity to store electrical energy. The energy lost during the activation by an EM wave is represented as permittivity loss. At 9.6 GHz, the maximum value of ε' was found to be 160. ε'' has a lower value than ε', fluctuating between 94 and 32 in the 8.2 to 12.4 GHz range. The dielectric tangent loss was found to be in the range of 0.85 to 0.36. SnO_2@RGO composites with high values of ε'' and tan δ have large dielectric losses. Surprisingly, two humps in dielectric tangent loss have been found, suggesting that dielectric losses are caused by two separate processes. These might include interfacial polarization between SnO_2 nanoparticles and RGO sheets, as well as the SnO_2@RGO composite's effective anisotropy energy. A schematic of the microwave shielding system is shown in Fig. (**15**) to provide a visual representation of the process mentioned above. Based on the foregoing, the results of the SnO_2@RGO composite show that such a structure may be utilized as a microwave shielding material.

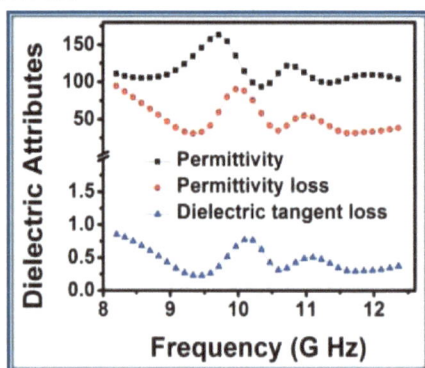

Fig. (14). Permittivity loss, permittivity loss, and dielectric tangent loss of the SnO_2@RGO composite are all frequency dependent. Reproduced from Ref. [63] with permission from the Royal Society of Chemistry.

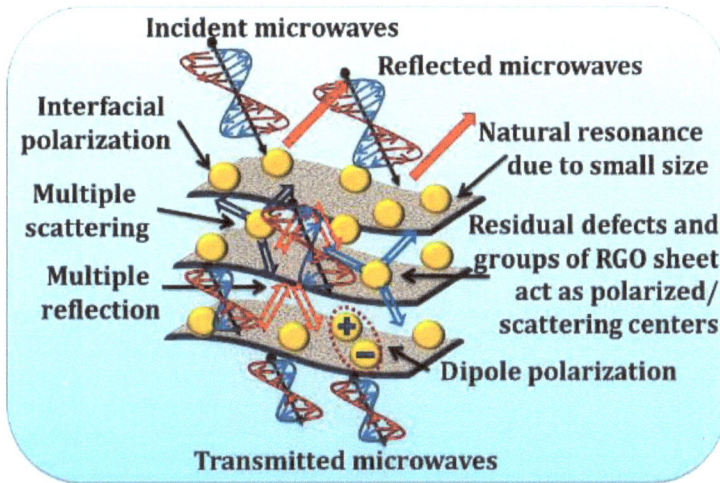

Fig. (15). Graphic representation of probable microwave shielding mechanisms in the SnO_2@RGO composite. Reproduced from Ref. [63] with permission from the Royal Society of Chemistry.

6.7. GRAPHENE-BASED NANOCOMPOSITES -3 - (GRAPHENE OXIDE/FERROFLUID/ CEMENT COMPOSITES)

To ensure uniform graphite and aqueous ferrofluid dispersion in the cement matrix, cement, graphene oxide, and aqueous ferrofluid were processed in a planetary ball mill for 5 hours. The resulting cement/graphene oxide/aqueous ferrofluid composite was crushed at 60MPa for 5 minutes in a piston-cylinder assembly into a 2.5 mm thick rectangular pellet with waveguide-specific dimensions, then treated with water. After being treated with water, powdery material forms hard pallets. At room temperature (25°C) and normal humidity, these pallets were cured for 35 days. Several cement/GO composites with varying weight ratios of cement to GO, cement: GO: 1:0(cg00), 1:0.1(cg 10), 1:0.2 (cg 20), 1:0.3 (cg 30) are produced in water medium, and to test the impact of FF, 10ml FF is employed in all above compositions and called cfg00, cfg10, cfg20, cfg30. Further loading of GO was avoided since the conductivity and hardness saturation levels were attained, and the shielding efficacy was more than 45 dB, which is necessary for commercial applications. For each concentration, seven pellets have been produced. The maximum variance in measured dc conductivity was found to be ± 2.5%, while the maximum variation in shielding efficiency was found to be ± 2% from the mean value. This guarantees that graphene oxide/aqueous ferrofluid is dispersed uniformly throughout the cement matrix. Fig. (**16a**) shows a schematic depiction of the incorporation of FF and GO into the cement matrix. X-ray diffraction patterns of the composites demonstrate the presence of magnetic FF particles and graphene oxide nanosheets in the cement matrix.

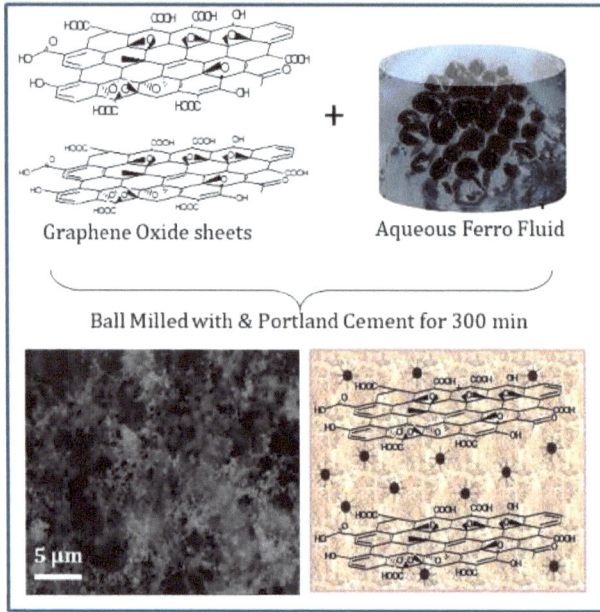

Fig. (16). Schematic illustration of graphene oxide nanosheets in ferrofluid cement matrix. and SEM images showing surface morphology cement having 30% of graphene oxide Reproduced from Ref. [59] with permission from the Institute of Physics.

Fig. (**16**) shows SEM micrographs of cement composites. Fig. (**16a**) shows an optical micrograph of ferrofluid. Spikes of ferrofluid can be seen in the optical micrograph, which are then integrated into the cement-graphene matrix. Fig. (**16b**), SEM micrograph (l), displays a cement-GO-FF (CGF) composite that demonstrates appropriate ferrofluid dispersion in cement and graphene oxide composite. The change of the SE_A and SE_R with frequency in the 8.2-12.4 GHz range is shown in Fig. (**16c and d**). According to the experimental results, the shielding effectiveness owing to absorption (SE_A) ranges from 10 to 40 dB as the GO concentration increases, whereas the SE_R varies from 11 to 2 dB. As a result, the cement-GO-FF composite achieves a total SE_T of 46dB (CFG30), which is much higher than pure cement. The shielding effectiveness (SE) of the conducting cement-GO-FF composite is dominated by absorption, whereas the shielding effectiveness owing to reflection (SE$_R$) is minimal and contributes very little. Furthermore, as seen in Fig. (**16d**), the shielding efficacy varies with sample thickness.

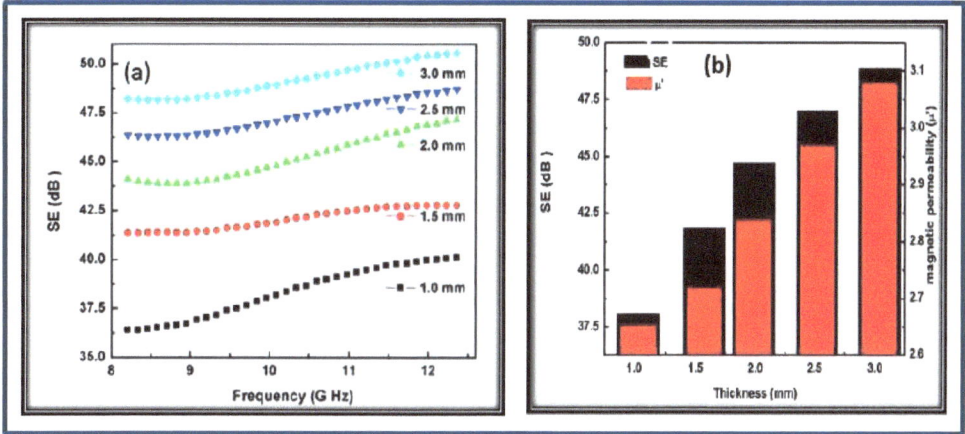

Fig. (17). (**a**) EMI Variation in the EMI shielding effectiveness, SE$_A$ and SE$_R$ of cement composites having different weight ratios of graphene oxide: cg10, cg20, cg30 and, cfg10, cfg20, cfg30 with frequency. (**b**) Effect of different thickness of composite on EMI shielding effectiveness Reproduced from Ref. [59] with permission from the Institute of Physics.

In the graphene oxide sheet, there are two types of charged motion: one that is mobile and free to travel along the chain, and another that has very limited mobility and accounts for the system's high polarization. The dipoles in the system are unable to reorient themselves rapidly enough to respond to the applied electric field when the frequency of the applied field is increased, resulting in a reduction in the dielectric constant. Because of the difference in dielectric constant and conductivity between GO and cement, some charge carriers in GO have been trapped, and space charge has formed on the surface of the cement particles and the GO. Ionic, electronic, orientational, and spatial charge polarization all affect a material's dielectric performance. The heterogeneity of the material appears to have a role in space charge polarization. Due to some space charge collecting at the interface, the presence of conducting GO and FF in the insulating matrix results in the creation of additional interfaces and a heterogeneous system, which leads to enhanced microwave absorption in the composites. Due to the existence of a bound charge, it contributes to orientational polarization.

This causes field distortion by generating a small amount of space charge at the heterogeneous contact. As the frequency of the applied field is reduced, the contribution of ionic conduction to total loss becomes more prominent. Interfacial polarization between GO and cement phase is predicted to diminish with increased frequency, resulting in a decrease in polarizability and loss factor. Furthermore, the dielectric losses caused by the SiO_2, $3Al_2O_3.2SiO_2$, and Fe_2O_3 components of cement, as well as multiple scattering, play a critical role in improving microwave absorption. Increased GO causes a decrease in skin depth and a rise in ac conductivity, as well as an improvement in input impedance. This improves the effective absorption capabilities as well as the quantity of electromagnetic radiation that can penetrate the shield.

CONCLUSION

Many studies have been conducted across the world to utilize graphene derivatives composites with dielectric and magnetic filler to absorb microwave radiation and understand the shielding process, but the quest for the ideal shield continues. As a result, there is much more to be discovered and utilized. Engineers, physicists, chemists, and materials scientists may investigate the features and characteristics of 2D systems because graphene has a unique 2D structure. Researchers from many disciplines should continue to work together to further reveal the potential of graphene composites in civic, commercial, and strategic applications for a more user-friendly final result. As a result of the current research, it can be stated that adding magnetic and dielectric fillers to the reduced graphene oxide matrix resulted in exceptional shielding capabilities. Graphene composites will be used as a next-generation building block material for EMI shielding and stealth technologies in the near future.

CONSENT FOR PUBLICATION

Not applicable.

CONFLICT OF INTEREST

The authors declare no conflict of interest, financial or otherwise.

ACKNOWLEDGEMENTS

Declared none.

REFERENCES

[1]   Wang, J.; Zhou, H.; Zhuang, J.; Liu, Q. *Influence of spatial configurations on electromagnetic interference shielding of ordered mesoporous carbon/ordered mesoporous silica/silica composites*, **2013**.
        http://dx.doi.org/10.1038/srep03252

[2]   Wen, B. *Reduced Graphene Oxides: Light-Weight and High-Efficiency Electromagnetic Interference Shielding at Elevated Temperatures*, **2014**.

[3]   Wilson, P.F.; Ma, M.T.; Adams-I, J. Techniques for measuring the electromagnetic shielding effectiveness of materials: Far-field source simulation. *IEEE Trans. Electromagn. Compat.*, **1998**, *30*(3), 11.

[4]   Chen, Z.; Xu, C.; Ma, C.; Ren, W.; Cheng, H.M. *Lightweight and flexible graphene foam composites for high-performance electromagnetic interference shielding*; Advanced Materials, **2013**, Vol. 25

[5]   Mishra, M.; Singh, A.P.; Dhawan, S.K. Expanded graphite-nanoferrite-fly ash composites for shielding of electromagnetic pollution. *J. Alloys Compd.*, **2013**, *557*, 244-251.
        http://dx.doi.org/10.1016/j.jallcom.2013.01.004

[6]   https://www.luckyvitamin.com/p-100185-cell-shield-cellular-phone-em-wave-radiation-blocker-1.

[7]   Volkow, N.D. *Effects of cell phone radiofrequency signal exposure on brain glucose metabolism*, **2011**.
        http://dx.doi.org/10.1001/jama.2011.186

[8]   Gultekin, D.H.; Moeller, L. *NMR imaging of cell phone radiation absorption in brain tissue*, **2013**.
        http://dx.doi.org/10.1073/pnas.1205598109

[9]   Engels, S. *Anthropogenic electromagnetic noise disrupts magnetic compass orientation in a migratory bird*, **2014**.
        http://dx.doi.org/10.1038/nature13290

[10]   Yang, M. *Mobile phone use and glioma risk: A systematic review and meta-analysis*, **2017**, *12*(5), e0175136.
        http://dx.doi.org/10.1371/journal.pone.0175136

[11]   Baan, R. *Carcinogenicity of radiofrequency electromagnetic fields*, **2011**.
        http://dx.doi.org/10.1016/S1470-2045(11)70147-4

[12]   http://www.who.int/peh-f/about/WhatisEMF/en/index1.html

[13]   Pavlík, M.; Kolcunová, I. Measuring of the shielding effectiveness of electromagnetic field of brick wall in the frequency range from 1 GHz to 9 GHz *Proceedings of the 2014 15th International Scientific Conference on Electric Power Engineering (EPE)*, **2014**, , pp. 261-265.
        http://dx.doi.org/10.1109/EPE.2014.6839488

[14]   Kim, M. *PET fabric/polypyrrole composite with high electrical conductivity for EMI shielding*, **2002**.
        http://dx.doi.org/10.1016/S0379-6779(01)00562-8

[15]   Liang, J. *Electromagnetic interference shielding of graphene/epoxy composites*, **2009**.
       http://dx.doi.org/10.1016/j.carbon.2008.12.038

[16]   Tzeng, S.-S. F.-Y. J. Chang M. S., and E. A, "EMI shielding effectiveness of metal-coated carbon
       fiber-reinforced ABS composites *302*(2), 258-267.**2001**,

[17]   Mishra, M.; Singh, A.P.; Gupta, V.; Chandra, A.; Dhawan, S.K. Tunable EMI shielding
       effectiveness using new exotic carbon: Polymer composites. *J. Alloys Compd.,* **2016**, *688*, 399-
       403.
       http://dx.doi.org/10.1016/j.jallcom.2016.07.190

[18]   Chaudhary, A. *Lightweight and easily foldable MCMB-MWCNTs composite paper with
       exceptional electromagnetic interference shielding*, **2016**.
       http://dx.doi.org/10.1021/acsami.5b12334

[19]   Chung, D. *Corrosion control of steel-reinforced concrete*, **2000**.
       http://dx.doi.org/10.1361/105994900770345737

[20]   Verma, M.; Singh, A.P.; Sambyal, P.; Singh, B.P.; Dhawan, S.K.; Choudhary, V. Barium ferrite
       decorated reduced graphene oxide nanocomposite for effective electromagnetic interference
       shielding. *Phys. Chem. Chem. Phys.,* **2015**, *17*(3), 1610-1618.
       http://dx.doi.org/10.1039/C4CP04284K PMID: 25437769

[21]   Singh, A.P.; Mishra, M.; Hashim, D.P.; Narayanan, T.N.; Hahm, M.G.; Kumar, P.; Dwivedi, J.;
       Kedawat, G.; Gupta, A.; Singh, B.P.; Chandra, A.; Vajtai, R.; Dhawan, S.K.; Ajayan, P.M.; Gupta,
       B.K. Probing the engineered sandwich network of vertically aligned carbon nanotube–reduced
       graphene oxide composites for high performance electromagnetic interference shielding
       applications. *Carbon,* **2015**, *85*, 79-88.
       http://dx.doi.org/10.1016/j.carbon.2014.12.065

[22]   Song, W. L. Magnetic and conductive graphene papers toward thin layers of effective
       electromagnetic shielding. *J. Mater. Chem. A,* **2015**, *3*
       http://dx.doi.org/10.1039/C4TA05939E

[23]   Sambyal, P.; Singh, A.P.; Verma, M.; Farukh, M.; Singh, B.P.; Dhawan, S.K. Tailored
       polyaniline/barium strontium titanate/expanded graphite multiphase composite for efficient radar
       absorption. *RSC Advances,* **2014**, *4*(24), 12614.
       http://dx.doi.org/10.1039/c3ra46479b

[24]   Huang, X.; Chen, Z.; Tong, L.; Feng, M.; Pu, Z.; Liu, X. Preparation and microwave absorption
       properties of BaTiO3@ MWCNTs core/shell heterostructure. *Mater. Lett.,* **2013**, *111*, 24-27.
       http://dx.doi.org/10.1016/j.matlet.2013.08.034

[25]   Li, Q.; Pang, J.; Wang, B.; Tao, D.; Xu, X.; Sun, L.; Zhai, J. Preparation, characterization and
       microwave absorption properties of barium-ferrite-coated fly-ash cenospheres. *Adv. Powder
       Technol.,* **2013**, *24*(1), 288-294.
       http://dx.doi.org/10.1016/j.apt.2012.07.004

[26]   Ashok Kumar, M. Conversion of Industrial Bio-Waste Into Useful Nanomaterials. *ACS Sustain.
       Chem.& Eng.,* **2013**.

[27]   Yousefi, N. *Highly aligned graphene/polymer nanocomposites with excellent dielectric properties
       for high-performance electromagnetic interference shielding*, **2014**.
       http://dx.doi.org/10.1002/adma.201305293

[28]   Singh, A.P.; Mishra, M.; Dhawan, S.K. *Conducting Multiphase Magnetic Nanocomposites for
       Microwave Shielding Application; "Nanomagnetism*; One Central Press UK, **2015**.

[29] Kumari, S.; Kumar, A.; Singh, A.P.; Garg, M.; Dutta, P.K.; Dhawan, S.K.; Mathur, R.B. Cu–Ni alloy decorated graphite layers for EMI suppression. *RSC Advances,* **2014**, *4*(44), 23202-23209.
http://dx.doi.org/10.1039/c4ra00567h

[30] Tripathi, P.; Prakash Patel, C.R.; Dixit, A.; Singh, A.P.; Kumar, P.; Shaz, M.A.; Srivastava, R.; Gupta, G.; Dhawan, S.K.; Gupta, B.K.; Srivastava, O.N. High yield synthesis of electrolyte heating assisted electrochemically exfoliated graphene for electromagnetic interference shielding applications. *RSC Advances,* **2015**, *5*(25), 19074-19081.
http://dx.doi.org/10.1039/C4RA17230B

[31] Shen, B. *Strong flexible polymer/graphene composite films with 3D saw-tooth folding for enhanced and tunable electromagnetic shielding*; Carbon, **2017**, Vol. 113, .

[32] Sudeep, P.M.; Vinayasree, S.; Mohanan, P.; Ajayan, P.M.; Narayanan, T.N.; Anantharaman, M.R. Fluorinated graphene oxide for enhanced S and X-band microwave absorption. *Appl. Phys. Lett.,* **2015**, *106*(22), 221603.
http://dx.doi.org/10.1063/1.4922209

[33] Hou, S.; Ma, W.; Li, G.; Zhang, Y.; Ji, Y.; Fan, F.; Huang, Y.J.J.M.S. Technology, Excellent Terahertz shielding performance of ultrathin flexible Cu/graphene nanolayered composites with high stability. *J. Mater. Sci. Technol.,* **2020**, *52*(0), 136-144.
http://dx.doi.org/10.1016/j.jmst.2020.04.007

[34] Kumar, A.; Singh, A.P.; Kumari, S.; Srivastava, A.; Bathula, S.; Dhawan, S.; Dutta, P.; Dhar, A. EM shielding effectiveness of Pd-CNT-Cu nanocomposite buckypaper. *J. Mater. Chem. A Mater. Energy Sustain.,* **2015**, *3*(26), 13986-13993.
http://dx.doi.org/10.1039/C4TA05749J

[35] Teotia, S.; Singh, B.P.; Elizabeth, I.; Singh, V.N.; Ravikumar, R.; Singh, A.P.; Gopukumar, S.; Dhawan, S.; Srivastava, A.; Mathur, R.J.R.A. Multifunctional, robust, light-weight, free-standing MWCNT/phenolic composite paper as anodes for lithium-ion batteries and EMI shielding material. *RSC Advances,* **2014**, *4*(63), 33168-33174.
http://dx.doi.org/10.1039/C4RA04183F

[36] Tripathi, P.; Prakash Patel, C.R.; Dixit, A.; Singh, A.P.; Kumar, P.; Shaz, M.A.; Srivastava, R.; Gupta, G.; Dhawan, S.K.; Gupta, B.K.; Srivastava, O.N. "High yield synthesis of electrolyte heating assisted electrochemically exfoliated graphene for electromagnetic interference shielding applications," vol. 5, no. 25. *RSC Advances,* **2015**, *5*(25), 19074-19081.
http://dx.doi.org/10.1039/C4RA17230B

[37] Wen, B. *Reduced graphene oxides: light-weight and high-efficiency electromagnetic interference shielding at elevated temperatures,* **2014**.

[38] Singh, A.P.; Mishra, M.; Sambyal, P.; Gupta, B.K.; Singh, B.P.; Chandra, A.; Dhawan, S.K. Encapsulation of γ-Fe_2O^3 decorated reduced graphene oxide in polyaniline core–shell tubes as an exceptional tracker for electromagnetic environmental pollution. *J. Mater. Chem. A Mater. Energy Sustain.,* **2014**, *2*(10), 3581-3593.
http://dx.doi.org/10.1039/C3TA14212D

[39] Sun, X.; He, J.; Li, G.; Tang, J.; Wang, T.; Guo, Y.; Xue, H. Laminated magnetic graphene with enhanced electromagnetic wave absorption properties. *J. Mater. Chem. C Mater. Opt. Electron. Devices,* **2013**, *1*(4), 765-777.
http://dx.doi.org/10.1039/C2TC00159D

[40]   Zhang, H-B.; Yan, Q.; Zheng, W-G.; He, Z.; Yu, Z-Z. Tough graphene-polymer microcellular foams for electromagnetic interference shielding. *ACS Appl. Mater. Interfaces,* **2011**, *3*(3), 918-924.

http://dx.doi.org/10.1021/am200021v PMID: 21366239

[41]   Chung, D. Carbon materials for structural self-sensing, electromagnetic shielding and thermal interfacing. *Carbon,* **2012**, *50*(9), 3342-3353.

http://dx.doi.org/10.1016/j.carbon.2012.01.031

[42]   Chen, S.; Zhu, J.; Wu, X.; Han, Q.; Wang, X. Graphene oxide--MnO_2 nanocomposites for supercapacitors. *ACS Nano,* **2010**, *4*(5), 2822-2830.

http://dx.doi.org/10.1021/nn901311t PMID: 20384318

[43]   Mishra, M.; Singh, A.P.; Sambyal, P.; Teotia, S.; Dhawan, S. Facile synthesis of phenolic resin sheets consisting expanded graphite/γ-Fe_2O_3/SiO_2 composite and its enhanced electromagnetic interference shielding properties. , **2014**; 52; pp. 478-485.

[44]   Reshi, H.A.; Singh, A.P.; Pillai, S.; Para, T.A.; Dhawan, S.; Shelke, V. X-band frequency response and electromagnetic interference shielding in multiferroic BiFeO3 nanomaterials. *Appl. Phys. Lett.,* **2016**, *109*(14), 142904.

http://dx.doi.org/10.1063/1.4964383

[45]   Eswaraiah, V.; Sankaranarayanan, V.; Ramaprabhu, S. Functionalized graphene–PVDF foam composites for EMI shielding. *Macromol. Mater. Eng.,* **2011**, *296*(10), 894-898.

http://dx.doi.org/10.1002/mame.201100035

[46]   Chen, D.; Wang, G-S.; He, S.; Liu, J.; Guo, L.; Cao, M-S. Controllable fabrication of mono-dispersed RGO–hematite nanocomposites and their enhanced wave absorption properties. *J. Mater. Chem. A Mater. Energy Sustain.,* **2013**, *1*(19), 5996-6003.

http://dx.doi.org/10.1039/c3ta10664k

[47]   Li, X.; Yi, H.; Zhang, J.; Feng, J.; Li, F.; Xue, D.; Zhang, H.; Peng, Y.; Mellors, N. Fe3O4–graphene hybrids: nanoscale characterization and their enhanced electromagnetic wave absorption in gigahertz range. *J. Nanopart. Res.,* **2013**, *15*(3), 1472.

http://dx.doi.org/10.1007/s11051-013-1472-1

[48]   Reshi, H.A.; Singh, A.P.; Pillai, S.; Yadav, R.S.; Dhawan, S.K.; Shelke, V. Nanostructured La 0.7 Sr 0.3 MnO3 compounds for effective electromagnetic interference shielding in the X-band frequency range. *J. Mater. Chem. C Mater. Opt. Electron. Devices,* **2015**, *3*(4), 820-827.

http://dx.doi.org/10.1039/C4TC02040E

[49]   Hu, C.; Mou, Z.; Lu, G.; Chen, N.; Dong, Z.; Hu, M.; Qu, L. 3D graphene–Fe3O4 nanocomposites with high-performance microwave absorption. *3D graphene–Fe3O4 nanocomposites with high-performance microwave absorption.,* **2013**, *15*(31), 13038-13043.

[50]   Gupta, T. K. MnO_2 decorated graphene nanoribbons with superior permittivity and excellent microwave shielding properties **2013**, *2*, 4256-4263.

[51]   Rawal, I.; Panwar, O.; Tripathi, R.; Singh, A.P.; Dhawan, S.; Srivastava, A. Effect of helium gas pressure on dc conduction mechanism and EMI shielding properties of nanocrystalline carbon thin films. *Mater. Chem. Phys.,* **2015**, *158*, 10-17.

http://dx.doi.org/10.1016/j.matchemphys.2015.03.024

[52]   Yang, J.; Liao, X.; Wang, G.; Chen, J.; Guo, F.; Tang, W.; Wang, W.; Yan, Z.; Li, G. Gradient structure design of lightweight and flexible silicone rubber nanocomposite foam for efficient electromagnetic interference shielding. *Chem. Eng. J.,* **2020**, *390*, 124589.

http://dx.doi.org/10.1016/j.cej.2020.124589

[53]   Han, D.; Zhao, Y-H.; Bai, S-L.; Ping, W.C. High shielding effectiveness of multilayer graphene oxide aerogel film/polymer composites. *RSC Advances,* **2016**, *6*(95), 92168-92174.
       http://dx.doi.org/10.1039/C6RA20976A

[54]   Che, R.C.; Peng, L.M.; Duan, X.F.; Chen, Q.; Liang, X.L. Microwave Absorption Enhancement and Complex Permittivity and Permeability of Fe Encapsulated within Carbon Nanotubes. *Adv. Mater.,* **2004**, *16*(5), 401-405.
       http://dx.doi.org/10.1002/adma.200306460

[55]   Shen, B.; Zhai, W.; Tao, M.; Ling, J.; Zheng, W. Lightweight, multifunctional polyetherimide/graphene@Fe3O4 composite foams for shielding of electromagnetic pollution. *ACS Appl. Mater. Interfaces,* **2013**, *5*(21), 11383-11391.
       http://dx.doi.org/10.1021/am4036527 PMID: 24134429

[56]   Ren, Y-L.; Wu, H-Y.; Lu, M-M.; Chen, Y-J.; Zhu, C-L.; Gao, P.; Cao, M-S.; Li, C-Y.; Ouyang, Q-Y. Quaternary nanocomposites consisting of graphene, Fe$_3$O4@Fe core@shell, and ZnO nanoparticles: synthesis and excellent electromagnetic absorption properties. *ACS Appl. Mater. Interfaces,* **2012**, *4*(12), 6436-6442.
       http://dx.doi.org/10.1021/am3021697 PMID: 23176086

[57]   Song, W.L. *Flexible graphene/polymer composite films in sandwich structures for effective electromagnetic interference shielding*; Carbon, **2014**, Vol. 66, .

[58]   Xu, H.-L.; Bi, H.; Yang, R.-B. Enhanced microwave absorption property of bowl-like Fe$_3$ O $_4$ hollow spheres/reduced graphene oxide composites. *Journal of Applied Physics,* **2012**, *111*(7), 07A522.

[59]   Singh, A.P.; Mishra, M.; Chandra, A.; Dhawan, S.K. Graphene oxide/ferrofluid/cement composites for electromagnetic interference shielding application. *Nanotechnology,* **2011**, *22*(46), 465701.
       http://dx.doi.org/10.1088/0957-4484/22/46/465701 PMID: 22024967

[60]   Yan, D.X.; Pang, H.; Li, B.; Vajtai, R.; Xu, L.; Ren, P.G.; Wang, J.H.; Li, Z.M. Structured reduced graphene oxide/polymer composites for ultra-efficient electromagnetic interference shielding. *Adv. Funct. Mater.,* **2015**, *25*(4), 559-566.
       http://dx.doi.org/10.1002/adfm.201403809

[61]   Hsiao, S.-T. Using a non-covalent modification to prepare a high electromagnetic interference shielding performance graphene nanosheet/water-borne polyurethane composite. *Carbon,* **2013**, *60*, 57-66.
       http://dx.doi.org/10.1016/j.carbon.2013.03.056

[62]   Mishra, M.; Singh, A.P.; Singh, B.P.; Singh, V.N.; Dhawan, S.K. Conducting ferrofluid: a high-performance microwave shielding material. *J. Mater. Chem. A Mater. Energy Sustain.,* **2014**, *2*(32), 13159-13168.
       http://dx.doi.org/10.1039/C4TA01681E

[63]   Mishra, M.; Singh, A.P.; Singh, B.P.; Dhawan, S. Performance of a nano architectured tin oxide@ reduced graphene oxide composite as a shield against electromagnetic polluting radiation. *RSC Advances,* **2014**, *4*(49), 25904-25911.
       http://dx.doi.org/10.1039/C4RA01860E

[64]   Gupta, A.; Singh, A.P.; Varshney, S.; Agrawal, N.; Sambyal, P.; Pandey, Y.; Singh, B.P.; Singh, V.; Gupta, B.K.; Dhawan, S. New insight into the shape-controlled synthesis and microwave

shielding properties of iron oxide covered with reduced graphene oxide. *RSC Advances,* **2014,** *4*(107), 62413-62422.

http://dx.doi.org/10.1039/C4RA10417J

[65]   Joon, S.; Kumar, R.; Singh, A.P.; Shukla, R.; Dhawan, S. Lightweight and solution processible thin sheets of poly (o-toluidine)-carbon fiber-novolac composite for EMI shielding. *RSC Advances,* **2015,** *5*(68), 55059-55065.

http://dx.doi.org/10.1039/C5RA07865B

[66]   Singh, K.; Ohlan, A.; Kotnala, R.K.; Bakhshi, A.K.; Dhawan, S.K. Dielectric and magnetic properties of conducting ferromagnetic composite of polyaniline with Fe2O3 nanoparticles. *Mater. Chem. Phys.,* **2008,** *112*(2), 651-658.

http://dx.doi.org/10.1016/j.matchemphys.2008.06.026

[67]   Ohlan, A.; Singh, K.; Chandra, A.; Singh, V.N.; Dhawan, S.K. Conjugated polymer nanocomposites: Synthesis, dielectric, and microwave absorption studies. *J. Appl. Phys.,* **2009,** *106*(4), 044305-044311.

http://dx.doi.org/10.1063/1.3200958

[68]   Ohlan, A.; Singh, K.; Chandra, A.; Dhawan, S.K. Microwave absorption properties of conducting polymer composite with barium ferrite nanoparticles in 12.4--18 GHz. *Appl. Phys. Lett.,* **2008,** *93*(5), 053114-053114.

http://dx.doi.org/10.1063/1.2969400

[69]   Sachdev, V.K.; Patel, K.; Bhattacharya, S.; Tandon, R.P. Electromagnetic interference shielding of graphite/acrylonitrile butadiene styrene composites. *J. Appl. Polym. Sci.,* **2011,** *120*(2), 1100-1105.

http://dx.doi.org/10.1002/app.33248

[70]   Lin, L. ZHANG, D., Research progress in electromagnetic shielding materials, J. *Journal of Functional Materials,* **2015,** *46*(03), 3016-3022.

[71]   Nicolson, A.; Ross, G. Measurement of the intrinsic properties of materials by time-domain techniques. *IEEE Trans. Instrum. Meas.,* **1970,** *19*(4), 19.

http://dx.doi.org/10.1109/TIM.1970.4313932

[72]   Weir, W.B. Automatic measurement of complex dielectric constant and permeability at microwave frequencies. *Proc. IEEE,* **1974,** *62*(1), 33-36.

http://dx.doi.org/10.1109/PROC.1974.9382

[73]   Wen, B.; Wang, X.X.; Cao, W.Q.; Shi, H.L.; Lu, M.M.; Wang, G.; Jin, H.B.; Wang, W.Z.; Yuan, J.; Cao, M.S. Reduced graphene oxides: the thinnest and most lightweight materials with highly efficient microwave attenuation performances of the carbon world. *Nanoscale,* **2014,** *6*(11), 5754-5761.

http://dx.doi.org/10.1039/C3NR06717C PMID: 24681667

[74]   Singh, A.P.; Gupta, B.K.; Mishra, M.; Govind, ; Chandra, A.; Mathur, R.B.; Dhawan, S.K. Govind; Chandra, A.; Mathur, R. B.; Dhawan, S. K., Multiwalled carbon nanotube/cement composites with exceptional electromagnetic interference shielding properties. *Carbon,* **2013,** *56,* 86-96.

http://dx.doi.org/10.1016/j.carbon.2012.12.081

[75]   Singh, A.P. S., A. K.; Chandra, A.; Dhawan, S. K., Conduction mechanism in Polyaniline-fly ash composite material for shielding against electromagnetic radiation in X-band & Ku band. *AIP Adv.,* **2011,** *1*(2)

http://dx.doi.org/10.1063/1.3608052

[76] Farukh, M.; Singh, A.P.; Dhawan, S. Enhanced electromagnetic shielding behavior of multi-walled carbon nanotube entrenched poly (3, 4-ethylenedioxythiophene) nanocomposites. *Compos. Sci. Technol.,* **2015**, *114*, 94-102.
http://dx.doi.org/10.1016/j.compscitech.2015.04.004

[77] Colaneri, N.F.; Schacklette, L. EMI shielding measurements of conductive polymer blends. *IEEE Trans. Instrum. Meas.,* **1992**, *41*(2), 291-297.
http://dx.doi.org/10.1109/19.137363

[78] Das, N.C.; Khastgir, D.; Chakia, T.K.; Chakraborty, A. Electromagnetic interference shielding effectiveness of carbon black and carbon fibre filled EVA and NR based composites. *Compos., Part A Appl. Sci. Manuf.,* **2000**, *31*(10), 1069-1081.
http://dx.doi.org/10.1016/S1359-835X(00)00064-6

[79] Idris, F.M.; Hashim, M.; Abbas, Z.; Ismail, I.; Nazlan, R.; Ibrahim, I.R. Recent developments of smart electromagnetic absorbers-based polymer-composites at gigahertz frequencies. *J. Magn. Magn. Mater.,* **2016**, *405*, 197-208.
http://dx.doi.org/10.1016/j.jmmm.2015.12.070

[80] Micheli, D.; Apollo, C.; Pastore, R.; Marchetti, M. X-Band microwave characterization of carbon-based nanocomposite material, absorption capability comparison and RAS design simulation. *Compos. Sci. Technol.,* **2010**, *70*(2), 400-409.
http://dx.doi.org/10.1016/j.compscitech.2009.11.015

[81] Wang, Y.Y.; Zhou, Z.H.; Zhou, C.G.; Sun, W.J.; Gao, J.F.; Dai, K.; Yan, D.X.; Li, Z.M. Lightweight and robust carbon nanotube/polyimide foam for efficient and heat-resistant electromagnetic interference shielding and microwave absorption. J. *ACS Appl. Mater. Interfaces,* **2020**, *12*(7), 8704-8712.
http://dx.doi.org/10.1021/acsami.9b21048 PMID: 31971778

[82] Yousefi, N.; Sun, X.; Lin, X.; Shen, X.; Jia, J.; Zhang, B.; Tang, B.; Chan, M.; Kim, J.K. Highly aligned graphene/polymer nanocomposites with excellent dielectric properties for high-performance electromagnetic interference shielding. *Adv. Mater.,* **2014**, *26*(31), 5480-5487.
http://dx.doi.org/10.1002/adma.201305293 PMID: 24715671

[83] Leslie-Pelecky, D.L.; Rieke, R.D. Magnetic Properties of Nanostructured Materials. *Chem. Mater.,* **1996**, *8*(8), 1770-1783.
http://dx.doi.org/10.1021/cm960077f

[84] Chen, Y-J.; Gao, P.; Wang, R-X.; Zhu, C-L.; Wang, L-J.; Cao, M-S.; Jin, H-B. Porous Fe_3O_4/SnO_2 Core/Shell Nanorods: Synthesis and Electromagnetic Properties. *J. Phys. Chem. C,* **2009**, *113*(23), 10061-10064.
http://dx.doi.org/10.1021/jp902296z

[85] Watts, P.C.; Hsu, W.K.; Barnes, A.; Chambers, B. High permittivity from defective multiwalled carbon nanotubes in the X-band. *Adv. Mater.,* **2003**, *15*(7-8), 600-603.
http://dx.doi.org/10.1002/adma.200304485

[86] Bhattacharyya, R.; Prakash, O.; Roy, S.; Singh, A.P.; Bhattacharya, T.K.; Maiti, P.; Bhattacharyya, S.; Das, S. Graphene oxide-ferrite hybrid framework as enhanced broadband absorption in gigahertz frequencies. *Sci. Rep.,* **2019**, *9*(1), 12111.
http://dx.doi.org/10.1038/s41598-019-48487-5 PMID: 31431643

[87]   Sambyal, P.; Singh, A.P.; Verma, M.; Gupta, A.; Singh, B.P.; Dhawan, S. Designing of MWCNT/ferrofluid/flyash multiphase composite as safeguard for electromagnetic radiation. *Adv. Mater. Lett.,* **2015**, *6*(7), 585-591.
        http://dx.doi.org/10.5185/amlett.2015.5807

[88]   Jia, X.; Shen, B.; Zhang, L.; Zheng, W. Construction of shape-memory carbon foam composites for adjustable EMI shielding under self-fixable mechanical deformation. *Chem. Eng. J.,* **2021**, *405*, 126927.
        http://dx.doi.org/10.1016/j.cej.2020.126927

[89]   Novoselov, K. S.; Geim, A. K.; Morozov, S. V.; Jiang, D.; Zhang, Y.; Dubonos, S. V.; Grigorieva, I. V.; Firsov, A. A. Electric field effect in atomically thin carbon films. science **2004**, *306*(5696), 666-669.

[90]   Hou, Y.; Cheng, L.; Zhang, Y.; Du, X.; Zhao, Y.; Yang, Z. High temperature electromagnetic interference shielding of lightweight and flexible ZrC/SiC nanofiber mats. *Chem. Eng. J.,* **2021**, *404*, 126521.
        http://dx.doi.org/10.1016/j.cej.2020.126521

[91]   Sutter, P.W.; Flege, J-I.; Sutter, E.A. Epitaxial graphene on ruthenium. *Nat. Mater.,* **2008**, *7*(5), 406-411.
        http://dx.doi.org/10.1038/nmat2166 PMID: 18391956

[92]   Nandamuri, G.; Roumimov, S.; Solanki, R. Chemical vapor deposition of graphene films. *Nanotechnology,* **2010**, *21*(14), 145604.
        http://dx.doi.org/10.1088/0957-4484/21/14/145604 PMID: 20215663

[93]   Wu, Z-S.; Ren, W.; Gao, L.; Zhao, J.; Chen, Z.; Liu, B.; Tang, D.; Yu, B.; Jiang, C.; Cheng, H-M.J. Synthesis of graphene sheets with high electrical conductivity and good thermal stability by hydrogen arc discharge exfoliation. *ACS Nano,* **2009**, *3*(2), 411-417.
        http://dx.doi.org/10.1021/nn900020u PMID: 19236079

[94]   Brodie, B.C. XIII. On the atomic weight of graphite. *Philos. Trans. R. Soc. Lond.,* **1859**, (149), 249-259.

[95]   Staudenmaier, L. Verfahren zur darstellung der graphitsäure. *Ber. Dtsch. Chem. Ges.,* **1898**, *31*(2), 1481-1487.
        http://dx.doi.org/10.1002/cber.18980310237

[96]   Hummers, W.S., Jr; Offeman, R.E. Preparation of graphitic oxide. *J. Am. Chem. Soc.,* **1958**, *80*(6), 1339-1339.
        http://dx.doi.org/10.1021/ja01539a017

[97]   Marcano, D.C.; Kosynkin, D.V.; Berlin, J.M.; Sinitskii, A.; Sun, Z.; Slesarev, A.; Alemany, L.B.; Lu, W.; Tour, J.M. Improved synthesis of graphene oxide. *ACS Nano,* **2010**, *4*(8), 4806-4814.
        http://dx.doi.org/10.1021/nn1006368 PMID: 20731455

[98]   Cravotto, G.; Cintas, P. Sonication-assisted fabrication and post-synthetic modifications of graphene-like materials. *Chemistry,* **2010**, *16*(18), 5246-5259.
        http://dx.doi.org/10.1002/chem.200903259 PMID: 20373309

[99]   Lopez, J.A.; González, F.; Bonilla, F.A.; Zambrano, G.; Gómez, M.E. Synthesis and characterization of Fe3O4 magnetic nanofluid. *Rev. Latinoam. Metal. Mater.,* **2010**, *30*(1), 60-66.

[100]   Chen, Q.; Zhang, Z.J. Size-dependent superparamagnetic properties of $MgFe_2O^4$ spinel ferrite nanocrystallites. J. *Appl. Phys. Lett.,* **1998**, *73*(21), 3156-3158.
        http://dx.doi.org/10.1063/1.122704

[101]   Zhang, J.; Xiong, Z.; Zhao, X. Graphene–metal–oxide composites for the degradation of dyes under visible light irradiation. *J. Mater. Chem. A Mater. Energy Sustain.,* **2011**, *21*(11), 3634-3640.

[102]   Gupta, T.K.; Singh, B.P.; Dhakate, S.R.; Singh, V.N.; Mathur, R.B. Improved nanoindentation and microwave shielding properties of modified MWCNT reinforced polyurethane composites. *J. Mater. Chem. A Mater. Energy Sustain.,* **2013**, *1*(32), 9138-9149.

http://dx.doi.org/10.1039/c3ta11611e

[103]   Ferrari, A.C.; Meyer, J.C.; Scardaci, V.; Casiraghi, C.; Lazzeri, M.; Mauri, F.; Piscanec, S.; Jiang, D.; Novoselov, K.S.; Roth, S.; Geim, A.K. Raman spectrum of graphene and graphene layers. *Phys. Rev. Lett.,* **2006**, *97*(18), 187401.

http://dx.doi.org/10.1103/PhysRevLett.97.187401 PMID: 17155573

[104]   Diéguez, A.; Romano-Rodríguez, A.; Vilà, A.; Morante, J.R. The complete Raman spectrum of nanometric SnO2 particles. *J. Appl. Phys.,* **2001**, *90*(3), 1550-1557.

http://dx.doi.org/10.1063/1.1385573

[105]   Gómez-Navarro, C.; Weitz, R.T.; Bittner, A.M.; Scolari, M.; Mews, A.; Burghard, M.; Kern, K. Electronic transport properties of individual chemically reduced graphene oxide sheets. *Nano Lett.,* **2007**, *7*(11), 3499-3503.

http://dx.doi.org/10.1021/nl072090c PMID: 17944526

[106]   Pathipati, S.R.; Pavlica, E.; Treossi, E.; Rizzoli, R.; Veronese, G.P.; Palermo, V.; Chen, L.; Beljonne, D.; Cai, J.; Fasel, R.; Ruffieux, P.; Bratina, G. Modulation of charge transport properties of reduced graphene oxide by submonolayer physisorption of an organic dye. *Org. Electron.,* **2013**, *14*(7), 1787-1792.

http://dx.doi.org/10.1016/j.orgel.2013.03.005

[107]   Lian, P.; Zhu, X.; Liang, S.; Li, Z.; Yang, W.; Wang, H. High reversible capacity of SnO2/graphene nanocomposite as an anode material for lithium-ion batteries. *Electrochim. Acta,* **2011**, *56*(12), 4532-4539.

http://dx.doi.org/10.1016/j.electacta.2011.01.126

[108]   Yan, D.X.; Ren, P.G.; Pang, H.; Fu, Q.; Yang, M.B.; Li, Z.M. Efficient electromagnetic interference shielding of lightweight graphene/polystyrene composite. *J. Mater. Chem. A Mater. Energy Sustain.,* **2012**, *22*(36), 18772-18774.

[109]   Lee, K.P.M.; Baum, T.; Shanks, R.; Daver, F. Graphene–polyamide-6 composite for additive manufacture of multifunctional electromagnetic interference shielding components. *J. Appl. Polym. Sci.,* **2021**, *138*(9), 49909.

http://dx.doi.org/10.1002/app.49909

CHAPTER 7

Utilization of Fly Ash Composites in Electromagnetic Shielding Applications

Swati Varshney[1]* and **S.K. Dhawan[2]**

[1] *Delhi Skill and Entrepreneurship University, Okhla - II, New Delhi 110020, India*

[2] *Advanced Materials Division, CSIR-National Physical Laboratory, New Delhi 110012, India*

Abstract: Electromagnetic interference (EMI) disturbs the working of electronic and electrical equipment used in aerospace, military, and many more areas. This disturbance leads to the complete failure of equipment and it is also very dangerous to human beings, especially radiation created by mobile phones. The considerable development in materials has been achieved with the fabrication of the shield in the form of composite sheets, paints, coatings, *etc*. Now, there is a need to fabricate flexible materials to achieve intricate shapes and structures to provide excellent EMI shielding in a wide frequency range. In addition to EM pollution, society is also dealing with pollution created by solid waste like fly ash. The disposal of fly ash has become a challenge due to its astounding amount produced in coal thermal power plants. This research work demonstrates the usage of fly ash to fabricate advanced composites for electromagnetic shielding applications. This chapter will throw some light on the electromagnetic shielding mechanism, EMI shielding measurement methods, fabrication of smart materials for shielding application in the designing of conducting polypyrrole nanocomposites, polyurethane composites, and cement paint composite using fly ash along with other magnetic/dielectric reinforcement to develop material offering optimized electromagnetic shielding properties. Moreover, these composites are further tested for other characterization techniques. These developed smart materials not only find a solution for the utilization of fly ash but also offer excellent shielding effectiveness in a wide frequency range.

Keywords: Cement paint, Chemical synthesis, Composites, Conductivity, Dielectric properties, Electromagnetic interference shielding, Electromagnetic radiation, Electron microscopy, Ferrite, Fly ash, Glass fibre, *In-situ* polymerization, Interfacial polarization, Magnetization, Microwave absorption, MWCNT, Polymers, Polypyrrole, Polyurethane, Solution casting.

***Corresponding author Swati Varshney:** Delhi Skill and Entrepreneurship University, Okhla – II, New Delhi 110020, India; Tel: +91 9311024810; E-mail: swati2581@yahoo.com

Sundeep K. Dhawan, Avanish Pratap Singh, Anil Ohlan, Kuldeep Singh Kakran and Pradeep Sambyal (Ed.)
All rights reserved-© 2022 Bentham Science Publishers

7.1. INTRODUCTION

Electric and magnetic fields subsist in all power lines and electrical equipment where the current flows. An electric field can be easily shielded by simple materials like wood, plastic, *etc*., but magnetic fields are not easily protected by these ordinary materials and but can simply pass-through building walls and human body parts. Electromagnetic radiations are of two types: ionizing and non-ionizing. The common source of non-ionizing radiation is telecommunication towers, wireless devices, mobile phones, microwave ovens, *etc*. These radiations are very harmful on long exposure and cause an adverse impact on human health. Hence, considerable work has been done over the past decade to ensure the adverse result of these radiations on human beings, mainly considering the effect of electrical appliances [1], exposure to Wi-Fi [2,3], exposure to mobile phone towers and base stations [4-6].

The constant increase of electromagnetic radiation exposure has raised questions about the potential adverse effect on living organisms. Mainly the usage of mobile phones has received much more attention due to their possible harm to human beings. Everyone is addicted to a mobile phone and keeps the mobile phone as close as possible. A large number of base station towers are required for the proper working of these mobile phones. Even in houses, people are exposed to electromagnetic radiation not only due to their phones but also due to nearby base stations and Wi-Fi connections, which leads to many health problems on long term exposure to radiations like infertility, neurological disorders, nervous system dysfunction, immune system dysfunction, and many others, while short term exposure leads to headache, insomnia and body pain [6-9].

WHO has recommended that long-term electromagnetic radiation studies should be conducted in the public interest [10]. Recently, IARC [11] has listed ELF-MFs under the 2B category due to its carcinogenic nature. Many researchers have studied the potentially harmful effect of electromagnetic radiation on human beings. Navarro *et al*. [12] have done a survey in the neighborhood of a cell phone tower, and it was observed that there is a considerable connection between the symptoms and power density. Some studies are based on the relationship between electromagnetic radiation exposure and tumours [13, 14]. Muscat *et al*. [15] studied the possibility of cancer with respect to cell phone usage. Singh *et al*. studied the relationship between electromagnetic radiations and the fitness of the human being living near propinquity to cell phone towers [16]. Many studies observed that radiation emitted by cell phones or base stations has a detrimental effect on the human body [17-20]. So, in order to avoid electromagnetic radiation, electromagnetic pollution, the development of electromagnetic shielding material is the only way. EMI shielding works excellently by providing hindrance for EM waves either by reflection or absorption of waves. This encumbrance of EM waves can be achieved with the help of some barrier made up of conductive or magnetic materials. Conventionally, metal shields made up of copper, aluminium, zinc were used for this purpose, but their size limitation, cost, heavyweight, corrosive nature, and difficult assembly discouraged their use as a shield. So, researchers pay more attention to the development of such alternative lighter materials that cost less but still, provide strong and tunable EMI shielding properties. Flexibility, corrosion resistance, and ease of fabrication are other requisite factors to be considered while designing an EMI shield. The type of material, size, and thickness of the shield are a few important parameters that decide the amount of radiation reduction or blocking or effectiveness of the shield. Thus, graphene, expanded graphite, multi-walled carbon nanotubes (MWCNT), and conducting polymers with some magnetic fillers like γ-Fe_2O_3, Fe_3O_4, and dielectric fillers like fly ash, TiO_2, *etc.*, have gained more attention for shielding application for offering the above-mentioned properties.

Moreover, there is a health hazard from electromagnetic pollution, as well as solid waste pollution from thermal power plant fly ash. The composition of fly ash can differ significantly depending on the type of coal burned, but all fly ash contains considerable amounts of silica as a major constituent with oxides of other metals and some traces of heavy metals. Fly ash particles are the source of air, water, and solid pollution [21, 22], as they contain small, lightweight, and airborne particles [23]. The health effects of this pollution are very serious. The smaller the particles, the greater will be the health risk. The smallest particle of fly ash can be easily inhaled and causes heart or lung diseases [24, 25], but fly ash is a precious resource with possible utilization in various applications [26-28]. In order to reduce health hazards caused by fly ash, the best way is to utilize it as filler material for engineering applications rather than disposing it off, which simultaneously leads to a reduction in landfilling problems.

In the past, several attempts have been made to utilize fly ash in many applications, especially for shielding purposes [29-31]. Tiwari *et al.* reported the usage of fly ash in place of conventional extenders in industrial coatings [32]. Cao *et al.* [33] observed an increase in the absorption-dominated shielding effectiveness of the cement paste having fly ash. Singh *et al.* [34, 35] has reported the use of lfy ash as. a functional filler in order to tune radiation shielding properties. Dou and co-workers [36] measured the EMI shielding effectiveness properties of Al alloy–fly ash composites in 30.0 kHz – 1.5 GHz frequency and obtained a maximum SE of 102.5±0.1dB. Budumuru *et al.* [37] studied shielding effectiveness and mechanical properties of aluminium composite reinforced with fly ash in X band.

This chapter provides oversight on the development of composites for microwave shielding applications by utilizing fly ash (as dielectric filler) along with some other fillers such as MWCNT, γ-Fe$_2$O$_3$ and ferrite encapsulated glass ifbers☐in a synergistic approach to get optimized shielding properties. In the present study, three different types of base material matrices have been chosen to get the best possible material for EMI shielding application. The first section of the chapter is focused on designing conducting nanocomposites, *i.e.*, polypyrrole nanocomposite with fly ash and γ-Fe$_2$O$_3$ and the second section of this chapter has described the fabrication of polyurethane composite sheet along with MWCNT and fly ash reinforcement, and the last section gives the overview on the development of the cement composite paints containing MWCNT, lfy ash and ferrite encapsulated glass. ifbers, which can be directly applied as wall paint to shield building walls. The motive of the work is to develop smart materials with balanced shielding effectiveness to avoid the hazardous health impact of electromagnetic radiation.

7.2. ELECTROMAGNETIC SHIELDING MECHANISM

Electromagnetic shielding is the process of decreasing the electromagnetic field (EMF) by obstructing the field with the help of blockade made of either conductive or magnetic materials, while shielding effectiveness (SE) of a material can be defined as a parameter that computes the efficacy to impede the electromagnetic (EM) radiation of a definite frequency when it penetrates on it.

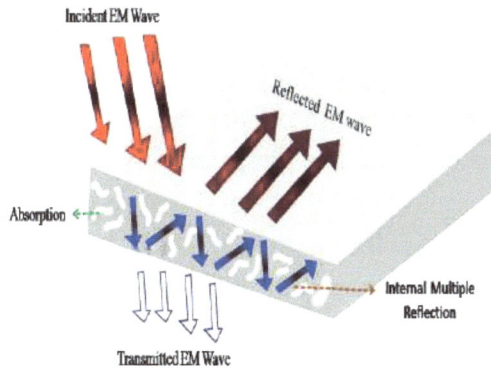

Fig. (1). Graphical representation of electromagnetic interference shielding mechanism.

All electromagnetic waves are composed of an oscillating magnetic field (H) and an electric field (E). The ratio of these two fields is called wave impedance, *i.e.*, E/H. The wave impedance in free space is decided by the radiation source. Specifically, the significant criterion is if the source is open-ended/dipole-like or closed/current loop and if the evaluation is made in the near or far-field shielding region [38]. The near and far-field shielding region is based on the distance between the radiation source and shielding material.

The electromagnetic plane wave theory is generally applied for EMI shielding in the far-field shielding region. Shielding effectiveness (SE) can be expressed as:

$$SE_T \text{ (dB)} = -10 \log (P_T/P_I) = -20 \log (E_T/E_I) \tag{1}$$

where P_T (E_T) and P_I (E_I) are the power (electric field intensity) of the transmitted and incident EM waves, respectively.

The shielding effectiveness of a material is governed by three mechanisms, namely, reflection, absorption, and multiple reflections. When a plane wave/radiation incident on a shield, a part of the wave is reflected from the outer surface of the shield, a part of the wave is absorbed within the shielding material, and a part of the wave is reflected at various interfaces within the shield, *i.e.*, internal multiple reflections. Therefore, total shielding effectiveness is the combination of these three phenomena. So, the total shielding effectiveness of a material can be expressed as [39, 40]:

$$SE \text{ (dB)} = SE_A + SE_R + SE_M \tag{2}$$

These three different phenomena named reflection (SE_R), absorption (SE_A), and multiple reflections (SE_M), contribute towards the total shielding effectiveness, and all are expressed in decibel (dB).

Fig. (**1**) shows the graphical representation of the EMI shielding mechanism in a conductive shield material. With the propagation of a transmitting wave, the amplitude of the wave decreases exponentially. Consequently, absorption loss results from ohmic losses and heating loss of the material due to the action between electric and/or magnetic dipoles in the shield with electromagnetic field and reflection loss of shield arises due to impedance imbalance between the shield and electromagnetic field. In the case of a plane wave, equation (2) can be represented as [41]:

$$SE \text{ (dB)} = SE_A + SE_R \tag{3}$$

Absorption loss and reflection loss of shielding material can be expressed as:

$$SE_A(dB) = 20\frac{d}{\delta}\log e = 20d\sqrt{\frac{\mu_r \omega \sigma_S}{2}} . \log e \qquad (4)$$

$$SE_R(dB) = 10\log\left(\frac{\sigma_S}{16\omega\mu_r \varepsilon_o}\right) \qquad (5)$$

where d is the thickness of the shield, μ_r is the magnetic permeability, $\delta = \sqrt{2/\mu_r \omega \sigma_S}$ is the skin depth, $\sigma_S = \omega \varepsilon_o \varepsilon''$ is the frequency-dependent microwave conductivity [42]. Here, ε_o is the permittivity of the free space, ω represents the angular frequency ($\omega = 2\pi f$), and ε'' is the imaginary part of permittivity. If the shield is thicker than the skin depth, the multiple reflections can be ignored. However, multiple reflections are generally considered when the shield is thinner than the skin depth and at frequencies below about 20 kHz.

7.3. EMI SHIELDING MEASUREMENT METHODS

Attenuation is the principal measure used to evaluate the effectiveness of EMI shielding. The principle is based on the measurement of the loss of electromagnetic signals, *i.e.*, the difference between the intensity of electromagnetic signals before shielding and after shielding. Attenuation is notified in decibels (dB) which corresponds to the ratio between field strength with and without a protective medium being present. EMI shield attenuation levels can be determined, although the testing can be a bit complex. The methods used to determine the outcome differ for different shielding applications. The four most common methods for testing shielding effectiveness include [43-45]:

1. The open field or free space method is used to measure the practical shielding of the finished product by assessing the practical shielding effectiveness of a complete electronic device. In this method, radiated emission and conducted emission are measured with the help of antennae placed at varying distances from the device. This method provides the normal operation condition for electronic devices as close as practicable. The device is kept 30 m away from the antenna, and radiated emissions are noted. A noise meter is used to document the results.

2. The shielded box method is based on the comparative study of specimens of dissimilar shielding materials. This technique uses a sealed metal box that has an opening. A coated conductive shielding unit is positioned over the opening of the box, and all transmitted and received emissions are measured. It assists in measuring electromagnetic signals both inside and outside the box. Shielding

efficiency reflects the ratio between the signals. This method has a constraint that it can be used for frequencies below 500 MHz.

3. In some situations, the amount of ambient sound in an area may not be significantly reduced, and a shielded room technique is required. Usually, the method involves at least two shielded rooms separated from a partition to avoid interference. The testing device and test equipment are placed in one room, and sensor arrays are placed in another. Shielding leads are frequently incorporated to decrease the potential for measuring errors created by external signals. This method is appropriate for evaluating a device's susceptibility.

4. In transmission-reflection (TR) procedures, the coaxial transmission line is the most commonly used measurement system. Shielding effectiveness (SE) of the planar shield has been measured by using coaxial transmission line systems [46, 47]. A vector network analyzer and a 50-ohm coaxial sample holder are used in this experiment to measure the S-parameters of the single-layered RF absorber [48]. Additionally, the coaxial transmission line can also be used to determine the data into reflected, absorbed, and transmitted elements. The measurement method depends on numerous aspects, like testing frequency, material, specimen size, and shape. The coaxial transmission line method is now frequently used as a standard method for the testing of planar material (ASTM D4935-99) [49].

7.4. SELECTION OF MATERIAL FOR THE DESIGNING OF MICROWAVE SHIELD

The absorption loss factor of shielding mechanism depends on the product of $\sigma_r \mu_r$, where σ_r is the relative conductivity of the shield referred to copper and μ_r is the relative permeability of the shield referred to as free space so high conductivity and high permeability are the two important requirements of material for a shield for maximum absorption loss [50]. However, the reflection loss depends on the ratio of σ_s/μ_r. Consequently, high reflection loss results due to high conductivity and low magnetic permeability. With due reason, metal shields were in use, but heavy weight, corrosion problems, difficulty in manufacturing, and more reflection loss have propagated the development of advanced material to overcome these drawbacks. Advances in research at this moment are now focused on improving absorption efficiency, expanding bandwidth, and improving the properties of microwave absorbing materials. Fabrication of conducting nanocomposites with the use of fillers to achieve a product with a wide range of electrical conductivity (σ) and/or electromagnetic attributes may be approached as a microwave absorbing shield. Several attempts have been made to build an efficient shield using different dielectric fillers like TiO_2, fly ash, $BaTiO_3$ [51-53] in order to efficiently increase the dielectric constant of microwave shields which is further responsible for high dielectric losses. In this concern, fly ash has drawn the attention of researchers. Development and innovation in new areas of fly ash usage have become essential for increasing its utilization in order to reduce waste pollution. This realization leads to the use of fly ash as dielectric filler in the production of composites for shielding application. In this sequence, Pattanaik *et al.* [54] has reported the fabrication of epoxy-fly ash composite, and research was concluded with the remark that fly ash increases the mechanical properties of composite and offers good dielectric strength to composite. Cao *et al.* [33] investigated the shielding effectiveness of cement paste along with fly ash as an admixture. He delineated that the usage of fly ash results in an increase in the shielding efficiency of the cement paste when tested at 1.0 and 1.5 GHz. It was reported that Fe_2O_3 present in the fly ash contributes to the shielding. Srivastava *et al.* [55] presented a brief review on fly ash utilization. In this paper, the use of fly ash for electromagnetic shielding interference has been explained. Bora *et al.* [56] has successfully prepared PANI-Co coated fly ash composite film by solution casting method. The absorption-dominated shielding effectiveness of ~ 30 dB was noted for the film in the Ku band. Cobalt ferrites, coal fly ash/cobalt ferrite particles, and TPU-based nanocomposites were successfully prepared by Gulzar *et al.*, and EMI shielding effectiveness of 37 dB was observed in the microwave region [57]. Poonguzhali *et al.* [58] have evaluated electrical conductivity studies of PANI-fly ash blends and reported that the presence of fly ash in PANI results in the improvement of shielding properties. AC and DC conductivity studies were studied for all samples having different concentrations of fly ash to aniline ratio. Dai *et al.* [59] studied the effect of the mineralogical phase and chemical composition of fly ash on electromagnetic wave-absorbing properties and concluded that fly ash can be used as a building material for electromagnetic protection. The wave absorbing properties of cement materials blended with high-iron fly ash has been investigated by Huang *et al.* [60], and the minimum reflectivity of 11 dB has been observed in the range of 9.5–18 GHz. whereas, Baoyi *et al.* [61] studied the usage of fly ash as cement replacement in order to improve the electromagnetic absorbing properties of cement composites. Bora *et al.* [62] synthesised PANI and NiOC composite by *in situ* synthesis, followed by the fabrication of thin-film by solution casting method. Total shielding effectiveness (SE) was foun, *i.e.*, ~24 dB, ~27-24 dB, ~21 dB in the J band, X band, and Ku band, respectively. Nam *et al.* [63] successfully fabricated MWNT, and lfy ash added cementitious material. The EMI SE recorded was -57.1 dB at 1 to 18 GHz. So, lots of research is going on to design lightweight, flexible, and adjustable effective microwave shields. The properties of microwave absorbers can be easily tuned by selecting apposite base material, fillers, synthesis method, and conditions.

7.5. SYNTHESIS OF POLYPYRROLE-Γ-FE₂O₃-FLY ASH (PFFA) NANOCOMPOSITES

In situ emulsion polymerization has been used for the fabrication of PFFA nanocomposites. Iron (III) chloride hexahydrate and iron (II) chloride tetrahydrate are used to produce γ-Fe_2O_3 nanoparticles (*via* co-precipitation method) [64]. 0.3 M sodium lauryl sulphate (SLS) has been utilized to form a uniform emulsion. 0.1

M of pyrrole and 0.2 M FeCl$_3$ has been added as monomer and dopant, respectively. Emulsion polymerization is completed in 4-5 hr, and after that resultant nanocomposite is washed with water to remove traces of impurities. Some compositions of the PFFA nanocomposites, designated as PFFA11, PFFA21, and PFFA12 with varying weight ratios of pyrrole, γ-Fe$_2$O$_3$, and fly ash, *i.e.*, 1:0.5:0.5, 1:1:0.5, and 1:0.5:1, have been synthesized. For comparison purposes, samples designated as pure polypyrrole (PPy) and PFA (pyrrole and fly ash in 1:1) have also been synthesized. Fig. (**2**) shows the illustration representing the development of polypyrrole–γ-Fe$_2$O$_3$-FA (PFFA) nano-sticks.

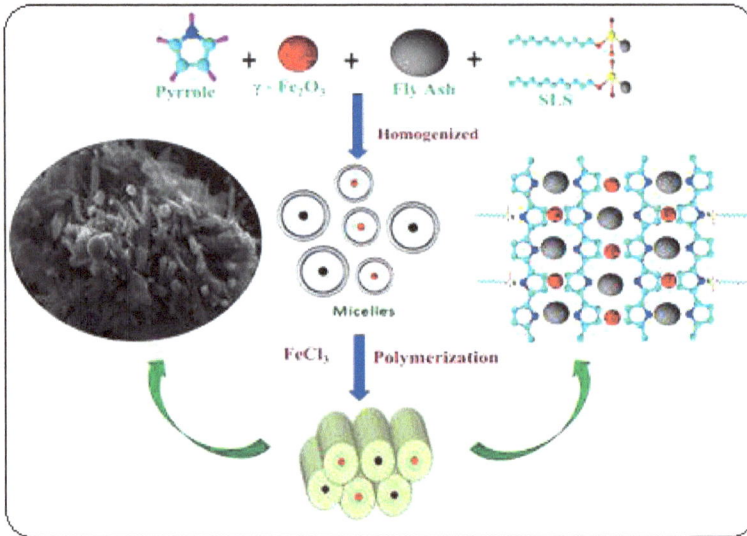

Fig. (2). Illustration of the development of PFFA composite nanostick. Reprinted with permission from ref. [53], Copyright 2014, American Chemical Society.

7.5.1. Characterization

7.5.1.1. X-ray Diffraction (XRD) Analysis

The XRD pattern of FA, γ-Fe$_2$O$_3$, PFA, and PFFA nanocomposites are depicted in Fig. (**3**), while the inset depicts the XRD pattern of polypyrrole. Fly ash (FA) has peaks at 2θ values of approximately 26.660°, 33.240°, 35.260°, 40.880°, 42.620°, 54.040°, 60.680°, and 64.500°. A typical XRD pattern for γ-Fe$_2$O$_3$ has been observed, showing peaks at 2θ = 30.28, 35.68, 43.34, 53.80, 57.34, 62.95°. All the peaks have a match with the standard pattern of γ-Fe_2O_3 (JCPDS No. 39-1346). The existence of γ-Fe_2O_3 and FA particles has also been confirmed by XRD results of PFFA nanocomposites. The broad peak observed at 21.5° confirms the semi-crystalline nature of polypyrrole. The crystallite size of γ-Fe_2O_3 particle was calculated by line broadening using Scherer's formula and estimated to be 9.17 nm.

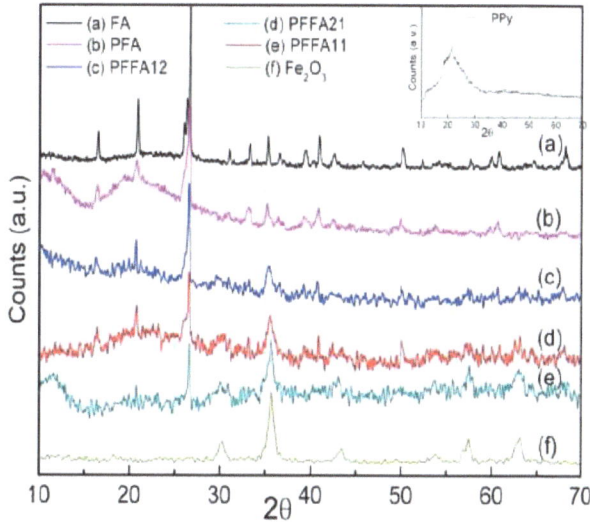

Fig. (3). X-ray diffraction Pattern. Reprinted with permission from ref. [53], Copyright 2014, American Chemical Society.

7.5.1.2. Micro Structural Analysis

The morphological structure of FA particles and PFFA nanocomposites has been studied by scanning electron microscope (SEM), which is further used to confirm the dispersal of γ-Fe_2O_3 and FA particles in PPy. SEM micrograph of fly ash particles confirms the sphere-shaped structures of particles from 400 nm to 3 μm in size (Fig. **4a**). In the same way, from Fig. (**4b**), rough structure formation of PFA particles has been observed, which validates the coating of sphere-shaped FA particles from the coarse – uneven structure of polypyrrole. While observing PFFA21 at low and high magnifications (Fig. **4c** and **4d**), complete alteration in structure has been observed from the uneven sphere-shape to nano-sticks of nanocomposites embellished with γ-Fe_2O_3 and FA particles. On further analysis, it has been found that the average length of PFFA21 particles is about ≈ 6μm with an average width of ≈ 50 nm. So, the findings reveal that some factors like monomer selection, choice of reinforcement, compatibility, reaction conditions, and synthesis method decide the morphology of polymer composite.

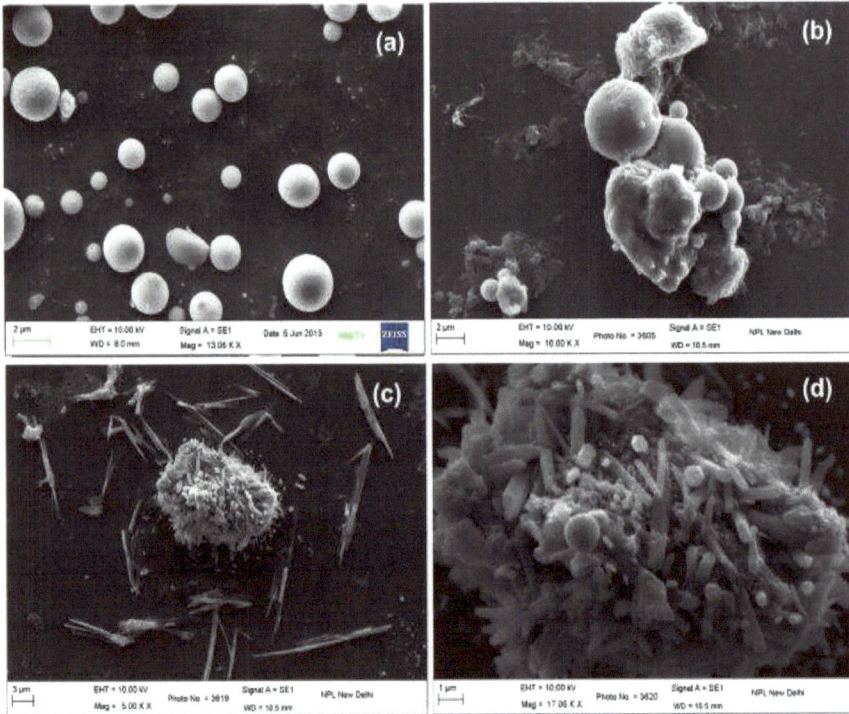

Fig. (4). Surface morphology of **(a)** fly ash **(b)** PFA **(c)** & **(d)** PFFA21. Reprinted with permission from ref. [53], Copyright 2014, American Chemical Society.

In Fig. (**5**), FA particles surrounded with PPy having γ-Fe_2O_3 have been clearly observed. The TEM image of FA shows spherical-shaped particles (inset of Fig. **5a**). TEM images indicate that the FA particles are covered with polypyrrole matrices having γ-Fe_2O_3 nanoparticles. Fig. (**5e**) displays the TEM image of the PFFA21 composite sample that illustrates the presence of γ-Fe_2O_3 nanoparticles in a composite structure. Results reveal that the existence of encapsulated magnetic and dielectric fillers into conducting matrices is very helpful in improving the absorption of the EM wave.

In Figs. (**5d** and **5f**), the HRTEM image of the PFFA12 & PFFA21 nanocomposite show the proper dispersion of γ-Fe$_2$O$_3$ nanoparticles in the polypyrrole matrix with d-spacing of 0.25 nm, which corresponds to (311) plane. XRD analysis reflects the same results, which are also observed by HRTEM.

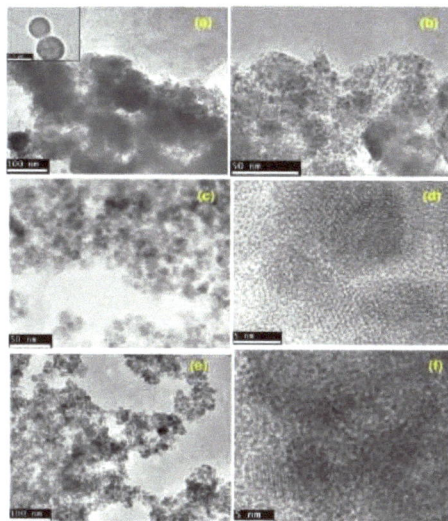

Fig. (5). HRTEM images of (**a** & **b**) PFFA11, (**c** & **d**) PFFA12 & (**e** & **f**) PFFA21 whereas inset displays TEM image for FA. Reprinted with permission from ref. [53], Copyright 2014, American Chemical Society.

7.5.1.3. Conductivity and Magnetic Measurements

Four probe method has been used to measure the room temperature conductivity of the PFFA composite sample, and a decreased trend in conductivity has been observed with the increase of γ-Fe$_2$O$_3$ and FA loading. Conductivity value has been found to be 0.062 S/cm and 0.046 S/cm for PFFA11 and PFFA21 composite samples, respectively. Transport hindrance may be the reason for the decrease in the value of conductivity which may be produced by the presence of insulating reinforcement in conducting PPy.

The magnetic properties of γ-Fe$_2$O$_3$ and PFFA nanocomposites have been studied by the magnetic hysteresis loop (Fig. **6**). The magnetic properties of composite samples have been studied under applied magnetic fields at room temperature. As a result, the composite sample shows magnetization under the applied magnetic field. This behaviour depends on the amount of ferrite content. γ-Fe$_2$O$_3$ nanoparticles show the saturation magnetization (M$_s$) value of the 61.24 emu/g at an external field of 5 kOe. In this case, no appreciable hysteresis loop has been observed with negligible coercivity and no remanence, which confirms its super paramagnetic behaviour. Nano size of magnetic particles is responsible for this super paramagnetic nature which gives fast relaxation behaviour in composite samples [65]. In the case of the PFFA11 and PFFA12 composite samples, M_s value of 7.7 emu/g has been observed at the external field of 5 kOe as both composite samples have the same number of γ-Fe_2O_3 nanoparticles but M_s value increases to 13.55 emu/g with the increase of γ-Fe_2O_3 particles in the case of PFFA21 composite.

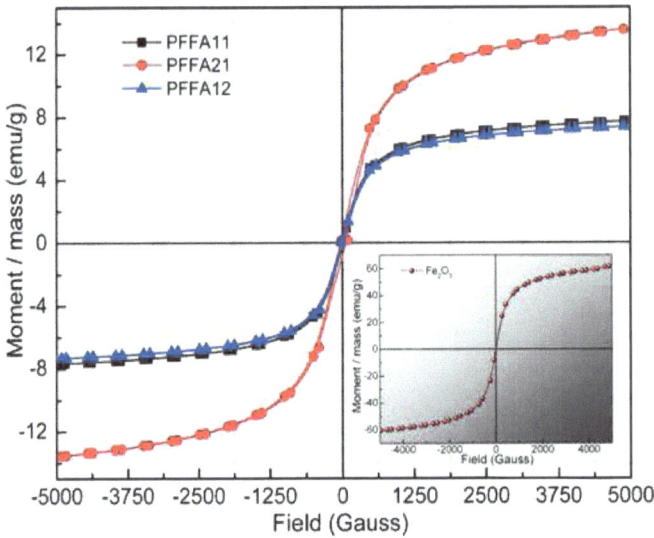

Fig. (6). Saturation magnetization (M_s) curve for PFFA composite. In comparison, inset shows the VSM of γ-Fe_2O_3. Reprinted with permission from ref. [53], Copyright 2014, American Chemical Society.

7.5.1.4. Dielectric, Permeability and Electromagnetic Interference Shielding Investigations

Complex permittivity ($\varepsilon_r = \varepsilon' - j\varepsilon''$) and complex permeability ($\mu_r = \mu' - j\mu''$) are the two main parameters that affect the shielding properties of polypyrrole composite. The real parts (ε') and (μ') is correlated with the polarization occurred and magnetic field induced within the material, respectively while imaginary parts (ε'' and μ'') represents dielectric and magnetic losses [66, 67]. These parameters are calculated from scattering parameters by using Nicholson, Ross, and Weir method [68-70], shown in Figs. (**7a** and **b**). Two electric currents, namely conduction, and displacement current, originate within the material under the influence of an electromagnetic field [71]. Mobile charge carriers are responsible for conduction current subsequently dielectric loss. However, the displacement current emerges due to localized charges (polarization inside the material) [72]. Space charge polymerization is dominant as compared to other polarizations in the case of the composite due to the heterogeneity of the material. Space charge polarization occurs in conjugated polymers due to a polaron/bipolaron system that is mobile and free to travel along the chain, and orientational polarization arises due to bound charges, *i.e.*, dipoles that have limited mobility.

A decreased trend in the ε' value has been observed for all the samples within the frequency range of 12.4-18 GHz. PFFA21 sample shows ε' value of 37.8–29.4, which is indicated in Fig. (**7a**). This value is much higher as compared to a pure PPy sample which has ε' value between 24.9 and 15.8. On account of interfacial polarization on the surface, PPy coating on reinforcement particles provides the elevated value for ε' and ε''. Fig. (**7a**) reveals that minimum variation in the value of ε'' has been observed for all the samples of PFFA composite. This may be related to the conductivity of the PFFA composite samples as all composite samples show the minimal difference in conductivity values.

The real part (μ') and imaginary part (μ'') of permeability of PFFA composites have been shown in Fig. (**7b**). It has been depicted from Fig. (**7b**) that an almost uniform value for the real part of the permeability has been noticed throughout the considered frequency, while the imaginary part shows a decreased trend with respect to the measured frequency range. This is ideal for a microwave surface impedance match [73].

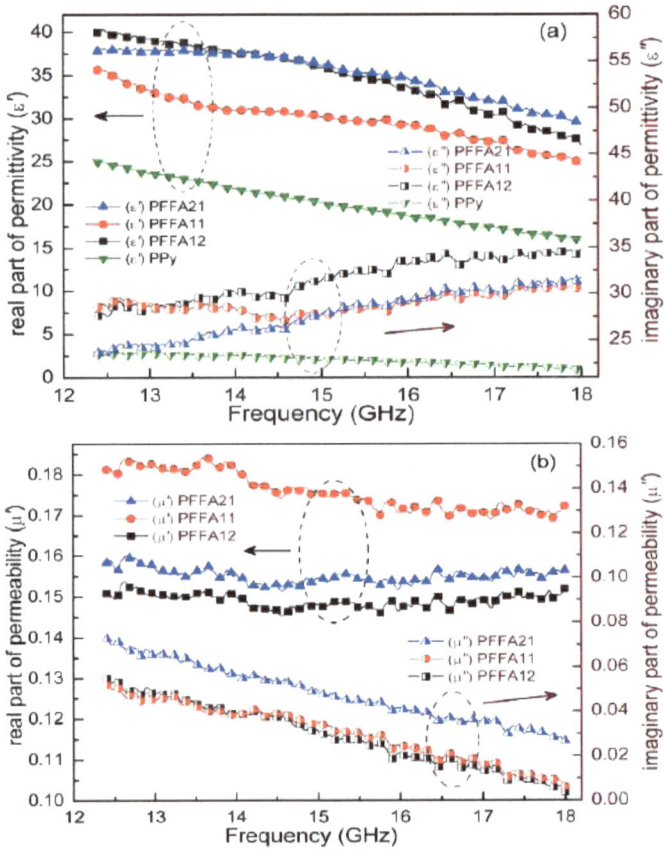

Fig. (7). (a) real (ε') and imaginary (ε'') part of complex permittivity **(b)** real (μ') and imaginary (μ'') part of complex permeability in the frequency range of 12.4-18 GHz. Reprinted with permission from ref. [53], Copyright 2014, American Chemical Society.

An absorber material should have optimal permeability (μ'), and the permeability of ferromagnetic materials can be stated as [74]:

$$\mu' = \frac{M_s^2}{akH_c M_s + b\lambda\xi} \qquad (6)$$

where a and b are two constants determined by the material composition, λ is the magnetostriction constant, and ξ is an elastic strain parameter of the crystal [75]. From this equation, it is clear that the value for μ' can be improved by either increasing M_s or by decreasing H_c. That, in effect, will increase the absorption of the microwave. It has been seen from Fig. **(7b)** that PFFA21 shows higher values of μ''

in comparison to that of PFFA12 & PFFA11 composite due to higher number of γ-Fe_2O_3 particles in the composite sample.

Dielectric tangent loss (tan $\delta_\varepsilon = \varepsilon''/\varepsilon'$) and magnetic tangent loss (tan $\delta_\mu = \mu''/\mu'$) have been calculated from the permittivity and permeability of the PFFA composite samples, as shown in Fig. (**8**). The samples show higher values of tan δ_ε (> 0.6) which reflects the lossy feature of the PFFA composite. It has also been observed from Fig. (**8**) that PFFA composites show a lesser value for dielectric loss (tan δ_ε) in comparison to pure PPy as a result of additional interfacial polarization. However, a high magnetic loss factor (tan δ_μ) has been observed for the PFFA composite, which increases with the addition of γ-Fe_2O_3. Domain-wall motion and eddy current loss are two vital aspects associated with magnetic loss in the material [72]. So, both dielectric and magnetic losses are responsible for the absorption of electromagnetic waves.

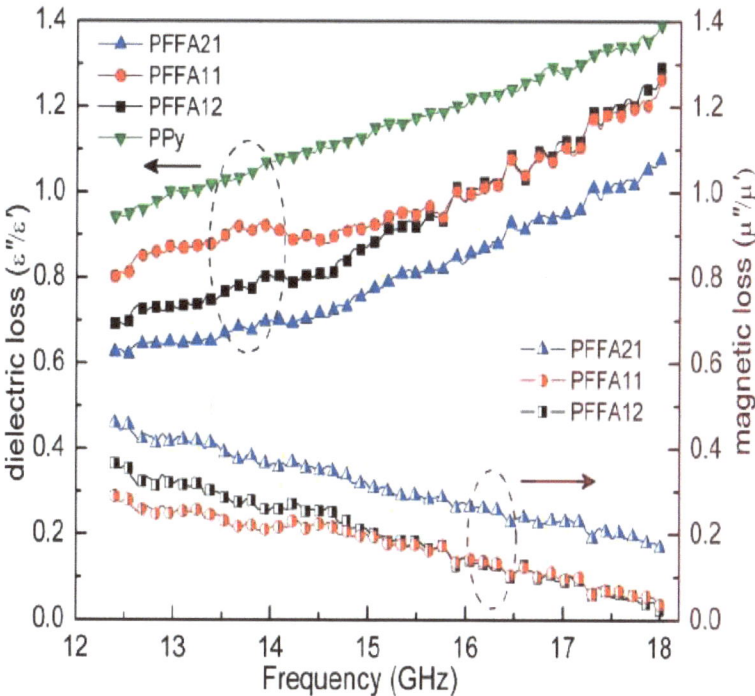

Fig. (8). Dielectric loss factor & magnetic loss factor in the frequency range of 12.4-18GHz. Reprinted with permission from ref. [53], Copyright 2014, American Chemical Society.

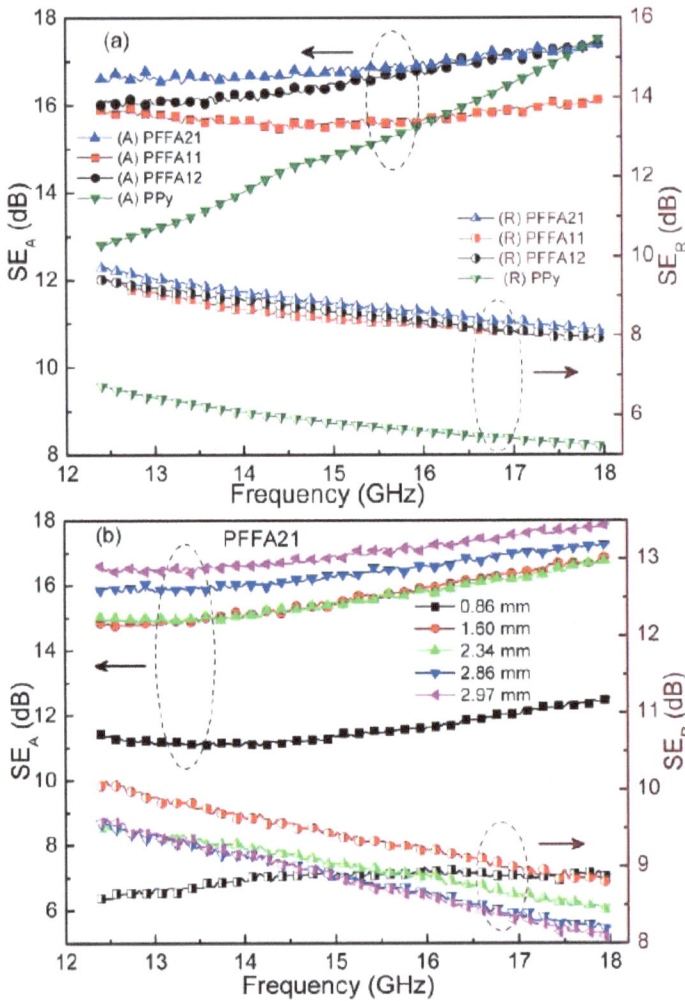

Fig. (9). (a) EMI shielding effectiveness due to absorption (SE$_A$) and reflection (SE$_R$) in the frequency range of 12.4-18GHz **(b)** Shielding effectiveness due to absorption (SE$_A$) and reflection (SE$_R$) of PFFA21 composite with respect to the thickness of the sample. Reprinted with permission from ref. [53], Copyright 2014, American Chemical Society.

It has been observed from Fig. **(9a)** that PFFA composites show SE$_A$ as the dominant shielding mechanism rather than SE$_R$. The SE$_A$ value also remains almost constant with the variation in frequency which is the requisite condition for an effective broadband absorber. As shown in Fig. **(9a)**, the SE value of the PFFA21 composite was found to be 25.5 dB (SE_A = 17.4 dB and SE_R= 8.1 dB) with a matching thickness of ~ 2.0 mm. The maximum value of SE_T, *i.e.*, 25.5 dB, corresponds to ~ 99.7 % attenuation, which is much better as compared to pure PPy. While considering Fig. (**9b**), it has been found that the absorption of the EM wave is directly proportional to the thickness of the composite sample, while a little fluctuation in the reflection part has been noticed. It is observed that shield thickness has a major effect on the absorption properties of the microwave.

The improved SE_A is due to the synergistic effect of dielectric and magnetic losses caused by the presence of FA and γ-Fe_2O_3 in the polymer matrix [76]. It is noteworthy to mention that dielectric loss mainly depends on the conductivity of materials, whereas magnetic loss is governed by both electrical conductivity and magnetic permeability [77]. Specifically, in comparison to dielectric loss materials, magnetic lossy materials show high absorption efficacy.

The dielectric and magnetic loss in the entire frequency range indicates that the enhanced radiation absorption ability is certainly the outcome of PPy, FA, and γ-Fe_2O_3 being mutually effective.

7.6. DESIGNING OF MULTIPHASE POLYURETHANE (PU) /FLY ASH/MWCNT/ COMPOSITE SHEET

PU/fly ash/MWCNT/composite sheets are fabricated by solution casting methods. For this work, the chemical vapour deposition (CVD) method has been used to synthesize MWCNTs on carbon substrate by using toluene (hydrocarbon source) and ferrocene (iron catalyst precursor**).** The deposited MWCNT (90 % pure) has a diameter range between 20 - 50 nm and length between 50 - 100 mm [78, 79]. After that, MWCNTs were dispersed in DMF, and PU was also dissolved separately in DMF solvent. Subsequently, fly ash particles are added to the dissolved PU solution for about 2 hours. Subsequently, both solutions mixed together. Then this solution was cast into a sheet at 75°C for 14 hours. This process ends up with the formation of a PU cast sheet having ~ 1.5 mm thickness [80, 81]. An illustration for the fabrication of a PU cast composite sheet has been shown in Fig. (**10**). Many preparations of composite have been designed by changing wt. % of PU, fly ash, and MWCNT, *i.e.*, 90/10/0, 90/0/5, 85/10/5, and 85/10/10, and designated as PUF, PUNT, PUFNT1, and PUFNT2, respectively. In addition to these samples, a pure PU sheet has also been cast for relative analysis.

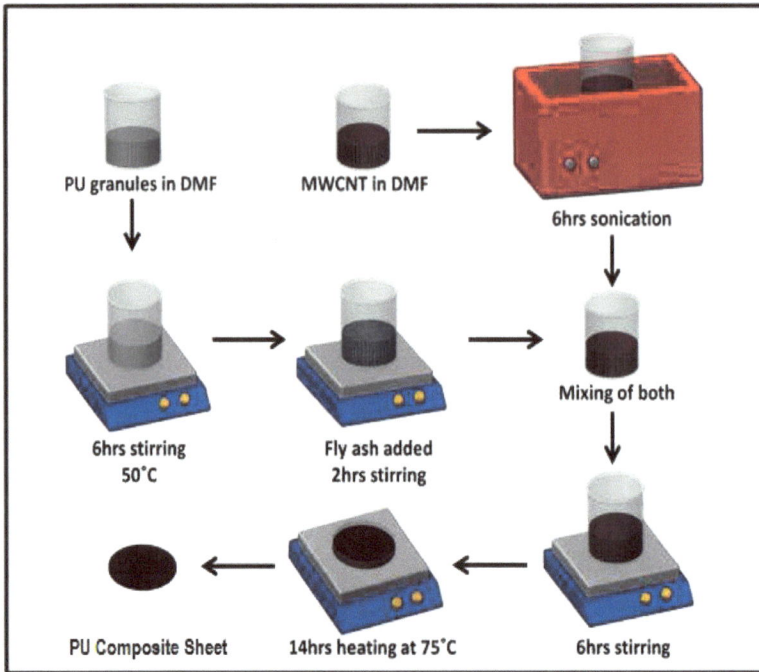

Fig. (10). Illustration showing fabrication of Polyurethane (PU) composite cast sheet.

7.6.1. Characterization

7.6.1.1. Micro Structural Analysis

A scanning electron microscope (SEM) has been used to analyse the morphological structure and distribution of filler particles in the PU matrix. Fig. (**11a**) indicates that the FA particles are of a spherical shape and particle size ranges between 50 nm and 3μm [82]. A micrograph of MWCNTs can be seen in Fig. (**11b**). Most MWCNTs look like entangled networks, while few of them can be identified individually. Fig (**11c**) depicts the morphology of the PUFNT2 composite, which confirms the excellent uniform distribution of filler particles in the PU matrix. Networks of MWCNTs are shown with the help of a red arrow, and fly ash particles are encircled in yellow. Fig. (**11d**) shows the morphological structure of the PUFNT2 composite under high magnification, which illustrates the mesh-like structures due to the presence of MWCNTs. These meshes may be responsible for the good electrical properties of PU cast sheets because of the possibility of additional charge transport paths formed throughout the sheet [78]. This morphological property results in the development of an excellent shield.

Fig. (11). Morphological structure of (a) FA, (b) MWCNT, (c) & (d) PUFNT2. Reprinted with permission from ref. [39], Copyright 2016, Springer Nature.

7.6.1.2. X-ray Diffraction (XRD) and Conductivity Analysis

X-ray diffraction method has been used to facilitate solid-state structural details of PU cast sheet (Fig. **12**). The characteristic diffraction peaks of MWCNT were noticed at $2\theta = 26.1°$, and $43.2°$, corresponding to (002) and (100) planes, respectively [83]. While the diffraction peaks observed at $2\theta = 26.660°$, $33.240°$, $35.260°$, $40.880°$, $42.620°$, and $54.040°$ proves the presence of FA in the PU composite sheet [53]. So, FA and MWCNT can be identified as the primary constituents of the PU cast sheet with the help of the XRD pattern. XRD pattern of PUF composite shows an intense peak at $2\theta = 26.660°$ which suggests the presence of a high concentration of FA. Likewise, PUNT samples show very prominent diffraction peaks of MWCNT, corresponding to the amount of reinforcement

(MWCNTs) in the polymer matrix. A wide diffraction peak has been observed for the pure PU cast sheet, which can correlate with its amorphous nature.

The conductivity of PU cast sheets has been calculated by the four probe method at room temperature. For PUFNT2 and PUFNT1 composite samples, room temperature conductivity has been found to be 7.2 and 2.9 S/cm, respectively. While for pure PU cast film, the conductivity of 10^{-14} S/cm has been observed. From the results, it is noted that improvement in conductivity has been reported with the increase of MWCNT wt. % loading. Thus, this multiscale reinforcement arrangement is expected to promote the good conducting properties in the composite sheet, which is a requisite property for shielding application.

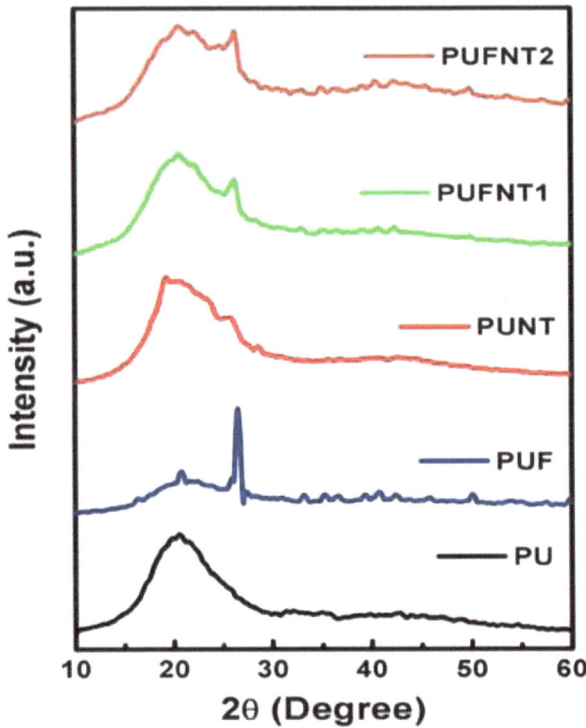

Fig. (12). X-ray diffraction pattern. Reprinted with permission from ref. [39], Copyright 2016, Springer Nature.

7.6.1.3. Thermogravimetric Analysis

Thermograms of PU, PUFNT1 and PUFNT2 composites are shown in Fig. (**13**). TGA was carried out to check the thermal stability of PU cast sheets in the temperature range of 25 to 600°C. Thermograms of all PU sheet samples show the same pattern, with three weight loss stages. The first step of weight loss refers to the residual moisture and solvent loss. Due to the presence of a solvent, the thermal degradation levels in PUFNT1 and PUFNT2 were initially higher as compared to PU. The noticeable degradation was reported between 250 and 400 ° C for all composites where maximum weight loss was observed. The left residue percentage for PU was 4.5 %, for PUFNT1 10.1 %, and for PUFNT2 16.8 %, respectively. Results reveal the higher thermal stability of composite cast sheets, having higher wt. % of fly ash and MWCNT reinforcement in PU matrix.

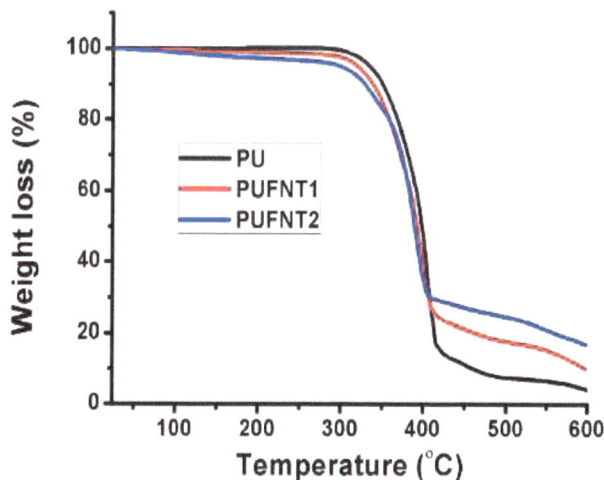

Fig. (13). TGA of pure PU, PUFNT composite cast sheet. Reprinted with permission from ref. [39], Copyright 2016, Springer Nature.

7.6.1.4. Electromagnetic interference shielding and dielectric investigations

An Agilent E8362B Vector Network Analyser has been used to calculate the SE of PU composite sheets in the frequency range of 12.4–18 GHz. Scattering parameters S_{11} (or S_{22}) and S_{21} (or S_{12}) obtained by measuring a two-port network analyser represents the reflection (R), transmission (T), and absorption (A) mechanism of

SE, where R = |S11|^2 and T = |S21|^2 and A = 1 - |S11|^2 - |S21|^2. SE$_M$ can be considered negligible in the case of SE$_T$ > 10 dB [84].

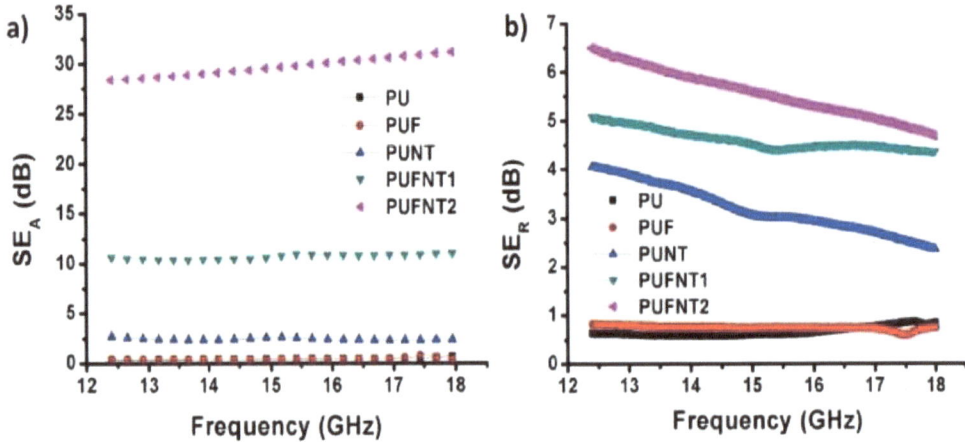

Fig. (14). Shielding effectiveness (a) SE$_A$ (b) SE$_R$ in the frequency range of 12.4 – 18 GHz. Reprinted with permission from ref. [39], Copyright 2016, Springer Nature.

SE$_A$ and SE$_R$ for PU cast sheets in the Ku band are shown in Figs. (**14a and b**). The SE$_{A (max)}$ for PUFNT2, PUFNT1, and PUNT cast sheets were found to be 30.5, 15.6, and 3.8 dB at 18 GHz, respectively (Fig. **14a**). The obtained SE$_R$ has been noted as 5.3, 4.8, and 2.4 dB for PUFNT2, PUFNT1, and PUNT samples at 18 GHz, respectively (Fig. **14b**). SE$_A$ and SE$_R$ values for PUF and PU are insignificant, but cast sheets having MWCNT show dominant SE$_A$ as compared to SE$_R$ in the Ku band.

As shown in Fig. (**15**), the curves for PU and PUF samples overlapped each other with almost negligible value for SE$_T$. In the case of the PUNT composite sheet, the obtained SE$_T$ value was 6.2 dB at18 GHz. It has also been noted that the combination of FA with MWCNT in PU improves SE$_T$ of PU composite sheet to 15.6 dB for PUFNT1, which further increases to 35.8 dB for PUFNT2 sample (> 99% attenuation of microwave) with the further addition of MWCNT in PU matrix.

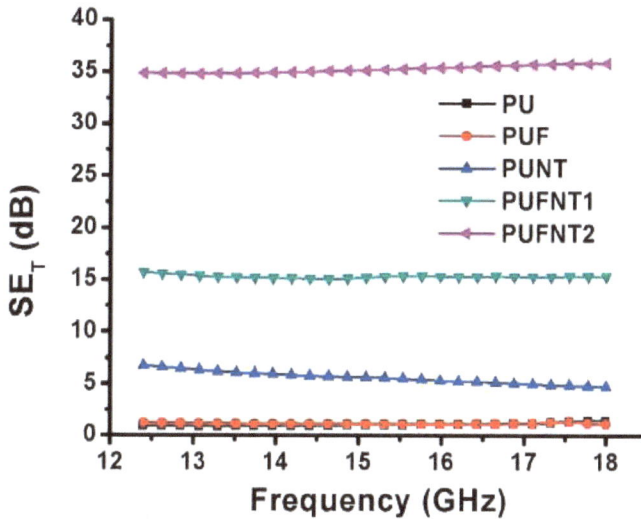

Fig. (15). Total shielding effectiveness (SE$_T$) in the frequency range of 12.4 – 18 GHz. Reprinted with permission from ref. [39], Copyright 2016, Springer Nature.

It is observed from the results that the individual effect of FA and MWCNT does not provide efficacious EMI SE property, but an amalgamation of both reinforcements into the PU matrix comes up with a considerable impact on the electromagnetic properties of the PU sheets and provide better performance. It should be mentioned here that the presence of Fe_2O_3 particles in FA compliments MWCNT's magnetic properties and increases the resultant magnetic permeability of the sheet, which evidently boosts the absorption of electromagnetic radiation. Thus, the shielding effectiveness of PU composite sheets was improved.

Theoretical calculations given in Nicholson, Ross, and Weir method were used to calculate complex permittivity (ε' & ε'') from scattering parameters (S_{11} and S_{21}) to further explore shielding mechanisms [70]. Figs. (**16a** and **b**) displays the variation of complex permittivity (ε' & ε'') with respect to frequency. Complex permittivity (ε' & ε'') for PUFNT2 and PUFNT1 decline with the frequency, whereas other composites show nearly steady values. It is clear from the graph that the inclusion of MWCNT and FA in the polymer matrix results in a higher dielectric constant, *i.e.*, 37.4 and dielectric loss value of 10.1 for the PUFNT2 sheet. Ionic, electronic, orientation, and space charge polarization occur in multiphase composite sheets

(PUFNT 1 and PUFNT2) under the influence of electromagnetic waves, but the heterogeneous nature of composite due to the presence of multi fillers leads to more space charge polarization as compared to other polarizations. In addition to this, there may be interfacial polarization between FA particles, MWCNT, and PU matrix. So, a synergistic combination of fillers provides an optimized exceptional combination of properties such as improved conductivity and better dielectric losses, which cannot be acquired by individual fillers. These findings suggest that these PU multiphase composite sheets could be a good shielding material for commercial applications.

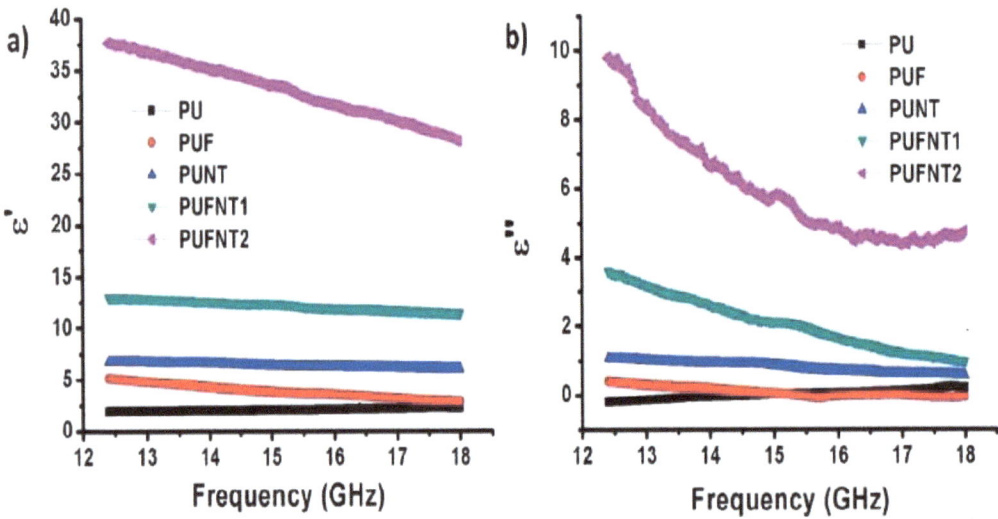

Fig. (16).(a) Dielectric constant (ε') **(b)** Dielectric loss (ε'') in the frequency range of 12.4 – 18 GHz. Reprinted with permission from ref. [39], Copyright 2016, Springer Nature.

7.7. DESIGNING OF CEMENT COMPOSITE PAINT

For this work, MWCNTs were fabricated by the chemical vapour deposition method (CVD), as described in the previous section [39]. The *in-situ* co-precipitation method has been used for the encapsulation of Fe_3O_4 particles onto glass fibres (GFF) by using $FeCl_3$ and $FeCl_2$. Cement paint (CP), fly ash (FA), and MWCNTs were combined together and ball-milled, and then GFF was mixed by using water as a dispersing medium. After that, the mixture was kept in the oven for drying, and finally, cement composite paint was obtained. By using the same procedure, many samples of cement composite paint have been prepared with varying ratios of constituents, *i.e.*, weight % ratio of cement paint to MWCNT. Prepared sample is abbreviated as CMGF1, CMGF2 and CMGF3 having 1:0.1, 1:0.2 and 1:0.3 weight % ratio of cement paint to MWCNT, respectively. The required amount of GFF and FA are also included in the above-mentioned samples to get the required effects.

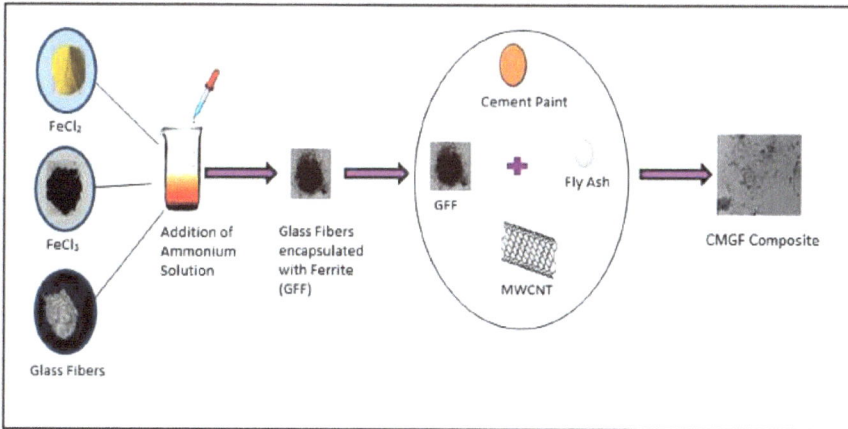

Fig. (17). Illustration for the preparation of Cement Composite paint (CMGF). Reprinted with permission from ref. [40], Copyright 2018, IOP Publishing, Ltd.

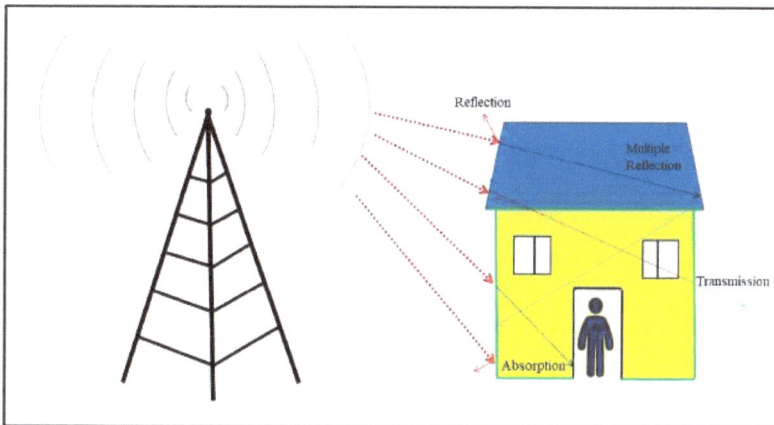

Fig. (18). Schematic representation showing shielding due to cement paint (CMGF) composite on building walls.

Fig. (**17**) displays the development of cement composite paint (CMGF) and schematic representation of shielding mechanism due to cement composite paint (CMGF) after applying on building walls has been illustrated in Fig. (**18**).

7.7.1. Characterization

7.7.1.1. X-Ray Diffraction (XRD)

XRD patterns for pure CP, MWCNT, FA, Fe_3O_4 particles, and CMGF composite paint samples have been shown in Fig. (**19**). MWCNTs show characteristic peaks at $2\theta = 26.07°$ and $40.16°$ corresponding to 002 and 100 diffraction planes [81, 90]. FA shows that all typical characteristic peaks [53] and diffraction peaks of ferrite particles match with the standard pattern of Fe_3O_4 (Powder Diffraction File, JCPDS No.88-0315) [76, 84]. Diffraction peaks of GFF, MWCNT and FA are apparent in all cement composite paint, *i.e.*, CMGF1, CMGF2 & CMGF3. Moreover, in the case of CMGF composite samples, a gradual change in the diffraction peaks has been noticed, which is the evidence of synergistic linkage amid the constituents present in the composite. The average size of Fe_3O_4 particles was calculated using Debye Scherer's formula and estimated as 11.2 nm for pure Fe_3O_4 particles encapsulated on glass fibers.

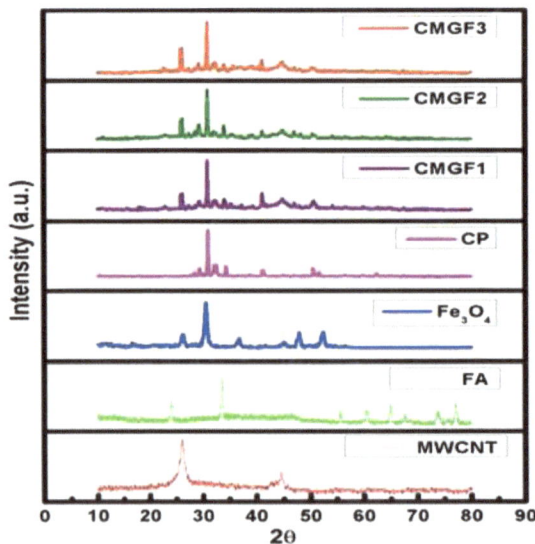

Fig. (19). X-ray diffraction pattern. Reprinted with permission from ref. [40], Copyright 2018, IOP Publishing, Ltd.

7.7.1.2. Micro Structural Analysis

To understand the morphology of composite paint samples, transmission electron microscopy (TEM) has been used. Fig. (**20a**) shows a TEM image of CP and displays the bulbous asymmetrical shape of CP particles. Several concentrically interlinked carbon nanotubes are clearly visible, with an interlayer spacing of 0.34 nm and a diameter of typically between 20-50nm (Fig. **20b**). The TEM and diffraction pattern of Fe_3O_4 have been shown in Figs. (**20c** and **d**), respectively. This shows that Fe_3O_4 nanoparticles of the 8-12 nm spherical shape are distributed uniformly on glass fibres. Similar results are also obtained from the XRD analysis. The crystalline nature of GFF is confirmed by the diffused SAED pattern (Fig. **20d**). TEM images of CMGF1 and CMGF3 under low and high magnification have been displayed in Figs. (**20e - 20g**). All reinforcements are observed distributed uniformly in the CP matrix, which endorses better interaction among them. The SAED pattern of CMFF3 composite paint informs about the crystallinity of composite paint (Fig. **20h).**

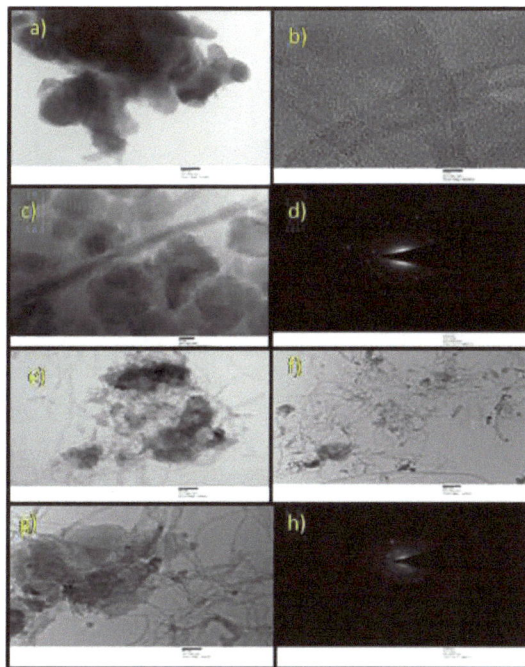

Fig. (20). TEM images of **(a)** CP, **(b)** MWCNT **(c)** GFF **(d)** SAED pattern of GFF **(e)** CMGF1 **(f)** & **(g)** CMGF3 under low and high magnification **(h)** SAED pattern of CMGF3. Reprinted with permission from ref. [40], Copyright 2018, IOP Publishing, Ltd.

7.7.1.3. Conductivity and Shore Hardness Test

Room temperature conductivity and shore hardness for CP and CMGF composite samples have been calculated in view of its commercial application as wall paint for the shielding of microwaves. The conductivity of CP cement paint composite varies from 0.02 S/cm to 0.72 S/cm from CMGF1 and CMGF3 composite, respectively, whereas the CP sample shows 1.02×10^{-5} S/cm conductivity. It is perceived from the results that conductivity increases sharply with the increase of the amount of MWCNT in the CP matrix. This may be due to the excellent conductivity of MWCNTs. Moreover, the development of conducting bridges into the CP matrix contributes to the improvement of conductivity.

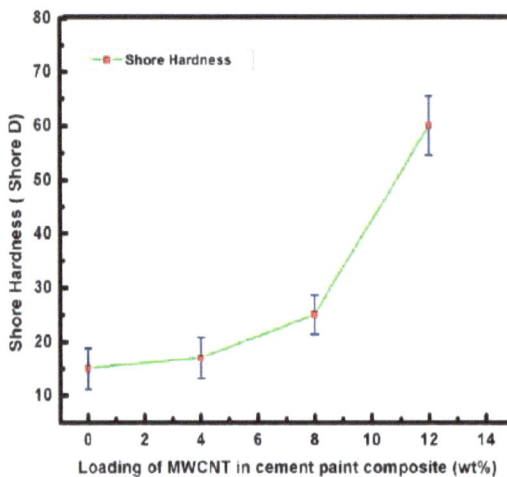

Fig. (21). Shore hardness *vs.* wt. % of MWCNT. Reprinted with permission from ref. [40], Copyright 2018, IOP Publishing, Ltd.

Shore scleroscope hardness tester has been used to calculate the hardness. It is noticeable from Fig. (**21**) that the hardness of the cement composite paint depends on wt. % of MWCNT in the composite, and an increasing trend has been noticed with respect to wt. % of MWCNT in CP. Maximum hardness of 60 ± 2 shore D is reported for CMGF3 composite sample while CP shows the hardness of 15 ± 1 shore D. It should be noted that the presence of MWCNT in all CMGF composite have ensured an excellent hardness value than that of conventional cement paint. Further, stronger affinity and homogeneous filler dispersion into the CP matrix may be an explanation for improved composite paint hardness.

7.7.1.4. Magnetic Measurements

A vibrating sample magnetometer (VSM) is used to measure the magnetic behaviour of Fe_3O_4 particles and composite paint having GFF at an external field of 5 kOe. Results are displayed as magnetic - hysteresis curve as shown in Fig. (**22**). GFF (Fe_3O_4 particles) shows a saturation magnetization (Ms) value of 35 emu/g, and a negligible value of coercivity and retentivity has been reported. Hysteresis loop has also not been noted in the GFF sample, which indicates its super paramagnetic nature. On the other hand, CMGF1, CMGF2, and CMGF3 composite paint also show magnetic behaviour with saturation magnetization of 0.60 emu/g, 0.61 emu/g, and 0.60 emu/g, respectively. The observed magnetic behaviour for all CMGF composite samples coincides with each other because of the addition of a similar quantity of GFF in the CP matrix.

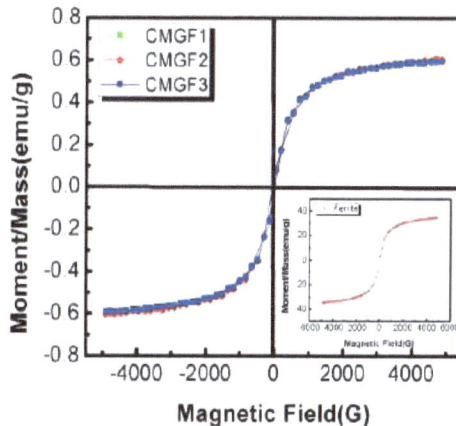

Fig. (22). Magnetization- hysteresis curve of CMGF cement composite paint with inset showing magnetization curve of Fe_3O_4. Reprinted with permission from ref. [40], Copyright 2018, IOP Publishing, Ltd.

7.7.1.5. Electromagnetic Interference Shielding Investigations

Fig. (**23**) shows the variation of SE_T, SE_A, and SE_R in the frequency range of 8-12.4 GHz (X-band). All CMGF composites under observation were fabricated with a thickness of about ~ 2.5 mm. The SE_T value of the CP sample is found to be almost negligible. CMGF1 sample shows SE_T value of 9.4 dB which increases to 66 dB for CMGF3 at 12.4 GHz, whereas CMGF2 sample shows an SE_T value of 34.7 dB, which is in between the SET values for CMGF1 and CMGF3.

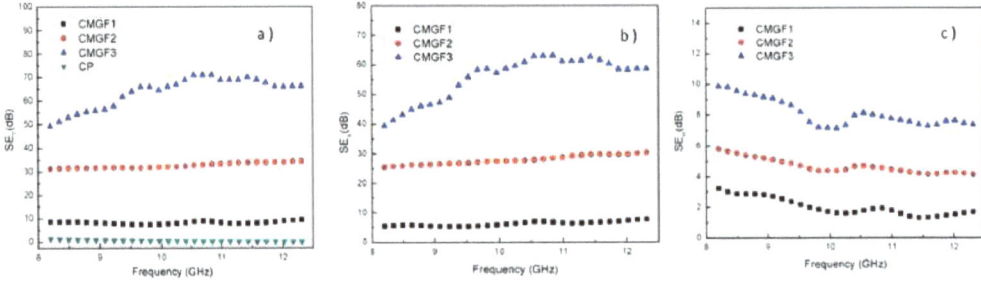

Fig. (23). Shielding effectiveness **(a)** SE_T **(b)** SE_A (c) SE_R for CMGF cement composite paint in X-Band (8-12.4GHz frequency). Reprinted with permission from ref. [40], Copyright 2018, IOP Publishing, Ltd.

It has been analyzed from Fig. (**23b** and **c**) that total shielding efficiency (SE_T) is primarily governed by SE_A while the contribution due to SE_R is very small. CMGF composite paint shows (*i.e.*, CMGF3) the maximum SE_A of 58.6 dB (Fig. **23b**), while maximum SE_R has been found to be 7.4 dB for CMGF composite paint (*i.e.*, for CMGF1) at 12.4 GHz (Fig. **23c**). The observed result confirms that the absorption-dominated SE is the prime mechanism in the case of CMGF composite paint and is much higher than that of CP. These results are attributable to the high proportion of MWCNTs in the CP matrix, which is also responsible for the improved conductivity of the composite paint. Referring to higher conductivity, polarization occurs within the composite material in the presence of radiation, resulting in superior shielding properties.

The variation of the skin depth with frequency has been shown in Fig. (**24**), and the plot shows that as the frequency increases, the skin depth (δ) gradually decreases. It should be noted that shielding properties depend on skin depth which can be understood by expression, $\delta = (\pi f \mu \sigma)^{-1/2}$ [85]. It is clear from this formula that the skin depth (δ) is a requisite parameter to determine the critical thickness of the material. Skin depth can also be related to the absorption loss by using equation $SE_A = 20(t/\delta) \log e = 8.69(t/\delta)$. It is very much clear from this equation that absorption loss (SE_A) is related to skin depth (δ). So, higher conductivity and good magnetic properties lead to low skin depth, and consequently, it is possible to get a thin shield with excellent SE_A [86]. It is reported that excellent EMI shielding can be attained at a thickness of the shield beyond skin depth [87]. It's worth noting that these cement composite paints could be a viable commercial shielding option for outside wall paint.

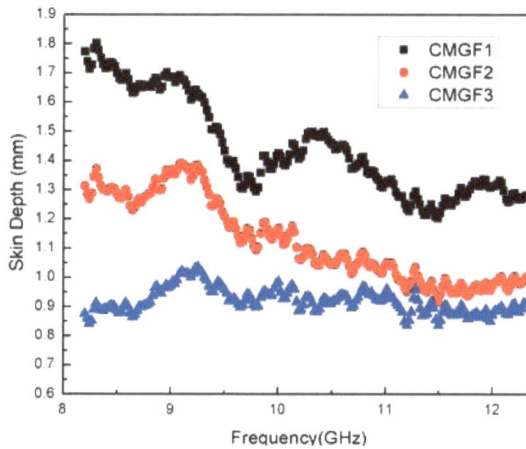

Fig. (24). Skin depth in the frequency range of 8-12.4 GHz frequency. Reprinted with permission from ref. [40], Copyright 2018, IOP Publishing, Ltd.

CONCLUDING REMARKS

Over the past decade, the dominancy of electronic items and the changes in lifestyle have driven EMI shield demand to new heights. Various formulations of composites have been prepared to meet this demand while removing all the flaws of conventionally used metal shields. All designed composites were characterized for morphological investigation, conductivity, magnetic and electromagnetic properties. Maximum total shielding effectiveness of polypyrrole nanocomposite along with γ-Fe_2O_3 and FA has been reported up to 25.5 dB in Ku Band (12.4-18 GHz). Multiphase PU /FA/MWCNT composite sheet shows SE_T of 35.8 dB at the frequency of 12.4–18 GHz. In comparison, cement composite paint having 12 wt. % of MWCNT with FA & GFF provides an efficient SE_T of 66 dB in the frequency range of 8-12.4 GHz. So, this is evident that absorption dominated shielding effectiveness of fabricated composites depends significantly on the incorporation of fillers into the base matrix material, and these advanced composite materials using waste fly ash particles along with other fillers could be a propitious material for an effective EMI shield.

CONSENT FOR PUBLICATION

Not applicable.

CONFLICT OF INTEREST

The authors declare no conflict of interest, financial or otherwise.

ACKNOWLEDGEMENTS

Declared none.

REFERENCES

[1]   Hatch, E.E.; Linet, M.S.; Kleinerman, R.A.; Tarone, R.E.; Severson, R.K.; Hartsock, C.T.; Haines, C.; Kaune, W.T.; Friedman, D.; Robison, L.L.; Wacholder, S. Association between childhood acute lymphoblastic leukemia and use of electrical appliances during pregnancy and childhood. *Epidemiology,* **1998**, *9*(3), 234-245.
       http://dx.doi.org/10.1097/00001648-199805000-00006 PMID: 9583414

[2]   Findlay, R.P.; Dimbylow, P.J. SAR in a child voxel phantom from exposure to wireless computer networks (Wi-Fi). *Phys. Med. Biol.,* **2010**, *55*(15), N405-N411.
       http://dx.doi.org/10.1088/0031-9155/55/15/N01 PMID: 20647607

[3]   Peyman, A.; Khalid, M.; Calderon, C.; Addison, D.; Mee, T.; Maslanyj, M.; Mann, S. Assessment of exposure to electromagnetic fields from wireless computer networks (wi-fi) in schools; results of laboratory measurements. *Health Phys.,* **2011**, *100*(6), 594-612.
       http://dx.doi.org/10.1097/HP.0b013e318200e203 PMID: 22004929

[4]   Ha, M.; Im, H.; Lee, M.; Kim, H.J.; Kim, B.C.; Gimm, Y.M.; Pack, J.K. Radio-frequency radiation exposure from AM radio transmitters and childhood leukemia and brain cancer. *Am. J. Epidemiol.,* **2007**, *166*(3), 270-279.
       http://dx.doi.org/10.1093/aje/kwm083 PMID: 17556764

[5]   Merzenich, H.; Schmiedel, S.; Bennack, S.; Brüggemeyer, H.; Philipp, J.; Blettner, M.; Schüz, J. Childhood leukemia in relation to radio frequency electromagnetic fields in the vicinity of TV and radio broadcast transmitters. *Am. J. Epidemiol.,* **2008**, *168*(10), 1169-1178.
       http://dx.doi.org/10.1093/aje/kwn230 PMID: 18835863

[6]   Hoang, A.S. Electrical conductivity and electromagnetic interference shielding characteristics of multiwalled carbon nanotube filled polyurethane composite films. *Adv. Nat. Sci.: Nanosci. Nanotech.,* **2011**, *2*(2), 025007.
       http://dx.doi.org/10.1088/2043-6262/2/2/025007

[7]   Kim, Y.Y.; Yun, J.; Lee, Y.S.; Kim, H.I. Preparation and Characteristics of Conducting Polymer-Coated MWCNTs as Electromagnetic Interference Shielding Materials. *Carbon Lett.,* **2011**, *12*(1), 48-52.
        http://dx.doi.org/10.5714/CL.2011.12.1.048

[8]   Lin, C.T.; Swanson, B.; Kolody, M.; Sizemore, C.; Bahns, J. Nanograin magnetoresistive manganite coatings for EMI shielding against directed energy pulses. *Prog. Org. Coat.,* **2003**, *47*(3-4), 190-197.
        http://dx.doi.org/10.1016/S0300-9440(03)00138-3

[9]   Wang, Y.; Jing, X. Intrinsically conducting polymers for electromagnetic interference shielding. *Polym. Adv. Technol.,* **2005**, *16*(4), 344-351.
        http://dx.doi.org/10.1002/pat.589

[10]   International EMF Project Progress Report. Geneva, Switzerland: World Health Organization. **2005**, 1-20.

[11]   IARC (International Agency for Research on Cancer). **2002**.

[12]   Navarro, E.A.; Segura, J.; Portoles, M.; Gomez-Perretta, C. The microwave syndrome: a preliminary study in Spain. *Electromagn. Biol. Med.,* **2003**, *22*(2-3), 161-169.
        http://dx.doi.org/10.1081/JBC-120024625

[13]   Khurana, V.G.; Teo, C.; Kundi, M.; Hardell, L.; Carlberg, M. Cell phones and brain tumors: a review including the long-term epidemiologic data. *Surg. Neurol.,* **2009**, *72*(3), 205-214.
        http://dx.doi.org/10.1016/j.surneu.2009.01.019 PMID: 19328536

[14]   Kohli, D.R.; Sachdev, A.; Vats, H.S. Cell phones and tumor: still in no man's land. *Indian J. Cancer,* **2009**, *46*(1), 5-12.
        http://dx.doi.org/10.4103/0019-509X.48589 PMID: 19282560

[15]   Muscat, J.E.; Malkin, M.G.; Thompson, S.; Shore, R.E.; Stellman, S.D.; McRee, D.; Neugut, A.I.; Wynder, E.L. Handheld cellular telephone use and risk of brain cancer. *JAMA,* **2000**, *284*(23), 3001-3007.
        http://dx.doi.org/10.1001/jama.284.23.3001 PMID: 11122586

[16]   Singh, K.; Nagaraj, A.; Yousuf, A.; Ganta, S.; Pareek, S.; Vishnani, P. Effect of electromagnetic radiations from mobile phone base stations on general health and salivary function. *J. Int. Soc. Prev. Community Dent.,* **2016**, *6*(1), 54-59.
        http://dx.doi.org/10.4103/2231-0762.175413 PMID: 27011934

[17]   Hermann, D.M.; Hossmann, K.A. Neurological effects of microwave exposure related to mobile communication. *J. Neurol. Sci.,* **1997**, *152*(1), 1-14.
        http://dx.doi.org/10.1016/S0022-510X(97)00140-8 PMID: 9395121

[18]   Braune, S.; Wrocklage, C.; Raczek, J.; Gailus, T.; Lücking, C.H. Resting blood pressure increase during exposure to a radio-frequency electromagnetic field. *Lancet,* **1998**, *351*(9119), 1857-1858.
        http://dx.doi.org/10.1016/S0140-6736(98)24025-6 PMID: 9652672

[19]   Al-Khlaiwi, T.; Meo, S.A. Association of mobile phone radiation with fatigue, headache, dizziness, tension and sleep disturbance in Saudi population. *Saudi Med. J.,* **2004**, *25*(6), 732-736.
        PMID: 15195201

[20]   Röösli, M.; Hug, K. Wireless communication fields and non-specific symptoms of ill health: a literature review. *Wien. Med. Wochenschr.,* **2011**, *161*(9-10), 240-250.
http://dx.doi.org/10.1007/s10354-011-0883-9 PMID: 21638215

[21]   Tiwari, M. Fly Ash Utilization: A Brief Review in Indian Context. *Int. Res. J. Engg. Tech.,* **2016**, *3*, 949-956.

[22]   Ashoka, D.; Saxena, M.; Asholekar, S.R. Coal Combustion Residue-Environmental Implication and Recycling Potential, Res. *Conserv. Recycl.,* **2005**, *3*, 1342-1355.

[23]   Iordanidis, A.; Buckman, J.; Triantafyllou, A.G.; Asvesta, A. Fly ash- airborne particles from Ptolemais-Kozani area, northern Greece, as determined by ESEM-EDX. *Int. J. Coal Geol.,* **2008**, *73*(1), 63-73.
http://dx.doi.org/10.1016/j.coal.2007.02.007

[24]   Gilmour, M.I.; O'Connor, S.; Dick, C.A.; Miller, C.A.; Linak, W.P. Differential pulmonary inflammation and *in vitro* cytotoxicity of size-fractionated fly ash particles from pulverized coal combustion. *J. Air Waste Manag. Assoc.,* **2004**, *54*(3), 286-295.
http://dx.doi.org/10.1080/10473289.2004.10470906 PMID: 15061611

[25]   Donaldson, K.; Tran, L.; Jimenez, L.; Duffin, R.; Newby, D.; Mills, N.; MacNee, W.; Stone, V. **2005**, Combustion-derived nanoparticles: A review of their toxicology following inhalation exposure. *Part. fib. tox, 2,* 1-14.
http://dx.doi.org/10.1186/1743-8977-2-10

[26]   Alam, J.; Akhtar, M. Fly ash utilization in different sectors in Indian scenario. *Int. J. Emerg. Trends Eng. Dev.,* **2011**, *1*, 1-14.

[27]   Loya, M.I.M.; Rawani, A.M. A review: promising applications for utilization of fly ash. *Int. J. Adv. Technol. Eng. Sci.,* **2014**, *2*, 143-149.

[28]   Basu, M.; Pande, M.; Bhadoria, P.B.S.; Mahapatra, S.C. Potential flyash utilization in agriculture in agriculture: a global review. *Prog. Nat. Sci.,* **2009**, *19*(10), 1173-1186.
http://dx.doi.org/10.1016/j.pnsc.2008.12.006

[29]   Amritphale, S.S.; Anshul, A.; Chandra, N.; Ramakrishnan, N. Development of Celsian Ceramics from Fly Ash Useful for X-Ray Radiation-Shielding Application. *J. Eur. Ceram. Soc.,* **2007**, *27*(16), 4639-4647.
http://dx.doi.org/10.1016/j.jeurceramsoc.2007.03.034

[30]   Bora, P.J.; Vinoy, K.J.; Ramamurthy, P.C.; Kishore, ; Madras, G. Kishore; Madras, G. Lightweight polyaniline-cobalt coated fly ash cenosphere composite film for electromagnetic interference shielding. *Electron. Mater. Lett.,* **2016**, *12*(5), 603-609.
http://dx.doi.org/10.1007/s13391-016-5447-0

[31]   Lu, N.N.; Wang, X.J.; Meng, L.L.; Ding, C.; Liu, W.Q.; Shi, H.L.; Hu, X.S.; Wu, K. Electromagnetic interference shielding effectiveness of magnesium alloy-fly ash composites. *J. Alloys Compd.,* **2015**, *650*, 871-877.
http://dx.doi.org/10.1016/j.jallcom.2015.08.019

[32]   Tiwari, S.; Saxena, M. Use of fly ash in high performance industrial coatings. *Br. Corros. J.,* **1999**, *3*(3), 184-191.
http://dx.doi.org/10.1179/000705999101500824

[33]   Cao, J.; Chung, D. Use of fly ash as an admixture for electromagnetic interference shielding. *Cement Concr. Res.,* **2004**, *34*(10), 1889-1892.
http://dx.doi.org/10.1016/j.cemconres.2004.02.003

[34]   Singh, K.; Singh, S.; Dhaliwal, A.S.; Singh, G. Gamma radiation shielding analysis of lead-flyash concretes. *Appl. Radiat. Isot.,* **2015**, *95*, 174-179.
http://dx.doi.org/10.1016/j.apradiso.2014.10.022 PMID: 25464195

[35]   Singh, K.; Singh, C.; Sidhu, G.S.; Singh, J.; Singh, P.S.; Mudahar, G.S. Flyash: a radiation shielding material. *Indian J. Phys.,* **2003**, *77A*, 41-45.

[36]   Dou, Z.; Wu, G.; Huang, X.; Sun, D.; Jiang, L. Electromagnetic shielding effectiveness of aluminum alloy-fly ash composites. *Compos., Part A Appl. Sci. Manuf.,* **2007**, *38*(1), 186-191.
http://dx.doi.org/10.1016/j.compositesa.2006.01.015

[37]   Budumuru, S.; Anuradha, M.S.; Avinash, B.S.C.; Raj, C.D. Analysis of Shielding Effectiveness on Al6061 Composite Material Reinforced with Fly Ash for Oblique Incidence. *Int. J. Innov. Technol. Explor. Eng.,* **2019**, *9*(2S3), 323-325.
http://dx.doi.org/10.35940/ijitee.B1073.1292S319

[38]   Young, C.S. *Countermeasures to Electromagnetic Signal Compromises, Information Security Science,* 1st ed; Elesvier, **2016**, pp. 185-202.

[39]   Gujral, P.; Varshney, S.; Dhawan, S.K. Designing of Multiphase Fly Ash/MWCNT/PU Composite Sheet against Electromagnetic Environmental Pollution. *J. Electron. Mater.,* **2016**, *45*(6), 3142-3148.
http://dx.doi.org/10.1007/s11664-016-4436-2

[40]   Kumari, D.; Dhawan, S.K.; Agrawal, N.; Varshney, S. Cement paint composite as pollution tracker for electromagnetic radiations. *Mater. Res. Express,* **2018**, *5*(12), 125602-125610.
http://dx.doi.org/10.1088/2053-1591/aae06a

[41]   Raagulan, K.; Kim, B.M.; Chai, K.Y. Recent Advancement of Electromagnetic Interference (EMI) Shielding of Two Dimensional (2D) MXene and Graphene Aerogel Composites. *Nanomaterials (Basel),* **2020**, *10*(4), 1-22.
http://dx.doi.org/10.3390/nano10040702 PMID: 32276331

[42]   Colaneri, N.F.; Shacklette, L.W. EMI Shielding Measurements of Conductive Polymer Blends. *IEEE Trans. Instrum. Meas.,* **1992**, *41*(2), 291-297.
http://dx.doi.org/10.1109/19.137363

[43]   Violette, J.L.N.; White, D.R.J.; Violette, M.F. *Electromagnetic Compatibility Handbook*; Van Nostrand Reinhold Company: New York, **1987**.
http://dx.doi.org/10.1007/978-94-017-7144-3

[44]   Geetha, S.; Kumar, K.K.S.; Rao, C.R.K.; Vijayan, M.; Trivedi, D.C. EMI Shielding: Methods and Materials—A Review. *J. Appl. Polym. Sci.,* **2009**, *112*(4), 2073-2086.
http://dx.doi.org/10.1002/app.29812

[45]   Daniel, S.; Thomas, S. *Shielding Efficiency Measuring Methods and Systems, Advanced Materials for Electromagnetic Shielding: Fundamentals, Properties and applications*; Wiley, **2018**, pp. 71-73.

[46]   ASTM International. Standard Test Method for Measuring the Electromagnetic Shielding Effectiveness of Planar Materials, ASTM D 4935-99, West Conshohocken: PA **1999**.

[47]   Vasquez, H.; Espinoza, L.; Lozano, K.; Foltz, H.; Yang, S. Simple Device for Electromagnetic Interference Shielding Effectiveness Measurement. *IEEE EMC Soc. Newslett.,* **2009**, *220*, 62-68.

[48]   Lin, H-N.; Chen, Y-Y.; Tsai, H-Y.; Lin, M-S. Characteristic Analysis and Applications of Electromagnetic Shielding Materials for Wireless Communications Device. *Open Mater. Sci. J.,* **2016**, *10*(1), 44-53.
http://dx.doi.org/10.2174/1874088X01610010044

[49]   Chen, C.; Lee, K.C.; Lin, J.H.; Koch, M. Comparison of electromagnetic shielding effectiveness properties of diverse conductive textiles *via* various measurement techniques. *J. Mater. Process. Technol.,* **2007**, *192–193*, 549-554.
http://dx.doi.org/10.1016/j.jmatprotec.2007.04.023

[50]   Sushmita, K.; Madras, G.; Bose, S. The journey of polycarbonate-based composites towards suppressing electromagnetic radiation. *Funct. Compost. Mat.,* **2021**, *2*(13), 1-38.
http://dx.doi.org/10.1186/s42252-021-00025-1

[51]   Huang, X.; Chen, Z.; Tong, L.; Feng, M.; Pu, Z.; Liu, X. Preparation and microwave absorption properties of $BaTiO_3$@MWCNTs core/shell heterostructure. *Mater. Lett.,* **2013**, *111*, 24-27.
http://dx.doi.org/10.1016/j.matlet.2013.08.034

[52]   Yang, Y.; Gupta, M.C.; Dudley, K.L.; Lawrence, R.W. Conductive carbon nanofiber-polymer foam structures. *Adv. Mater.,* **2005**, *17*(16), 1999-2003.
http://dx.doi.org/10.1002/adma.200500615

[53]   Varshney, S.; Ohlan, A.; Jain, V.K.; Dutta, V.P.; Dhawan, S.K. *In situ* synthesis of polypyrrole–γ-Fe_2O_3-fly ash nanocomposites for protection against EMI pollution. *Ind. Eng. Chem. Res.,* **2014**, *53*(37), 14282-14290.
http://dx.doi.org/10.1021/ie500512d

[54]   Pattanaik, A.; Bhuyan, S.K.; Samal, S.K.; Behera, A.; Mishra, S.C. Dielectric properties of epoxy resin fly ash composite. *IOP Conf. Series Mater. Sci. Eng.,* **2016**, *115*, 012003.
http://dx.doi.org/10.1088/1757-899X/115/1/012003

[55]   Srivastava, A.; Krishan, A. A Brief Review on Fly Ash Utilization. *Int. J. Sci. Res. Sci. Eng. Technol.,* **2017**, *3*, 388-396.

[56]   Bora, P.J.; Vinoy, K.J.; Ramamurthy, P.C.; Madras, G. Lightweight polyaniline-cobalt coated fly ash cenosphere composite film for electromagnetic interference shielding. *Electron. Mater. Lett.,* **2016**, *12*(5), 603-609.
http://dx.doi.org/10.1007/s13391-016-5447-0

[57]   Gulzar, G.; Zubair, K.; Shakir, M.F.; Zahid, M.; Nawab, Y.; Rehan, Z.A. Effect on the EMI Shielding Properties of Cobalt Ferrites and Coal-Fly-Ash Based Polymer Nanocomposites. *J. Supercond. Nov. Magn.,* **2020**, *33*(11), 3519-3524.
http://dx.doi.org/10.1007/s10948-020-05608-w

[58]   Poonguzhalia, S.; Gopalakrishnanb, K.; Balajiprasadc, M.; Chandrasekarand, J.; Babu, B. Electrical conductivity studies of PANI-flyash blends. *J. Elec. Elec. Eng.,* **2014**, *9*, 67-70.

[59]   Yinsuo, D.; Jianhua, W.; Derong, W.; Chunhua, L.; Zhongzi, X. Effect of Mineralogical Phase and Chemical Composition of Fly Ash on Electromagnetic Wave-Absorbing Properties. *Mater. Trans.,* **2018**, *5*, 876-882.

[60]   Yubin, H.; Jueshi, Q.; Jianye, Z. Influence of high-iron fly ash on absorption property of cement based materials. *J. Funct. Mater.,* **2009**, *11*, 1787-1790. [In Chinese].

[61]   Baoyi, L.; Yuping, D.; Shunhua, L. The electromagnetic characteristics of fly ash and absorbing properties of cement-based composites using fly ash as cement replacement. *Constr. Build. Mater.,* **2012**, *27*(1), 184-188.
       http://dx.doi.org/10.1016/j.conbuildmat.2011.07.062

[62]   Bora, P.J.; Vinoy, K.J.; Ramamurthy, P.C.; Kishore, ; Madras, G. Kishore; Madras, G. Electromagnetic interference shielding effectiveness of polyaniline-nickel oxide coated cenosphere composite film. *Compos. Commun.,* **2017**, *4*, 37-42.
       http://dx.doi.org/10.1016/j.coco.2017.04.002

[63]   Nam, I.W.; Lee, H.K. Synergistic effect of MWNT/fly ash incorporation on the EMI shielding/absorbing characteristics of cementitious materials. *Constr. Build. Mater.,* **2016**, *115*, 651-661.
       http://dx.doi.org/10.1016/j.conbuildmat.2016.04.082

[64]   Cornell, R.M.; Schertmann, U. *Iron Oxides in the Laboratory Preparation and Characterization,* 2nd ed; Wiley: New York, **2000**.

[65]   Qiao, R.; Yang, C.; Gao, M. Superparamagnetic iron oxide nanoparticles: from preparations to *in vivo* MRI applications. *J. Mater. Chem.,* **2009**, *19*(35), 6274-6293.
       http://dx.doi.org/10.1039/b902394a

[66]   Li, Y.B.; Chen, G.; Li, Q.H.; Qiu, G.Z.; Liu, X.H. Facile synthesis, magnetic and microwave absorption properties of Fe_3O_4/polypyrrole core/shell nanocomposite. *J. Alloys Compd.,* **2011**, *509*(10), 4104-4107.
       http://dx.doi.org/10.1016/j.jallcom.2010.12.100

[67]   Cui, C.; Du, Y.; Li, T.; Zheng, X.; Wang, X.; Han, X.; Xu, P. Synthesis of electromagnetic functionalized Fe_3O_4 microspheres/polyaniline composites by two-step oxidative polymerization. *J. Phys. Chem. B,* **2012**, *116*(31), 9523-9531.
       http://dx.doi.org/10.1021/jp3024099 PMID: 22800337

[68]   Nicolson, A.M.; Ross, G.F. Measurement of the intrinsic properties of materials by time domain techniques. *IEEE Trans. Instrum. Meas.,* **1970**, *19*(4), 377-382.
       http://dx.doi.org/10.1109/TIM.1970.4313932

[69]   Weir, W.B. Automatic measurement of complex dielectric constant and permeability at microwave frequencies. *Proc. IEEE,* **1974**, *6*(1), 33-36.
       http://dx.doi.org/10.1109/PROC.1974.9382

[70]   Shukla, V. Review of electromagnetic interference shielding materials fabricated by iron ingredients. *Nanoscale Adv.,* **2019**, *1*(5), 1640-1671.
       http://dx.doi.org/10.1039/C9NA00108E

[71]   Li, M.; Huang, X.; Wu, C.; Xu, H.; Jiang, P.; Tanaka, T. Fabrication of two-dimensional hybrid sheets by decorating insulating PANI on reduced graphene oxide for polymer

nanocomposites with low dielectric loss and high dielectric constant. *J. Mater. Chem.,* **2012**, *22*(44), 23477-23484.

http://dx.doi.org/10.1039/c2jm34683d

[72]   Ameli, A.; Nofar, M.; Park, C.B.; Pötschke, P.; Rizvi, G. Polypropylene/carbon nanotube nano/microcellular structures with high dielectric permittivity, low dielectric loss, and low percolation threshold. *Carbon,* **2014**, *71*, 206-217.

http://dx.doi.org/10.1016/j.carbon.2014.01.031

[73]   Yang, Y.; Zhang, B.; Xu, W.; Shi, Y.; Zhou, N.; Lu, H. Microwave absorption studies of W-hexaferrite prepared by co-precipitation/mechanical milling. *J. Magn. Magn. Mater.,* **2003**, *265*(2), 119-122.

http://dx.doi.org/10.1016/S0304-8853(03)00237-3

[74]   Stonier, R.A. Stealth aircraft and technology from world war- II the gulf. *Sampe J.,* **1991**, *27*, 9-17.

[75]   Wang, C.; Han, X.; Xu, P.; Wang, J.; Du, Y.; Wang, X.; Qin, W.; Zhang, T. Controlled Synthesis of Hierarchical Nickel and Morphology-Dependent Electromagnetic Properties. *J. Phys. Chem. C,* **2010**, *114*(7), 3196-3203.

http://dx.doi.org/10.1021/jp908839r

[76]   Varshney, S.; Singh, K.; Ohlan, A.; Jain, V.K.; Dutta, V.P.; Dhawan, S.K. Synthesis of ferrofluid based nanoarchitectured polypyrrole composites and its application for electromagnetic shielding. *Mater. Chem. Phys.,* **2014**, *143*(2), 806-813.

http://dx.doi.org/10.1016/j.matchemphys.2013.10.018

[77]   Wan, M. *Conducting Polymers with Micro or Nanometer Structure*; Springer Berlin Heidelberg: New York, **2008**.

[78]   Gupta, T.K.; Singh, B.P.; Teotia, S.; Katyal, V.; Dhakate, S.R.; Mathur, R.B. Designing of multiwalled carbon nanotubes reinforced polyurethane composites as electromagnetic interference shielding materials. *J. Polym. Res.,* **2013**, *20*(6), 1-7.

http://dx.doi.org/10.1007/s10965-013-0169-6

[79]   Mathur, R.B.; Chatterjee, S.; Singh, B.P. Growth of carbon nanotubes on carbon fibre substrates to produce hybrid/phenolic composites with improved mechanical properties. *Compos. Sci. Technol.,* **2008**, *67*(7-8), 1608-1615.

http://dx.doi.org/10.1016/j.compscitech.2008.02.020

[80]   Wang, X.; Du, Z.; Zhang, C.; Li, C.; Yang, X.; Li, H. Multi-walled carbon nanotubes encapsulated with polyurethane and its nanocomposites. J. Polym. Sci: Part A. *J. Polym. Sci. A Polym. Chem.,* **2008**, *46*(14), 4857-4865.

http://dx.doi.org/10.1002/pola.22818

[81]   Gupta, T.K.; Singh, B.P.; Dhakate, S.R.; Singh, V.N.; Mathur, R.B. Improved nanoindentation and microwave shielding properties of modified MWCNT reinforced polyurethane composites. *J. Mater. Chem. A Mater. Energy Sustain.,* **2013**, *1*(32), 9138-9149.

http://dx.doi.org/10.1039/c3ta11611e

[82]   Kutchko, B.G.; Kim, A.G. Fly ash characterization by SEM-EDS. *Fuel,* **2006**, *85*(17-18), 2537-2544.

http://dx.doi.org/10.1016/j.fuel.2006.05.016

[83]   Rojas, J.V.; Toro-Gonzalez, M.; Molina-Higgins, M.C.; Castano, C.E. Facile radiolytic synthesis of ruthenium nanoparticles on graphene oxide and carbon nanotubes. *Mater. Sci. Eng. B,* **2016**, *205*, 28-35.
        http://dx.doi.org/10.1016/j.mseb.2015.12.005

[84]   Jianhong, W.; Liu, Z.; Shi, J.; Pan, C. Synthesis of polyaniline-Fe_3O_4 nanocomposites and their conductivity and magnetic properties. J. Wuhan. *Univ. Technol.,* **2010**, *25*, 760-764.

[85]   Spies, B.R. Depth of investigation in electromagnetic sounding methods. *Geophysics,* **1989**, *54*(7), 872-888.
        http://dx.doi.org/10.1190/1.1442716

[86]   Gill, N.; Gupta, V.; Tomar, M.; Sharma, A.L.; Pandey, O.P.; Singh, D.P. Tomar, Sharma, A. L.; Pandey, O.P.; Singh, D. P. Improved electromagnetic shielding behaviour of graphene encapsulated polypyrrole-graphene nanocomposite in X-band. *Compos. Sci. Technol.,* **2020**, *192*, 108113.
        http://dx.doi.org/10.1016/j.compscitech.2020.108113

[87]   Chen, W.; Wang, J.; Zhang, B.; Wu, Q.; Su, X. Enhanced electromagnetic interference shielding properties of carbon fiber veil/Fe_3O_4 nanoparticles/epoxy multiscale composites. *Mater. Res. Express,* **2017**, *4*(12), 126303.
        http://dx.doi.org/10.1088/2053-1591/aa9af9

CHAPTER 8

Fabrication and Microwave Shielding Properties of Free-Standing Conducting Polymer-Carbon Fiber Thin Sheets

Rakesh Kumar[1]* and S K Dhawan[2]

[1] *Department of Applied Sciences, Maharaja Surajmal Institute of Technology (Affiliated to GGSIP University), Janakpuri, Delhi – 110058, India*

[2] *Advanced Materials & Devices Metrology Division, CSIR-National Physical Laboratory, New Delhi – 110012, India*

Abstract: EMI is a 20th-century radiation pollution that not only results in various health hazards but also weakens the electronic system's performance. With the rapid global development in various fields, this problem is increasing consistently. To ensure the uninterrupted performance of electronic gadgets and avoid the effects on human health, EMI shielding has become a necessity. In the recent past, a large number of materials having a wide range of conductivity and good electromagnetic attributes have been exploited for EMI shielding applications. Initially used metallic shields, due to their high cost & weight, corrosion propensity, and reflection-based shielding, have been replaced by various types of materials. Among them, intrinsically conducting polymers (ICPs) like polyaniline, polythiophene, polypyrrole, *etc.*, and their composites with various types of conductive and/or magnetic fillers have played a significant role. Among all the conducting polymers, polyaniline has been studied the most due to its special properties like moderately high conductivity, ease of synthesis, proton doping, low cost, and high environmental stability. Most of the developments related to EMI shielding have been focused on the synthesis of new materials with high shielding effectiveness (SE). For this purpose, polyaniline and its composites have been widely explored due to its appropriate properties. But the commercial use of polyaniline for EMI shielding applications has always been hampered due to its infusibility and limited

*Corresponding author Rakesh Kumar:** Department of Applied Sciences, Maharaja Surajmal Institute of Technology (Affiliated to GGSIP University), Janakpuri, Delhi–110058, India; Tel: +91 9416270673; E-mails: rakeshchikara@gmail.com, rakeshkumar@msit.in

Sundeep K. Dhawan, Avanish Pratap Singh, Anil Ohlan, Kuldeep Singh Kakran and Pradeep Sambyal (Ed.)
All rights reserved-© 2022 Bentham Science Publishers

processability. Also, limited work has been done for the fabrication of polyaniline composites in the form of sheets that have sufficient SE along with improved thermal and mechanical stability. The work presented in this chapter is based on the fabrication of lightweight, thin sheets of polyaniline composites for EMI shielding application in the X-band of microwave range (8.2-12.4 GHz). The polyaniline-CF-novolac (PACN) composite sheets thus obtained were finally tested for EMI shielding applications using vector network analyzer (VNA) in the X-band of microwave range. Characterization of all the composites and/or their sheets was done by UV-*vis*, FT-IR, SEM, TGA, electrical conductivity (standard four-probe method), flexural strength, and flexural modulus measurements.

Keywords: Carbon fibers, Conducting polymers, EMI shielding, Polyaniline, Shielding Effective ness.

8.1. INTRODUCTION

Polymers are high molecular weight compounds formed by the interlinking of monomers during polymerization. The properties of the polymers depend on the type of chemical bonding, type of elements, and functional groups present in the polymer chain. For instance, polymers may be saturated or unsaturated, which affects the properties, including the conductivity. Due to their better and excellent properties, polymers have replaced traditional materials such as wood, metals, ceramics, *etc.* Polymers are versatile materials, with uses in several areas, as often remarked 'from buckets to rockets.' Formerly plastics/polymers were considered insulators. Later, the plastics under certain conditions were made to act like inorganic semiconductors and even as conductors or superconductors, like metals, which led to the discovery of conducting polymers [1]. Apart from domestic applications like simple utensils, polymers find their uses in electronics and communication systems, optical devices, energy storage devices, smart conjugated coatings, sensors, microwave shielding, *etc.* Before knowing about the synthesis, characterization, conduction mechanism, and applications of conducting polymers, it is first necessary to know about the journey of these materials from insulators to conductors.

8.1.1.Conducting Polymers

All the polymers in which charge can propagate while exhibiting conductivity are known as electrochemically active polymers. Depending on the type of charge carriers and their propagation, electrochemically active polymers can be categorized as (a) ionically conducting polymers, (b) redox polymers, and (c) intrinsically conducting polymers (ICPs).

When we talk about conducting polymers, it is assumed that we are discussing ICPs, because the majority of the conducting polymers are ICPs. There are very few examples of redox polymers or ionically conducting polymers. So, normally conducting polymers means ICPs. ICPs are conjugated polymers having alternate single and double bonds in their chain backbone [2]. ICPs consist of π-electrons on alternate carbon atoms in their chain backbone. Overlapping their wave function leads to conjugation and delocalization of unpaired electrons along the polymeric chain [3]. Hence, all the ICPs have π-conjugated delocalized systems. The π-conjugation of the polymer chain generates high-energy-occupied molecular orbitals and low-energy unoccupied molecular orbitals, leading to a system that can be readily oxidized or reduced [4], so the system has low ionization potentials and high electron affinities. The combination of all these factors impart special electrical properties to ICPs and provide them the capability to support positive and negative charge carriers with high mobility along the entire polymer chain [5-7].

The band gap of undoped conjugated polymers lies between 1 to 4 eV, so they are either insulators or semiconductors having low conductivity (10^{-10} to 10^{-5} S/cm) at room temperature. After doping, the conductivity of ICPs can increase many folds [8]. The conductivity of ICPs after doping reaches the 'metallic' regime (1 to 10^4 S/cm). Doping converts the polymer into highly delocalized polycations or polyanion by the introduction of an oxidizing agent, a reducing agent, or a protonic acid [9-13]. The conductivity of ICPs can be tuned within a wide range by chemical manipulation because it depends on different factors like synthesis method & conditions, type of oxidizing/reducing agent or protonic acid used, type and nature of the dopant used, the level of doping, *etc*. Fig. (**1**) shows the chemical structure of some ICPs in their neutral insulating state.

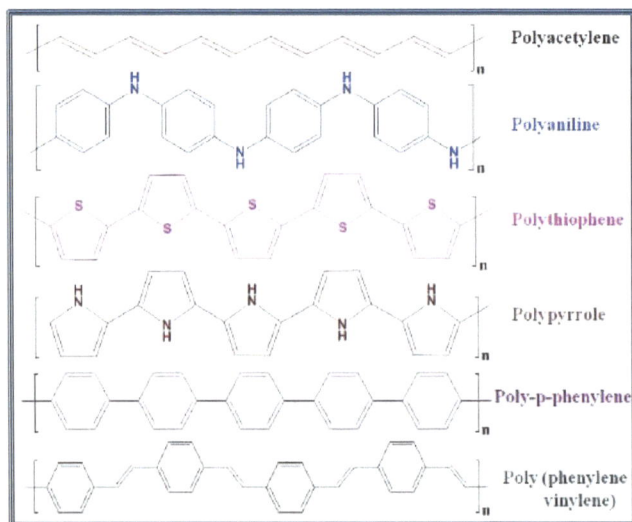

Fig. (1). Chemical structure of some undoped ICPs.

8.1.2. Polyaniline and its Applications

Polyaniline, which is the oxidation product of aniline, has been known as aniline black since 1962 [14]. Later, the proton exchange and redox attributes with the influence of water [15] and H_2SO_4 [16] were also reported. However, much curiosity in PA and other ICPs was developed after 1977 with the fundamental discovery of iodine-doped polyacetylene having metal-like conductivity [1, 17]. Polyaniline, which has –NH– group on both sides of the phenylene ring, is made up of oxidized (-B-N=Q=N-) and reduced (-B-NH-B-NH-) repeated units, where Q denotes quinoid ring & B denotes benzenoid ring. Hence, different forms of PA obtained with a number of imine and amine segments on the polymer chain due to oxidation and reduction take place on the –NH– group. Depending upon the ratio of amine to imine, different forms of PA (Fig. **2**) are formed in the reaction mixture, such as leucoemeraldine (reduced form, y = 1), emeraldine base (50 % oxidized form, y = 0.5), and pernigraniline (fully oxidized form, y = 0). Emeraldine base is considered the most useful form of polyaniline because upon acid doping, it turns into an emeraldine salt (Fig. **2**), having high conductivity and high stability at room temperature [18]. The pernigraniline and leucoemeraldine forms of polyaniline are poor conductors of electricity even after doping with an acid. The oxidizing agent plays an important role in achieving the desired structure and electrical properties.

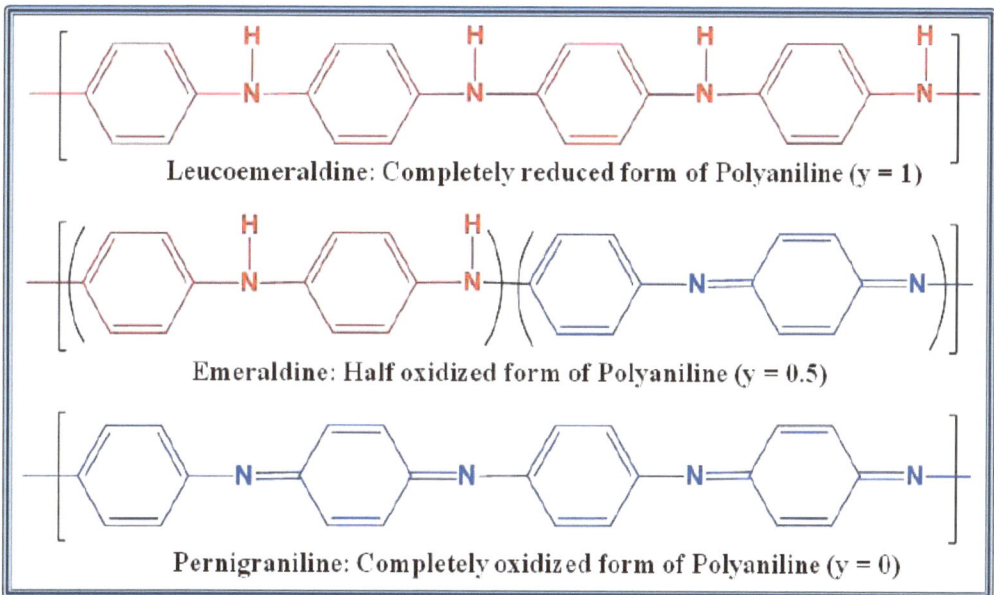

Fig. (2). Different forms of polyaniline.

Polyaniline is the most widely used conducting polymer due to the combination of its superior properties such as low cost, easy synthesis [19], special proton doping mechanism [3], moderately high conductivity after doping, and excellent environmental stability [20-30]. Among all developments, the techniques found to be more advantageous and which are in common practice are derivatization, copolymerization, and preparations of blends and composites of ICPs with insulating polymers. Substitution can be done with hydrogen attached to the benzene ring (ring substitution) or at nitrogen (N-substitution) by an alkyl, aryl, alkoxy, hydroxy, amino, sulfonic acid, or halogen group [25, 31-35] Alkyl and alkoxy ring-substituted polyanilines show better optical and electrochemical properties and have attracted much attention [35, 36]. The general structure of substituted anilines along with some of their derivatives are shown in Fig. (**3**).

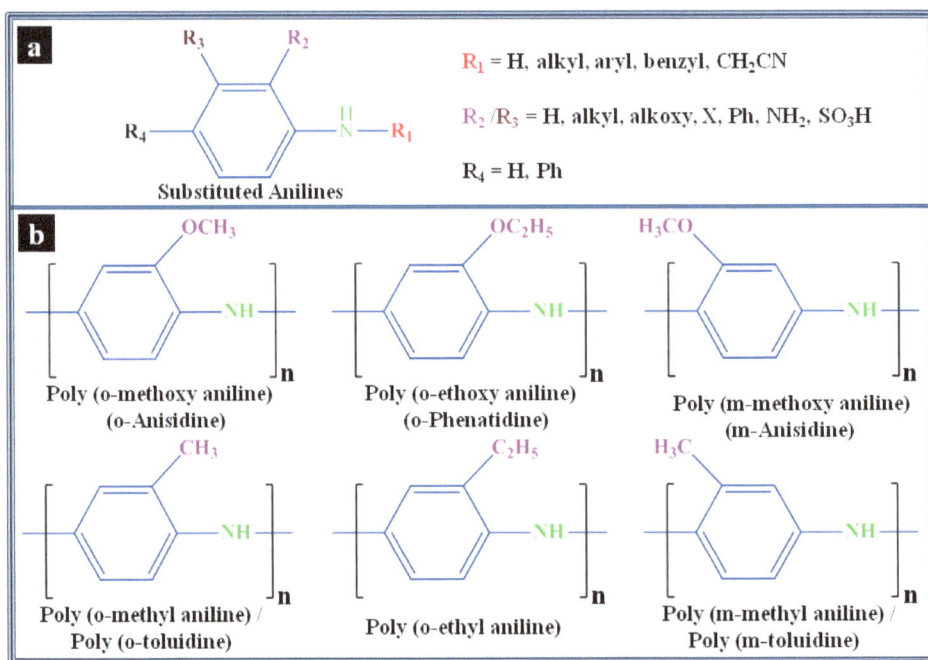

Fig. (3). Structure of (**a**) substituted anilines, (**b**) alkoxy and alkyl-substituted polyanilines.

Incorporation of substituents improves solubility, processability, and electrochemical stability of the polymer, but the electrical conductivity decreases due to a reduction in π-conjugation [37, 38]. Therefore, research work is now progressing to solve the processability and solubility issue without compromising with the electrical and physio-chemical properties. Compared to the derivatives of polyaniline, copolymers of polyaniline with some ICPs (including the substituted aniline) are more advantageous in terms of improvement in physical and mechanical properties. However, polyaniline copolymers and blends/composites of polyaniline with insulting polymers do not have enough conductivity to be useful in applications that need high conductivity, such as EMI shielding, microwave absorption, conductive fabrics, *etc.*To meet the desired properties for the targeted application, some other parameters have also been taken into consideration, *e.g.*, synthesis methods & conditions, incorporation of conductive fillers into PA matrix and fabrication of the resultant composites in the form of EMI shields. A lot of research has been carried out to improve the properties of polyanilines prepared by emulsion polymerization using different surfactants [39-72]. Various reports have shown that DBSA, NSA, and SLS/SDS acts as good surfactants because all these form nanoparticles of the polymer have good morphology [61,73, 74] and the good electrical conductivity and good solubility, as well as processability of PA, have been observed. Palaniappan *et al*. [75] have also reported that the efficiency of APS is increased by the use of acid and surfactant and observed that improved conductivity of PA is obtained when the polymer is prepared by emulsion polymerization pathway as compared to simple aqueous polymerization. The kinetics, particle size, and conductivity of PA synthesized by conventional solution polymerization method and by emulsion polymerization using SDS as a surfactant were compared by Kim *et al*. [76]. It is reported that the PA particles prepared *via* aqueous polymerization showed irregular shapes, whereas these particles showed spherical shape (particle size of 25–60 nm in diameter) when prepared via emulsion polymerization pathway using SDS as a surfactant [76].

ICPs have some unique properties like nonlinear optical properties, tunable conductivity within the range of conductor to the metallic regime, special doping/de-doping mechanism, electrical redox reversibility, *etc*. Due to these special properties conducting polymers are used for technological applications. Due to their high conductivity, conducting polymers are excellent candidates for EMI shielding, microwave absorption, and as conductive textiles. Due to semiconducting behaviour, they are used in electronic devices like Schottky rectifiers, field-effect transistors, organic light-emitting diodes (OLED) and solar cells. They find their applications in rechargeable batteries and supercapacitors due to high conductivity in combination with the reversible redox potential. Due to moderate conductivity with reversible doping/de-doping process, these can be used in chemical or biochemical sensors, gas separation membranes, and drug delivery agents. Electrochromism provoked by the electrochemical doping/de-doping facilitates their use in electrochromic smart windows and electrochromic displays. properties and corresponding applications of ICPs are illustrated in Fig. (**4**).

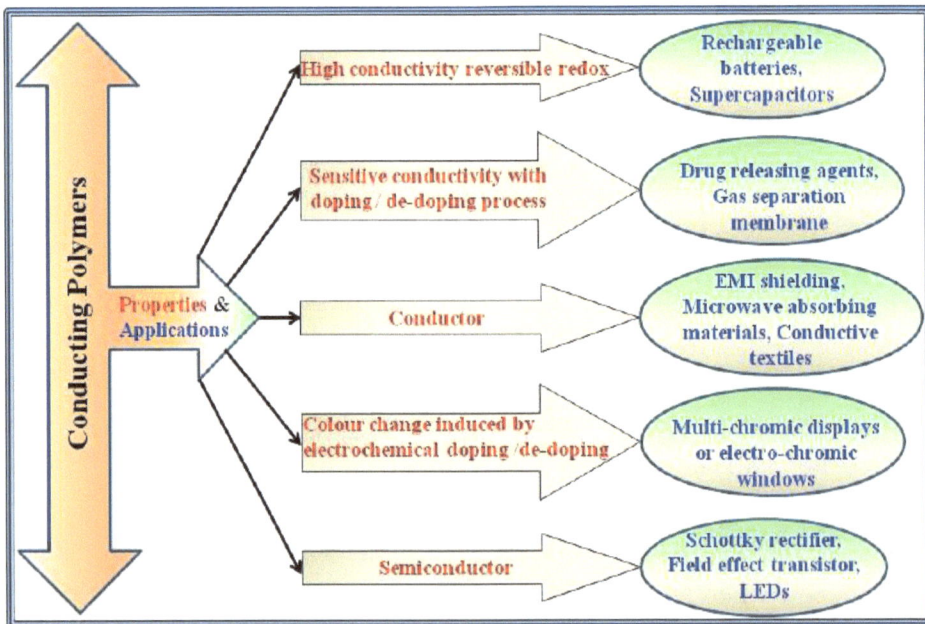

Fig. (4). Properties and corresponding applications of ICPs.

PA and its derivatives have been widely used in rechargeable batteries [77], EMI shielding [68, 78], microwave and radar absorbing materials [30, 79], OLED [80], Schottky diodes [81], electron field emitters [82], field-effect transistors [83], sensors and indicators [84-86], erasable optical information storage devices [87], digital memory devices [88], supercapacitors [89], electrochromic devices [90], electromechanical actuators [91], catalysts [92], asymmetric films [93], membranes [94], antistatic and anticorrosion coatings [95, 96], solar cells [97] *etc.*

8.2. ELECTROMAGNETIC INTERFERENCE (EMI) SHIELDING

Electromagnetic interference (EMI) is a modern type of environmental radiation pollution called electromagnetic pollution. EMI is the consequence of extensive emission of electromagnetic (EM) radiations in the environment due to the rapid growth in the field of electronics and telecommunication, wireless systems, navigation systems, military and aircraft technology [98]. Lots of redundant and unnecessary radiated signals are discharged from an external source carrying transient currents, which create disturbance in the circuits of various electronic devices [99]. Every electronic gadget is prone to EMI, and hence the electronic equipment becomes incompatible with each other. The most common occurrence of EMI is in electronic equipment working in high energy radiations (radio frequency and microwave), *e.g.*, mobile communications, digital devices, remote sensing, satellite communications, and military equipment and radar surveillance systems [100, 101], which disturb and affects their safety operations. Consequences of EMI can be experienced all around us, *e.g.*, "ghosts" in TV picture reception, radio interference of taxicab with police cab, power line transient interference with personal computers, disturbance or interruption in mobile communication equipment, medical equipment, military and aircraft devices, *etc*. EMI can perturb the working of sensitive components, degradation of equipment performance and loss of data, explosions, and accidents [102]. EMI is also harmful to human health as it can cause nervousness, languidness, insomnia, headaches, and even cancers [98]. Studies have shown that extreme use of mobiles can cause brain cancer [103]. To avoid the problem of EMI, the electronic equipment is made to be compatible with each other as well as with the environment. This compatibility is called electromagnetic compatibility (EMC). So, EMC is actually the solution to this novel kind of pollution (EMI), which can be achieved by the shielding of both electronic and radiation sources. This method of controlling electromagnetic interference shielding is called EMI shielding. EMI shielding is of critical use not only due to the reliability of electronics [104] but also due to health concerns. Shielding can be attained either by confining the radiated energy within the specific region or by preventing the entry of radiated energy into a specific region. EMI shielding may be pursued by any of the following main approaches or by a combination of them:

i.   Interposition of a shield between the source and the receiver.

ii. Diversion of the EM field away from the area of interest.

iii. Reduction of the EM field level of the original source by the introduction of an additional source.

Among all the approaches, the first one is the most efficient approach to reduce the EM field levels. A shield is a barrier to control the transmission of the EM wave across its bulk. A shield prevents the equipment by controlling the penetration of EM waves coming from an outside source and equally prevents the outside electronic items or living organisms that are susceptible to the EM waves emitted from the equipment [105]. A certain level of attenuation can be extended by using a deliberately devised EM shield. Hence, a shield acts as an enclosure for equipment or for a system/particular area that regulates the transmission of EM radiation.

8.2.1. Theory of EMI Shielding and Microwave Absorption

When a thin, conductive, and spherical shell/shield encloses any sensitive equipment which is placed in an electric field (E-field), it will be shielded because the current of electromagnetic waves cannot be conducted inside of the shell. This is primarily due to the development of electronic charges of different polarity by the E-field along with the spherical shell, and the electrical field produced by these charges has a tendency to abandon the effect of the original field inside the shield [106]. In this case, a very thin shield can be very effective if the wave frequency is high enough. Here, electromagnetic current acquires a least resistive path and flows through the periphery of the conductive shield. But when it comes to H-fields, then the H-field intensity will be reduced by a sufficiently thick shield having magnetic material of good permeability. Because the magnetic material provides a low-reluctance path, the H-field tends to stay in the layer of magnetic material [62, 64]. However, a thin shield made up of conductive material with low permeability can also perform effective shielding for H-fields at high frequencies. Because the shield has sufficient conductivity, it will itself induce eddy currents in the shield screen, which will generate an alternating H-field of the opposite orientation inside the shield. This effect increases with the increase of frequency of EM waves, so the shielding effectiveness will be high at high frequencies. Thus, shielding against low-frequency H-fields is relatively difficult and requires thick shields made up of expensive magnetic materials. However, conductive shields based on the principle of induced eddy currents may be rationally effective at power line frequencies [106].

8.2.1.1. Shielding Effectiveness

Shielding effectiveness (SE) measures the shielding efficiency of a shield which is the ratio of the magnitude of the transmitted electric (magnetic) field to the magnitude of the incident electric (magnetic) field for a barrier. The SE of a shield relies on frequency, EMI attenuation, the distance between shield and source, thickness, design, and nature of the shield. SE is usually expressed in decibels (dB) as a function of the logarithm of the ratio of the intensity of incident and transmitted plane wave (P), electric (E), or magnetic (H) field as follows:

$$SE\ (dB) = -10\ log\ \{P_T/P_I\} = -20\ log\ \{E_T/E_I\} = -20\ log\ \{H_T/H_I\} \qquad (8.3)$$

In any type of EMI, the total SE of a shield is due to three mechanisms. A part of the incident radiation is absorbed inside the shield material, a part is reflected from the front surface of the shield, and a part is reflected from the rear surface of the shield to the front surface, as shown in Fig. (5). The third type of phenomenon, depending on its phase relationship with the incident wave, can assist or hamper the effectiveness of the shield. Hence, the total SE is the sum of the SE due to absorption factor (SE_A), the reflection factor (SE_R), and the correction factor to account for multiple reflections (SE_M) in thin shields [107];

$$SE = SE_A + SE_R + SE_M \qquad (8.4)$$

The multiple reflection factor SE_M can be neglected if the absorption loss (SE_A) is greater than 10 dB. In practical calculations, SE_M can also be neglected for electric fields and plane waves, where the shield is thicker than the skin depth (δ). For a shielding material, the skin depth (δ) is the distance up to where the intensity of the EM wave decreases by 1/e of its original strength.

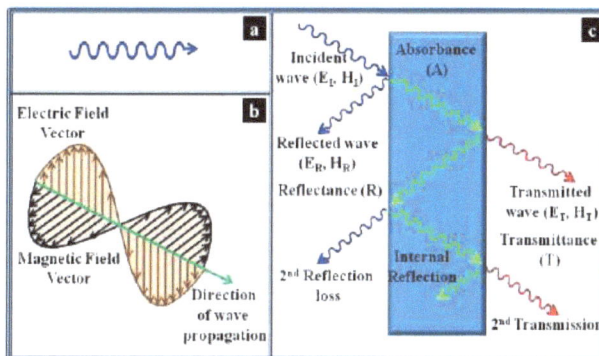

Fig. (5). Graphical representation of (**a**) EM wave, (**b**) Electric & magnetic field vector perpendicular to the direction of wave propagation, and (**c**) Mechanism of EMI shielding.

8.2.1.2. Absorption Loss (SE$_A$)

SE$_A$ is due to the physical characteristics of the shield and is not dependent on the field of the source, so it is the same for all three waves [108]. The amplitude of an EM wave decreases exponentially when it passes through a medium. This is called decay or absorption loss, which occurs due to the ohmic losses and heating of the material carried out by the currents induced in the medium. E$_T$ and H$_T$ can be expressed as $E_T = E_I e^{-t/\delta}$ and $H_T = H_I e^{-t/\delta}$ [109]. When the wave is attenuated up to the distance of 1/e or 37 %, then it is called the skin depth. Hence, the SE$_A$ in dB is given as:

$$SE_A = 20 \, (t/\delta) \log e = 8.69 \, (t/\delta) = 131 \, t\sqrt{f\mu\sigma} \tag{8.5}$$

where t is the thickness of the shield in mm; f is the frequency of the wave in MHz; μ is the relative permeability (1 for copper); σ is conductivity relative to copper, and δ is the skin depth.

$$\delta = 1/\sqrt{\pi f \mu \sigma} \tag{8.6}$$

About 9 dB of absorption loss is due to skin depth in a shield. Skin effect is particularly important at low frequencies, where the H-fields are predominant having a lower wave impedance than 377 Ω. From the viewpoint of absorption loss, the shield should have high conductivity, high permeability, and sufficient thickness to attain the required number of skin depths at the lowest frequency [108].

8.2.1.3. Reflection Loss (SE$_R$)

SE$_R$ is associated with the relative mismatch of the incident wave and the surface impedance of the shield. The computation of SE$_R$ can mainly be simplified by considering SE for incident electric fields as a separate problem from that of electric, magnetic, or plane waves. These three principle fields are expressed as [108, 109]:

$$R_E = K_1 \, 10 log \left({\sigma}/{f^3 r^2 \mu} \right) \tag{8.7}$$

$$R_H = K_2 \, 10 log \left({f r^2 \sigma}/{\mu} \right) \tag{8.8}$$

$$R_P = K_3 \, 10log\left(\frac{f\mu}{\sigma}\right) \tag{8.9}$$

where R_E, R_H, and R_P in dB are the reflection losses for the electric, magnetic, and plane wave fields, respectively, σ is the relative conductivity referred to copper, f is the frequency in Hz, μ is the relative permeability referred to free space, r is the distance from the source to the shielding in meter.

8.2.1.4. Multiple Reflection Correction Factors (SE$_M$)

Mathematically the SE$_M$ can be positive or negative (in practice, it is always negative) and becomes insignificant when SE$_A$> 6 dB [108]. It is only important when metals are thin and at low frequencies (*i.e.*, below approximately 20 kHz). SE$_M$ can be expressed as:

$$SE_M = -20 \, log \, (1 - e^{-2t/\delta}) \tag{8.10}$$

From the shielding theory, it is clear that in order to attain efficient shielding, the shield should possess proper physical geometry, sufficient electrical conductivity (σ), and/or good dielectric properties [110]. For reflection, which is the primary mechanism of EMI shielding, the shield must possess mobile charge carriers (electrons or holes) that can interact with the electromagnetic fields to cause ohmic (heating) losses in the shield, although only moderate conductivity (10^{-3} to 1 S/cm) is sufficient [30, 111]. For absorption, which is the secondary EMI shielding mechanism, the shield should possess electric and/or magnetic dipoles which can interact with the electromagnetic fields in the radiation.

8.3. I.CPs and their Composites as Shielding Materials

Initially, the metallic shields in the forms of metallic sheets, meshes, coatings, *etc.*, were used as conventional shielding materials and showed excellent SE [112]. But now, metallic shields have become obsolete due to their high cost, weight penalty, corrosion susceptibility, and difficult processability [113]. Besides this, the shielding process by metal shields occurred through reflection, so the electromagnetic pollution is not completely mitigated. Hence, in the last two decades, lots of research has been focused on the design and development of shields that work by absorption phenomenon. In this direction, the shields based on polymeric materials are highly admired due to their lightweight, low cost, easy shaping, and corrosion resistance. Initially, the metallic shields were replaced by shields made up of conductive polymer composites. These composites are a physical mixture of conventional insulating polymers and conductive fillers such as metallic fillers, metal-coated glass fiber, carbon powder, carbon fiber (CF), graphene, carbon nanotubes (CNTs), *etc*. [29, 114]. But the conductivity of these composites is not sufficient (as high as 10^{-1} S/cm at percolation threshold) for shielding applications, especially at high frequencies [29].

ICPs and their composites offered an attractive solution for EMI shielding because of their tunable electrical conductivity, good dielectric properties, controllable electromagnetic attributes, corrosion resistance, facile processing as compared to metals, better compatibility with filler materials and other polymeric matrices. Moreover, the shielding mechanism of ICPs by reflection plus absorption is unique. Many scientific ideas to adopt them for microwave shielding applications were inspired by their intrinsic conductivity in the field of microwave (100-20MHz) and the dependence of their conductivity on frequency [110, 111]. The exclusive properties of ICPs promote their use in techno-commercial applications and high-tech areas like space, defence, navigation/communication control, and as a radar absorbing material (RAM) in stealth technology [111]. The reversible electrical properties of ICPs, possible by redox doping/de-doping mechanism, provide them the dynamic (switchable) microwave absorption capability, and that is the reason that the conducting polymeric materials are considered as "intelligent stealth materials."

A large number of shielding materials are available, but no single-phase material can take care of all the aspects of a shield-like absorption coefficient, thickness, volume, broadband response, *etc*., to give a desired performance under different environments and for different application areas. Therefore, various attempts have also been made to make admixtures, blends, and composites of these polymers with other conducting/insulating materials and exploit them for shielding applications [22, 30, 68, 79, 98, 104, 110]. Among all these options, composites of ICPs with conducting fillers as well as with some insulating polymers have paid maximum attention due to their mesmerizing properties and prevalent applications [115-118]. For shielding applications, the use of conductive fillers (guest) in conducting polymer matrix (host) was always found to be advantageous. Among the several conductive fillers used, carbon fiber (CF) and other carbon materials have detained the utmost attention. Although PA alone can be used for shielding, the use of conductive filler increases the conductivity and shielding property of PA, and it may be used for techno-commercial applications. Although the electrical conductivity is not the only criteria for shielding, it improves with the conductivity, and the composites formed by the combination of a conducting polymer matrix and highly conductive filler shows excellent results. Here, the electrically conducting polymer matrix has the extra advantage of being able to electrically connect the filler units that do not interact with one another, thereby enhancing the connectivity and improving the conductivity. Although shielding does not require connectivity yet, it is found that it is improved by connectivity [104].

For high efficiency, the conductive filler should have a small unit size, a high conductivity, and a high aspect ratio (for better connectivity). Different types of carbon materials such as carbon black, CF, carbon nanotubes (CNTs), graphene, *etc.*, are commonly used for EMI shielding due to their high conductivity. Due to the high aspect ratio, ibres are preferred as compared to particles. Particularly, the CF, due to its high strength, high modulus, low density and chemical inertness find their use in various technological applications. However, these properties of CF are pretty much useless unless embedded in a polymer matrix. EMI shielding is one of the main applications of conventional short carbon fiber (SCF) [119]. Recently, carbon particles with a very high aspect ratio, *i.e.*, CNTs and graphene, are also used for shielding applications. The CF used as fillers in lightweight polymer composites is usually prepared by the pyrolysis of polyacrylonitrile (PAN) and petroleum pitch [120]. A long thin strand of SCF is about 7-15 μm in diameter and is mainly composed of carbon atoms (90-95 % carbon).

Like graphite, the structure of CF consists of sheets of C-atoms stacked in a regular hexagonal pattern. The microscopic crystals having bonded carbon atoms are more or less aligned parallel to the long axis of the fiber. Such alignment of the crystals makes the fiber exceptionally strong for its size. The mechanical properties and weight-to-strength ratio of CF (having density ~ 1.8 g/cc) are extremely good. The tensile strength (T.S.) of CF is high, *i.e.*, 2-7 GPa (T.S. of

"CF Toray-T-300 Japan" is ~4GPa), and the modulus of elasticity is also very high, *i.e.*, more than 450 GPa, which is greater than steel.

8.4. DESIGN AND DIFFERENT FORMS OF EMI SHIELDS

Materials for shielding can be devised and used in different forms like foils, films, tapes, coatings/adhesives, laminates, EMI shielded vent panels/windows, conductive elastomeric EMI gaskets, metal sheets, polymer composite sheets (thickness 0.5 mm to 32mm), *etc.* [121] . The polymeric composites can be used for EMI shielding in liquid emulsion form to make conductive coatings or adhesive and/or in solid powder form to make sheets, films, laminates, conductive gaskets, *etc.* Among the various forms, the polymeric composite sheets find practical applications on a commercial scale to make enclosures and cabinets for equipment. If the thickness of the shield is less than 0.5 mm, it is regarded as a film, and even lesser thickness, *i.e.*, in μm range, is regarded as a foil. If the thickness of the shield is more than 0.5 mm, then it is considered as a sheet. Although the thicker sheets are easy to fabricate and show high efficiency, the material required is in high quantity. Moreover, the thick sheets have to pay the weight penalties and cannot be suitable for future gadgets which are going to be lighter. Fabrication of free-standing thin sheets of conducting polymer composites itself is a typical task. Particularly, polyanilines composite, due to their better thermal, mechanical and chemical stability in combination with moderately high conductivity, are used to make efficient and durable enclosures/cabinets for EMI shielding of electronics and avionics.

8.5. BINDERS FOR THE FABRICATION OF POLYANILINE COMPOSITE SHEETS

Binders or bonding agents are used to bind the discrete particles in any heterogeneous system and make the structure compact and solid. Different types of polymer resins like epoxy resins, polyurethane resins, polyester resins, acrylic resins, phenolic resins, *etc.*, are used as binders for wood, textiles, papers, concrete, explosives, polymer composites, *etc.* In polymer composites, binders are used to bind the polymer molecules/chains and the filler units very tightly. So, the resultant structure becomes compact and dense, thereby increasing the overall mechanical strength of the resultant composite. Among all the strongest binders used, phenolic resins are one of the important classes. Phenolic resins are of two types: novolac and resol. Resol has reactive sites in the form of free methylol groups; hence, it can be crosslinked simply by heating; it can be converted into a crosslinked thermosetting phenol formaldehyde (PF) resin. So, it is called a single-stage phenolic resin. Due to the lack of a reactive site (free methylol groups), novolac/phenolic/ novolac resin is a linear and thermoplastic resin in its initial stage. For crosslinking, to make it as hard & thermosetting phenolic resin, it has to be processed with curing agents (usually hexamethylenetetramine, commercially called hexamine) in its second stage. So novolac is also called a double stage phenolic resin. Cross-linking occurs at a sufficient rate at 140 °C – 160 °C. Synthesis and crosslinking reactions for novolac and/or resol are shown in Scheme **8.1**. During the processing of conducting polymer composites, the phenolic resins effortlessly penetrate and adhere to the structure of fillers or reinforcements used. The unique ability of phenolic resins to "wet out" and to cross-link throughout the fillers provides desired properties, like high mechanical strength, good thermal stability, and chemical inertness, to the composite structure.

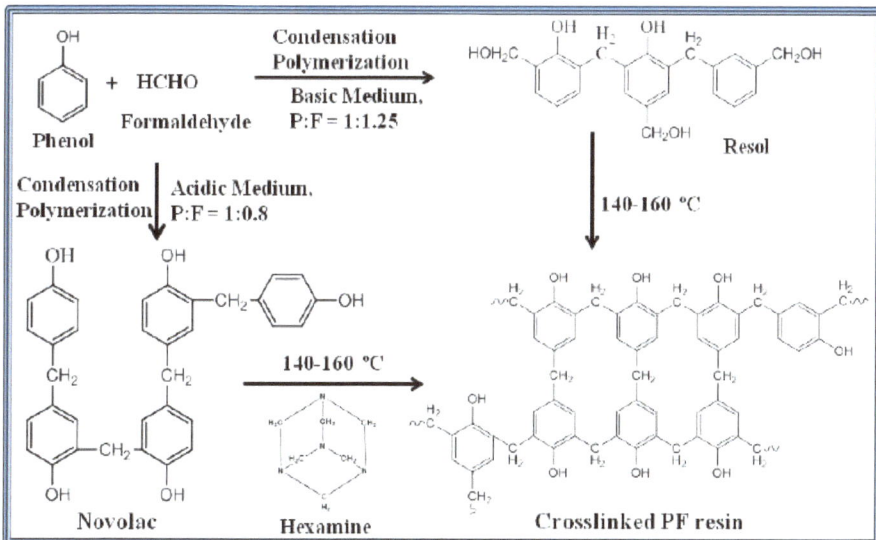

Scheme. (8.1). Synthesis and crosslinking of novolac and/or resol resin.

8.6. SCOPE AND OBJECTIVE OF THE CHAPTER

Literature review on different aspects of conducting polymers and their composites presented in the previous sections illustrates that there is still a lot to be done in order to realize the full commercial potential of the highly promising semiconducting polymer composites. Since the demand for conducting plastics is growing, further research is needed to keep up the pace and come up with new and more improved materials for electronic applications. Initial studies performed on such materials have shown that combinations of conductive fillers with insulating as well as conducting polymers (ICPs) increase the conductivity of these materials, while the incorporation of magnetic particles leads to the magnetic properties at the cost of conductivity. These conducting polymer composites provide cost-effective and compatible alternatives for EMI shielding of electronic components. Although the ICP composites have been widely used for various applications, including EMI shielding, a lot of work is needed to explore their use in EMI shielding up to their commercial level. Also, there is a growing demand for thermally as well as mechanically stable polymer composites. Conventional conducting polymers are thermally stable up to 150°C-180°C and require modification in order to make them more stable. In this regard, a lot of research is needed (1) to search the simple, efficient, and cost-effective methods to synthesize ICPs composites; (2) to improve processability, mechanical and thermal properties of these composites without affecting the electrical and shielding properties significantly; (3) to fabricate or process these composites by simple solution and/or melt processing techniques to achieve their end-use at a commercial scale; (4) to study the morphological aspects of these composites regarding the interaction and compatibility of conductive fillers with the conducting polymer matrix.

The main objective of the present chapter is to fabricate conducting polyaniline-carbon fiber composite thin sheets for EMI shielding application. The polymer composites are synthesized in such a way so as to get the maximum output of all the properties. For improving the shielding, electrical, thermal, and mechanical properties together with the processability of the polymers, various parameters have been considered for emulsion polymerization, use of strong oxidizing agent

(APS), use of protonic salt (SDS) as surfactant as well as dopant and reinforcement by a highly conductive and thermally as well as a mechanically stable filler (carbon fiber). Further, to optimize the required properties for the desired shielding, different amounts of novolac resin are used as binders for the fabrication of thin and light weight sheets.

8.7. SYNTHESIS OF POLYANILINE AND POLYANILINE-CARBON FIBER (PA-CF) COMPOSITES

Pristine PA is usually synthesized by chemical oxidation of aniline by solution polymerization technique in aqueous acidic media, using APS as an oxidant [13]. However, for *In-situ* incorporation of fillers (such as carbon fiber) into the polymer matrix to synthesize polymer carbon fiber composites, the emulsion polymerization technique is appropriate because, in this situation, the conventional method for polymerization of aniline with fillers in the presence of inorganic acid has some drawbacks like inappropriate distribution of fillers into the polymer matrix and poor wettability of fillers in aqueous acidic medium.

In-situ chemical oxidative emulsion polymerization was employed for the synthesis of PA and its composites with CF using APS as an oxidant. In all the syntheses, a salt of protonic acid, *.ie.*, SLS/SDS, was used as a surfactant, which also performed as a dopant. SDS is first converted to its acid form, *i.e.*, dodecyl hydrogen sulphate (DHS), by the addition of equimolar HCl solution until pH was maintained between 2 to 3. The synthesis was carried out in a stainless-steel triple wall reactor at -5 to 0 °C. The entire setup used for the chemical oxidative emulsion polymerization of PA and PA-CF composites is shown in Fig. (**6 b**).

Fig. (**6a**) shows the emulsion of the aniline before polymerization, and Fig. (**9c**) shows the emulsion of the resultant polymer/polymer composites after polymerization.

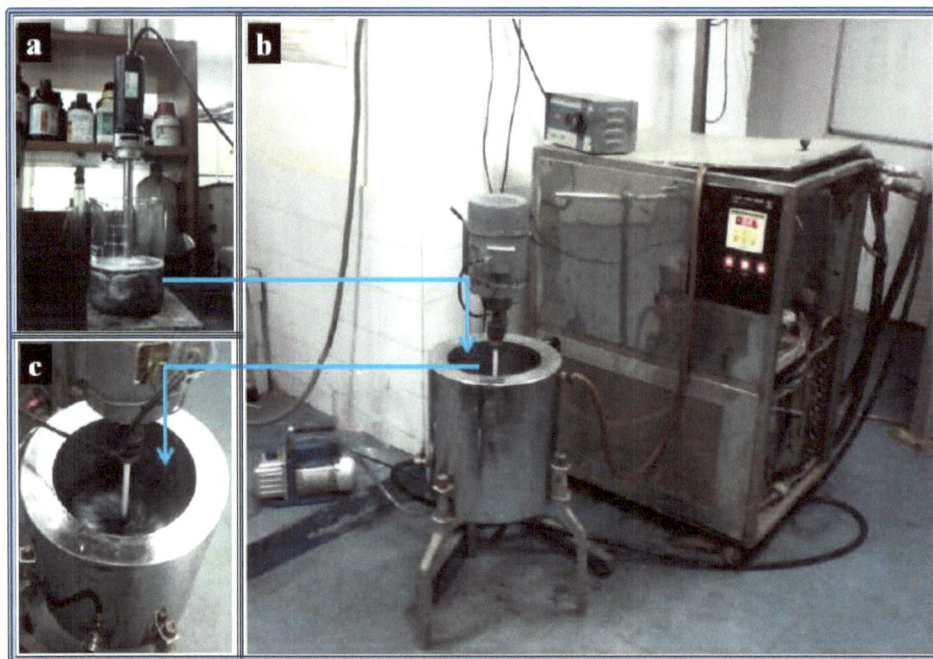

Fig. (6). (**a**) Emulsion of monomer (aniline) in DHS before polymerization, (**b**) setup for chemical oxidative emulsion polymerization of polyanilines and polyanilines-CF composites, and (**c**) Emulsion of the resultant polymer/polymer composites after polymerization.

In a typical synthesis process, DHS is used, which, being a bulky compound, acts both as a surfactant as well as a dopant. In a beaker, equimolar DHS and monomer were homogenized (~ 17500 rpm) with CF (20 wt. %) in distilled water (DW) until the critical micelle concentration (CMC) was reached (as shown in Scheme **8.2**). For surfactants, the CMC values are different which also depend on the medium and other additives used in the reaction. The CMC value for SLS/SDS in pure DW at 20-25 °C is reported at 8.0-8.2 mM [122-124]. Micellar size for DHS at this CMC is around 6 nm in diameter, and aggregation number is about 62 at 20 °C [124], and at this CMC, the micelles are assumed to be spherical in shape. However, the micellar shape for different surfactants may be different, like spherical, cylindrical, tubular, hexagonal, lamellar, *etc.*, and may change with a change in surfactant concentration. Surfactants are amphiphilic in nature containing hydrophilic heads (polar) and hydrophobic tails (non-polar) [125]. At CMC, micelles are formed, which are actually the aggregates of surfactant molecules. In a micelle, the hydrophobic tails of surfactant molecules are aligned towards the interior of the micelle, whereas the hydrophilic heads come in contact with the aqueous medium, *i.e.*, oriented towards the exterior of the micelle [126]. In emulsion polymerization, the position of the monomers in the micelles is significant because it dictates the reaction mechanism and the properties of the final polymer formed [127]. The aniline monomers and carbon fibers assemble themselves inside the micelles. The semi-polar aniline monomer is located at the palisade layer (the region between the hydrophilic groups and the first few carbon atoms of the hydrophobic groups) of the micelle. So, the interior of the micelle provides the site necessary for polymerization. APS (a water-soluble free radical initiator/oxidant) is then added dropwise, which travels into micelle and reacts with a monomer molecule to form anilinium cations, thereby initiation taking place. Anilinium cations are further polymerized within the micelle to form high molecular weight polyaniline. As CF is already used during the reaction, hence some of the polymers formed are deposited on the surface of the CF. The polyaniline-CF composite so formed was not isolable directly because the polyaniline salt remained embarked in the emulsion along with the by-products of the reaction. In most of the cases, the product was isolated by breaking the emulsion (by adding a particular destabilizer, *e.g.*, acetone) [128]. So, the PA-CF precipitates were collected, filtered, washed repeatedly with DW, and dried. Finally, it was crushed with pixels and mortar to get PA-CF composite powder. By a similar route, pristine PA doped with DHS was also synthesized for comparative study. It is certain that the PA and PA-CF composites synthesized by emulsion polymerization have good solubility, good processability, reasonably high molecular weight, and good electrical conductivity [127].

8.7.1. Chemistry of *In-situ* Chemical Oxidative Emulsion Polymerization

The overall reaction pathway for the emulsion polymerization of PA-CF composite using SDS as a surfactant as well as a dopant in the presence of CF as a filler is shown in Scheme **8.2**. In the first step, the Na^+ ions of SDS are replaced by H^+ ions by treating it with equimolar HCl (0.1M) solution hence SDS converted to its acid form, *i.e.*, DHS. In the second step, chopped CF are added to the DHS solution and homogenized, so the emulsion of DHS is formed in the water system at CMC [124, 129, 130]. The micelles formed at CMC also contain CF. The polar head groups of the DHS oriented at the outer surface of the micelles stabilize the emulsion system, and some micro micelles are also absorbed on the polar surface of the CF. Although the CF incorporated into DHS emulsion has a strong influence on micelle aggregation number and second CMC. The third step involves the addition of aniline monomer into the emulsion system, and it is again homogenized. The fourth step is the polymerization step which took place with the slow addition of the APS initiator (oxidant). Here, a spherical micellar phase in water has been created *via* mutual interaction of aqueous APS solution and DHS. Aniline and APS molecules, due to their hydrophobic nature, diffused into the micelles and were surrounded by hydrophilic sulfonate (–SO$_3$H) units. Here, aniline was polymerized to form PA by oxidation with APS. Some of the PA was also deposited on the surface of CF. Polyaniline doped with DHS shows that the surfactant molecules have become a part of the resulting polymeric composite material due to ionic interaction between the polymer and the surfactant. In this way, PA-CF composite, showing spherical shapes of PA and reinforced with CF, has been prepared. The morphology of the resultant composite is confirmed by SEM images.

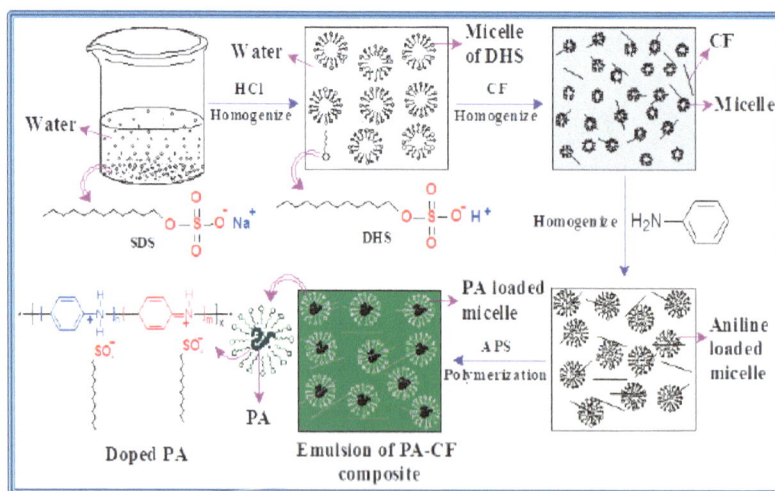

Scheme. (8.2). Schematic representation for the synthesis of PA-CF composite by *In-situ* chemical oxidative emulsion polymerization. Reproduced from American Journal of Polymer Science 2015, 5(1A):28-39.

8.7.2. Mechanism of Polymerization

Although the mechanism of oxidative polymerization of aniline is still dubious, the most generalized mechanism suggested by Wei *et al.* [131-133] is shown in Scheme **8.3**. This mechanism is primarily based on kinetic studies of the electrochemical polymerization of aniline.

The process of aniline polymerization is referred to as 'non-classical chain polymerization' or somewhat between a classical chain-growth polymerization and a classical step-growth polymerization [133]. According to the proposed mechanism, "the first step in which oxidation of aniline occurs to form dimeric species is the slowest step because of the higher oxidation potential of aniline than those of dimers. The dication molecule is rapidly deprotonated to yield a nitrenium cation; it is because of the significant positive charge on dication. The dimers are immediately oxidized and then reacted with an aniline monomer *via* an electrophilic aromatic substitution, followed by further oxidation and deprotonation to form the trimers. In the successive steps, this process is repeated and leads ultimately to the formation of PA with a linear structure [134, 135].

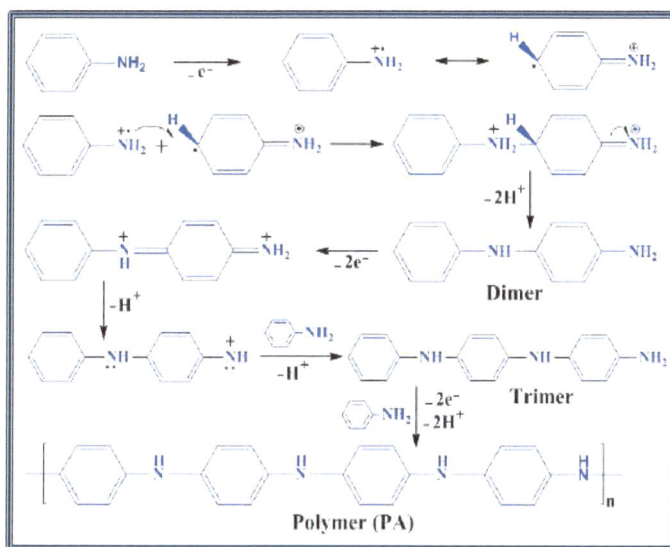

Scheme. (8.3). Mechanism of oxidative polymerization of aniline; proposed by Wei *et al.* [131-133].

8.8. BLENDING AND MIXING OF PA-CF COMPOSITES WITH NOVOLAC RESIN TO PREPARE POLYANILINE-CARBON FIBER-NOVOLAC (PACN) COMPOSITES

Conducting polymer composites can be mixed and blended with some conventional polymeric resins by various techniques like a mechanical mixer, extrusion, ball mills, roll mills, *etc*. In this chapter, PA-CF composite was blended with novolac resin by mixing in a ball mill, followed by ultrasonication. Specific proportions of PA-CF composite powder and novolac powder were taken with stainless steel balls in tungsten carbide jars of Retsch "PM-400" planetary ball mill (Fig. **7**). Then, the two components were mixed and ground in the ball mill for 2 hours at a speed of 200 rpm. A homogeneous mixture of these two components was thus formed by the coaxial movement of the carbide jars of the ball mill with high speed. The homogeneity was further increased by dissolving it in ethanol (10 wt. %) and ultra-sonicated at room temperature for 3 hours. Finally, the ethanol from the solution was evaporated and the PACN composite powder thus prepared was then dried in a vacuum oven at 60-70 °C. By this route, different compositions of PACN were prepared by taking different amounts of novolac resin, such as 5, 15, and 25 wt. %. of PA-CF composite. The different powdered samples of PACN composites thus formed were abbreviated as PACN5p, PACN 15p & PACN25p, respectively (here 'p' stands for powder form).

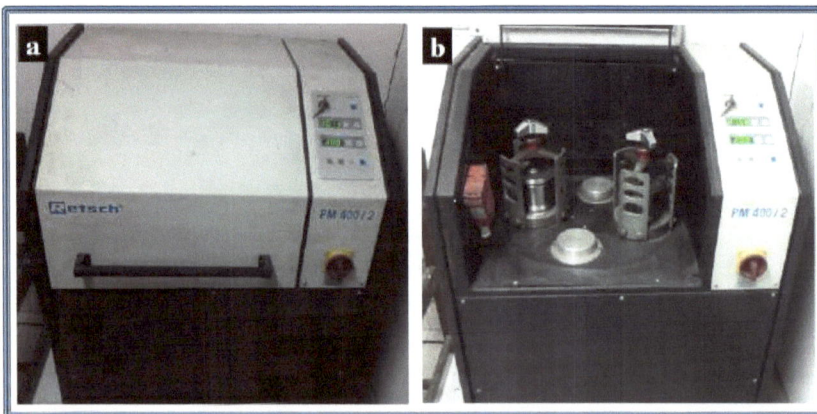

Fig. (7). (**a**) Retsch "PM-400" planetary Ball Mill and (**b**) Carbide Jars placed in Ball Mill.

8.9. FABRICATION OF FREE-STANDING SELF-SUPPORTED THIN SHEETS OF PACN COMPOSITES

Various techniques like hand lay casting, solution casting, compression moulding using an injector, compression moulding using hydraulic hot/cold press, *etc.*, may be used for the fabrication of the conducting polymer composite sheets or films. Due to limited solubility, poor processability, and infusibility of most of the conducting polymers, the solution and hand lay casting techniques are not suitable for them. Thus, the compression moulding technique is usually preferred for their processing and fabrication. Among the various compression techniques, the compression moulding techniques using the hydraulic hot press are appropriate for the powdered samples. Thus, the PACN composites were fabricated into the form of free-standing thin sheets by hydraulic hot pressing.

8.9.1. Hydraulic Hot Press & Fabrication Process

Hydraulic hot press is generally used to fabricate plywood boards, flush doors, coir mattresses, PC boards and laminates, films, or sheets of fiber-reinforced plastic (FRP) composites. Hot pressing is a high-pressure powder metallurgy method that converts a powder into a compact solid at high temperature and pressure [136]. Hot pressing is mainly preferred to obtain hard and brittle articles. The technique is widely exploited for different types of polymers and polymeric composites. Fig. (**8**) (left) shows a hydraulic hot press. In the process, the loose powdered sample of PACN composites was filled into an iron mould that allows induction or resistance heating up to high temperatures. Here, a three-piece iron mould/die was used to fabricate sheets having dimensions 70 mm × 20 mm. Prior to its use, the mould was cleaned properly with acetone, and then silicone wax was applied inside the mould for easy release of the sheet to be prepared. A calculated amount of PACN composite powder was taken into the mould for getting a sheet having a thickness of less than 1 mm. Then, the mould is placed between the hot plates (Fig. **8**, right) of the hydraulic hot press at 100 °C and contact pressure (~ 100 psi). At this temperature, crosslinking of novolac resin occurred in the polymer matrix, and thus the three phases (PA, CF, and novolac) of the heterogeneous PACN composite were condensed. The excess resin was squeezed out after performing its action. Further curing of the resultant sheet was carried out at 150 °C for about 2 hours. Finally, the sheet was ejected from the mould after slow and gradual cooling up to room temperature.

Fig. (8). Hydraulic hot press (left) & three-piece mould placed between hot plates (right).

Free standing self-supported thin sheets of PACN composites were thus prepared by taking different amounts of novolac resin as a binder to see the effect of novolac resin on mechanical, thermal, electrical, and shielding properties of PA-CF composites. Sheets containing different proportions of novolac resin from 5 wt. %, 15 wt. % and 25 wt. %, are abbreviated as PACN5s, PACN15s, and PACN25s, respectively (here 's' stands for sheet).

A pictorial view of the process of preparation of PACN composites and fabrication of their thin sheets is shown in Scheme **8.4**.

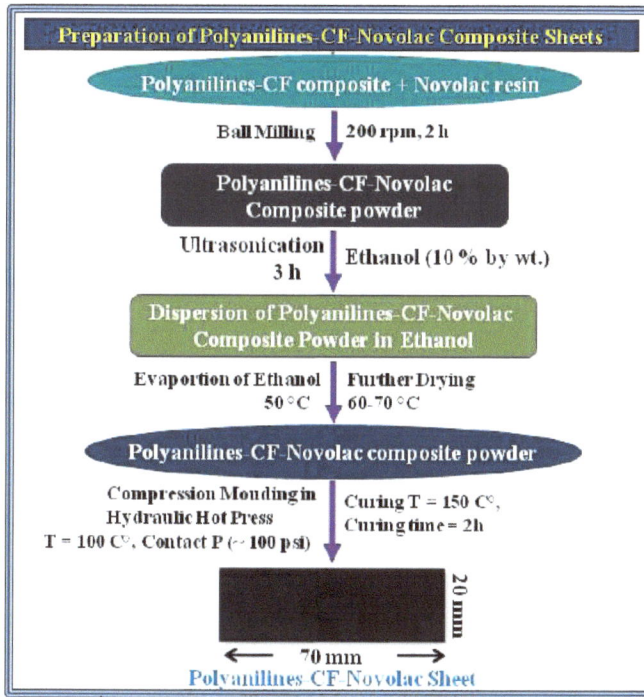

Scheme. (8.4). Pictorial representation for the fabrication of thin sheets of PACN composites.

8.10. CHARACTERIZATION AND ANALYSIS

8.10.1. UV-*vis* Spectroscopy

In an acidic medium, the green protonated (doped) polyaniline shows three characteristic absorption peaks/bands at 325-360, 400-430, and 780-825 nm [137]. The first peak arises due to π-π* electronic transition within benzenoid segments. The second peak and third peak depict the doping level and formation of polaron and bipolaron, respectively. However, the position of these bands may vary slightly according to the type of dopant used for doping and solvent used.

UV-*vis* absorption studies of the powdered samples of pristine PA, PA-CF composite, and PACN composites were carried out on Shimadzu 1601 Spectrophotometer in the wavelength range of 200-1100 nm. The solutions of all the samples were prepared in N-methyl pyrrolidone (NMP). But NMP converts the polymer to its undoped form (blue coloured), so a sufficient amount of HCl (up to the change of colour from blue to green) is thereafter added for doping of the polymer again. Fig. (**9**) shows the UV-*vis* spectra of PA, PA-CF composite, and PACN composites. Pristine PA doped with DHS showed three usual peaks in its spectrum. The first one is at 303 nm, which is due to the π- π* electron transition that occurred within the benzenoid structure. The second peak is at 439 nm, which signifies the intermediate state formed during the oxidation of the leucoemraldine form of PA. The third peak appeared at 877 nm, which illustrated that the emeraldine form of PA changed to a fully oxidized pernigraniline form. Hence, the second and third peaks confirmed the doping level of PA and the development of polarons. As per Fig. (**9**), comparison of UV-*vis* spectrum of PA with the spectra of PA-CF and PACN composites (PACN5p, PACN15p & PACN25p) displayed a hypsochromic shifting (blue shift) of the first peak, *i.e.*, from 303 to 282 nm in the case of PA-CF and all the PACN composites. Shifting of this peak shows an interaction of the PA matrix with CF. However, no unusual shifting in the peak position corresponding to polaronic transitions indicates that the introduction of CF in the polymer matrix did not affect the doping mechanism of PA in any way. The UV-*vis* study of the PA and PA-CF composite confirms their oxidation and doping levels.

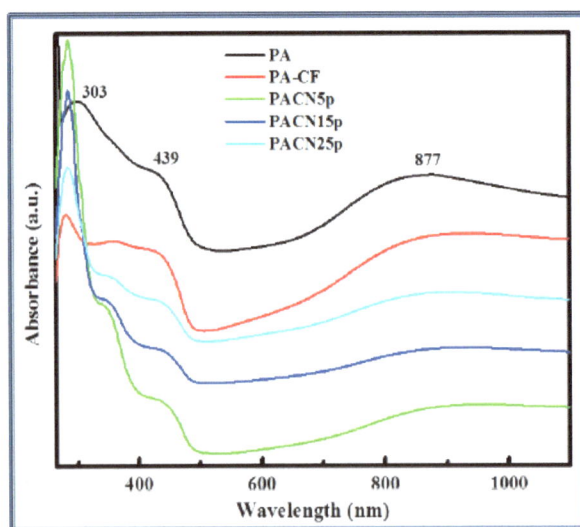

Fig. (9). UV-*vis* spectra of PA, PA-CF, PACN5p, PACN15p & PACN25p samples. Reproduced from the American Journal of Polymer Science 2015, 5(1A):28-39.

8.10.2. FTIR Spectroscopy

Fourier transform infrared (FTIR) spectra of the samples were recorded on Nicolet 5700 within the range of 4000-500 cm^{-1} wavenumber. FTIR spectra of PA, PA-CF, and PACN powder samples are shown in Fig. (**10**). In the spectrum of PA, the major peaks (vibrational bands) appeared at 1560, 1470, 1290, 1110, and 791 cm^{-1}, which are similar to those reported in the standard spectrum of PA [138]. The peaks at 1560 and 1470 cm^{-1} appeared due to the stretching of quinoid and benzenoid rings, respectively. These bands are assigned to nitrogen quinone (Q) and benzenoid ring (B) [139, 140]. Tang *et al.* [26] observed that the relative intensity of these bands affects the electrical conductivity of polyaniline. The peak at 1290 cm^{-1} signifies the C-N stretching of an aromatic amine which usually appears within the range of 1300-1200 cm^{-1}. A strong band appeared at 1110 cm^{-1}, is termed as the electronic band or a vibrational band of nitrogen quinine. According to Han and co-workers [129], the peak around 1120 cm^{-1} can be assigned to C–H in-plane-bending vibration modes (N=Q=N, Q=N$^+$H–B, and B–N$^+$H–B; Q = quinoid ring, B = benzenoid ring), which occurs during the protonation. The small shifting of peak and variation of intensity might be due to the different dopant levels of the resulting PA. The band appearing at 791 cm^{-1} (due to C-H out-of-plane bending mode) revealed the para-substituted benzene ring in PA, so confirms the polymer formation [141]. In addition to these characteristic bands of PA, the presence of bands at 2920 and 2850 cm^{-1} are due to the asymmetric and symmetric aliphatic C-H stretching vibrations, respectively [142] and indicates the existence of alkyl substituent of DHS (used as dopant) in the PA chain. Furthermore, the appearance of a band at 570 cm^{-1}, which is attributed to the degenerate bending mode of the SO$_3^-$ group, confirming the presence of the SO$_3^-$ group of DHS [143]. Hence, the FTIR spectrum confirmed the structure, polymerization, and doping of the PA synthesized.

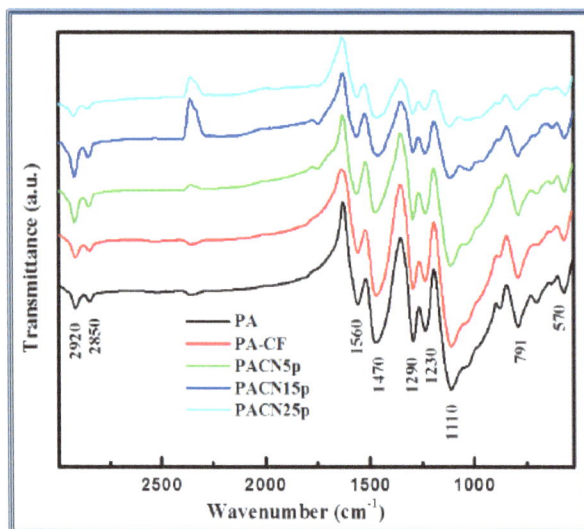

Fig. (10). FTIR spectra of PA, PA-CF, PACN5p, PACN15p & PACN25p samples. Reproduced from the American Journal of Polymer Science 2015, 5(1A):28-39.

On comparing the IR spectrum of PA with PA-CF, PACN5p, PACN15p, and PACN25p composites (Fig. **10**), it is found that there is no appearance of extra bands in the FTIR spectra of these composites, rather only some slight shifting in the main peaks is observed. It indicates that there is no net chemical reaction occurring between PA, CF, and novolac resin, but some ionic interaction might happen between these components. Also, in the case of PA-CF and PACN composites, the benzenoid/quinoid intensity ratio is reduced. This may reveal that the CF and novolac resin in the PA matrix promote and stabilize the quinoid ring structure. These observations suggest that PA-CF &PACN composites must have a higher value of electrical conductivity than pristine PA. The comparison of the IR spectrum of PA with the spectra of the composites confirms the presence of CF and/or novolac resin in PA-CF & PACN composites.

8.10.3. SEM Analysis

The morphology of PA and its composites was examined by scanning electron microscope (SEM, Zeiss EVO MA-10). Fig. (**11**) displays the SEM images of PA, PA-CF, PACN15p, and PACN15s samples. Fig. (**11a**) demonstrates that the PA, synthesized by emulsion polymerization using DHS as surfactant as well as dopant, has a spherical structure with a porous surface. The observed morphology of PA is similar to the earlier reports [129, 143]. Fig. (**11b**) showed that CF are uniformly distributed in the PA matrix, and most of the CF are longitudinally elongated, which is a good sign for the betterment of the mechanical strength and the conducting path in the composite sheet. CF being conductive fillers connected even the discrete particles of the conducting polymer matrix and offered an uninterrupted path for electron movement. This image also indicates that some polymers are also deposited on the surface of CF. This implies that the interaction between polymer molecules and CF conquers the van der Waals interaction between CF, with the effective interaction between the π-bonds in the aromatic ring of PA, hence CF should strongly facilitate the charge transfer reaction between the two components [144]. The SEM image of PACN15p (Fig. **11c**) shows that the size of the CF reduced during blending of PA-CF composite with novolac resin.

Fig. (**11d**) displays the SEM micrograph of PACN15s. This image demonstrates that the structure of composite became solid and dense after blending with novolac resin and upon subsequent hot pressing. Hence, novolac resin used as a binder in the PACN sheet increased the connectivity. Fig. (**12**) is the high-resolution optical image of the synthesized PACN15 sheet with its dimensions.

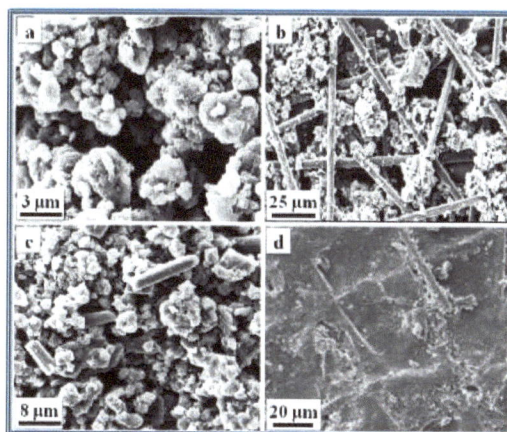

Fig. (11). SEM images of (**a**) PA, (**b**) PA-CF composite, (**c**) PACN15p and (**d**) PACN15s. Reproduced from the American Journal of Polymer Science 2015, 5(1A):28-39.

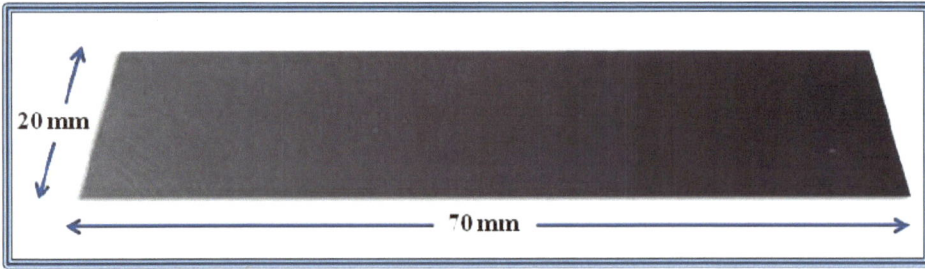

Fig. (12). High-resolution optical image of PACN15s.

8.10.4. Thermogravimetric Analysis

Thermogravimetric analysis (TGA) was performed to find the effect of CF and novolac resin on the thermal stability of PA. TGA of all the samples was carried out by thermogravimetric analyzer (Mettler Toledo TGA/SDTA 851^e) in an inert atmosphere in the temperature range of 25-900 $^\circ$C at a uniform heating rate of 8 $^\circ$C/min. The thermograms of PA, PA-CF, PACN composite powder samples, along with CF and novolac are shown in Fig. (**13a**). The thermal stability of CF, as well as novolac resin, is high, which can be observed in Fig. (**13a**). TGA curve of PA is characterized by three weight-loss steps, which is similar to previous reports [145]. In the first step, weight loss observed up to 110°C is due to the loss of residual water molecules/moisture entrapped in the polymer moiety [145, 146]. The second weight loss observed in the range of 182–300 °C is due to the removal of dopant molecules from the polymer structure [147]. The weight loss observed after this step corresponds to the complete degradation and decomposition of the polymer main chain [145, 148]. From this thermogram of PA (Fig. **13a**), it is clear that PA is thermally stable up to 182 °C. Comparison of thermal behaviour of PA with PA-CF composite reveals that the weight loss that occurred in the case of PA-CF composite is gradual as compared to weight loss that occurred in PA. Also, the total weight loss that occurred in the case of PA-CF composite is less as compared to the total weight loss that occurred in pristine PA, and the PA-CF composite is found thermally stable up to 210 °C. The obtained results confirmed that CF has enormously increased the thermal stability of the polymer. On further comparison of the thermogram of PA-CF with the thermograms of PACN15p and PACN15s (Fig. **13a**), it is observed that the novolac resin used as binder increased the thermal stability of the PA-CF composite, which further increased significantly when the PACN composite powder is moulded into the form of thin sheets by hydraulic hot press.

Fig. (**13b** and **c**) indicate that the thermal stability of the composites/sheets increased with an increase in the novolac percentage (5 to 25 %) in the PA-CF composite. From the thermogram of PACN25s, it is clear that the initial loss in weight due to the loss of water molecules, in this case, is extremely less because the sheet was already thermally cured at 150 °C. The weight loss due to partial leaching of the dopant starts at 230 °C. Hence, the thermal stability of PACN25s is found to be highest (230 °C). The weight loss due to the removal of dopant here is very less (as compared to other samples), which signifies that the novolac resin might have developed some kind of ionic interaction with the PA structure. Moreover, the weight loss that occurs due to degradation of the polymer backbone (300-900°C) is gradual; it means that the polymer degradation is slow. The weight loss observed at 480-650 °C is due to partial leaching of novolac resin from the polymer composite.

From all these observations, it is concluded that the incorporation of CF (as fillers) into the PA matrix improved its thermal stability, which further increased upon blending the PA-CF composite with novolac resin. This happened primarily due to the high thermal stability of CF and novolac both. The binder might also have increased the interaction of dopant and CF with the polymer molecules, hence increasing the thermal stability. Moreover, during hot pressing at 150 °C, the novolac resin in the PACN composite got crosslinked and bound the CF and PA matrix very tightly, thereby increasing the thermal stability significantly.

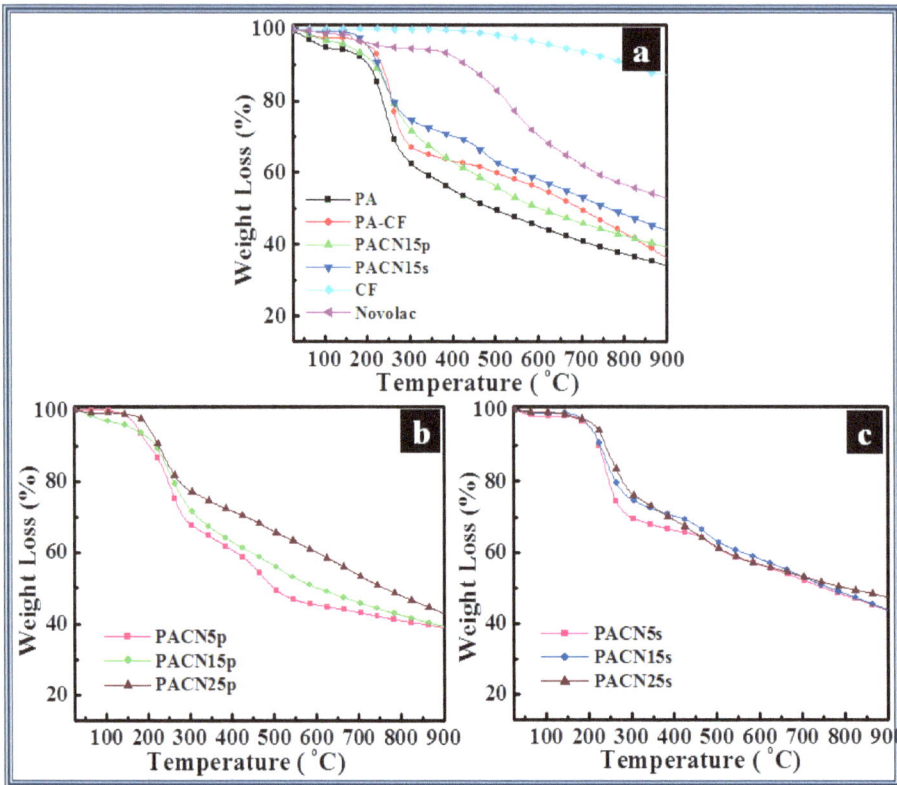

Fig. (13). (**a**) TGA of PA, PA-CF, PACN15p, PACN15s, CF & Novolac, (**b**) TGA of PACN5p, PACN15p & PACN25p and (**c**) TGA of PACN5s, PACN15s & PACN25s. Reproduced from the American Journal of Polymer Science 2015, 5(1A):28-39.

8.10.5. Flexural Strength and Flexural Modulus

The mechanical strength of the PACN composite sheets is elucidated in terms of measurement of flexural strength (FS) and flexural modulus (FM). FM refers to a material's resistance to breaking when bending forces are applied perpendicular to its longitudinal axis. FS and FM values were measured by the three-point bending test based on ASTM D790 [149], using an Instron Universal Testing Machine (model 4411) at a crosshead speed of 0.5 mm/min. The dimension of the specimens used is 70×20 mm^2 with thickness ~ 0.6 mm. The support span length (L) between the two supports of the specimen was 40 mm. Variation of FS and FM of PACN sheets with varying amounts of novolac resin is shown in Fig. (**14**), and their values are given in Table **2**. Fig. (**14**) shows that FS and FM increased with an increase in the amount of novolac resin in the PA-CF composite. The interfacial ionic interaction that might have occurred between novolac resin and CF enabled the load to be transferred between the resin and the reinforcement (CF units). The novolac resin increases the dispersion and wettability of CF in conducting polymer matrices. So, adhesion bonding between three phases of the composite (resin, fillers, and polymer matrix) is enhanced, which imparts more strength to the composite by enabling the stress to be transferred around the stiff reinforcement phase. Moreover, the novolac resin crosslinked after thermal curing of the sheets at 150 $^\circ$C, so the sheets become stiffer and stronger. Excess resin leached out of the sheet during this process. The effect of novolac resin on mechanical and thermal properties was investigated in this work up to 25 wt. % loading of the resin, because up to this loading, the conductivity and shielding properties of PACN sheets have been markedly decreased because of the insulating nature of the resin. Although higher loading of resin has also been reported for other composites [119], multiple factors are responsible for the variation of electrical and shielding properties.

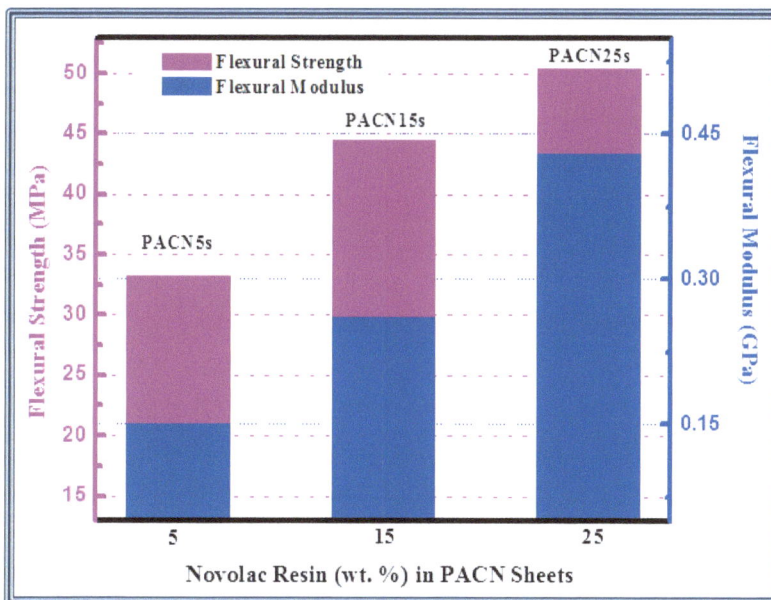

Fig. (14). FS and FM of composite sheets (PACN5s, PACN15s & PACN25s).

8.10.6. Electrical Conductivity (EC)

Standard four-probe method using Keithley programmable current source (model 6221) is used to determine room temperature conductivity of all the samples. EC values of PA, PA-CF, and PACN composites/sheets are given in Table **1** and compared with the help of the bar graph in Fig. (**15**). On comparing these values, it is found that EC of PA increased amazingly after the incorporation of CF as filler into the polymer matrix. The introduction of CF actually facilitates the charge transfer process between the two components of the composite and thereby increases the conductivity by the formation of conductive channels. Some reports also presented the improved mechanical and electrical properties by using CNTs and/or CF as fillers in the conjugated polymer matrix [150, 151]. Here, CF loading in the polymer matrix was fixed (20 % by wt.) to get the desired conductivity for shielding applications, in accordance with the previous reports [151-153]. In contrast, the amount of novolac resin in PA-CF composite is varied to optimize mechanical, electrical, and shielding properties of the prepared PACN thin sheets. By comparing the absolute values of conductivity (Table **1**) of different samples, it is found that EC of PA-CF composite decreases after blending it with novolac resin, which further keeps on decreasing with an increase in the percentage of novolac resin in PACN composite powder. The reason behind this is that the insulating novolac resin hinders the free flow of electrons in the conducting polymer composite. When powdered PACN composites were converted to thin sheets by compression in hydraulic hot press, EC again increased because the binder then binds the filler units tightly with the polymer matrix and makes the structure very compact, thereby enhancing the connectivity of filler and matrix. Moreover, after the thermal curing of thin sheets at 150 °C for 2 hours, the excess of the insulating component (novolac resin) from the sheets squeezed out after performing its action. From Fig. (**15**), it is concluded that the EC decreases beyond 15 % loading of novolac resin in the PACN sheets, and at 25 % loading, the EC decreased extensively because up to this limit, the insulating nature of the novolac resin dominated.

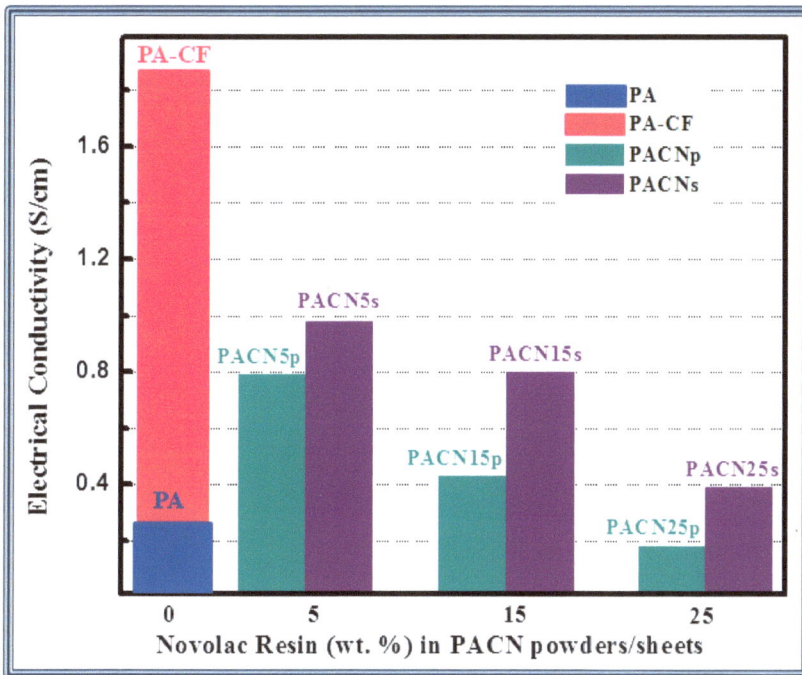

Fig. (15). EC of PA, PA-CF, PACN composites and PACN sheets and variation of EC with novolac percentage in the composites/sheets.

Table 1. Comparison of FS, FM, EC, and SE of PA and PA composite/sheet samples.

Sample name	CF (wt. %)	Novolac resin (wt. %)	Flexural strength (MPa)	Flexural modulus (GPa)	Electrical conductivity (S/cm)	SE (dB) at 8.2 GHz	Sample thickness (mm)
PA	0	0	-	-	0.29	14.5	2.25
PA-CF	20	0	-	-	1.87	31.6	2.23
PACN5p	20	5	-	-	0.79	34.6	2.26
PACN15p	20	15	-	-	0.43	35.0	2.23
PACN25p	20	25	-	-	0.18	29.0	2.17

(Table 1) cont.....

PACN5s	20	5	33.28	0.15	0.98	15.2	0.61
PACN15s	20	15	44.53	0.26	0.80	15.8	0.59
PACN15s	20	15	-	-	-	18.6	1.2
PACN15s	20	15	-	-	-	23.0	1.8
PACN15s	20	15	-	-	-	32.4	2.4
PACN15s	20	15	-	-	-	38.5	3.0
PACN 25s	20	25	50.48	0.41	0.39	10.6	0.62

8.10.7. EMI Shielding Measurements

The EMI shielding effectiveness (SE) of a material is the ratio of the transmitted power to the incident power [154-156];

$$SE(dB) = -10log\left(\frac{P_T}{P_I}\right) \tag{8.11}$$

Here, P_T and P_I are transmitted and incident EM powers, respectively. The total EMI shielding effectiveness (SE) is the sum of SE_A, SE_R, and SE_M;

$$SE = SE_A + SE_R + SE_M \tag{8.12}$$

where SE_A, SE_R, and SE_M are shielding effectiveness due to absorption, reflection, and multiple reflections, respectively.

In a two-port network, S-parameters S_{11} (S_{22}) and S_{12} (S_{21}) represent the reflection and transmission coefficients given as [157]:

$$T = \left|\frac{E_T}{E_I}\right|^2 = |S_{21}|^2 = |S_{12}|^2 \tag{8.13}$$

$$R = \left|\frac{E_R}{E_I}\right|^2 = |S_{11}|^2 = |S_{22}|^2 \tag{8.14}$$

and absorption coefficient, A is,

$$(A) = 1 - R - T \qquad (8.15)$$

Here, it is noted that A is given with respect to the power of the incident EM wave. If the effect of multiple reflections between both interfaces of the material is negligible, then the relative intensity of the effective incident EM wave inside the material after reflection is based on the quantity $(1 - R)$. Therefore, the effective absorbance (A_{eff}) can be described as $A_{eff} = (1 - R - T)/(1 - R)$ with respect to the power of the effective incident EM wave inside the shielding material. So, it is convenient to express the reflectance and effective absorbance in the form of $-10 \log (1 - R)$ and $-10 \log (1 - A_{eff})$ in decibel (dB), respectively, which give SE_R and SE_A as [158]:

$$SE_R = -10 \log(1 - R) \qquad (8.16)$$

$$SE_A = -10 \log(1 - A_{eff}) = -10 \log\left(\frac{T}{1-R}\right) \qquad (8.17)$$

EMI shielding measurements were carried out by Agilent E8362B Vector Network Analyzer (Fig. **16b**) in the frequency range of 8.2-12.4 GHz (X band) of the microwave region. To check the EMI shielding of the synthesized PACN sheets (thickness ~ 0.6 mm), the sheets were cut to make rectangular samples of dimensions 25.7×13 mm^2 to fit into the copper sample holder (Fig. **16c**) of dimensions $22.86 \times 10.14 \times 6$ mm^3 connected between the wave-guide flanges of network analyzer. For comparing the EMI shielding of thin sheets with the powdered composite samples, the PA, PA-CF, and PACN powdered samples were compressed by a hydraulic press to get rectangular pellets of the same dimensions and thickness ~ 2.2 mm.

A network analyzer is one of the most important tools for analysing analogue circuits. By measuring the amplitudes and phases of transmission and reflection coefficients of an analogue circuit, a network analyzer reveals all the network characteristics of the circuit. In microwave engineering, network analysers are used to analyse a wide variety of materials, components, circuits, and systems.

A measurement of the reflection and/or transmission through a material, along with knowledge of its physical dimensions, provides the information to characterize the permittivity and permeability of the material. The PNA, PNA-L, ENA, and ENA-L are vector network analysers that make high frequency stimulus-response measurements from 300 kHz to 110 GHz or even 325 GHz. A vector network analyzer consists of a signal source, a receiver, and a display (Fig. **16a**). The source launches a signal at a single frequency to the material under test. The receiver is tuned to that frequency so as to detect the reflected and transmitted signals from the material. The measured response produces the magnitude and phase data at that frequency. The source is then stepped to the next frequency, and the measurement is repeated to display the reflection and transmission measurement response as a function of frequency.

Simple components and connecting wires that perform well at low frequencies behave differently at high frequencies. At microwave frequencies, wavelengths become small compared to the physical dimensions of the devices such that two closely spaced points can have a significant phase difference. Low-frequency lumped-circuit element techniques must be replaced by transmission line theory to analyse the behaviour of devices at higher frequencies. Additional high-frequency effects such as radiation loss, dielectric loss, and capacitive coupling make microwave circuits complex and expensive. It is time-consuming and costly to design a perfect microwave network analyzer.

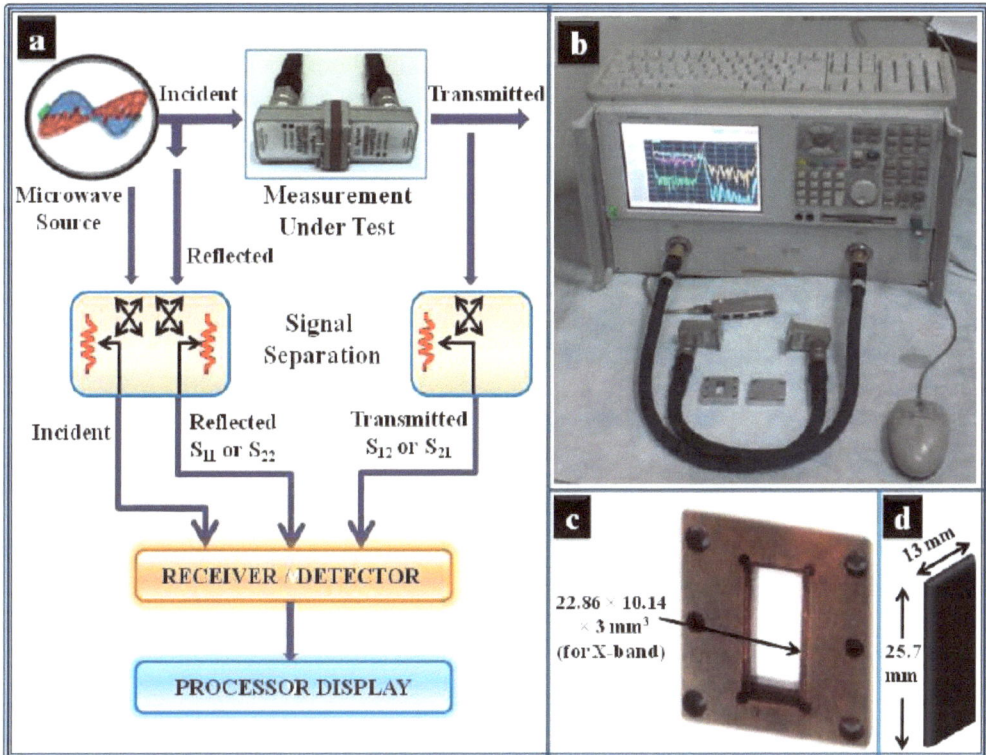

Fig. (16). (**a**) Block diagram of Vector Network Analyzer, (**b**) Agilent E8362B Vector Network Analyzer, (**c**) Sample holder & (**d**) Rectangular sample.

Fig. (**17**) shows the variation of SE_A, SE_R & SE_T with frequency for PA, PA-CF, and PACN composite powder (PACN5p, PACN15p & PACN25p) samples compressed in a rectangular die at a pressure of 5 tons. In the microwave range, the contribution of SE_A becomes more as compared to SE_R. From the experimental results, the SE due to absorption (SE_A) has been found to vary from 8.5-29.7 dB for composite powder pellets of comparable thicknesses of ~ 2.2 mm (Fig. **17a**). The addition of CF in PA increased the SE_A of pristine PA from 8.5 to 22 dB, which further increased up to 29.7 dB upon the addition of novolac resin in PA-CF composite but only up to 15 % loading of novolac resin, then decreased to 24.9 dB at 25 % loading of novolac. Fig. (**17b**) shows that the SE due to reflection (SE_R) varies from 6.2-10.1 dB, so there is a small increase in SE_R upon the addition of CF and novolac in PA. Moreover, SE_R values decreased with an increase in novolac percentage in the PA-CF composite. Fig. (**20c**) demonstrates that the total shielding effectiveness (SE_T) of pristine PA increased from 14.7 to 35 dB upon the addition of CF and novolac resin. These results suggest that the microwave absorption loss of the PACN composites is better than the pristine PA. This is due to the combined contribution of CF and novolac resin. Due to the high conductivity of CF, it improved the electrical properties of PA and thereby enhanced the shielding behaviour. In spite of having insulating behaviour, novolac resin improved the SE of PA-CF composite, but up to a certain amount of loading (15 wt. %). Actually, novolac resin used as binder makes a unique interlocking arrangement with PA-CF composite and makes the system compact and dense, which results in better connectivity and improvement in EMI shielding. But at higher loading (25 wt. %) of resin, the insulating character of resin dominates, leading to a decrease in shielding effectiveness value.

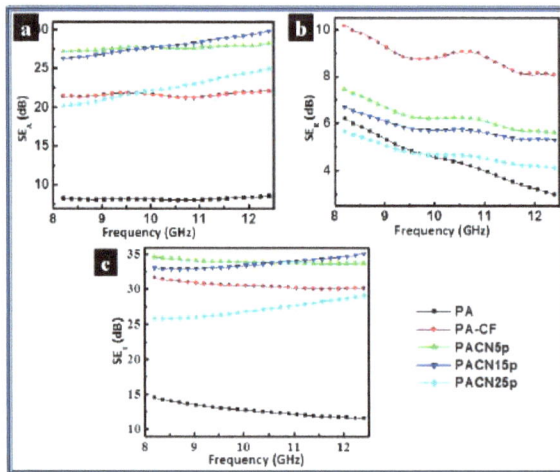

Fig. (17). Variation of (**a**) SE_A, (**b**) SE_R, and (**c**) SE_T with frequency for PA, PA-CF, PACN composites in powder form before transforming into sheets. Reproduced from the American Journal of Polymer Science 2015, 5(1A):28-39.

From the scattering parameters obtained from VNA, the values of transmission coefficient (T) and reflection coefficient (R) for PACN sheets were calculated by Equations 8.13 and 8.14, respectively. The value of their absorption coefficient (A) is then calculated by using equation 8.15. Fig. (**18**) shows the calculated reflection coefficient (R), absorption coefficient (A), transmission coefficient (T), and absorption efficiency of the PACN sheets with a comparable thickness of ~

0.6 mm in the frequency range of 8.2–12.4 GHz. The transmittance coefficient (T) value among all PACN sheets is maximum for the PACN25s sample. After 15 wt. % loading of novolac resin, the transmittance coefficient is observed to be more than 0.09, as shown in Fig. (**18c**). In other words, due to higher values of A and R, more EM waves are consumed by the PACN sheets, which leads to a more significant decrease of the T values. According to the plots, as exhibited in Fig. (**18b**), the A values of the PACN sheets were in the range of 0.2~0.3. The R values of the PACN sheets at each wt. % loading of novolac resin were found to be between 0.6-0.8 (Fig. **18a**). These plots show that the trend of the R depends on the amount of novolac resin in the sheets and the frequency. Moreover, PACN15s composite shows an absorption efficiency of more than 85%.

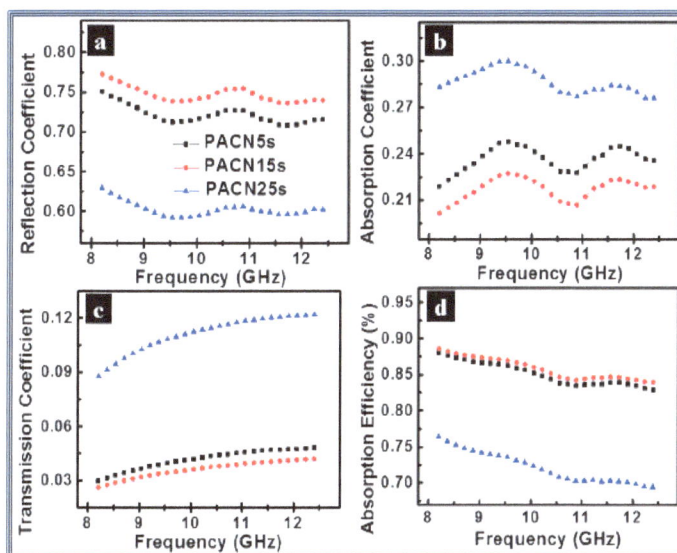

Fig. (18). (**a**) Reflection coefficient, (**b**) Absorption coefficient, (**c**) Transmission coefficient & (d) Absorption efficiency of the synthesized PACN sheets. Reproduced from the American Journal of Polymer Science 2015, 5(1A):28-39.

After calculating the value of A, R and T, the values of SE_R & SE_A for PACN sheets were calculated using Equations 8.16 and 8.17, respectively. Fig. (**19**) shows the variation of SE with frequency for PACN sheets in the 8.2-12.4 GHz range of microwave. When the PACN composite powders were compression moulded in the form of thin sheets, it was observed that these sheets have a sufficient value of SE_T (10.6-15.8 dB) even at very less thickness (~ 0.6 mm) (Fig. **19c**). The SE_T achieved for the PACN5s, PACN15s, and PACN25s is 15.25, 15.83, and 10.58 dB, respectively. It has been observed that for conducting PACN sheets, SE is contributed by absorption and reflection (Figs. **8.19a** and **8.19b**). The sufficient SE_T of PACN sheets is due to the improved interfacial interaction between resin and reinforcement phase of composite upon thermal curing (during hot pressing) of the sheets at 150 °C temperature. The above observation is associated with two unique features of the PACN sheets, namely (i) the CF present in polymer matrix contributed positively to shielding the microwaves, mainly by absorption phenomenon and (ii) the highly conducting PACN sheets had high charge storage capacities, capable of absorbing the incident EM waves by polarization in the electric field. In contrast, the SE of PACN sheets was found very low beyond 25 wt. % loading of novolac resin, about 5-6 dB. It is marginally higher than the value of SE of the neat novolac resin sheet (~2 dB) [127], where the insulating nature of novolac resin results in the formation of insulating sheets with little potential for EMI shielding. Earlier works have reported the electromagnetic shielding and microwave absorption properties of conducting polymer filled with CF and some other magnetic materials, but thick samples are required to achieve high SE. SE is dependent on multiple factors, especially SE_A, which is directly proportional to thickness (from Equation 8.17). Therefore, it is observed that the thickness of the shield has a great influence on the microwave shielding properties, as shown in Fig. (**19d**), which demonstrates the SE_T for PACN15s samples at different thicknesses (also shown in Table **1**). When the thickness of PACN15s shield is 1.8 mm (multiple PACN15s samples), the total SE achieved is more than 20 dB which is greater than the SE required for techno-commercial applications [159]. Therefore, the sheets prepared in the present work can find their application as a futuristic microwave absorbing material.

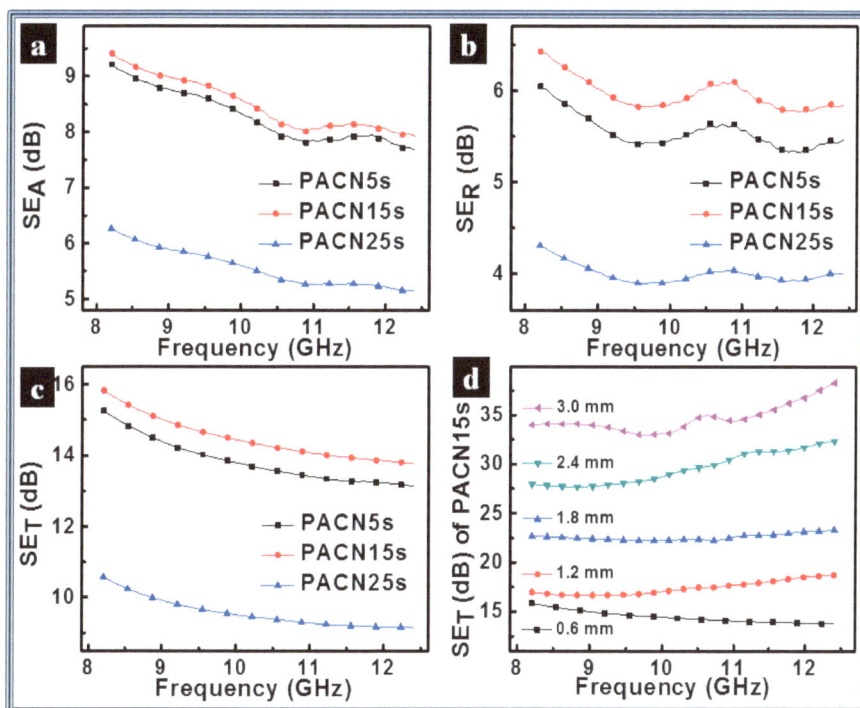

Fig. (19). Variation of (**a**) SE_A, (**b**) SE_R, & (**c**) SE_T with frequency in 8.2-12.4 GHz, showing the effect of novolac concentration in PACN sheets and (**d**) variation of SE_T of PACN15s with thickness. Reproduced from the American Journal of Polymer Science 2015, 5(1A):28-39.

CONCLUSION

The work presented in this chapter is about the fabrication of polyaniline-carbon fiber composites in the form of self-supported thin sheets. Thin sheets of PA-CF have been successfully prepared by compression moulding technique in hydraulic hot press, using novolac resin as a binder. Dispersion of conductive filler (CF) in the PA matrix improved the mechanical as well as electrical properties of the prepared sheets. By improving the filler matrix interaction, novolac resin has improved the mechanical properties of PACN sheets without affecting the electrical and shielding properties significantly. Novolac resin and CF jointly enhance the flexural strength of PACN sheets up to 50.48 MPa and flexural modulus up to 0.41 GPa. The thermal stability of the PACN sheet has been found to be 230 ^{0}C at 25 % loading of novolac resin. Self-supported thin sheets of PA-

CF composite with 15 % loading of novolac (PACN15s) have greater shielding effectiveness (15.8 dB at ~ 0.6 mm thickness) in 8.2-12.4 GHz frequency range. In addition, the PACN15s at a critical thickness of 3.0 mm (multiple PACN15s samples) had shown SE up to 35 dB. It is believed that these composite sheets could be promising candidates for next-generation building block material in microwave shielding with vast utility in aerospace applications due to their sufficient shielding even at low thickness, lightweight, good thermal stability, and mechanical strength.

CONSENT FOR PUBLICATION

Not applicable.

CONFLICT OF INTEREST

The author declares no conflict of interest, financial or otherwise.

ACKNOWLEDGEMENTS

Declared none.

REFERENCES

[1]   Shirakawa, H.; Louis, E.J.; MacDiarmid, A.G.; Chiang, C.K.; Heeger, A.J. Synthesis of electrically conducting organic polymers: halogen derivatives of polyacetylene, (CH). *J. Chem. Soc. Chem. Commun.,* **1977**, (16), 578-580.
       http://dx.doi.org/10.1039/c39770000578
[2]   Nalwa, H.S. *Handbook of Organic Conductive Molecules and Polymers, Volume 4, Conductive Polymers: Transport, Photophysics and Applications.,* **1997**.
[3]   Chiang, J-C.; MacDiarmid, A.G. 'Polyaniline': Protonic acid doping of the emeraldine form to the metallic regime. *Synth. Met.,* **1986**, *13*(1–3), 193-205.
       http://dx.doi.org/10.1016/0379-6779(86)90070-6
[4]   Diaz, A.F.; Rubinson, J.F.; Mark, H.B., Jr Electrochemistry and electrode applications of electroactive/conductive polymers.*Electronic Applications*; Springer Berlin Heidelberg, **1988**, pp. 113-139.
       http://dx.doi.org/10.1007/BFb0025905
[5]   Stenger-Smith, J.D. Intrinsically electrically conducting polymers. Synthesis, characterization, and their applications. *Prog. Polym. Sci.,* **1998**, *23*(1), 57-79.
       http://dx.doi.org/10.1016/S0079-6700(97)00024-5
[6]   Banerji, A.; Tausch, M.W.; Scherf, U. Classroom Experiments and Teaching Materials on OLEDs with Semiconducting Polymers. *Educ. Quím.,* **2013**, *24*(1), 17-22.
       http://dx.doi.org/10.1016/S0187-893X(13)73190-2

[7] Rehahn, M. Elektrisch leitfähige Kunststoffe: Der Weg zu einer neuen Materialklasse. *Chem. Unserer Zeit*, **2003**, *37*(1), 18-30.
http://dx.doi.org/10.1002/ciuz.200390000

[8] Skotheim, T.A.; Elsenbaumer, R.; Reynolds, J. *Handbook of conducting polymers*; New York, **1998**.

[9] Skotheim, T.A. *Handbook of conducting polymers*; CRC press, **1997**.

[10] Bishop, A.R.; Campbell, D.K.; Fesser, K. Polyacetylene and Relativistic Field Theory Models. *Mol. Cryst. Liq. Cryst. (Phila. Pa.)*, **1981**, *77*(1-4), 253-264.
http://dx.doi.org/10.1080/00268948108075245

[11] Bredas, J.L.; Street, G.B. Polarons, bipolarons, and solitons in conducting polymers. *Acc. Chem. Res.*, **1985**, *18*(10), 309-315.
http://dx.doi.org/10.1021/ar00118a005

[12] Su, W.P.; Schrieffer, J.R.; Heeger, A.J. Solitons in Polyacetylene. *Phys. Rev. Lett.*, **1979**, *42*(25), 1698-1701.
http://dx.doi.org/10.1103/PhysRevLett.42.1698

[13] Street, G.B.; Skotheim, T. *Handbook of conducting polymers*, **1986**.

[14] Letheby, H. XXIX.-On the production of a blue substance by the electrolysis of sulphate of aniline. *J. Chem. Soc.*, **1862**, *15*(0), 161-163.
http://dx.doi.org/10.1039/JS8621500161

[15] De Surville, R.; Jozefowicz, M.; Yu, L.T.; Pepichon, J.; Buvet, R. Electrochemical chains using protolytic organic semiconductors. *Electrochim. Acta*, **1968**, *13*(6), 1451-1458.
http://dx.doi.org/10.1016/0013-4686(68)80071-4

[16] Doriomedoff, M. F. HautiBre-Cristofini, and S.-V. DE, *R., JOZEFOWICZ, M., YU, LT and BUVET, R. J. Chim. Phys.*, **1971**, *68*, 1055.
http://dx.doi.org/10.1051/jcp/1971681055

[17] Chiang, C.K.; Fincher, C.R.; Park, Y.W.; Heeger, A.J.; Shirakawa, H.; Louis, E.J.; Gau, S.C.; MacDiarmid, A.G. Electrical conductivity in doped polyacetylene. *Phys. Rev. Lett.*, **1977**, *39*(17), 1098-1101.
http://dx.doi.org/10.1103/PhysRevLett.39.1098

[18] MacDiarmid, A.G. "Synthetic Metals": A Novel Role for Organic Polymers (Nobel Lecture). *Angew. Chem. Int. Ed.*, **2001**, *40*(14), 2581-2590.
http://dx.doi.org/10.1002/1521-3773(20010716)40:14<2581::AID-ANIE2581>3.0.CO;2-2

[19] Stejskal, J.; Gilbert, R.G. Polyaniline. Preparation of a conducting polymer(IUPAC Technical Report), in Pure and Applied Chemistry , **2002**; p. 857.

[20] Camalet, J.L.; Lacroix, J.C.; Aeiyach, S.; Chane-Ching, K.; Lacaze, P.C. Electrosynthesis of adherent polyaniline films on iron and mild steel in aqueous oxalic acid medium. *Synth. Met.*, **1998**, *93*(2), 133-142.
http://dx.doi.org/10.1016/S0379-6779(97)04099-X

[21] Jaymand, M. Recent progress in chemical modification of polyaniline. *Prog. Polym. Sci.*, **2013**, *38*(9), 1287-1306.
http://dx.doi.org/10.1016/j.progpolymsci.2013.05.015

[22] Cao, Y.; Smith, P.; Heeger, A.J. Counter-ion induced processibility of conducting polyaniline and of conducting polyblends of polyaniline in bulk polymers. *Synth. Met.*, **1992**, *48*(1), 91-97.
http://dx.doi.org/10.1016/0379-6779(92)90053-L

[23] Palaniappan, S.; John, A. Polyaniline materials by emulsion polymerization pathway. *Prog. Polym. Sci.*, **2008**, *33*(7), 732-758.

http://dx.doi.org/10.1016/j.progpolymsci.2008.02.002

[24]   Li, X-G.; Zhou, H-J.; Huang, M-R. Synthesis and properties of a functional copolymer from N-ethylaniline and aniline by an emulsion polymerization. *Polymer (Guildf.)*, **2005**, *46*(5), 1523-1533.

http://dx.doi.org/10.1016/j.polymer.2004.12.021

[25]   Gök, A.; Sarı, B.; Talu, M. Synthesis and characterization of conducting substituted polyanilines. *Synth. Met.*, **2004**, *142*(1–3), 41-48.

http://dx.doi.org/10.1016/j.synthmet.2003.07.002

[26]   Holze, R. Copolymers—A refined way to tailor intrinsically conducting polymers. *Electrochim. Acta*, **2011**, *56*(28), 10479-10492.

http://dx.doi.org/10.1016/j.electacta.2011.04.013

[27]   Huang, L-M.; Wen, T-C.; Gopalan, A. Synthesis and characterization of soluble conducting poly(aniline-co-2, 5-dimethoxyaniline). *Mater. Lett.*, **2003**, *57*(12), 1765-1774.

http://dx.doi.org/10.1016/S0167-577X(02)01066-2

[28]   Tsotra, P.; Friedrich, K. Short carbon fiber reinforced epoxy resin/polyaniline blends: their electrical and mechanical properties. *Compos. Sci. Technol.*, **2004**, *64*(15), 2385-2391.

http://dx.doi.org/10.1016/j.compscitech.2004.05.003

[29]   Huang, J-C. Carbon black filled conducting polymers and polymer blends. *Adv. Polym. Technol.*, **2002**, *21*(4), 299-313.

http://dx.doi.org/10.1002/adv.10025

[30]   Saini, P.; Choudhary, V.; Singh, B.P.; Mathur, R.B.; Dhawan, S.K. Enhanced microwave absorption behavior of polyaniline-CNT/polystyrene blend in 12.4–18.0 GHz range. *Synth. Met.*, **2011**, *161*(15–16), 1522-1526.

http://dx.doi.org/10.1016/j.synthmet.2011.04.033

[31]   Storrier, G.D.; Colbran, S.B.; Hibbert, D.B. Chemical and electrochemical syntheses, and characterization of poly (2, 5-dimethoxyaniline)(PDMA): a novel, soluble, conducting polymer. *Synth. Met.*, **1994**, *62*(2), 179-186.

http://dx.doi.org/10.1016/0379-6779(94)90309-3

[32]   Yue, J.; Wang, Z.H.; Cromack, K.R.; Epstein, A.J.; MacDiarmid, A.G. Effect of sulfonic acid group on polyaniline backbone. *J. Am. Chem. Soc.*, **1991**, *113*(7), 2665-2671.

http://dx.doi.org/10.1021/ja00007a046

[33]   Raghunathan, A.; Kahol, P.K.; McCormick, B.J. Electron localization studies of alkoxy polyanilines. *Synth. Met.*, **1999**, *100*(2), 205-216.

http://dx.doi.org/10.1016/S0379-6779(99)00008-9

[34]   Malinauskas, A.; Holze, R. An *in situ* UV—vis spectroelectrochemical investigation of the initial stages in the electrooxidation of selected ring- and nitrogen-alkylsubstituted anilines. *Electrochim. Acta*, **1999**, *44*(15), 2613-2623.

http://dx.doi.org/10.1016/S0013-4686(98)00390-9

[35]   Borole, D.D.; Kapadi, U.R.; Mahulikar, P.P.; Hundiwale, D.G. Electrochemical behaviour of polyaniline, poly(o-toluidine) and their copolymer in organic sulphonic acids. *Mater. Lett.*, **2004**, *58*(29), 3816-3822.

http://dx.doi.org/10.1016/j.matlet.2004.07.035

[36]   Kumar, D. Electrochemical and optical behaviour of conducting polymer: poly(o-toluidine). *Eur. Polym. J.*, **1999**, *35*(10), 1919-1923.

http://dx.doi.org/10.1016/S0014-3057(98)00178-5

[37] Pinto, N.J.; Kahol, P.K.; McCormick, B.J.; Dalal, N.S.; Wan, H. Charge transport and electron localization in polyaniline derivatives. *Phys. Rev. B Condens. Matter,* **1994**, *49*(19), 13983-13986.
http://dx.doi.org/10.1103/PhysRevB.49.13983 PMID: 10010348

[38] Kahol, P.K.; Pinto, N.J.; McCormick, B.J. Charge transport and electron localization in alkyl ring-substituted polyanilines. *Solid State Commun.,* **1994**, *91*(1), 21-24.
http://dx.doi.org/10.1016/0038-1098(94)90835-4

[39] Segawa, H.; Shimidzu, T.; Honda, K. A novel photo-sensitized polymerization of pyrrole. *J. Chem. Soc. Chem. Commun.,* **1989**, (2), 132-133.
http://dx.doi.org/10.1039/c39890000132

[40] Della Pina, C.; Falletta, E.; Rossi, M. Conductive materials by metal catalyzed polymerization. *Catal. Today,* **2011**, *160*(1), 11-27.
http://dx.doi.org/10.1016/j.cattod.2010.05.023

[41] Carter, G.M.; Thakur, M.K.; Chen, Y.J.; Hryniewicz, J.V. Time and wavelength resolved nonlinear optical spectroscopy of a polydiacetylene in the solid state using picosecond dye laser pulses. *Appl. Phys. Lett.,* **1985**, *47*(5), 457-459.
http://dx.doi.org/10.1063/1.96146

[42] Wang, J.; Neoh, K.G.; Zhao, L.; Kang, E.T. Plasma polymerization of aniline on different surface functionalized substrates. *J. Colloid Interface Sci.,* **2002**, *251*(1), 214-224.
http://dx.doi.org/10.1006/jcis.2002.8389 PMID: 16290721

[43] Yamamoto, T.; Hayashi, Y.; Yamamoto, A. A Novel Type of Polycondensation Utilizing Transition Metal-Catalyzed C–C Coupling. I. Preparation of Thermostable Polyphenylene Type Polymers. *Bull. Chem. Soc. Jpn.,* **1978**, *51*(7), 2091-2097.
http://dx.doi.org/10.1246/bcsj.51.2091

[44] Snow, A.W. Vapour deposition polymerization of butadiyne. *Nature,* **1981**, *292*(5818), 40-41.
http://dx.doi.org/10.1038/292040a0

[45] Karasz, F.E.; Capistran, J.D.; Gagnon, D.R.; Lenz, R.W. High Molecular Weight Polyphenylene Vinylene. *Mol. Cryst. Liq. Cryst. (Phila. Pa.),* **1985**, *118*(1), 327-332.
http://dx.doi.org/10.1080/00268948508076234

[46] Macdiarmid, A.G.; Chiang, J-C.; Halpern, M.; Huang, W-S.; Mu, S-L.; Nanaxakkara, L.D.; Wu, S.W.; Yaniger, S.I. "Polyaniline": Interconversion of Metallic and Insulating Forms. *Mol. Cryst. Liq. Cryst. (Phila. Pa.),* **1985**, *121*(1-4), 173-180.
http://dx.doi.org/10.1080/00268948508074857

[47] Sapurina, I.; Stejskal, J. The mechanism of the oxidative polymerization of aniline and the formation of supramolecular polyaniline structures. *Polym. Int.,* **2008**, *57*(12), 1295-1325.
http://dx.doi.org/10.1002/pi.2476

[48] Ćirić-Marjanović, G.; Konyushenko, E.N.; Trchová, M.; Stejskal, J. Chemical oxidative polymerization of anilinium sulfate versus aniline: Theory and experiment. *Synth. Met.,* **2008**, *158*(5), 200-211.
http://dx.doi.org/10.1016/j.synthmet.2008.01.005

[49] Kogan, I. An oxidizing agent for aniline polymerization. *Synth. Met.,* **1999**, *100*, 303.

[50] Palaniappan, S. Benzoyl peroxide oxidation route to polyaniline salts—Part I. *Polym. Adv. Technol.,* **2004**, *15*(3), 111-117.
http://dx.doi.org/10.1002/pat.424

[51] Yasuda, A.; Shimidzu, T. Chemical Oxidative Polymerization of Aniline with Ferric Chloride. *Polym. J.,* **1993**, *25*(4), 329-338.
http://dx.doi.org/10.1295/polymj.25.329

[52]   Chowdhury, P.; Saha, B. Potassium dichromate initiated polymerization of aniline. *Indian J. Chem. Technol.,* **2005**, *12*(6), 671-675.

[53]   Rodrigues, M.A.; de Paoli, M-A. Electrochemical properties of chemically prepared poly(aniline). *Synth. Met.,* **1991**, *43*(1–2), 2957-2962.
         http://dx.doi.org/10.1016/0379-6779(91)91215-V

[54]   Wang, Y.; Liu, Z.; Han, B.; Sun, Z.; Huang, Y.; Yang, G. Facile synthesis of polyaniline nanofibers using chloroaurate acid as the oxidant. *Langmuir,* **2005**, *21*(3), 833-836.
         http://dx.doi.org/10.1021/la047442z PMID: 15667157

[55]   Kim, B-K.; Kim, Y.H.; Won, K.; Chang, H.; Choi, Y.; Kong, K.; Rhyu, B.W.; Kim, J-J.; Lee, J-O. Electrical properties of polyaniline nanofibre synthesized with biocatalyst. *Nanotechnology,* **2005**, *16*(8), 1177-1181.
         http://dx.doi.org/10.1088/0957-4484/16/8/033

[56]   Erdem, E.; Karakışla, M.; Saçak, M. The chemical synthesis of conductive polyaniline doped with dicarboxylic acids. *Eur. Polym. J.,* **2004**, *40*(4), 785-791.
         http://dx.doi.org/10.1016/j.eurpolymj.2003.12.007

[57]   Tan, S.; Tieu, J.H.; Bélanger, D. Chemical polymerization of aniline on a poly(styrene sulfonic acid) membrane: Controlling the polymerization site using different oxidants. *J. Phys. Chem. B,* **2005**, *109*(29), 14085-14092.
         http://dx.doi.org/10.1021/jp051278m PMID: 16852769

[58]   Cao, Y.; Andreatta, A.; Heeger, A.J.; Smith, P. Influence of chemical polymerization conditions on the properties of polyaniline. *Polymer (Guildf.),* **1989**, *30*(12), 2305-2311.
         http://dx.doi.org/10.1016/0032-3861(89)90266-8

[59]   Ding, H.; Wan, M.; Wei, Y. Controlling the diameter of polyaniline nanofibers by adjusting the oxidant redox potential. *Adv. Mater.,* **2007**, *19*(3), 465-469.
         http://dx.doi.org/10.1002/adma.200600831

[60]   Dyachkova, T. *Effects of the nature of oxidant and synthesis conditions on properties of nanocomposites of polyaniline/carbon nanotubes,*

[61]   Kim, J.; Kwon, S.; Ihm, D. Synthesis and characterization of organic soluble polyaniline prepared by one-step emulsion polymerization. *Curr. Appl. Phys.,* **2007**, *7*(2), 205-210.
         http://dx.doi.org/10.1016/j.cap.2006.05.001

[62]   Kinlen, P.J.; Frushour, B.G.; Ding, Y.; Menon, V. Synthesis and Characterization of Organically Soluble Polyaniline and Polyaniline Block Copolymers. *Synth. Met.,* **1999**, *101*(1–3), 758-761.
         http://dx.doi.org/10.1016/S0379-6779(98)00280-X

[63]   Ichinohe, D.; Aral, T.; Kise, H. Synthesis of soluble polyaniline in reversed micellar systems. *Synth. Met.,* **1997**, *84*(1–3), 75-76.
         http://dx.doi.org/10.1016/S0379-6779(96)03843-X

[64]   Stejskal, J.; Omastová, M.; Fedorova, S.; Prokeš, J.; Trchová, M. Polyaniline and polypyrrole prepared in the presence of surfactants: a comparative conductivity study. *Polymer (Guildf.),* **2003**, *44*(5), 1353-1358.
         http://dx.doi.org/10.1016/S0032-3861(02)00906-0

[65]   Kudoh, Y. Properties of polypyrrole prepared by chemical polymerization using aqueous solution containing $Fe_2(SO_4)_3$ and anionic surfactant. *Synth. Met.,* **1996**, *79*(1), 17-22.
         http://dx.doi.org/10.1016/0379-6779(96)80124-X

[66]   Lee, Y.H.; Lee, J.Y.; Lee, D.S. A novel conducting soluble polypyrrole composite with a polymeric co-dopant. *Synth. Met.,* **2000**, *114*(3), 347-353.

http://dx.doi.org/10.1016/S0379-6779(00)00268-X

[67]   Kobayashi, A.; Xu, X.; Ishikawa, H.; Satoh, M.; Hasegawa, E. Electrical conduction in polyaniline compressed pellets doped with alkylbenzenesulfonic acids. *J. Appl. Phys.,* **1992**, *72*(12), 5702-5705.
       http://dx.doi.org/10.1063/1.351921

[68]   Dhawan, S.K.; Singh, N.; Rodrigues, D. Electromagnetic shielding behaviour of conducting polyaniline composites. *Sci. Technol. Adv. Mater.,* **2003**, *4*(2), 105-113.
       http://dx.doi.org/10.1016/S1468-6996(02)00053-0

[69]   Shannon, K.; Fernandez, J.E. Preparation and properties of water-soluble, poly(styrenesulfonic acid)-doped polyaniline. *J. Chem. Soc. Chem. Commun.,* **1994**, (5), 643-644.
       http://dx.doi.org/10.1039/c39940000643

[70]   Proń, A.; Laska, J.; Österholm, J-E.; Smith, P. Processable conducting polymers obtained *via* protonation of polyaniline with phosphoric acid esters. *Polymer (Guildf.),* **1993**, *34*(20), 4235-4240.
       http://dx.doi.org/10.1016/0032-3861(93)90182-A

[71]   Eisazadeh, H.; Khorshidi, H.R. Preparation and characterization of polyaniline-DBSNa/Fe2O3 and polyaniline-DBSNa/CoO nanocomposites using surfactive dopant sodium dodecylbenzenesulfonate (DBSNa). *Journal of Vinyl and Additive Technology,* **2010**, *16*(1), 105-110.
       http://dx.doi.org/10.1002/vnl.20214

[72]   Leng, W.; Zhou, S.; Gu, G.; Wu, L. Wettability switching of SDS-doped polyaniline from hydrophobic to hydrophilic induced by alkaline/reduction reactions. *J. Colloid Interface Sci.,* **2012**, *369*(1), 411-418.
       http://dx.doi.org/10.1016/j.jcis.2011.11.080 PMID: 22196348

[73]   Rao, P.S.; Sathyanarayana, D.N.; Palaniappan, S. Polymerization of Aniline in an Organic Peroxide System by the Inverted Emulsion Process. *Macromolecules,* **2002**, *35*(13), 4988-4996.
       http://dx.doi.org/10.1021/ma0114638

[74]   Wan, M.; Li, J. Tubular poly(ortho-toluidine) synthesized by a template-free method. *Polym. Adv. Technol.,* **2003**, *14*(3-5), 320-325.
       http://dx.doi.org/10.1002/pat.314

[75]   Palaniappan, S.; Nivasu, V. Emulsion polymerization pathway for preparation of organically soluble polyaniline sulfate. *New J. Chem.,* **2002**, *26*(10), 1490-1494.
       http://dx.doi.org/10.1039/b106222k

[76]   Kim, B-J.; Oh, S-G.; Han, M-G.; Im, S-S. Preparation of Polyaniline Nanoparticles in Micellar Solutions as Polymerization Medium. *Langmuir,* **2000**, *16*(14), 5841-5845.
       http://dx.doi.org/10.1021/la9915320

[77]   Xiao, L.; Cao, Y.; Xiao, J.; Schwenzer, B.; Engelhard, M.H.; Saraf, L.V.; Nie, Z.; Exarhos, G.J.; Liu, J. A soft approach to encapsulate sulfur: polyaniline nanotubes for lithium-sulfur batteries with long cycle life. *Adv. Mater.,* **2012**, *24*(9), 1176-1181.
       http://dx.doi.org/10.1002/adma.201103392 PMID: 22278978

[78]   Dhawan, S.K.; Singh, N.; Venkatachalam, S. *Shielding behaviour of conducting polymer-coated fabrics in X-band, W-band and radio frequency range.,* **2002**, *129*(3), 261–267-261–267.
       http://dx.doi.org/10.1016/S0379-6779(02)00079-6

[79]   Dhawan, S.K.; Singh, K.; Bakhshi, A.K.; Ohlan, A. Conducting polymer embedded with nanoferrite and titanium dioxide nanoparticles for microwave absorption. *Synth. Met.,* **2009**, *159*(21–22), 2259-2262.
       http://dx.doi.org/10.1016/j.synthmet.2009.08.031

[80]   Rafiqi, F.A.; Majid, K. Synthesis, characterization, luminescence properties and thermal studies of polyaniline and polythiophene composites with rare earth terbium(III) complex. *Synth. Met.,* **2015**, *202,* 147-156.

http://dx.doi.org/10.1016/j.synthmet.2015.01.032

[81]   Rivera, R.; Pinto, N.J. Schottky diodes based on electrospun polyaniline nanofibers: Effects of varying fiber diameter and doping level on device performance. *Physica E,* **2009**, *41*(3), 423-426.

http://dx.doi.org/10.1016/j.physe.2008.09.002

[82]   Patil, S.S. *Enhanced field emission from chemically synthesized cadmium sulphide-polyaniline (CdS-PANI) nanotube composite. Vacuum Electronics Conference (IVEC),* **2013**.

http://dx.doi.org/10.1109/IVEC.2013.6571110

[83]   Jussila, S.; Puustinen, M.; Hassinen, T.; Olkkonen, J.; Sandberg, H.G.O.; Solehmainen, K. Self-aligned patterning method of poly(aniline) for organic field-effect transistor gate electrode. *Org. Electron.,* **2012**, *13*(8), 1308-1314.

http://dx.doi.org/10.1016/j.orgel.2012.04.004

[84]   Ansari, R.; Emsakpour, F.; Mohammad-Khah, A.; Arvand, M. Application of Polyaniline Conducting Polymer as a New Indicator Electrode for Potentiometric Titration of Halide Ions. *Curr. Phys. Chem.,* **2012**, *2*(3), 218-223.

http://dx.doi.org/10.2174/1877946811202030218

[85]   Steffens, C.; Corazza, M.L.; Franceschi, E.; Castilhos, F.; Herrmann, P.S.P., Jr; Oliveira, J.V. Development of gas sensors coatings by polyaniline using pressurized fluid. *Sens. Actuators B Chem.,* **2012**, *171–172,* 627-633.

http://dx.doi.org/10.1016/j.snb.2012.05.044

[86]   Sangamithirai, D.; Narayanan, V.; Muthuraaman, B.; Stephen, A. Investigations on the performance of poly(o-anisidine)/graphene nanocomposites for the electrochemical detection of NADH. *Mater. Sci. Eng. C,* **2015**, *55,* 579-591.

http://dx.doi.org/10.1016/j.msec.2015.05.066 PMID: 26117792

[87]   McCall, R.; Ginder, J.M.; Leng, J.M.; Coplin, K.A.; Ye, H.J.; Epstein, A.J.; Asturias, G.E.; Manohar, S.K.; Masters, J.G.; Scherr, E.M.; Sun, Y.; Macdiarmid, A.G. Photoinduced absorption and erasable optical information storage in polyanilines. *Synth. Met.,* **1991**, *41*(3), 1329-1332.

http://dx.doi.org/10.1016/0379-6779(91)91618-K

[88]   Tseng, R.J.; Huang, J.; Ouyang, J.; Kaner, R.B.; Yang, Y. Polyaniline nanofiber/gold nanoparticle nonvolatile memory. *Nano Lett.,* **2005**, *5*(6), 1077-1080.

http://dx.doi.org/10.1021/nl050587l PMID: 15943446

[89]   Khosrozadeh, A.; Xing, M.; Wang, Q. A high-capacitance solid-state supercapacitor based on free-standing film of polyaniline and carbon particles. *Appl. Energy,* **2015**, *153,* 87-93.

http://dx.doi.org/10.1016/j.apenergy.2014.08.046

[90]   Ji, Y.; Qin, C.; Niu, H.; Sun, L.; Jin, Z.; Bai, X. Electrochemical and electrochromic behaviors of polyaniline-graphene oxide composites on the glass substrate/Ag nano-film electrodes prepared by vertical target pulsed laser deposition. *Dyes Pigments,* **2015**, *117,* 72-82.

http://dx.doi.org/10.1016/j.dyepig.2015.01.026

[91]   Ji, Y. *Review.* Dyes and Pigments **2015**, *117*(complete), 72-82.

[92]   Drelinkiewicz, A.; Waksmundzka-Góra, A.; Sobczak, J.W.; Stejskal, J. Hydrogenation of 2-ethyl-9,10-anthraquinone on Pd-polyaniline(SiO2) composite catalyst: The effect of humidity. *Appl. Catal. A Gen.,* **2007**, *333*(2), 219-228.

http://dx.doi.org/10.1016/j.apcata.2007.09.011

[93]   Huang, J.; Kaner, R.B. Flash welding of conducting polymer nanofibres. *Nat. Mater.*, **2004**, *3*(11), 783-786.
        http://dx.doi.org/10.1038/nmat1242 PMID: 15502832

[94]   Blinova, N.V.; Stejskal, J.; Trchová, M.; Ćirić-Marjanović, G.; Sapurina, I. Polymerization of aniline on polyaniline membranes. *J. Phys. Chem. B*, **2007**, *111*(10), 2440-2448.
        http://dx.doi.org/10.1021/jp067370f PMID: 17311453

[95]   Bhandari, H.; Sathiyanaranayan, S.; Choudhary, V.; Dhawan, S.K. Synthesis and characterization of proccessible polyaniline derivatives for corrosion inhibition. *J. Appl. Polym. Sci.*, **2009**, *111*(5), 2328-2339.
        http://dx.doi.org/10.1002/app.29283

[96]   Li, S. Preparation of bamboo-like PPy nanotubes and their application for removal of Cr (VI) ions in aqueous solution. **2012**, *378*(1), 30-35.

[97]   Thomas, S.; Deepak, T.G.; Anjusree, G.S.; Arun, T.A.; Nair, S.V.; Nair, A.S. A review on counter electrode materials in dye-sensitized solar cells. *J. Mater. Chem. A Mater. Energy Sustain.*, **2014**, *2*(13), 4474-4490.
        http://dx.doi.org/10.1039/C3TA13374E

[98]   Sambyal, P.; Singh, A.P.; Verma, M.; Farukh, M.; Singh, B.P.; Dhawan, S.K. Tailored polyaniline/barium strontium titanate/expanded graphite multiphase composite for efficient radar absorption. *RSC Advances*, **2014**, *4*(24), 12614-12624.
        http://dx.doi.org/10.1039/c3ra46479b

[99]   Kim, H.K.; Kim, M.S.; Song, K.; Park, Y.H.; Kim, S.H.; Joo, J.; Lee, J.Y. EMI shielding intrinsically conducting polymer/PET textile composites. *Synth. Met.*, **2003**, *135–136*, 105-106.
        http://dx.doi.org/10.1016/S0379-6779(02)00876-7

[100]   Al-Saleh, M.H.; Sundararaj, U. Electromagnetic interference shielding mechanisms of CNT/polymer composites. *Carbon*, **2009**, *47*(7), 1738-1746.
        http://dx.doi.org/10.1016/j.carbon.2009.02.030

[101]   Singh, A.P.; Mishra, M.; Sambyal, P.; Gupta, B.K.; Singh, B.P.; Chandra, A.; Dhawan, S.K. Encapsulation of [gamma]-Fe_2O_3 decorated reduced graphene oxide in polyaniline core-shell tubes as an exceptional tracker for electromagnetic environmental pollution. *J. Mater. Chem. A Mater. Energy Sustain.*, **2014**, *2*(10), 3581-3593.
        http://dx.doi.org/10.1039/C3TA14212D

[102]   Vasquez, H. *Simple device for electromagnetic interference shielding effectiveness measurement.*, **2009**.

[103]   Kumar, S.; Pathak, P. Effect of electromagnetic radiation from mobile phones towers on human body. *Indian J. Radio Space Phys.*, **2011**, *40*(6), 340-342.

[104]   Chung, D.D.L. *Electromagnetic interference shielding effectiveness of carbon materials.*, **2001**, *39*(2), 279285-279285. Carbon.
        http://dx.doi.org/10.1016/S0008-6223(00)00184-6

[105]   Saini, P.; Arora, M. *Microwave Absorption and EMI Shielding Behavior of Nanocomposites Based on 2 Intrinsically Conducting Polymers, 3 Graphene and Carbon Nanotubes 4.*, **2012**.
        http://dx.doi.org/10.5772/48779

[106]   Bjorklof, D. *Shielding for EMC*; Compliance Engineering, **1999**.

[107]   Anoop Kumar, S.; Singh, A.P.; Saini, P.; Khatoon, F.; Dhawan, S.K. Synthesis, charge transport studies, and microwave shielding behavior of nanocomposites of polyaniline with Ti-doped γ-Fe2O3. *J. Mater. Sci.*, **2012**, *47*(5), 2461-2471.
        http://dx.doi.org/10.1007/s10853-011-6068-5

[108]   Reddy, B. *Advances in nanocomposites-Synthesis, characterization and industrial applications*; InTech, **2011**.

http://dx.doi.org/10.5772/604

[109]   Ott, H.W.; Ott, H.W. *Noise reduction techniques in electronic systems*; Wiley New York, **1988**, Vol. 442, .

[110]   Saini, P.; Choudhary, V.; Singh, B.P.; Mathur, R.B.; Dhawan, S.K. Polyaniline–MWCNT nanocomposites for microwave absorption and EMI shielding. *Mater. Chem. Phys.,* **2009**, *113*(2–3), 919-926.

http://dx.doi.org/10.1016/j.matchemphys.2008.08.065

[111]   Olmedo, L.; Hourquebie, P.; Jousse, F. *Handbook of organic conductive molecules and polymers*; John Wiley and Sons Ltd: New York, **1997**.

[112]   Geetha, S.; Satheesh Kumar, K.K.; Rao, C.R.K.; Vijayan, M.; Trivedi, D.C. EMI shielding: Methods and materialsâ€"A review. *J. Appl. Polym. Sci.,* **2009**, *112*(4), 2073-2086.

http://dx.doi.org/10.1002/app.29812

[113]   Ott, H.W. *Electromagnetic compatibility*; Electromagnetic Compatibility Engineering, **2009**, pp. 1-43.

http://dx.doi.org/10.1002/9780470508510

[114]   Zhang, R.; Agar, J.C.; Wong, C.P. *Conductive Polymer Composites*, **2002**.

[115]   *Conductive Polymers: Synthesis and Electrical Properties*; Nalwa, H.S. Handbook of Organic Conductive Molecules and PolymersJohn Wiley and Sons: Chichester, **1997**, 2, .

[116]   Heeger, A.J. Semiconducting and Metallic Polymers: The Fourth Generation of Polymeric Materials (Nobel Lecture). *Angew. Chem. Int. Ed.,* **2001**, *40*(14), 2591-2611.

http://dx.doi.org/10.1002/1521-3773(20010716)40:14<2591::AID-ANIE2591>3.0.CO;2-0

[117]   Shirakawa, H. The Discovery of Polyacetylene Film: The Dawning of an Era of Conducting Polymers (Nobel Lecture). *Angew. Chem. Int. Ed.,* **2001**, *40*(14), 2574-2580.

http://dx.doi.org/10.1002/1521-3773(20010716)40:14<2574::AID-ANIE2574>3.0.CO;2-N

[118]   Freund, M.S.; Deore, B.A. *Self-doped conducting polymers*; John Wiley & Sons, **2007**.

http://dx.doi.org/10.1002/9780470061725

[119]   Singh, A.P.; Garg, P.; Alam, F.; Singh, K.; Mathur, R.B.; Tandon, R.P.; Chandra, A.; Dhawan, S.K. Phenolic resin-based composite sheets filled with mixtures of reduced graphene oxide, γ-Fe$_2$O$_3$ and carbon fibers for excellent electromagnetic interference shielding in the X-band. *Carbon,* **2012**, *50*(10), 3868-3875.

http://dx.doi.org/10.1016/j.carbon.2012.04.030

[120]   Subramoney, S. Carbon Nanotubes--A Status Report. *Electrochem. Soc. Interface,* **1999**, *8*(4), 34-41.

http://dx.doi.org/10.1149/2.F06994IF

[121]   *A review of EMI shielding and suppression materials. Proceedings of the International Conference on,* **1997**.

[122]   Mukerjee, P.; Mysels, K.J. *Critical micelle concentrations of aqueous surfactant systems*; DTIC Document, **1971**.

http://dx.doi.org/10.6028/NBS.NSRDS.36

[123]   Dominguez, A.; Fernandez, A.; Gonzalez, N.; Iglesias, E.; Montenegro, L. Determination of Critical Micelle Concentration of Some Surfactants by Three Techniques. *J. Chem. Educ.,* **1997**, *74*(10), 1227.

http://dx.doi.org/10.1021/ed074p1227

[124]   van Os, N.M.; Haak, J.R.; Rupert, L.A.M. *Physico-chemical properties of selected anionic, cationic and nonionic surfactants*; Elsevier, **2012**.

[125]   Huang, H.-L.; Lee, W-M.G. Enhanced naphthalene solubility in the presence of sodium dodecyl sulfate: effect of critical micelle concentration. *Chemosphere,* **2001**, *44*(5), 963-972.
        http://dx.doi.org/10.1016/S0045-6535(00)00367-2 PMID: 11513430

[126]   Shaw, D.J.; Costello, B. , "Introduction to colloid and surface chemistry: Butterworth-Heinemann", Oxford, 1991, ISBN 0 7506 1182 0, 306 pp 14-95: Elsevier, **1993**.

[127]   Gogoi, J.P.; Bhattacharyya, N.S. *Microwave Characterization of Expanded Graphite/Phenolic Resin Composite for Strategic Applications* Session 4P6 Plasmas, Composite Media, Materials Science: , 896.

[128]   Yan, F.; Xue, G. Synthesis and characterization of electrically conducting polyaniline in water-oil microemulsion. *J. Mater. Chem.,* **1999**, *9*(12), 3035-3039.
        http://dx.doi.org/10.1039/a905146e

[129]   Kim, B-J.; Oh, S-G.; Han, M-G.; Im, S-S. Synthesis and characterization of polyaniline nanoparticles in SDS micellar solutions. *Synth. Met.,* **2001**, *122*(2), 297-304.
        http://dx.doi.org/10.1016/S0379-6779(00)00304-0

[130]   Oh, S.G.; Shah, D.O. The effect of micellar lifetime on the rate of solubilization and detergency in sodium dodecyl sulfate solutions. *J. Am. Oil Chem. Soc.,* **1993**, *70*(7), 673-678.
        http://dx.doi.org/10.1007/BF02641002

[131]   Wei, Y.; Jang, G.W.; Chan, C.C.; Hsueh, K.F.; Hariharan, R.; Patel, S.A.; Whitecar, C.K. Polymerization of aniline and alkyl ring-substituted anilines in the presence of aromatic additives. *J. Phys. Chem.,* **1990**, *94*(19), 7716-7721.
        http://dx.doi.org/10.1021/j100382a073

[132]   Wei, Y.; Hariharan, R.; Patel, S.A. Chemical and electrochemical copolymerization of aniline with alkyl ring-substituted anilines. *Macromolecules,* **1990**, *23*(3), 758-764.
        http://dx.doi.org/10.1021/ma00205a011

[133]   Wei, Y.; Hsueh, K.F.; Jang, G-W. Monitoring the chemical polymerization of aniline by open-circuit-potential measurements. *Polymer (Guildf.),* **1994**, *35*(16), 3572-3575.
        http://dx.doi.org/10.1016/0032-3861(94)90927-X

[134]   Ahmed, S.M. Mechanistic investigation of the oxidative polymerization of aniline hydrochloride in different media. *Polym. Degrad. Stabil.,* **2004**, *85*(1), 605-614.
        http://dx.doi.org/10.1016/j.polymdegradstab.2004.01.003

[135]   Han, D.; Chu, Y.; Yang, L.; Liu, Y.; Lv, Z. Reversed micelle polymerization: a new route for the synthesis of DBSA–polyaniline nanoparticles. *Colloids Surf. A Physicochem. Eng. Asp.,* **2005**, *259*(1–3), 179-187.
        http://dx.doi.org/10.1016/j.colsurfa.2005.02.017

[136]   German, R.M. *AZ of powder metallurgy*; Elsevier Science Limited, **2005**.

[137]   Sai Ram, M.; Palaniappan, S. Benzoyl peroxide oxidation route to polyaniline salt and its use as catalyst in the esterification reaction. *J. Mol. Catal. Chem.,* **2003**, *201*(1–2), 289-296.
        http://dx.doi.org/10.1016/S1381-1169(03)00157-2

[138]   Palaniappan, S.; Lakshmi Devi, S. Thermal stability and structure of electroactive polyaniline–fluoroboric acid–dodecylhydrogensulfate salt. *Polym. Degrad. Stabil.,* **2006**, *91*(10), 2415-2422.
        http://dx.doi.org/10.1016/j.polymdegradstab.2006.03.016

[139]   Tang, J.; Jing, X.; Wang, B.; Wang, F. Infrared spectra of soluble polyaniline. *Synth. Met.,* **1988**, *24*(3), 231-238.
        http://dx.doi.org/10.1016/0379-6779(88)90261-5

[140]   Cao, Y.; Li, S.; Xue, Z.; Guo, D. Spectroscopic and electrical characterization of some aniline oligomers and polyaniline. *Synth. Met.,* **1986**, *16*(3), 305-315.
        http://dx.doi.org/10.1016/0379-6779(86)90167-0

[141]   Palaniappan, S.; Amarnath, C. Polyaniline-dodecylhydrogensulfate-acid salt: synthesis and characterization. *Mater. Chem. Phys.,* **2005**, *92*(1), 82-88.
http://dx.doi.org/10.1016/j.matchemphys.2004.12.033

[142]   Hwang, G-W.; Wu, K-Y.; Hua, M-Y.; Lee, H-T.; Chen, S-A. Structures and properties of the soluble polyanilines, N-alkylated emeraldine bases. *Synth. Met.,* **1998**, *92*(1), 39-46.
http://dx.doi.org/10.1016/S0379-6779(98)80020-9

[143]   Hino, T.; Namiki, T.; Kuramoto, N. Synthesis and characterization of novel conducting composites of polyaniline prepared in the presence of sodium dodecylsulfonate and several water soluble polymers. *Synth. Met.,* **2006**, *156*(21–24), 1327-1332.
http://dx.doi.org/10.1016/j.synthmet.2006.10.001

[144]   Fonseca, C.P.; Almeida, D.A.L.; Baldan, M.R.; Ferreira, N.G. Influence of the PAni morphology deposited on the carbon fiber: An analysis of the capacitive behavior of this hybrid composite. *Chem. Phys. Lett.,* **2011**, *511*(1), 73-76.
http://dx.doi.org/10.1016/j.cplett.2011.05.042

[145]   Stejskal, J.; Omastová, M.; Fedorova, S.; Prokeš, J.; Trchová, M. Polyaniline and polypyrrole prepared in the presence of surfactants: a comparative conductivity study. *Polymer (Guildf.),* **2003**, *44*(5), 1353-1358.
http://dx.doi.org/10.1016/S0032-3861(02)00906-0

[146]   Singh, A.P. *Conduction mechanism in Polyaniline-flyash composite material for shielding against electromagnetic radiation in X-band & Ku band.,* **2011**, *1*(2)

[147]   Kanungo, M.; Kumar, A.; Contractor, A.Q. Studies on electropolymerization of aniline in the presence of sodium dodecyl sulfate and its application in sensing urea. *J. Electroanal. Chem. (Lausanne),* **2002**, *528*(1-2), 46-56.
http://dx.doi.org/10.1016/S0022-0728(02)00770-2

[148]   Kuramoto, N.; Geniès, E.M. Micellar chemical polymerization of aniline. *Synth. Met.,* **1995**, *68*(2), 191-194.
http://dx.doi.org/10.1016/0379-6779(94)02284-6

[149]   Rao, P.S.; Subrahmanya, S.; Sathyanarayana, D.N. Inverse emulsion polymerization: a new route for the synthesis of conducting polyaniline. *Synth. Met.,* **2002**, *128*(3), 311-316.
http://dx.doi.org/10.1016/S0379-6779(02)00016-4

[150]   Al-Saleh, M.H.; Saadeh, W.H.; Sundararaj, U. EMI shielding effectiveness of carbon based nanostructured polymeric materials: A comparative study. *Carbon,* **2013**, *60*(0), 146-156.
http://dx.doi.org/10.1016/j.carbon.2013.04.008

[151]   Varshney, S.; Ohlan, A.; Singh, K.; Jain, V.K.; Dutta, V.P.; Dhawan, S.K. Robust Multifunctional Free Standing Polypyrrole Sheet for Electromagnetic Shielding. *Sci. Adv. Mater.,* **2013**, *5*(7), 881-890.
http://dx.doi.org/10.1166/sam.2013.1534

[152]   Li, L.; Chung, D.D.L. Electrical and mechanical properties of electrically conductive polyethersulfone composites. *Composites,* **1994**, *25*(3), 215-224.
http://dx.doi.org/10.1016/0010-4361(94)90019-1

[153]   Thomassin, J-M.; Jérôme, C.; Pardoen, T.; Bailly, C.; Huynen, I.; Detrembleur, C. Polymer/carbon based composites as electromagnetic interference (EMI) shielding materials. *Mater. Sci. Eng. Rep.,* **2013**, *74*(7), 211-232.
http://dx.doi.org/10.1016/j.mser.2013.06.001

[154]   Tripathi, P.; Prakash Patel, C.R.; Dixit, A.; Singh, A.P.; Kumar, P.; Shaz, M.A.; Srivastava, R.; Gupta, G.; Dhawan, S.K.; Gupta, B.K.; Srivastava, O.N. High yield synthesis of electrolyte heating assisted

electrochemically exfoliated graphene for electromagnetic interference shielding applications. *RSC Advances,* **2015**, *5*(25), 19074-19081.

http://dx.doi.org/10.1039/C4RA17230B

[155]   Singh, A.P. *Probing the engineered sandwich network of vertically aligned carbon nanotube–reduced graphene oxide composites for high performance electromagnetic interference shielding applications,* **2015**.

http://dx.doi.org/10.1016/j.carbon.2014.12.065

[156]   Reshi, H.A. Nanostructured La0.7Sr0.3MnO3 compound for effective Electromagnetic Interference shielding in X-band frequency range. *J. Mater. Chem. C Mater. Opt. Electron. Devices,* **2014**.

[157]   Singh, A.P.; Mishra, M.; Chandra, A.; Dhawan, S.K. Graphene oxide/ferrofluid/cement composites for electromagnetic interference shielding application. *Nanotechnology,* **2011**, *22*(46), 465701.

http://dx.doi.org/10.1088/0957-4484/22/46/465701 PMID: 22024967

[158]   Mishra, M.; Singh, A.P.; Dhawan, S.K. Expanded graphite-nanoferrite-fly ash composites for shielding of electromagnetic pollution. *J. Alloys Compd.,* **2013**, *557*, 244-251.

http://dx.doi.org/10.1016/j.jallcom.2013.01.004

[159]   Verma, M.; Verma, P.; Dhawan, S.K.; Choudhary, V. Tailored graphene based polyurethane composites for efficient electrostatic dissipation and electromagnetic interference shielding applications. *RSC Advances,* **2015**, *5*(118), 97349-97358.

http://dx.doi.org/10.1039/C5RA17276D

CHAPTER 9

EMI Shielding Properties of Conducting Poly (aniline-co-o-toluidine)-CF-Novolac Composites

Seema Joon[1,*] and S.K. Dhawan[2]

[1] *D.T.E.A. Sr. Sec. School, Pusa Road, New Delhi – 11005, India*

[2]*CSIR-National Physical Laboratory, New Delhi 110012, India*

Abstract: A study was made to design a copolymer of aniline and o-toluidine and its composite with carbon fiber (CF) in making PANIoTCFN sheets for controlling electromagnetic interference. PANIoTTCFN composite synthesized by emulsion polymerization was physically blended with different proportions of novolac resin to prepare a composite sheet by hot press compression moulding. *In-situ* incorporation of carbon fiber into the copolymer during the synthesis leads to the formation of composites with improved mechanical, thermal, electrical, and shielding properties. Structural and morphological studies were carried out by FTIR, XRD, and SEM. PANIoTCFN composite sheets with 50 % loading of novolac resin have a flexural strength of 52.4 MPa and exhibited shielding effectiveness of 26 dB at a thickness of 1.48 mm of the composite sheet, which reveals that these composite sheets can be used for EMI shielding applications.

Keywords: Carbon Fiber, Conducting Copolymer Composites, Conducting Polymer, EMI Shielding, Polyaniline, Poly (o-Toluidine) Copolymer.

9.1. INTRODUCTION

Among all conducting polymers, polyaniline (PANI) is one of the most widely used polymers for EMI shielding application because of its easy synthesis, good environmental stability, and moderately high conductivity upon doping [1, 2]. Unfortunately, its insolubility in common organic solvents, except for N-methyl-2-pyrrolidone (NMP) and dimethyl sulfoxide (DMSO), results in difficult processability, which restricts its use in various technological applications [3, 4]. Solutions to this problem have been found by the use of substituted polyanilines

*Corresponding author Seema Joon: D.T.E.A. Sec. Sec. School, New Delhi-110005, India; Tel: 91 9416568509; E-mail: seemajoon03@gmail.com

Sundeep K. Dhawan, Avanish Pratap Singh, Anil Ohlan, Kuldeep Singh Kakran and Pradeep Sambyal (Ed.)
All rights reserved-© 2022 Bentham Science Publishers

such as alkyl & alkoxy substituted polyanilines, sulfonated polyaniline, halogenated polyaniline, *etc*. Due to good optical & electrochemical properties along with better solubility [5] and processability [6] than polyaniline [7, 8], poly (o-toluidine) has attracted the attention of researchers. Poly (o-toluidine) is a PANI derivative that contains the methyl group at the ortho position of the aromatic ring of the aniline monomer. But, the progress insolubility of poly (o-toluidine) facilitated its processability at the cost of conductivity, thereby restricting its applications where moderate conductivity is of paramount importance. To improve the solubility of PANI and combine the advantages of poly (o-toluidine) with polyaniline, copolymerization might be the best method [9-11]. It is one of the finest alternate methods for the improvement of processability of PANI without compromising the electrical properties much. Properties of copolymers differ significantly from those of respective homopolymers; therefore, copolymerization extends the capability of the polymer scientists to tailor-make a material with specific desired properties. Besides this, copolymerization from a pair of monomers also leads to an increase in the number of conductive polymers obtainable from the same set of monomers. It is well known from the literature that the properties of the copolymer are intermediary to those of homopolymers [12, 13]. Further, the properties of copolymers depend on the synthesis conditions, types, and ratio of their co-monomers [13, 14]. Studies on conducting copolymers are meagre as compared to those of conductive homopolymers. Among the copolymers of polyanilines, poly (aniline-co-o-toluidine) has attracted much attention due to its excellent optical and electrochemical properties [15, 16]. The synthesis of poly (aniline-co-o-toluidine) was initiated in 1990 [17], and to date, a lot of reports have been published on the synthesis of poly (aniline-co-o-toluidine) using different synthetic routes [18-22]. A wide range of electrical conductivity (0.1–10 S/cm) of this copolymer was established by Wei *et al.* [17].

The present work reported in this chapter aims to extract a combination of desired properties of a copolymer of aniline with o-toluidine. In the present chapter, the effect of copolymerization on various properties of PANI and PoT is reported in terms of measurement of thermal, mechanical, electrical, and shielding properties. The copolymer of aniline with o-toluidine, *i.e.*, poly (aniline-co-o-toluidine), (PANIoT), and its composite with CF were prepared by chemical oxidative emulsion polymerization for the preparation of its homopolymers and their composites. Further, thin sheets of poly (aniline-co-o-toluidine)-carbon fiber (PANIoT-CF) composite were also fabricated by solution casting cum hot pressing technique, using varying amounts of novolac resin as a binder. Light weight thin sheets of poly (aniline-co-o-toluidine)-carbon fiber-novolac (PANIoTCFN) composites were prepared and used for EMI shielding application for the first time.

9.2. SYNTHESIS OF MATERIALS

The monomers used for the synthesis of copolymer (PANIoT) and its composites/sheets are aniline (Merck, India) and o-toluidine (Merck, India). Both were freshly distilled prior to use. The oxidant taken was ammonium peroxydisulfate (APS, Merck, India), and other chemicals were of reagent grade and used as received. β-naphthalene sulphonic acid (β-NSA, Hi-media, India) was used as a surfactant as well as a dopant. Before use, PAN based carbon fiber (Toray-T-300-Japan) was chopped to get an average size of 5-6mm. Novolac resin (Pheno-organic, India) used as a binder has been used as received. Aqueous solutions were prepared in double distilled water having specific resistivity of $10^6 \, \Omega$-cm.

9.2.1. Synthesis of Poly (Aniline-co-o-Toluidine)-Carbon Fiber (PANIoT-CF) Composite

PANIoT-CF composite was synthesized by *in-situ* chemical oxidative emulsion polymerization. The preparation process for PANIoT-CF composite was as follows: 0.1M solution of NSA and chopped CF (20% by wt. of monomer) of 5mm average length were homogenized for 1h so that carbon fibers were uniformly dispersed. Aniline and o-toluidine (equimolar, 0.1M) were then added to the above dispersion and again homogenized for 1h to form an emulsion (micelles) of aniline and o-toluidine along with CF in NSA. This emulsion was then transferred to a triple walled stainless steel reactor and polymerized by *in-situ* chemical oxidative emulsion polymerization. The reaction was carried out at -3 to 0°C temperature with drop wise addition of 0.1M APS aqueous solution along with continuous stirring for 5-6h. The copolymer composite (PANIoT-CF) so prepared was filtered, washed, dried at 60-65°C in a vacuum oven, and finally crushed to get PANIoT-CF composite powder. In addition, PANIoT without CF (pristine PANIoT) was also synthesized by the same route for comparative study.

9.2.2. Fabrication of Thin Sheets of Poly (Aniline-co-o-Toluidine)-Carbon Fiber-Novolac (PANIoTCFN) Composites

PANIoTCFN thin sheets were fabricated using solution casting cum compression moulding (using a hydraulic hot press) techniques. In the preparation of these sheets, novolac resin was used as a binder and the concentration of novolac resin was varied to optimize maximum shielding effectiveness along with thermal and mechanical properties. CF here acts as a reinforcement and filler material. In order to optimize the concentration of novolac resin, 25, 50, and 75 wt. % of novolac resin was added into PANIoT-CF composite. The fabrication procedure of PANIoTCFN thin sheets was following: first of all, a measured amount of PANIoT-CF was dispersed in ethanol by means of magnetic stirring and thereafter, an adequate amount of novolac resin was added (varying from 25 to 75 wt. % keeping the PANIoT-CF plus novolac resin wt. % as 100). The above dispersion was then ultrasonicated for 2h. Further, the solution was homogenized for 30 min and then poured into a specially designed vessel attached with vacuum filtration. In this way, thin sheets were prepared and the excess amount of novolac resin underwent into the solvent. These obtained sheets were further compressed in a hydraulic hot press at 100°C and contact pressure. The curing of sheets was carried out at 150° C for 2h. The compression moulded sheets were ejected out from the mould after cooling at room temperature. In this way, thin sheets of PANIoT-CF with 25, 50, and 75wt. % loading of novolac resin were prepared and abbreviated as PANIoTCFN25, PANIoTCFN50, and PANIoTCFN75 sheets, respectively. Before casting the sheets, one part of the PANIoTCFN composites (PANIoTCFN25, PANIoTCFN50, and PANIoTCFN75) was dried in powder form. The resultant composites and sheets of PANIoTCFN25, PANIoTCFN50, and PANIoTCFN75 were further characterized and tested for EMI shielding application.

9.3. ANALYSIS OF MATERIALS

The structural and morphological analysis of the materials was carried out by XRD, FTIR, and SEM. Besides these characterizations, to acquire information about the thermal, electrical, mechanical properties of the samples, TGA, conductivity and flexural strength measurements were performed. Furthermore, the samples were tested for EMI shielding using VNA in the frequency range of 8-12GHz (X-band).

9.3.1. XRD Analysis

The X-ray diffraction (XRD) studies were carried out on a D8 advance X-ray diffractometer (Bruker) using CuKα radiation (λ= 1.540598A°) in scattering range(2θ) of 3°- 90° with a scan rate of 0.02°sec^{-1} and slit width of 1 mm. Fig. (**1**) illustrates the X-ray diffraction patterns of PANIoT, PANIoT–CF composite, and PANIoTCFN sheets with different compositions of novolac. The diffractogram of PANIoT shows some strong peaks at 2θ =8.74° (d= 10.11A°), 18.78° (d= 4.72A°), 19.78° (d= 4.48A°) and 23.96° (d= 3.71A°) while some less intense peaks at 2θ =

6.76°, 20.36°, 25.63°, 27.06°. The sharp peaks indicate that PANIoT exhibits semi-crystalline nature.

The diffractogram of PANIoT-CF shows peaks at $2\theta = 8.8°$, 18.61° but with less intensity, at the same time, all other PANIoT peaks are not seen clearly; instead, a broad halo is seen at $2\theta = 23°$-27° range. These changes are due to the incorporation of CF into the copolymer matrix, which explains that there might be some interaction of CF with PANIoT, thereby decreasing the crystallinity of PANIoT. In the case of PANIoTCFN sheets, the sharp peaks are not present at all; instead, the broad peaks are seen clearly showing the amorphous nature of sheets. This might be due to the presence of novolac in the sheets, which is amorphous in nature. The appearance of peaks during the formation of the copolymer suggests that some interaction occurred when the composite was created with novolac, as seen by the XRD pattern of the copolymer composite.

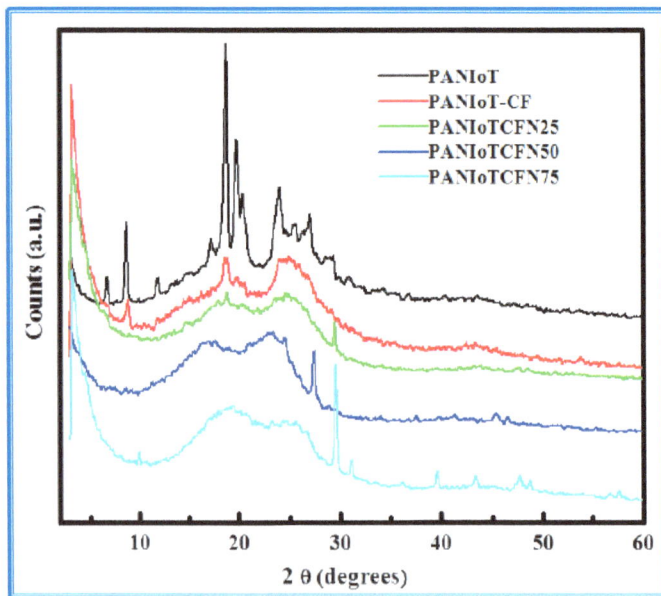

Fig. (1). XRD spectra of PANIoT, PANIoT-CF and PANIoTCFN25, PANIoTCFN50 and PANIoTCFN75 sheets.

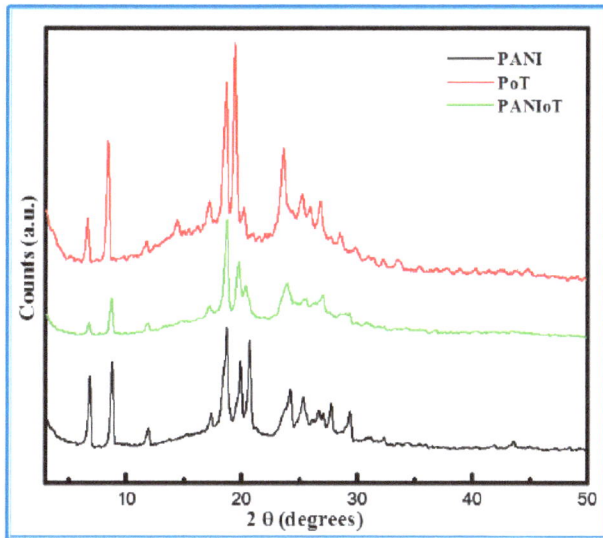

Fig. (2). Comparison of the XRD spectra of PANI, PoT, and PANIoT.

Fig. (**2**) shows the diffractogram of PANIoT along with PANI and PoT. The figure shows that all three diffractograms have semi-crystalline nature. However, in the case of a copolymer, the intensity and sharpness of the peak decrease, showing less crystallinity than the homopolymers. Moreover, there is no major change in the peaks; only slight shifting of peaks is observed in PoT & PANIoT in comparison to PANI.

9.3.2 FTIR Spectral Study

FTIR analysis of different polymer and polymer composite samples was carried out to see how the presence of methyl groups at the ortho position shifts the band position of functional groups when copolymerization is carried out. The instrument used for this purpose was Nicolet 5700, and the spectra were recorded in the wavenumber range of 4000-500 cm^{-1}. Fig. (**3**) demonstrates the Fourier-transform infrared spectroscopy (FTIR) spectra of pristine PANIoT, PANIoT-CF, and PANIoTCFN composites. The spectrum of the PANIoT copolymer is consistent with the previously reported spectrum of copolymer [19]. The bands at 1580 cm^{-1} and 1492 cm^{-1} correspond to the stretching of quinoid and benzenoid rings of PANIoT; the position of these peaks are in the range as reported in the earlier reports of polyanilines copolymers [23, 24]. The peak at 1303 cm^{-1} is because of the C–N asymmetric stretching of the benzenoid and quinoid ring [25, 26], while a band at

1157 cm^{-1} in conductive PANIoT exhibits the degree of electron delocalization over the polymer backbone, thus the electrical conductivity [15]. The peak at 1018 cm^{-1} is due to the vibrations of dopant ion [27], hence confirming the presence of the SO$_3^-$ group of NSA. A peak at 879 cm^{-1} appears due to the methyl group attached to the benzene ring [15, 28]. The band at 809 cm^{-1} attributes to para coupled phenyl rings indicating the polymer formation [15, 29]. A peak at 744 cm^{-1} may be assigned to 1,2 substituted benzene ring of ortho-toluidine unit present in the copolymer and indicates the presence of substituent group on ortho position. A similar sequential structure is observed in the spectra of PANIoT-CF and PANIoTCFN composites. There is no appearance of a new band in FTIR spectra of composites which indicates that there is no net chemical reaction occurring between the three components of the composite. Only a slight shifting in the main peaks of PANIoT is observed. This shifting of peaks may be due to the incorporation of CF and novolac resin into the PANIoT matrix, which indicates that some interaction of CF & novolac takes place with the polymer chains. Moreover, the intensity ratio of the peak due to quinoid ring stretching (1580 cm^{-1}) to the peak due to benzenoid ring stretching (1492 cm^{-1}) increased in the spectrum of PANIoT-CF as compared to the spectrum of PANIoT.

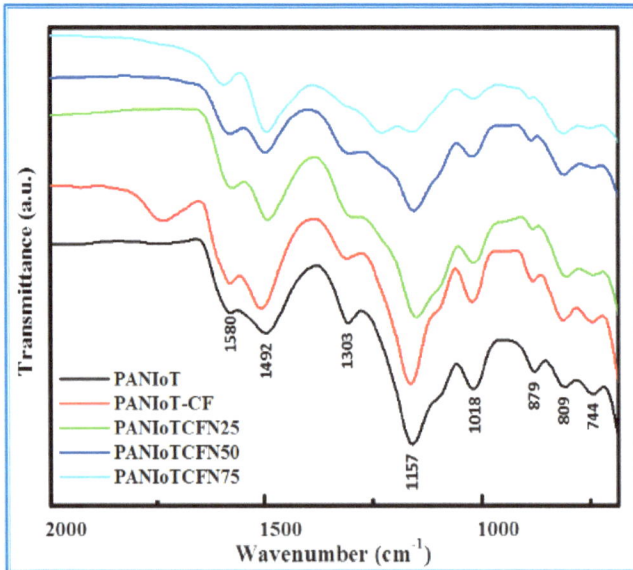

Fig. (3). FTIR spectra of PANIoT, PANIoT-CF and PANIoTCFN composites.

This may suggest that the *in-situ* incorporation of CF into polymer matrix favours the formation of quinoid structure, and this might be a sign of higher conductivity of PANIoT-CF composites as compared to pristine PANIoT. The FTIR spectra of PANIoT, PANIoT-CF, and PANIoTCFN confirm the functional groups present in the copolymer.

Fig. (**4**) shows the comparison of the FTIR spectrum of PANIoT with the spectra of PANI and PoT. The positions and characteristics of the peaks shown by PANI, PoT, and PANIoT are mentioned in Table **1**. Here, it is to be noted that PANI, PoT, and PANIoT showed all the characteristic peaks according to the previous reports of polyanilines [30, 31].

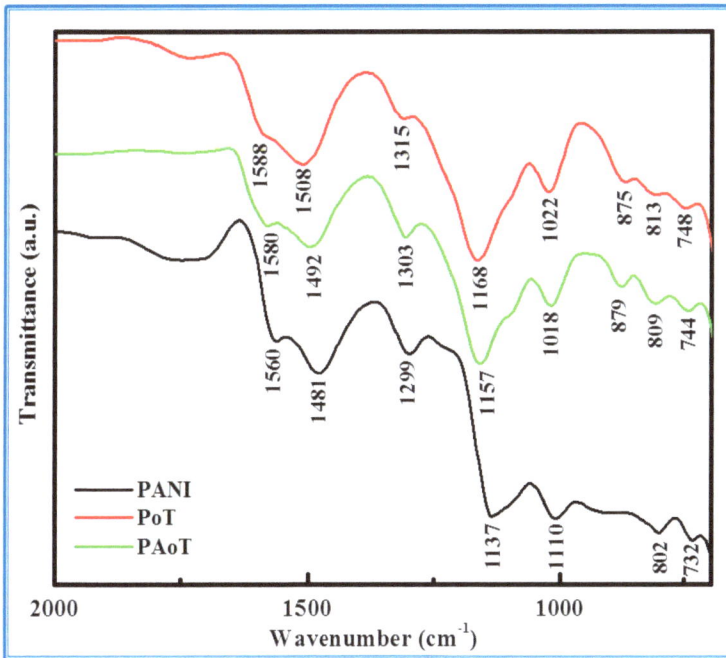

Fig. (4). Comparison of the FTIR spectra of PANI, PoT and PANIoT.

From the comparison of PANI with the PoT spectrum, it is revealed that PoT not only showed all the characteristic peaks like PANI but also showed one extra peak (875 cm^{-1}), which is due to the presence of substituent methyl group. In addition, the positions and intensity of some characteristic peaks of PANI are also changed in the case of PoT (Table **1**) which maybe because of the steric effect of the methyl group and difference in protonation ability of PoT and PANI. In the case of PoT, a major decrease in the relative intensity ratio of quinoid (1588 cm^{-1}) to benzenoid (1508 cm^{-1}) structures is observed, indicating a less favourable quinoid structure in PoT than in PANI; hence, PoT may have less conductivity as compared to PANI. On comparing the spectrum of PANIoT with its homopolymers (Fig. **9.4**), it is found that all the characteristic peaks of homopolymers are present in the PANIoT spectrum, but shifting in the position and intensity of the major peaks is observed, which may be due to the effect of copolymerization. A peak at 879 cm^{-1}, which is due to the substituent methyl group, indicates the presence of an o-toluidine unit in the copolymer chains. The position of most of the PANIoT peaks is between the peaks of PANI and PoT. It suggests that the synthesized polymer is a copolymer and not a mixture of homopolymers. Moreover, the relative intensity ratio of quinoid (1580 cm^{-1}) to benzenoid (1492 cm^{-1}) structures is observed between those of PANI and PoT, which indicates that the electrical conductivity of the synthesized copolymer may lie in between the conductivities of its homopolymers. For the same reason, FTIR characteristic peaks of PANIoT composites are in between the composites of PANI and PoT.

Table 1. Comparison of the FTIR characteristics of PANI, PoT, and PANIoT.

Peak position (cm^{-1})			Band Characteristics
PANI	PANIoT	PoT	
732	744	748	1,2 substitutions on benzene ring
802	809	813	Para coupled phenyl ring indicating the polymer formation
–	879	875	1,2,4 trisubstituted benzene ring in the o-toluidine unit (confirm the presence of methyl group on the ortho position
1010	1018	1022	Vibration band of dopant ion (SO$_3^-$ of NSA)
1137	1157	1168	C-H in-plane bending vibrations of nitrogen quinine (electronic band)
1299	1303	1315	C-N stretching of aromatic amine
1481	1492	1508	Benzenoid ring stretching
1560	1580	1588	Quinoid ring stretching

9.3.3. SEM Analysis

The surface morphology of the pristine PANIoT, PANIoT-CF composite, and PANIoTCFN sheets was determined by scanning electron microscope (SEM, Zeiss EVO MA-10). Fig. (**5**) shows the SEM micrographs of the pristine PANIoT (a)PANIoT-CF (b). The SEM micrograph of the pristine PANIoT (displayed in Fig. **6.5a**) shows a different morphology than its homopolymers. Here, the tubular morphology is not seen clearly; instead, small granules of PANIoT were gathered to form flakes or slices. Few of the microtubules of the copolymer (PANIoT) are seen in between the flakes. Fig. (**5b**) shows the micrograph of the PANIoT-CF composite, which confirms the insertion of CF into the copolymer matrix. It is seen from this micrograph that the CF is well dispersed, and some of the PANIoT molecules are also deposited on the surface of CF. So, CF makes an interwoven fibrous network into the conducting copolymer matrix and improves the conducting channels for the free flow of electrons. SEM image of PANIoTCFN50 sheet sample (Fig. **9.5c**) shows the signatures of CF into the solid thin sheet. This image indicates that the PANIoT-CF composite, when blended with novolac resin and transformed into thin sheets, the structure of the resultant three-phase (copolymer-CF-novolac) system became compact and dense. It seems that novolac resin binds the filler unit (CF) very tightly with the copolymer matrix.

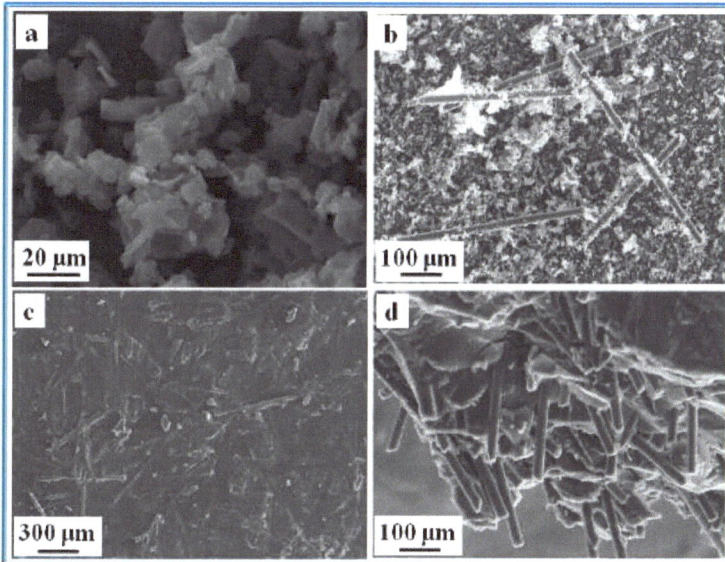

Fig. (5). SEM image of (**a**) PANIoT, (**b**) PANIoT-CF, (**c**) PANIoTCFN50 sheet and (**d**) Fractured PANIoTCFN50 sheet.

Fig. (**5d**) shows the SEM image of the fractured PANIoTCFN50 sheet. The fractured morphology of the PANIoTCFN50 sheet displays that some of the CF were pulled out, creating empty holes on the surface, while most others were broken. The broken-out CF indicates that during hot pressing of the sheets, strong interfacial interaction occurred between CF and novolac, which makes the structure of composite more compact.

9.3.4. Thermogravimetric Analysis

The thermal behaviour of the copolymer PANIoT, PANIoT-CF composite, and PANIoTCFN sheets has been investigated through thermogravimetric analysis (TGA), as shown in Fig. (**6**). It was carried out on Mettler Toledo TGA 851e. The thermogram of the copolymer and its composite samples show three major weight losses analogous to thermograms obtained for PANI and PoT.

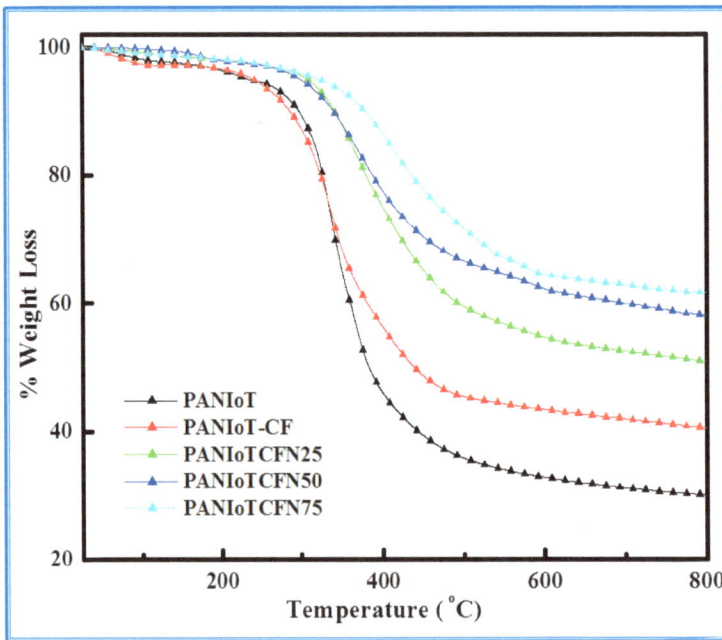

Fig. (6). Thermograms of PANIoT, PANIoT-CF and PANIoTCFN25, PANIoTCFN50 and PANIoTCFN75 sheets.

First weight loss at 110°C is due to the loss of water/moisture entrapped inside the copolymer moiety, second weight loss, starting from ~ 205°C and lasting up to ~ 425°C, is due to the loss of dopant (NSA) and third weight loss from 425 - 800°C is due to the degradation of the polymer backbone. Figure 6.6 demonstrates the effect of CF on the thermal stability of PANIoT. On comparing the thermograms of PANIoT-CF composite with PANIoT, the thermal stability of the composite has been found to increase up to 220°C. It is observed that CF increases the thermal stability of the copolymer significantly in a similar way as in the case of its homopolymers (PANI & PoT). This increase in thermal stability can be accounted for by the interaction of CF (acting as reinforcement) with the copolymer matrix. Further, it is clear from the thermograms of PANIoTCFN sheets that there is a tremendous increase in the thermal stability of PANIoT-CF composite when it is transformed into the form of thin sheets by using different amounts of novolac resin (as a binder). This enhancement in thermal stability is due to the presence of novolac, which is cross-linked during the curing of the composite sheets at 150°C. This increase in thermal stability is sustained with the % increase of novolac in the sheets (25% to 75%). Thus, the thermal stability of the PANIoTCFN75 sheet is maximum, which is ~ 305°C.

Fig. (**7**) shows the comparison of thermal stability of copolymer PANIoT with its homopolymers (PANI and PoT) to establish the effect of derivatization and copolymerization on the thermal stability of PANI. The thermogram of the PANIoT shows a similar three-step thermal degradation as that of PANI and PoT but at a lower temperature than PANI & at a higher temperature than that of PoT. This suggests that PoT and PANIoT have less thermal stability as compared to PANI. This may be attributed to the attached methyl group at the ortho position, which due to steric hindrance, distorts the geometry of the polymer, thereby reducing the inter-chain interaction and decreasing the thermal stability. This effect is less in PANIoT as compared to PoT because of the smaller number of substituent groups in PANIoT as compared to pure PoT. So, PANIoT showed thermal stability intermediate to its homopolymers. In the same pattern, the thermal stability of PANIoTCFN sheets is found in between the thermal stability of PANICFN and PoTCFN sheets.

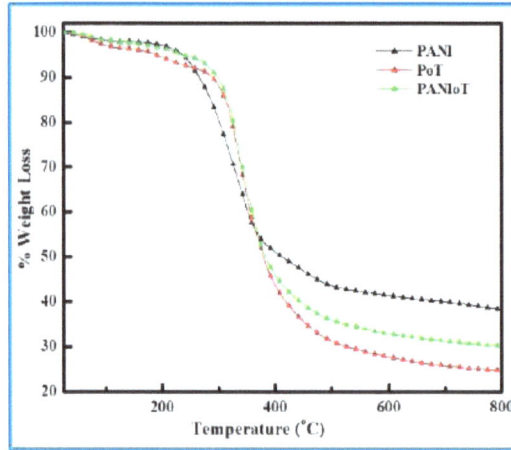

Fig. (7). Comparison of the thermograms of PANI, PoT, and PANIoT.

9.3.5. Flexural Strength

Flexural strength is one of the important properties of composite materials to determine mechanical strength. Thus, in the present work, the flexural strength of PANIoTCFN sheets was measured by the three-point bending test based on ASTM D790 [32] using an Instron Universal Testing Machine (model 4411) at a crosshead speed of 0.5mm/min. Flexural strength values of PANIoTCFN sheets (PANIoTCFN25, PANIoTCFN50 & PANIoTCFN75) are given in Table **2** and also compared with the help of bar diagram in Fig. (**8**).

Table 2. Comparison of flexural strength, electrical conductivity, and shielding effectiveness of PANIoT, PANIoT-CF, and PANIoTCFN sheet samples.

Sample Name	Carbon Fiber (wt. %)	Novolac Resin (wt. %)	Flexural Strength (MPa)	Electrical Conductivity (S/cm)	Shielding Effectiveness (dB)	Thickness of Sample (mm)
PANIoT	0	0	--	6.7×10^{-2}	14.3	1.10
PANIoT-CF	20	0	--	8.9×10^{-1}	21.2	1.00
PANIoTCFN 25	20	25	24.9	6.6×10^{-1}	15.0	0.80
PANIoTCFN 50	20	50	40.0	3.8×10^{-1}	19.0	0.68

(Table 2) cont.....

PANIoTCFN 75	20	75	52.4	6.4×10^{-2}	13.1	0.73
PANIoTCFN 50	20	50	--	--	26.4	1.48

* Sample with higher thickness.

It is believed that CF and novolac resin jointly increase the mechanical strength of the polymers [33, 34]. In the present work loading of CF in PANIoT is fixed (20% by weight), but the amount of novolac resin in PANIoT-CF composite is varied (25, 50 & 75% by wt. of PANIoT-CF composite) to optimize the electrical and shielding properties along with flexural strength of the resultant PANIoTCFN sheets. Novolac being a strong binder improved the adhesion bonding between the copolymer matrix and the filler units. Moreover, during the sheet preparation in a hydraulic hot press at 150°C, novolac resin (having hexamine as hardener) caused cross-linking and made the structure of the composite more compact and stiffer. These characteristics of novolac resin increase the flexural strength of the resultant PANIoTCFN sheets, which further increases significantly with an increase in the percentage of novolac resin in the composite sheets. This trend of flexural strength for PANIoTCFN sheets, as shown in Fig. **(8)**, is similar to the PANICFN and PoTCFN sheets.

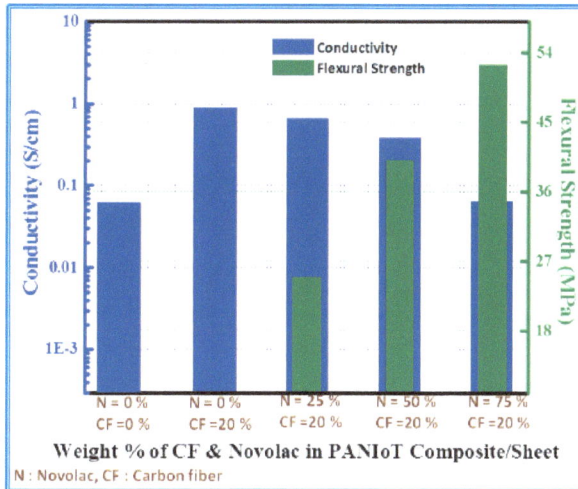

Fig. (8). Variation of conductivity in PANIoT, PANIoT-CF, PANIoTCFN sheets and variation of flexural strength in PANIoTCFN sheets as a function of wt. % of novolac resin.

When the flexural strength of PANIoTCFN sheets is compared with the flexural strength of the respective PANICFN and PoTCFN sheets having an equal amount of novolac resin (Figure 9.9 & Table 3), it is noticed that the flexural strength of PANIoTCFN sheets has an intermediate value to PANICFN and PoTCFN sheets. The PoTCFN sheets have the lowest value of flexural strength. It is due to the presence of the methyl group as a substituent in PoT and PANIoT. The methyl group being electron donating in nature increases the reactivity of o-toluidine units, and the resultant polymer (PoT) has shorter chains [35] and lower molecular weight. Moreover, the steric hindrance caused by the incorporation of methyl groups distorts the geometry of the polymer [1], which further decreases the regularity and rigidity of the polymer chains. All these factors are responsible for the lowest flexural strength of PoTCFN sheets. The effect of the methyl group in the copolymer (PANIoT) is lower than that of PoT; hence the flexural strength of the resultant copolymer composite (PANIoTCFN) is intermediary to the respective PANICFN and PoTCFN sheets.

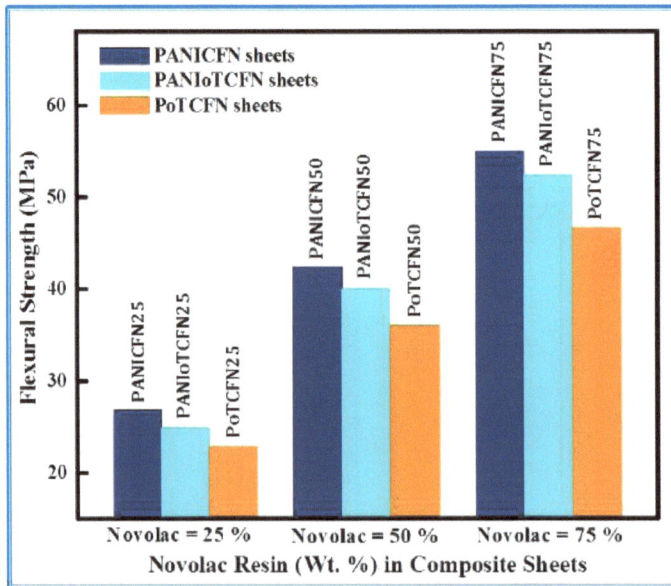

Fig. (9). Comparison of the flexural strength of PANICFN, PoTCFN and PANIoTCFN sheets.

9.3.6. Conductivity

The electrical conductivity of the synthesized conducting polymer and its composite plays an important role while testing the materials for EMI shielding application. Thus, the room temperature conductivity of the different samples was measured by the four-probe technique using Keithley programmable current source (model 6221). For conductivity measurement, pellets of PANIoT and PANIoT-CF was prepared in a hydraulic press, and PANIoTCFN sheets were cut with a dimension of 13×7 mm. Four contacts of the samples (pellet/sheet) were made on each end using silver paste. The measured values of electrical conductivity for PANIoT, PANIoT-CF composite, and PANIoTCFN sheets are given in Table **2** and are also compared with the help of the bar diagram in Fig. (**8**). The conductivity of pristine PANIoT increases by the addition of CF into the PANIoT matrix. This is attributed to two reasons; first, the CF itself possesses good conductivity. Second, the conducting network is improved with the loading of CF, resulting in enhancement of the electrical conductivity of the composite. The conductivity of PANIoT-CF composite decreases after blending it with novolac resin, which further decreases with an increase in the novolac resin in the resultant PANIoTCFN sheets. This is due to the insulating character of novolac resin. But, in spite of the insulating nature of novolac resin, the decrease in conductivity is not pronounced up to 50 wt. % loading of novolac into PANIoTCFN sheets. This is due to the fact that during the fabrication of sheets by hot pressing, novolac resin being a strong binder increases the connectivity between CF and the conducting copolymer matrix. But, at 75 wt. % loading of novolac resin, the insulating character of novolac dominates, thereby decreasing the conductivity of PANIoTCFN75 sheet is significant. The conductivity of PANIoTCFN sheets is found on the same line as PANICFN and PoTCFN sheets. Though the conductivity of sheets decreases, PANIoTCFN25 and PANIoTCFN50 sheets possess the optimum value of conductivity and polarization, which is desired for exhibiting good microwave shielding responses.

On comparing the conductivity of PANIoTCFN sheets with the conductivity values of its homopolymers composite sheets, the conductivity of PANIoTCFN sheets is found to have an intermediary value of conductivity to the values of the respective PANICFN and PoTCFN sheets having the same amount of novolac resin. This is because of the copolymerization of highly conducting aniline monomers with less conducting o-toluidine monomers. The low conductivity of the conducting polymer poly (o-toluidine) (PoT) is due to the introduction of a substituent group (methyl group), as discussed in the previous chapter. The substituent methyl group increases the torsional angle between the two adjacent phenylene rings and facilitates better solvation at the–NH group on the polymer backbone. This leads to the reduction in the conjugation of the polymer and hence a reduction in the conductivity [36, 37]. Also, the steric effect of the methyl group increases the interchain distance and, as interchain hopping becomes difficult, the conductivity level is lowered [38].

Hence, the conductivity of PoT is very less as compared to PANI. When the copolymer of aniline and o-toluidine is prepared, a part of the o-toluidine units in the copolymer chains (with respect to PoT chains) is replaced with highly conducting aniline units. The introduction of aniline units permits the steric effects of the methyl groups to decrease and, therefore, increasing the conjugation length. Hence, the conductivity of the copolymer is higher than PoT but less than PANI.

Table 3. Comparison of flexural strength, electrical conductivity and shielding effectiveness of PANI, PoT, PANIoT and their composites/sheets.

Sample Name	Flexural Strength (MPa)	Electrical Conductivity (S/cm)	Shielding Effectiveness (dB)	Sample Thickness (mm)
PANI	--	0.37	17.9	0.97
PANIoT	--	6.7×10^{-2}	14.3	1.10
PoT	--	2.62×10^{-3}	7.0	1.03
PANI-CF	--	1.46	25.2	1.02
PANIoT-CF	--	8.9×10^{-1}	21.2	1.00
PoT-CF	--	5.40×10^{-1}	13.0	0.90
PANICFN25	26.8	1.02	17.5	0.69
PANIoTCFN25	24.9	6.6×10^{-1}	15.0	0.80
PoTCFN25	22.9	3.10×10^{-1}	9.1	0.81
PANICFN50	42.4	0.71	20.2	0.73
PANICFN50*	--	--	31.9	1.47
PANIoTCFN50	40.0	3.8×10^{-1}	19.0	0.68
PANIoTCFN50*	--	--	26.4	1.48
PoTCFN50	36.0	9.80×10^{-2}	16.0	0.85
PoTCFN50*	--	--	23.9	2.11
PANICFN75	55.0	0.24	16.1	0.82
PANIoTCFN75	52.4	6.4×10^{-2}	13.1	0.73
PoTCFN75	46.6	4.68×10^{-4}	8.1	0.90

* Sample with higher thickness

The conductivity of the PANIoT-CF composite is also found intermediary to the conductivities of the PANI-CF and PoT-CF composite because the amount of CF is the same (20% by weight of monomers). Also, the conductivity of PANIoTCFN sheets is found intermediary to the conductivities of the PANICFN and PoTCFN sheets having the same amount of novolac resin. It is because of the fact that at constant composition, the effect of CF and novolac is found to be similar for

PANIoT, PoT, and PANI. The comparison of the conductivity values of PANI, PoT, PANIoT, their composites with CF, and their resultant sheets having similar amounts of novolac resin is given in Table **3** and shown in Fig. (**10**).

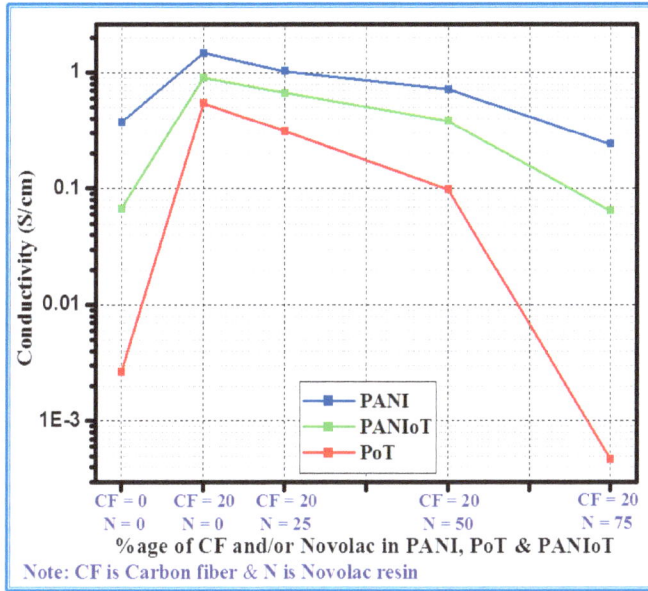

Fig. (10). Comparison of the electrical conductivity of PANI, PoT, PANIoT & their composites and sheets.

9.4. EMI SHIELDING AND DIELECTRIC MEASUREMENTS

After the confirmation of the formation of the copolymer and its composites and measurement of thermal, mechanical, and electrical properties, shielding measurements were carried out on an Agilent E8362B vector network analyzer in a microwave range of 8.2 to 12.4 GHz (X band). The PANIoTCFN sheet samples (thickness ~0.75mm) are prepared by cutting the sheets according to the dimensions of the sample holder and pellet samples (thickness ~ 1.0mm) of PANIoT and PANIoT-CF, with the same dimensions, is prepared by hydraulic press. Measurements were carried out using a 22.86 x 10.14 x 6mm^3 copper sample holder connected between the waveguide flanges of the network analyzer. To avoid air gaps, the above sample holder was modified with a groove of 1.5 mm on each side with 3mm depth.

Shielding effectiveness of a material is defined as the ratio of the transmitted power to incident power; hence, SE measured in decibel (dB) is given as:

$$SE(dB) = -10log\left(\frac{P_T}{P_I}\right) \tag{9.1}$$

where P_I and P_T are the power of incident and transmitted EM waves, respectively.

For a single layer of shielding material, SE is the sum of contribution due reflection (SE$_R$), absorption (SE$_A$), and multiple reflections (SE$_M$) and can be expressed by the following equation:

$$SE = SE_A + SE_R + SE_M \tag{9.2}$$

The S_{11} (or S_{22}) and S_{12} (or S_{21}) parameters of the two-port network system represent the reflection and transmission coefficients, respectively. According to the analysis of S parameters, transmittance (T), reflectance (R), and absorbance (A) through the shielding material can be described as

$$R = \left|\frac{E_R}{E_I}\right|^2 = |S_{11}|^2 = |S_{22}|^2 \tag{9.3}$$

$$T = \left|\frac{E_T}{E_I}\right|^2 = |S_{21}|^2 = |S_{12}|^2 \tag{9.4}$$

$$A = 1 - R - T \tag{9.5}$$

Here, it is noted that A is given with respect to the power of the incident EM wave. If the effect of multiple reflections between both interfaces of the material is negligible, the relative intensity of the effectively incident EM wave inside the material after reflection is based on the quantity as$(1 - R)$. Therefore, effective absorbance (A$_{eff}$) can be described as $A_{eff} = (1 - R - T)/(1 - R)$ with respect to the power of the effectively incident EM wave inside the shielding material. Convenience reflectance and effective absorbance are expressed in the form of $-10log(1 - R)$ and$-10\left(1 - A_{eff}\right)$ in decibel (dB) [39, 40], respectively, which provide the SE$_R$ & SE$_A$ as:

$$SE_R = -10\log(1 - R) \tag{9.6}$$

$$SE_A = -10\log(1 - A_{eff}) = -10\log\left(\frac{T}{1-R}\right) \tag{9.7}$$

EMI shielding effectiveness of PANIoT, PANIoT-CF composite pellets (~1.0 mm thickness), and PANIoTCFN sheets (~ 0.75 mm thickness) have been evaluated using Equations 6.6 and 6.7. Fig. (**9.11a**) shows the variation of SE_R, SE_A & SE_T of PANIoT and PANIoT-CF composite with frequency in 8.2-12.4 GHz range.

The value of SE for PANIoT and PANIoT-CF composite is 14.3 and 21.2 dB, at 8.2 GHz, respectively. The loading of CF within the PANIoT matrix increases the conductivity of PANIoT, which is already proved by conductivity measurements. The establishment and enhancement of electrical conductivity are of paramount importance as it leads to parallel enhancement of overall shielding effectiveness [39]. Interestingly, absorption loss increases by a larger magnitude compared to corresponding reflection loss.

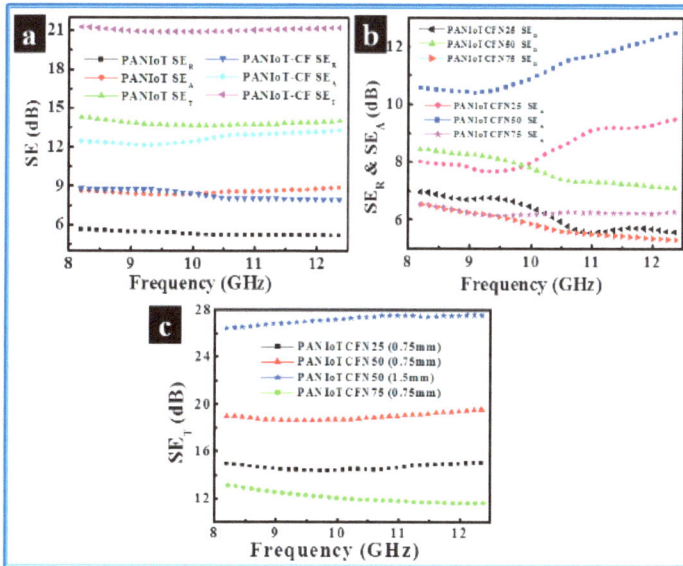

Fig. (11). Variation of SE with frequency, showing the effect of CF and/or novolac concentration on the shielding of PANIoT; (**a**) SE_A, SE_R & SE_T of PANIoT and PANIoT-CF, (**b**) SE_A, & SE_R of PANIoTCFN25, PANIoTCFN50 and PANIoTCFN75 sheets and (**c**) SE_T of PANIoTCFN25, PANIoTCFN50 and PANIoTCFN75 sheets at a comparable thickness of 0.75mm & SE_T of PANIoTCFN50 sheets at a higher thickness (1.5mm).

In addition to pristine PANIoT and PANIoT-CF composite, the SE_R, SE_A & SE_T of PANIoTCFN sheets have also been plotted against frequency in Fig. (**11b** & **c**). Fig. (**11b**) shows the variation of SE_R and SE_A of PANIoTCFN25, PANIoTCFN50, and PANIoTCFN75 sheets with frequency in the X-band of the frequency range. Fig. (**11c**) shows the variation of SE_T of PANIoTCFN25, PANIoTCFN50, and PANIoTCFN75 sheets in the same frequency range (X-band); also the SE_T of PANIoTCFN50 sheet at thickness 1.5mm is plotted in the same figure to see the effect of thickness. The SE_T values for PANIoTCFN25, PANIoTCFN50 and PANIoTCFN75 sheets were found to be 15.0 dB, 19.0 dB and 13.1dB, respectively at 8.2GHz. It is clear from the values that the SE_T value is found to be higher for the PANIoTCFN50 sheet though its conductivity is less than PANIoTCFN25. The presence of carbon fiber in the novolac present in the PANIoT-CF composite helps to electrically connect the CF that doesnot interact with each another in the absence of novolac. Novolac performs its action during the curing of sheets and enhances the connectivity. It has also been found that electrical conductivity is not the sole scientific criteria for exhibiting high shielding effectiveness [41], and good attenuations were also extended by moderate conductors with good dielectric properties. Further, the effect of increasing the thickness of samples on shielding effectiveness was also observed for the PANIoTCFN50 sheet. EMI performance increases with increasing material thickness. Here, for copolymer composite sheets, the effect of thickness is the same (shown in Figure 9.11c); the SE_T for PANIoTCFN50 sheet increased from 19.0dB at 0.75 mm thickness to 26.4dB at 1.5mm thickness.

Further, the electromagnetic absorption behaviour of a material also depends on the dielectric properties, represented by complex permittivity (ε' and ε'') and permeability (μ). In our case, all the samples are nonmagnetic; therefore, only complex permittivity (ε' and ε'') contributes to the microwave absorption properties. The complex permittivity has been obtained through scattering parameters S_{11} and S_{21} using Nicolson and Ross technique [42, 43]. The dielectric properties of PANIoT-CF composite are enhanced as CF present in the matrix as a filler also induces interfacial polarization, which may contribute towards dielectric properties, which are similar as observed for PANI-CF and PoT-CF composites. The variation of the real and imaginary parts of permittivity as a function of frequency is plotted in Figs. (**9.12a** & **b**). It is clear from the figure that both ε' and ε'' decrease with frequency. Fig. (**12c**) shows the variation of tangent loss with frequency. The PANIoT–CF composite shows a higher value of tangent loss showing more dissipation of energy by PANIoT-CF composite.

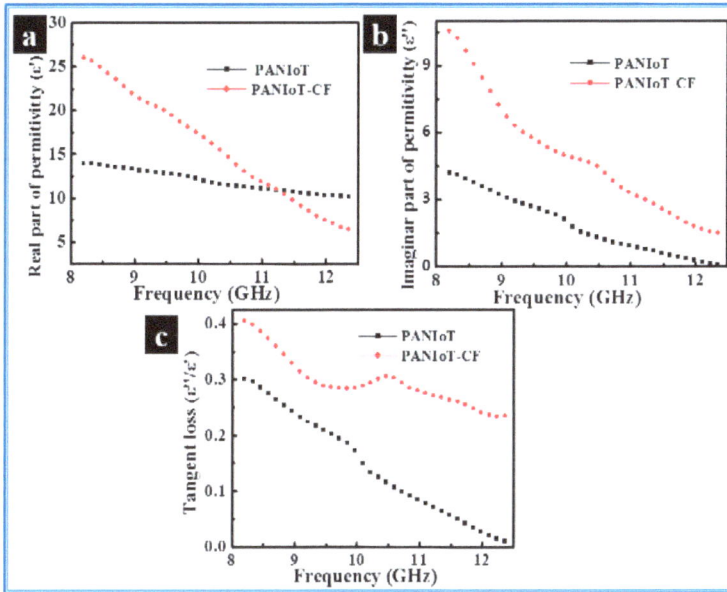

Fig. (12). Variation of **(a)** Real part of the permittivity (ε'), **(b)** imaginary part of the permittivity (ε'') and **(c)** tangent loss (tan δ_E) with frequency for PANIoT & PANIoT-CF.

Together with the complex permittivity of PANIoT and PANIoT-CF composite, the complex permittivity for PANIoTCFN sheets has also been measured by the same method. The real part of complex permittivity (ε') and imaginary part of complex permittivity (ε'') plot *vs.* frequency for PANIoTCFN sheets are shown in Figs. (**13a & b**). The real part of complex permittivity (ε') and the imaginary part of complex permittivity (ε'') depend on the polarizability of the material, which in turn depends on the dipole density and its orientation. In the case of conducting polymers, there are two types of charged species, a polaron/bipolaron system, which is mobile and free to move along the chain and the other one is bound charges (dipoles), which only have restricted mobility and account for strong polarization in the system. When the frequency of the applied field is increased, the dipoles present in the system cannot reorient themselves fast enough to respond to applied electric fields, and as a result, the dielectric constant decreases. With the increase in the novolac loading, besides the polarization due to mobile charges and the bound charges, the filler induced interfacial polarization may also contribute towards dielectric properties. Due to the presence of conducting CF and insulating novolac in the copolymer matrix, the improvement in the dielectric properties was observed. Such a polarization occurs due to electrical conductivity differences between novolac and PANIoT-CF interfaces, leading to charge localization at interfaces *via* Maxwell-Wagner-Sillars interfacial polarization phenomenon. Such polarization and related relaxation phenomena contribute towards energy storage and losses. The PANIoTCFN50 sheet sample fabricated with 50 wt. % of novolac has the highest dielectric constant (ε'=95 to 83), while the PANIoTCFN75 has the lowest dielectric constant (ε' = 30 to 21). The actual losses can be computed by the normalization of these losses with storage terms (*i.e.*, by the ratio of dielectric losses/imaginary permittivity (ε'') with dielectric constant/real permittivity (ε') to quantity loss tangent tan δ_E. The tan δ_E for PANIoTCFN sheets has been plotted in Fig. (**13c**), which explains the lossy character of the sheets. The much-enhanced imaginary and real parts of permittivity of these PANIoTCFN sheets indicate that they are suitable for use as EMI materials in the measured frequency region. In order to see the effect of copolymerization, the shielding effectiveness of PANIoT, PANIoT-CF composite, and PANIoTCFN sheets is compared with respective homo polymers (Table **3**). These values indicate that in the case of all the pristine samples, *i.e.*, PANI, PoT and PANIoT, the effect of addition of CF is the same. This means that the shielding effectiveness enhances the composites PANI-CF, PoT-CF, and PANIoT-CF. Moreover, the SE values of PANIoT and PANIoT-CF (Table **2**) are found to be more inclined towards PANI and PANI-CF SE values. In the case of PANIoTCFN sheets also, the effect of the addition of CF and novolac is the same. The SE value is found to be maximum for the PANIoTCFN50 sheet. The same trend was observed for the PANICFN and PoTCFN sheets. Moreover, the effect of increasing the thickness is found to be the same, *i.e.*, SE increases with the increase in thickness.

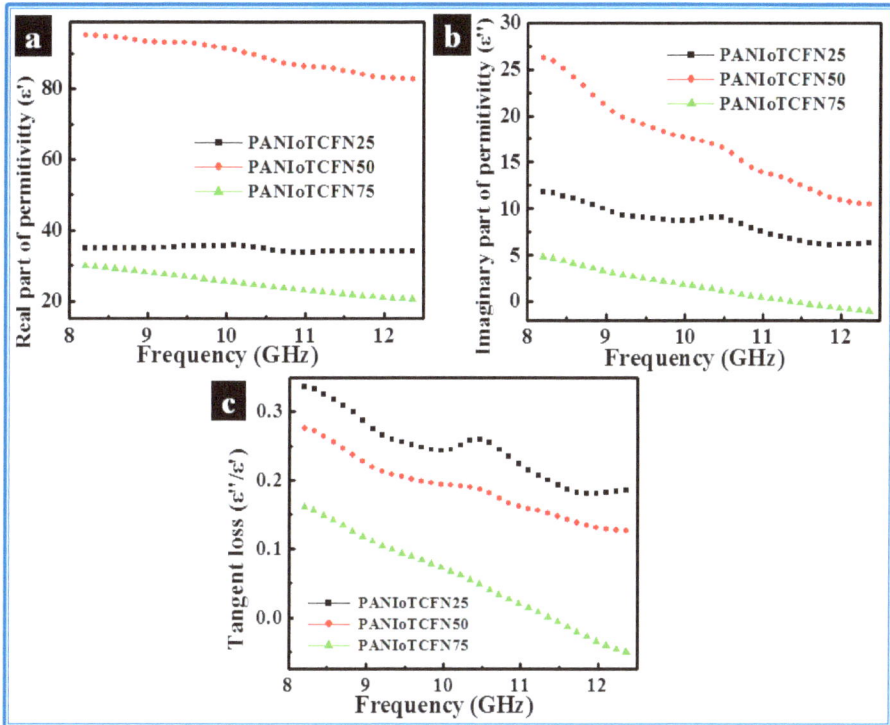

Fig. (13). Variation of **(a)** Real part of the permittivity (ε'), **(b)** imaginary part of the permittivity (ε''), and **(c)** tangent loss (tan δ_E) with frequency for PANIoTCFN25, PANIoTCFN50 and PANIoTCFN75 sheets.

CONCLUSION

The copolymer of aniline and o-toluidine and its composite with CF was successfully prepared. PANIoTCFN sheets were fabricated by solution casting cum hot pressing techniques. Fabrication of PANIoTCFN sheets was found to be easier than PANICFN sheets due to the better processability of the copolymer composite because of the flexible PoT units in PANIoT. FTIR and XRD studies suggested the formation of copolymers, which was further supported by the measured results of different properties of the copolymer composites. The thermal, mechanical, electrical, and shielding properties of PANIoT, PANIoT-CF composite, and PANIoTCFN sheets were measured to see the effect of CF and novolac resin. CF improved the thermal and mechanical properties together with the enhancement of the electrical and shielding properties. Moreover, novolac resin also improved the mechanical and thermal properties of PANIoTCFN sheets without affecting the shielding properties significantly. The flexural strength of PANIoTCFN sheets increased up to 52.4MPa for the PANIoTCFN75 sheet, while the thermal stability of the PANIoTCFN75 sheet is up to 305°C. The thin sheet of PANIoTCFN50 was found to have maximum shielding effectiveness of 19.0 dB at 0.68 mm thickness at 8.2 GHz frequency, and the same sheets at a higher thickness (1.48mm) showed SE up to 26 dB. The comparison of PANIoTCFN sheets was made with the sheets of PANICFN & PoTCFN, and it was found that the copolymer composite sheets have intermediary properties to the composite sheets of its homopolymers. So, the thermal, mechanical, electrical, and shielding properties are optimized in the copolymer together with the benefits of easy processability due to copolymerization. Hence, PANIoTCFN sheets have the potential to be used in techno-commercial as well as aerospace applications for microwave shielding due to their lightweight, ease of processing, and sufficient shielding even at low thickness.

CONSENT FOR PUBLICATION

Not applicable.

CONFLICT OF INTEREST

The authors declare no conflict of interest, financial or otherwise.

ACKNOWLEDGEMENTS

Declared none.

REFERENCES

[1]   Nguyen, M.T.; Kasai, P.; Miller, J.L.; Diaz, A.F. Synthesis and properties of novel water-soluble conducting polyaniline copolymers. *Macromolecules,* **1994**, *27*(13), 3625-3631.
         http://dx.doi.org/10.1021/ma00091a026

[2]   Prevost, V.; Petit, A.; Pla, F. Studies on chemical oxidative copolymerization of aniline and o-alkoxysulfonated anilines: I. Synthesis and characterization of novel self-doped polyanilines. *Synth. Met.,* **1999**, *104*(2), 79-87.
         http://dx.doi.org/10.1016/S0379-6779(99)00009-0

[3]   Abdiryim, T.; Xiao-Gang, Z.; Jamal, R. Comparative studies of solid-state synthesized polyaniline doped with inorganic acids. *Mater. Chem. Phys.,* **2005**, *90*(2), 367-372.
         http://dx.doi.org/10.1016/j.matchemphys.2004.10.036

[4]   Kathirgamanathan, P.; Adams, P.N.; Quill, K.; Underhill, A.E. Novel conducting soluble co-polymers of aniline. *J. Mater. Chem.,* **1991**, *1*(1), 141-142.
         http://dx.doi.org/10.1039/jm9910100141

[5]   Kuo, C-T.; Weng, S-Z.; Huang, R-L. Field-effect transistor with polyaniline and poly (2-alkylaniline) thin film as semiconductor. *Synth. Met.,* **1997**, *88*(2), 101-107.
        http://dx.doi.org/10.1016/S0379-6779(97)80887-9

[6]   Wang, Y.; Joo, J.; Hsu, C-H.; Pouget, J.; Epstein, A. Charge transport of hydrochloric acid doped polyaniline and poly (o-toluidine) fibers: role of processing. *Macromolecules,* **1994**, *27*(20), 5871-5876.
        http://dx.doi.org/10.1021/ma00098a047

[7]   Wei, Y.; Focke, W.W.; Wnek, G.E.; Ray, A.; MacDiarmid, A.G. Synthesis and electrochemistry of alkyl ring-substituted polyanilines. *J. Phys. Chem.,* **1989**, *93*(1), 495-499.
        http://dx.doi.org/10.1021/j100338a095

[8]   Leclerc, M.; Guay, J.; Dao, L.H. Synthesis and characterization of poly (alkylanilines). *Macromolecules,* **1989**, *22*(2), 649-653.
        http://dx.doi.org/10.1021/ma00192a024

[9]   Faulques, E.C.; Perry, D.L.; Yeremenko, A.V. Spectroscopy of Emerging Materials **2004**.

[10]   Griffiths, P.R. *Chemical infrared Fourier transform spectroscopy*; Wiley, **1975**.

[11]   Koenig, J.L. *Infrared and Raman Spectroscopy of Polymers.* **2001**.

[12]   Thomassin, J-M.; Jérôme, C.; Pardoen, T.; Bailly, C.; Huynen, I.; Detrembleur, C. Polymer/carbon based composites as electromagnetic interference (EMI) shielding materials. *Mater. Sci. Eng. Rep.,* **2013**, *74*(7), 211-232.
        http://dx.doi.org/10.1016/j.mser.2013.06.001

[13]   Dhakate, S.R.; Subhedar, K.M.; Singh, B.P. Polymer nanocomposite foam filled with carbon nanomaterials as an efficient electromagnetic interference shielding material. *RSC Advances,* **2015**, *5*(54), 43036-43057.
        http://dx.doi.org/10.1039/C5RA03409D

[14]   Huang, H-L.; Lee, W-M.G. Enhanced naphthalene solubility in the presence of sodium dodecyl sulfate: effect of critical micelle concentration. *Chemosphere,* **2001**, *44*(5), 963-972.
        http://dx.doi.org/10.1016/S0045-6535(00)00367-2 PMID: 11513430

[15]   Kumar, D. Synthesis and characterization of poly (aniline-co-o-toluidine) copolymer. *Synth. Met.,* **2000**, *114*(3), 369-372.
        http://dx.doi.org/10.1016/S0379-6779(00)00270-8

[16]   Elmansouri, A.; Hadik, N.; Outzourhit, A.; Lachkar, A.; Abouelaoualim, A.; Achour, M.; Oueriagli, A.; Ameziane, E. Schottky diodes and thin films based on copolymer: poly (aniline-co-toluidine). *Active and Passive Electronic Components,* **2009**.

[17]   Wei, Y.; Hariharan, R.; Patel, S.A. Chemical and electrochemical copolymerization of aniline with alkyl ring-substituted anilines. *Macromolecules,* **1990**, *23*(3), 758-764.
        http://dx.doi.org/10.1021/ma00205a011

[18]   Probst, M.; Holze, R. A systematic spectroelectrochemical investigation of alkyl-substituted anilines and their polymers. *Macromol. Chem. Phys.,* **1997**, *198*(5), 1499-1509.
        http://dx.doi.org/10.1002/macp.1997.021980515

[19]   Huang, M.R.; Li, X.G.; Yang, Y.L.; Wang, X.S.; Yan, D. Oxidative copolymers of aniline with o-toluidine: Their structure and thermal properties. *J. Appl. Polym. Sci.,* **2001**, *81*(8), 1838-1847.
        http://dx.doi.org/10.1002/app.1617

[20] Borole, D.; Kapadi, U.; Mahulikar, P.; Hundiwale, D. Electrochemical behaviour of polyaniline, poly (o-toluidine) and their copolymer in organic sulphonic acids. *Mater. Lett.,* **2004**, *58*(29), 3816-3822.

http://dx.doi.org/10.1016/j.matlet.2004.07.035

[21] Yang, C-H.; Yang, T.C.; Chih, Y.K. Mixture design applied to electrochemical polymerization of ternary aniline derivatives on ITO electrodes. *J. Electrochem. Soc.,* **2005**, *152*(9), E273-E281.

http://dx.doi.org/10.1149/1.1990499

[22] Savitha, P.; Sathyanarayana, D. Copolymers of aniline with o-and m-toluidine: synthesis and characterization. *Polym. Int.,* **2004**, *53*(1), 106-112.

http://dx.doi.org/10.1002/pi.1316

[23] Chan, H.; Ng, S.; Sim, W.; Tan, K.; Tan, B. Preparation and characterization of electrically conducting copolymers of aniline and anthranilic acid: evidence for self-doping by x-ray photoelectron spectroscopy. *Macromolecules,* **1992**, *25*(22), 6029-6034.

http://dx.doi.org/10.1021/ma00048a026

[24] Chan, H.; Ng, S.; Sim, W.; Seow, S.; Tan, K.; Tan, B. Synthesis and characterization of conducting poly (o-aminobenzyl alcohol) and its copolymers with aniline. *Macromolecules,* **1993**, *26*(1), 144-150.

http://dx.doi.org/10.1021/ma00053a022

[25] Nguyen, M.T.; Diaz, A.F. Water-soluble conducting copolymers of o-aminobenzyl alcohol and diphenylamine-4-sulfonic acid. *Macromolecules,* **1994**, *27*(23), 7003-7005.

http://dx.doi.org/10.1021/ma00101a048

[26] Chen, S-A.; Hwang, G-W. Synthesis of water-soluble self-acid-doped polyaniline. *J. Am. Chem. Soc.,* **1994**, *116*(17), 7939-7940.

http://dx.doi.org/10.1021/ja00096a078

[27] Huang, J.; Wan, M. In situ doping polymerization of polyaniline microtubules in the presence of β-naphthalenesulfonic acid. *J. Polym. Sci. A Polym. Chem.,* **1999**, *37*(2), 151-157.

http://dx.doi.org/10.1002/(SICI)1099-0518(19990115)37:2<151::AID-POLA5>3.0.CO;2-R

[28] Kulkarni, M.V.; Viswanath, A.K.; Marimuthu, R.; Mulik, U. Investigation of effect of protonic acid media on the optical and thermal properties of chemically synthesized poly (o-toluidine). *J. Mater. Sci. Mater. Electron.,* **2004**, *15*(12), 781-785.

http://dx.doi.org/10.1023/B:JMSE.0000045299.21124.ec

[29] Anand, J.; Palaniappan, S.; Sathyanarayana, D. Spectral, thermal, and electrical properties of poly (o-and m-toluidine)-polystyrene blends prepared by emulsion pathway. *J. Polym. Sci. A Polym. Chem.,* **1998**, *36*(13), 2291-2299.

http://dx.doi.org/10.1002/(SICI)1099-0518(19980930)36:13<2291::AID-POLA16>3.0.CO;2-5

[30] Palaniappan, S.; Lakshmi Devi, S. Thermal stability and structure of electroactive polyaniline fluoroboric acid dodecylhydrogensulfate salt. *Polym. Degrad. Stabil.,* **2006**, *91*(10), 2415-2422.

http://dx.doi.org/10.1016/j.polymdegradstab.2006.03.016

[31] Shreepathi, S.; Holze, R. Benzoyl-Peroxide-Initiated Inverse Emulsion Copolymerization of Aniline and o-Toluidine: Effect of Dodecylbenzenesulfonic Acid on the Physicochemical Properties of the Copolymers. *Macromol. Chem. Phys.,* **2007**, *208*(6), 609-621.

http://dx.doi.org/10.1002/macp.200600491

[32] Rao, P. S.; Subrahmanya, S.; Sathyanarayana, D. N. Inverse emulsion polymerization: a new route for the synthesis of conducting polyaniline. *Synthetic Metals,* **2002**, *128*(3), 311â€"316-311â€"316.

http://dx.doi.org/10.1016/S0379-6779(02)00016-4

[33] Singh, A.P.; Garg, P.; Alam, F.; Singh, K.; Mathur, R.B.; Tandon, R.P.; Chandra, A.; Dhawan, S.K. Phenolic resin-based composite sheets filled with mixtures of reduced graphene oxide, γ-Fe_2O_3 and carbon fibers for excellent electromagnetic interference shielding in the X-band. *Carbon,* **2012**, *50*(10), 3868-3875.

http://dx.doi.org/10.1016/j.carbon.2012.04.030

[34] Varshney, S.; Ohlan, A.; Singh, K.; Jain, V.K.; Dutta, V.P.; Dhawan, S.K. Robust Multifunctional Free Standing Polypyrrole Sheet for Electromagnetic Shielding. *Sci. Adv. Mater.,* **2013**, *5*(7), 881-890.

http://dx.doi.org/10.1166/sam.2013.1534

[35] Rao, P.S.; Sathyanarayana, D.; Palaniappan, S. Polymerization of aniline in an organic peroxide system by the inverted emulsion process. *Macromolecules,* **2002**, *35*(13), 4988-4996.

http://dx.doi.org/10.1021/ma0114638

[36] Clark, R.L.; Yang, S.C. **1989**.

[37] Gomes, M.A.B.; Gonçalves, D.; Pereira de Souza, E.C.; Valla, B.; Aegerter, M.A.; Bulhões, L.O.S. Solid state electrochromic display based on polymer electrode-polymer electrolyte interface. *Electrochim. Acta,* **1992**, *37*(9), 1653-1656.

http://dx.doi.org/10.1016/0013-4686(92)80131-5

[38] John, H.; Thomas, R.M.; Mathew, K.T.; Joseph, R. Studies on the dielectric properties of poly(o-toluidine) and poly(o-toluidine-aniline) copolymer. *J. Appl. Polym. Sci.,* **2004**, *92*(1), 592-598.

http://dx.doi.org/10.1002/app.20044

[39] Saini, P.; Choudhary, V.; Singh, B.P.; Mathur, R.B.; Dhawan, S.K. Polyaniline–MWCNT nanocomposites for microwave absorption and EMI shielding. *Mater. Chem. Phys.,* **2009**, *113*(2–3), 919-926.

http://dx.doi.org/10.1016/j.matchemphys.2008.08.065

[40] Hong, Y.K.; Lee, C.Y.; Jeong, C.K.; Lee, D.E.; Kim, K.; Joo, J. Method and apparatus to measure electromagnetic interference shielding efficiency and its shielding characteristics in broadband frequency ranges. *Rev. Sci. Instrum.,* **2003**, *74*(2), 1098-1102.

http://dx.doi.org/10.1063/1.1532540

[41] Joo, J.; Epstein, A.J. Electromagnetic radiation shielding by intrinsically conducting polymers. *Appl. Phys. Lett.,* **1994**, *65*(18), 2278-2280.

http://dx.doi.org/10.1063/1.112717

[42] Nicolson, A.M.; Ross, G.F. Measurement of the Intrinsic Properties of Materials by Time-Domain Techniques. *IEEE Trans. Instrum. Meas.,* **1970**, *19*(4), 377-382.

http://dx.doi.org/10.1109/TIM.1970.4313932

[43] Weir, W.B. Automatic measurement of complex dielectric constant and permeability at microwave frequencies. *Proc. IEEE,* **1974**, *62*(1), 33-36.

http://dx.doi.org/10.1109/PROC.1974.9382

CHAPTER 10

Porous 2D MXenes for EMI Shielding

Pradeep Sambyal[1,*,] Chong Min Koo[1,2,3*] and S.K. Dhawan[4]

[1] *Materials Architecturing Research Center, Korea Institute of Science and Technology, Hwarangno 14-gil 5, Seongbuk Gu, Seoul 02792, Republic of Korea*

[2] *KU-KIST Graduate School of Converging Science and Technology, Korea University, Seoul 02841, Korea*

[3] *Division of Nano & Information Technology, KIST School, University of Science and Technology, Seoul 02792, Republic of Korea*

[4] *Materials Physics & Engineering Division, CSIR-National Physical Laboratory, New Delhi – 110012, India*

Abstract: Advancement in modern electronic devices needs special requirements such as compact size, lightweight, and easy processing ability for the new innovative systems. This chapter describes the fundamentals of porous MXene composites (foams and aerogels) with the aim of inhibiting electromagnetic (EM) pollution. The first article that elucidated the EM shielding capabilities of MXene composites demonstrated superior performances to those of the existing materials, owing to their metallic conductivity, large surface area, surface modifiability, and ease of processability. Various approaches have been used to attenuate EM waves, including the application of laminate, porous, and hybrid structures. Among these, the porous morphology can contribute to the design of the absorption-dominant EM shield. Herein, the variations in electrical conductivity, mechanical stability, and electromagnetic interference shielding effectiveness (EMI SE) were explored with the use of a porous morphology. Subsequently, the theoretical and experimental results were analyzed to obtain new insights into the shielding mechanisms. This chapter will provide an overview of porous MXene composite materials and future challenges and strategies to design hybrid materials for next-generation EMI shielding applications.

Keywords: Aerogels, Electromagnetic Shielding, Foams, MXenes, Porous Structure.

---

*Corresponding authors Pradeep Sambyal, Chong Min Koo: Materials Architecturing Research Center, Korea Institute of Science and Technology, Hwarangno 14-gil 5, Seongbuk Gu, Seoul 02792, Republic of Korea; Tel: +82-10-5171-0507; E-mails: Pradeepsambyal23@gmail.com, Koo@kist.re.kr

Sundeep K. Dhawan, Avanish Pratap Singh, Anil Ohlan, Kuldeep Singh Kakran and Pradeep Sambyal (Ed.)
All rights reserved-© 2022 Bentham Science Publishers

10.1. INTRODUCTION

Advancements in modern innovative technologies have resulted in new challenges for material scientists and researchers. The use of miniaturized and lightweight systems represents the future of present-day systems. These small, lightweight systems generate electromagnetic (EM) pollution, which causes adverse effects on the smooth functioning of adjacent systems and results in hazardous consequences for human health [1-3]. Several health problems have been reported, such as headaches, nausea [4], eye problems [5], and abnormal organ development [6, 7] in infants. Moreover, EM waves cause the malfunctioning of certain medical instruments such as hearing aids and cardiac pacemakers [8, 9]. The existing solutions are not efficient in inhibiting secondary EM pollution; therefore, designing a highly absorption dominant shield is important. Specific solutions are required for a particular application to address explicit problems. Conventionally, metals, metal foils, and sheets were primarily used as electromagnetic interference (EMI) shields, owing to their high electrical conductivity; they performed exceptionally well against EM pollution. However, high density, low processability, and high vulnerability to corrosive environments limit their applications [10, 11]. Intrinsic conducting polymers also show good EMI shielding performances; nevertheless, their limited electrical conductivities and high sample thicknesses are major shortcomings [3, 12]. In the last decade, 2D materials have become popular in the area of EMI shielding. At very low thicknesses, remarkable EMI shielding properties have been observed. However, inadequate mechanical properties also affect their prospects. In 2016, novel 2D MXenes were reported to exhibit a shielding effectiveness performance of 92 dB EMI with 45 μm thickness [2]. Thus far, MXenes have attained significant authority among EMI shielding materials, owing to their metallic conductivity, high surface area, low density, and tunable surface chemistry [13, 14]. However, despite its excellent EMI shielding performance, it is unable to mitigate secondary EM pollution. Thus, EMI shields that are highly adsorptive and able to inhibit secondary EM pollution are required. Recently, porous composite materials have gained considerable attention, owing to their low densities, porous structures, superior mechanical properties, and superior EMI shielding performances [15, 16]. Different structural approaches have been employed to mitigate EM pollution, such as porous films, foams, and aerogels. Porous films were fabricated using vacuum filtration techniques, while porous foams were prepared through *in-situ* synthesis or *ex-situ* dipping methods. Porous aerogels were prepared using direct freeze-drying or bi-directional freeze techniques. The porous structure enhances the shielding ability by introducing additional interfaces within the shield material. These interfaces, with their mismatching impedance characteristics, tend to increase the internal scattering or multiple reflections and subsequently augment the absorption loss. Therefore, porous EMI shielding materials with low density, ultra-light weight, and good mechanical properties can be considered ideal candidates for applications in the aerospace, military, and mobile electronics industries. This chapter discusses the advancements in EMI shielding materials and porous EMI shielding materials with different structural morphologies while providing insights into the future challenges and possible solutions for next-generation smart electronic devices.

10.2 MXENES

2D MXenes are a new addition to the family of 2D materials. Further, MXenes belong to the family of transition metal carbides, nitrides, and carbonitrides with the characteristic formula $M_{n+1}X_nT_x$, where M is an early transition metal (*e.g.*, Ti, Zr, V, Nb, Ta, or Mo), X is carbon and/or nitrogen, and T_x represents the functional groups on the surface of MXene [17]. MXenes are young members of the 2D material family, which witnessed a rapid increase in the number of publications from its first reported article, as shown in Fig. (**1a**). MXenes are synthesized by the top-down synthesis approach in an acidic medium. The presence of an additional terminal group (-OH, -O, -F) at the surface of MXene results in hydrophilicity, which increases the ease of processing and synthesis of hybrid composites in aqueous media [18]. They have demonstrated superior electrical conductivity, large specific surface area, lightweight, tunable surface chemistry, and ease of processability. To date, the synthesis of more than twenty MXene compounds has been reported [2]. MXenes are generally synthesized from the MAX phase ($M_{n+1}AX_n$), where A represents elements of groups 13-16 of the periodic table. Ti_3AlC_2 is the most studied member among the group of 100+ layered ternary carbides and nitrides, which are also known as MAX phases. MXenes can be obtained from all the respective MAX phase precursors, thereby facilitating the ease of designing materials, especially for systems based on highly conductive 2D materials [6]. The etching process is used to etch the "A" group elements in the MAX phase using a combination of acids and salts, such as hydrofluoric acid (HF), NH_4F, NaF, CaF, HCl + LiF mixture (forming *in-situ* HF), and HF + HCl [19, 20]. Certain surface termination groups such as $-T_x$ (*e.g.*, –O, –OH, and –F) were formed on the surface of transition metal carbides, which are bonded to the outer M layers [17]. MXenes offer stable dispersions in various solvents, owing to their surface chemistry. Thus, MXenes can be used in different processing methods such as spray-coating, spin casting, dip-coating, inkjet printing, and interfacial assembly [21, 22]. Furthermore, the MXene family comprises three atomic structures, namely, M_2X, M_3X_2, and M_4X_3. These enhance the processability and design of materials based on essential requirements. Therefore, superior properties such as large surface areas, multi-layered structures, abundant functional groups, defects, and excellent electrical conductivities render it a unique candidate for EMI shielding applications. Moreover, it finds various applications such as water purification, energy storage, sensors, antibacterial activity and catalysis and many more, as shown in Fig. (**1b**) and Table **1**.

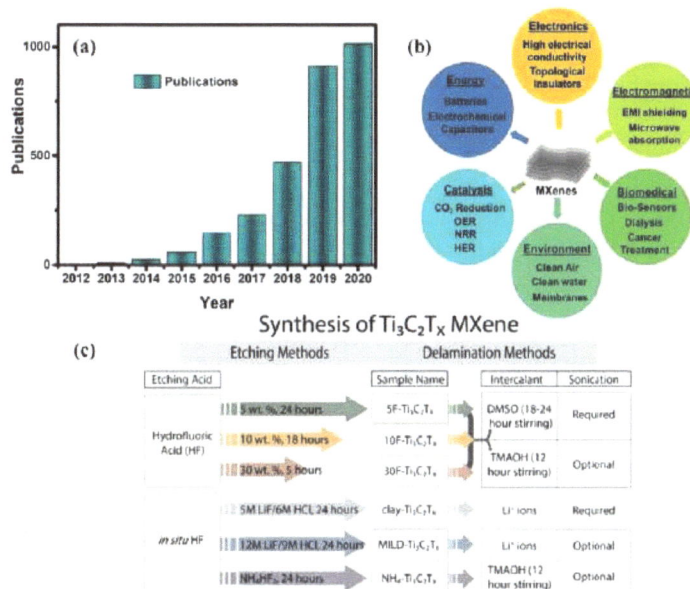

Fig. (1). Summary of publications of MXenes from 2012 to 2020 (a). (Result is obtained *via Web of Science,* Sept., 2020.). (b) Explored applications of MXenes. (c) Etching of $Ti_3C_2T_x$ from Ti_3AlC_2 through HF and *in-situ* HF processes. Reproduced with permission from a study [17] Copyright 2017, American Chemical Society.

10.3. SYNTHESIS OF MXENES

Various approaches have been reported in the literature for the synthesis of MXene materials. The selection of the synthesis route depends purely on the desired properties, end-product, and application. Thus, not all types of MXenes can be obtained from one universal route. In a typical synthesis, the A-element atomic layers in the MAX phase were etched from the $M_{n+1}X_n$ layers adjacent to the A-element layers. The etching process yields terminated multi-layered MXene with a 2D layer structure comprising hydrogen and Van der Waals bonds. The etched product was washed to remove the residual acid and impurities until a pH of 6 was attained; subsequently, repetitive washing was carried out. Fig. (**1c**) shows a chart of the synthesis procedures of MXenes using hydrofluoric acid and *in-situ* HF etching (as shown in Fig. **1c**) [17]. Gogotsi *et al.* have reported the exfoliation of the MAX phase through the HF with ultrasonication to yield 2D Mxenes for the first time. During the etching process, M-A bonds get broken easily due to weaker bond strength in comparison to M-X bonds. The following chemical reactions explain the selective etching process using HF:

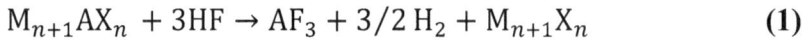

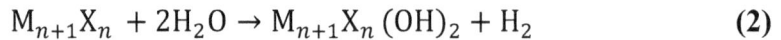

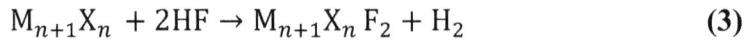

$$M_{n+1}AX_n + 3HF \rightarrow AF_3 + 3/2\,H_2 + M_{n+1}X_n \tag{1}$$
$$M_{n+1}X_n + 2H_2O \rightarrow M_{n+1}X_n\,(OH)_2 + H_2 \tag{2}$$
$$M_{n+1}X_n + 2HF \rightarrow M_{n+1}X_n\,F_2 + H_2 \tag{3}$$

The extraction of 'A' element from the MAX phase yielding $M_{n+1}X_n$ is depicted by equation 1. While equation 2 and 3 show the $M_{n+1}X_n$ functionalization with OH and F elements, respectively, resulting in $M_{n+1}X_nT_x$ (T= OH or F). Moreover, M in $M_{n+1}X_n$ has electrophilic nature that facilitates the surface functionalization with electron-donating groups (T). The physical and chemical properties of MXenes highly rely on these surface terminated groups. Fig. (**1c**) shows, that MXenes can be obtained from the MAX phase using the *in-situ* HF process, where HF is formed through the hydrochloric acid-lithium fluoride solution, or ammonium fluoride and ammonium hydrogen bifluoride [23]. The etching process depends upon various factors like particle size, structure, and atomic bonding. Thus, processes need to be modified as per the MAX phase or desired MXene, as a higher atomic number M in the MAX phase requires strong etchant due to the strong M-A bonds [23].

10.4. STRUCTURAL PROPERTIES OF MXENES

The crystal structure arrangement, surface functionalization (terminal groups), and compositions control the final electronic properties. The metallic nature of MXenes can be explained by analyzing the structure-property relationship. Various theoretical models are used to understand the core mechanism, which is then authenticated by the experimental outcomes. In this segment, the crystal structures of MXenes are discussed.

10.5. CRYSTAL STRUCTURE

MXenes are synthesized through the selective etching of the parent ternary carbide (MAX phase) $M_{n+1}AX_n$ Phase ($n = 1$ for 211, $n = 2$ for 312, and $n = 3$ for 413) [19].

The MAX phase shows a crystal structure comprising a layered hexagonal crystal structure with a $P6_3/mmc$ space group. The characteristic lattice parameters for $Ti_{n+1}AlC_n$ MAX exhibit a value of approximately 3Å, while the c values are approximately 13, 18, and 23–24 Å for the 211, 312, and 413 phases, respectively [24]. Moreover, the c/a ratio ranges from 4 to 8. The "M" layers are closely packed in the MAX phase, and the octahedral sites are filled by "X" atoms. The "A" layer atoms are interleaved within the $M_{n+1}X_n$ layers, and a metallic bond (M–A) is formed with "M" atoms, while the M–X bond shows a mixed ionic/covalent/metallic character as it is stronger than the M–A bond. It has an edge-sharing M_6X octahedron interwoven with layers of "A" elements [25]. The basic difference among the MAX phases (211, 312, and 413) is the different number of "M" layers separating the "A" layers (Fig. **2**). $M_{n+1}AX_nT_x$MXenes (M_2XF_2, $M_2X(OH)_2$, and M_2XO_2) [26] are formed due to the formation of six chemical bonds between the "M" group elements as well as the adjacent "X" group atoms and the surface terminations (T_x: OH, O, F). A large variety of MAX phases can be made by varying compositions or by using solid solutions of M, A, and X elements [27]. MXene synthesis from the various MAX phases can be classified as 1) mono-metal, 2) double-metal (M' and M''), and 3) double-X solid solution MXenes (C and N, also known as carbonitrides) (Fig. **2**). Double-metal MXenes are classified into two different categories; 1) a solid solution, in which the M-atom sites are arbitrarily occupied by M' and M'' metals; 2) ordered forms where M elements can be present in in-plane (*i*-MAX) or out-of-plane (*o*-MAX) arrangement [28, 29]. A quaternary phase is formed upon the addition of a fourth element in the MAX phase, including Ti_3AlCN and $Ti_3(Si, Al)C_2$. The MXenes obtained from the *o*-MAX phases are known as ordered double-metal transition metal MXenes, including $Mo_2TiC_2T_x$, $Cr_2TiC_2T_x$, and $Mo_2TiC_2T_x$, which exhibit considerably different electronic properties in comparison to those of the normal Mxenes, *i.e.*, $Ti_3C_2T_x$.

Table 1. Various applications of MXenes other than EMI shielding.

S.No.	Material	Application	Reference
1.	$Ti_3C_2T_x$, Ti_2CT_x, Nb_2CT_x	Energy storage	[31-33]
2.	TiO_2-C/TiC, $Ti_3C_2T_x$-Fe_2O_3, $Ti_3C_2(OH)_2$	Water purification	[34-36]
3.	$Ti_3C_2T_x$	Water desalination	[37]
4.	V_2C–PDMAEMA, Ti_2CO_2	Sensors (CO_2, NH_3)	[38, 39]
5.	Ti_3C_2	Biosensors	[40]
6.	$Ti_3C_2T_x$–TiO_2	Photocatalysis	[41, 42]
7.	$Ti_3C_2(OH)_2$, V_2CT_x, TiC_2T_x	Nuclear waste management	[43-45]

(Table 1) cont.....

8.	Ti$_3$C$_2$Ti$_{,2}$,CT i$_3$C$_2$-chitosan	Antibacterial activity	[46-48]
9.	Ti$_3$C$_2$, Ti$_3$C$_2$-Co$_3$O$_4$	Dye adsorption	[49, 50]
10.	RuNi-Ti$_3$C$_2$T$_x$, NiS$_2$/V$_2$CT$_X$	H$_2$ generation	[51, 52]

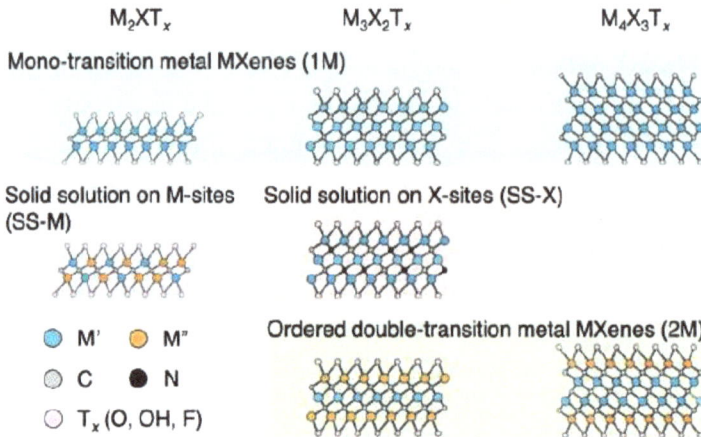

Fig. (2). Basic structure of MXenes reported in the literature. MXenes are classified into three different formulas: M$_2$X, M$_3$X$_2$, and M$_4$X$_3$, where M represents an early transition metal and X denotes carbon and/or nitrogen. Reproduced with permission from ref [30]. Copyright 2020, American Chemical Society.

10.6. MXENES AS EMI SHIELDING MATERIALS

The recent advancements in the fifth-generation (5G) of modern electronic and electrical systems have led to high-end technologies that require lightweight and compact systems with high processing speeds. However, these systems also yield undesirable EM pollution. MXene has received much attention after Shahzad *et al.* reported MXene as an EMI shielding material [2]. A 2D MXene µm thick film with an electrical conductivity of 4600 S cm^{-1} was obtained through vacuum filtration. Moreover, electrical conductivity is the prime factor for the excellent EMI shielding performance of Ti$_3$C$_2$T$_x$, Mo$_2$TiC$_2$T$_x$, and Mo$_2$Ti$_2$C$_3$T$_x$. They reported electromagnetic interference shielding effectiveness (EMI SE) of 54, 20, and 24 dB at 2.5 µm thickness for Ti$_3$C$_2$T$_x$, Mo$_2$TiC$_2$T$_x$, and Mo$_2$Ti$_2$C$_3$T$_x$ in the X-band frequency range, respectively (as shown in Fig. **3a**). Interestingly, as the thickness of the film is increased, an EMI SE value of 92 dB is achieved for 45 µm thickness. Even the introduction of sodium alginate to Ti$_3$C$_2$T$_x$ improves the mechanical strength, oxidation resistance, and EMI shielding effectiveness. Fig. (**3b**) demonstrates the EMI SE variation with frequency. The 90 wt.% loading composite film showed a 57 dB EMI SE value with a corresponding SSE/t of more than 30000 dB cm^2 g^{-1}. The present Ti$_3$C$_2$T$_x$ film outperforms all literature material values at equivalent thickness. This superior shielding performance is attributed to the synergistic combination of the 2D laminar structure and excellent electrical conductivity. Fig. (**3c**) shows the EMI SE value of MXenes outperforming the reported literature materials and metals. The proposed EMI shielding mechanism for the MXene films, where the impedance mismatch between air and the MXene surface leads to the reflection of the incident EM wave. The remaining part of the EM wave absorbs within the lamellar structure and gets attenuated through the ohmic losses and eddy current losses.

Yun *et al.* fabricated nanometer-sized Ti$_3$C$_2$T$_x$MXene films through the interfacial self-assembly of 2D MXene flakes floating on the water-air interface [53]. The continuous stacking of monolayers yields thick bilayer films. The sheet resistance decreased linearly from 1.056 to 15Ω cm^{-2} with an increase in the number of MXene layers, which yields outstanding shielding as a result of the thickness. Moreover, the EMI SE value varied from 1 to 20 dB for a thickness of 2.3 to 55 nm (as depicted in Fig. **3d**). Here, two theoretical models (Simon's formula and the transfer matrix method) were used to obtain the role of multiple reflections in the EMI shielding performance of the assembled Ti$_3$C$_2$T$_x$MXene films [54, 55]. The transfer matrix methods substantiated the experimental results and fit well with the contribution of the multiple reflections to the overall shielding performance, as shown in Fig. (**3e**). This research study achieved the highest SSE/t value of 3.89×10^6 dB cm^2 g^{-1} for a film annealed at 400 °C (Fig. **3f**). A novel material with an ultrathin size and ultralight characteristics was fabricated, which is a promising candidate for advanced electronic systems.

He *et al.* demonstrated the EMI shielding performance of Ti$_3$C$_2$T$_x$MXene synthesized compounds in two different ways [56]. In both the routes, Ti$_3$AlC$_2$ MAX powder was etched using 40% hydrofluoric acid (HF), and then LiF and HCl were incorporated for 16 h at 40 °C to attain ultrathin MXene sheets. The reaction product (M-Ti$_3$C$_2$T$_x$ composite) possessed more F terminations, whereas the second reaction product (U-Ti$_3$C$_2$T$_x$ composite) yield comprised of high =O terminations. SiO$_2$ nanoparticles were introduced in different concentrations (20, 40, 60, and 80 wt.%) of Ti$_3$C$_2$T$_x$ matrix in the cold press under a pressure of 5 MPa. The U-Ti$_3$C$_2$T$_x$ and M-Ti$_3$C$_2$T$_x$ composite at 60 wt.% MXenes demonstrate a two-order difference in electrical conductivity (4.2×10^{-3} S cm^{-1} and 6.3×10^{-5} Scm^{1-}). The U-Ti$_3$C$_2$T$_x$ 80 wt.% composite exhibited an EMI SE value of 58 dB at 1 mm thickness in the X-band frequency range. M-Ti$_3$C$_2$T$_x$MXene composites displayed excellent EMI shielding performance, higher electrical conductivity, larger surface area, and highly conductive networks as compared to those of U-Ti₃C₂Tₓ. Moreover, the proposed EMI shielding mechanism showed that the U- Ti₃C₂Tₓ MXene composite consisted of a large number of surface terminations and point defects, leading to a better dipolar polarization loss and high attenuation.

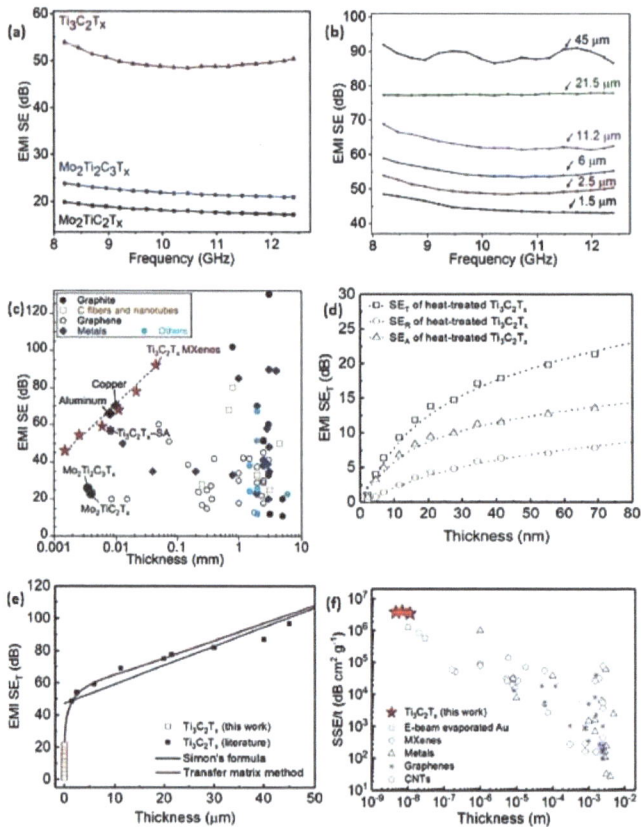

Fig. (3). (a) EMI SE of $Ti_3C_2T_x$, $Mo_2TiC_2T_x$, and $Mo_2Ti_2C_3T_x$MXenes. (b) EMI SE of $Ti_3C_2T_x$MXene films at different thicknesses. (c) Comparison of EMI SE *vs.* thickness with reported literature [2]. Copyright 2016, Science. (d) Comparison of EMI SE_T, SE_R, and SE_A values of electrically thin $Ti_3C_2T_x$ MXene films between calculation using transfer matrix method and experiment. (e) Comparison between simulation and experimental values of EMI SE_T. (f) SSE/t as a function of thickness for various superior shielding materials. Reproduced with permission from a study [53] Copyright 2020, Wiley-VCH.

The EMI shielding performances of 16 different MXene laminates with variable compositions (*i.e.*, M₂X, M₃X₂, and M₄X₃) at different thicknesses were studied by Han *et al.* (Fig. **4a**) [30]. The provided outcome suggested contributions of electrical conductivity, which improved shielding performance. Nb₂CTₓMXenes exhibited an electrical conductivity of 5 S cm^{-1}, while Ti$_3$C$_2$T$_x$ showed approximately 8,500 S cm^{-1}, as depicted in Fig. (**4b**). Electrical conductivity of approximately 15,000 S cm^{-1} was achieved by maintaining the synthesis parameters, which were further increased by controlling the processing and investigational conditions. This outstanding EMI shielding performance of MXenes is attributed to their excellent electrical conductivity (Fig. **4c**). It is important to note that the electrical conductivity and EMI shielding performance are dependent on the synthesis route, precursor quality, and chemical composition. It is significant that this competent EMI shielding performance fulfills the commercial requirement scale of 20 dB at a thickness of 5 μm. Moreover, theoretical and computational studies using the transfer matrix method also support the experimental results and thus, confirm the synergetic contribution of the multiple reflection mechanisms to the EMI shielding performance of an electrically thin MXene film.

10.7. MXENE POROUS FOAMS FOR EMI SHIELDING

Modern electronic systems need to exhibit certain prerequisites for applications in aerospace, defense, and portable electronic industries, which include lightweight properties, compactness, and high processing speeds. Porous lightweight materials with good electrical conductivity and robust mechanical stability are potential candidates for EMI shielding. The ultralow density and excellent conductivity of porous MXene foams and aerogels make them promising candidates for these applications. Liu *et al.* synthesized flexible, porous, and lightweight Ti$_3$C$_2$T$_x$MXene foams from compacted MXene films after immersion in a hydrazine solution at 90 °C [57]. MXene sheets showed a volume expansion when the diffusion of hydrazine solution into MXene occurred, which led to the formation of a continuous cellular network morphology. Ti$_3$C$_2$T$_x$MXene films with 1, 3, and 6 μm thickness) were converted into Ti$_3$C$_2$T$_x$ MXene foams with thicknesses of 6.2, 18, and 60 μm, respectively. The porosity is induced in the MXene structure as a result of the generation of CO or CO$_2$ gases in the chemical reaction between the hydrazine molecules and the hydroxyl (–OH) groups on the MXene surface. The fractional elimination of oxygen moieties from the MXene foams was confirmed through the higher Ti/O and C/O ratios as compared to those of the compact films. This introduced the hydrophobic nature to the surface of the MXene foam and consecutively improved the oxidation resistance. The hydrazine treatment increased the porosity, causing a reduction in the electrical conductivity of the three compacted Ti$_3$C$_2$T$_x$ films from 4000 S cm^{-1} to 588, 625, and 580 S cm^{-1}. The foam structure showed a reduction in the electrical conductivity but exhibited higher EMI SE values (32, 50, and 70 dB) as compared to those of the compacted films (28, 46, and 53 dB) (Fig. **5d–e**). The superior EMI shielding effectiveness was attributed to the high porosity, which enhanced the internal scattering, leading to augmented internal multiple reflections in the MXene layers.

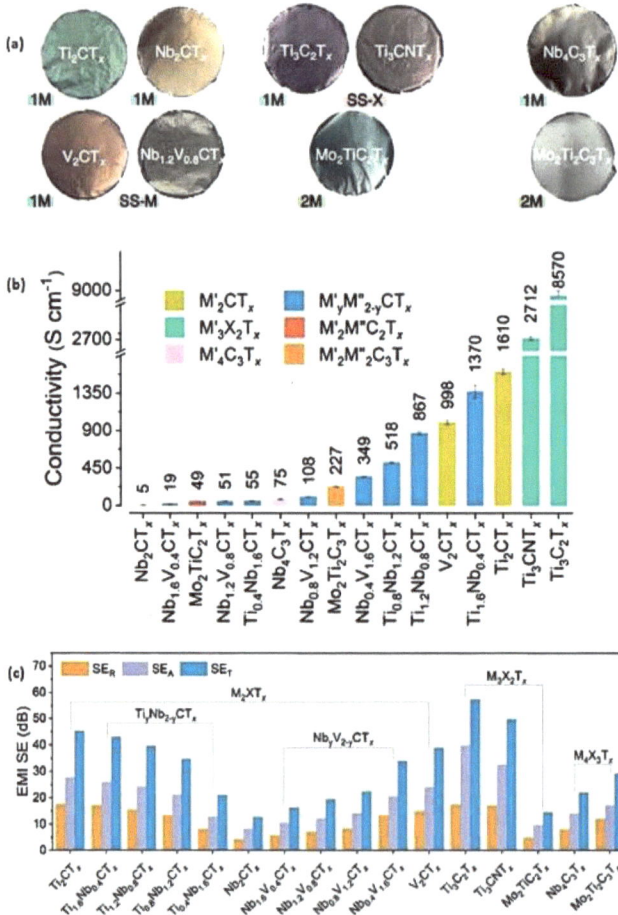

Fig. (4). (a) Images of different MXenes composed of single or double transition metals and solid solutions. **(b)** Electrical conductivity of 16 different MXenes. **(c)** Average EMI SE (SE_R, SE_A, and SE_T) of various MXene films at a thickness of 5 ± 0.3 µm in 8.2–12.4 GHz, representing reflection and absorption contributions. Adapted with permission from a study [30]. Copyright 2020, American Chemical Society.

$Ti_3C_2T_x$ MXene/graphene hybrid foams have been developed for EMI shielding applications [58]. Different compositions of hybrid foams were synthesized by changing the MXene-to-GO ratio, which was achieved by using vacuum drying of frozen solutions at -65 °C and thermal reduction at 300 °C for 1 h. An increase in density was observed as pure rGO showed a density of 3.1 mg cm^{-3}, while the introduction of MXene (MX: rGO:: 1:1) increased the density to 7.2 mg cm^{-3}. Thus, an increase in the density of the hybrid foams is attributed to the higher density of Ti$_3$C$_2$T$_x$MXene as compared to that of graphene. Moreover, the density values decreased to 4.6 and 3.7 mg cm^{-3} on increasing the graphene content to 1:2 and 1:3, respectively. Correspondingly, a significant improvement in the electrical conductivity was observed with the addition of MXene, and the value increased from 140 S m^{-1} to 1250 S m^{-1}. The enhanced electrical conductivity and brilliant EMI shielding efficiency were obtained for the porous structure foams, which increased with an increase in MXene content. The pure rGO foam showed an EMI SE value of 15 dB at a thickness of 1.5 mm, while the 1:1 MXene hybrid exhibited double the value at the same thickness. The outstanding EMI shielding performance of the hybrid foam was ascribed to internal scattering in the conductive network of MXene. The hybrid foam with a ratio of 1:2 (MXene: rGO) yields an EMI SE value that exceeds 50 dB with an increase in thickness (3 mm).

The sol–gel and thermal reduction method was employed to fabricate the Ti$_3$C$_2$T$_x$MXene/carbon hybrid foam using a mechanically strong epoxy [16]. In this facile synthesis method, 10 mL of Ti$_3$C$_2$T$_x$MXene dispersion was added to a mixture of resorcinol, formaldehyde, and a sodium carbonate catalyst, which was subsequently sonicated and cured at 90 °C for 5 h under nitrogen and then freeze-dried. The resultant porous foam was annealed at 400 °C for 2 h to yield the MXene/carbon hybrid foam. Consequently, the epoxy solution was infused in the hybrid foam under vacuum. The MXene/carbon foam composed of an epoxy with 1.64 wt. % MXene and 2.61 wt. % carbon showed an electrical conductivity of 1.84 S cm^{-1}. The hybrid foam exhibited an EMI SE value of 46 dB at 2 mm thickness in the X-band (8.2-12.4 GHz) frequency range. The increment in the MXene content in the hybrid foam caused an increase in the hardness (from 0.28 to 0.31 GPa) and Young's modulus (from 3.51 to 3.96 GPa), which were higher as compared to those of the pure carbon/epoxy composites. The electrical conductivity, mechanical properties, and EMI shielding performance of epoxy-based composites and hybrid foams are highly dependent on the conductive 2D MXenes.

A low-reflection Ti$_2$CT$_x$MXene/PVA composite foam was assessed for its EMI shielding application [59]. The free-drying method was employed to synthesize the Ti$_2$CT$_x$/PVA foam. The Ti$_2$CT$_x$/PVA foam showed excellent flexibility and robust mechanical strength due to the formation of strong hydrogen bonding between the surface terminations of MXene and the molecular chains of PVA (as depicted in Fig. **5a**). The Ti$_2$CT$_x$/PVA composite foam demonstrated a lower electrical conductivity than that of the porous MXene composite. However, for Ti$_3$C$_2$T$_x$, it is higher than those of the other carbon-based materials. Ti_2CT_x/PVA foam with a density of 10.9 mg cm^{-3} showed an EMI SE value of 33 dB at a thickness of 5 mm. Moreover, the compressed foam structure exhibited an SSE/t value of 5,136 dB cm^2 g^{-1} due to the low density (as shown in Fig. **5c-d**). The EMI shielding of the porous foam structure was more efficient than that of the compact film structure, where absorbed EM waves dissipated in the form of heat owing to the strong internal scattering within the porous architecture. The low reflection and high absorption behavior of the MXene/PVA foam were due to the contribution of dielectric and interfacial polarizations.

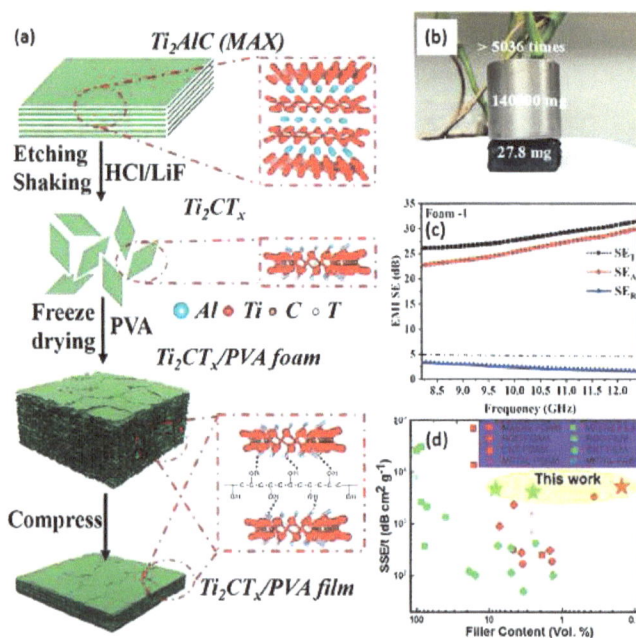

Fig. (5). (a) Schematic representation of the synthesis Ti_2CT_X/PVA foam, **(b)** photograph of f-Ti_2CT_x/PVA foam that supports more than 5000 times its own weight, **(c)** EMI shielding performances of f-Ti_2CT_x/PVA foam-1, and **(d)** SSE/t *versus* filler ratio (vol %) of different materials. Adapted with permission from a study [59] Copyright 2019, American Chemical Society.

A general freeze-drying method was used to prepare a compressible and durable $Ti_3C_2T_x$MXene/sodium alginate/polydimethylsiloxane (MXene/SA/PDMS) hybrid foam [60]. MXene/SA lightweight composed of 95% MXene exhibits a density of approximately 20 mg cm^{-3}, with an electrical conductivity of 22.11 S cm^{-1}, and exhibits an EMI SE value of 72 dB. The introduction of a thin layer of PDMS into MXene/SA foam through vacuum filtration resulted in a decrease in the EMI shielding efficiency up to 50 dB. Moreover, the structural strength and robustness were enhanced by the PDMS coating, as MXene/SA/PDMS hybrid foam sustained its EMI shielding performance in over 500 cycles of bending and compression. The role and mechanism of the porous architecture in the attenuation of EM waves were evaluated by comparing the simple blend of MXene/SA/PDMS and solid PDMS-coated MXene/SA foams that filled all of the pores. The MXene/SA/PDMS hybrid foam with a thin uniform PDMS coating, solid PDMS coating, and blended composite demonstrated EMI SE values of 54, 50, and 9 dB, respectively.

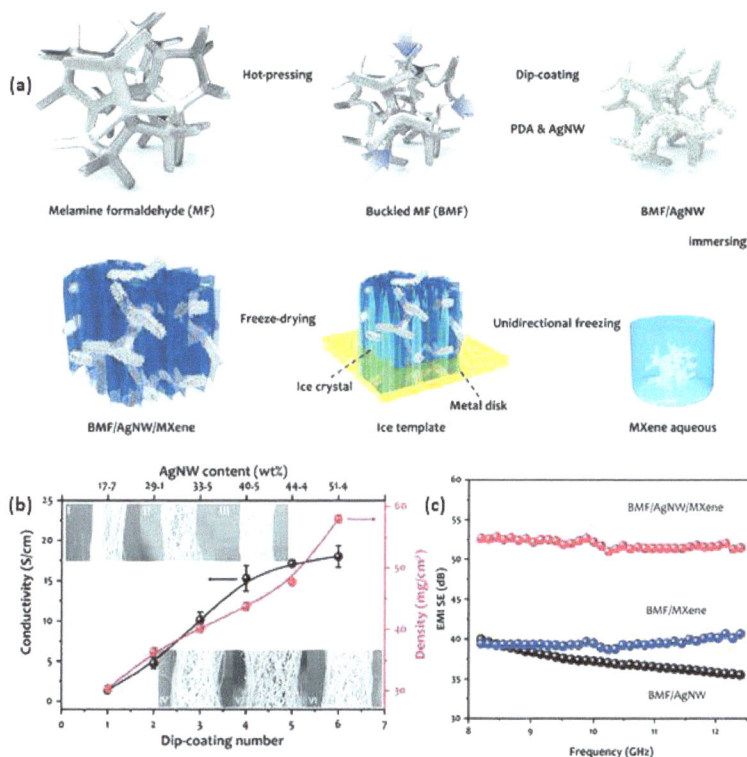

Fig. (6). (a) Schematic representation of the fabrication process of the BMF/AgNW/MXene hybrid sponge. **(b)** Conductivity and density of the BMF/AgNW foams at different dip-coating cycles. Inset: SEM images of AgNW coated backbone of BMF sponge; **(c)** The comparison of EMI SE between BMF/MXene, BMF/AgNW and BMF/AgNW/MXene sponges. Adapted with permission [61]. Copyright 2019, The Royal Society of Chemistry.

In another study, hybrid foams of a silver nanowire AgNW/Ti$_3$C$_2$T$_x$MXene were synthesized and evaluated for EMI shielding effectiveness [61] (as depicted in Fig. **6a**). A buckled melamine-formaldehyde (BMF) foam was fabricated through tri-axial compression of a melamine-formaldehyde (MF) porous substrate foam. The dip-coating technique was used to grow AgNWs uniformly on the substrate, and the electrical conductivity increased with an increasing number of dip-coating cycles. An irregular honeycomb MXene structure was obtained by dipping the conductive foams in $Ti_3C_2T_x$MXene solution and was then dried using the freeze-drying method. Moreover, the BMF/AgNW/$Ti_3C_2T_x$ hybrid composite demonstrated higher electrical conductivity and EMI shielding performance than those of the BMF/AgNW and/or BMF/MXene composites. The hybrid porous architecture of electrical conductivity and EMI shielding efficiency is attributed to the synergistic combination of AgNW and $Ti_3C_2T_x$MXene (Fig. **6b-c**). The BMF/AgNW/$Ti_3C_2T_x$ hybrid foam shows an EMI SE value of 52.6 dB at 2 mm thickness, while both BMF/AgNW and BMF/MXene foams exhibit EMI SE values of 40 dB. The outstanding shielding efficiency was ascribed to the enhanced electrical conductivity, which resulted in an impedance mismatch between the composite and the empty spaces. Furthermore, the absorption-dominant EMI shielding performance is attributed to the internal scattering of EM waves from conductive porous interfaces.

10.8. MXENE AEROGELS FOR EMI SHIELDING

A $Ti_3C_2T_x$MXene aerogel was fabricated using a unidirectional freeze-drying method using various concentrations of MXene (Fig. **7a**) [62]. In the unidirectional freeze-drying method, the density of the aerogels was controlled through the concentration of the MXene solution. The pore morphology depends on the formation of ice crystals during freezing. The $Ti_3C_2T_x$MXene aerogels are light weight and possess good structural integrity and an extremely low density of 20.7 mg cm^{-3}. The aerogel demonstrated an electrical conductivity of 22 S cm^{-1} and an EMI SE value of 75 dB at 2 mm thickness, corresponding to an SSE/t value of 18,116 dB cm^2 g^{-1} (as shown in Fig. **7d**). Composite aerogel films of $Ti_3C_2T_x$/calcium alginate were also fabricated through vacuum filtration and the freeze-drying process [63]. MXene/CA aerogel with 90 wt.% MXene concentration showed shielding effectiveness and a specific SE of 54.3 dB and 17,586 dB cm^2 g^{-1} at a thickness of 26 µm, respectively. The porous morphology enhanced the EMI shielding performance due to the increment in the internal reflections by multiple scattering surfaces and interfaces. Moreover, this caused high absorption dominance within the material.

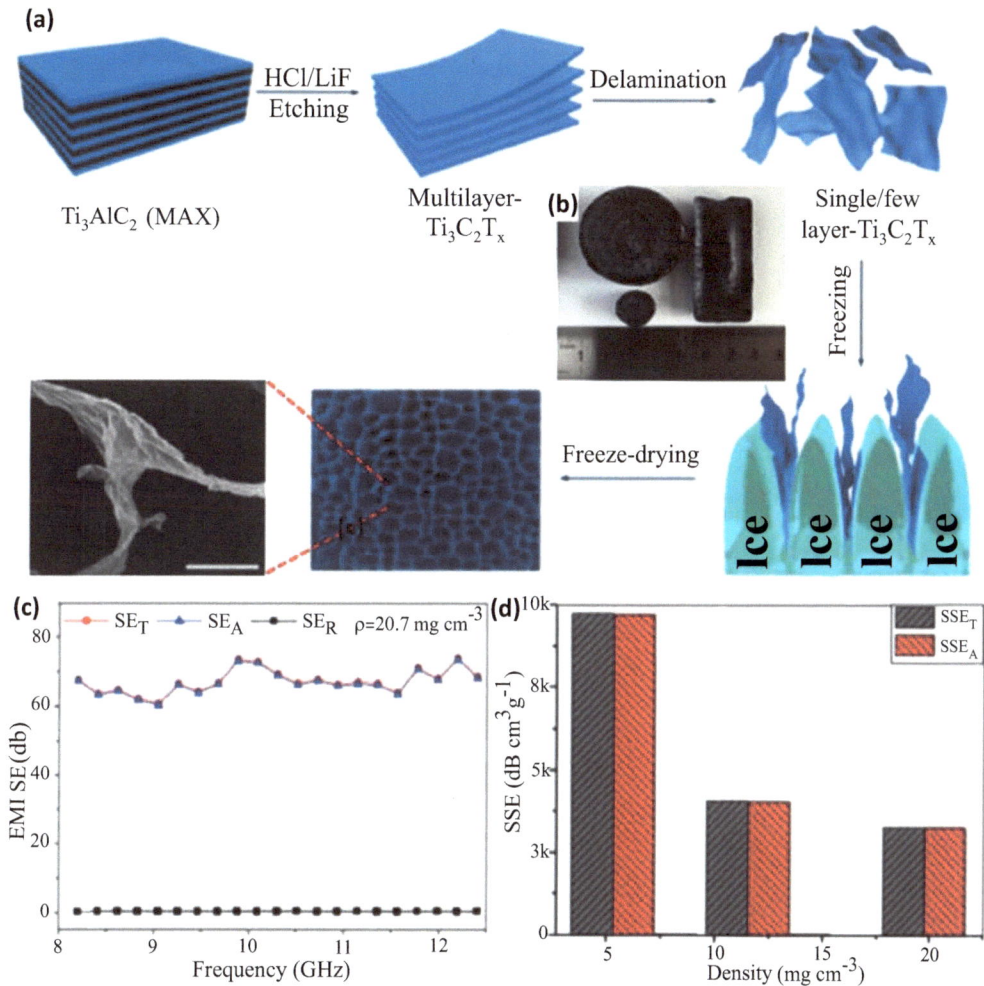

Fig. (7). (a) Fabrication of the $Ti_3C_2T_x$MXene aerogel. **(b)** Digital photograph of the MXene aerogel and **(c)** the EMI of the MXene aerogel with a density of 20.7 mg cm^{-3}. **(d)** The SSE$_T$ and SSE$_A$ of the MXene aerogel at 8.2 GHz. Adapted with permission [62], Copyright 2019, The Royal Society of Chemistry.

Han *et al*. designed aerogels using three different MXenes ($Ti_3C_2T_x$, Ti_3CNT_x, and Ti_2CT_x) for EMI shielding applications (Fig. **8a-c**) [64].

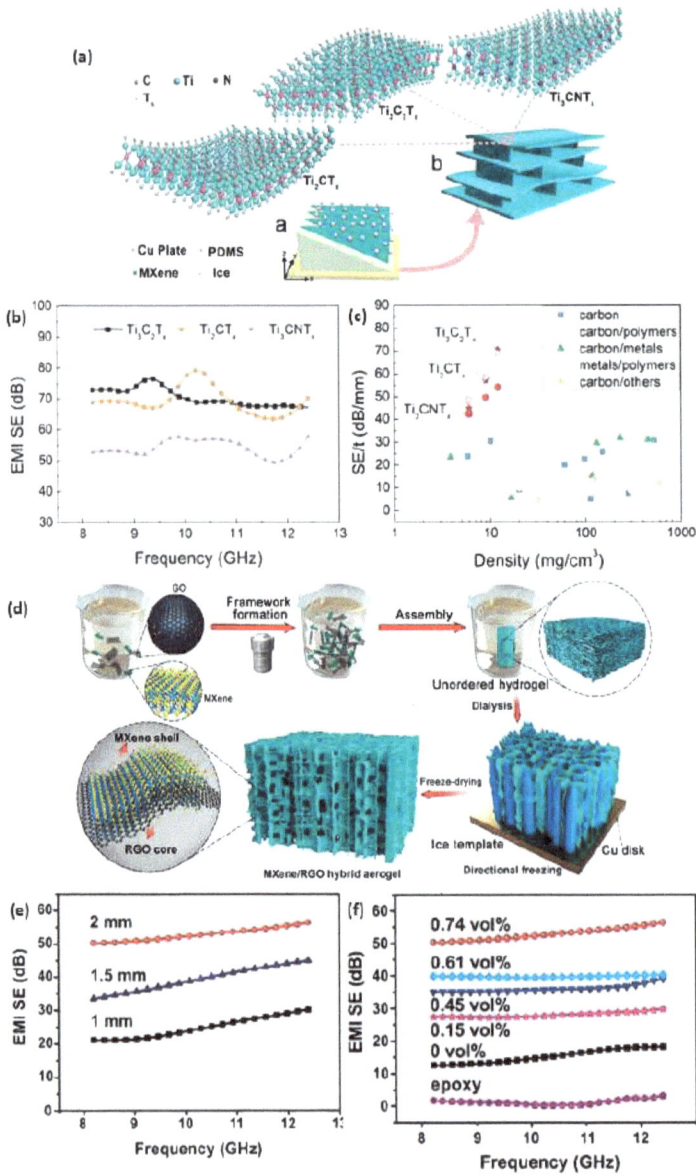

Fig. (8). (a) Bidirectional freeze-casting mechanism; **(b)** EMI SE of $Ti_3C_2T_x$, Ti_2CT_x, and Ti_3CNT_x. **(c)** Comparison plot of SE/t *vs.* density of various reported materials. Adapted with permission [64]. Copyright 2019, Wiley-VCH. **(d)** Schematic for the synthesis of MXene/RGO Hybrid aerogel and the EMI SE of hybrid aerogels with varying **(e)** thicknesses and **(f)** MXene contents. Adapted with permission [65]. Copyright 2018, American Chemical Society.

Bidirectional freeze-drying methods were employed to fabricate flexible aerogels by varying the concentration of the MXene solutions. A uniform lamellar-structured aerogel was fabricated as a result of the dual temperature gradient, which cannot be attained *via* a conventional freeze-drying method. As the concentration of the MXene solution increased, the formation of a lamellar bridge structure was observed. The uniform porous lamellar-bridged structure results in a highly conductive MXene-based framework. Thus, a robust and conductive aerogel was obtained from the MXene solution at a concentration of 11 mg mL^{-1}. The low density and excellent electrical conductivities of the $Ti_3C_2T_x$, Ti_3CNT_x, and Ti_2CT_x aerogels outperform those of the other carbon materials and subsequently reveal outstanding EMI SE and SE/t values (Fig. **8c**). The $Ti_3C_2T_x$, Ti_3CNT_x, and Ti_2CT_x aerogels exhibit the average total EMI SE (SE$_T$) values of 70.5, 69.2, and 54.1 dB, respectively, at a thickness of 1 mm in the X-band frequency range. The EMI shielding performance of aerogels is ascribed to the uniform porous lamellar structure, high electrical conductivity, and multiple reflections within the porous structure. MXene foams and aerogels with low densities and robust mechanical properties are promising contenders as new-generation lightweight EMI-shielding materials for utilization in modern electronics, defense, and industrial applications. Zhao *et al.* synthesized $Ti_3C_2T_x$/reduced graphene oxide (MXene/rGO) aerogels *via* a freeze-drying method wherein rGO was added to the MXene solution in order to improve the structural stability of MXene aerogels (Fig. **9d**) [65]. This freeze-drying method yielded compacted aligned porous structures, where the graphene sheet shell was covered by conductive $Ti_3C_2T_x$ sheets. The mechanically stable aerogel porous framework was obtained owing to the larger sheet size of rGO, while the electrical conductivity is attributed to the conductive MXene sheet. On the addition of 0.74 vol % MXene, electrical conductivity of 10.85 S cm^{-1} was observed with an EMI SE of 50 dB at a thickness of 2 mm (Fig. **9e, f**). The proposed mechanisms revealed that when an EM wave enters the porous structure, it is subjected to internal scattering through multiple interfaces, which leads to the dissipation of absorbed energy in the form of heat. Further absorption loss was attributed to the dielectric polarization losses arising from the surface terminations and the structural defects.

Sambyal *et al.* synthesized MXene/CNT hybrid (MXCNT) aerogels by reinforcing mechanically strong CNTs in $Ti_3C_2T_x$MXene (Fig. **9a-g**) [15]. The bidirectional freeze-drying process fabricated a uniform lamellar structure that was supported by MXene bridges. The introduction of CNTs yielded outstanding electrical conductivity and robust mechanical properties, owing to the reinforcement of CNTs to the MXene bridges in the alternating layers. Moreover, MXCNT aerogels showed an anisotropic behavior for electrical conductivity. The compression strength analysis of the MXCNT aerogels revealed higher compressive strengths as compared to those of the pristine $Ti_3C_2T_x$ MXene aerogel. Three different compositions with varying CNT concentrations of MXene (1:1, 1:2, and 1:3) demonstrated an increment in the compression moduli by 3898, 7796, and 9661%, respectively, as compared to that of the pure MXene aerogel. The MXCNT13 aerogel with an MXene-to-CNT ratio of 1:3 exhibited shielding effectiveness of 103 dB at 3 mm, and a corresponding SSE/t value of 8,253.9 dB cm^2 g^{-1} was achieved (Fig. **9f**). The MXCNT aerogels exhibited an absorption-dominant EMI shielding performance, which was attributed to the porous lamellar structure. This shielding performance was achieved by three factors, electron migration, dipolar polarization, and internal scattering within the porous architecture.

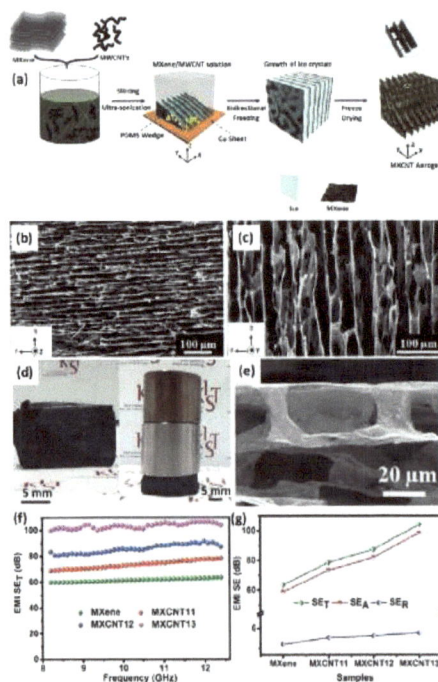

Fig. (9). (a) Schematic representation of a bidirectional freezing method for the fabrication of MXCNT Hybrid aerogel, **(b-d)** the SEM images of $Ti_3C_2T_x$MXene aerogel, the MXCNT13 aerogel **(e)** before and during compression; **(f)** the total EMI SE of the $Ti_3C_2T_x$/CNTs aerogels, and **(g)** the absolute EMI SET, SER and SEA of the $Ti_3C_2T_x$/CNTs aerogels at 12.4 GHz. Adapted with permission [15]. Copyright 2019, American Chemical Society.

In an interesting research work, Liang *et al.* fabricated ultralight 3D MXene aerogel/wood-derived porous carbon (WPC) composites for EMI shielding applications [66]. Here, the WPC skeleton obtained from natural wood acts as a template and provides a uniform honeycomb cell structure. A higher degree of graphitization of natural wood was observed at a high carbonization temperature. The carbonization at three different temperatures, WPC-500, WPC-1000, and WPC-1500, exhibited water contact angles of 75.4°, 104.1°, and 112.7°, respectively. MXene aerogel/WPC composite with a density of 0.197 g/cm^3 showed an EMI SE of 71.3 dB. Interestingly, the composite showed a wall-like structure composed of the MXene aerogel (brick) and WPC skeleton (mortar), which stabilized the unstable MXene aerogel structure. Furthermore, this structure increases the transmission path of the EM wave and attenuates the EM wave in the form of heat and ohmic losses, thereby yielding outstanding EMI shielding efficiency. The MXene aerogel/WPC composite showed excellent thermal insulation, flame-retardant properties, and good anisotropic compressive strength.

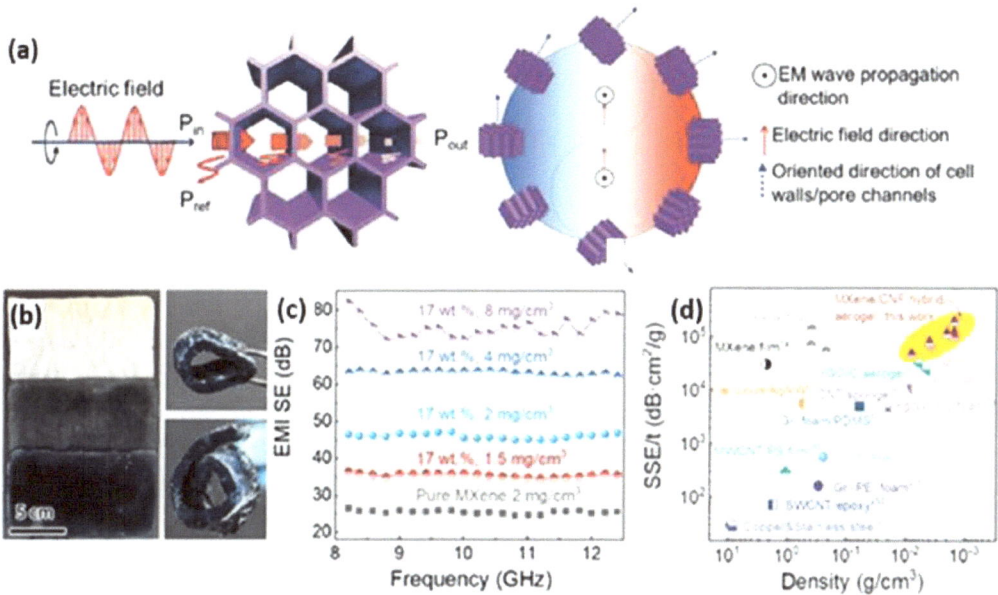

Fig. (10). (a) Oriented cell walls-induced EMI shielding mechanism of the MXene/CNF hybrid aerogels. (b) Digital images of the MXene-based hybrid aerogels, showing large-area MXene/CNF aerogels (12 × 6 cm^2) with various CNF contents (from top to bottom, corresponding to the pure CNF aerogels, the MXene/CNF with 50 wt.% CNF, and MXene/CNF with 17 wt.% CNF), and an image showing the flexible performance of the MXene/CNF aerogels. (c) EMI SE in the X-band and (d) Comparison of the MXene/CNF hybrid aerogels' shielding performance with reported literature. Adapted with permission [67], copyright 2020, Wiley-VCH.

Zeng *et al.* designed nanocellulose-MXene aerogels with orientation-tunable EMI shielding efficiency [67]. Ultrathin cellulose nanofibrils (CNFs) were introduced to MXene to fabricate ultralow-density, robust, and highly flexible aerogels with oriented biomimetic cell walls. They discovered that the EMI shielding effectiveness strongly relies on the angle between the electric field direction of the incident EM wave and the oriented cell walls of the aerogel; here, the parallel configuration yielded a maximized EMI SE, in which the oriented cell walls were parallel to the electric field direction (Fig. **10a**). Thus, a wide range of governable EMI SE values can be attained without changing the frame materials, and they provide a great opportunity for the design of oriented and structured aerogels or systems for brilliant EMI shielding efficiencies. Subsequently, MXene/CNF hybrid aerogel showed EMI SE values of 35.5 and 74.6 dB at densities of 1.5 and 8.0 mg cm^{-3}, respectively (Fig. **10c**). Furthermore, the ultralight aerogel exhibits outstanding SSE and SSE/t values of 30,660 and 189,400 dB cm^2 g^{-1}, respectively, and significantly outperforms all other EMI shielding materials that have been reported thus far (Fig. **10d**). The brilliant EMI shielding performance was obtained from the synergistic combination of MXene/CNF hybrid cell walls that allow ultralight aerogels to eliminate EM pollution.

CONCLUSION

Herein, we have compiled a porous MXene structural approach to attenuate EM pollution. Since the EMI shielding properties of MXenes were first reported, MXenes have been extensively explored as EMI shield materials, owing to their metallic conductivity, large specific area, and ease of surface modification. Therefore, excellent EMI shielding is achieved, and numerous studies on EMI shielding with MXene have been carried out. However, there are still many aspects and challenges that need to be addressed. MXenes have shown reflection-dominant EMI shielding efficiency, which can lead to secondary pollution. Thus, a highly absorption-dominant MXene shield is required against EM waves. As discussed above, the porous structure approach subsequently reduces the reflection contribution to the total shielding effectiveness. Various architectures, such as foam and aerogel, are employed against EM pollution because of their lightweight, outstanding electrical properties, and excellent EMI shielding performance. Moreover, the porous architecture increases the impedance matching, which leads to the penetration of EM waves into the shield. The absorbed EM wave is subjected to multiple internal reflections due to the presence of different interfaces within the shield material, which results in EM wave attenuation.

Thus, various structural approaches, such as the use of laminate, layer-by-layer, segregated, and porous structures, require exploration of their EMI shielding properties. Modern electronic systems have certain prerequisites, such as lightweight, portability, high-speed processing, and excellent shielding performance. From the above literature, we can conclude that MXenes are promising candidates for futuristic smart electronics, medical equipment, defense systems, and analogous applications.

CONSENT FOR PUBLICATION

Not applicable.

CONFLICT OF INTEREST

The authors declare no conflict of interest, financial or otherwise.

ACKNOWLEDGEMENTS

Declared none.

REFERENCES

[1]   Chen, Z.; Xu, C.; Ma, C.; Ren, W.; Cheng, H. M. J. A. m. *Lightweight and flexible graphene foam composites for high-performance electromagnetic interference shielding.*, **2013**, *25*(9), 1296-1300.
        http://dx.doi.org/10.1002/adma.201204196

[2]   Shahzad, F.; Alhabeb, M.; Hatter, C.B.; Anasori, B.; Man Hong, S.; Koo, C.M.; Gogotsi, Y. Electromagnetic interference shielding with 2D transition metal carbides (MXenes). *Science,* **2016**, *353*(6304), 1137-1140.
        http://dx.doi.org/10.1126/science.aag2421 PMID: 27609888

[3]   Singh, A.P.; Mishra, M.; Sambyal, P.; Gupta, B.K.; Singh, B.P.; Chandra, A.; Dhawan, S. Encapsulation of γ-Fe₂O₃ decorated reduced graphene oxide in polyaniline core–shell tubes as an exceptional tracker for electromagnetic environmental pollution. *J. Mater. Chem. A Mater. Energy Sustain.,* **2014**, *2*(10), 3581-3593.
        http://dx.doi.org/10.1039/C3TA14212D

[4]   Stam, R.; Yamaguchi-Sekino, S. J. I. h. *Occupational exposure to electromagnetic fields from medical sources.*, **2017**.

[5]   Redlarski, G.; Lewczuk, B.; Żak, A.; Koncicki, A.; Krawczuk, M.; Piechocki, J.; Jakubiuk, K.; Tojza, P.; Jaworski, J.; Ambroziak, D. J. B. r. i. The influence of electromagnetic pollution on living organisms: historical trends and forecasting changes. **2015**.

[6]   Christ, A.; Douglas, M.; Nadakuduti, J.; Kuster, N. J. P. o. t. I. *Assessing human exposure to electromagnetic fields from wireless power transmission systems.*, **2013**, *101*(6), 1482-1493.
        http://dx.doi.org/10.1109/JPROC.2013.2245851

[7]   Carpenter, D. O. J. R. o. e. h. *Human disease resulting from exposure to electromagnetic fields.*, **2013**, *28*(4), 159-172.

[8]   Hocking, B.; Mild, K. H. J. I. J. o. O. S. Ergonomics, Guidance note: risk management of workers with medical electronic devices and metallic implants in electromagnetic fields. **2008**, *14*(2), 217-222.

[9]   Irnich, W.; De Bakker, J.; Bisping, H. J. J. P.; Electrophysiology, C. *Electromagnetic interference in implantable pacemakers.*, **1978**, *1*(1), 52-61.
        http://dx.doi.org/10.1111/j.1540-8159.1978.tb03441.x

[10] Choi, Y.-S.; Yoo, Y.-H.; Kim, J.-G.; Kim, S.-H. J. S.; Technology, C. *A comparison of the corrosion resistance of Cu–Ni–stainless steel multilayers used for EMI shielding.*, **2006**, *201*(6), 3775-3782.
http://dx.doi.org/10.1016/j.surfcoat.2006.03.040

[11] Luo, X.; Chung, D. J. C. P. B. E. *Electromagnetic interference shielding using continuous carbon-fiber carbon-matrix and polymer-matrix composites.*, **1999**, *30*(3), 227-231.
http://dx.doi.org/10.1016/S1359-8368(98)00065-1

[12] Ohlan, A.; Singh, K.; Chandra, A.; Dhawan, S. Microwave absorption properties of conducting polymer composite with barium ferrite nanoparticles in 12.4–18 GHz. *Appl. Phys. Lett.,* **2008**, *93*(5), 053114.
http://dx.doi.org/10.1063/1.2969400

[13] Li, X.; Yin, X.; Han, M.; Song, C.; Xu, H.; Hou, Z.; Zhang, L.; Cheng, L. Ti3C2 MXenes modified with *in situ* grown carbon nanotubes for enhanced electromagnetic wave absorption properties. *J. Mater. Chem. C Mater. Opt. Electron. Devices,* **2017**, *5*(16), 4068-4074.
http://dx.doi.org/10.1039/C6TC05226F

[14] Iqbal, A.; Shahzad, F.; Hantanasirisakul, K.; Kim, M-K.; Kwon, J.; Hong, J.; Kim, H.; Kim, D.; Gogotsi, Y.; Koo, C.M. Anomalous absorption of electromagnetic waves by 2D transition metal carbonitride Ti$_3$CNT$_x$ (MXene). *Science,* **2020**, *369*(6502), 446.
http://dx.doi.org/10.1126/science.aba7977 PMID: 32703878

[15] Sambyal, P.; Iqbal, A.; Hong, J.; Kim, H.; Kim, M-K.; Hong, S.M.; Han, M.; Gogotsi, Y.; Koo, C.M. Ultralight and Mechanically Robust Ti$_3$C$_2$T$_x$ Hybrid Aerogel Reinforced by Carbon Nanotubes for Electromagnetic Interference Shielding. *ACS Appl. Mater. Interfaces,* **2019**, *11*(41), 38046-38054.
http://dx.doi.org/10.1021/acsami.9b12550 PMID: 31509378

[16] Wang, L.; Qiu, H.; Song, P.; Zhang, Y.; Lu, Y.; Liang, C.; Kong, J.; Chen, L.; Gu, J. 3D Ti$_3$C$_2$Tx MXene/C hybrid foam/epoxy nanocomposites with superior electromagnetic interference shielding performances and robust mechanical properties. *Compos., Part A Appl. Sci. Manuf.,* **2019**, *123*, 293-300.
http://dx.doi.org/10.1016/j.compositesa.2019.05.030

[17] Alhabeb, M.; Maleski, K.; Anasori, B.; Lelyukh, P.; Clark, L.; Sin, S.; Gogotsi, Y. J. C. o. M. Guidelines for synthesis and processing of two-dimensional titanium carbide (Ti3C2T x MXene). **2017**, *29*(19), 7633-7644.

[18] Naguib, M.; Kurtoglu, M.; Presser, V.; Lu, J.; Niu, J.; Heon, M.; Hultman, L.; Gogotsi, Y.; Barsoum, M. W. J. A. m. *Two-dimensional nanocrystals produced by exfoliation of Ti3AlC2.*, **2011**, *23*(37), 4248-4253.
http://dx.doi.org/10.1002/adma.201102306

[19] Anasori, B.; Lukatskaya, M. R.; Gogotsi, Y. J. N. R. M. **2017**.

[20] Liu, F.; Zhou, A.; Chen, J.; Jia, J.; Zhou, W.; Wang, L.; Hu, Q. J. A. S. S. Preparation of Ti3C2 and Ti2C MXenes by fluoride salts etching and methane adsorptive properties. **2017**, *416*, 781-789.

[21] Kim, D.; Ko, T. Y.; Kim, H.; Lee, G. H.; Cho, S.; Koo, C. M. J. A. n. Nonpolar Organic Dispersion of 2D Ti3C2T x MXene Flakes via Simultaneous Interfacial Chemical Grafting and Phase Transfer Method. **2019**, *13*(12), 13818-13828.

[22] Zhao, S.; Li, L.; Zhang, H.-B.; Qian, B.; Luo, J.-Q.; Deng, Z.; Shi, S.; Russell, T. P.; Yu, Z.-Z. J. M. C. F. *Janus MXene nanosheets for macroscopic assemblies.*, **2020**, *4*(3), 910-917.
http://dx.doi.org/10.1039/C9QM00681H

[23] Chaudhari, N.K.; Jin, H.; Kim, B.; San Baek, D.; Joo, S.H.; Lee, K. MXene: an emerging two-dimensional material for future energy conversion and storage applications. *J. Mater. Chem. A Mater. Energy Sustain.,* **2017**, *5*(47), 24564-24579.
http://dx.doi.org/10.1039/C7TA09094C

[24]   Sokol, M.; Natu, V.; Kota, S.; Barsoum, M. W. J. T. i. C. *On the chemical diversity of the MAX phases.*, **2019**, *1*(2), 210-223.
http://dx.doi.org/10.1016/j.trechm.2019.02.016

[25]   Barsoum, M. W. J. P. i. s. s. c. The MN+ 1AXN phases: A new class of solids: Thermodynamically stable nanolaminates. **2000**, *28*(1-4), 201-281.

[26]   Khazaei, M.; Ranjbar, A.; Arai, M.; Sasaki, T.; Yunoki, S. J. J. o. M. C. C. *Electronic properties and applications of MXenes: a theoretical review.*, **2017**, *5*(10), 2488-2503.
http://dx.doi.org/10.1039/C7TC00140A

[27]   Naguib, M.; Mochalin, V. N.; Barsoum, M. W.; Gogotsi, Y. J. A. m. 25th anniversary article: MXenes: a new family of two-dimensional materials. **2014**, *26*(7), 992-1005.

[28]   Rosen, J.; Dahlqvist, M.; Tao, Q.; Hultman, L. *In-and Out-of-Plane Ordered MAX Phases and Their MXene Derivatives. In 2D Metal Carbides and Nitrides (MXenes)*; Springer, **2019**, pp. 37-52.
http://dx.doi.org/10.1007/978-3-030-19026-2_3

[29]   Anasori, B.; Xie, Y.; Beidaghi, M.; Lu, J.; Hosler, B.; Hultman, L.; Kent, P.; Gogotsi, Y.; Barsoum, M. J. N.; Mashtalir, O.; Carle, J.; Presser, V.; Lu, J.; Hultman, L.; Gogotsi, Y.; Barsoum, MW **2015**, *9*, 9507.

[30]   Han, M.; Shuck, C.E.; Rakhmanov, R.; Parchment, D.; Anasori, B.; Koo, C.M.; Friedman, G.; Gogotsi, Y. Beyond $Ti_3C_2T_x$: MXenes for Electromagnetic Interference Shielding. *ACS Nano,* **2020**, *14*(4), 5008-5016.
http://dx.doi.org/10.1021/acsnano.0c01312 PMID: 32163265

[31]   Liang, X.; Garsuch, A.; Nazar, L. F. J. A. C. *Sulfur cathodes based on conductive MXene nanosheets for high-performance lithium–sulfur batteries.*, **2015**, *127*(13), 3979-3983.

[32]   Zhao, X.; Liu, M.; Chen, Y.; Hou, B.; Zhang, N.; Chen, B.; Yang, N.; Chen, K.; Li, J.; An, L. J. J. o. M. C. A. Fabrication of layered Ti_3C_2 with an accordion-like structure as a potential cathode material for high performance lithium–sulfur batteries. **2015**, *3*(15), 7870-7876.

[33]   Naguib, M.; Halim, J.; Lu, J.; Cook, K. M.; Hultman, L.; Gogotsi, Y.; Barsoum, M. W. J. o. t. A. C. S. *New two-dimensional niobium and vanadium carbides as promising materials for Li-ion batteries.*, **2013**, *135*(43), 15966-15969.
http://dx.doi.org/10.1021/ja405735d

[34]   Zou, G.; Guo, J.; Peng, Q.; Zhou, A.; Zhang, Q.; Liu, B. J. J. o. M. C. A. *Synthesis of urchin-like rutile titania carbon nanocomposites by iron-facilitated phase transformation of MXene for environmental remediation.*, **2016**, *4*(2), 489-499.
http://dx.doi.org/10.1039/C5TA07343J

[35]   Zhang, Q.; Teng, J.; Zou, G.; Peng, Q.; Du, Q.; Jiao, T.; Xiang, J. J. N. Efficient phosphate sequestration for water purification by unique sandwich-like MXene/magnetic iron oxide nanocomposites. **2016**, *8*(13), 7085-7093.

[36]   Guo, J.; Peng, Q.; Fu, H.; Zou, G.; Zhang, Q. J. T. J. o. P. C. C. *Heavy-metal adsorption behavior of two-dimensional alkalization-intercalated MXene by first-principles calculations.*, **2015**, *119*(36), 20923-20930.
http://dx.doi.org/10.1021/acs.jpcc.5b05426

[37]   Ren, C. E.; Hatzell, K. B.; Alhabeb, M.; Ling, Z.; Mahmoud, K. A.; Gogotsi, Y. J. T. j. o. p. c. l. Charge- and size-selective ion sieving through $Ti_3C_2T_x$ MXene membranes. **2015**, *6*(20), 4026-4031.

[38]   Chen, J.; Chen, K.; Tong, D.; Huang, Y.; Zhang, J.; Xue, J.; Huang, Q.; Chen, T. J. C. C. CO2 and temperature dual responsive "Smart" MXene phases. **2015**, *51*(2), 314-317.

[39]   Yu, X.-f.; Li, Y.-c.; Cheng, J.-b.; Liu, Z.-b.; Li, Q.-z.; Li, W.-z.; Yang, X.; Xiao, B. J. A. a. m. interfaces, Monolayer Ti_2CO_2: a promising candidate for NH3 sensor or capturer with high sensitivity and selectivity **2015**, *7*(24), 13707-13713.

[40]   Liu, H.; Duan, C.; Yang, C.; Shen, W.; Wang, F.; Zhu, Z. J. S.; Chemical, A. B. A novel nitrite biosensor based on the direct electrochemistry of hemoglobin immobilized on MXene-Ti_3C_2. **2015**, *218*, 60-66.

[41] Peng, C.; Yang, X.; Li, Y.; Yu, H.; Wang, H.; Peng, F. J. A. a. m. interfaces, Hybrids of two-dimensional Ti3C2 and TiO2 exposing {001} facets toward enhanced photocatalytic activity. **2016**, *8*(9), 6051-6060.

[42] Cao, S.; Shen, B.; Tong, T.; Fu, J.; Yu, J. J. A. F. M. 2D/2D heterojunction of ultrathin MXene/Bi2WO6 nanosheets for improved photocatalytic CO2 reduction. **2018**, *28*(21), 1800136.

[43] Zhang, Y.-J.; Lan, J.-H.; Wang, L.; Wu, Q.-Y.; Wang, C.-Z.; Bo, T.; Chai, Z.-F.; Shi, W.-Q. J. J. o. H. M. *Adsorption of uranyl species on hydroxylated titanium carbide nanosheet: A first-principles study.*, **2016**, *308*, 402-410.
http://dx.doi.org/10.1016/j.jhazmat.2016.01.053

[44] Wang, L.; Yuan, L.; Chen, K.; Zhang, Y.; Deng, Q.; Du, S.; Huang, Q.; Zheng, L.; Zhang, J.; Chai, Z. J. A. A. M. Interfaces, Loading actinides in multilayered structures for nuclear waste treatment: the first case study of uranium capture with vanadium carbide MXene. **2016**, *8*(25), 16396-16403.

[45] Zhang, P.; Wang, L.; Yuan, L-Y.; Lan, J-H.; Chai, Z-F.; Shi, W-Q. Sorption of Eu(III) on MXene-derived titanate structures: The effect of nano-confined space. *Chem. Eng. J.*, **2019**, *370*, 1200-1209.
http://dx.doi.org/10.1016/j.cej.2019.03.286

[46] Rasool, K.; Helal, M.; Ali, A.; Ren, C. E.; Gogotsi, Y.; Mahmoud, K. A. J. A. n. Antibacterial activity of Ti3C2T x MXene. **2016**, *10*(3), 3674-3684.

[47] Jastrzębska, A. M.; Karwowska, E.; Wojciechowski, T.; Ziemkowska, W.; Rozmysłowska, A.; Chlubny, L.; Olszyna, A. J. J. o. M. E. Performance, The atomic structure of Ti2C and Ti3C2 MXenes is responsible for their antibacterial activity toward E. coli bacteria. **2019**, *28*(3), 1272-1277.

[48] Mayerberger, E. A.; Street, R. M.; McDaniel, R. M.; Barsoum, M. W.; Schauer, C. L. J. R. a. Antibacterial properties of electrospun Ti 3 C 2 T z (MXene)/chitosan nanofibers. **2018**, *8*(62), 35386-35394.

[49] Cai, C.; Wang, R.; Liu, S.; Yan, X.; Zhang, L.; Wang, M.; Tong, Q.; Jiao, T. J. C.; Physicochemical, S. A.; Aspects, E. *Synthesis of self-assembled phytic acid-MXene nanocomposites via a facile hydrothermal approach with elevated dye adsorption capacities.*, **2020**, *589*, 124468.
http://dx.doi.org/10.1016/j.colsurfa.2020.124468

[50] Luo, S.; Wang, R.; Yin, J.; Jiao, T.; Chen, K.; Zou, G.; Zhang, L.; Zhou, J.; Zhang, L.; Peng, Q. J. A. o. Preparation and dye degradation performances of self-assembled MXene-Co3O4 nanocomposites synthesized via solvothermal approach. **2019**, *4*(2), 3946-3953.

[51] Li, X.; Zeng, C.; Fan, G. J. I. J. o. H. E. Ultrafast hydrogen generation from the hydrolysis of ammonia borane catalyzed by highly efficient bimetallic RuNi nanoparticles stabilized on Ti3C2X2 (X= OH and/or F). **2015**, *40*(10), 3883-3891.

[52] Kuang, P.; He, M.; Zhu, B.; Yu, J.; Fan, K.; Jaroniec, M. 0D/2D NiS2/V-MXene composite for electrocatalytic H2 evolution. *J. Catal.*, **2019**, *375*, 8-20.
http://dx.doi.org/10.1016/j.jcat.2019.05.019

[53] Yun, T.; Kim, H.; Iqbal, A.; Cho, Y. S.; Lee, G. S.; Kim, M. K.; Kim, S. J.; Kim, D.; Gogotsi, Y.; Kim, S. O. J. A. M. *Electromagnetic shielding of monolayer MXene assemblies.*, **2020**, *32*(9), 1906769.
http://dx.doi.org/10.1002/adma.201906769

[54] Simon, R.M. Emi Shielding Through Conductive Plastics. *Polym. Plast. Technol. Eng.*, **1981**, *17*(1), 1-10.
http://dx.doi.org/10.1080/03602558108067695

[55] Moore, R. *Electromagnetic composites handbook*; McGraw Hill Professional, **2016**.

[56] He, P.; Wang, X.-X.; Cai, Y.-Z.; Shu, J.-C.; Zhao, Q.-L.; Yuan, J.; Cao, M.-S. J. N. Tailoring Ti3 C2T x nanosheets to tune local conductive network as an environmentally friendly material for highly efficient electromagnetic interference shielding. **2019**, *11*(13), 6080-6088.

[57] Liu, J.; Zhang, H. B.; Sun, R.; Liu, Y.; Liu, Z.; Zhou, A.; Yu, Z. Z. J. A. M. *Hydrophobic, flexible, and lightweight MXene foams for high-performance electromagnetic-interference shielding.*, **2017**, *29*(38), 1702367.

http://dx.doi.org/10.1002/adma.201702367

[58]   Fan, Z.; Wang, D.; Yuan, Y.; Wang, Y.; Cheng, Z.; Liu, Y.; Xie, Z. J. C. E. J. A lightweight and conductive MXene/graphene hybrid foam for superior electromagnetic interference shielding. **2020**, *381*, 122696.

[59]   Xu, H.; Yin, X.; Li, X.; Li, M.; Liang, S.; Zhang, L.; Cheng, L. Lightweight Ti$_2$CT $_x$ MXene/Poly(vinyl alcohol) Composite Foams for Electromagnetic Wave Shielding with Absorption-Dominated Feature. *ACS Appl. Mater. Interfaces,* **2019**, *11*(10), 10198-10207.

   http://dx.doi.org/10.1021/acsami.8b21671 PMID: 30689343

[60]   Wu, X.; Han, B.; Zhang, H.-B.; Xie, X.; Tu, T.; Zhang, Y.; Dai, Y.; Yang, R.; Yu, Z.-Z. J. C. E. J. *Compressible, durable and conductive polydimethylsiloxane-coated MXene foams for high-performance electromagnetic interference shielding.*, **2020**, *381*, 122622.

   http://dx.doi.org/10.1016/j.cej.2019.122622

[61]   Weng, C.; Wang, G.; Dai, Z.; Pei, Y.; Liu, L.; Zhang, Z. J. N. Buckled AgNW/MXene hybrid hierarchical sponges for high-performance electromagnetic interference shielding. **2019**, *11*(47), 22804-22812.

[62]   Bian, R.; He, G.; Zhi, W.; Xiang, S.; Wang, T.; Cai, D. J. J. o. M. C. C. *Ultralight MXene-based aerogels with high electromagnetic interference shielding performance.*, **2019**, *7*(3), 474-478.

   http://dx.doi.org/10.1039/C8TC04795B

[63]   Zhou, Z.; Liu, J.; Zhang, X.; Tian, D.; Zhan, Z.; Lu, C. Ultrathin MXene/Calcium Alginate Aerogel Film for High-Performance Electromagnetic Interference Shielding. *Adv. Mater. Interfaces,* **2019**, *6*(6), 1802040.

   http://dx.doi.org/10.1002/admi.201802040

[64]   Han, M.; Yin, X.; Hantanasirisakul, K.; Li, X.; Iqbal, A.; Hatter, C. B.; Anasori, B.; Koo, C. M.; Torita, T.; Soda, Y. J. A. O. M. *Anisotropic MXene aerogels with a mechanically tunable ratio of electromagnetic wave reflection to absorption.*, **2019**, *7*(10), 1900267.

   http://dx.doi.org/10.1002/adom.201900267

[65]   Zhao, S.; Zhang, H-B.; Luo, J-Q.; Wang, Q-W.; Xu, B.; Hong, S.; Yu, Z-Z. Highly Electrically Conductive Three-Dimensional Ti$_3$C$_2$T $_x$ MXene/Reduced Graphene Oxide Hybrid Aerogels with Excellent Electromagnetic Interference Shielding Performances. *ACS Nano,* **2018**, *12*(11), 11193-11202.

   http://dx.doi.org/10.1021/acsnano.8b05739 PMID: 30339357

[66]   Liang, C.; Qiu, H.; Song, P.; Shi, X.; Kong, J.; Gu, J. Ultra-light MXene aerogel/wood-derived porous carbon composites with wall-like "mortar/brick" structures for electromagnetic interference shielding. *Sci. Bull. (Beijing),* **2020**, *65*(8), 616-622.

   http://dx.doi.org/10.1016/j.scib.2020.02.009

[67]   Zeng, Z.; Wang, C.; Siqueira, G.; Han, D.; Huch, A.; Abdolhosseinzadeh, S.; Heier, J.; Nüesch, F.; Zhang, C.; Nyström, G. *Nanocellulose-MXene Biomimetic Aerogels with Orientation-Tunable Electromagnetic Interference Shielding Performance.*, **2020**, *7*(15), 2000979.

CHAPTER 11

Nanostructured Two-Dimensional (2D) Materials as Potential Candidates for EMI Shielding

Ayushi Saini[1], Anil Ohlan[2], S. K. Dhawan[3], and Kuldeep Singh[1]*

[1] *CSIR-Central Electrochemical Research Institute (CECRI) Chennai Unit, CSIR Madras Complex, Taramani, Chennai – 600113, India*

[2] *Department of Physics, Maharshi Dayanand University, Rohtak124001, India*

[3] *Materials Physics & Engineering Division, CSIR-National Physical Laboratory, New Delhi – 110012, India*

Abstract: For an effective EMI shielding, materials should have high electrical conductivity as EMI attenuation is a sum of relfection, absorption, and multiple relfections which requires the existence of mobile charge carriers (electrons or holes), electric and/or magnetic dipoles, usually provided by materials having high dielectric constants (ε) or magnetic permeability (μ) and the large surface area or interface area. Until now, a metal shroud was the material of choice as an EMI shield. However, metal fillers add additional weight and are susceptible to corrosion, making them less desirable. Therefore, we have focused on new emerging two-dimensional 2D nanomaterials that are light in weight and have a low cost. Here, the focus is to address the challenges in their synthesis especially transition metal carbides (MXenes), MoS_2, functionalized graphene/ferromagnetic conducting polymer composites, and their fabrication for EMI reductions. These articles also evaluate and explain the recent progress explicitly and underline the complex interplay of its intrinsic properties of 2D nanostructured materials (MXene, MoS_2, Graphene/ferromagnetic polymer composite) as a potential candidate for EMI shielding and evaluate their electromagnetic compatibility. The chapter will cover the facets related to a newly emerging area of EMI shields in the automotive industry, especially lithium-ion battery-operated electric vehicles and self-driving cars, high-speed wireless communication devices, and next-generation mobile phones with 4G and 5G technology.

Keywords: Dielectric measurements, Graphene, Graphene/ferromagnetic Polymer composites, Hydrothermal reactions, MXene, MoS_2.

***Corresponding author Kuldeep Singh:** CSIR-Central Electrochemical Research Institute (CECRI) Chennai Unit, CSIR Madras Complex, Taramani, Chennai – 600113, India; Tel: +918586939107; E-mail: kuldeep.kakran@gmail.com

Sundeep K. Dhawan, Avanish Pratap Singh, Anil Ohlan, Kuldeep Singh Kakran and Pradeep Sambyal (Ed.)
All rights reserved-© 2022 Bentham Science Publishers

11.1. INTRODUCTION

The technology developments driven by the advancement in nano-engineering of materials have led to smarter, smaller, and lighter electrical & electronic devices. They are denser, fast, efficient, and operate at a higher frequency. However, during the operation of electronic equipment, electromagnetic radiation will be generated continuously and radiated outward [1, 2]. The more use of high frequency and consumption of electronic devices has resulted in a significant amount of electromagnetic (EM) radiation sources in the human environment. EM interference (EMI) can occur when EM radiation is emitted by a specific electrical circuit and interferes with the operation of nearby devices [3, 4]. This affects the working accuracy of electronic devices and can cause malfunction to medical apparatus, industry robots, and communication devices. It may also cause data theft and the vulnerability of personal security in electronic components. While electronic devices have provided convenience to humans, they have also introduced health hazards such as mutation, insomnia, headache, and leukaemia, as well as organ damage, thermal injuries, and cancer due to continuous exposure of living organisms to EM fields. This continues to be a serious concern in the society [5–7]. Therefore, a lot of attention has been paid by researchers, technocrats, and engineers to develop effective microwave absorption materials that are cost-efficient, lightweight, and effective over a broad frequency range. Most of the research in EMI shielding science and technology was focused on metals and polymer-based composite having both electrical and magnetic properties. For EMI shielding, materials should have electrical conductivity as EMI attenuation is a sum of relfection, absorption, and multiple relfections, which requires the existence of mobile charge carriers (electrons or holes), electric and/or magnetic dipoles, usually provided by materials having high dielectric constants (ε) or magnetic permeability (μ) and the large surface area or interface area [8, 9]. However, metal-based EMI shielding materials are dense and corrosive, which limits their further development. Compared with metal materials, polymer-based EMI shielding materials are corrosion resistant, light, easy to process, and adjustable, but their limited electrical conductivity restricts their use. Therefore, more efforts have been made to improve their electric conductivity or magnetic permeability by adding conductive fillers or additives such as graphite, carbon black, and metal particles in various shapes and sizes ranging zero–dimension (0D), one–dimension (1D), two–dimension (2D), and three–dimension (3D) materials or combination of these materials, to give different structural feature for excellent EMI shielding behaviour [10–18]. The MXene [19], MAX phase [20], graphene (GN), graphite, single-wall carbon nanotube (SWCNT), multi-walled carbon nanotube (MWCNT), magnetic nanoparticle, metal oxides, core-shell nanomaterials, and conductive ferromagnetic polymer composites are being utilized for EMI shielding application [9].

In this chapter, more in-depth insight is given on new emerging two-dimensional 2D nanomaterials, which are light in weight and of low cost. Here, the focus is to address the challenges in their synthesis, especially graphene, and beyond graphene 2D materials like MXenes, MoS_2, boron nitride (h-BN) and graphitic carbon nitride (g-C_3N_4), functionalized graphene/ferromagnetic conducting polymer composites, and their fabrication in the form of conductive coatings, paint and shield for EMI reductions. The chapter will cover the facets related to a newly emerging area of EMI shields in the automotive industry, especially lithium-ion battery-operated electric vehicles and self-driving cars, fast-charging stations, along with underlined basic standard testing protocols as per ISO/ IEC standards.

11.2. EMI SHIELDING REQUIREMENT AND MATERIAL APPLICATIONS

The requirement for miniaturization of electrical & electronic devices in civil and military applications requires lightweight smart materials offering the same level of effectiveness as the traditional metallic shields. Metals are prone to corrosion and have high density. Their processing and forming often require high temperature and high pressure, making them expensive. Therefore, extensive research focus has been directed towards replacing metals with polymer matrix composites to develop lightweight EMI shielding materials. Although polymers and polymer matrix composites are inherently poor conductors and ineffective in EMI shielding, novel methods have been developed to impart EMI shielding properties in them in order to benefit from their low density for weight reduction of the part. In recent years, due to the exceptional structural and electrical properties of two-dimensional(2D) nanomaterials like graphene, and beyond graphene 2D materials like MXenes [21], boron nitride (h-BN) [22], and graphitic carbon nitride (g-C_3N_4) [23], transition metal dichalcogenides (TMDs), and black phosphorus (BP) [24] have emerged as potential candidates for EMI shielding materials.

With the development of sophisticated and advanced artificial intelligence tools, electronic warfare has utilized electromagnetic radiation as a jamming tool or weapon on battlefields to destroy electronic components of ships, radars, and flights and destroy the security of the enemy countries. Hence, electromagnetic interference shielding is the unavoidable choice around the globe [25]. Secondly, there is a growing concern for a clean and green environment to meet global goals on climate change. EVs offer the potential for large-scale reduction in carbon dioxide (CO_2) emissions from the transport sector. Therefore, electric vehicles (EVs) are getting popular, and the EVs automobile industry is growing at a CAGR of 21.1% [26]. The advancement of electronics and the use of artificial intelligence makes use of high-density electronics in EVs along with high voltage Lithium-ion batteries. The existing automobiles are based on IC engines, a 12 V battery powers the cars, and 24 V powers the bus, truck & HCVs. While in lithium-ion battery operated electric Vehicles (EVs), the voltage can go up to 400 V for the mid to high-end cars & 900 V for the HCVs. The electric currents can also go up to 200 A. Like other electronic devices, Lithium-ion batteries also emit electromagnetic radiation and lead to EMI (Fig. **1**).

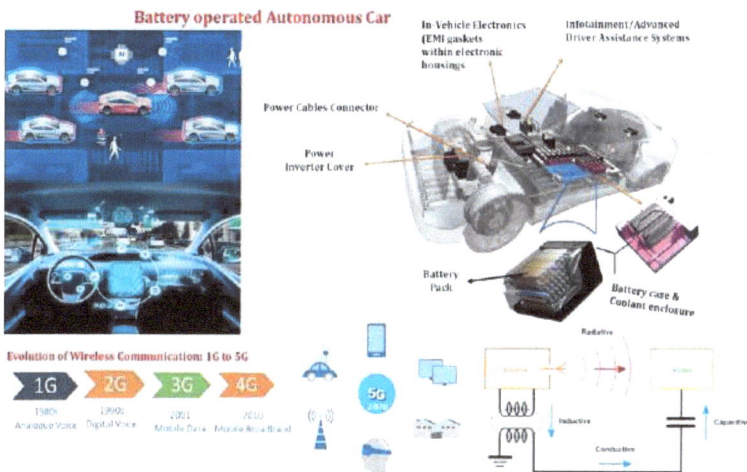

Fig. (1). New emerging sources of EMI: Autonomous car, EVs having high voltage Lithium-ion battery and progressive growth in telecommunications from the first generation 1G to 5 G.

These EMI emissions are a problem as they will disrupt electronic equipment and can cause malfunctions. To counter this problem, shielding is used. Typically shielding is metal; however, due to potential weight savings, a new polymer composite shield is suggested; they are lighter and save fuel consumption while helping in the reduction of carbon footprint. In a high voltage lithium-ion battery system, a battery management system (BMS) has a charging module for controlling the charging current, protection circuits against overheating and thermal management systems are essential; therefore BMS needs to go through the EMC tests. Parallel to this, the requirement of low-cost EMI shield in high-speed wireless communication devices, the new generation of mobile phones 4G and 5G technology have revolutionised the telecommunications industry. Due to take-off, wireless communication makes it possible to receive and send information *via* mobile devices anytime and anywhere; since the first handheld mobile phone demonstrated by Motorola in 1973, mobile wireless communication has experienced a tremendous change in nearly five decades, as shown in Fig. (**1**). The first generation (1G) mobile communication was developed in the 1980s. Afterward, mobile communication has undergone a generation of technology revolution and evolution about every 10 years, namely from 1G to 4G communications. A large number of organizations around the world have been established to promote 5G research to make more advanced higher speed data communication for a variety of smart applications and breed the Internet of Things (IoT) [27]. The seamless and in-depth coverage of base stations in the 5G era, meanwhile, is bound to bring on a boost in electromagnetic interference (EMI) and radiation on an unprecedented scale. On the one hand, various electronic devices have to insure themselves against the growing interference of external electromagnetic waves. The increasingly worsening electromagnetic pollution produced by electronics would be part of a public health challenge, which could hardly be overlooked in the near future. As a result, EMI shielding materials will embrace unprecedented new opportunities and become a critical component of electronic materials and even new materials of the future. Intelligent and connected EVs are becoming the most sophisticated mobile clients in the Internet of Things (IoT). However, this confluence of various technologies also means that system designers are challenged with the task of proper system integration in an evolving and shifting technology and regulatory environment to enable high-speed communication speeds to reach 5 Gb/s, with a digital transmission reference frequency of 2.5 GHz. This can radiate as interference noise in the 2.4-GHz band unless the device has the appropriate design, compliance evaluation, and leakage countermeasure implementations.

The sources of EMI may present in all environments, including land, sea, air, and space activities, including systems installed on vehicles, in the buildings, or portables. Hence, a wide range of equipment needs to be protected from electromagnetic radiation. A breakdown of the EMI materials application is presented in Fig. (**2**).

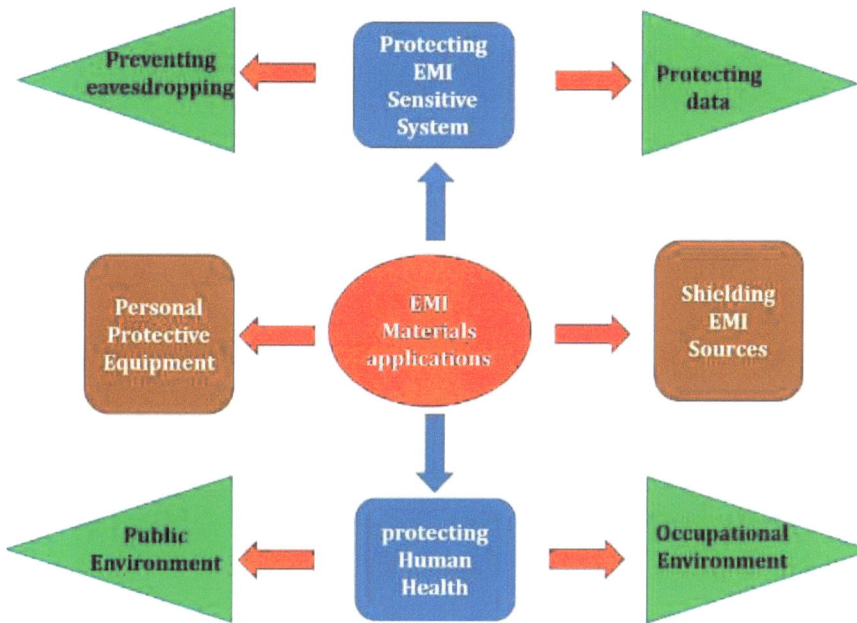

Fig. (2). EMI Materials application on the bases of their type sector.

Humans may be exposed to electromagnetic radiation in private, public, and working places. Several technologies may result in high occupational exposure, including induction heating, radiofrequency welding and sealing, microwave heating, working with plastic sealers and glue dyers, operating or servicing radio/TV transmission equipment, and radar equipment [28]. To facilitate ever-growing technological progress and the resulting EMR, the European Union member states in 2016 issued new safety limits to both guarantee the safety of humans and also the operability of technological systems, which sets new challenges in managing EMI issues. With the new legislative requirements, the employer is required to pay more attention to the risk management of EM fields and EMI. In an occupational environment, there are risk groups such as female workers who are pregnant or workers with medical implants [29]. The functionality of active medical implants, such as cardiac pacemakers, insulin pumps, *etc.*, may be compromised when exposed to strong EM fields [30].

11.3. EMI TEST METHODS AND STANDARDS

Electromagnetic interference problems can be solved using different tactics like grounding, bonding, filtering, isolating and shielding [31]. In particular, EMI shielding acts as a barrier to unwanted electrical noise and protects the equipment. EMI shielding is strongly influenced by the design of the enclosure serving as shield, mounting of the enclosure, gaskets used, environmental exposure, continuity of the shield, among others, and ultimately by the material used to build the part serving as the EMI shield. Different types of shielding solutions are used to protect these devices from the effects of EMI. Not all EMI shields or enclosures work at equal efficiency levels. Therefore, there are specific set of tests and measurement standards that can be used to evaluate the effectiveness of an EMI shield or enclosure. The level of attenuation (the gradual loss of signal) is used as one of the principal indicators for such evaluation. However, determining attenuation can be complex. Different methods are used for different shielding applications. The most common techniques for testing shielding strength include:

11.3.1. Open Field Test

Meant for finished electronic products, this test is generally conducted on an open site, free from any metallic equipment. Radiated field strength and conductive emissions are measured with the help of some antennae that are placed at varying distances from the device.

11.3.2. Coaxial Transmission Line Test

It's a kind of comparative technique that seeks to measure the shielding effectiveness of planar material. A reference-testing device is placed near it, and the voltage it receives at multiple frequencies is recorded. It's then replaced by a load-device, and voltage readings are taken in the same manner. The shielding effectiveness is measured by comparing the reading of reference and load devices.

11.3.3. Shielded Box Test

This method is effective for frequencies of less than 500 megahertz. It involves the use of a sealed box with an opening. A shielding unit is inserted through the box's opening. This helps record electromagnetic signals from both inside and outside the box. The ratio between the signals represents shielding effectiveness.

11.3.4. Shielded Room Test

This method involves the use of two shielded rooms with a separating wall between them. Testing equipment is placed in one room, and sensors in the other. Shielding leads are used to minimize errors caused by external signals. This process is ideal for evaluating a device's susceptibility.

Several EMI shielding effectiveness (SE) tests methods have been developed in order to quantify the shielding effectiveness capabilities of materials; the two most common methods are ASTM D4935-10 and IEEE Std 299-2006 standards [32]. Other existing standards available to characterize the SE of materials are ASTM D4935-99, ASTM ES7-83, MIL-STD-188-125A, IEEE-STD-299-1991, MIL-STD-461C, and MIL-STD-462. It has been reported that using the ASTM D4935-99 standard coaxial holder to measure SE is convenient because of the relatively small specimens required for testing in comparison with military standards, which require 46 cm square samples. The EMI shielding test method of enclosures was first introduced by the MIL-STD-285 standard and later replaced by an improved version, the IEEE Std 299-2006 standard. However due to large number of potential test conditions, the electrical nature of the materials to be tested, and setup related errors the EMI SE values for the same material may differ between different test methods. Thus, a more robust test setup to measure EMI shielding would be highly desirable [33–37].

A reverberation chamber technique for measuring the SE of enclosures in the order of 0.1 m cube and smaller is proposed. This sort of input led to the origin of the IEEE Std 299.1-2013 standard that covers EMI shielding tests of enclosures with dimensions between 0.1 m and 2 m not covered in the IEEE Std 299-2006 standard [34, 38].

The transmission line method has two distinct approaches to measure the EMI SE, the continuous conductor (CC) and the split conductor (SC). The CC configuration, based on the ASTM ES7-83 standard withdrawn in 1988, consisted of a 50 ohms continuous line that could be disassembled to insert an annular testing sample. The SC approach, defined by the standard ASTM D4935-10, consists of a sample holder, also known as test fixture, transmission line holder, tester and flanged coaxial sample holder, that works as a coaxial transmission line with geometrical characteristics such that the impedance at the connection points and throughout the line is maintained at 50 ohms, the characteristic impedance. The holder is made of two equal halves with a pair of flanges in the middle to hold the specimens allowing capacitive coupling of energy into insulating materials through displacement current [33]. The gap discontinuity due to the sample thickness is taken in account by using a reference sample that consists of a ring and a disc of the same material. The difference between the measurement without material (reference specimen) and with material (load specimen) gives the SE that is the result of two distinct shielding mechanisms, the shielding due to reflection and shielding due to absorption. The ASTM D4935-10 standard covers a frequency range from 30 MHz to 1.5 GHz, however various works [39] have used with more or less success a modified version of this method to perform SE tests in a wider range of frequencies. Other insightful work based also on the ASTM D4935 standard report on a tester developed to operate at frequencies up to 8.0 GHz, [40] and up to 18 GHz [41] reported in a more recent work. These testers are dedicated to the EMI SE measurements of thin films.

Despite the fact that CC and SC methods are widely used, the appearance of higher modes inside the sample, due to the reduction of wavelength in conductive materials considered electrically thick, may result in SE errors. This is justified by the fact that at high frequencies the shielding contributions due to absorption is considerable for the case of electrically thick samples. Dual-TEM cells, TEM-t, nested reverberation chambers, and flanged dual-ridged waveguides or H-t cells are examples of other SE test methods used with planar materials.

11.4. MARKET FOR EMI SHIELDING MATERIALS

Surging demand for consumer electronics and increasing electromagnetic pollution are the driving factors for the EMI shielding market. The global EMI shielding materials market has been evaluated and estimated by different indicators. The growth of this market is propelled by factors such as ongoing demonstrations of field trials and pilot projects to develop the viability of 5G technology, automobiles, autopilot cars, lithium-ion battery operated electric vehicles (EVs) and stringent environmental and EMC regulations across industries. The global EMI shielding market size is projected to grow from USD 6.8 billion in 2020 to USD 9.2 billion by 2025; it is expected to grow at a CAGR of 6.3 % from 2020 to 2025 [42]. While Transparency Market Research forecasted EMI Shielding Market of USD 6.21 Billion in 2019 and expected to reach USD 8.91 Billion by 2026 with a CAGR of 5.30 %. The Global Industry Analysts forecasts this market to reach USD 9 billion by 2024 [43]. The actual market volume in the coming years could be significantly higher due to many factors. European regulation on guaranteeing workers' safety from exposure to EM fields has obligated employers to minimize the workers' exposure to such radiation. Regulations like this can increase the market size and accelerate the adoption of more expensive next-generation materials. Some of the key players in the global EMI shielding materials market are listed here in Table **1.**

Table 1. Global players in EMI shielding materials.

RTP Company (US)	RTP Company (US) provides thermoplastic elastomers and conductive compounds. These specialty compounds give designers and processors immense flexibility and provide significant benefits over metals, resins, or coatings. For EMI shielding solutions, materials such as carbon fiber, stainless steel fiber, or nickel-coated carbon fiber are used in the thermoplastic matrix to meet specific application requirements in automotive, electrical and electronics, energy, healthcare, and industrial markets.

(Table 1) cont.....

Greene Rubber Company (US)	Greene Rubber Company offers solutions to gasketing, sealing, insulating and vibration isolation and create innovative solutions to EMI shielding problems.
Hillas.com (US)	Provide shielding 3M tapes and adhesives
Leader Tech Inc. (US)	Global leads in variety of EMI shielding materials
Parker Hannifin Corp (US)	Parker-Hannifin (US) offers EMI Shielding products through its Chomerics division. Chomerics is a global leader in the development and application of electrically conductive and thermal interface materials. The company designs, develops, and manufactures EMI shielding solutions, thermal interface materials, integrated display solutions, engineered plastic solutions, and test compliance equipment for EMC compliance and safety
MAJR Products (US)	Experts in Shielding Gaskets & Tapes, EMI-Gaskets-Honeycomb-Ventilation-Panels, Shielded Ventilation Panels EMI Windows & Conductive Coatings, RF Absorber & Ferrite Materials
Laird Plc. (UK)	Laird Performance Materials (UK)is one of the market leaders, providing EMI shielding solutions with a wide range of products for customers across various industry verticals. Laird delivers effective EMI shielding and gasket solutions ranging from board-level shields and fabric-over-foam gaskets to finger stock and foam-in-place
Less EMF(US)	They produce high quality shielding fabrics for different application ranges from curtains to covers
Polymer Science, Inc. (US)	Polymer Science Inc offers a large range of Polymer based EMI shielding and grounding materials. These materials are used in electronics assemblies to prevent or restrict electromagnetic interference. The product offering includes conductive fabric, foam, foil and film tapes. Other products include Silicone gel, acrylics, hydrogels, hydrophilic coatings, urethanes, gap fillers, heat spreaders, phase change, dielectric pads, electromagnetic wave absorbers, fabric tapes, foam tapes, film tapes, foil tapes, and transfer adhesives.
Schaffner Holding AG (Switzerland)	Global company to provide EMC, power quality and power magnetics solutions, helping to improve the quality of electricity and the availability of electrical supplies.
Schlegel Electronic Materials (Belgium)	Originator of the fabric-clad foam EMI shielding technology, shielding clips, fasteners and adhesives
Henkel AG & Co. KGaA Germany)	Henkel (Germany) – The company is a global leader in the adhesives business, offering an unparalleled breadth

(Table 1) cont.....

	of technologies serving all markets and industries. The company manufactures and distributes home and beauty care products and also provides adhesive solutions. Its industrial product portfolio is organized into technology cluster brands -- Loctite, Bondertie, Tehcnomelt, Teroson, and Aquence. For consumer and professional markets, the company focuses on global brands -- Pritt, Pattex, Loctite, and Ceresit.
Other Major companies	Most of the available product ranges are oriented towards electronic components and cable protection. However, personnel protection products are also widely presented in a form of metalized fabrics. Most reflective products use metalized (Al, Cu, Ag) fabric and films, metal- and ferrite-loaded plastics, or metal sheets or films. A major shortcoming of most of the products is their inability to absorb EM waves because they are based on high reflectivity
HEICO Corporation (US)	
Tech-Etch Inc. (US)	
Premix Group (US)	
Sealing Devices (US)	
Spira Manufacturing Corporation (US)	
Chomerics (US)	
Parafix, RSI Inc. (UK)	
Statex Productions GmbH (Germany)	
Holland Shielding (Netherlands)	
Kitagawa Industries Co., Ltd. (Japan)	

Here in this chapter, we have focused on a different synthetic route of 2D materials which are used for the mass production and special attention has been pain on graphene (GN) and its Graphene polymer composite, MXene, and its composites, and beyond graphene 2D materials like transition metal dichalcogenides (TMDs)MoS_2, boron nitride (h-BN) and graphitic carbon nitride (g-C_3N_4), black phosphorus (BP) and MXene. Special attention has been paid on MXene structural features and synthesis in detail.

11.5. GRAPHENE AND ITS PROPERTIES

Graphene is a conjugated sp^2 hybridized planar structure of carbon atoms tightly packed into a two-dimensional (2D) hexagonal or honeycomb lattice [44]. It is the basic building block or mother of all graphitic materials. It can be wrapped up into 0 D fullerenes, rolled into a 1D nanotube, or stacked into 3D graphite. Originally, it was believed that this 2 D material could not exist because it would be thermodynamically unstable to exist until 2004, when Andre Geim and Kostya

Novoselov used a rather simple technique to separate the graphene layer from graphite [45–47]. These two great physicists used a common adhesive tape to mechanically exfoliate the layer of graphite into just a few layers of graphene repetitively. The successful identiifed single layers of graphene added a revolutionary discovery in the ifeld of nanoscience and nanotechnology. Graphene shows many interesting phenomena and properties due to which the interest of researchers increased dramatically. It is a zero-band gap 2D semiconductor exhibiting a strong ambipolar electric field effect having charge carrier concentrations of 1013 cm^{-2} and mobility of ~10000 $cm^{-2}s^{-1}$. Graphene is highly transparent, with the absorption of ~ 2.3 % towards visible light having a high thermal conductivity with a value of ~ 5000 WmK−1 for a single-layer sheet at room temperature [45, 47].

For real use in applications, the mass production of 2D materials meeting high yield and efficiency, low cost, and expansion to industrial-grade are prerequisites for their scalability and commercialization. In the last few years, numerous methods of preparation have been explored which can be classified into two broadways *i.e.,* "top-down" exfoliation and "bottom-up" synthesis. The 2D sheets exist in their raw bulk materials and can be exfoliated by "top-down" methods by breaking or weakening the interlayer interactions of the layered materials by external forces and facilitate the large-scale production of 2D materials at a low cost. Besides those top-down exfoliation techniques, direct synthesis of 2D material powders or foams on the templates by "bottom-up" synthetic methods paves another way to prepare high-quality 2D materials in large quantities. The top-down methods, including mechanical exfoliation and chemical exfoliation [48,49] and (2) bottom-up methods, such as chemical vapor deposition and chemical synthesis [50,51].

Graphene has been synthesized by various techniques namely mechanical exfoliation, chemical synthesis, epitaxial growth on silicon carbide (SiC), chemical vapor deposition (CVD), and other various methods [44,52] like unzipping carbon nanotubes *etc*. Among these methods, the most popular and promising way to synthesize graphene is CVD because it can produce high-quality graphene on a large scale [52]. In Table **2** different synthetic methods of graphene preparation are shown below:

Table 2. Different synthetic method for the preparation of graphene.

Methods	Synthetic Methods	Properties
Bottom-Up	Epitaxial growth of GN on SiC	SiC precursor, lacks homogeneity and quality, expensive due to energy consumption, have an environmental concern because of tetrafluoroethylene (C_2F_4)
	Dried ice method	Produced by complete burning of Mg ribbon inside the dry ice bowl.
	Chemical Vapour deposition (CVD)	It types of the deposition process, gas-phase precursors (CH_4, C_2H_4, C_2H_2, and C_6H_{14}) are used, elevated temperature (450-1000 ∘C), metallic catalyst (Cu, Ni), low defective GN, and excellent electrical and optical property.
	Template route	Good quality and well-defined structure can get high yield by using pyrrole under mild conditions, and a less desirable method due to the damage during purification.
	Total Organic synthesis	Synthesis from polycyclic aromatic hydrocarbons, high-quality GN with high yield, and limited size.
	Substrate-free gas-phase synthesis	New method, gas phase precursor (isopropyl alcohol and dimethyl ether and ethanol), clean and high-quality GN.
Top-Down	Arch discharge	The conventional method used to synthesis fullerene, CNT and GN, high-temperature plasma reaction (3727–5727 ∘C) in inert and air condition, and affordable cost.
	Liquid phase exfoliation	Common synthetic method, exfoliation occurs in aqueous and non-aqueous medium
	Graphite intercalation	Intercalation of chemical species into graphite interlayers and improves electrical conductivity.
	Radiation based methods	Short processing time, High quality, financially not viable, and radiation sources are UV and laser.
	Pyrolysis method	Solvothermal process, can be scaled up, good yield, and speed method.
	Un-zipping of CNT- GN nanoribbon (GNR)	Cutting the cylindrical CNT by various methods (metal-catalyzed cutting, chemical unzipping, plasma etching, intercalation and exfoliation), low yield, and expensive precursors and chemicals.
	Mechanical exfoliation	Use normal force (roll milling) and shear force (ball milling), high production cost, large processing time (24-48 h), low yield, and undesirable for large scale production.
	Sonication	Ultrasonic energy needs large amounts of energy, difficult to remove impurities, surfactants are used for sonication, and electrical conductivity.

	Thermal or hydrothermal reduction	Reduced to rGO, high temperature, greenhouse gas effect, and high operational cost.
	Chemical reduction	Reduced to rGO and GN, many reducing agent are used (hydrazine (N_2H_4), zinc/hydrochloric acid, Aluminium hydride, borohydrides, nitrogen-based reagents, sulphur-based reagents, sodium borohydride, microorganisms, and caffeic acid), lengthy synthesis time, additional chemical cost, environmental pollution, and toxicity.
	Electro-chemical reduction	Cost effective, less toxic, environmentally friendly, and rapid process.
	Other reduction methods	photo thermal, laser, microwave, photo catalytic, sono-chemical, and plasma treatment

11.5.1. Synthesis of Graphene by CVD Method

Numerous methods have been established for graphene synthesis. Among these methods, CVD, chemical synthesis and mechanical exfoliation are the most commonly used methods. However, CVD is regarded as having the most potential as a promising way to synthesize high-quality Graphene for large-scale production of mono or few layers graphene films on metal substrate. The CVD technique applies the decomposition of the variety of precursors, including solid, liquid and gas precursor's carbon source molecules to synthesize graphene film. The first report on planar few layers graphene (PFLG) was found in 2006 where Somania *et al.* [53] have synthesized graphene film on nickel foil using camphor (a natural, eco-friendly, and low-cost precursor), where one-meter-long quartz tube (diameter 50 mm) serves as a CVD reactor and kept horizontally inside two horizontal furnaces. Camphor (0.1–0.5g) is evaporated in the first furnace (180 °C) and pyrolyzed in the second furnace (700–850 °C) with argon as carrier gas. In each experiment, 3–4 samples of Ni sheets (2 x 2 cm^2) are kept on the Alumina boat in the centre of the second furnace. The substrates are used as received and were cleaned ultrasonically with acetone and methanol. The material collected on the Ni substrates is scrapped using a sharp blade and the powder collected is studied using a high-resolution transmission electron microscope. Graphene, thus produced, was found to have multiple folds and estimated to have approximately 35 layers of graphene sheets. This study opened up a new processing route for Graphene synthesis, though several issues like controlling the number of layers, minimizing the folds, *etc*, were yet to be solved.

Besides camphor, poly (methyl methacrylate) (PMMA), sucrose ($C_{12}H_{22}O_{11}$), and fluorene ($C_{13}H_{10}$) are used as solid carbon precursors to produce a high-quality single-layer Graphene film, where no D peak was observed. Besides, a lower decomposition temperature was required for the process when polystyrene was utilized as the carbon precursor because the C–H bonds in polystyrene are comparably weaker, thus less energy is needed to decompose polystyrene. Hence, this renders a simpler and more convenient choice for the production of single-layer graphene.

Hydrocarbon gas precursors, such as methane, ethylene (reaction at 1000 °C), and acetylene (reaction at 650 °C) are among the most popular carbon sources used for synthesizing graphene [54–58]. Among the gaseous carbon precursors, methane (CH_4) is commonly used to synthesize graphene layers. For example, Lewis's group employed diluted CH_4 gas to synthesize graphene on nickel film deposited over complete Si/SiO$_2$ wafers, which was a great benefit for device fabrication. Ruoff group [57] reported a CVD method for large-area synthesis of high-quality and uniform GNS films on copper foils using a mixture of methane and hydrogen as precursors. As obtained, films are predominantly single-layer GNS with a small percentage (less than 5 %) of the area having few layers, and continuous across copper surface steps and grain boundaries. Particularly, one of the major benefits of their process is that it could be used to grow GNS on 300 mm copper films on Si substrate and this GNS film could also be easily transferred to alternative substrates, such as SiO$_2$/Si or glass.

Along with solid and gas precursors, liquid precursors have attracted the interest of many researchers because of better availability and low cost compared to hydrogen gas precursors. Liquid carbon sources, such as benzene, methanol, and ethanol, have been employed as carbon sources to synthesize good-quality graphene film using CVD [58]. Gadipelli *et al*. [61] utilized benzene and methanol to grow Graphene films on copper foil. It was shown that by using methanol or benzene liquid carbon precursors, the usage of explosive gas (methane and hydrogen) can be omitted, since hydrogen, CO and methane can be produced during the catalytic decomposition of methanol. Hence, in this way, the safety of the personnel carrying out the experiment is guaranteed. Besides, methanol is also found to be an inhibitor of amorphous carbon growth. Guermoune's group [62] demonstrated various alcohols as liquid carbon precursors to produce good-quality monolayer Graphene on copper foils by CVD.

11.5.2. Epitaxial Growth of Graphene

The Epitaxial growth of graphene on silicon carbide (SiC) is a very promising method for the synthesis of uniform, wafer-size graphene nano layers, in which single crystal SiC substrates are heated in vacuum to high temperatures in the range of 1200–1600 °C. Since the sublimation rate of silicon is higher than that of carbon, excess carbon is left behind on the surface, which rearranges to form GNS [59]. More recently, Bao *et al*. has reported an interesting route for the preparation of GNS that employed commercial polycrystalline SiC granules instead of single-crystal SiC to formulate high-quality freestanding single-layer GNS [60]. This is a very efficient method and can be easily scaled up since the throughput is only limited by the size of the oven.

11.5.3. Chemically Modified Graphene (CMG)

For large scale production of graphene at affordable cost is thought to be chemical conversion of graphite to graphene oxide followed by successive reduction. It is usually referred to as chemically modified graphene (CMG). In 1859, Brodie was first to prepared graphite oxide by the oxidation of graphite with fuming nitric acid and potassium chlorate under cooling [61], while in 1898, Staudenmaier improved this protocol by using concentrated sulfuric acid as well as fuming nitric acid and adding the chlorate in multiple aliquots over the course of the reaction. This small change in the procedure made the production of highly oxidized GO in a single reaction vessel. In 1958, Hummers reported the method most commonly used today in which graphite is oxidized by treatment with $KMnO_4$ and $NaNO_3$ in concentrated sulfuric acid [62]. These three methods comprise the primary routes for forming GO. In continuation of this, Tour, J. M. group reported an improved method where they have used $KMnO_4$ as the only oxidant and an acid mixture of concentrated sulfuric and phosphoric acid in the 9:1 volume ratio as the acidic medium. This technique greatly increased the efficiency of oxidizing graphite to GO and also prevented the formation of toxic gases, such as NO_2 and N_2O_4. The graphene oxide prepared by this method is more oxidized than that prepared by Hummer's method and also possesses a more regular structure [63]. Compared to pristine graphite, GO is heavily oxygenated bearing hydroxyl and epoxy groups on sp^3 hybridized carbon on the basal plane, in addition to carbonyl and carboxyl groups located on the sheet edges on sp^2 hybridized carbon. Hence, GO is highly hydrophilic and readily exfoliated in water, yielding stable dispersion consisting mostly of single layered sheets (graphene oxide). The chemical composition of graphite oxide and graphene oxide are the same but they have different structures. Graphene oxide (1nm) is a monolayer material produced by the exfoliation of graphite oxide. For exfoliation of GON, N-dimethylformamide (DMF), tetrahydrofuran (THF), N-methyl-2-pyrrolidone (NMP) and ethylene glycol are commonly used solvent. The exfoliated sheets of GO have many hydrophilic functional groups like –OH, ─COOH, ─C─O─C─, C=O which keep them highly dispersible and it can be reduced to graphene-like sheets by removing the oxygen-containing groups with the recovery of a conjugated structure. The reduced GO (RGO) sheets are usually considered as one kind of chemically modified graphene (CMG). Functionalization breaks the conjugated structure and localizes p-electrons, which results in a decrease of both carrier mobility and carrier concentration therefore, it is not appropriate to refer to RGO/CMG, simply as graphene since the properties are considerably different. Several reducing agents have been used to reduce graphene oxide, such as hydrazine, sodium borohydride, hydroiodic acid, sulphur-containing compounds, ascorbic acid, and vitamin C [64,65]. Since a reduction process can dramatically improve the electrical conductivity of GO, it can be a direct criterion to judge the effect of different reduction methods. Another important change is the C/O ratio obtained by the elemental analysis measurements and X-ray photo-electron spectrometry (XPS) analysis. Depending on the preparation method, GO with chemical compositions ranging from $C_8O_2H_3$ to $C_8O_4H_5$.

11.6 ELECTRICAL PROPERTIES AND PERMITTIVITY OF GRAPHENE COMPOSITES FOR EMI

It is observed that conductivity and permittivity play a vital role in the shielding of the EM wave. In graphene, the conductivity decreases with increasing graphene layer, and finally approaches the conductivity of graphite. The conductivity of graphene prepared by different methods are listed in Table **3**.

Table 3. Electric conductivity of Graphene prepared by different methods.

S. No.	Material	Method	Σ (S cm^{-1})	Refs.
1	Graphite	Natural	1.1×10^5	[66]
2	Few-layer Graphene	Chemical vapor deposition. (CVD)	$\sim 10^6$	[67]
3	Single-layer Graphene.	Chemical vapor deposition. (CVD)	10^8	[67]
4	Chemically modified graphene (CMG)	Chemical reduction by acetic acid–hydriodic acid.	2.98×10^4	[68]
5	Chemically modified graphene (CMG)	Chemical reduction by hydrazine hydrate	7.2×10^3	[69]
6	Chemically modified graphene (CMG)	Chemical reduction by hydrazine hydrate	3.51×10^4	[70]
7	Chemically modified graphene (CMG)	Thermal reduction.	2.8×10^4	[71]
8	Chemically modified graphene (CMG)	Thermal reduction.	1.39×10^5	[72]

The conductivity determines both the reflection and conduction loss, meanwhile the functional groups and defects enhance the polarization loss. Therefore, although the conductivity of chemically fabricated graphene is lower, it is usually used in EMI shielding, owing to its induced functional groups and defects. For best performance, material not only requires conductivity but also absorption towards penetrating EM waves. The lightweight with strong absorption and weak secondary reflection are preferred material to achieve high EMI shielding efficiency (SE) [73]. Generally, the reflection (reflection from material surfaces and interface scattering in propagation) increases with the increase of conductivity of materials, while the absorption refers to the dielectric properties. The real permittivity (ε') corresponds to the polarization capacity, and the imaginary permittivity (ε''), a dielectric loss factor, presents the attenuation towards the penetrating EM wave [9]. The attenuation arises from the dielectric loss and magnetic loss of dielectric-magnetic materials. Dielectric loss includes the relaxation and conduction loss, influenced by polarization and conductivity, respectively. It is interesting that polarization is derived from functional groups, defects and interfaces, while the conductivity comes from the intrinsic conduction and hopping conduction, where the hopping conduction dominates the conductivity even though it is really low. Actually, the above mechanisms not only have impacts on the dielectric properties but also influence the S parameters which determine EMI SE. The total EMI SE can be explained as the sum of the absorption shielding efficiency (SE$_A$) and the reflection shielding efficiency (SE$_R$). The SE is a measure of the material's ability to attenuate the intensity of EM waves and can be expressed as [9]:

$$SE(dB) = -10\log\frac{P_t}{P_0} = SE_R + SE_A + SE_M$$

(1)

where, P_t and P_0 are the transmitted and incident electromagnetic power, respectively. SE$_R$ and SE$_A$ are the shielding effectiveness due to reflection and absorption, respectively. SE$_M$ is multiple reflection effectiveness inside the material, which can be negligible when SE >10 dB. It has been observed that a minimum SE of 20 dB (indicates 99 % of the EM radiation has been blocked by the shield) is required for commercial applications in electronic appliances like laptops and desktop computers and for military application the required SE is more than 30 dB.

The dependence of SE on complex permittivity and permeability can be expressed as [74]:

$$SE_A(dB) = 20\frac{d}{\delta}\log e = 20d\sqrt{\frac{\mu_r \omega \sigma_S}{2}}.\log e \qquad (2)$$

$$SE_R(dB) = 10\log\left(\frac{\sigma_S}{16\omega\mu_r\varepsilon_0}\right) \qquad (3)$$

where, d is the thickness of the shield, μ_r is the magnetic permeability, δ is the skin depth, $\sigma_S = \omega\varepsilon_0\varepsilon''$ is the frequency dependent conductivity [75], ε'' is the imaginary part of permittivity (dielectric loss factor), ω is the angular frequency ($\omega = 2\pi f$) and ε_0 is the permittivity of the free space. From equations 2 and 3, it is observed that with the increase in frequency, the SE_A values increase while the contribution of the reflection decreases. Dependence of SE_A and SE_R on conductivity and permeability reveal that the material having higher conductivity and magnetic permeability can achieve better absorption properties. To relate σ_S with the shielding parameters of the material, σ_S and skin depth were plotted against the frequency for the composites. The skin depth of the samples was calculated using the relation:

$$\delta = \sqrt{2/\omega\mu\sigma_S}$$

Dependence of SE_A and SE_R on conductivity and permeability reveal that the material having higher conductivity and magnetic permeability can achieve better absorption properties. The complex permittivity and the permeability of the materials can be calculated using scattering parameters (S_{11} and S_{21}) based on the theoretical calculations given in Nicholson, Ross and Weir method [76]. The real part (ε') is mainly associated with the amount of polarization occurring in the material and the imaginary part (ε'') is related with the dissipation of energy. The dielectric performance of the material depends on ionic, electronic, orientational (arises due to presence of bound charges) and space charge polarization (due to the heterogeneity in the system). According to the theory of EMI SE based on good conductor approximation, the SE is a function of thickness d of the composites. Song *et al.* demonstrated that the permittivity values of graphene nanosheet (GN) composites were higher than those of expanded graphite composites [77]. Generally, from expanded graphite to GN and graphene, the thickness decreases, while the permittivity and EMI SE increase with the same mass ratio. Wen *et al.* [77] presented the first interpreters to the higher loss tangent. It arises from three aspects: graphene with abundant functional groups and defects enhances polarization loss; the decreased thickness increases the conductive paths, which enhances conduction loss; the thin, flexible and corrugated graphene enhances the propagation paths for the EM wave.

Electromagnetic shielding efficiency of graphene can be as high as 135 dB, whereas the commercial application's requirement is only 20 dB. The EMI shielding properties of a 2D material were first reported for graphene-based materials in which an EMI SE ≈ 21 dB was achieved for a 15 wt. % loading in graphene/epoxy composites. Table **4** shows the selected EMI shielding of graphene and its polymer composite.

Table 4. EMI shielding properties of selected graphene and graphene polymer composites.

Material	Type	Wt. %	Thickness (mm)	EMI Shielding SE (dB)	Ref.
Graphene/wax	Bulk	20	2	29.68	[78]
Graphene/PDMS	foam	0.8	1	20.0	[2]
Graphene/epoxy	Bulk	15	—	21.0	[79]
rGO	Bulk	20	2	38.0	[80]
rGO/SiO$_2$	film	20	1.5	38.0	[80]
rGO/γ-Fe$_2$O$_3$/ PANI	film	75	2.5	51.0	[81]
S-doped rGO	film	100	0.15	38.6	[82]
Graphene/Fe$_3$O$_4$/ PANI	Bulk	100	2	26.0	[9]
GO/epoxy	film	15	n/a	21.0	[79]
GO	foam	100	0.3	26.3	[83]

11.7. TWO-DIMENSIONAL MATERIALS BEYOND GRAPHENE

Beyond Graphene, transition metal dichalcogenides (TMDs) are the most studied 2D materials. In broader terms, most of the 2D layered materials in their bulk state are quite similar to graphite. They exist in vertically aligned stakes of individual sheets, attached by weak van der Waals forces. Layered TMD single sheets have the general formula MX_2, with an M^{4+} transition metal (M= Ti, Zr, Hf, V, Nb, Ta, Mo, W, Tc, Re, Co, Rh, Ir, Ni, Pd, P or V), and X is a chalcogen (S, Se, Te), it is a large family of materials and their electronic characters could be semiconducting, metallic and superconducting TMDs [84] exhibit unique quantum confinement and surface effects that occur through the transition of an indirect to direct bandgap when bulk materials are exfoliated to monolayers. The TMD monolayer is not truly one-atom-thick as there is a metal atom sandwiched between the chalcogen atoms. Among the TMDs, MoS_2 displays versatile chemistry and tunable bandgap that is much more beneficial to design practical applications. Molybdenum disulphide (MoS_2) is a widely recognized TMD and naturally occurs as molybdenite, an anisotropic semiconductor material. The highly specific surfaces and polymorphism of MoS_2 can endow its critical electronic properties. Monolayer MoS_2 is a semiconductor with a direct bandgap of 1.8 eV and can be tuned depending on the number of layers in the stakes, along with tunable band gap [85]. Very similarly to graphene, MoS_2 is mechanically flexible with a Young's modulus of 0.33 ± 0.07 TPa, excellent mobility ($\sim$700 cm^2 V^{-1} s^{-1}), high current on/off ratio of $\sim$107$-$108, and broad optical absorption ($\sim$107m^{-1} in the visible range) thus making it scientific and industrially important [86]. The known polymorphs of MoS_2 are the 1T, 2H, and 3R phases. The letters signify the type of symmetry in which T stands for tetragonal, H for hexagonal, and R for rhombohedral. The number designates the number of layers in the crystallographic unit cell. MoS_2 shifts from semiconductor (H phase) to metallic (T and R phases) with distinct electronic properties. Chemically exfoliated 1T MoS_2 phase is considerably more conductive than the 2H semiconducting polymorph [87].

11.7.1. Synthesis of MoS$_2$ and its Composite Application in Electromagnetic Absorption

11.7.1.1. Mechanical Exfoliation

The mechanical exfoliation method utilizing the Scotch tape was reported by Novoselov and Geim *et al.* [45] to obtain graphene sheets, a commonly used method to produce high-quality 2D materials monolayers. However, it is not appropriate for large-scale production because of its low production yield and limited ability to control flake thickness, shape, and size. For single layer MoS$_2$, the bulk phase of MoS$_2$ is attached to an adhesive tape, repeatedly exfoliated, followed by attaching to a substrate. This results in high quality ultra-thin layered MoS$_2$ transferred onto the substrate.

11.7.1.2. Liquid Exfoliation (LE)

It is another technique for generating large quantities of single- and few-layered MoS$_2$ nanoflakes, including an ion intercalation method and solvent exfoliation [88,89]. Here, 2D martials nanosheets are produced by direct ultrasonication of their bulk materials in liquid media. Here, two parameters playa very important role, first ultrasonication (bath sonication or probe sonication), ultrasonic waves generate shear forces or cavitation bubbles to provide high energy in liquids to break the layered structure and produce single- or few-layered nanosheets. Second, the liquid media (organic solvent or the aqueous solutions of stabilizers, ionic liquids, or salts solution). The main challenge with liquid-generated MoS$_2$ flakes is their subsequent separation, which is dependent on the layer size and thickness. Foad Ghasemi *et al.* reports the preparation of Molybdenum disulphide (MoS$_2$) flakes with controlled size and thickness utilizing the sequential solvent exchange method. Here, they did sonication in dimethylformamide (DMF) and N-methyl-2-pyrrolidone (NMP) solvents. The NMP solvent acts more effectively in reducing the thickness of flakes while DMF shows better potential in conserving the lateral size of nanosheets. The distribution of size and thickness of nanoflakes is a function of sonication time and power of sonication probes.

11.7.1.3. Chemical vapor deposition (CVD)

This method has been utilized to grow MoS_2 thin layers on insulating substrate (SiO_2 or sapphire) by using Mo, $(NH_4)_2MoS_4$, or MoO_3 as precursor. Here, the precursors are firstly deposited on supports as thin layers then sulfurization or thermal decomposition of the precursors at elevated working temperature is executed to control the structure of MoS_2. In this method, it is very difficult to control the number of layers.

To address these issues, Xinsheng Wang *et al*. [90] reported a process where MoO_3 powder is firstly evaporated and reduced by sulphur vapor to generate MoO_2, which then nucleates on SiO_2/Si substrates and grows into rhomboidal micro sheets in a CVD furnace. The MoO_2 micro sheets are then annealed in sulphur vapor. In the process of annealing, varied numbers of MoS_2 layers are obtained by sulfurizing the surface of MoO_2 micro sheets, according to the annealing duration. During the high temperature annealing, the surface of MoO_2 microplates were sulfurized to MoS_2 with a varied number of layers depending on the annealing duration.

After that, a PMMA thin film was spin-coated on the MoS_2/MoO_2 microplates to peel-off MoS_2 thin layers from MoO_2 and transfer them onto other substrates using PMMA mediated Nano transfer printing technique. The obtained MoS_2 flakes exhibited rhomboidal shape with lengths up to tens of micrometres, having high crystallinity with a crystal domain size of ~10 μm and mobility comparable with mechanically exfoliated MoS_2 flakes. This simple and reliable approach opens up a new way for producing highly crystalline MoS_2 atomic layers in a controlled manner.

11.7.1.4. Hydrothermal or Solvothermal Methods

In this method low toxic organic sulphur source like thiourea or thioacetamide are allowed to react with a molybdenum salt especially ammonium molybdate ($(NH_4)_6Mo_7O_2.4H_2O$) or sodium molybdate ($Na_2MoO_4\cdot2H_2O$) in a Teflon-lined stainless-steel autoclave at various temperatures [91, 92] to achieve high-surface-area and high crystallinity MoS_2 nanomaterials. The major challenge lies in the limited utilization and high selectivity of suitable metal and chalcogen sources, as well as in mastering how to slow down vertical growth while simultaneously accelerating lateral growth.

11.7.1.5. EMI shielding properties of MoS$_2$

For the new generation of EMI shielding materials, the shielding mechanism should mainly be attributed to absorption rather than reflection. Pure MoS$_2$, graphene or RGO can't provide a sufficient bandwidth and intensity of electromagnetic absorption (EA). Therefore, the electron transport ability of MoS$_2$ can be distinctly improved by combining it with RGO, Metal oxides, or conducting polymers. Some research groups have reported MoS$_2$/RGO nano-composites prepared by simple heating the mixture of graphene oxide and molybdate under sulfide gas [93]. However, this synthetic process involves sour gas treatment and high temperature calcination, so it is hazardous and not energy efficient. This makes it unfavourable, especially for large-scale preparation. Aming Xie *et al.* [93] have reported a simple hydrothermal route, which is much simpler and safer and no hazardous tail gas or reagent was utilized. This improved method involves the hydrothermal growth of RGO into layered MoS$_2$ to form RGO/MoS$_2$ nano-sheets to form more regular sheet microstructures. Here, Graphene oxide along with thiourea and $(NH_4)_6Mo_7O_{24}H_2O$ were the first ultra-sonication to form a homogenous solution. The resulting solution was transferred to a Teflon-lined stainless-steel autoclave and kept at 200 °C for 20 h. Once the reaction completed the final product was filtered, washed with deionized water and absolute ethanol for several times and dried at 50 °C under vacuum. Fig. (**3A**) shows the morphology of pure MoS$_2$ and MoS$_2$/RGO nanosheets while Fig. (**3B**) shows the Raman spectra of GO, RGO and RGO/MoS$_2$ composite. Typical D band at 1347 cm^{-1} is due to the breathing mode of κ-point photons of A1g symmetry band while G bant at 1581cm^{-1}arises from the first-order scattering of E$_{2g}$ phonons by sp^2 carbon atoms. The defects of carbon lattice structures were aroused during the reduction process, the I$_D$/I$_G$ was increased from 0.77 in GO to 1.03 in RGO and 0.91 in RGO/MoS$_2$ composite. In addition, two strong peaks at 377 and 404 cm^{-1} respectively stand for the E$_{2g}$ and A$_{1g}$ peaks of MoS$_2$ (Fig. **3B (ii)**). Here, dielectric properties of MoS$_2$ based hybrids can be effectively regulated by altering the amount of RGO sheets. This composite with optimized compositions showed strong E$_A$ in a wide bandwidth (≤ -10 dB) at low filler loading ratio and thin thickness. In one composition (20 wt. % of RGO/MoS2) in a wax matrix, the composite exhibited an effective E_A bandwidth of 5.7 GHz and the minimum reflection loss (R_L) of − 60 dB was observed (Fig. **3C**).

In another study Wang *et al.* [94], reports high performance RGO/MOS2 nanocomposite prepared by combination of freeze dry and thermal heating in a tubular furnace in an inert atmosphere. Here, they mixed the calculated amount of GO solution with Sodium molybdate dihydrate ($Na_2MoO_4 \cdot 2H_2O$) and sonicated to form a homogeneous dispersion followed by freeze-drying. The spongy powder obtained was transferred to a tube furnace, and heated at 650 °C (10 °C/min rise in temperature) for 2 hours in an inert atmosphere as shown in the schematic of Fig. (**4**). Here, a thin sample of 2.0 mm having 10 wt. % MoS2/RGO hybrid in the wax matrix exhibited an effective microwave absorption bandwidth of 5.72 GHz while highest reflection loss of −50.9 dB was observed at 11.68 GHz for a sample with a thickness of 2.3 mm.

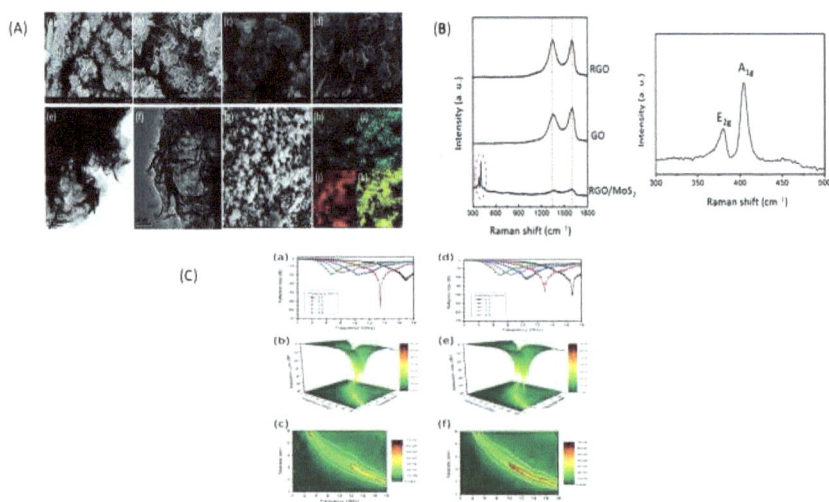

Fig. (3). (A) SEM images of pure MoS2 (**a, b**) and MoS2/RGO nanosheets (**c, d**); TEM images of MoS2/RGO nano-sheets (**e, f**); EDS mapping area of MoS2/RGO nano-sheets (**g**) and the element signals of C (**h**), O (**i**), S (**j**) and Mo (**k**). (B) (**a**) Raman spectra of RGO/MoS2, GO and RGO; (**b**) specific Raman spectrum of RGO/MoS2 from 300 to 500 cm^{-1}. (C) RL curves and 3D RL plots of MoS2/RGO with the filler loading ratio of 20 wt. % (**a-c**) and 30 wt. % (**d-f**) in wax composites, the test frequency range is from 2 to 18 GHz (white line: region of − 10 dB; yellow line: region of −15 dB; red line: region of −20 dB). Reproduced with permission from ref. [93], Copyright 2013, Royal Society of Chemistry.

Similarly, Ao-Ping Guo *et al.* [95] have reported the loose structure of reduced graphene oxides (rGO)@MoS$_2$ composites in poly (vinylidene fluoride) (PVDF) to form the composites which shows superior microwave absorption and excellent electromagnetic interference shielding performances. The maximum reflection loss of the rGO@MoS$_2$/PVDF composites with a low filling rate (only 5.0 wt. %) can reach - 43.1 dB at 14.48 GHz, and the frequency bandwidth less than -10 dB is from 3.6 to 17.8 GHz (in the frequency range of 2 ~ 18 GHz) with the thickness of 1-5 mm. Furthermore, the rGO@MoS$_2$/PVDF composites with a higher filling rate (25 wt. %) also exhibit outstanding electromagnetic interference (EMI) shielding effectiveness (SE) reaching a maximum at 27.9 dB.

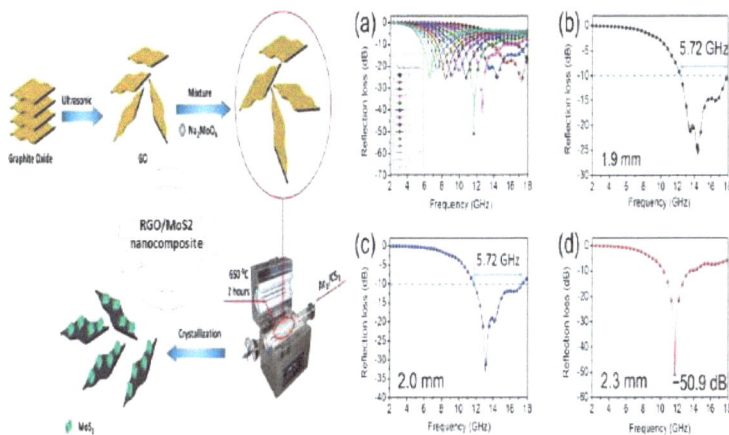

Fig. (4). Synthetic Process of the MoS$_2$/RGO Hybrid preparation and reflection losses (**a**) RL curves of MoS$_2$/RGO−wax composites (10 wt. %) at thicknesses of 1.9 mm (**b**) and 2.0 mm (**c**), 2.3 mm (**d**), Reproduced with permission from ref [94], Copyright 2015 American Chemical Society.

11.7.2. 2D Nitrides: Hexagonal Boron Nitride (h-BN)

Hexagonal boron nitride (h-BN) is isoelectronic and structurally similar to graphite having almost the same crystal lattice parameters with a composition of alternating B and N atoms in a hexagonal layer. As a result, h-BN Nanomaterials show many similar properties as graphitic materials, such as low density, high mechanical strength, good chemical corrosive resistance, excellent oxidation resistance, good lubricating property and outstanding thermal conductivity [96]. The first synthesis of boron nitride was in 1842 by Balmain (1842) using molten boric acid and potassium cyanide. It can synthesise in amorphous or crystalline form produced in amorphous and crystalline forms. Hexagonal boron nitride (h-BN), sphalerite boron nitride (β-BN), and urtzite boron nitride (γ-BN) are crystalline forms of BN. Boron nitride also exists in BN fullerenes, 1D BN nanotubes (BNNTs), isoelectric to CNTs in terms of chirality, tube diameters, and number of walls. Out of these different phases, h-BN is the most common stable form of BN and most of the interest started after the isolation of graphene sheets in 2004. Although h-BN has a similar structure to graphene, it is a wide bandgap material with an intrinsic band gap (Eg) of 5.9 eV when compared to the highly conductive graphene [97]. H-BN nanosheets can be obtained *via* top-down (typical exfoliation-type approaches) or bottom-up approaches (usually CVD or other deposition techniques). The most common methods that have been used are mechanical exfoliation, chemical exfoliation, CVD, and pulsed laser deposition (PLD). Similar to graphene synthesis by mechanical exfoliation in this method, BN are peeled off with an adhesive tape, attached to a substrate (*e.g.*, Si/SiO₂). Nanosheets obtained with this method have a good combination of thickness and lateral size, making them suitable for fundamental studies in physics and optoelectronics. In chemical exfoliation, Vigorous sonication of BN crystal or powder in a solution of 1,2-dichloroethane solution, a strong polar solvent N, N-dimethylformamide (DMF), N-methyl-2-pyrrolidinone (NMP) or poly (m-phenylenevinylene co-2,5-dictoxy-p-phenylenevinylene) have been used to form BN sheets. The solvents could be optimized by choosing a proper media whose surface energy matches the energy per unit area required to overcome the van der Waals forces while peeling BN nanoparticles away [98]. Very recently, molten hydroxides have been used to exfoliate BN [99]. Briefly in this method, sodium hydroxide (NaOH), potassium hydroxide (KOH) and BN powders are ground, then transferred to a poly(tetrafluoroethylene) (PTFE)-lined stainless-steel autoclave and heated at 180°C for 2 hours. As a result, BN nanosheets and nano scrolls are obtained. The exfoliation process involves the following sequence: (1) self-curling of the sheets at the edges due to the adsorption of cations (Na⁺ or K+) on the outmost BN surface; (2) anions and cations entering the interlayer space and the adsorption of anions (OH) on the positively curved surface which drives continuous curling of the BN layer; (3) direct peeling away from the parent materials, or cutting by the reaction of BN surface with hydroxides. This method has the advantages of being simple, one-step, low-cost, and the product can be easily transferred to any substrate by redispersion in common solvents such as water and ethanol [99].

In chemical vapour deposition (CVD) method, NH_3 and diborane (B_2H_6) were used as precursors for the deposition of amorphous BN thin films (less than 600 nm thick) on silicon (Si) and metallic substrates, such as tantalum (Ta), molybdenum (Mo) and germanium (Ge), in the temperature range of 600–1000 1C. Apart from this various chemical precursors such as BF_3–NH_3, BCl_3–NH_3, B_2H_6–NH_3 or the pyrolysis of a single precursor such as borazine ($B_3N_3H_6$), trichloro borazine ($B_3N_3H_3Cl_3$), or hexachlorobutadiene ($B_3N_3Cl_6$) [100].

In most cases, the BN layers are grown on a substrate; however, there are a few substrate-free methods too. Rui Gao *et al.* report the first successful synthesis of BN nanoflakes synthesized *via* a simple template and catalyst-free chemical vapour deposition process at 1100-1300 °C [101]. In a typical procedure, the precursors B_2O_3 and melamine were first mechanically mixed, and put into a graphite crucible. The graphite crucible containing the precursors was placed at the centre of a high-purity graphite induction-heated cylinder. Argon (Ar) was introduced as a carrier gas, and N_2 as the reaction gas. The thickness of the BN sheets could be controlled in the range of 25–50 nm by tuning the synthesis temperature. Jie Yu, *et al.* has grown Vertically-aligned BN nanosheets on a silicon (Si) substrate were grown at 800 °C from a gaseous mixture of BF_3–N_2–H_2 *via* microwave plasma CVD (MPCVD) technique [102]. Jie Yu *et al.* were able to control size, shape, thickness, density, and alignment of the BNNSs by appropriately changing the growth conditions.

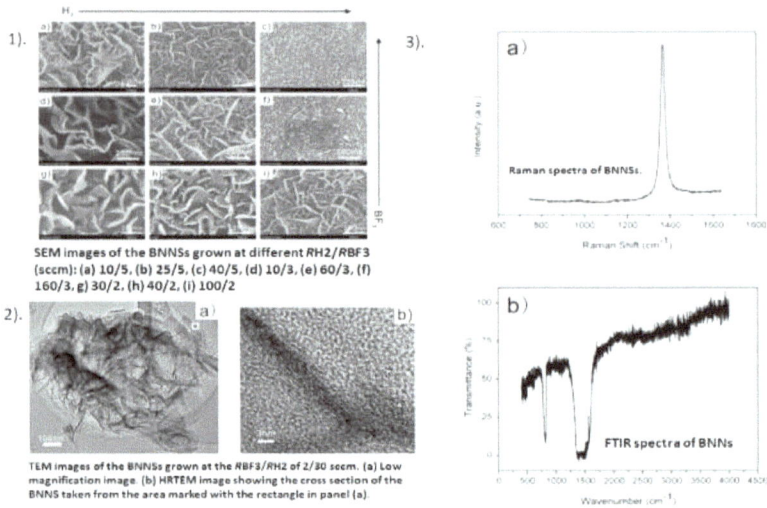

Fig. (5). SEM images of the BNNSs grown at different RH2/RBF3 (sccm): (**a**) 10/5, (**b**) 25/5, (**c**) 40/5, (**d**) 10/3, (**e**) 60/3, (**f**) 160/3, (**g**) 30/2, (**h**) 40/2, (**i**) 100/2 2) TEM images of the BNNSs grown at the RBF3/RH2 of 2/30 sccm. (**a**) Low magnification image. (**b**) HRTEM image showing the cross section of the BNNS taken from the area marked with the rectangle in panel (**a**) 3) Typical Raman (**a**) and FTIR (**b**) spectra of the BNNSs. Reproduced with permission from ref [102], Copyright 2015 American Chemical Society.

11.7.2.1 Solid State Reaction

Gang Lian *et al.* [103] prepared flower-like BN nano-flakes by a template-free solid phase reaction. Here, $NaBF_4$, NH_4Cl and NaN_3 powders were mixed, pressed into pellets at room temperature, and then heated in an autoclave at 300°C for 20 hours. The BN production reaction is:

$$NaBF_4 + 3NH_4Cl + 3NaN_3 \longrightarrow 3HF\ BN + NaF + 3NaCl + 4N_2 + 3NH_3^+$$

The as-prepared BN Nano flowers were composed of vertically standing BN nano flakes. The nanoflakes composing the nanoflowers can be easily exfoliated into few layered graphene-like BN, which is readily dispersible in strong polar organic solvents.

11.7.2.2. Graphene/hexagonal Boron Nitride Nanoparticle Hybrids for Microwave Absorption

The superior electrical resistivity and wide band gap of hBN nanomaterials are beneficial to tune the dielectric property of graphene through hybridizing of them. The formation of the graphene/h-BN nanomaterial hybrids can keep the low density, high mechanical strength, good chemical the interface effect appeared between graphene and h-BN nanomaterials will cause polarization to generate magnetism, which is also critical for enhancing the microwave absorption property of graphene. Yongqing Bai *et al.* [96] demonstrated a simple approach to fabricate multilayer graphene in a kilo-mass/hour (≥ 2.5 kg/h) scale through an oxidation-thermal expansion-air convection shearing process.

Here, the subsequent incorporation of h-BN nanoparticles (h-BNNPs) tailored the dielectric and magnetic properties to significantly boost the microwave absorption performance. The as obtained multilayer graphene/h-BNNP hybrid with 40 wt. % of h-BNNPs, exhibits extremely low reflection loss value of -67.35 dB at 8.04 GHz when the absorber thickness is 3.29 mm, ranking it as one of the most attractive absorbers reported to date. Moreover, the multilayer graphene/h-BNNP hybrids showed very low densities (0.45 g/cm^3), making it a very attractive material for practical microwave absorption application. Similarly, in another report Lv *et al.* [104] developed a graphene/g-C$_3$N$_4$ hybrid by loading g-C$_3$N$_4$ nanosheets (15 wt. %) on graphene through a simple liquid-phase approach and then homogeneously mixing in wax. A 1.5 mm thickness composite having filled ratio of 10 wt. %, delivered a reflection loss (R$_L$) of 29.6 dB in a bandwidth of 5.2 GHz (12.8 to18 GHz) [9]. Yue Kang *et al.* [105] demonstrated a modified assembling and annealing route to make the sandwich like rGO/h-BN hybrids in large scale. Here they used GO as a starting template; h-BN precursor (ammonia borane) was formed and attached to rGO by annealing the self-assembled hybrids of GO and AB. The heat-treated rGO-h-BN sandwiches are lightweight and thermally stable. Semiconducting filler like rGO can improve the thermal conductivity and the impedance matching of the hybrid composition. It was observed that the complex permittivity and MWA are tunable by varying the h-BN content and that the layered structure ensures a high dielectric loss and superior impedance matching. Therefore, in a 1.6 mm thick wax composite loaded with rGO/h-BN 25 wt. %, a minimum RL of - 40.5 dB was obtained at 15.3 GHz. The effective absorption bandwidth achieved was 5 GHz. Zhang *et al.* fabricated a flexible film composed of multilayer GO/polymer and amino-functionalized h-BN/polymer using a layer-by-layer (LbL) casting method. An ordered architecture with 11 layers and 235.2 μm thickness maintained the electrical insulation, enhancing the EMI shielding and thermal conductivity performances. An EMI SE value of 37.9 dB, electrical insulation of 1.5 MV/m, and high thermal conductivity of 12.6 W/mK were reached. The EMI shielding properties were ascribed to the GO because the h-BN layers blocked the conductive paths throughout the multilayer film. The possibility to tune the film thickness and the layered interfacial architecture, in order to reflect and absorb EM waves, thus increasing the polarization and conduction loss, are advantages in the preparation of EMI shielding materials [106]. The layer-by layer structure of the hybrid makes great contributions to the increased approaches and possibilities of electron migrating and hopping, which has both highly efficient dielectric loss and excellent impedance matching for microwave consumption.

11.7.3. 2D Black Phosphorus

Black phosphorus (BP) is an emerging two-dimensional(2D) material with a natural bandgap, which has unique anisotropy and extraordinary physical properties [107]. Among all the allotropes of Phosphorus such as red, violet and white phosphorus, BP has more stability and exhibits in-plane anisotropy due to its puckered structure [108]. BP is a semiconductor with a predicted direct fundamental bandgap of ≈ 0.3 eV for bulk and ≈ 2 eV for a single layer [109,110], which is identified by high-resolution scanning tunnelling spectroscopy (STS) [111]. Furthermore, comparing numerous calculated and experimental results we can deduce that the optical bandgap of single layer phosphorene should be around 1.45 eV due to the huge exciton binding energy [109,112,113], BP distinguishes from other widely investigated 2D layered materials because of its unique crystal structure. It has a puckered structure along the armchair (AM) direction, and this structural anisotropy is due to its local bonding configurations. The bond angle along the zigzag (ZZ) direction is 94.3° and the adjacent P-P bond length is 2.253 Å. These values are smaller than the corresponding dihedral angle along the AM direction (103.3°) and the connecting bond length (2.287 Å). The lattice constants along the two perpendicular directions are different, at 3.30 Å and 4.53 Å, respectively. This kind of unique structural arrangement, resembling a network of connected hinges, is the origin of the anisotropic physical properties.

Phosphorene (single layer of BP) can also be exfoliated by bulk crystals similar to Graphene and its band gap can be tuned making its wide application in ultrafast electronics and high frequency optoelectronic applications, telecommunication, field effect transistors, cathode anode materials in batteries, photo detectors, thermoelectric applications [109,114].

11.7.3.1. Preparation of Black Phosphorus

Very few methods are available for the synthesis of BP single crystals such as high-pressure synthesis, chemical vapour transport (CVT), mercury catalysis, and liquid bismuth. In 1914, Bridgman first reported that white phosphorus (WP) can be converted into BP single crystals under a hydrostatic pressure of 1.2 GPa at 200 °C within 5–30 min^{-1}. Typically, high purity BP are synthesis by using high pressure process where red phosphorus (RP) is heated to 1000 °C followed by slow cooling down to 600 °C at a cooling rate of 100 °C per hour under a constant pressure of 10 kbar to give high quality BP crystals with highest purity [115]. The bottom-up approach is not scalable due to extremely low yield and challenging to test for volume applications. Liquid phase exfoliation (LPE) is one of the most promising large-scale techniques, where direct conversions of bulk layered crystals into thin nanosheets are stabilized by electrostatic repulsion. In aqueous solvents and surface energy matching in organic solvent-based systems. Hanlon *et al.* reported high quality, few-layer black phosphorus nanosheets produced in large quantities by liquid phase exfoliation in the solvent N-cyclohexyl-2-pyrrolidone (CHP) [116]. Unfortunately, phosphorene is not fully explored in preparation partially due to the complicated chemistry of phosphorus. It is also important to develop methods that can synthesize large-area single-crystal thin films (at least in the order of square millimetres) in which the anisotropic properties of BP may be explored at a larger scale [117]. Li *et al.* reported a scalable approach to synthesize a large-area (up to 4 mm) thin black phosphorus (BP) film on a flexible substrate. Here, first deposit a red phosphorus thin-film on a flexible polyester substrate, followed by its conversion to BP in a high-pressure multi-anvil cell at room temperature [118].

Sonication of ground BP crystals yields a brown dispersion of exfoliated layers successfully without introducing any defects like other 2D materials with this method. Up to date there are few articles published on making BP nanosheets from single to few layers in solvents such as N-methyl pyrrolidone (NMP), dimethyl sulfoxide (DMSO), and di-methyl acetamide (DMA) [119] *etc*. Fig. (**6**) shows solvent exfoliation of BP in various solvents *via* tip ultrasonication and dispersion in N-methyl pyrrolidone after ultra-sonication at different rpm (5000 rpm centrifugation and 15000 rpm) centrifugation. The AFM height image shows that the thickness values range from 16 to 128 nm. No bubbles, droplets, or other signs of BP degradation are present. Further, the SEM image of Fig. (**7**) indicates that the lateral dimensions of the BP flakes agree well with the AFM images. Functionalization of BP can open new ways of exfoliation that can be highly stable in water. Shear exfoliation technique has been reported for making thin nanosheets of almost any 2D materials [120].

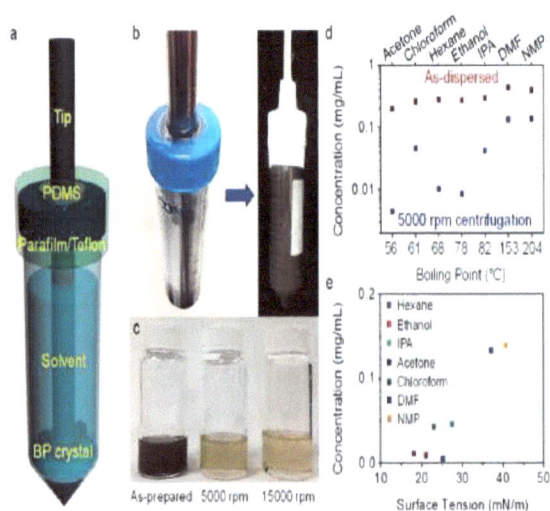

Fig. (6). Solvent exfoliation of BP in various solvents *via* tip ultrasonication. (**a**) Schematic and (**b**) photograph of the custom tip ultra-sonication setup that minimized exposure to ambient air during processing. (**c**) Photograph of BP dispersion in N-methyl pyrrolidone after ultra-sonication, 5000 rpm centrifugation, and 15000 rpm centrifugation (left to right). (**d**) BP concentration plot for various solvents with different boiling points before and after 5000 rpm centrifugation and (**e**) with different surface tensions after 5000 rpm centrifugation. Reproduced from reference [119], Copyright 2015 American Chemical Society.

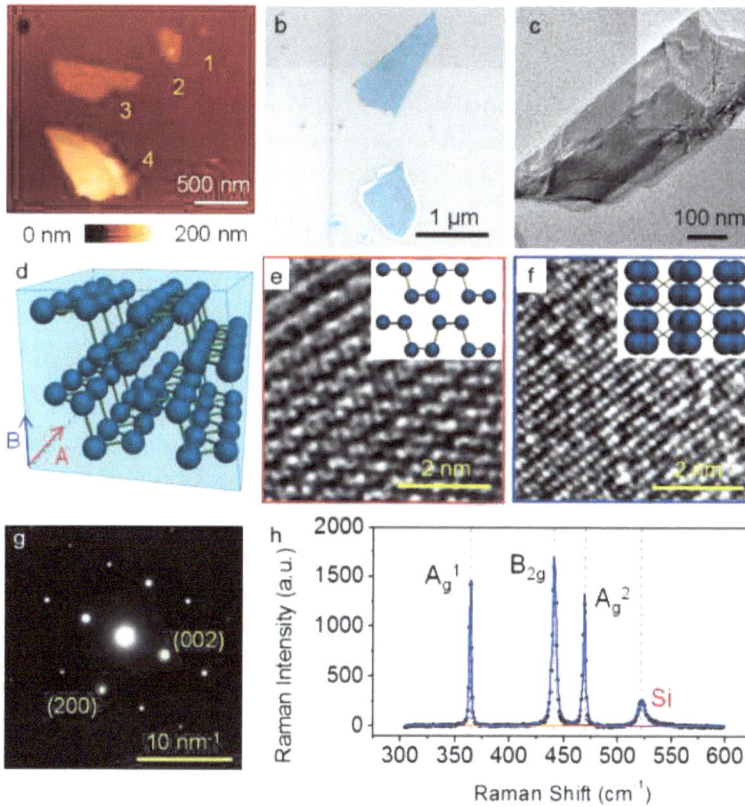

Fig. (7). Characterization of solvent-exfoliated BP nanosheets. (**a**) AFM height image of solvent-exfoliated BP nanosheets that were deposited onto a 300 nm SiO_2/Si substrate with different heights (1:16, 2:40, 3:29, and 4:128 nm). No bubbles or other evidence of degradation are apparent from the solvent exfoliation process. The height data were taken in a N_2 environment. (**b**) False-coloured SEM image of solvent-exfoliated BP nanosheets. (**c**) Low-resolution TEM image of solvent exfoliated BP nanosheets. (**d**) Schematic showing the atomic structure of BP. High-resolution TEM images of solvent exfoliated BP nanosheets along direction (**e**) A and (**f**) B. (**g**) Selected area electron diffraction pattern of solvent-exfoliated BP nanosheets. (**h**) Raman spectrum of solvent-exfoliated BP nanosheets. Reproduced from reference [119], Copyright 2015 American Chemical Society.

11.7.4. Uses of 2D Black Phosphorus in EMI Shielding

Semiconducting properties of BP make it a promising candidate for EM wave attenuation. However, to the best of our knowledge, very few reports are available on BP-based materials for EM absorption [121–123]. BP degradation and challenges in the synthesis strategies make this material challenging to manipulate. Wu *et al.* prepared few-layer BP (Fig. **6a**) by liquid-phase exfoliation and then mixed 30 and 50 wt. % in a wax matrix. The composite loaded with 30 wt. % of few-layer BP, with a 2.5 mm thickness reached an effective EM absorption up to 6.20 GHz [121].The fabrication of an rGO/BP composite aerogel was reported by Hao *et al.* through a simple self-assembling process [122]. The BP nanosheets were uniformly distributed into an rGO framework, resulting in a porous structure possessing excellent performance as a microwave absorber. It achieved a minimum reflection loss (RL) of -46 dB and a broad absorption band of 6.1 GHz (RL < -10 dB) at 2.5 mm. The existence of multiple interfaces between rGO network and BP nanosheets allowed for several dielectric relaxations causing dielectric losses, which improved the EM wave absorption. More recently, hybrid films of GO and BP were proposed by Zhou *et al.* [123] to fabricate tough BP-graphene films which exhibited high ambient stability and mechanical strength. With the improvement of σ from 8.5 ± 0.6 up to 22.7 S cm-1 due to the increasing BP content, the graphene-based films showed excellent EMI shielding properties. At 8 GHz, the 5 μm thick films reached an EMI SE value of 29.7 dB, twice as high as the starting rGO film (14.8 dB). This improvement is attributed to the layered and micro-/nanoscale structure and synergetic effects between BP and rGO sheets [123].

11.7.5. MXene

Among the 2D materials (graphene, metal dichalcogenides (TMDs,) MoS_2, boron nitride, black phosphorus), the newly discovered early transition metal carbides and/or nitrides, named '**MXenes**', are exploited enormously for EMI shielding in the past three-four years. The first report on MXenes (Ti_3C_2) was published by the Gogotsi research group in 2011 and due to their unique combination of high electrical conductivity and mechanical properties, MXene became one of the hottest researchers' topics [124]. It has been extensively explored and applied in the field of chemistry, materials science and nanotechnology. Today, the family of MXenes is the largest class of 2D materials and is still growing rapidly, expanding both in size and scope in the range of potential applications [125]. The high electrical conductivity and mechanical properties, hydrophilic nature due to functionalized surfaces, high negative zeta-potential, enabling stable colloidal solutions in water;

and efficient absorption of electromagnetic waves have led to a large number of applications (Fig. **8**) namely in energy storage, which remains a large proportion of MXene activities; biomedical field for photo thermal therapy of cancer, theranostics, biosensors, dialysis, and neural electrodes and electromagnetic applications, including electromagnetic interference shielding and printable antennas is getting more popularity among the research group [126].

11.7.5.1. Structure, Synthesis of MXene and its Composite

The MXene is a fastest flourishing 2D material, derived from its corresponding 3DMAX phase by an etching process. The general formula of MXene is $M_{n+1}X_nT_x$ where M is an early transition element, n = 1–3, X is a carbon or nitrogen and T_x is a surface functional groups like -F, -OH, =O, and -Cl, which are directly attached to the M. The $M_{n+1}AX_n$ is used to denote the MAX phase, where generally is group A 13/14 element, but, several other elements also utilize for A layer (Fig. **8**) [125]. The different combination of elements has been used for both MAX and MXene synthesis shown below in Figs. **8(a)** and **8(b)**. The number of layers in MXene is determined by n where the n+1 layer of MXene is formed which is true for the MAX phase as well [126,127]. All known MAX phases are layered hexagonal with $P6_3$/mmc symmetry, where the M layers are nearly closed packed, and the X atoms fill the octahedral sites. The $M_{n+1}X_n$ layers are, in turn, interleaved with layers of A atoms. In other words, the MAX phase structure can be described as 2D layers of early transition metal carbides and/or nitrides "glued" together with an A element [128]. The strong M–X bond has a mixed covalent/metallic/ionic character, whereas the M–A bond is metallic. So, in contrast to other layered materials, such as graphite and TMDs, where weak van der Waals interactions hold the structure together, the bonds between the layers in the MAX phases are too strong to be broken by shear or any similar mechanical means. However, by taking advantage of relative strengths of the M–A compared with the M–X bonds, the A layers can be selectively etched by chemical means without disrupting the M–X bonds. M–A bonds are weaker than the M–X bonds, heating of MAX phases under vacuum, in molten salts, or in certain molten metals at high temperatures results in the selective loss of the element [132]. The first MXene ($Ti_3C_2T_x$) was discovered in 2011 and since then 30 MXene compositions have been published and (marked in blue in Fig. **9**), and dozens more have been explored by computational methods [125].

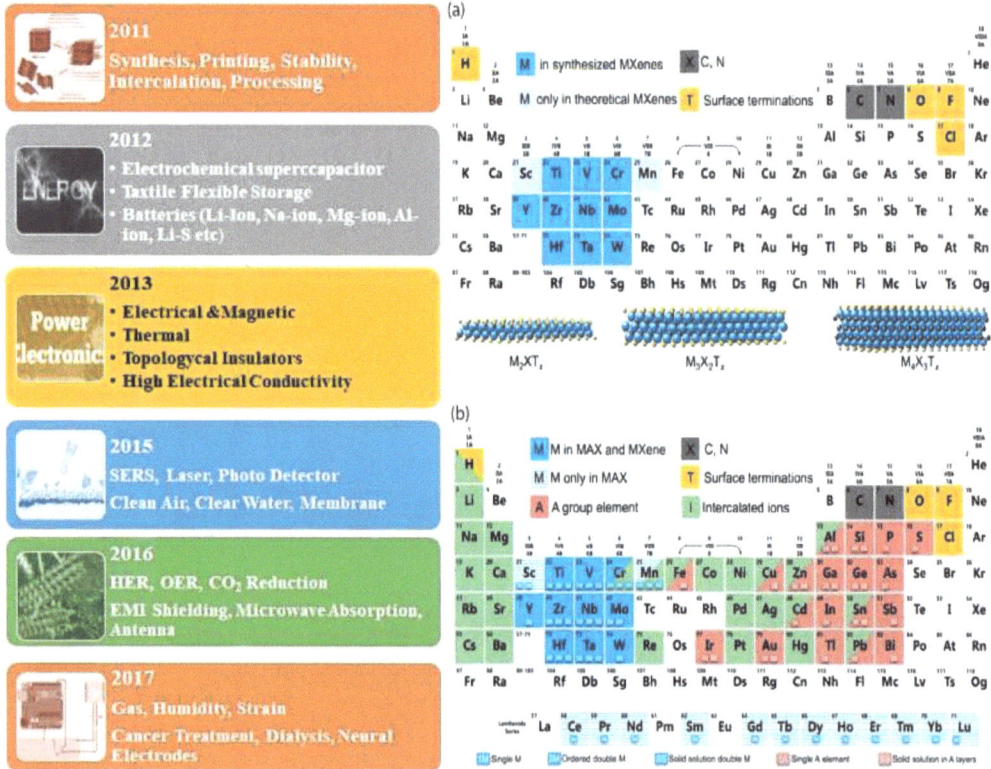

Fig. (8). Year wise application of MXene in different sectors. Periodic tables showing compositions of MXenes and MAX phases. (**a**) Elements used to create MAX phases, MXenes, and their intercalated ions. The A elements are denoted by a red background and are used to synthesize MAX phases that can possibly be utilized to make MXenes. The elements denoted by a green background, have been intercalated into MXenes (to date) and the symbols are at the bottom, 1M and 1A designate the formation of a single (pure) transition metal and A element MAX phase (and MXene). Solid solutions are indicated by an SS in transition metal atomic planes (blue) or A element planes (red); and 2M indicates the formation of an ordered double-transition metal MAX phase or MXene (in-plane or out-of-plane). The MAX phase elements denoted by blue striped background have not yet been used to synthesize MXene Reproduced from reference [125], Copyright 2019 American Chemical Society.

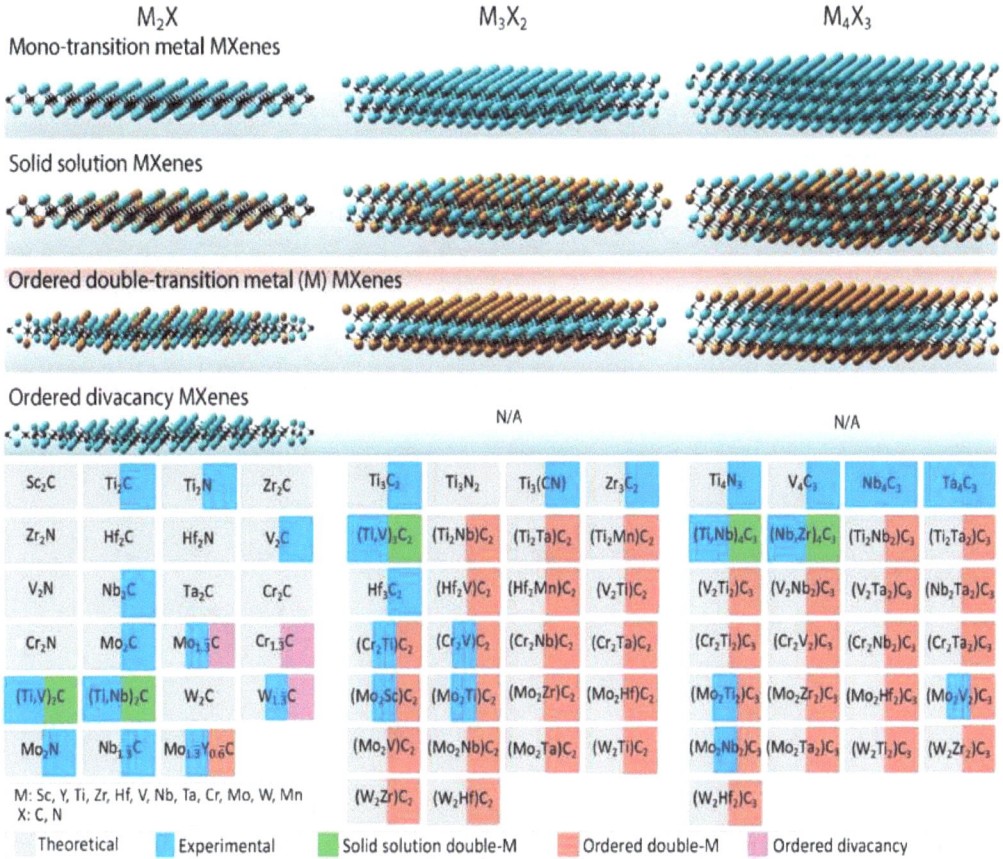

Fig. (9). MXenes synthesized up to date. The top row illustrates structures of (top−down) mono−M MXenes, double-M solid solutions (SS) (marked in green), ordered double-M MXenes (marked in red), and ordered divacancy structure (only for the M_2C MXenes), respectively. This table shows the MXene reported both experimentally (blue) and theoretically (grey) so far [reproduced from reference [125], Copyright 2019, American Chemical Society.

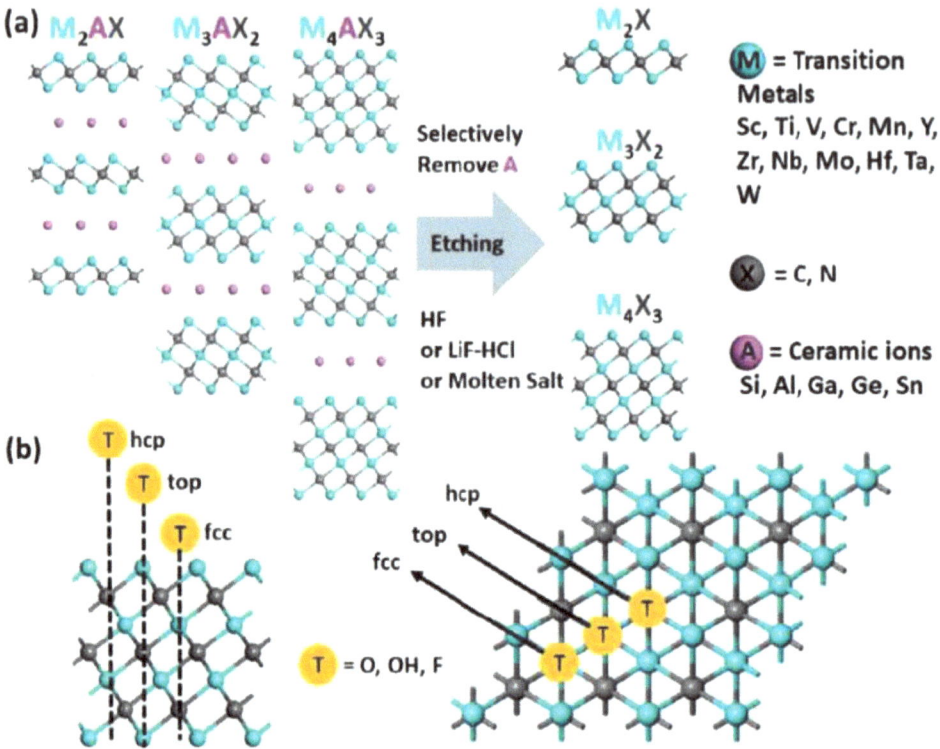

Fig. (10). (a) Experimental steps of synthesizing MXenes by chemical etching the MAX phases. **(b)** Surface terminal groups on MXenes with fcc, hcp, and top sites. Reproduced from reference [129]. Copyright 2019, American Chemical Society.

11.7.5.2. Synthesis of MXene

Small scale preparation method of MXene in the laboratory scale is now well-developed, where MXene is synthesised by chemically and selectively etch out the A-elements, usually group IIIA and IVA elements, such as Al^{3+} and Si^{4+}, from the parent bulk MAX phase precursor [124], where the M stands for the early transition metals usually from group IIIB to VIIB, and X is either C or N. The fluoride-containing acidic solution such as hydrofluoric acid (HF) solution or LiF−HCl mixture solution or ammonium bifluoride ((NH_4)HF_2) was the most widely used as etchant for the preparation of MXenes. Fig. (**10**) shows the schematic of synthesizing MXenes by chemical etching of MAX phases. Using the most common acid etchant. The main challenges for MXene synthesis are (1) to find non-hazardous synthesis routes; and (2) to establish a broader range of MAX-phase precursors [129].

In recent development in the etching method, the use of molten salts enables the A-element to be removed at high temperature [130]. The obtained MXenes are subsequently found to be purer than MXene obtained by normal HF etching method [131,132]. Li *et al*. reported the synthesis of Ti_3ZnC_2 MAX phase from the reaction of Ti_3AlC_2 in $ZnCl_2$ Lewis acidic molten salt at 550 °C, *via* a replacement reaction mechanism [133]. Ti_3ZnC_2 could be further transformed into $Ti_3C_2Cl_2$MXene by increasing the MAX: $ZnCl_2$ ratio. However, the formation mechanism of $Ti_3C_2Cl_2$MXene was not fully understood in terms of the chemistry; since the molten salt and the MAX phase shared the same element (Zn), a reaction mechanism assuming the existence of a low valence Zn^{2+}cation was proposed. In one of the reports by Youbing Li *et al*. [130], proposed a generic method to etch MAX phases by direct redox coupling between the A element and the cation of the Lewis acid molten salt, which allows us to predict the reactivity of the MAX in the molten salt and drastically increases the number of MXenes. MXenes obtained by chemical etching are usually terminated by O, F, and OH groups on the surface [134]. Alternatively, other than wet etching, the chemical vapor deposition (CVD) is also a possible means for MXene synthesis [135] and one of the advantages of this approach is to leave the 2D MXenes unterminated.

11.7.6. EMI Shielding Properties of MXene and MXene/polymer Composite

2D materials with high electrical conductivity and large aspect ratios are the choice for fabricating high performance EMI shielding and their properties can be tailored by varying the layer spacing. Many paths have been developed that exfoliate MXene flakes to meet the requirements of EMI shield and avoid electromagnetic pollution. The high electrical conductivity of 2D $Ti_3C_2T_x$ leads to the reflection of radiation by the interaction between charge carriers and electromagnetic (EM) fields, as well as absorption of EM radiation by electric and/or magnetic dipoles interacting with it. At the same time, the contribution of multiple internal reflections by interfaces and defects of the material is also demonstrated to scatter and absorb the EM waves. Faisal Shahzad *et al*. reported the EMI SE of $Ti_3C_2T_x$ (1.5-µm-thickness) exceeded 40 dB at the width of 8–13 GHz and scaled with the thickness

(t) to be 92 dB (45-µm-thickness), which is the highest among synthetic materials of comparable thickness produced to date [136]. Therefore, MXene with varying thickness and abundant terminating groups helped usefully tune the shielding performance, which surpasses that of carbon nanomaterials and metals with the comparable thickness [137].

To date, MXenes have been successfully integrated with various types of materials, such as metals, carbon nanotubes, high-quality graphene, metal sulphides, nanospheres, and polymers. Among them, polymers have been extensively used to prepare MXene-based composites because of their advantages of low cost, easy to fabricate, and tunable functionalities [138].The key interest in EMI shielding by MXene rises due to its unique intrinsic properties such as conductivity and surface-rich functional groups, high surface area arising due to inter-layers structure providing a tortuous path for incident electromagnetic waves [131]. In addition, MXene has a high dielectric loss in a solution processable polymer matrix at a low percolation limit showing a promising ultra-light EMI shield. When forming composites with polymers, MXenes with excellent mechanical properties, hydrophilic surfaces, and metallic conductivity can improve the mechanical and thermal properties of polymers [139]. In contrast to multi-layered MXenes, single-layer MXenes have higher accessible surface hydrophobicity and better compatibility with polymers. Thus, MXenes are usually delaminated before combining with polymers. Naguib *et al*. synthesize a $Ti_3C_2T_x$–polyacrylamide (PAM)composite by using a two-step method [140]. Here, DMSO is introduced in the interlayers of Ti_3C_2Tx to increase the layer spacing of $Ti_3C_2T_x$ and to achieve full delamination of individual $Ti_3C_2T_x$ layers. Then the prepared $Ti_3C_2T_x$ and PAM solutions are uniformly mixed and dried at ambient temperature for 4–5 days. Different $Ti_3C_2T_x$–PAM composites are prepared by varying the mass ratio of $Ti_3C_2T_x$ to PAM. When$Ti_3C_2T_x$ accounts for 6 wt. %, the composite has the best mechanical properties and the highest electrical conductivity of 3.3×10^{-2} S / m^{-1}. 2D transition metal carbides (MXenes) and their impregnation into sodium alginate (SA) polymer were reported for EMI shielding with highest EMI SE ~ 90 dB for thickness 45 µm [141]. The hydrophilicity of the $Ti_3C_2T_x$MXene was advantageous and polymer composites with bio-compatible sodium alginate (SA) polymer. A 90 wt. % $Ti_3C_2T_x$–SA composite film with a thickness of 9 µm exhibited an excellent

EMI SE of 57 dB with the highest absolute shielding effectiveness (SSE/t) value exceeding 30000 dB cm^{2}g^{-1}. Liu *et al.* reported lightweight MXenes foam for EMI SE [142]. Similarly, Bian *et al.* reported the MXene aerogel for EMI shielding [142]. Ti$_3$C$_2$T$_x$MXene loaded polymer nanocomposites are proposed as an absorption predominant EMI shield by different research groups. Wang *et al.* studied the epoxy-Ti$_3$C$_2$T$_x$MXene nanocomposites for EMI shielding, according to their study, maximum EMI SE of ~ 30 dB could be achieved for 2 mm. However, SE increased to ~ 40 dB after heat treatment for the same thickness. Modern digital systems require coat-able thin polymer nanocomposites (film) with excellent absorption predominant EMI SE value. Intrinsically conducting polymers (ICP) such as polyaniline (PANI) nanocomposite films are well-known for EMI shielding as the matrix itself could provide absorption predominant EMI shielding. Zhang *et al.* has synthesized co-doped PANI-Ti$_3$C$_2$T$_x$MXene nanocomposite and investigated EMI shielding properties [143]. They have reported EMI SE value ~37 dB for the 40 μm thicker PANI–Ti$_3$C$_2$T$_x$MXene nanocomposite film. However, maintaining conductivity and stability of PANI is a challenge due to its de-doping over a period.

Among conducting polymers, poly(3,4-ethylenedioxythiophene): poly (styrene sulfonate) (PEDOT: PSS) has drawn special attraction due to its facile process ability and tuneable conductivity [144]. Yet, the understanding of EMI shielding property of PEDOT: PSS film is very limited, which is believed to be due to its water-soluble nature. Li *et al.* studied the EMI shielding properties of polyurethane-PEDOT: PSS blend, and reported EMI SE of ~ 70 dB achievable for 0.15 mm thicker PU-PEDOT: PSS blend [144]. Liu *et al.* has studied the EMI shielding properties of PEDOT: PSS Ti$_3$C$_2$T$_x$MXene film. The EMI SE ~ 40 dB was reported for PEDOT: PSS-Ti$_3$C$_2$T$_x$MXene film (10 μm) in X-band (8.2-12.4 GHz) [145]. Pritom J. Bora *et al.* [146] proposed a facile method of preparation of MXene interlayered crosslinked conducting polymer, poly(3,4-ethylenedioxythiophene): poly (styrene sulfonate) (PEDOT: PSS) PEDOT: PSS-Ti$_3$C$_2$T$_x$MXene (XPM50) nanocomposite film using divinyl sulfone (DVS) as a crosslinker and demonstrated its EMI shielding properties ((SE) ~ -41dB for 6 μm thick coated film) in the X-band (8.2-12.4 GHz). Here, it was interesting to note that for ~ 6 μm thicker XPM50 film, total EMI SE was ~ 41 dB and corresponding SE$_A$ value was ~ 37 dB. The crosslinked conducting polymer (PEDOT: PSS) film possesses some extraordinary properties, such as better conductivity, better mechanical strength and most importantly, it changes from hydrophilic to hydrophobic. R. Liu *et al.* [143] reported a maximum EMI SE of 9 μm thick PEDOT: PSS-MXene film to be ~ 42 dB and the corresponding SE_A and SE_R values were ~30 dB and ~ 12 dB, respectively. Thus, it indicates that the crosslinked conducting film (PEDOT: PSS-MXene film) has better absorption compared to simple PEDOT: PSS-MXene film at lower thickness. R. Liu at el. first developed ultrathin, flexible, and nacre-like composite films with a "brick-and-mortar" structure using $Ti_3C_2T_x$ MXene and conductive PEDOT: PSS *via* a vacuum assisted filtration process. The mechanical properties of Ti_3C_2Tx/PEDOT: PSS composite films were significantly enhanced according to the heterogeneous interpermeating of $Ti_3C_2T_x$MXene and PEDOT: PSS. The pure $Ti_3C_2T_x$ film (shown inside the dashed area in Fig. **12.2a**) with high flexibility exhibited good metallic lustre on the outer surface. Surface SEM photographs for pure $Ti_3C_2T_x$ film and polymeric composite film with the $Ti_3C_2T_x$ to PEDOT:PSS ratio of 3:1. With the addition of PEDOT: PSS, $Ti_3C_2T_x$MXenes embedded within the polymeric matrix (Fig. **12a**), leading to layered "brick-and-mortar" structures.

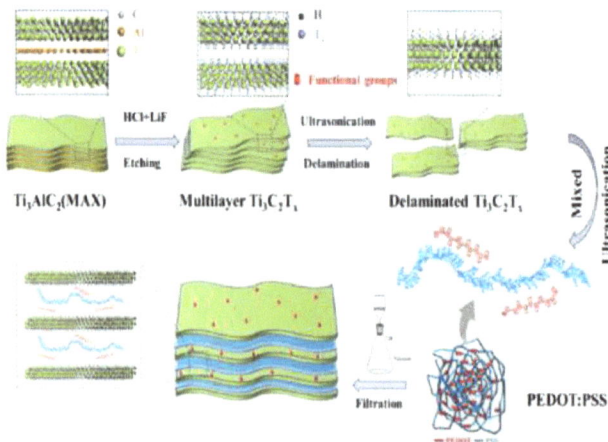

Fabrication of $Ti_3C_2T_x$ MXene/PEDOT:PSS composite film via a vacuum-assisted filtration process.

Fig. (11). Schematic Illustration of the Assembly Process for $Ti_3C_2T_x$ MXene/PEDOT: PSS Composite Films Reproduced from reference [145], Copyright 2018, American Chemical Society.

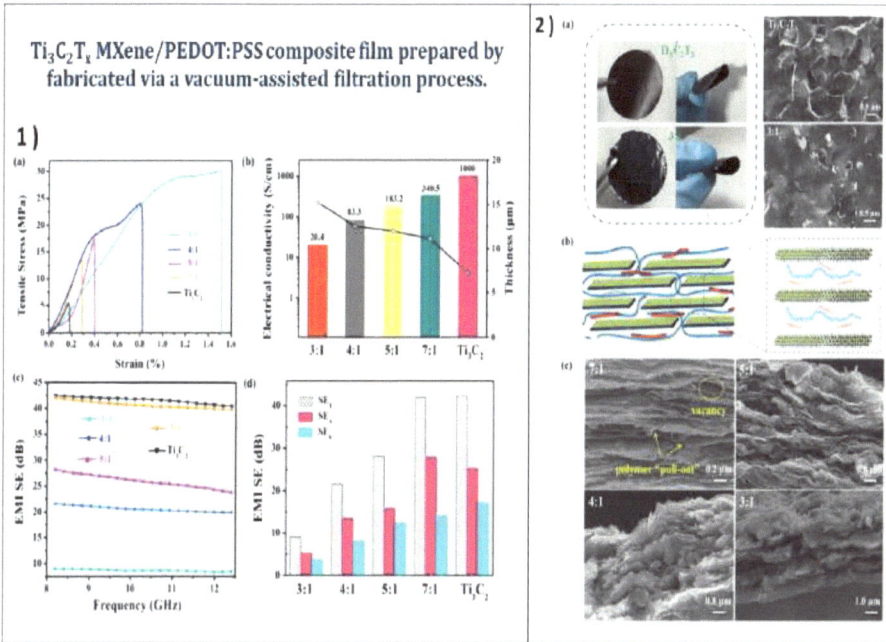

Fig. (12). **(a)** Tensile stress−strain curves of pure $Ti_3C_2T_x$ MXene film and the polymeric composite film with different ratios ($Ti_3C_2T_x$ to PEDOT: PSS). **(b)** Electrical conductivity and thickness of pure Ti_3C_2Tx MXene film and the polymeric composite film with different ratios. **(c)** EMI SE of pure Ti_3C_2Tx MXene film and the polymeric composite film with different ratios. **(d)** Total EMI SE and its absorption (SEA) and reflection (SER) of pure $Ti_3C_2T_x$ MXene film and the polymeric composite film with different ratios. Digital images and top-view SEM images of the as prepared pure Ti_3C_2Tx MXene film and the polymeric composite film with 3:1 ratio ($Ti_3C_2T_x$ to PEDOT: PSS). **(b)** Schematic illustration of the Ti_3C_2Tx MXene/PEDOT: PSS composite film with a "brick-and-mortar" structure. **(c)** Cross-sectional SEM images of pure Ti_3C_2Tx MXene film and the Ti_3C_2Tx MXene/PEDOT: PSS composite film with different ratios. Reproduced from reference [145], Copyright 2018, American Chemical Society.

Electrical conductivity is of great importance for EMI shielding materials as it is directly related to reflecting electromagnetic waves for high EMI SE. With increasing PEDOT: PSS content in the polymeric composite films, the electrical conductivity markedly decreases as shown in Fig. (**12b**) and pure $Ti_3C_2T_x$ MXene film (7.2 μm, 2.62 g cm^{-3}) displayed the highest electrical conductivity of 1000 S/cm. Relatively, the better conductivity endows the pure $Ti_3C_2T_x$ film with an excellent EMI SE of 42.48 dB (Fig. **12a**). By increasing the $Ti_3C_2T_x$ to PEDOT:PSS ratio from 7:1 to 3:1, the corresponding electrical conductivity decreased gradually from 340.5 to 20.4 S/cm, which greatly meet the requirements (0.01S/cm) for EMI shielding materials in practical applications [147]. Apart from the polymeric composite film with the 3:1 ratio ($Ti_3C_2T_x$ to PEDOT:PSS), all other polymeric composite films with excellent EMI performance and mechanical properties are satisfactory for industrial uses (EMI SE > 20 dB). As shown in Fig. (**12c**), the overall EMI shielding performance in X-band declined with the increment of PEDOT:PSS content, following a trend similar to the variation of electrical conductivities. For $Ti_3C_2T_x$ to PEDOT:PSS ratio from 7:1 to 3:1, the EMI SE of the polymeric composite films corresponds to 42.10, 28.22, 21.61, and 8.99 dB, respectively. Interestingly, the EMI SE of the composite film with the 7:1 ratio (11.1 μm, 1.94 g cm^{-3}) is very close to that of pure $Ti_3C_2T_x$ film (42.48 dB), but the strength increases by 2.4 times, showing a significant difference from the previously reported insulating polymer matrices, such as SA, chitosan, and PVA.

Table 5. EMI shielding performances of MXene/polymer membranes.

MXene/polymer	Method	Wt. %	Thickness (μm)	SE (dB/μm)	Range (GHz)	Ref.
Ti_3C_2/wax	Cold Press (CP)	90	1000	76.1	8.2–12.4	[148]
Ti_3C_2/ SA	Vacuum Assisted Filtration (VAF)	90	~45	92	8.2–12.4	[136]
Ti_3C_2–CNT /paraffin	Cold Press (CP)	35	1550	52.9	2.0–18.0	[149]
Ti_3C_2-rGO/epoxy	Hot press (HP)	0.74	2000	56.4	8.2–12.4	[150]
Ti_3C_2/PANI	Cold Press (CP)	50	1500	23@10.9	8.2–12.4	[151]
Ti_3C_2/PPy/PET	Dip Coating (DIP)		1300	90	8.2–12.4	[152]

(Table 5) cont.....

Ti$_3$C$_2$/PANI/paraffin	Cold press (CP)	26.5	1800	56.3@13.8	2.0–18.0	[153]
Ti$_2$C/PVA-film	Press Rolling (PR)	7.6	100	26	8.2–12.4	[154]
Ti$_3$C$_2$/PVA/ MWCNT/PSS	LBL	4.2	0.17	2.81	8.2–12.4	[155]

11.7.7. Future Direction of MXene Research

Because of the various functionalities of MXenes and polymers, MXene/polymer membranes exhibit outstanding and/or unique properties compared with their precursors. Despite the enhanced performances compared with that of the commercial devices, numerous unprecedented challenges restrict the commercialization of current MXene/polymer membranes, such as scale-up fabrication, cost, and design of devices. The deployment of large-scale fabrication for MXene/polymer membrane is the key for its commercialization. This finding further demonstrates that the addition of an optimal amount of conductive polymers will not only maintain prominent EMI shielding performance for the polymeric composite films but dramatically improve their mechanical property. Keeping the above issues in mind, there is high need for following Future direction for research in 2D MXene area [132]:

1. Controlling and modifying MXene' surfaces.

2. Establishing the exact structure of M$_{n+1}$X$_n$T$_x$ as a function of T and x.

3. Detailing and understanding the structure and properties of MXenes intercalated with various compounds/ions.

4. Determining the chemical and thermal stabilities of MXenes in different environments.

5. Large-scale delamination of MXenes other than Ti$_3$C$_2$T$_x$.

6. Finding alternative, robust, and safe routes of MAX phase exfoliation and MXene delamination.

7. Synthesizing MXenes without surface functional groups.

8. Direct gas phase synthesis of single-layer MXene films and characterizing single-layer MXenes, including electronic magnetic, optical, thermal, and mechanical properties.

9. Exploring MXenes in various applications, such as composite reinforcement, catalysis, transparent electronic conductors, sensors, *etc*.

10. Expanding the family through synthesis of new MXenes.

CONCLUDING REMARKS

The rise in demand for small, smart and advanced electronic devices has led to the rapid growth of research on EMI shielding materials, issues related to EMI pollution, and health hazards for humans and other living organisms make this a socially relevant topic. Several materials have been used to solve the problem of EMI, wherein 2D materials demonstrate potential due to their unique conductive properties and high surface area. Among them, graphene and MXenes have been widely highlighted and this chapter summarizes the current progress on the EMI shielding properties of graphene, and beyond graphene 2D materials MoS_2, black phosphorus, h-BH and MXene-based materials. The 2D materials synthesized by various methods in which radiative and chemical vapor deposition gives rise to good quality graphene and utilization of other methods is dependent on the types of applications. MXene synthesis also adapts various techniques and lithium fluoride/hydrochloric acid etching being the most desirable etching method. 2D materials have demonstrated excellent EMI shielding performance and many stoichiometric compositions and surface groups of MXenes graphene and their various hybrids were obtained by tuning the properties. Therefore, the electrical conductivity, filler content, thickness, and structure of MXene-based materials have been investigated in order to understand their influence on EMI shielding. Outstanding achievements have been reported; however, further investigations related to the MXene-form and its stability, combined with materials such as polymer matrices and structure–composition optimizations are needed. Moreover, besides the performances already reached in academia and regardless of MXenes or other emerging 2D materials, knowing the products already available on the market and the current trends in the industry may inspire new investigative strategies on the EMI shielding properties of 2D materials.

CONSENT FOR PUBLICATION

Not applicable.

CONFLICT OF INTEREST

The authors declare no conflict of interest, financial or otherwise.

ACKNOWLEDGEMENTS

The authors are grateful to the Director, CSIR- Central Electrochemical research Institute, CSIR-Madras Complex, Chennai Tamilnadu 600113 India, for his kind support and encouragement. Author's thanks are also extended to the technicians of the laboratory for their help in offering the resources in running the program.

REFERENCES

[1]   Yang, Y.; Gupta, M. C.; Dudley, K. L.; Lawrence, R. W. Novel Carbon Nanotube - Polystyrene Foam Composites for Electromagnetic Interference Shielding. *Nano Letters* **2005**, *5* (11), 2131–2134. https://doi.org/10.1021/nl051375r.

[2]   Chen, Z.; Xu, C.; Ma, C.; Ren, W.; Cheng, H. M. Lightweight and Flexible Graphene Foam Composites for High-Performance Electromagnetic Interference Shielding. *Advanced Materials* **2013**, *25* (9), 1296–1300. https://doi.org/10.1002/adma.201204196.

[3]   Hanada, E.; Antoku, Y.; Tani, S.; Kimura, M.; Hasegawa, A.; Urano, S.; Ohe, K.; Yamaki, M.; Nose, Y. Electromagnetic Interference on Medical Equipment by Low-Power Mobile Telecommunication Systems. *IEEE Transactions on Electromagnetic Compatibility* **2000**, *42* (4), 470–476. https://doi.org/10.1109/15.902316.

[4]   Architectural Electromagnetic Shielding Handbook: A Design and Specification Guide | Wiley https://www.wiley.com/en-us/Architectural+Electromagnetic+Shielding+ Handbook %3A +A +Design +and+Specification+Guide-p-9780780360242 (accessed 2021 -01 -10).

[5]   Jia, Z.; Zhang, M.; Liu, B.; Wang, F.; Wei, G.; Su, Z. Graphene Foams for Electromagnetic Interference Shielding: A Review. *ACS Applied Nano Materials*. American Chemical Society July 24, 2020, pp 6140–6155. https://doi.org/10.1021/acsanm.0c00835.

[6]   Song, W. L.; Guan, X. T.; Fan, L. Z.; Cao, W. Q.; Wang, C. Y.; Cao, M. S. Tuning Three-Dimensional Textures with Graphene Aerogels for Ultra-Light Flexible Graphene/Texture Composites of Effective Electromagnetic Shielding. *Carbon* **2015**, *93*, 151–160. https://doi.org/10.1016/j.carbon.2015.05.033.

[7]   Zeng, Z.; Jin, H.; Chen, M.; Li, W.; Zhou, L.; Zhang, Z. Lightweight and Anisotropic Porous MWCNT/WPU Composites for Ultrahigh Performance Electromagnetic Interference Shielding. *Advanced Functional Materials* **2016**, *26* (2), 303–310. https://doi.org/10.1002/adfm.201503579.

[8]   Zhang, X.; Liu, W.; Wang, H.; Zhao, X.; Zhang, Z.; Nienhaus, G. U.; Shang, L.; Su, Z. Self-Assembled Thermosensitive Luminescent Nanoparticles with Peptide-Au Conjugates for Cellular Imaging and Drug Delivery. *Chinese Chemical Letters* **2020**, *31* (3), 859–864. https://doi.org/10.1016/j.cclet.2019.06.032.

[9]   Singh, K.; Ohlan, A.; Pham, V. H.; Balasubramaniyan, R. B.; Varshney, S.; Jang, J.; Hur, S. H.; Choi, W. M.; Kumar, M.; Dhawan, S. K.; Kong, B. S.; Chung, J. S. Nanostructured Graphene/Fe3O4 Incorporated Polyaniline as a High-Performance Shield against Electromagnetic Pollution. *Nanoscale* **2013**, *5* (6), 2411–2420. https://doi.org/10.1039/c3nr33962a.

[10]   Wang, Y.; Wang, H.; Ye, J.; Shi, L.; Feng, X. Magnetic CoFe Alloy@C Nanocomposites Derived from ZnCo-MOF for Electromagnetic Wave Absorption. *Chemical Engineering Journal* **2020**, *383*, 123096. https://doi.org/10.1016/j.cej.2019.123096.

[11]   Singh, A. K.; Yadav, A. N.; Srivastava, A.; Haldar, K. K.; Tomar, M.; Alaferdov, A. v.; Moshkalev, S. A.; Gupta, V.; Singh, K. CdSe/V2O5 Core/Shell Quantum Dots Decorated Reduced Graphene Oxide Nanocomposite for High-Performance Electromagnetic Interference Shielding Application. *Nanotechnology* **2019**, *30* (50), 505704. https://doi.org/10.1088/1361-6528/ab4290.

[12]   Hawkins, S. A.; Yao, H.; Wang, H.; Sue, H. J. Tensile Properties and Electrical Conductivity of Epoxy Composite Thin Films Containing Zinc Oxide Quantum Dots and Multi-Walled Carbon Nanotubes. *Carbon* **2017**, *115*, 18–27. https://doi.org/10.1016/j.carbon.2016.12.058.

[13]   El-Shamy, A. G. Novel Conducting PVA/Carbon Quantum Dots (CQDs) Nanocomposite for High Anti-Electromagnetic Wave Performance. *Journal of Alloys and Compounds* **2019**, *810*, 151940. https://doi.org/10.1016/j.jallcom.2019.151940.

[14]   Ge, C.; Zou, J.; Yan, M.; Bi, H. C-Dots Induced Microwave Absorption Enhancement of PANI/Ferrocene/C-Dots. *Materials Letters* **2014**, *137*, 41–44. https://doi.org/10.1016/j.matlet.2014.08.111.

[15]   Lakshmi, N. v.; Tambe, P. EMI Shielding Effectiveness of Graphene Decorated with Graphene Quantum Dots and Silver Nanoparticles Reinforced PVDF Nanocomposites. *Composite Interfaces* **2017**, *24* (9), 861–882. https://doi.org/10.1080/09276440.2017.1302202.

[16]   Ji, B.; Giovanelli, E.; Habert, B.; Spinicelli, P.; Nasilowski, M.; Xu, X.; Lequeux, N.; Hugonin, J. P.; Marquier, F.; Greffet, J. J.; Dubertret, B. Non-Blinking Quantum Dot with a Plasmonic Nanoshell Resonator. *Nature Nanotechnology* **2015**, *10* (2), 170–175. https://doi.org/10.1038/nnano.2014.298.

[17]   Wang, Y.; Wang, W.; Qi, Q.; Xu, N.; Yu, D. Layer-by-Layer Assembly of PDMS-Coated Nickel Ferrite/Multiwalled Carbon Nanotubes/Cotton Fabrics for Robust and Durable

Electromagnetic Interference Shielding. *Cellulose* **2020**, *27* (5), 2829–2845. https://doi.org/10.1007/s10570-019-02949-1.

[18]   Sangwan, V. K.; Hersam, M. C. Electronic Transport in Two-Dimensional Materials. *Annual Review of Physical Chemistry* **2018**, *69* (1), 299–325. https://doi.org/10.1146/annurev-physchem-050317-021353.

[19]   Gao, L.; Li, C.; Huang, W.; Mei, S.; Lin, H.; Ou, Q.; Zhang, Y.; Guo, J.; Zhang, F.; Xu, S.; Zhang, H. MXene/Polymer Membranes: Synthesis, Properties, and Emerging Applications. *Chemistry of Materials* **2020**, *32* (5), 1703–1747. https://doi.org/10.1021/acs.chemmater.9b04408.

[20]   Tan, Y.; Luo, H.; Zhou, X.; Peng, S.; Zhang, H. Dependences of Microstructure on Electromagnetic Interference Shielding Properties of Nano-Layered Ti3AlC2 Ceramics. *Scientific Reports* **2018**, *8* (1), 1–8. https://doi.org/10.1038/s41598-018-26256-0.

[21]   Iqbal, A.; Kwon, J.; Kim, M. K.; Koo, C. M. MXenes for Electromagnetic Interference Shielding: Experimental and Theoretical Perspectives. *Materials Today Advances* **2021**, *9*, 100124. https://doi.org/10.1016/J.MTADV.2020.100124.

[22]   Huang, L.; Chen, C.; Li, Z.; Zhang, Y.; Zhang, H.; Lu, J.; Ruan, S.; Zeng, Y. J. Challenges and Future Perspectives on Microwave Absorption Based on Two-Dimensional Materials and Structures. *Nanotechnology*. Institute of Physics Publishing February 4, 2020, p 162001. https://doi.org/10.1088/1361-6528/ab50af.

[23]   Kumar, P. Ultrathin 2D Nanomaterials for Electromagnetic Interference Shielding. *Advanced Materials Interfaces* **2019**, *6* (24), 1901454. https://doi.org/10.1002/admi.201901454.

[24]   Hu, M.; Zhang, N.; Shan, G.; Gao, J.; Liu, J.; Li, R. K. Y. Two-Dimensional Materials: Emerging Toolkit for Construction of Ultrathin High-Efficiency Microwave Shield and Absorber. *Frontiers of Physics*. Higher Education Press August 1, 2018, pp 1–39. https://doi.org/10.1007/s11467-018-0809-8.

[25]   Fundamentals of electromagnetic compatibility (EMC) https://www.slideshare.net/ BrunoDeWachter1/fundamentals-of-electromagnetic-compatibility-emc (accessed 2021 -01 -11).

[26]   Electric Vehicle Market Growth, Industry Trends, and Statistics by 2030 | COVID-19 Impact Analysis | Marketsand Markets https://www.marketsandmarkets.com/Market-Reports/electric-vehicle-market-209371461.html (accessed 2021 -01 -11).

[27]   Liu, C.; Wang, L.; Liu, S.; Tong, L.; Liu, X. Fabrication Strategies of Polymer-Based Electromagnetic Interference Shielding Materials. *Advanced Industrial and Engineering Polymer Research* **2020**, *3* (4), 149–159. https://doi.org/10.1016/j.aiepr.2020.10.002.

[28]   Mild, K. H.; Alanko, T.; Hietanen, M.; Decat, G.; Falsaperla, R.; Rossi, P.; Gryz, K.; Karpowicz, J.; Sandström, M. Exposure of Workers to Electromagnetic Fields. A Review of Open Questions on Exposure Assessment Techniques. *International Journal of Occupational Safety and Ergonomics* **2009**, *15* (1), 3–33. https://doi.org/10.1080/10803548.2009.11076785.

[29]   Gryz, K.; Karpowicz, J.; Leszko, W.; Zradziński, P. Evaluation of Exposure to Electromagnetic Radio frequency Radiation in the Indoor Workplace Accessible to the Public by the Use of Frequency-Selective Exposimeters. *International Journal of Occupational Medicine and Environmental Health* **2014**, *27* (6), 1043–1054. https://doi.org/10.2478/s13382-014-0334-0.

[30]   Hocking, B.; Mild, K. H. Guidance Note: Risk Management of Workers with Medical Electronic Devices and Metallic Implants in Electromagnetic Fields. *International Journal of Occupational Safety and Ergonomics* **2008**, *14* (2), 217–222. https://doi.org/10.1080/10803548.2008.11076763.

[31]   Tips on shielding and grounding in Industrial Automation | SMAR Technology Company https://www.smar.com/en/technical-article/tips-on-shielding-and-grounding-in-industrial-automation (accessed 2021 -09 -20).

[32]   Koohestani, M.; Perdriau, R.; Levant, J. L.; Ramdani, M. A Novel Passive Cost-Effective Technique to Improve Radiated Immunity on PCBs. *IEEE Transactions on Electromagnetic Compatibility* **2019**, *61* (6), 1733–1739. https://doi.org/10.1109/TEMC.2018.2882732.

[33]   ASTM D4935 - Standard Test Method for Measuring the Electromagnetic Shielding Effectiveness of Planar Materials | Engineering360 https://standards.globalspec.com/std/10382691/ ASTM% 20D4935 (accessed 2021 -01 -11).

[34]   299.1-2013 - 299.1-2013 - IEEE Standard Method for Measuring the Shielding Effectiveness of Enclosures and Boxes Having all Dimensions between 0.1 m and 2 m - IEEE Standard https://ieeexplore.ieee.org/document/6712029 (accessed 2021 -01 -12).

[35]   Wilson, P. F.; Ma, M. T.; Adams, J. W. Techniques for Measuring the Electromagnetic Shielding Effectiveness of Materials: Part I: Far-Field Source Simulation. *IEEE Transactions on Electromagnetic Compatibility* **1988**, *30* (3), 239–250. https://doi.org/10.1109/15.3302.

[36]   Wilson, P. F.; Ma, M. T. Techniques for Measuring the Electromagnetic Shielding Effectiveness of Materials: Part II—Near Field Source Simulation. *IEEE Transactions on Electromagnetic Compatibility* **1988**, *30* (3), 251–259. https://doi.org/10.1109/15.3303.

[37]   Shan, Y.; Li, P.; Deng, J. Planar Material Sample Fixture Characterization and Application for EMI Shielding Effectiveness Evaluations. In *cccc2012 Asia-Pacific Symposium on Electromagnetic Compatibility, APEMC 2012 - Proceedings*; 2012; pp 181–184. https://doi.org/10.1109/APEMC.2012.6237804.

[38]   Holloway, C. L.; Ladbury, J.; Coder, J.; Koepke, G.; Hill, D. A. Measuring the Shielding Effectiveness of Small Enclosures/Cavities with a Reverberation Chamber. In *IEEE International Symposium on Electromagnetic Compatibility*; 2007. https://doi.org/10.1109/ISEMC.2007.241.

[39]   Vasquez, H.; Espinoza, L.; Lozano, K.; Foltz, H.; Yang, S. *Simple Device for Electromagnetic Interference Shielding Effectiveness Measurement.*

[40]   Sarto, M. S.; Tamburrano, A. Innovative Test Method for the Shielding Effectiveness Measurement of Conductive Thin Films in a Wide Frequency Range. *IEEE Transactions on Electromagnetic Compatibility* **2006**, *48* (2), 331–341. https://doi.org/10.1109/TEMC.2006.874664.

[41]   Tamburrano, A.; Desideri, D.; Maschio, A.; Sabrina Sarto, M. Coaxial Waveguide Methods for Shielding Effectiveness Measurement of Planar Materials Up to 18 GHz. *IEEE Transactions on Electromagnetic Compatibility* **2014**, *56* (6), 1386–1395. https://doi.org/10.1109/TEMC.2014.2329238.

[42]   EMI Shielding Market | Size, Share, Industry Analysis and Market Forecast to 2025 | Marketsand Markets[TM] https://www.marketsandmarkets.com/Market-Reports/emi-shielding-market-105681800.html?gclid=Cj0KCQiA6Or_BRC_ARIsAPzuer8xtfSEEyQPJVL3ngPHo6zKrd29Q2st8brsn-vO56ZDoJrAyzrw8kwaAvyXEALw_wcB (accessed 2021 -01 -11).

[43]   Increase in Electromagnetic Radiation from Electronic Components & the Need to Comply with Stringent EMC Regulations to Support Demand for EMI/RFI Shielding Materials & Technologies   https://strategyr.blogspot.com/2018/11/increase-in-electromagnetic-radiation.html (accessed 2021 -01 -11).

[44]   Singh, K.; Ohlan, A.; Dhawan, S. K. Polymer-Graphene Nanocomposites: Preparation, Characterization, Properties, and Applications. In *Nanocomposites - New Trends and Developments*; InTech, 2012. https://doi.org/10.5772/50408.

[45]   Novoselov, K. S.; Geim, A. K.; Morozov, S. v.; Jiang, D.; Zhang, Y.; Dubonos, S. v.; Grigorieva, I. v.; Firsov, A. A. Electric Field in Atomically Thin Carbon Films. *Science* **2004**, *306* (5696), 666–669. https://doi.org/10.1126/science.1102896.

[46]   Oliveira, F. M.; Gusmaõ, R. Recent Advances in the Electromagnetic Interference Shielding of 2D Materials beyond Graphene. *ACS Applied Electronic Materials*. American Chemical Society October 27, 2020, pp 3048–3071. https://doi.org/10.1021/acsaelm.0c00545.

[47]   Zhang, H. bin; Yan, Q.; Zheng, W. G.; He, Z.; Yu, Z. Z. Tough Graphene-Polymer Microcellular Foams for Electromagnetic Interference Shielding. *ACS Applied Materials and Interfaces* **2011**, *3* (3), 918–924. https://doi.org/10.1021/am200021v.

[48]   Novoselov, K. S.; Jiang, D.; Schedin, F.; Booth, T. J.; Khotkevich, V. v.; Morozov, S. v.; Geim, A. K. Two-Dimensional Atomic Crystals. *Proceedings of the National Academy of Sciences of the United States of America* **2005**, *102* (30), 10451–10453. https://doi.org/10.1073/pnas.0502848102.

[49]   Li, H.; Yin, Z.; He, Q.; Li, H.; Huang, X.; Lu, G.; Fam, D. W. H.; Tok, A. I. Y.; Zhang, Q.; Zhang, H. Fabrication of Single- and Multilayer MoS 2 Film-Based Field-Effect Transistors for Sensing NO at Room Temperature. *Small* **2012**, *8* (1), 63–67. https://doi.org/10.1002/smll.201101016.

[50]   Li, Q.; Newberg, J. T.; Walter, E. C.; Hemminger, J. C.; Penner, R. M. Polycrystalline Molybdenum Disulfide (2H-MoS2) Nano- and Microribbons by Electrochemical/Chemical Synthesis. *Nano Letters* **2004**, *4* (2), 277–281. https://doi.org/10.1021/nl035011f.

[51]   Altavilla, C.; Sarno, M.; Ciambelli, P. A Novel Wet Chemistry Approach for the Synthesis of Hybrid 2D Free-Floating Single or Multilayer Nanosheets of MS2@oleylamine (M=Mo, W). *Chemistry of Materials* **2011**, *23* (17), 3879–3885. https://doi.org/10.1021/cm200837g.

[52]   Lee, H. C.; Liu, W. W.; Chai, S. P.; Mohamed, A. R.; Aziz, A.; Khe, C. S.; Hidayah, N. M. S.; Hashim, U. Review of the Synthesis, Transfer, Characterization and Growth Mechanisms of Single and Multilayer Graphene. *RSC Advances*. Royal Society of Chemistry March 9, 2017, pp 15644–15693. https://doi.org/10.1039/C7RA00392G.

[53]   Somani, P. R.; Somani, S. P.; Umeno, M. Planer Nano-Graphenes from Camphor by CVD. *Chemical Physics Letters* **2006**, *430* (1–3), 56–59. https://doi.org/10.1016/j.cplett.2006.06.081.

[54]   Nandamuri, G.; Roumimov, S.; Solanki, R. Chemical Vapor Deposition of Graphene Films. *Nanotechnology* **2010**, *21* (14), 145604. https://doi.org/10.1088/0957-4484/21/14/145604.

[55]   Guermoune, A.; Chari, T.; Popescu, F.; Sabri, S. S.; Guillemette, J.; Skulason, H. S.; Szkopek, T.; Siaj, M. Chemical Vapor Deposition Synthesis of Graphene on Copper with Methanol, Ethanol, and Propanol Precursors. *Carbon* **2011**, *49* (13), 4204–4210. https://doi.org/10.1016/j.carbon.2011.05.054.

[56]   Chen, Z.; Ren, W.; Liu, B.; Gao, L.; Pei, S.; Wu, Z. S.; Zhao, J.; Cheng, H. M. Bulk Growth of Mono- to Few-Layer Graphene on Nickel Particles by Chemical Vapor Deposition from Methane. *Carbon* **2010**, *48* (12), 3543–3550. https://doi.org/10.1016/j.carbon.2010.05.052.

[57]   Li, X.; Cai, W.; An, J.; Kim, S.; Nah, J.; Yang, D.; Piner, R.; Velamakanni, A.; Jung, I.; Tutuc, E.; Banerjee, S. K.; Colombo, L.; Ruoff, R. S. Large-Area Synthesis of High-Quality and Uniform Graphene Films on Copper Foils. *Science* **2009**, *324* (5932).

[58]   Gadipelli, S.; Calizo, I.; Ford, J.; Cheng, G.; Hight Walker, A. R.; Yildirim, T. A Highly Practical Route for Large-Area, Single Layer Graphene from Liquid Carbon Sources Such as Benzene and Methanol. *Journal of Materials Chemistry* **2011**, *21* (40), 16057–16065. https://doi.org/10.1039/c1jm12938d.

[59]   Aristov, V. Y.; Urbanik, G.; Kummer, K.; Vyalikh, D. v.; Molodtsova, O. v.; Preobrajenski, A. B.; Zakharov, A. A.; Hess, C.; Hänke, T.; Büchner, B.; Vobornik, I.; Fujii, J.; Panaccione, G.; Ossipyan, Y. A.; Knupfer, M. Graphene Synthesis on Cubic SiC/Si Wafers. Perspectives for Mass Production of Graphene-Based Electronic Devices. *Nano Letters* **2010**, *10* (3), 992–995. https://doi.org/10.1021/nl904115h.

[60]   Deng, D.; Pan, X.; Zhang, H.; Fu, Q.; Tan, D.; Bao, X. Freestanding Graphene by Thermal Splitting of Silicon Carbide Granules. *Advanced Materials* **2010**, *22* (19), 2168–2171. https://doi.org/10.1002/adma.200903519.

[61]   XIII. On the Atomic Weight of Graphite. *Philosophical Transactions of the Royal Society of London* **1859**, *149*, 249–259. https://doi.org/10.1098/rstl.1859.0013.

[62]   Hummers, W. S.; Offeman, R. E. Preparation of Graphitic Oxide. *Journal of the American Chemical Society* **1958**, *80* (6), 1339. https://doi.org/10.1021/ja01539a017.

[63]   Marcano, D. C.; Kosynkin, D. v.; Berlin, J. M.; Sinitskii, A.; Sun, Z.; Slesarev, A.; Alemany, L. B.; Lu, W.; Tour, J. M. Improved Synthesis of Graphene Oxide. *ACS Nano* **2010**, *4* (8), 4806–4814. https://doi.org/10.1021/nn1006368.

[64]   Stankovich, S.; Dikin, D. A.; Piner, R. D.; Kohlhaas, K. A.; Kleinhammes, A.; Jia, Y.; Wu, Y.; Nguyen, S. B. T.; Ruoff, R. S. Synthesis of Graphene-Based Nanosheets *via* Chemical Reduction of Exfoliated Graphite Oxide. *Carbon* **2007**, *45* (7), 1558–1565. https://doi.org/10.1016/j.carbon.2007.02.034.

[65]   Pham, H. D.; Pham, V. H.; Cuong, T. V.; Nguyen-Phan, T. D.; Chung, J. S.; Shin, E. W.; Kim, S. Synthesis of the Chemically Converted Graphene Xerogel with Superior Electrical Conductivity. *Chemical Communications* **2011**, *47* (34), 9672–9674. https://doi.org/10.1039/c1cc13329b.

[66]   American Institute of Physics Handbook, 3rd ed. [1972].pdf | Atomic Mass Unit | Electronvolt https://www.scribd.com/document/437090631/American-Institute-of-Physics-Handbook-3rd-ed-1972-pdf (accessed 2021 -01 -11).

[67]   Nirmalraj, P. N.; Lutz, T.; Kumar, S.; Duesberg, G. S.; Boland, J. J. Nanoscale Mapping of Electrical Resistivity and Connectivity in Graphene Strips and Networks. *Nano Letters* **2011**, *11* (1), 16–22. https://doi.org/10.1021/nl101469d.

[68]   Pei, S.; Zhao, J.; Du, J.; Ren, W.; Cheng, H. M. Direct Reduction of Graphene Oxide Films into Highly Conductive and Flexible Graphene Films by Hydrohalic Acids. *Carbon* **2010**, *48* (15), 4466–4474. https://doi.org/10.1016/j.carbon.2010.08.006.

[69]   Li, D.; Müller, M. B.; Gilje, S.; Kaner, R. B.; Wallace, G. G. Processable Aqueous Dispersions of Graphene Nanosheets. *Nature Nanotechnology* **2008**, *3* (2), 101–105. https://doi.org/10.1038/nnano.2007.451.

[70]   Chen, H.; Müller, M. B.; Gilmore, K. J.; Wallace, G. G.; Li, D. Mechanically Strong, Electrically Conductive, and Biocompatible Graphene Paper. *Advanced Materials* **2008**, *20* (18), 3557–3561. https://doi.org/10.1002/adma.200800757.

[71]   Zhao, X.; Hayner, C. M.; Kung, M. C.; Kung, H. H. Flexible Holey Graphene Paper Electrodes with Enhanced Rate Capability for Energy Storage Applications. *ACS Nano* **2011**, *5* (11), 8739–8749. https://doi.org/10.1021/nn202710s.

[72]   Lin, X.; Shen, X.; Zheng, Q.; Yousefi, N.; Ye, L.; Mai, Y. W.; Kim, J. K. Fabrication of Highly-Aligned, Conductive, and Strong Graphene Papers Using Ultralarge Graphene Oxide Sheets. *ACS Nano* **2012**, *6* (12), 10708–10719. https://doi.org/10.1021/nn303904z.

[73]   Cao, M.-S.; Wang, X.-X.; Cao, W.-Q.; Yuan, J. Ultrathin Graphene: Electrical Properties and Highly Efficient Electromagnetic Interference Shielding. *J. Mater. Chem. C* **2015**, *3*, 6589. https://doi.org/10.1039/c5tc01354b.

[74]   Colaneri, N. F.; Shacklette, L. W. EMI Shielding Measurements of Conductive Polymer Blends. *IEEE Transactions on Instrumentation and Measurement* **1992**, *41* (2), 291–297. https://doi.org/10.1109/19.137363.

[75]   Ohlan, A.; Singh, K.; Chandra, A.; Dhawan, S. K. Microwave Absorption Properties of Conducting Polymer Composite with Barium Ferrite Nanoparticles in 12.4-18 GHz. *Applied Physics Letters* **2008**, *93* (5), 053114. https://doi.org/10.1063/1.2969400.

[76]   Nicolson, A. M.; Ross, G. F. Measurement of the Intrinsic Properties Of Materials by Time-Domain Techniques. *IEEE Transactions on Instrumentation and Measurement* **1970**, *19* (4), 377–382. https://doi.org/10.1109/TIM.1970.4313932.

[77]   Wen, B.; Cao, M. S.; Hou, Z. L.; Song, W. L.; Zhang, L.; Lu, M. M.; Jin, H. B.; Fang, X. Y.; Wang, W. Z.; Yuan, J. Temperature Dependent Microwave Attenuation Behavior for Carbon-Nanotube/Silica Composites. *Carbon* **2013**, *65*, 124–139. https://doi.org/10.1016/j.carbon.2013.07.110.

[78]   Wen, B.; Wang, X. X.; Cao, W. Q.; Shi, H. L.; Lu, M. M.; Wang, G.; Jin, H. B.; Wang, W. Z.; Yuan, J.; Cao, M. S. Reduced Graphene Oxides: The Thinnest and Most Lightweight Materials with Highly Efficient Microwave Attenuation Performances of the Carbon World. *Nanoscale* **2014**, *6* (11), 5754–5761. https://doi.org/10.1039/c3nr06717c.

[79]   Liang, J.; Wang, Y.; Huang, Y.; Ma, Y.; Liu, Z.; Cai, J.; Zhang, C.; Gao, H.; Chen, Y. Electromagnetic Interference Shielding of Graphene/Epoxy Composites. *Carbon*. Pergamon March 1, 2009, pp 922–925. https://doi.org/10.1016/j.carbon.2008.12.038.

[80]   Wen, B.; Cao, M.; Lu, M.; Cao, W.; Shi, H.; Liu, J.; Wang, X.; Jin, H.; Fang, X.; Wang, W.; Yuan, J. Reduced Graphene Oxides: Light-Weight and High-Efficiency Electromagnetic Interference Shielding at Elevated Temperatures. *Advanced Materials* **2014**, *26* (21), 3484–3489. https://doi.org/10.1002/adma.201400108.

[81]   Singh, A. P.; Mishra, M.; Sambyal, P.; Gupta, B. K.; Singh, B. P.; Chandra, A.; Dhawan, S. K. Encapsulation of γ-Fe_2O_3 Decorated Reduced Graphene Oxide in Polyaniline Core-Shell Tubes as an Exceptional Tracker for Electromagnetic Environmental Pollution. *Journal of Materials Chemistry A* **2014**, *2* (10), 3581–3593. https://doi.org/10.1039/c3ta14212d.

[82]   Shahzad, F.; Kumar, P.; Kim, Y.-H.; Hong, S. M.; Koo, C. M. Biomass-Derived Thermally Annealed Interconnected Sulfur-Doped Graphene as a Shield against Electromagnetic Interference. *ACS Applied Materials & Interfaces* **2016**, *8* (14), 9361–9369. https://doi.org/10.1021/acsami.6b00418.

[83]   Shen, B.; Li, Y.; Yi, D.; Zhai, W.; Wei, X.; Zheng, W. Microcellular Graphene Foam for Improved Broadband Electromagnetic Interference Shielding. *Carbon* **2016**, *102*, 154–160. https://doi.org/10.1016/j.carbon.2016.02.040.

[84]   Li, X.; Zhu, H. Two-Dimensional MoS2: Properties, Preparation, and Applications. *Journal of Materiomics* **2015**, *1* (1), 33–44. https://doi.org/10.1016/j.jmat.2015.03.003.

[85]   Chhowalla, M.; Shin, H. S.; Eda, G.; Li, L. J.; Loh, K. P.; Zhang, H. The Chemistry of Two-Dimensional Layered Transition Metal Dichalcogenide Nanosheets. *Nature Chemistry*. Nat Chem April 2013, pp 263–275. https://doi.org/10.1038/nchem.1589.

[86]   Zhang, X.; Lai, Z.; Tan, C.; Zhang, H. Solution-Processed Two-Dimensional MoS2Nanosheets: Preparation, Hybridization, and Applications. *Angewandte Chemie - International Edition*. Wiley-VCH Verlag July 25, 2016, pp 8816–8838. https://doi.org/10.1002/anie.201509933.

[87]   Hirsch, A.; Hauke, F. Post-Graphene 2D Chemistry: The Emerging Field of Molybdenum Disulfide and Black Phosphorus Functionalization. *Angewandte Chemie International Edition* **2018**, *57* (16), 4338–4354. https://doi.org/10.1002/ANIE.201708211@10.1002/(ISSN)1521-3773.MERCK_350.

[88]   Ghasemi, F.; Mohajerzadeh, S. Sequential Solvent Exchange Method for Controlled Exfoliation of MoS$_2$ Suitable for Phototransistor Fabrication. *ACS Applied Materials and Interfaces* **2016**, *8* (45), 31179–31191. https://doi.org/10.1021/acsami.6b07211.

[89]   Niu, L.; Coleman, J. N.; Zhang, H.; Shin, H.; Chhowalla, M.; Zheng, Z. Production of Two-Dimensional Nanomaterials via Liquid-Based Direct Exfoliation. *Small* **2016**, *12* (3), 272–293. https://doi.org/10.1002/smll.201502207.

[90]   Wang, X.; Feng, H.; Wu, Y.; Jiao, L. Controlled Synthesis of Highly Crystalline MoS2 Flakes by Chemical Vapor Deposition. *Journal of the American Chemical Society* **2013**, *135* (14), 5304–5307. https://doi.org/10.1021/ja4013485.

[91]   Yin, X. L.; Li, L. L.; Jiang, W. J.; Zhang, Y.; Zhang, X.; Wan, L. J.; Hu, J. S. MoS2/CdS Nanosheets-on-Nanorod Heterostructure for Highly Efficient Photocatalytic H2 Generation under Visible Light Irradiation. *ACS Applied Materials and Interfaces* **2016**, *8* (24), 15258–15266. https://doi.org/10.1021/acsami.6b02687.

[92]   Xie, J.; Zhang, J.; Li, S.; Grote, F.; Zhang, X.; Zhang, H.; Wang, R.; Lei, Y.; Pan, B.; Xie, Y. Controllable Disorder Engineering in Oxygen-Incorporated MoS2 Ultrathin Nanosheets for Efficient Hydrogen Evolution. *Journal of the American Chemical Society* **2013**, *135* (47), 17881–17888. https://doi.org/10.1021/ja408329q.

[93]   Xie, A.; Sun, M.; Zhang, K.; Jiang, W.; Wu, F.; He, M. In Situ Growth of MoS2 Nanosheets on Reduced Graphene Oxide (RGO) Surfaces: Interfacial Enhancement of Absorbing Performance against Electromagnetic Pollution. *Physical Chemistry Chemical Physics* **2016**, *18* (36), 24931–24936. https://doi.org/10.1039/c6cp04600b.

[94]   Wang, Y.; Chen, D.; Yin, X.; Xu, P.; Wu, F.; He, M. Hybrid of MoS2 and Reduced Graphene Oxide: A Lightweight and Broadband Electromagnetic Wave Absorber. *ACS Applied Materials and Interfaces* **2015**, *7* (47), 26226–26234. https://doi.org/10.1021/acsami.5b08410.

[95] Guo, A.-P.; Zhang, X.-J.; Wang, S.-W.; Zhu, J.-Q.; Yang, L.; Wang, G.-S. Excellent Microwave Absorption and Electromagnetic Interference Shielding Based on Reduced Graphene Oxide@MoS$_2$/Poly(Vinylidene Fluoride) Composites. *ChemPlusChem* **2016**, *81* (12), 1305–1311. https://doi.org/10.1002/cplu.201600370.

[96] Bai, Y.; Zhong, B.; Yu, Y.; Wang, M.; Zhang, J.; Zhang, B.; Gao, K.; Liang, A.; Wang, C.; Zhang, J. Mass Fabrication and Superior Microwave Absorption Property of Multilayer Graphene/Hexagonal Boron Nitride Nanoparticle Hybrids. *npj 2D Materials and Applications* **2019**, *3* (1), 1–10. https://doi.org/10.1038/s41699-019-0115-5.

[97] Pakdel, A.; Bando, Y.; Golberg, D. Nano Boron Nitride Flatland. *Chemical Society Reviews.* Royal Society of Chemistry February 7, 2014, pp 934–959. https://doi.org/10.1039/c3cs60260e.

[98] Zhi, C.; Bando, Y.; Tang, C.; Kuwahara, H.; Golberg, D. Large-Scale Fabrication of Boron Nitride Nanosheets and Their Utilization in Polymeric Composites with Improved Thermal and Mechanical Properties. *Advanced Materials* **2009**, *21* (28), 2889–2893. https://doi.org/10.1002/adma.200900323.

[99] Li, X.; Hao, X.; Zhao, M.; Wu, Y.; Yang, J.; Tian, Y.; Qian, G. Exfoliation of Hexagonal Boron Nitride by Molten Hydroxides. *Advanced Materials* **2013**, *25* (15), 2200–2204. https://doi.org/10.1002/adma.201204031.

[100] Shi, Y.; Hamsen, C.; Jia, X.; Kim, K. K.; Reina, A.; Hofmann, M.; Hsu, A. L.; Zhang, K.; Li, H.; Juang, Z. Y.; Dresselhaus, M. S.; Li, L. J.; Kong, J. Synthesis of Few-Layer Hexagonal Boron Nitride Thin Film by Chemical Vapor Deposition. *Nano Letters* **2010**, *10* (10), 4134–4139. https://doi.org/10.1021/nl1023707.

[101] Gao, R.; Yin, L.; Wang, C.; Qi, Y.; Lun, N.; Zhang, L.; Liu, Y. X.; Kang, L.; Wang, X. High-Yield Synthesis of Boron Nitride Nanosheets with Strong Ultraviolet. *Journal of Physical Chemistry C* **2009**, *113* (34), 15160–15165. https://doi.org/10.1021/jp904246j.

[102] Yu, J.; Qin, L.; Hao, Y.; Kuang, S.; Bai, X.; Chong, Y. M.; Zhang, W.; Wang, E. Vertically Aligned Boron Nitride Nanosheets: Chemical Vapor Synthesis, Ultraviolet Light Emission, and Superhydrophobicity. *ACS Nano* **2010**, *4* (1), 414–422. https://doi.org/10.1021/nn901204c.

[103] Lian, G.; Zhang, X.; Tan, M.; Zhang, S.; Cui, D.; Wang, Q. Facile Synthesis of 3D Boron Nitride Nanoflowers Composed of Vertically Aligned Nanoflakes and Fabrication of Graphene-like BN by Exfoliation. *Journal of Materials Chemistry* **2011**, *21* (25), 9201–9207. https://doi.org/10.1039/c0jm04503a.

[104] Lv, H.; Zhang, H.; Ji, G. Development of Novel Graphene/g-C$_3$N$_4$ Composite with Broad-Frequency and Light-Weight Features. *Particle & Particle Systems Characterization* **2016**, *33* (9), 656–663. https://doi.org/10.1002/ppsc.201600065.

[105] Kang, Y.; Jiang, Z.; Ma, T.; Chu, Z.; Li, G. Hybrids of Reduced Graphene Oxide and Hexagonal Boron Nitride: Lightweight Absorbers with Tunable and Highly Efficient Microwave Attenuation Properties. *ACS Applied Materials and Interfaces* **2016**, *8* (47), 32468–32476. https://doi.org/10.1021/acsami.6b11843.

[106] Zhang, X.; Zhang, X.; Yang, M.; Yang, S.; Wu, H.; Guo, S.; Wang, Y. Ordered Multilayer Film of (Graphene Oxide/Polymer and Boron Nitride/Polymer) Nanocomposites: An Ideal EMI Shielding Material with Excellent Electrical Insulation and High Thermal Conductivity. *Composites Science and Technology* **2016**, *136*, 104–110. https://doi.org/10.1016/j.compscitech.2016.10.008.

[107] Bridgman, P. W. Two New Modifications of Phosphorus. *Journal of the American Chemical Society* **1914**, *36* (7), 1344–1363. https://doi.org/10.1021/ja02184a002.

[108] Keyes, R. W. The Electrical Properties of Black Phosphorus. *Physical Review* **1953**, *92* (3), 580–584. https://doi.org/10.1103/PhysRev.92.580.

[109] Tran, V.; Soklaski, R.; Liang, Y.; Yang, L. Layer-Controlled Band Gap and Anisotropic Excitons in Few-Layer Black Phosphorus. *Physical Review B - Condensed Matter and Materials Physics* **2014**, *89* (23), 235319. https://doi.org/10.1103/PhysRevB.89.235319.

[110] Takao, Y.; Asahina, H.; Morita, A. Electronic Structure of Black Phosphorus in Tight Binding Approach. *Journal of the Physical Society of Japan* **1981**, *50* (10), 3362–3369. https://doi.org/10.1143/JPSJ.50.3362.

[111] Liang, L.; Wang, J.; Lin, W.; Sumpter, B. G.; Meunier, V.; Pan, M. Electronic Bandgap and Edge Reconstruction in Phosphorene Materials. *Nano Letters* **2014**, *14* (11), 6400–6406. https://doi.org/10.1021/nl502892t.

[112] Cai, Y.; Zhang, G.; Zhang, Y. W. Layer-Dependent Band Alignment and Work Function of Few-Layer Phosphorene. *Scientific Reports* **2014**, *4*. https://doi.org/10.1038/srep06677.

[113] Wang, X.; Jones, A. M.; Seyler, K. L.; Tran, V.; Jia, Y.; Zhao, H.; Wang, H.; Yang, L.; Xu, X.; Xia, F. Highly Anisotropic and Robust Excitons in Monolayer Black Phosphorus. *Nature Nanotechnology* **2015**, *10* (6), 517–521. https://doi.org/10.1038/nnano.2015.71.

[114] Castellanos-Gomez, A. Black Phosphorus: Narrow Gap, Wide Applications. *Journal of Physical Chemistry Letters*. American Chemical Society October 8, 2015, pp 4280–4291. https://doi.org/10.1021/acs.jpclett.5b01686.

[115] Li, L.; Yu, Y.; Ye, G. J.; Ge, Q.; Ou, X.; Wu, H.; Feng, D.; Chen, X. H.; Zhang, Y. Black Phosphorus Field-Effect Transistors. *Nature Nanotechnology* **2014**, *9* (5), 372–377. https://doi.org/10.1038/nnano.2014.35.

[116] Hanlon, D.; Backes, C.; Doherty, E.; Cucinotta, C. S.; Berner, N. C.; Boland, C.; Lee, K.; Harvey, A.; Lynch, P.; Gholamvand, Z.; Zhang, S.; Wang, K.; Moynihan, G.; Pokle, A.; Ramasse, Q. M.; McEvoy, N.; Blau, W. J.; Wang, J.; Abellan, G.; Hauke, F.; Hirsch, A.; Sanvito, S.; O'Regan, D. D.; Duesberg, G. S.; Nicolosi, V.; Coleman, J. N. Liquid Exfoliation of Solvent-Stabilized Few-Layer Black Phosphorus for Applications beyond Electronics. *Nature Communications* **2015**, *6* (1), 8563. https://doi.org/10.1038/ncomms9563.

[117] Ling, X.; Wang, H.; Huang, S.; Xia, F.; Dresselhaus, M. S. The Renaissance of Black Phosphorus. *Proceedings of the National Academy of Sciences of the United States of America*. National Academy of Sciences April 14, 2015, pp 4523–4530. https://doi.org/10.1073/pnas.1416581112.

[118] Li, X.; Deng, B.; Wang, X.; Chen, S.; Vaisman, M.; Karato, S. I.; Pan, G.; Lee, M. L.; Cha, J.; Wang, H.; Xia, F. Synthesis of Thin-Film Black Phosphorus on a Flexible Substrate. *2D Materials* **2015**, *2* (3), 031002. https://doi.org/10.1088/2053-1583/2/3/031002.

[119] Kang, J.; Wood, J. D.; Wells, S. A.; Lee, J. H.; Liu, X.; Chen, K. S.; Hersam, M. C. Solvent Exfoliation of Electronic-Grade, Two-Dimensional Black Phosphorus. *ACS Nano* **2015**, *9* (4), 3596–3604. https://doi.org/10.1021/acsnano.5b01143.

[120] Paton, K. R.; Varrla, E.; Backes, C.; Smith, R. J.; Khan, U.; O'Neill, A.; Boland, C.; Lotya, M.; Istrate, O. M.; King, P.; Higgins, T.; Barwich, S.; May, P.; Puczkarski, P.; Ahmed, I.; Moebius, M.; Pettersson, H.; Long, E.; Coelho, J.; O'Brien, S. E.; McGuire, E. K.; Sanchez, B. M.; Duesberg, G. S.; McEvoy, N.; Pennycook, T. J.; Downing, C.; Crossley, A.; Nicolosi,

V.; Coleman, J. N. Scalable Production of Large Quantities of Defect-Free Few-Layer Graphene by Shear Exfoliation in Liquids. *Nature Materials* **2014**, *13* (6), 624–630. https://doi.org/10.1038/nmat3944.

[121] Wu, F.; Xie, A.; Sun, M.; Jiang, W.; Zhang, K. Few-Layer Black Phosphorus: A Bright Future in Electromagnetic Absorption. *Materials Letters* **2017**, *193*, 30–33. https://doi.org/10.1016/j.matlet.2017.01.052.

[122] Hao, C.; Wang, B.; Wen, F.; Mu, C.; Xiang, J.; Li, L.; Liu, Z. Superior Microwave Absorption Properties of Ultralight Reduced Graphene Oxide/Black Phosphorus Aerogel. *Nanotechnology* **2018**, *29* (23), 235604. https://doi.org/10.1088/1361-6528/aab83a.

[123] Zhou, T.; Ni, H.; Wang, Y.; Wu, C.; Zhang, H.; Zhang, J.; Tomsia, A. P.; Jiang, L.; Cheng, Q. Ultra- tough Graphene–Black Phosphorus Films. *Proceedings of the National Academy of Sciences of the United States of America* **2020**, *117* (16), 8727–8735. https://doi.org/10.1073/pnas.1916610117.

[124] Naguib, M.; Kurtoglu, M.; Presser, V.; Lu, J.; Niu, J.; Heon, M.; Hultman, L.; Gogotsi, Y.; Barsoum, M. W. Two-Dimensional Nanocrystals Produced by Exfoliation of Ti 3AlC 2. *Advanced Materials* **2011**, *23* (37), 4248–4253. https://doi.org/10.1002/adma.201102306.

[125] Gogotsi, Y.; Anasori, B. The Rise of MXenes. *ACS Nano*. American Chemical Society August 27, 2019, pp 8491–8494. https://doi.org/10.1021/acsnano.9b06394.

[126] Raagulan, K.; Kim, B. M.; Chai, K. Y. Recent Advancement of Electromagnetic Interference (EMI) Shielding of Two Dimensional (2D) MXene and Graphene Aerogel Composites. *Nanomaterials* **2020**, *10* (4), 702. https://doi.org/10.3390/nano10040702.

[127] Alhabeb, M.; Maleski, K.; Anasori, B.; Lelyukh, P.; Clark, L.; Sin, S.; Gogotsi, Y. Guidelines for Synthesis and Processing of Two-Dimensional Titanium Carbide (Ti3C2Tx MXene). *Chemistry of Materials* **2017**, *29* (18), 7633–7644. https://doi.org/10.1021/acs.chemmater.7b02847.

[128] Naguib, M.; Mochalin, V. N.; Barsoum, M. W.; Gogotsi, Y. 25th Anniversary Article: MXenes: A New Family of Two-Dimensional Materials. *Advanced Materials* **2014**, *26* (7), 992–1005. https://doi.org/10.1002/adma.201304138.

[129] Zhan, C.; Sun, W.; Xie, Y.; Jiang, D. E.; Kent, P. R. C. Computational Discovery and Design of MXenes for Energy Applications: Status, Successes, and Opportunities. *ACS Applied Materials and Interfaces*. American Chemical Society July 17, 2019, pp 24885–24905. https://doi.org/10.1021/acsami.9b00439.

[130] Li, Y.; Shao, H.; Lin, Z.; Lu, J.; Liu, L.; Duployer, B.; Persson, P. O. Å.; Eklund, P.; Hultman, L.; Li, M.; Chen, K.; Zha, X. H.; Du, S.; Rozier, P.; Chai, Z.; Raymundo-Piñero, E.; Taberna, P. L.; Simon, P.; Huang, Q. A General Lewis Acidic Etching Route for Preparing MXenes with Enhanced Electrochemical Performance in Non-Aqueous Electrolyte. *Nature Materials* **2020**, *19* (8), 894–899. https://doi.org/10.1038/s41563-020-0657-0.

[131] Iqbal, A.; Sambyal, P.; Koo, C. M. 2D MXenes for Electromagnetic Shielding: A Review. *Advanced Functional Materials* **2020**, *30* (47), 2000883. https://doi.org/10.1002/adfm.202000883.

[132] Urbankowski, P.; Anasori, B.; Makaryan, T.; Er, D.; Kota, S.; Walsh, P. L.; Zhao, M.; Shenoy, V. B.; Barsoum, M. W.; Gogotsi, Y. Synthesis of Two-Dimensional Titanium Nitride Ti4N3 (MXene). *Nanoscale* **2016**, *8* (22), 11385–11391. https://doi.org/10.1039/c6nr02253g.

[133] Li, M.; Lu, J.; Luo, K.; Li, Y.; Chang, K.; Chen, K.; Zhou, J.; Rosen, J.; Hultman, L.; Eklund, P.; Persson, P. O. Å.; Du, S.; Chai, Z.; Huang, Z.; Huang, Q. Element Replacement Approach by Reaction with Lewis Acidic Molten Salts to Synthesize Nano laminated MAX Phases and MXenes. *Journal of the American Chemical Society* **2019**, *141* (11), 4730–4737. https://doi.org/10.1021/jacs.9b00574.

[134] Hope, M. A.; Forse, A. C.; Griffith, K. J.; Lukatskaya, M. R.; Ghidiu, M.; Gogotsi, Y.; Grey, C. P. NMR Reveals the Surface Functionalisation of Ti3C2 MXene. *Physical Chemistry Chemical Physics* **2016**, *18* (7), 5099–5102. https://doi.org/10.1039/c6cp00330c.

[135] Gogotsi, Y. Chemical Vapour Deposition: Transition Metal Carbides Go 2D. *Nature Materials.* Nature Publishing Group November 1, 2015, pp 1079–1080. https://doi.org/10.1038/nmat4386.

[136] Shahzad, F.; Alhabeb, M.; Hatter, C. B.; Anasori, B.; Hong, S. M.; Koo, C. M.; Gogotsi, Y. Electromagnetic Interference Shielding with 2D Transition Metal Carbides (MXenes). *Science* **2016**, *353* (6304), 1137–1140. https://doi.org/10.1126/science.aag2421.

[137] Yang, L.; Chen, W.; Yu, Q.; Liu, B. Mass Production of Two-Dimensional Materials beyond Graphene and Their Applications. *Nano Research.* Tsinghua University Press June 21, 2020, pp 1–15. https://doi.org/10.1007/s12274-020-2897-3.

[138] Gao, L.; Li, C.; Huang, W.; Mei, S.; Lin, H.; Ou, Q.; Zhang, Y.; Guo, J.; Zhang, F.; Xu, S.; Zhang, H. MXene/Polymer Membranes: Synthesis, Properties, and Emerging Applications. *Chemistry of Materials* **2020**, *32* (5), 1703–1747. https://doi.org/10.1021/acs.chemmater.9b04408.

[139] Zhan, X.; Si, C.; Zhou, J.; Sun, Z. MXene and MXene-Based Composites: Synthesis, Properties and Environment-Related Applications. *Nanoscale Horizons.* Royal Society of Chemistry February 1, 2020, pp 235–258. https://doi.org/10.1039/c9nh00571d.

[140] Naguib, M.; Saito, T.; Lai, S.; Rager, M. S.; Aytug, T.; Parans Paranthaman, M.; Zhao, M. Q.; Gogotsi, Y. Ti3C2TX (MXene)-Polyacrylamide Nanocomposite Films. *RSC Advances* **2016**, *6* (76), 72069–72073. https://doi.org/10.1039/c6ra10384g.

[141] Liu, J.; Zhang, H. bin; Sun, R.; Liu, Y.; Liu, Z.; Zhou, A.; Yu, Z. Z. Hydrophobic, Flexible, and Lightweight MXene Foams for High-Performance Electromagnetic-Interference Shielding. *Advanced Materials* **2017**, *29* (38). https://doi.org/10.1002/adma.201702367.

[142] Bian, R.; He, G.; Zhi, W.; Xiang, S.; Wang, T.; Cai, D. Ultralight MXene-Based Aerogels with High Electromagnetic Interference Shielding Performance. *Journal of Materials Chemistry C* **2019**, *7* (3), 474–478. https://doi.org/10.1039/c8tc04795b.

[143] Zhang, Y.; Wang, L.; Zhang, J.; Song, P.; Xiao, Z.; Liang, C.; Qiu, H.; Kong, J.; Gu, J. Fabrication and Investigation on the Ultra-Thin and Flexible Ti3C2Tx/Co-Doped Polyaniline Electromagnetic Interference Shielding Composite Films. *Composites Science and Technology* **2019**, *183*, 107833. https://doi.org/10.1016/j.compscitech.2019.107833.

[144] Li, P.; Du, D.; Guo, L.; Guo, Y.; Ouyang, J. Stretchable and Conductive Polymer Films for High-Performance Electromagnetic Interference Shielding. *Journal of Materials Chemistry C* **2016**, *4* (27), 6525–6532. https://doi.org/10.1039/c6tc01619g.

[145] Liu, R.; Miao, M.; Li, Y.; Zhang, J.; Cao, S.; Feng, X. Ultrathin Biomimetic Polymeric Ti3C2Tx MXene Composite Films for Electromagnetic Interference Shielding. *ACS Applied Materials and Interfaces* **2018**, *10* (51), 44787–44795. https://doi.org/10.1021/acsami.8b18347.

[146] Bora, P. J.; Anil, A. G.; Ramamurthy, P. C.; Tan, D. Q. MXene Interlayered Crosslinked Conducting Polymer Film for Highly Specific Absorption and Electromagnetic Interference Shielding. *Materials Advances* **2020**, *1* (2), 177–183. https://doi.org/10.1039/d0ma00005a.

[147] Yang, W.; Zhao, Z.; Wu, K.; Huang, R.; Liu, T.; Jiang, H.; Chen, F.; Fu, Q. Ultrathin Flexible Reduced Graphene Oxide/Cellulose Nanofiber Composite Films with Strongly Anisotropic Thermal Conductivity and Efficient Electromagnetic Interference Shielding. *Journal of Materials Chemistry C* **2017**, *5* (15), 3748–3756. https://doi.org/10.1039/C7TC00400A.

[148] Han, M.; Yin, X.; Wu, H.; Hou, Z.; Song, C.; Li, X.; Zhang, L.; Cheng, L. Ti_3C_2 MXenes with Modified Surface for High-Performance Electromagnetic Absorption and Shielding in the X-Band. *ACS Applied Materials and Interfaces* **2016**, *8* (32), 21011–21019. https://doi.org/10.1021/acsami.6b06455.

[149] Li, X.; Yin, X.; Han, M.; Song, C.; Xu, H.; Hou, Z.; Zhang, L.; Cheng, L. Ti_3C_2 MXenes Modified with: In Situ Grown Carbon Nanotubes for Enhanced Electromagnetic Wave Absorption Properties. *Journal of Materials Chemistry C* **2017**, *5* (16), 4068–4074. https://doi.org/10.1039/c6tc05226f.

[150] Zhao, S.; Zhang, H. bin; Luo, J. Q.; Wang, Q. W.; Xu, B.; Hong, S.; Yu, Z. Z. Highly Electrically Conductive Three-Dimensional $Ti_3C_2T_x$ MXene/Reduced Graphene Oxide Hybrid Aerogels with Excellent Electromagnetic Interference Shielding Performances. *ACS Nano* **2018**, *12* (11), 11193–11202. https://doi.org/10.1021/acsnano.8b05739.

[151] Kumar, S.; Arti; Kumar, P.; Singh, N.; Verma, V. Steady Microwave Absorption Behavior of Two-Dimensional Metal Carbide MXene and Polyaniline Composite in X-Band. *Journal of Magnetism and Magnetic Materials* **2019**, *488*, 165364. https://doi.org/10.1016/j.jmmm.2019.165364.

[152] Wang, Q. W.; Zhang, H. bin; Liu, J.; Zhao, S.; Xie, X.; Liu, L.; Yang, R.; Koratkar, N.; Yu, Z. Z. Multifunctional and Water-Resistant MXene-Decorated Polyester Textiles with Outstanding Electromagnetic Interference Shielding and Joule Heating Performances. *Advanced Functional Materials* **2019**, *29* (7). https://doi.org/10.1002/adfm.201806819.

[153] Wei, H.; Dong, J.; Fang, X.; Zheng, W.; Sun, Y.; Qian, Y.; Jiang, Z.; Huang, Y. Ti3C2Tx MXene/Polyaniline (PANI) Sandwich Intercalation Structure Composites Constructed for Microwave Absorption. *Composites Science and Technology* **2019**, *169*, 52–59. https://doi.org/10.1016/j.compscitech.2018.10.016.

[154] Xu, H.; Yin, X.; Li, X.; Li, M.; Liang, S.; Zhang, L.; Cheng, L. Lightweight Ti_2CT_x MXene/Poly (Vinyl Alcohol) Composite Foams for Electromagnetic Wave Shielding with Absorption-Dominated Feature. *ACS Applied Materials and Interfaces* **2019**, *11* (10), 10198–10207. https://doi.org/10.1021/acsami.8b21671.

[155] Weng, G. M.; Li, J.; Alhabeb, M.; Karpovich, C.; Wang, H.; Lipton, J.; Maleski, K.; Kong, J.; Shaulsky, E.; Elimelech, M.; Gogotsi, Y.; Taylor, A. D. Layer-by-Layer Assembly of Cross-Functional Semi-Transparent MXene-Carbon Nanotubes Composite Films for Next-Generation Electromagnetic Interference Shielding. *Advanced Functional Materials* **2018**, *28* (44). https://doi.org/10.1002/adfm.201803360.

CHAPTER 12

Novel Radiation Shielding Concrete Utilizing Industrial Waste for Gamma-Ray Shielding

Manish Mudgal[1*] and **Er R.K. Chouhan**[1]

[1] *Centre for Advanced Radiation Shielding & Geo-polymeric Materials, CSIR-Advanced Materials and Processes Research Institute (AMPRI) Bhopal, India*

Abstract: For the first time, the capability of red mud waste has been explored for the development of advanced synthetic radiation shielding aggregate and radiation shielding concrete. Red mud, an aluminium industry waste, consists of multi-component and multi-elemental characteristics. Approximately two tons of red mud are generated for every ton of aluminium production. There are about 85 alumina plants all over the world, thus leading to the generation of about 77 million tons of highly alkaline waste annually. The major mineral content of red mud waste includes hematite, anatase, and cancrinite, thus making red mud waste the most suitable multi-component resource material for developing multi phases containing shielding aggregate. Further, these multi-elements in the red mud are present in the form of oxide, oxy-hydroxide, and hydroxides, having low as well as high atomic number elements, namely sodium, iron, titanium compounds, respectively, and are non-toxic in nature. The concrete possessing specific gravity higher than 2600 kg/m³ is known as heavyweight concrete, and aggregate with specific gravities higher than 3000 kg/m³ is called heavyweight aggregate as per TS EN 206-1 (2002). The shielding aggregate contains both naturally occurring as well as some of the artificial aggregate. The natural aggregate includes hematite, magnetite, limonite barite, *etc.*, which are non-replenishable and are useful for many other important applications, and the artificial aggregate includes the use of iron shots and steel filing and in some cases, lead shots, *etc.* The use of lead shots makes the material toxic in nature, therefore, there is a need to avoid the use of lead-based materials for shielding applications, as it ranks second in the list of hazardous materials. Apart from toxicity associated with lead, the low melting point of lead is also prohibitive as the shielding concrete should be preferably heat and fire-resistant. Further, all the natural minerals inherently contain only a single shielding phase, therefore, conventionally shielding concretes are developed by a combination of various natural minerals, which leads to an inhomogeneous radiation shielding matrix in the developed conventional radiation shielding concrete. In view of the above, there is an urgent need to develop advanced non-toxic synthetic shielding aggregate capable of providing homogeneous radiation shielding matrix

---

*Corresponding authors Manish Mudgal:** Centre for Advanced Radiation Shielding & Geo-polymeric Materials, CSIR-Advanced Materials and Processes Research Institute (AMPRI) Bhopal, India; Tel: +91 9425019217; E-mail: mmudgal1969@rediffmail.com

Sundeep K. Dhawan, Avanish Pratap Singh, Anil Ohlan, Kuldeep Singh Kakran and Pradeep Sambyal (Ed.)
All rights reserved-© 2022 Bentham Science Publishers

preferably obviating the use of toxic lead and conventional non-replenishable natural minerals resources. In this chapter, aluminium industrial waste, *i.e.*, red mud, has been utilized. Chemical formulation and mineralogical designing of the red mud has been done by ceramic processing using appropriate reducing agents and additives. The chemical analysis, SEM microphotographs, and XRD analysis confirm the presence of multi-component, multi shielding, and multi-layered phases in developed advanced synthetic radiation shielding aggregate. The maximum density of developed synthetic aggregate is found to be 4.16 g/cc. The mechanical properties, namely aggregate impact value, aggregate crushing value, and aggregate abrasion value, have been evaluated and was compared with hematite ore aggregate and found to be an excellent material useful for making advanced radiation shielding concrete for the construction of nuclear power plants and other radiation installations.

For the first time, the development and design mix of novel radiation shielding concrete using innovative red mud-based synthetic shielding aggregates have been carried out in which the heavy density shielding aggregates are developed using red mud and are basically ceramic materials consisting of shielding phases, namely barium silicate (san-bornite), barium iron titanium silicate (bafertisite), barium aluminium silicate, iron titanium oxide (pseudorutile), barium titanate, barium iron titanium oxide, barium aluminium oxide, and magnetite, which are multi-elemental, multi phases, multi-layered crystal structures, therefore, they are excellent shielding materials.

The radiation shielding concrete was made using developed synthetic shielding aggregates adopting IS 10262-2009 standard for grade designation of M-30 concrete. The reference hematite ore concrete and developed concrete tested for radiation shielding attenuation properties for gamma rays using ^{137}Cs (of photon energy 662 keV) and ^{241}Am (of photon energy 60 keV) were found to possess highly effective shielding properties. The developed novel design mix concrete achieved an attenuation factor of 5.8 as compared to 5.1 attenuation factor for reference hematite ore concrete. The developed radiation shielding concrete using red mud-based synthetic shielding aggregates possess a broad application spectrum ranging from the construction of diagnostic X-ray, CT scanner rooms, and storing radioactive waste to nuclear power plants.

Keywords: Gamma-ray shielding, Radiation shielding concrete, Red mud.

12.1. INTRODUCTION

Radiation shields are usually fabricated using materials like lead, graphite, steel, polyethylene, concrete, *etc.*, depending on the radiation type and energy. Of all the materials, concrete is considered to be the most suitable material that can be gainfully employed in fabricating shields that possess not only adequate attenuation property but also the required mechanical strength. Further, concrete can be cast into moulds of any required shape, and by varying the constituents, high densities can be achieved. Since concrete shields require almost no maintenance, they have the additional advantage of being economical too.

Concrete is the most widely used material for reactor radiation shielding due to its cost-effectiveness and satisfactory mechanical properties. It is usually a mixture of hydrogen and other light nuclei and has a high atomic number [1]. The aggregate of concrete containing many heavy elements plays an important role in improving concrete shielding properties, therefore, it has good shielding properties for the attenuation of photons and neutrons [2, 3]. The density of heavyweight concrete is based on the specific gravity of the aggregate and the properties of the other components of concrete. Concretes with specific gravities higher than 2600 kg/m^3 are called heavyweight concrete, and aggregates with specific gravities higher than 3000 kg/m^3 are called heavyweight aggregate according to TS EN 206-1 [4]. The aggregates and other components are based upon the exact application of the high-density concrete. Some of the natural minerals used as aggregates in high-density concrete are hematite, magnetite, limonite, barite and some of the artificial aggregates include materials like steel punching and iron shot. Bauxite, hydrous iron ore, or serpentine, all slightly heavier than normal weight concrete, can be used in case of high fixed water content. It is essential that heavy-weight aggregates are inert with respect to alkalis and free of oil as well as foreign coatings, which may have undesired effects on the bonding of the paste to the aggregate particles or on cement hydration. Presently, heavyweight concrete is extensively used as a shield in nuclear plants, radio therapy rooms, and for transporting as well as storing radioactive wastes. For this purpose, concrete must have high strength and density. Heavyweight and high-strength concrete can be used for shielding purposes. Such concrete with magnetite aggregates can have a density in the range of 3.2–4 t/m^3, which is significantly higher than that with normal aggregates [5, 6]. Concrete specimens prepared with magnetite, datolite-galena, magnetite-steel, limonite-steel, and serpentine were simulated. Researchers [7] used heavyweight aggregates of different minerals (limonite and siderite) in order to prepare different series for the radiation shielding of this concrete. It was reported that the concrete prepared with heavy-weight aggregates of different minerals are useful radiation absorbents. The heart of a nuclear power project is the 'Calandria,' and it is housed in a reactor concrete building typically with a double containment system, a primary (or inner) containment structure (PCS), and a secondary (or outer) containment structure (SCS). This reactor containment structure is the most significant concrete structure in a nuclear power plant [8].

12.2. INDUSTRIAL WASTE - RED MUD

India is one of the top ten mineral-producing countries around the world and is rich in mineral resources and mining. The Indian economy also depends on the value of mineral processing. It shows that the major portion of the mining raw materials in India is used for industrial activities like bauxite used for aluminium production, *etc.* Aluminium is the most usable metal after steel around the world. Day by day, the demand for aluminium is increasing due to its various applications in many sectors like automobile, aerospace, defence, *etc.* The aluminium is being produced through two process steps of bauxite processing. In the 1st step, alumina is refined through the Bayer process (developed by Carl Josef Bayer), and in the 2nd step, aluminium is produced by dissolving alumina using the Hall-Heroult process [9]. However, the processing of alumina from bauxite using 1st step involved serious environmental concern because residue generated during the Bayer process is an insoluble product called red mud (bauxite residue). It is the major waste product depending on the quality of bauxite processing. However, ~ 55 to 65 % of residue was generated during bauxite processing, *i.e.*, 1 ton of alumina produces around 1.25 to 1.50 tonnes of red mud [10]. The annual production of metallurgical alumina in 2017 was approximately ~130 million tons in the whole world, which means that ~182 million tons of red mud has been generated globally every year [11]. In India, the annual production of metallurgical alumina in 2017-18 was ~6.2 million tons, which means that ~8.6 million tons red Mud were generated every year [12], which is a huge amount. However, aluminium has been used in many applications from home to space, therefore, demand for this lightweight material will increase in the near future, in turn increasing the red mud generation every year. Around the world, the disposal of this residue (red mud) is either through dry disposal on land or slurry disposal in sea/pond. Red mud generation involves a highly alkaline reaction, so red mud is highly alkaline in nature (pH 11 to 13) [13]. However, the red mud is harmful to water, land, and the surrounding atmosphere of the disposal area. The higher cost and huge area of land are also involved with red mud disposal dams [14].

Recycling and potential utilization of this insoluble by-product (red mud) have become a major issue in the 21st century for aluminium-producing industries all over the world, especially for environmental pollution and human health concerns. However, many researchers have been working on recycling, potential utilization, and treatment of this insoluble residue (red mud) for the last two decades. According to a search result from the Web of Science based on the term 'red mud,' from 2001 to 2018, the total number of publications was 997 [15]. The Indian researchers are also more concerned about this serious environmental issue and paying greater attention to the recycling, bulk utilization, and treatment of red mud, *e.g.*, valuable element recovery, storage as red mud ponds and reclamation, construction industry, pollution control and development of new materials.

12.2.1. Red Mud Generation

The production of alumina in the world is around 130 million metric tons. India produced around 6.7 million metric tons in the year 2019 [16]. China is the leading alumina producer with a share of 55-56 %, followed by Australia 16%, Brazil 7%, and India 5 %. The alumina production in India is shown in Figure 1. The four major producers, National Aluminium Co. Ltd. (NALCO), Hindalco Industries Ltd., Bharat Aluminium Co. Ltd., and Vedanta Aluminium Ltd. (VAL), are at the forefront in alumina and aluminium production in India. However, around the world, bauxite is the primary raw material for producing alumina through two different processes known as sintering and the Bayer process [17]. The Bayer process is widely used in industry for refining bauxite as well as alumina production because this process is the most economical process for producing alumina from bauxite. In this process, raw bauxite and caustic soda are mixed together in the presence of lime to complete the pre-desilication stage (~ 100°C, ATM Pressure). Further, hot sodium hydroxide solution (NaOH) is used for processing the pre-desilication stage mixture; hence alumina present in bauxite is converted to aluminium hydroxide $Al(OH)_3$. The aluminium hydroxide $Al(OH)_3$ decomposes alumina. A large amount of alumina is dissolved to produce aluminium using the smelting Hall-Heroult process. The other elements present in bauxite are not dissolved in hot sodium hydroxide solution (NaOH); then this undissolved solution is washed and filtered for obtaining the bauxite residue [18]. This residue is called red mud. Fig. (**3**) shows the line diagram of the Bayer process. It is a strong solid waste, high alkaline in nature, and very fine-grained material. This bauxite residue is also called red mud because of its red colour due to being rich in iron (III). Approximately 1.5-2.5 tons of red mud is discharged during per ton of alumina production [19], depending upon the bauxite source and alumina generation process efficiencies.

The worldwide reserve of red mud was forecasted to reach about 4 billion tons in 2019, with a production rate of 0.18 billion tons/year [20]. The amount of red mud generation can be calculated by using a mean ratio of 1.4 to alumina production data. Hence ~182 million tonnes of red mud were generated globally in 2019. The red mud generation is expected to increase in the future; due to the initial exploitation of the highest quality bauxite reserves, the grade of bauxite ore is decreasing, resulting in an increased ratio of red mud production to bauxite [21]. In India, red mud generation is ~ 9 million tonnes per year in 2019, which is around 5% of the worldwide generation. The red mud generation is also continuously increasing in India because of the increasing per capita consumption of alumina/aluminium. The generation of red mud in India is shown in Fig. (**2**).

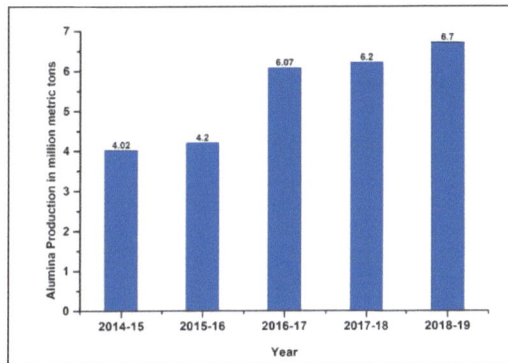

Fig. (1). Almina Production in India.

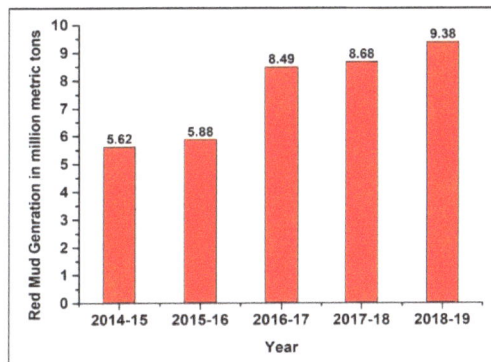

Fig. (2). Red Mud Generation in India.

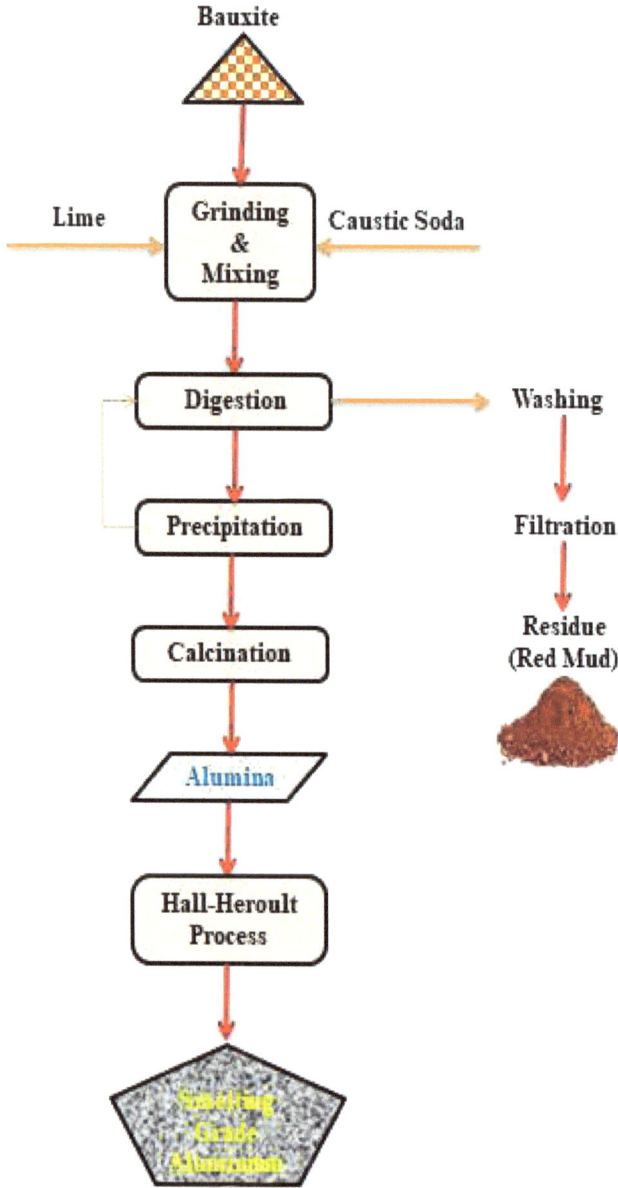

Fig. (3). Line diagram of the Bayer Process.

12.2.2. Properties and Characterization of Red Mud

The chemical compositions, physical properties, and characterization of red mud depend on the location of mines, type of bauxite, and different parameters of the production process. The main chemical constituents present in red mud are Fe_2O_3, Al_2O_3, SiO_2, TiO_2, CaO, Na_2O, as well as minor elements like K, Cr, V, Ba, Cu, Mn, Pb, Zn, P, *etc.* The worldwide variation of chemical constituents of red mud is high. The typical chemical composition of red mud is given in Table **1**. The red mud generated from the Bayer process has much higher contents of Fe_2O_3 and Al_2O_3 compared to the sintering process. However, the sintering process has higher contents of CaO and SiO_2 as compared to the Bayer process. The major valuable chemical constituents available in red mud in various countries are given in Table **2**. The typical chemical compositions of red mud produced from Indian alumina/aluminium plants are given in Table **3**.

Table 1. Typical chemical composition of red mud [22-24].

Composition	Weight %
Fe_2O_3	30-60%
Al_2O_3	10-20%
SiO_2	3-50%
TiO_2	Trace-25%
CaO	2-8%
Na_2O	2-10%

Table 2. Main chemical constituents of red mud in various countries.

Country	Plant	Chemical Constituents (wt. %)						Refs.
		Fe_2O_3	Al_2O_3	TiO_2	SiO_2	Na_2O	CaO	
Italy	Eurallumina	35.2	20	9.2	11.6	7.5	6.7	[25]
Turkey	Seydisehir	36.94	20.39	4.98	15.74	10.10	2.23	[26]
China	Guizhou	26.41	18.94	7.40	8.52	4.75	21.84	[27]
Australia	Kwinana	28.5	24.0	3.11	18.8	3.4	5.26	[28]
Brazil	Alunorte	45.6	15.1	4.29	15.6	7.5	1.16	[28]
Germany	AOSG	44.8	16.2	12.33	5.4	4.0	5.22	[28]
USA	RMC	35.5	18.4	6.31	8.5	6.1	7.73	[28]
UK	ALCAN	46.0	20.0	6.0	5.0	8.0	1.0	[29]
France	Gardanne	26.62	15.0	15.76	4.98	1.02	22.21	[29]
Vietnam	Tanrai	30.8	15.6	2.58	31.7	3.14	3.51	[30]
Russia	Uralsky	36.9	11.8	3.54	8.71	0.27	23.8	[31]
Greece	Boeotia	42.34	16.26	4.27	6.97	3.83	11.64	[32]
Spain	Alcoa	47.85	20.20	9.91	7.50	8.40	6.22	[33]

Table 3. Chemical constituents of red mud generated from Indian Alumina/Aluminium Plant.

Ref.	Company	Chemical Compositions in Wt.%						
		Fe_2O_3	Al_2O_3	TiO_2	SiO_2	Na_2O	CaO	LOI

(Table 3) cont.....

[34]	NALCO, Damanjodi	51-57	16-18	3-5	8-12	4-6	1-2.5	11-13
[35]	HINDALCO, Renukoot	34 -40	17-19	15 -16	7 - 8	5 – 6	1 – 5	8 – 10
[36]	HINDALCO, Muri	44-46	19-21	17-19	5-7	3-4	1 – 2	12-14
[37]	HINDALCO, Belgaum	44 -47	17-20	8-11	7-9	3-5	1 – 3	10-14
[37]	MALCO, Metturdam	40-26	18-22	2-4	12-16	4-5	1-3	11-15
[37]	BALCO, Korba	35-37	18-21	17-19	6-7	5-6	1-2	11-14

The red mud is generated during the alumina extraction process, so it is a very fine-grained material. The particle size distribution of Indian red mud is usually finer than 75μm for 90-95% of total particles [38]. The red mud is alkaline, thixotropic, and has a specific surface area in the range of 10-30 m^2/g with a specific gravity in the range of 2.85 to 3.34 [39], depending on the degree of bauxite grinding. A zeta potential value of red mud is - 45mV [39].

The mineralogy of red mud consists of more compounds because the composition is not uniform. The majority are iron oxides, mainly hematite (Fe_2O_3) and goethite (FeOOH), together with boehmite (AlOOH), gibbsite (Al(OH)$_3$), rutile (TiO_2), Calcite ($CaCO_3$), sodium aluminium silicate (Na(AlSiO$_4$)), dicalcium silicate (Ca_2SiO_4) and quartz (SiO_2). The detailed characterization of red mud has been analysed by using XRD, FE-SEM & EDX, FT-IR, TGA, DTG, TEM, and UV–vis-NIR spectroscopy [40-42].

12.3. CONVENTIONAL SHIELDING AGGREGATES

The shielding aggregate contains both naturally occurring as well as some of the artificial aggregate. The natural aggregate includes hematite, magnetite, limonite, and barite, *etc.*, which are non-replenishable and are useful for many other important applications, and the artificial aggregate includes the use of iron shots and steel filing and even in some cases, lead shots [3,43,44], *etc.* The use of lead shots makes the material toxic in nature, therefore, there is a need to avoid the use of lead-based materials for shielding applications, as it ranks second in the list of hazardous materials. Apart from toxicity associated with lead, the low melting point of lead is also prohibitive as the shielding concrete should be preferably heat and fire-resistant [45, 46]. Further, all the natural minerals inherently contain only a single shielding phase, therefore, conventionally, the shielding concrete is developed by a combination of various natural minerals, which leads to an inhomogeneous radiation shielding matrix in the developed conventional radiation shielding concrete. In view of the above, there is an urgent need to develop advanced non-toxic synthetic shielding aggregate, having (a) non-toxic nature, (b) being capable of providing homogeneous radiation shielding matrix and (c) fire resistance and preferably obviating the use of toxic lead and conventional non-replenishable natural minerals resources.

12.4. NOVEL DEVELOPED SYNTHETIC RADIATION SHIELDING AGGREGATES

In the present chapter, for the first time in the world, a novel process for making advanced non-toxic synthetic radiation shielding aggregate by ceramic processing of red mud has been developed utilizing industrial waste, namely red mud and barium-containing additives. The developed advanced non-toxic synthetic shielding aggregate possesses unique characteristic features like (1) non-toxicity, (2) fire resistance, and (3) containing multi-component, multi shielding, and multi-layered phases, therefore being capable of providing homogeneous radiation shielding matrix. The use of industrial waste red mud for making synthetic shielding aggregate obviates the use of toxic lead and conventional non-replenishable natural minerals resources. Red mud is an aluminium industry waste, and approximately two tons of red mud are generated for every ton of aluminium production. There is a total of about 85 alumina plants all over the world, thus leading to the generation of about 77 million tons of highly alkaline waste annually [47–51]. The major mineral content of red mud waste includes hematite, anatase, and cancrinite, thus making red mud waste the most suitable multi-component resource material for developing multi phases containing shielding aggregate. Further, these multi-elements in the red mud are present in the form of oxide, oxy-hydroxide, and hydroxides, having low as well as high atomic numbers elements, namely sodium, iron compounds, and non-toxicity in nature.

In the present study, the red mud, a waste generated during aluminium extraction from bauxite, has been converted into useful resource materials for making gamma radiation shielding materials by ceramic processing of red mud.

12.5. CHARACTERIZATION STUDIES

12.5.1. Chemical Analysis

The red mud waste from the aluminium industry and developed advanced non-toxic synthetic radiation shielding aggregates were analyzed in the laboratory, and the results are shown in Figs. (**4a** and **b**).

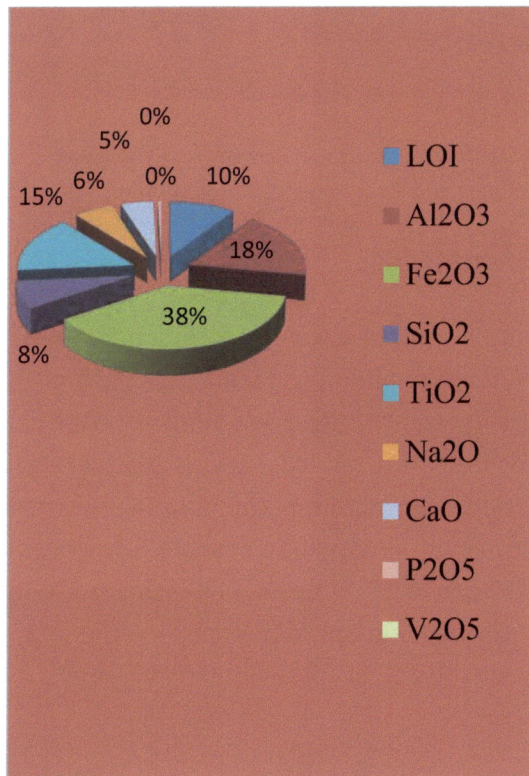

(Fig. 4) cont.....

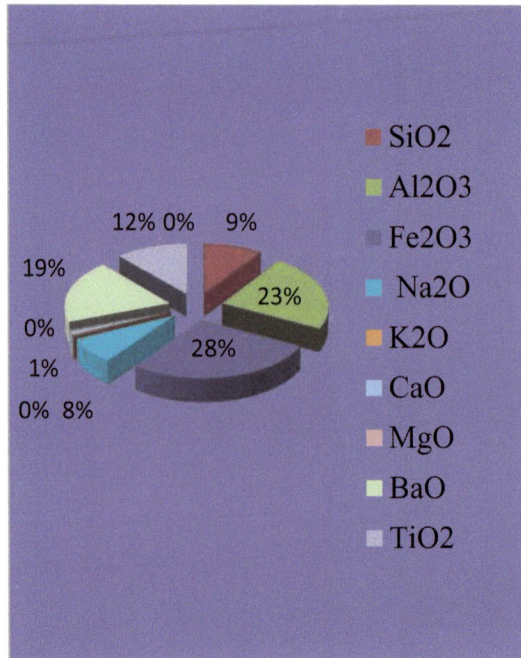

Fig. (4). (a) Chemical constituents present in red mud; 4 **(b)** Chemical constituents present in developed synthetic radiation shielding aggregates.

12.5.2. XRD Phase Identification

X-ray diffraction pattern of red mud was obtained on D-8 advanced X-ray diffractometer using Cu Kα radiation to identify the various different phases. The X-ray diffraction intensity was recorded as a function of Bragg's 2θ in the angular range of 5° - 70°. Identification of the various phases present in red mud and developed advanced synthetic radiation shielding aggregates was carried out by comparing the experimental interplanar spacing (d values) with those of the respective likely substances listed in the JCPDS standard X-ray diffraction (XRD) data files [53]. XRD of red mud and developed synthetic aggregate is shown in Figs. (**5a** and **b**), respectively.

12.5.3. X-ray Diffraction Analysis of Red Mud

X-ray diffraction pattern of red mud indicated the presence of anatase (tetragonal, 21-1272), rutile (tetragonal, JCPDS file Number 21-1276), quartz (hexagonal, JCPDS file Number 33-1161), hematite (cubic, JCPDS file Number 33-664,) boehmite (orthorhombic, JCPDS file Number 21-1307), gibbsite (JCPDS file Number 7-0324), cancrinite (hexagonal, JCPDS file Number 25- 776), chantalite (tetragonal, JCPDS file Number 29-1410) phases.

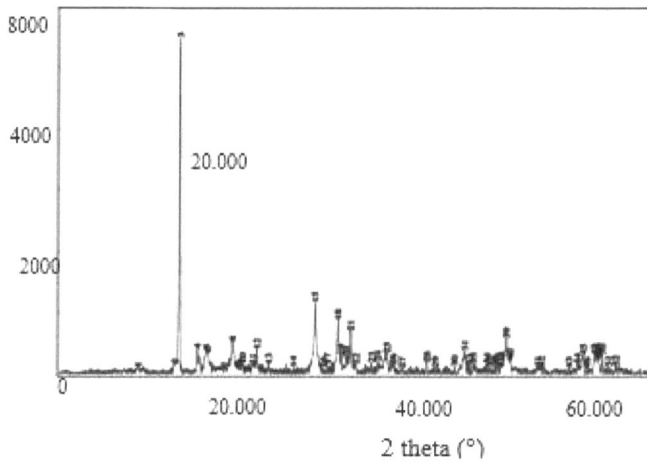

Fig. (5). (a) XRD of developed synthetic radiation shielding aggregates.

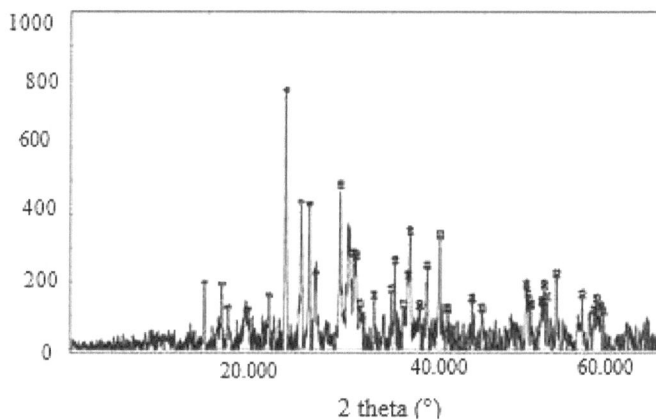

Fig. (5). (b) XRD of developed synthetic radiation shielding aggregates.

12.5.4. X-ray Diffraction Analysis of Developed Synthetic Shielding Aggregates

X-ray diffraction pattern of developed shielding aggregates contain barium silicate (sanbornite, JCPDS file Number 26-176, orthorhombic), barium aluminium silicate (JCPDS file Number 28-125, hexagonal), iron titanium oxide (pseudorutile, JCPDS file Number 13-270 and JCPDS file Number 19-635, tetragonal and hexagonal), barium titanate (JCPDS file Number 8-368, Monoclinic), barium iron titanium silicate (Bafertisite, JCPDS file Number 14-541, orthorhombic), barium iron titanium oxide (JCPDS file Number 26-1032, hexagonal), barium aluminium oxide (JCPDS file Number 28-121, tetragonal to cubic), magnetite (JCPDS file Number 19-629, cubic), iron sulphide (troilite, JCPDS file Number 11-151, hexagonal) phases.

12.5.5. Scanning Electron Microphotographs

The scanning electron micrographs of red mud and advanced non-toxic synthetic radiation shielding aggregates were examined using a JEOL model JEM-35-CF scanning electron microscope. The samples were mounted on aluminium stubs using carbon tape and then coated with a thin layer of platinum to prevent charging before the observation.

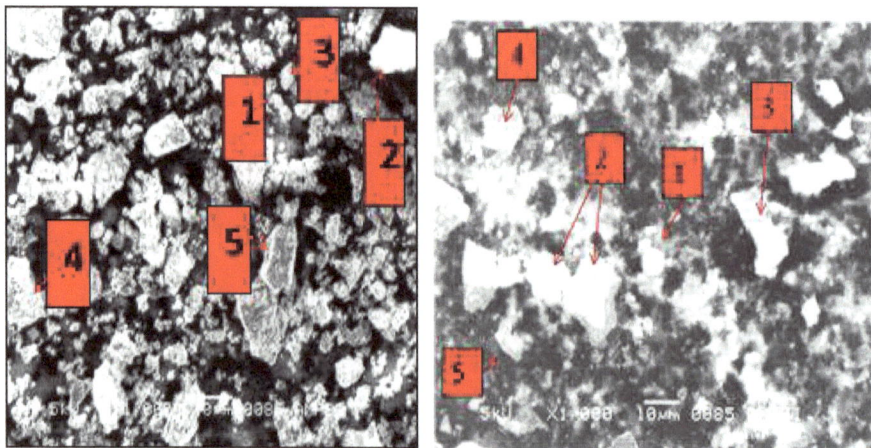

Fig. (6). (a) Scanning electron microphotographs of red mud; **(b)**: Scanning electron microphotographs of synthetic radiation shielding aggregates.

The scanning electron microphotographs exhibiting microstructure of red mud and developed advanced non-toxic synthetic radiation shielding aggregates utilizing red mud are given in Figs. (**6a** and **b**), respectively. In Fig. (**6a**), the presence of varying heavy multi-phased, multi-elemental constituents are seen in scattered morphology of red mud, and also aluminium silicates are distributed with varying phases like 1) anatase and rutile - tetragonal, 2) cancrinite - hexagonal, 3) hematite - cubic 4) boehmite – orthorhombic, which can be easily identified in the microphotograph. Further, in Fig. (**6b**) SEM micrograph of developed advanced non-toxic synthetic radiation shielding aggregates utilizing red mud confirms its multi-component, multi-phases, and multi-layered structure. Magnetite - FC cubic, barium silicate (Sanbornite), and barium iron titanium silicate - Bafertisite) - orthorhombic, iron titanium oxide (Pseudorutile) - tetragonal, barium iron titanium oxide – hexagonal and barium titanate – monoclinic, *etc.*, were some of the phases observed

12.5.6. EDXA Analysis

The EDX of advanced non-toxic synthetic radiation shielding aggregates utilizing red mud was studied for its chemical composition at one of the points in the EDX graph and shown in Fig. (**7**). The EDX pattern of the developed non-toxic radiation shielding aggregate was found to be containing carbon, oxygen, aluminum, silicon, sodium, calcium, titanium, and iron with their composition (weight %) 20.72 %, 23.98% 9.00%, 7.20%, 1.68%, 6.11%, 14.16%, and 17.14%, respectively. The strong peak around 4.5 to 5.5keV can correspond to the characteristic of barium metal. The EDX spectra revealed the predominance of multi-component, multi-elements like carbon, oxygen, sodium, aluminum, silicon, calcium, titanium and iron.

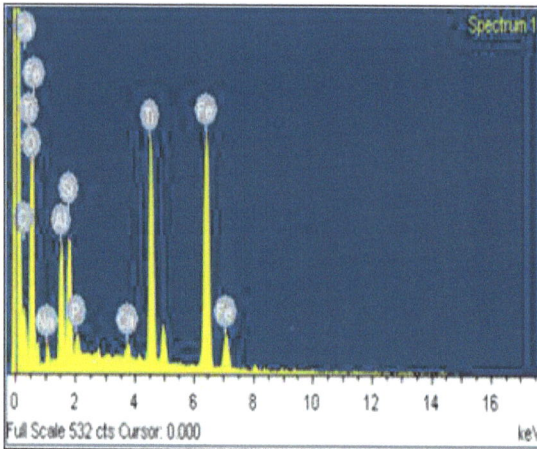

Element	Weight%	Atomic%
C K	9.47	19.07
O K	33.41	50.51
Na K	1.04	1.09
Al K	3.99	3.58
Si K	3.99	3.43
P K	0.85	0.67
Ca K	0.63	0.38
Ti K	14.88	7.52
Fe K	31.75	13.75
Totals	100.00	

Fig. (7). EDXA outcomes of the selected area and (b) elemental compositions (weight%).

12.5.7. Engineering Properties of Advanced non-toxic Synthetic Radiation Shielding Aggregates

The engineering properties, namely aggregate impact value, aggregate crushing value, and aggregate abrasion value, have been evaluated, and the results are given in Tables **4-6,** respectively.

Table 4. Aggregate Impact Value as per IS:2386 Part IV 1963 (Reaffirmed:1997).

S. No.	Type of Aggregate	Impact Value	Requirement as per IS 383:1970 (Reaffirmed:1997)
1	Conventional Basaltic Aggregates	24.28%	a) For aggregates to be used in concrete for wearing surfaces, the impact value shall not exceed 30% by weight.
2	Hematite Ore Aggregates	17.14%	

(Table 4) cont.....

3	Advanced Synthetic Shielding Aggregates	10.42%	b) For aggregates to be used in concrete for other than wearing surfaces, the impact value shall not exceed 45% by weight.

Table 5: Aggregate crushing value as per IS: 2386 Part IV 1963 (Reaffirmed: 1997).

S. No.	Type of Aggregate	Crushing Value	Requirement as per IS 383:1970 (Reaffirmed:1997)
1	Conventional Basaltic Aggregates	17.4%	a) For aggregates to be used in concrete for wearing surfaces, the crushing value shall not exceed 30% by weight.
2	Hematite Ore Aggregates	23.6%	
3	Advanced Synthetic Shielding Aggregates	16.9%	b) For Aggregates to be used in concrete for other than wearing surfaces, the crushing value shall not exceed 45% by weight.

Table 6. Aggregate abrasion value as per IS: 2386 Part IV 1963 (Reaffirmed: 1997).

S. No.	Type of Aggregate	Abrasion Value	Requirement as per IS 383:1970 (Reaffirmed: 1997)
1	Conventional Basaltic Aggregates	22.9%	a)For Aggregates to be used in concrete for wearing surfaces, the abrasion value shall not exceed 30% by weight.
2	Hematite Ore Aggregates	19.8%	b)For aggregates to be used in concrete for other than wearing surfaces, the abrasion value shall not exceed 50% by weight.
3	Advanced Synthetic Shielding Aggregates)	19.3%	

12.6. INNOVATIVE RADIATION SHIELDING CONCRETE

Conventionally, shielding concrete is based on merely physical mixtures of iron metal shots, hematite ores, and cement, *etc.* Further, the large variations in the densities of these constituents require very special care in obtaining a homogeneous shielding matrix.

Advanced chemically formulated and mineralogically designed multi-component-multi phases containing synthetic high-density shielding aggregates were developed by ceramic processing of aluminium industry waste, *i.e.*, red mud.

The ceramic processing of the red mud with appropriate additives enables the formation of varieties of ceramic phases with multi-elemental compositions and multi-layered crystal structures, namely, barium aluminates, called celsian and silicates of barium, iron, titanium, namely bafertisite, possessing a hexagonal with a trigonal bipyramidal crystal structure. The red mud waste inherently contains varieties of elements, namely iron, titanium, aluminum, silicon, calcium, magnesium, and sodium, *etc.*, thus making red mud waste the most suitable multi-component resource material for developing multi phases containing shielding aggregates. Further, these multi-elements in the red mud are present in the form of oxide, oxy-hydroxide, and hydroxides, having low as well as high atomic numbers elements, namely sodium and iron compounds, ,and non-toxicity in nature.

The chemical analysis, SEM microphotographs, and XRD analysis of developed synthetic shielding aggregates confirm the presence of multi-component, multi shielding, and multi-layered phases in developed advanced synthetic radiation shielding aggregate. The maximum density of developed synthetic aggregate is found to be 4.16 g/cc. The mechanical properties, namely aggregate impact value, aggregate crushing value, and aggregate abrasion value, have been evaluated and was compared with hematite ore aggregate and found to be superior; hence it is an excellent material useful for making advanced radiation shielding concrete for the construction of nuclear power plants and other radiation installations.

For the first time, the synthetic shielding aggregates for radiation shielding concrete were developed adopting IS 10262-2009 standards, as shown in Fig. **(8)**, which is used for grade designation of M-30 concrete. The reference hematite ore concrete and developed concrete tested for radiation shielding attenuation properties for gamma rays using ^{137}Cs (of photon energy 662 keV) and ^{241}Am (of photon energy 60 keV) were found to possess highly effective shielding properties. The developed novel design mix concrete achieved an attenuation factor of 5.8 as compared to 5.1 attenuation factor for reference hematite ore concrete. The developed radiation shielding concrete using red mud-based synthetic shielding aggregates possess a broad application spectrum ranging from the construction of diagnostic X-ray, CT scanner rooms, and storing radioactive waste to nuclear power plants [53, 54].

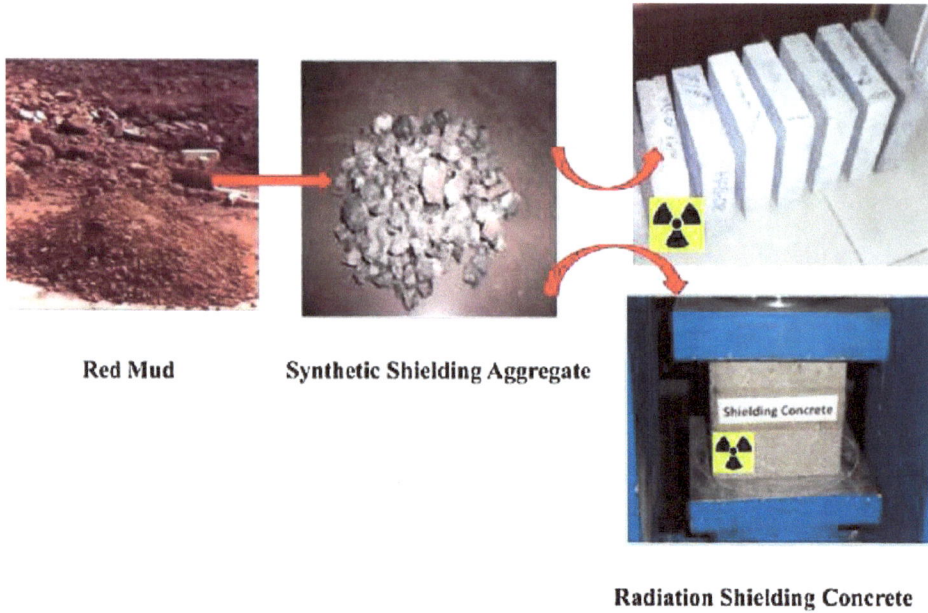

Red Mud   Synthetic Shielding Aggregate

Radiation Shielding Concrete

Fig. (8). Radiation shielding concrete made from red mud.

12.7. GAMMA ATTENUATION CHARACTERISTICS

The concrete mixes were tested for gamma attenuation factors using two varying sources: (a) Cesium 137 at energy 662 keV and (b) Americium 241 at energy 60 keV. The test results are depicted in Table **7** and Fig.s **9a, b**. Further, the trial runs for a design mix of heavy cement concrete using developed innovative advanced synthetic aggregates were carried out. Each design mix cubes of 15 cm × 15 cm × 15 cm were tested for compressive strength after 3, 14, and 28 days of curing, and the results are reported in Table **8**. Further, the heavy concrete slabs of size 30 cm × 30 cm × 7.2 cm were developed for each design mix proportion. The developed concrete mix slabs were taken to Radiological Physics and Advisory Division (RPAD), BARC, Mumbai, and tested for the attenuation factor of ^{137}Cs, and the results are given in Table **8**.

Table 7. Radiation shielding concrete slab samples.

Sr No.	Sample details	X-ray Attenuation Factor (using 300 kVP)	Gamma Attenuation Factor (using Cs-137 source) at energy 662 keV	Gamma Attenuation Factor (using Cs-137 source) at energy 662 keV
1.	AMPRI-BCC-1	45.8	3.2	169.3
2.	AMPRI-OSC-1	79.8	4.7	1646.9
3.	AMPRI-OSC-2	62.6	3.9	1167.9
4.	AMPRI-HSC-1	87.0	5.1	1454.6

Table 8. Results of compressive strength using standard 150 mm x 150 mm x 150 mm cubes after different days.

Sr No.	Sample Details	3-Days Compressive strength (MPa)	14 Days Compressive Strength (MPa)	28-Days Compressive Strength (MPa)	Density (Kg/m^3)
1.	AMPRI-HAS (01)	19.2	24.5	35.0	3242
2	AMPRI-HAS (02)	21.7	25.6	35.2	3083
3.	AMPRI-HAS (03)	18.2	30.7	33.8	3185
4.	AMPRI-HAS (04)	21.2	27.8	30.5	3151
5.	AMPRI-HAS (05)	23.2	33.3	35.3	3199

(Table 8) cont.....

6.	AMPRI-HAS (06)	25.4	32.8	39.0	3395
7.	AMPRI-HAS (07)	25.6	30.2	38.5	3586
8.	AMPRI-HAS (08)	21.5	27.4	32.5	3484

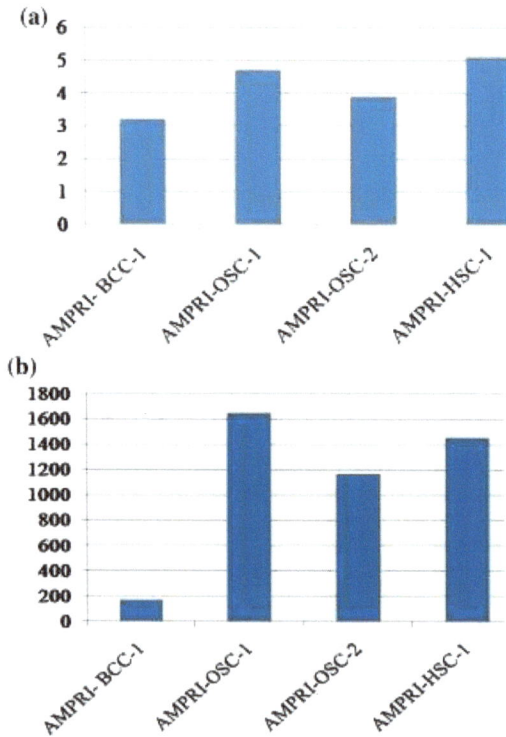

Fig. (9). (a) Gamma attenuation factor using Cs-137 source at energy 662 keV for developed design mix concrete of varying sources. **(b)** Gamma attenuation factor using Americium 241 source at energy 60 keV for developed design mix concrete of varying sources (Reproduced from J. Inorg. Organomet. Poly (2017)).

CONCLUSION

This chapter reports a novel method of fabricating radiation shielding concrete using an industrial waste from the alumina industry, namely red mud, which can be used in designing shielding gamma radiation. This novel radiation shielding material has been successfully tested for radiation shielding attenuation using ^{137}Cs, ^{241}Am, and 300 kVp X-rays. The developed design mix will overcome the challenge posed by using conventional radiation shielding material, which uses toxic materials like Pb containing compositions. This novel material achieved target parameters such as compressive strength of M-30.

CONSENT FOR PUBLICATION

Not applicable.

CONFLICT OF INTEREST

The authors declare no conflict of interest, financial or otherwise.

ACKNOWLEDGEMENTS

The authors of this chapter gratefully acknowledge the guidance and support of Director CSIR-AMPRI, Bhopal, for providing necessary institutional facilities and encouragement for R&D work.

The authors are thankful to BRNS, NPCIL & BARC Mumbai, and DST, New Delhi, for their support in radiation attenuation studies for radiation shielding and radiation protection.

REFERENCES

[1]   Ikraiam, F.A.; Abd El-Latif, A. A. Abd ELAzziz, J.M. Ali, Effect of steel fiber addition on mechanical properties and Gamma-ray attenuation for ordinary concrete used in El-Gabal El-Akhdar area in Libya for radiation shielding purposes. *Arab J. Nucl. Sci. Appl.,* **2009**, *42*, 287-295.

[2]   El-Sayed, A. Calculation of the cross-sections for fast neutrons and gamma-rays in concrete shields. *Ann. Nucl. Energy,* **2002**, *29*(16), 1977-1988.
       http://dx.doi.org/10.1016/S0306-4549(02)00019-1

[3]   Akkurt, I.; Basyig̃it, C.; Akkas, A.; Kilinc̨arsian, S.; Mavi, B.; Gu̇nog̃lu, K. Determination of some heavyweight aggregate half value layer thickness used for radiation shielding. *Acta Phys. Pol. A,* **2012**, *121*(1), 138-140.
       http://dx.doi.org/10.12693/APhysPolA.121.138

[4]   TS EN 206–1. *Concrete-Part 1: Specification, Performance, Production and conformity TSE.,* **2002**,

[5]   Gencel, O.; Bozkurt, A.; Kam, E.; Korkut, T. Determination and calculation of gamma and neutron shielding characteristics of concretes containing different hematite proportions. *Ann. Nucl. Energy,* **2011**, *38*(12), 2719-2723.
        http://dx.doi.org/10.1016/j.anucene.2011.08.010

[6]   Gencel, O.; Koksal, F.; Ozel, C.; Brostow, W. Combined effect of fly ash and waste ferrochromium on properties of concrete. *Constr. Build. Mater.,* **2012**, *29*, 633-640.
        http://dx.doi.org/10.1016/j.conbuildmat.2011.11.026

[7]   Basyig˘it, C.; Uysal, V.; Kilinc¸arsian, S.; Mavi, B.; Gu¨nog˘lu, K.; Akkurt, I.; Akkas, A. Investigating radiation shielding properties of different mineral origin heavyweight concretes. *AIP Conf. Proc.,* **2011**, *1400*(1), 232-235.
        http://dx.doi.org/10.1063/1.3663119

[8]   } Ahmed S. Ouda, Development of high-performance heavy density concrete using different aggregates for gamma-ray shielding. *Housing and Building National Research Center HBRC Journal,* **2015**, *11*, 328-338.

[9]   Habashi, F. *A Hundred Years of the Bayer Process for Alumina Production. Essential Readings in Light Metals.,* **2016**, , 85-93.
        http://dx.doi.org/10.1007/978-3-319-48176-0_12

[10]   Zhang, R.; Zheng, S.; Ma, S.; Zhang, Y. Recovery of alumina and alkali in Bayer red mud by the formation of andradite-grossular hydrogarnet in hydrothermal process. *J. Hazard. Mater.,* **2011**, *189*(3), 827-835.
        http://dx.doi.org/10.1016/j.jhazmat.2011.03.004 PMID: 21444152

[11]   survey USg. *Mineral commodity summaries U.S. geological survey,* **2018**.

[12]   Part-II:Metals and Alloys. Indian Minerals Yearbook 2018 INDIAN BUREAU OF MINES. 15 ed. Indira Bhavan, Civil Lines. *NAGPUR: GOVERNMENT OF INDIA,* **2019**, (November), 1-15.

[13]   Liu, Y.; Lin, C.; Wu, Y. Characterization of red mud derived from a combined Bayer Process and bauxite calcination method. *J. Hazard. Mater.,* **2007**, *146*(1-2), 255-261.
        http://dx.doi.org/10.1016/j.jhazmat.2006.12.015 PMID: 17208370

[14]   Mahadevan, H.; Chandwani, H.; Prasad, P. *An appraisal of the methods for red mud disposal under Indian conditions. Proc of Nat Sem on Bauxites and Alumina (BAUXAL-96).,* **1996**, , 337-347.

[15]   Alam, M.; Zanganeh, J.; Moghtaderi, B. *The composition, recycling and utilisation of Bayer red mud,* **2018**.

[16]   *Mineral commodity summaries,* Report. Reston, VA. **2020**.

[17]   Barbosa, F.M.; Bergerman, M.G.; Horta, D.G. M.G. Bergerman, D.G. Horta, Removal of iron-bearing minerals from gibbsitic bauxite by direct froth flotation. *Tecnol. Metal. Mater. Min.,* **2016**, *13*(1), 106-112.
        http://dx.doi.org/10.4322/2176-1523.0924

[18]   Klauber, C.; Gräfe, M.; Power, G. Bauxite residue issues: II. options for residue utilization. *Hydrometallurgy,* **2011**, *108*(1), 11-32.

http://dx.doi.org/10.1016/j.hydromet.2011.02.007

[19]   Paramguru, R.K.; Rath, P.C.; Misra, V.N. Trends in Red Mud Utilization – A Review. *Miner. Process. Extr. Metall. Rev.,* **2004**, *26*(1), 1-29.
http://dx.doi.org/10.1080/08827500490477603

[20]   Wang, L.; Hu, G.; Lyu, F.; Yue, T.; Tang, H.; Han, H.; Yang, Y.; Liu, R.; Sun, W. Application of red mud in wastewater treatment. *Minerals (Basel),* **2019**, *9*(5), 281.
http://dx.doi.org/10.3390/min9050281

[21]   Hua, Y.; Heal, K.V.; Friesl-Hanl, W. The use of red mud as an immobiliser for metal/metalloid-contaminated soil: A review. *J. Hazard. Mater.,* **2017**, *325*, 17-30.
http://dx.doi.org/10.1016/j.jhazmat.2016.11.073 PMID: 27914288

[22]   Sutar, H.; Mishra, S.C.; Sahoo, S.K.; Maharana, H.S. *Progress of red mud utilization: An overview.,* **2014**, , 255-279.

[23]   Khairul, M.; Zanganeh, J.; Moghtaderi, B. The composition, recycling and utilisation of Bayer red mud. *Resour. Conserv. Recycling,* **2019**, *141*, 483-498.
http://dx.doi.org/10.1016/j.resconrec.2018.11.006

[24]   Sglavo, V.M.; Campostrini, R.; Maurina, S.; Carturan, G.; Monagheddu, M.; Budroni, G.; Cocco, G. Bauxite 'red mud'in the ceramic industry. Part 1: thermal behaviour. *J. Eur. Ceram. Soc.,* **2000**, *20*(3), 235-244.
http://dx.doi.org/10.1016/S0955-2219(99)00088-6

[25]   Altundoğan, H.S.; Altundoğan, S.; Tümen, F.; Bildik, M. Arsenic adsorption from aqueous solutions by activated red mud. *Waste Manag.,* **2002**, *22*(3), 357-363.
http://dx.doi.org/10.1016/S0956-053X(01)00041-1 PMID: 11952183

[26]   Wang, P.; Liu, D-Y. Physical and chemical properties of sintering red mud and bayer red mud and the implications for beneficial utilization. *Materials (Basel),* **2012**, *5*(10), 1800-1810.
http://dx.doi.org/10.3390/ma5101800

[27]   Snars, K.; Gilkes, R. Evaluation of bauxite residues (red muds) of different origins for environmental applications. *Appl. Clay Sci.,* **2009**, *46*(1), 13-20.
http://dx.doi.org/10.1016/j.clay.2009.06.014

[28]   Srikanth, S.; Ray, A.K.; Bandopadhyay, A.; Ravikumar, B.; Jha, A. Phase constitution during sintering of red mud and red mud–fly ash mixtures. *J. Am. Ceram. Soc.,* **2005**, *88*(9), 2396-2401.
http://dx.doi.org/10.1111/j.1551-2916.2005.00471.x

[29]   Pera, J.; Boumaza, R.; Ambroise, J. Development of a pozzolanic pigment from red mud. *Cement Concr. Res.,* **1997**, *27*(10), 1513-1522.
http://dx.doi.org/10.1016/S0008-8846(97)00162-2

[30]   Hai, L.D.; Khai, N.M.; Quy, T.V.; Huan, N.X. *Material composition and properties of red mud coming from alumina processing plant Tanrai.,* **2014**, , 2311-2484.

[31]   Valeev, D.; Zinoveev, D.; Kondratiev, A.; Lubyanoi, D.; Pankratov, D. Reductive smelting of neutralized red mud for iron recovery and produced pig iron for heat-resistant castings. *Metals (Basel),* **2020**, *10*(1), 32.

http://dx.doi.org/10.3390/met10010032

[32]   Alkan, G.; Schier, C.; Gronen, L.; Stopic, S.; Friedrich, B. A mineralogical assessment on residues after acidic leaching of bauxite residue (red mud) for titanium recovery. *Metals (Basel),* **2017**, *7*(11), 458.
       http://dx.doi.org/10.3390/met7110458

[33]   Pascual, J.; Corpas, F.; López-Beceiro, J.; Benítez-Guerrero, M.; Artiaga, R. Thermal characterization of a Spanish red mud. *J. Therm. Anal. Calorim.,* **2009**, *96*(2), 407-412.
       http://dx.doi.org/10.1007/s10973-008-9230-9

[34]   Meher, S. Thermal analysis of nalco red mud. *Int. J. Chem. Stud.,* **2014**, *1*(5), 1-9.

[35]   Verma, S.; Amritphale, S.S.; Das, S. Synthesis and characterization of advanced red mud and MWCNTs based EMI shielding material via ceramic processing. *Mater. Sci. Appl.,* **2016**, *7*(4), 192-201.
       http://dx.doi.org/10.4236/msa.2016.74019

[36]   Rai, S.; Wasewar, K.; Mukhopadhyay, J.; Yoo, C.K.; Uslu, H. Neutralization and utilization of red mud for its better waste management. *WORLD,* **2012**, *6*, 5410.

[37]   Reddy, N.G.; Rao, B.H. Characterization of settled particles of the red mud waste exposed to different aqueous environmental conditions. *Indian Geotechnical Journal.,* **2018**, *48*(3), 405-419.
       http://dx.doi.org/10.1007/s40098-018-0300-z

[38]   Sundaram, R.; Gupta, S. Constructing foundations on red mud. 6th International Congress on environmental Geotechnics; p. 1172-5

[39]   Rao, B.H.; Reddy, N.G. *Zeta potential and particle size characteristics of red mud waste. Geoenvironmental Practices and Sustainability.,* **2017**, 69-89.

[40]   Nath, H.; Sahoo, A. A study on the characterization of red mud. *Int. J. Appl. Bioeng.,* **2014**, *8*(1), 8.
       http://dx.doi.org/10.18000/ijabeg.10118

[41]   Antunes, M.L.P.; Couperthwaite, S.J.; da Conceição, F.T.; Costa de Jesus, C.P.; Kiyohara, P.K.; Coelho, A.C.V.; Frost, R.L. Red Mud from Brazil: Thermal Behavior and Physical Properties. *Ind. Eng. Chem. Res.,* **2012**, *51*(2), 775-779.
       http://dx.doi.org/10.1021/ie201700k

[42]   Palmer, S.J.; Reddy, B.J.; Frost, R.L. Characterisation of red mud by UV-vis-NIR spectroscopy. *Spectrochim. Acta A Mol. Biomol. Spectrosc.,* **2009**, *71*(5), 1814-1818.
       http://dx.doi.org/10.1016/j.saa.2008.06.038 PMID: 18693065

[43]   Gencel, O.; Brostow, W.; Özel, C.; Filiz, M. Concretes containing hematite for use as shielding barriers materials science. *Materials Science (Medžiagotyra),* **2010**, *16*, 8. ISSN 1392–1320

[44]   Akkurt, I.; Akyildirim, H.; Mavi, B.; Kilinçarsian, S.; Basyigit, C. Photon attenuation coefficients of concrete includes barite in different rate. *Ann. Nucl. Energy,* **2010**, *37*(7), 910-914.
       http://dx.doi.org/10.1016/j.anucene.2010.04.001

[45]   H. Shen, B. Dowton, E. Mainegra-Hing, Radiation attenuation by lead and non-lead materials used in radiation shielding garments. *Med. Phys.,* **2007**, *3*, 530.

[46]   Kaplan, M.F. Concrete radiation shielding. England (1989), Longman Scientific and Technical.

[47]   Thakur, R.S.; Das, S.N. International series on environment. *Red mud –analysis and utilization.,* **1994,**

[48]   Ayres, R.U.; John, H.; Bojorn, A. Utilization of the wastes in the new millennium. *MRSI Bull.,* **2010,** *7,* 477.

[49]   Mann Singh, K.; Kaur, B.; Singh Sidhu, G. Investigations of some building materials for g-rays shielding effectiveness. *Radiat. Phys. Chem.,* **2013,** *87,* 16-25.

http://dx.doi.org/10.1016/j.radphyschem.2013.02.012

[50]   Amritphale, S.S.; Anshul, A.; Chandra, N.; Ramakrishnan, N. A novel process for making radiopaque materials using bauxite red mud. *J. Eur. Ceram. Soc.,* **2007,** *27*(4), 1945-1951.

http://dx.doi.org/10.1016/j.jeurceramsoc.2006.05.106

[51]   Amritphale, S.S.; Navin, C.; Narayanrao, R. Low temperature process for making radiopac materials utilizing industrial/agricultural waste as raw material US 7524452 B2 **2009.**

[52]   Powder Diffraction File, Alphabetical Index Inorganic Phases 1984) Published By JCPDS International Centre for Diffraction Data 1601, Park Lane Swarthmore, Pennsylvania 19081 USA. **1984.**

[53]   Mudgal, Manish; Chouhan, Ramesh Kumar; Verma, Sarika; Amritphale, Sudhir Sitaram; Das, Satyabrata; Shrivastva, Arvind Development of advanced, non-toxic, synthetic radiation shielding aggregate, Radiochim. *Acta,* **2017.**Doi.org/10.1515/ract-2016-2715

[54]   Chauhan, R.K. Manish Mudgal, Sarika Verma, S. S. Amritphale, Satyabrata Das. *J. Inorg. Organomet. Polym.,*

http://dx.doi.org/10.1007/s10904-017-0531-y

SUBJECT INDEX

A

Absorbers, noise 39
Absorbing coatings 51
Absorption 31, 35, 46, 60, 62, 74, 76, 82, 87, 90, 92, 93, 112, 156, 157, 187, 255, 259, 272, 273, 290, 305, 484
  capabilities, effective 305
  radiation 272, 273
Acid 75, 136, 137, 138, 173, 190, 215, 218, 219, 220, 282, 287, 358, 357, 360, 372, 438, 441, 443, 446, 479, 482, 483, 493, 505
  acetic 483
  ascorbic 482
  caffeic 479
  chloroauric 218
  fatty 220
  hydriodic 483
  hydrofluoric 441, 443, 446, 505
  hydroiodic 482
  inorganic 372
  molten boric 493
  nitric 136, 282
  para toluene sulphonic (PTSA) 219
  phosphoric 482
  polystyrene sulfonic 215
  protonic 357, 372
Acrylonitrile butadiene styrene (ABS) 79, 116, 134
Aerogel(s) 76, 86, 88, 173, 439, 440, 448, 453, 455, 456, 458, 459
  flexible 456, 458
  hybrid 86, 88, 173, 455, 458, 459
  porous framework, stable 456
  ultralight 459
  uniform lamellar-structured 456
Aerogel-like 69, 505
  carbon (ALC) 69
  sugarcanes (ALSs) 69, 505
Aerospace industries 137

Agglomerated graphene-based nanosheets 285
Agglomeration 74, 122, 287
Aggregates 185, 373, 528, 529, 543, 544, 545
  artificial 529
  heavy density shielding 528
Aggregation 138, 166, 173, 180, 190
Aircraft 2, 84, 181, 193, 274, 362
  efficient 193
  technology 362
Alumina production 531
Aluminium 317, 528, 530, 531, 538, 542
  extraction 538
  industry 538
  production 530, 531
  silicates 542
Aluminum 2, 63, 64, 36, 60, 71, 156, 542, 545
  industry 2
  nitrate 63, 64
Amalgamation 338
Ammonium 63, 64, 413, 443, 505
  bicarbonate 63, 64
  bifluoride 505
  fluoride 443
  hydrogen bifluoride 443
  peroxydisulfate 413
Amperometric sensor 217
Analysis 213, 223, 227, 241, 296, 414
  elemental composition 241
  morphological 296, 414
  thermo gravimetric 213, 223, 227
Aniline 374, 376
  monomer, semi-polar 374
  polymerization 376
Anionic sulfate radicals 220
Anisotropic 498
  properties 498
Anisotropy 29, 33, 40, 44, 252, 253, 280, 291, 292
  magneto-crystalline 40, 252, 253
  energy 44, 280, 291, 292
Anticorrosion properties 37
Applications 41, 367, 368, 371, 397, 498, 502

Sundeep K. Dhawan, Avanish Pratap Singh, Anil Ohlan, Kuldeep Singh Kakran and Pradeep Sambyal (Eds.)
All rights reserved-© 2022 Bentham Science Publishers

electromagnetic 502
electronic 371
techno-commercial 41, 367, 368, 397
thermoelectric 498
Applied energy 188
Arc discharging methods 282
Argon atmosphere 66, 67, 68
Aromatic 384, 382, 412, 419
   amine 382, 419
   thiophene rings 225, 240
Assembly, electronic 18
ATR accessory 235
Autoclave 65, 117, 493, 495
  high-temperature high-pressure 65
  lined stainless-steel 493
Automobile applications 117

B

Bands, disorder-induced 288, 298
Bandwidth 491, 497
  effective absorption 497
  effective microwave absorption 491
Battery management system (BMS) 469
Behavior, anisotropic 457
Bidirectional freeze-drying methods 456
Bi-directional freeze techniques 440
Buckled melamine-formaldehyde (BMF) 452

C

Carbonaceous sugarcane monoliths 69
Carbon 46, 60, 63, 65, 66, 67, 90, 109, 110,
     111, 116, 117, 120, 123, 124, 125, 126,
     129, 130, 132, 133, 134, 141, 142, 143,
     144, 132, 163, 175, 178, 182, 193, 216,
     368, 458, 466, 480
  based nanostructures (CBNS) 109, 110,
     111, 116, 117, 120, 123, 124, 125, 126,
     129, 130, 132, 143, 144
  black (CB) 60, 117, 132, 163, 178, 193,
     216, 368, 466
  dioxide foaming process 90

nanofiber(s) (CNF) 46, 111, 117, 123, 132,
     133, 134, 141, 142, 182, 193, 458
nanostructures-based polymer composites
     175
  precursor 63, 65, 66, 67, 480
  sources 67, 480
Carbon fibers 38, 45, 155, 163, 367, 372, 374,
     411, 412, 413, 423, 431, 474
  nickel-coated 38, 474
Carbon foam 63, 67, 70, 75, 86, 87, 94
  coated 94
  display 75
  fabricated 87
  heat-treated pristine 75
  nickel nanoparticles 86
  synthesis of 63, 70
  synthesized 67
Carbonization 63, 64, 65, 66, 67, 74, 82, 86,
     87, 458
  sintering 66
Carbonization process 69
  hydrothermal 69
Carbon nanotube(s) (CNTs) 37, 45, 46, 48, 59,
     60, 72, 73, 117, 118, 119, 122, 134, 163,
     172, 173, 186, 187, 216, 234, 368
  charged 186
  concentric 234
Carrier mobility 138, 482
Casting 179, 378
  film process 179
  techniques 378
Catalysis 442, 498, 513
  mercury 498
Catalytic 61, 117, 119, 120, 481
  chemical vapour deposition (CCVD) 117,
     119, 120
  decomposition 119, 481
  processes 61
Chemical conversion process 291
Chemical vapor deposition 75, 165, 170, 171,
     172, 173, 178, 186, 477, 483, 489, 506,
     513
  method 165
  technique 75, 186
Chemical vapour transport (CVT) 498

Chitosan 76, 511
Coal thermal power plants 315
Coatings 116, 132, 178, 192, 194, 215, 230,
    232, 315, 317, 324, 361, 366, 474
  anticorrosion 361
  antistatic 192, 194, 215
  industrial 317
  resistant 178
Coaxial transmission line technique 21
Colloidal 166, 169, 219
    dispersions 169, 219
    graphene suspensions 166
Commercial applications 46, 72, 134, 137,
    193, 239, 302, 339, 343, 484
Communication devices 60, 465, 466, 469
  high-speed wireless 465, 469
Communication systems 112, 217, 356
Community 47, 117
  nanoscience 117
Composite foams 59, 60, 72, 73, 75, 76, 77,
    79, 80, 81, 82, 83, 84, 85, 88, 90, 91, 92,
    94, 95
  flexible 84
  reinforced 72
Composites 34, 44, 116, 131, 163, 167, 177,
    179, 180, 184, 187, 188, 193, 227, 254,
    255, 257, 262, 274, 281, 282, 290, 293,
    294, 305, 346, 355, 366, 367, 371, 372,
    383, 398, 413, 428, 441, 486
  conductive 163, 193
  fabricated 346
  graphene-based 167, 179
  graphene derivatives 274, 305
  high-performance 116
  hybrid 187, 441
  phenolic resin 131
  polyaniline-carbon fiber 398
  polymer-based 177, 180
  polymer-ferromagnetic 44
  polyurethane-based 188
  synthesize ferrite-graphene 282
Compositions 231, 290, 317, 323, 449, 457,
    491, 492, 496, 513, 534, 536, 542
  hybrid 496
  stoichiometric 513

Compression moulding techniques 378, 398
Conducting ferrofluid 276, 283, 284, 285,
    286, 287, 288, 290
Conducting 44, 45, 46, 90, 118, 181, 214, 217,
    283, 284, 283, 284, 287, 292, 293, 294,
    321, 367
  ferrofluid composites 283, 284, 287, 292,
    293, 294
  magnetic fillers 90
  nanocomposites 181, 321
  polymeric materials 367
  polymers (CP) 44, 45, 214, 217, 342, 343,
    345, 355, 356, 357, 360, 371, 508, 511,
    512
  properties 118
  silicone 46
Conduction, intrinsic 484
Conductive 73, 187, 293, 360, 447, 450, 471,
    475
  emissions 471
  fabrics 360, 475
  network 73, 187, 447, 450
  pathways 293
Conductivity 8, 29, 46, 181, 239, 249, 250,
    256, 257, 259, 281, 294, 335, 343, 356,
    367, 368, 389, 426, 427, 439, 440, 459,
    466, 483, 484, 485, 507
  boosting 294
  electric 46, 281, 466
  intrinsic 367
  metallic 439, 440, 459, 507
Conductors 1, 35, 36, 38, 51, 154, 155, 356,
    358, 360, 467, 513
  inner core 154
  transparent electronic 513
Conjugated polymers 35, 37, 217, 218, 249,
    250, 328, 357
  composites 37, 217
Conversion techniques 23, 27, 161, 162
Copolymer 360, 411, 412, 413, 415, 416, 417,
    418, 419, 420, 421, 422, 427, 428, 434,
    435
  composites 434
  polyanilines 360, 412, 416
  synthesized 419

Corrosion 36, 60, 116, 132, 153, 162, 163,
    178, 214, 274, 317, 355, 366, 367 465,
    466, 467
  propensity 355
  protection 214
  resistance 132, 317, 366, 367
  resistive 274
  susceptibility 153, 162, 366
Crosslinked conducting polymer 509
Crosslinking reactions 370
Crystals, microscopic 368

D

Debye relaxation laws 281
Deposition 120, 170, 174, 332, 339, 478, 494
  chemical vapour 120, 332, 339, 478, 494
Deposition process 478, 494
  catalyst-free chemical vapour 494
Deposition 173, 493
    reactions 173
    techniques 493
Deprotonation 376
Developed shielding aggregates 541
Devices 2, 22, 47, 59, 60, 110, 111, 112, 153,
    155, 178, 193, 214, 320, 356, 361, 362,
    393, 470, 471, 512
  biomedical 193
  commercial 512
  digital memory 361
  electrochromic 361
  energy storage 178, 356
  microwave-frequency 2
  photovoltaic 214
  telecommunication 60
DHS emulsion 374, 375
Dielectric 1,2, 9, 10, 23, 27, 35, 37, 38, 40, 44,
    46, 51, 60, 62, 71, 87, 89, 90, 92, 143,
    144, 160, 184, 186, 188, 191, 249, 250,
    261, 262, 271, 272, 280, 293, 301, 304,
    327, 330, 332, 431, 432, 484, 490, 496
  balanced 35
  attributes 1, 2, 271, 272, 301
  composites 35, 37

  conducting 46
  constant 44, 71, 144, 160, 186, 188, 191,
    249, 250, 280, 304
  measurement techniques 1, 27, 51
  mechanisms 9, 10
  properties 23, 27, 37, 38, 184, 293, 431,
    432, 484, 490, 496
Dielectric fillers 44, 46, 111, 122, 299, 305,
    317, 318, 321, 322, 325
  composite 44
Dielectric loss 185, 250, 279, 292, 301, 330,
    484, 485
  factor 250, 330, 484, 485
  mechanism 279, 301
  microwave absorbers 43
  tangents 185, 292
Discovery 72, 356 ,477
  revolutionary 477
Dispersion 123, 125, 126, 166, 168, 170, 174,
    176, 177, 179, 214, 238, 240, 344, 413,
    414
  homogeneous filler 344
  techniques 179
Dissipation, electrostatic 179, 181
Distilled water (DW) 221, 223, 247, 283, 295,
    373, 374
Dodecyl 215, 219, 221, 222, 247, 360, 215,
    219, 221, 222, 247, 360
  benzene sulfonic acid (DBSA) 215, 219,
    221, 222, 247, 360
  hydrogen sulphate (DHS) 215, 219, 221,
    222, 247, 360
Dopant degradation 227
Dried ice method 478
Dye adsorption 445

E

Effect, hysteresis 253
Elastic modulus 180
Electrical 1, 40, 117, 134, 179, 181, 316, 358,
    360, 389, 395, 398, 412, 428, 497
  appliances 316
  insulation 497

properties 1, 40, 117, 179, 181, 358, 389,
    395, 398, 412, 428
  redox reversibility 360
  resistivity 134
Electrical conductivity (EC) 60, 82, 83, 85,
    114, 135, 136, 142, 180, 181, 184, 188,
    194, 389, 390, 448, 450, 453, 456, 510
  hybrid composite demonstrated higher 453
Electric energy 115
Electric vehicles (EVs) 465, 467, 468, 474
  battery-operated 465, 467
Electrochemical polymerization methods 215
Electrochromism 361
Electromagnetic 1, 31, 36, 110, 111, 112, 113,
    114, 115, 116, 117, 118, 119, 120, 121,
    122, 123, 153, 193, 218, 281, 319, 320,
    322, 338, 362, 346, 439, 465
  absorber 36
  inhibiting 439
  compatibility 116, 362, 465
  contamination 193
  energy 281
  environment 111
  plane wave theory 319
  properties 1, 338, 346
  protection 322
  signals 31, 153, 320
  spectrum 111, 112
  theory 31, 218
Electromagnetic absorption (EA) 488, 490
  behaviour 431
Electromagnetic interference 11, 318, 502
  amplitude 11
  and efficient absorption 502
  shielding mechanism 318
Electromagnetic noise 59, 272
  suppressing 59
Electromagnetic radiation(s) 2, 5, 43, 44, 45,
    47, 48, 111, 153, 154, 155, 276, 315,
    316, 318, 338, 470
  absorption of 36, 338
  exposure 316
  traditional 111
  shielding theory 276

Electromagnetic waves 2, 3, 4, 5, 29, 31, 32,
    112, 154, 155, 253, 254, 255, 256, 260,
    279, 330
  absorption of 31, 260, 330
Electronic(s) 2, 5, 47, 50, 59, 60, 154, 172,
    179, 181, 193, 215, 356, 362, 369, 444,
    468, 469, 474
  commerce 47
  conductivity 215
  equipment, sensitive 5
  flexible 193
  instruments 179
  pollutions 59
  properties 172, 215, 444
  systems 2, 60, 154
  warfare 468
Electronic devices 2, 3, 47, 48, 59, 60, 61, 84,
    153, 154, 360, 362, 466, 467, 468, 469
  flexible portable 84
  portable 60
Electron(s) 11, 40, 155, 156, 259, 280, 291,
    366, 389, 420, 465, 466, 497
  hopping 259
  migrating 497
Electron microscopy 133, 315
  emission scanning 133
Electron transport 75, 490
  ability 490
  properties 75
Electro spun nanofibers 181
EMI shielding 8, 41, 49, 60, 72, 74, 86, 88,
    90, 95, 112, 113, 115, 122, 132, 136,
    142, 156, 282, 182, 184, 446, 447, 465,
    475
  and electrical conductivity 142
  applications, high-performance 41, 95
  behaviour of composites 184
  capabilities 49, 182
  effective 112, 465, 475
  futuristic 136
  mechanism 8, 60, 74, 88, 90, 113, 115, 122,
    156, 446, 447
  of carbon foam 72, 74, 86
  of filler 181
  polymer nanocomposites 49

theory 132
EMI shielding measurement 63, 141, 315,
    320, 391, 392
    methods 315, 320
EMI shielding properties 46, 49, 59, 60, 71,
    72, 132, 134, 137, 140, 143, 290, 508,
    514
    of carbon composite foams 59, 72
    of carbon foam 71
    of composites 290
Emissions18, 218, 320, 362, 468
    conducted 320
    radiated 18, 320
EMI theory 114, 137, 486
Emulsification 219, 247
Emulsion 373, 375
    of monomer 373
    system 375
Emulsion polymerization 213, 219, 220, 221,
    222, 323, 360, 371, 372, 373, 374, 375,
    384, 411, 412, 413
    chain-growth 220
    chemical oxidative 372, 373, 375, 412, 413
Emulsion polymerization 220, 221, 360, 372
    pathway 360
    process 220
    techniques 221, 372
Energy 362, 478, 528, 546
    consumption 478
    photon 528, 546
    radiated 362
Epoxy nanocomposites 141, 142, 143
    reinforced 141, 142
Equipment, test compliance 475
Etching 83, 441, 442, 443, 478, 502, 505, 506,
    513
    chemical 505, 506
    lithium fluoride/hydrochloric acid 513
    method 506, 513
    plasma 478
    process 441, 442, 443, 502
Exfoliation, liquid-phase 166, 168, 501

F

Fabricating 41, 123, 506
Fabrication 63, 130, 162, 414, 452
    methods 130, 162
    procedure 414
    processes 63, 452
Fabric tapes 475
Facile solvothermal method 93
Faraday's law 32
Fe metals 122
Ferrites 2, 39, 40, 41, 43, 47, 276, 315, 318
    commercial 40
    nanocomposites 276
Ferro fluid 140, 141
Ferrofluid 286, 287, 303
    dispersion 303
    nanoparticles 286, 287
Fiber(s) 128, 129, 131, 155, 368, 371, 378
    conducting polyaniline-carbon 371
    conductive metal 155
    reinforced plastic (FRP) 378
Fiber fabric 128, 130
    coated 130
    reinforcing 130
Field(s) 4, 6, 11, 13, 33, 34, 38, 115, 131, 132,
    155, 157, 158, 159, 160, 214, 252, 260,
    318, 362, 363, 471, 498, 502
    anisotropic 33, 34, 252
    biomedical 502
    effect transistors 498
    emission 214
    energy storage 260
    magnetizing 38
    periodic 115
    vector 11, 13
Fillers 111, 214, 230, 274, 305, 367, 465
    dielectric nanomaterial 214
    magnetic and dielectric 111, 274, 305
    metal 465
    metallic 367
    reinforcement 230
Films 128, 163, 215, 479
    photographic 215

plastic 128
synthesized graphene 479
transparent conductive 163
Flexural modulus (FM) 387, 388, 390, 398
Fly ash (FA) 45, 46, 92, 141, 315, 317, 318,
    321, 322, 323, 325, 332, 334, 338, 339,
    341
  coated 322
  reinforcement 318
  thermal power plant 317
Freeze-drying 70, 76, 77, 88, 451, 453, 456
  method 70, 76, 77, 88, 451, 453, 456
  process 453
Frequency dependence of shielding 42
  effectiveness 42
FTIR 382, 416
  analysis 416
  Spectroscopy 382

G

Gadgets, electronic 109, 110, 271, 355, 362
Galvanic corrosion 218
Gas 65, 481
  explosive 481
  volatile 65
Glucose metabolism 272
Graphene 69, 70, 163, 166, 167, 172, 176,
    182, 187, 276, 467, 476, 478, 479, 484,
    486, 493
  carbon nanotubes hybrid 172, 187
  conductive 493
  crystallites 166
  deposit 167
  exfoliate 166
  fabricated 484
  honeycomb-induced 69, 70
  monolayer 163
  nanocomposites 176
  nanosheet (GN) 176, 182, 276, 467, 476,
    478, 479, 486
  synthesis 164, 165, 479, 480, 493
Graphene foam 68, 80, 83, 93
  networks 93

Graphene nanoplatelets 176, 186, 187, 188,
    190
  hybridization of 186, 188
Graphene oxide 70, 72, 120, 121, 168, 169,
    178, 282, 283, 295, 302, 303, 304, 481,
    482, 490
  chemical reduction of 282, 295
Graphite 179, 181, 282
  nanoparticles 179
  oxidation 181, 282
Graphite oxide 121, 167, 168, 170, 274, 482
  dry 170
  exfoliate 121
  oxidized 167
Growth 109, 116, 119, 174, 271, 474, 481,
    490
  amorphous carbon 481
  augmented 271
  hydrothermal 490
  plasma-enhanced CVD 174

H

Hall-Heroult process 530
Headaches 272, 316, 362, 440, 466
Health 2, 48, 47, 112, 316, 317, 355, 362, 440,
    466, 469, 513
  effects 317
  electronic 47
  hazards 2, 317, 355, 466, 513
  problems 316, 440
  public 48, 469
  risk 317
Heat 66, 88, 184, 187, 192, 282, 508
  energy 184, 187
  loss 282
  release 192
  treatment 66, 88, 508
Heterogeneity 34, 239, 249, 250, 251, 304,
    328, 486
High 60, 71, 116, 136, 139, 141, 142, 235,
    242, 378, 456, 465, 466, 475, 498, 501,
    506
  dielectric constants 465, 466

electrical conductivity 60, 71, 116, 131, 136, 139, 141, 142, 143, 456, 465, 501, 506
EMI reflection 60
frequency optoelectronic applications 498
magnification image 235, 242
pressure powder metallurgy method 378
quality shielding fabrics 475
speed homogenization 131, 143
High absorption 74, 93, 262, 451,
  ability 262
  behavior 451
High EMI shielding 112, 484
  capability 112
  efficiency 484
Higher conductivity 30, 35, 36, 37, 215, 216, 294, 345, 346, 418, 485
High-resolution 284, 384, 385, 497, 500
  scanning tunnelling spectroscopy 497
High-resolution transmission electron 186, 223, 224, 326
  microscopy (HRTEM) 186, 223, 224, 326
Hindalco Industries Ltd 531
Homogenization 124, 131
Homopolymers 412, 416, 419, 420, 422, 426, 435
  conductive 412
Hot 130, 131
  press compression 131
  press method 130
Hot pressing 130, 412, 434
  machine 130
  method 130
  techniques 412, 434
Hummer's method 120, 282, 283, 482
  modified 282
Hybrid nanocomposite 139, 140
Hydraulic press machine 131
Hydrazine 167, 169, 283, 479, 482, 483, 448
  hydrate 169, 283, 483
  molecules 448
  treatment 448
Hydrocarbon(s) 119, 134, 170, 478, 480
  gases 170
  gas precursors 480

polycyclic aromatic 478
Hydrochloric acid-lithium fluoride solution 443
Hydrogen 218, 480
  gas precursors 480
  peroxide 218
Hydrophilic 167, 168, 219, 220, 375, 475, 482, 509
  coatings 475
  sulfonate 375
Hydrophilicity 441, 507
Hydrophobic 217, 219, 220, 509
  forces 217
Hydrothermal 45, 465, 489
  reactions 45, 465
Hydroxyls 167, 448, 482
Hysteresis loop 38, 41, 287, 344

I

Immune system dysfunction 316
Impedance 259, 446, 453
  matching properties 259
  mismatch 446, 453
Impinging energy 155
Impregnation forces 128
Incident electromagnetic 30, 34, 260, 507
  radiations 260
  waves 30, 34, 260, 507
Incident microwave radiation 182
Industrial 96, 110, 123, 154, 456
  applications 110, 123, 154, 456
  waste materials 96
Industries 49, 85, 110, 117, 179, 274, 441, 465, 467, 474, 476, 514, 530, 531
  aluminium-producing 530
  automobile 117, 274
  automotive 117, 465, 467
  electronic 85, 179
  mobile electronics 441
Infrared spectroscopy 213, 416
Input impedance mismatch 194
Insomnia 316, 362, 466
Instron universal testing machine 387, 423

Insulating 163, 239, 388, 389
  nature 239, 388, 389
  novolac resin hinders 389
  polymer resins 163
Interfaces 33, 34, 109, 157, 249, 277, 280,
    304, 429, 433, 440, 453, 459
  conductive porous 453
  electronic 109
Interfacial 287, 388, 441
  assembly 441
  forces 287
  ionic interaction 388
Interfacial polarization 34, 36, 88, 90, 92, 140,
    141, 249, 250, 251, 301, 328, 330
  enhanced 249
Interference, radiofrequency 59
International agency for research on cancer
    (IARC) 272, 316
Internet of things (IoT) 469
Intrinsically conducting polymers (ICPs) 355,
    356, 357, 358, 360, 367, 371, 508
Intrinsic 32, 34, 126, 440, 465
  conducting polymers 440
  properties 32, 34, 126, 465
Ionic transmission 271
Ion intercalation method 488
Iron 86, 87, 536
  nanoparticles 86, 87
  oxides 536
Isocyanates 63

K

Ku band 46, 75, 141, 143, 182, 186, 253, 322,
    337, 346
  frequency range 141, 143
  of frequency 253

L

Lattice 167, 476
  contiguous aromatic 167
  honeycomb 476

Light 26, 32, 116, 117, 163, 193, 260, 315,
    459, 465, 466, 467
  weight shielding materials 163, 193
Light-emitting diodes 167, 215, 361
  organic 361
Lightweight 60, 69, 368
  EMI shielding enclosures 60
  polymer composites 368
  porous carbon 69
Liquid 480, 488
  exfoliation 488
  precursors 480
Liquid carbon 480, 481
  precursors 481
  sources 480
Liquid phase exfoliation (LPE) 478, 498
  methods 166, 167
Lithium-ion batteries 120, 186, 468, 474
Losses 8, 9, 44, 46, 159, 252, 253, 272, 319,
    365, 446, 458
  hysteresis 252, 253
  joule-heating 46
  memory 272
  ohmic 8, 44, 159, 319, 365, 446, 458
  refection 9
  reflective 46
Low 235, 393, 495
  frequency lumped-circuit element
    techniques 393
  magnification image 235, 495
  percolation thresholds 163
LTE applications 47
Lung diseases 317

M

Magnetic 1, 6, 7, 8, 9, 27, 31, 35, 40, 43, 60,
    62, 87, 89, 95, 109, 110, 111, 154, 156,
    158, 160, 214, 252, 273, 274, 276, 280,
    287, 290, 303, 305, 326, 363, 364, 365,
    366, 371, 465, 466, 467, 485, 513
  electronic 513
  anisotropy 40
  crust 6

hysteresis loop 252, 326
nanoparticles 60, 154, 274, 276, 290, 467
permeability 1, 43, 60, 62, 87, 89, 154, 156, 465, 466, 485
polarization 31
properties 27, 35, 40, 95, 110, 273, 280, 287, 326, 371, 466
storage 160
Magnetic energy 30, 31, 112, 249, 280
dissipation 30, 249
storage 31
Magnetic fields 4, 6, 7, 13, 30, 155, 158, 159, 249, 272, 316, 318, 327
oscillating 318
vector 13
Magnetic loss 43, 218, 291
absorbers 43
microwave 43
property 218
tangent 291
Magnetization 29, 31, 33, 38, 41, 112, 116, 218, 252, 272, 315, 326, 344
high saturated 116
hysteresis curve of CMGF cement 344
processes 112
Malfunctions 466, 469
Materials 1, 6, 36, 37, 38, 39, 43, 46, 48, 59, 72, 85, 88, 109, 112, 122, 132, 141, 143, 156, 167, 172, 173, 174, 185, 186, 192, 218, 273, 363, 366, 368, 439, 456, 469, 482, 484, 528, 550
biodegradable 59
carbon 36, 37, 43, 72, 85, 368, 456
ceramic 528
conductive 6, 112, 363
dielectric 1, 37, 38, 46, 132, 141, 156, 218, 273
dielectric-magnetic 484
electronic 48, 469
electrostatic discharge 192
ferri-magnetic 39
ferrite 282
fiber 128
flame retardant 192

hybrid 109, 122, 143, 172, 173, 174, 185, 439
hydrogen storage 186
monolayer 482
oxidized graphene 167
polymeric 182, 366
synthetic 88, 507
toxic 550
Maxwell's equations 11, 12
Measuring electromagnetic signals 320
Mechanical 140, 179, 180, 317, 356, 439, 368, 371, 414, 501, 509, 511, 512, 513, 528, 529
mixing and melt treatment 140
properties 179, 180, 317, 368, 371, 414, 501, 509, 511, 512, 513, 528, 529
stability 356, 439
Mechanical exfoliation 165, 166, 282, 477, 478, 479, 488, 493
method 488
Mechanism 61, 85, 89, 110, 115, 157, 186, 221, 222, 293, 364, 376, 452, 455, 456
absorbing 293
bidirectional freeze-casting 455
Melamine-formaldehyde (MF) 452
Melt 125
mixing process 125
Melt processing 125, 371
method 125
techniques 371
Metal 48, 49, 85, 216 , 468, 501
dichalcogenides 468, 501
nanoparticles 48, 49, 85, 216
Metallic 38, 49, 487, 502
catalyst 478
nanowires and nanoparticles 49
oxide 38
Metallic shields 37, 218, 355, 366
traditional 467
Methods 59, 63, 72, 121, 126, 172, 181, 450, 478, 506
evaporation 126
foaming 59, 63, 72
free-drying 450
generic 506

hydrothermal 172, 181
pyrolysis 478
thermal-mediated 121
Micrograph 137, 541
scanning electron 541
transmission electron 137
Micromechanical exfoliation 165
Microwave 1, 2, 21, 22, 40, 43, 44, 45, 253,
254, 259, 260, 281, 282, 293, 294, 321,
361, 362, 396
absorbers 39, 43, 257, 322, 501
absorbing material (MAM) 43, 44, 140,
188, 262, 321, 397
circuits 393
consumption 497
devices 40
electronics 2
engineers 21
futuristic 397
exfoliated graphite oxide (MEGO) 180
exfoliation 120
frequencies 22, 43, 252, 279, 393
heating 470
radiations 88, 132, 305
shields 1, 274, 321
Microwave absorption 41, 44, 87, 89, 90, 253,
257, 259, 260, 292, 294, 360, 363
application 496
materials 43
properties 45, 71, 140, 141, 397, 431, 496
Microwave shielding
mechanisms 302
properties 194, 397
system 301
Mobile communication 48, 49, 362, 469
equipment 362
Mobile wireless communication 469
Molybdate 490
ammonium 489
Monomers 127, 177, 216, 218, 219, 220, 221,
373, 374, 412, 413
hydrophobic 219
Morphology 29, 34, 111, 120, 219, 223, 325,
333, 342, 375, 383
agglomerated 285

Multi-component 527, 528, 542, 546
ferrites 41
Multifunctional, developing 143
Multilayer graphene 138, 496

N

Nanocomposites 51, 109, 132, 134, 136, 138,
140, 141, 143, 144, 179, 180, 181, 187,
188
coated graphene 139
polymers-based 144
Nanomaterials 178, 214, 465, 467
carbon-based 177
Nanotubes networks 173
Natural
graphite 119, 178, 182
resonances 44, 113, 140, 141, 280, 291,
292, 300
Nature 256, 257, 357, 364, 373, 527, 530, 531,
537, 538, 546
amorphous 224, 234, 335, 415
carcinogenic 316
corrosive 317
crystalline 342
electrical 472
electrophilic 443
graphitic 288
hydrophobic 375, 448
metallic 443
Nervous system dysfunction 316
Network(s) 41, 49, 78, 113, 498
analyzer 21, 24, 28, 392, 428
communication 50, 193
fibrous 420
Neurological disorders 316
Newton-Raphson NIST Iterative technique 26
Next-generation
mobile phones 465
shielding applications 1
Nicholson-Ross-Weir 1, 24, 51, 248
algorithm 248
method 24
technique 1, 51

NIR spectroscopy 536
NIST Iterative Technique 26, 161
Nitrogen 359, 441, 445, 450, 502
    atmosphere 227, 245
Novolac 370, 385, 386, 387, 394, 414, 415,
    417, 421, 422, 424, 426, 427, 431, 432,
    433
    insulating 432
    powder 377
Novolac resin 377, 383, 384, 385, 386, 388,
    389, 395, 398, 413, 414, 420, 424, 426,
    434
    proportions of 379, 411
NRW
    method 24, 26
    technique 26, 161
Nuclear
    magnetic resonance (NMR) 272
    power plants 528, 529, 546

O

Oil spillage 68
Open
    ended coaxial probe technique 27
    field test 471
Operation 154, 272, 466
    electronic 3
Optical
    image 384, 385
    transparency 215
Optical properties 214, 271, 478
    bio-functional 186
    nonlinear 360
Organic light-emitting diodes (OLED) 361
Orientation polarization 1, 10, 115, 251, 291
Oxidation 121, 137, 168, 169, 217, 358, 375,
    376, 381, 481
    chemical 372
    reaction 121
    resistance 445, 448
    stabilization 63, 66
Oxidation-thermal expansion-air convection
    shearing process 496

Oxidative polymerization of aniline 376
Oxides 39, 41, 167, 274, 317, 527, 537, 546
    iron titanium 528, 541, 542
Oxyethylene deformation 225, 244
Oxygen 170, 542
    reactive 170
    reduction processes 186
    residual 170

P

PACN sheets, synthesized 392, 396
PAM composites 507
Paths 7, 109, 506
    charge transport 334
    electrical conduction 73
PDMS, developed graphene 94
PEDOT polymer 37, 221, 237, 249, 257
PEDOT's nanocomposites 248, 262
    synthesis 248
Percolation threshold 143, 180, 181, 184, 239,
    367
    electrical 134, 177
Permeability properties 31, 228
PET-coated single-walled carbon nanotubes
    45
Phenol formaldehyde (PF) 117, 370
    crosslinked thermosetting 370
Phenolic resins 63, 81, 82, 128, 130, 131, 369,
    370
    thermosetting 370
Photocatalysis 444
Photons 529
Photo thermal therapy 502
Photovoltaics 167
Planar few layers graphene (PFLG) 479
Planes 11, 12, 13, 121, 155, 224, 234, 280,
    285, 326, 334
    basal 168, 482
    transition metal atomic 503
Plane waves
    transmitted 364
    uniform 11
Plasma treatment 479

Plastics 155, 316, 356
  ferrite-loaded 476
  fiber-reinforced 378
Polarizability 9, 305, 432
Polarization 9, 10, 44, 249, 250, 251, 261,
    262, 280, 281, 291, 293, 328, 432, 484
  behavior 37
  electric 184
  electron 141
  electronic 10, 11, 280
  interface 291
  ionic 249
  loss 484, 486
  saturation 38
  space-charge 251
  spatial charge 304
Polaronic transitions 381
Pollution 109, 110, 154, 315, 317, 362, 439,
    440, 445, 459
  environmental 2, 479, 530
Poly 78, 79, 176, 177, 186, 213, 215, 411,
    412, 413, 492, 493, 508
  biodegradable 75
  conducting polymer 426
Polyacrylamide 507
Polyacrylonitrile 368
Polyaniline 355, 356, 358, 359, 360, 369, 372,
    373, 375, 380, 382, 411, 412
  alkyl-substituted 359
  halogenated 412
  reinforced hybrid 138
  sulfonated 412
Polyaniline-carbon fiber 372
  novolac 377
Polyaniline
  CF-novolac 356
Polyaniline composite(s) 356
  sheets 369
Polydopamine 182, 190
Polymer(s) 37, 59, 124, 125, 126, 127, 141,
    176, 193, 215, 221, 250, 252, 356, 360,
    385, 416, 507
  based composite foams 60
  carbonized 228
  cyanoethyl pullulan 186

  dispersion 219
  organic 127
  resins 155, 369
  synthesized conducting 425
Polymer composite(s) 35, 49, 109, 132, 163,
    177, 186, 192, 213, 369, 371
  carbon-based conductive 163
  conductive 180, 238, 366
  conductive ferromagnetic 467
Polymer nanocomposites 48, 110, 117, 153,
    154, 176, 178, 181
  coat-able thin 508
  reinforced 123
Polymer matrices 109, 111, 117, 122, 123,
    126, 144, 173, 192, 282, 514
  thermoset 109, 153
Polymeric 369, 509, 510, 511, 512
  film 171
  matrix 509
  resins 377
Polymerization 126, 127, 219, 220, 221, 247,
    372, 373, 374, 376, 382
  electrochemical 376
  of conjugated polymers 218
  process 214, 262
  reaction of EDOT 221
Polymerized polymer globules 127
Polymethyl methacrylate, synthesized 81
Polymorphism 487
Polymorphs 487
Polypyrrole
  matrices 325
  matrix 326
  nanocomposite 318, 346
Polystyrene sulfonate 217
  charged saturated 216
Polythiophene 214, 215, 355
Polyurethane 63, 66, 77, 79, 85, 176, 179,
    186, 230, 231, 315, 318
  coatings 180
  composites 315
  loaded 178
  matrix 153, 179, 182
  pure insulating 194
  reinforced 178

resins 369
Polyurethane nanocomposites 193
  based thermoplastic 183
Powder moulding process 87
Power 7, 50, 61, 62, 276, 319, 391, 392, 429,
    468, 488
  battery 468
  density 316
  flux-multiplying 40
  incident electromagnetic 484
  magnetics solutions 475
Pressure 66, 128, 129, 170, 378, 394, 446
  hydrostatic 498
  release method 59, 65, 87
  sensors 125
Problem, electromagnetic radiation pollution
    274
Process 82, 87, 88, 126, 127, 128, 130, 131,
    174, 218, 301, 376, 480, 489, 531
  complicated 300
  deoxygenating 169
  filtration 509
  high-temperature 144
  oxidation 167, 281
  synthetic 490, 492
Processing 110, 111, 174, 176, 182, 370, 435,
    441, 448, 530, 531
  ceramic 528, 545
  facile 367
  mineral 530
  of CBNS reinforced thermoplastic PNCS
    123
  of CBNS reinforced thermoset PNCS 128
Production 49, 65, 117, 119, 167, 170, 177,
    193, 194, 480, 481
  commercial 218
  industrial 166
  mass 476, 477
Production 172, 534
  methods 172
  process 534
Products 18, 38, 51, 156, 194, 218, 256, 321,
    374, 475, 476
  etched 442
  toxic 214

Propagation 31, 32, 319, 356, 484
  constant 26, 31
Properties 117, 118, 119, 122, 123, 124, 177,
    179, 180, 321, 322, 338, 339, 356, 412
  absorbing 282, 322
  absorptive 188
  catalytical 192
  electrochemical 359, 412
  engineering 543
  evaporation 124
  flame-retardant 458
  mesmerizing 367
  nanocomposite 190
  nanomechanical 137
  sustainable 69
  synergistic 214
  wave-absorbing 322
Protection, environmental 92
Pulsed laser deposition (PLD) 493

Q

Quarter wave principle 279

R

Radar 41, 43, 135, 140, 367
  absorbing material (RAM) 43, 140, 367
  absorption materials 41, 140
  wave absorption 135
Radiation pollution 272
  environmental 362
Radiations 3, 109, 110, 112, 132, 133, 218,
    262, 272, 273, 315, 316, 362, 363, 506
  designing shielding gamma 550
  emitting 154
  non-ionizing 271, 272, 316
Radiation shielding 527, 528, 529, 537, 546,
    547, 548
  advanced synthetic 527
  developed 528, 546
  developed non-toxic 542
  fabricating 550
  making advanced non-toxic synthetic 537

Radiation shields 528
  selective frequency 272
Radiological physics and advisory division
  (RPAD) 547
Raman 213, 225, 243, 285, 287, 298, 490,
  491, 500
  spectra 225, 285, 287, 298, 490
  spectroscopy 213, 287, 298
  spectrum 243, 287, 491, 500
Range 48, 50, 188, 193, 194, 289, 291, 382,
  385, 415, 416, 501, 536
  microwave frequency 44, 193, 218, 272
Reaction 167, 219, 247, 282, 283, 373, 374,
  480, 481, 490, 493
  alkaline 530
  charge transfer 384
  high-temperature plasma 478
Reaction pathway 374
Reactor 223, 247, 529
  containment structure 529
  walled stainless steel 413
Rectangular pallets 299
Red mud 92, 93, 527, 528, 530, 531, 532, 534,
  536, 537, 539, 540, 541, 542, 545
  ceramic processing of 537, 538
  generation 530, 531, 532
  production 532
Red phosphorus (RP) 9, 158, 366, 498
Reduced graphene oxide (RGO) 79, 82, 83,
  121, 169, 170, 183, 250, 274, 281, 285,
  286, 287, 288, 291, 294, 297, 298, 299,
  490
  composites 275
Reduction 135, 137, 167, 168, 169, 170, 172,
  173, 182, 183, 358, 359, 426, 448, 481
  chemical 165, 169, 178, 181, 281, 479
  chemical vapor 173
  electrochemical 169, 173
  hydrothermal 479
Reflectance 162, 392, 429
  convenience 429
Reflection
  mechanism 245, 254
  method 28
  phenomena 34, 157

  power 112
Reflection coefficient 21, 24, 132, 161, 162,
  276, 289, 392, 395, 396
  calculated 395
Reflection loss (RL) 9, 45, 46, 141, 142, 156,
  157, 158, 256, 273, 319, 320, 321, 496,
  501
  gross 46
Reflection shielding 85, 484
  efficiency 484
  materials 85
Research, neurophysiological 217
Resin 59, 66, 78, 128, 130, 155, 370, 388,
  395, 397, 474
  phenolformaldehyde 67
  polyester 117, 369
  transfer moulding 128
Resonance frequency 252
  natural 33
Resonant cavity perturbation technique 71
Resources 154
  mineral 530
  non-replenishable natural minerals 528
  transient power 271
Reversible doping/de-doping process 361
Ross technique 431
RTM process 129

S

Salts 372, 441
  emeraldine 358
  molybdenum 489
  protonic 372
Sanbornite 541, 542
Satellite communication 48, 362
Scale preparation method 505
Scaling law 184
Scanning electron 285, 324, 333, 383, 420,
  541
  microphotographs of red mud 541
  microscope 285, 324, 333, 383, 420
Scanning tunnelling spectroscopy (STS) 497
Scherer's formula 324

Scherrer equation 298
Semiconducting 487
  polymorph 487
Semiconductors 1, 35, 36, 51, 357, 477, 487,
    497
  inorganic 356
Sensors 117, 178, 194, 214, 356, 361, 442,
    472, 513
  biochemical 361
Separation 18, 119, 488
  micro-phase 230
Shear exfoliation technique 499
Shearing force 126
Sheets, metallic 366
Shell, spherical magnetic 6
Shield 7, 61, 112, 155, 182, 193, 364, 440,
    459
  electromagnetic energy 193
  material 7, 61, 112, 155, 182, 364, 440, 459
Shielded box 17, 19, 21, 63, 320
  method 17, 19, 63, 320
  techniques 21
Shielded enclosure application 3
Shielded room 19, 63, 321, 472
  method 19, 63
  technique 321
  test 472
Shielding 7, 9, 20, 23, 29, 30, 31, 33, 34, 37,
    41, 94, 112, 178, 184, 213, 228, 259,
    261, 262, 274, 302, 304, 318, 397, 428,
    439, 457, 458, 475
  applications, sensing 94
  capabilities 184, 439
  clips 475
  components 112
  effectiveness measurement 20
  efficacy 7, 9, 41, 302, 304
  electromagnetic 23, 34, 178, 213, 228, 261,
    318, 397, 439
  measurements 428
  performance 29, 30, 31, 33, 34, 37, 259,
    262, 274, 457, 458
  theory 300, 366

Shielding effectiveness 7, 15, 21, 29, 31, 42,
    50, 62, 155, 184, 187, 255, 303, 319,
    363, 427, 472
  absorption-dominated 317, 322
  capabilities 472
  electromagnetic interference 439
  measurement 17, 18, 19, 50, 62, 299
Shielding effectiveness performance 440
  higher electromagnetic 261
Shielding efficiency 3, 7, 16, 20, 21, 112, 187,
    302, 322, 363
  orientation-tunable EMI 458
  outstanding EMI 458
Shielding materials 7, 8, 36, 37, 60, 61, 62,
    71, 72, 112, 153, 272, 319, 366, 429
  polymer-based 49
Shielding mechanism 1, 5, 32, 111, 154, 238,
    245, 254, 273, 338, 341
  secondary EMI 366
  walls-induced EMI 458
Shielding properties 254, 255, 256, 322, 327,
    368, 371, 388, 389, 411, 412, 434, 435
  electrical conductivity and EMI 59, 177
  tune radiation 317
Shifting 225, 381, 382, 419
  hypsochromic 381
  technology 469
Shore hardness test 343
Short carbon fiber (SCF) 368
Short circuit line (SCL) 23, 26, 161, 162
  technique 1, 51
Signal(s) 22, 27, 162, 362, 393
  frequency 393
  radiated 362
  transmitted 22, 27, 162, 393
Sintering process 534
Skin 7, 157, 485
  depth, shallow 162, 278
  irritations 272
Snoek's law 33, 252
Sodium lauryl sulphate (SLS) 322
Solar cells 120, 164, 194, 214, 361
  dye-sensitized 217
Sol-gel auto-combustion method 45
Solid 315, 495

state reaction 495
  waste 315
Solution casting 177, 315, 378, 412, 413, 434
  methods 322, 332
  technique 230, 231
Solution mixing technique 143
Solution polymerization
  method 360
  technique 372
Solution processing 123, 124, 175, 176
  method 124
  technique 124, 216
Solvothermal 478, 489
  methods 489
  process 478
Sonication 168, 176, 285, 297, 478, 488, 499
Specific shielding effectiveness (SSE) 76
Spectrophotometer 380
Spin coating technique 45
Spira manufacturing corporation 476
Staudenmaier's method 172
Stored energy 115
Stretching 226, 236, 382, 419
  asymmetric 416
Styrene 130
  acrylonitrile butadiene 79, 116
Substrate-free gas-phase synthesis 478
Sucrose 64, 74, 480
  graphene composites foam 80
  molten 74
  MWCNTs Composite 78
Sugarcane 69, 78
  carbonaceous 69
Sulfonic acid 359
  sodium dodecyl benzene 220
Superconductors 356
Supercritical carbon dioxide foaming 75
Surface 128, 156, 157, 164, 165, 171, 183,
    184, 240, 241, 242, 285, 375, 489, 509
  catalytic 170
  coatings 179
  conductive 114
  crystal 170
  fiber 181
  hydrophilic 507

modification 144, 459
  tensions 499
Surfactant coating 287
Surfactants 166, 168, 176, 214, 215, 219, 220,
    240, 360, 372, 373, 374, 375
Susceptibility 36, 218
  device's 321, 472
Suspension 217, 240, 283, 295
  bluish-green 221
Synthesis 88, 119, 172, 218, 219, 221, 222,
    299, 355, 356, 372, 411, 441, 442, 498
  chemical 315, 477, 479
  facile 153
  fullerene 478
  high-pressure 498
  of dodecyl benzene sulfonic acid 221
  of MXENES 442
  of PEDOT grafted MWCNT composites 222
  of tin oxide 295
  solvothermal 36
Synthesis methods 109, 119, 172, 322, 325,
    357, 360
  facile 450
  straight forward 170
Synthetic shielding aggregates 546
  innovative red mud-based 528
  red mud-based 528, 546
Systems 17, 21, 125, 126, 304, 305, 357, 392,
    395, 432, 440, 441, 445
  coaxial transmission line 321
  defense 49, 460
  electrical 445
  electric machinery 112
  hybrid nanofiller 173
  organic solvent-based 498
  thermal management 469
  vacuum filtration 131
  wireless 362

T

Technique 23, 24, 26, 27, 124, 126, 160, 161,
    162, 166, 176, 177, 282, 377, 378
  blending 140, 179

compression 378
dip-coating 452
evaporation 135, 136
foaming 63
spray 216
Technocrats 466
Technology 2, 47, 43, 44, 60, 71, 144, 216,
218, 262, 305, 367, 475
fabric-clad foam EMI shielding 475
mobile networking 2, 47
silicon-based CMOS 216
stealth 43, 44, 60, 71, 144, 218, 262, 305,
367
Telecommunications 2, 272, 362, 468, 498
industry 469
towers 316
Temperature, polymerization reaction 221
Tetrafluoroethylene 478, 493
Tetrahydrofuran 133, 168, 482
Thermal 35, 60, 72, 164, 169, 172, 182, 192,
165, 171, 186, 187, 220, 274, 466, 489,
491, 496
conductivity 35, 60, 72, 164, 172, 182, 274,
496
contact resistance 192
decomposition 165, 171, 220, 489
degradation temperature 186
energy 187
heating 491
injuries 466
treatment 169
Thermal reduction 170, 172, 174, 178, 449,
483
methods 170, 450
Thermal stability 60, 144, 178, 180, 227, 336,
385, 386, 398, 422, 435
properties 60
Thermo gravimetric analysis (TGA) 213, 223,
227, 245, 336, 356, 385, 387, 414, 421,
536
Thermoplastic 109, 110, 116, 132, 144, 153
reinforced 132
elastomers 230, 474
polymeric materials 132
polyurethanes 116, 135, 153, 179, 186

resin 370
Thermoplastic polymer(s) 116, 153, 179
nanocomposites 110, 123
Thermoset polymer nanocomposites 109, 110,
142
Thermosetting polymers 116
Thin-film transistors 167
Thioacetamide 489
Titanium 63, 542, 545
hydride 63
Tools 21, 112, 287
advanced artificial intelligence 468
Top-down exfoliation techniques 477
Toxic gases 121, 167, 482
TPU 116, 135, 153, 179, 181, 182, 184, 191,
322
based nanocomposites 322
electrical conductivity of 184, 191
TPU nanocomposites 136, 137, 139
filled 135, 137
Transfer 185, 282, 446, 447, 448, 475
adhesives 475
electron 185
heat energy 282
matrix method 446, 447, 448
Transferred impedance data 17
Transformers 7
Transition 467, 487
electron 381
electronic 380
Transition metal(s) 40, 85, 170, 171, 487, 503
carbides 441, 465, 507
dichalcogenides (TMDs) 468, 476, 487,
501, 502
Transmission 21, 28, 112, 114, 156, 157, 272,
336, 362, 363, 393
amplitudes and phases of 21, 392
coefficient calculation 24
reflection (TR) 321
reflection line technique 27
Transmission coefficients 13, 24, 26, 161,
391, 395, 396, 429
synthesized conducting ferrofluid's 289
Transmission electron 138
microscopy 186, 213, 342

Transmission line 27, 63, 321, 393, 473
  coaxial 21, 63, 321, 473
  holder 473
  measurement calibrations use 27
  theory 11, 22, 393
Transmission line method 17, 21, 63, 321, 473
  coaxial 17, 21, 63, 321
Transmittance 162, 276, 288, 289, 396, 429
  coefficient 276, 288, 289, 396

U

Ultra-high vacuum condition 171
Ultrasonication 166, 168, 172, 173, 174, 176, 178, 180, 186, 377, 443
  infiltration 186
Ultrathin cellulose nanofibrils 458
UV-vis Spectroscopy 380

V

Vacuum assisted resin 128, 129
  infusion moulding (VARIM) 128
  transfer moulding (VARTM) 128, 129
  drying 449
  filtration 414, 445, 451, 453
  filtration techniques 141, 440
  infiltration technique 240, 241
  oven 131, 221, 223, 241, 295, 296, 377, 413
  pump 127, 130
Vapor-induced phase separation process 80, 90
Variation 191, 258
  in EMI shielding effectiveness 191
  of microwave conductivity 258
VARTM method 129
Vector network analyzer (VNA) 21, 22, 27, 28, 61, 63, 74, 162, 276, 356, 393, 394, 395
Vedanta aluminium Ltd. (VAL) 531
Vehicles 470
  automotive 135, 137

Vibrating sample magnetometer (VSM) 287, 344
Vibrational bands 382
Vibrations, anti-symmetric 225
Voltage 11, 468, 472

W

Waals 119, 122, 165, 170, 384, 442, 487, 493, 502
  bonds 442
  forces 119, 122, 170, 487, 493
  interactions 165, 384, 502
Wastes 96, 538
  aluminium industry 527, 537, 545
  industrial 2, 60, 92, 528, 530, 537, 550
  pollution 321
  radioactive 528, 529, 546
Water 66, 67, 166, 168, 169, 170, 180, 181, 182, 215, 219, 220, 221, 222, 283, 284, 295, 302, 442, 444, 499
  deionized 283, 295, 490
  desalination 444
  emulsion 219, 220
  purifications 68, 442, 444
  reinforced 180
  soluble polyelectrolyte 215
  synthesized 182
Waterborne polyurethane 78
Wave(s) 7, 12, 13, 35, 60, 61, 62, 63, 76, 89, 92, 184, 229, 277, 316, 319, 363, 365, 392, 429, 459, 484
  absorption 229, 316, 501
  energy 294
  outgoing 61
  radio 154
  reflected 160
  transmitted 13, 61, 157, 162
  transmitting 319
  ultrasonic 488
  ultrathin silk curtain 184
Wavelength(s) 12, 18, 22, 24, 25, 26, 48, 111, 112, 115, 161, 164
  laser 287

propagating 279
Wave propagation 155, 364
   pathways 293
Weight 170, 227, 336, 385, 386, 421, 467
   loss 170, 227, 336, 385, 386, 421
   reduction 467
Weir method 229, 328, 338, 485
Wet layup process 128
White phosphorus (WP) 497, 498
Wi-fi 153, 316
   connections 316
   fi routers 153
Wind turbines 193
Wireless 2, 47, 110
   communication 153, 469
   devices 316
World health organization (WHO) 272, 316

X

X-ray diffraction (XRD) 213, 286, 297, 323,
   324, 334, 341, 411, 414, 536, 539

Y

Young's modulus 119, 164, 450, 487

Sundeep K. Dhawan

Dr. S.K. Dhawan is working as Emeritus Scientist in CSIR-National Physical Laboratory, New Delhi, India. Dr. Dhawan has been nominated in the top 2% scientists in the world as published by Stanford University and Elsevier Foundation. Dr. Dhawan has worked in the area of conducting polymers nano ferromagnetic composites for EMI shielding, self-healing smart conducting polymer coatings for corrosion protection, and waste plastic management. Dr. Dhawan has published 180 papers and filed 16 US & Indian patents. He has received DST-Lockheed Innovation Award in 2014 for Smart Self-Healing Coatings. His technology on "Utilization of Waste Plastics for designing Tiles for Societal Usage" has been selected in the top Smart 50 Innovations by DST & IIM in 2018. In 2019, Dr. Dhawan received NRDC Societal Innovation Award for his contribution to Energy Storage Devices.

Avanish Pratap Singh

Dr. Avanish Pratap Singh is working as an assistant professor at the Department of Physics, ARSD College, University of Delhi, India. He obtained his Ph.D. in Physics from the University of Delhi and the National Physical Laboratory, New Delhi. His current area of research interests includes EMI shielding, exotic carbon forms, conducting polymers and magnetic materials. He has authored or co-authored more than 40 articles in peer-reviewed journals.

Anil Ohlan

Dr. Anil Ohlan is working as an Assistant Professor at the Department of Physics. Maharshi Dayanand University, Rohtak, India. His main research interests are in EMI shielding, multiferroics, super capacitors, ferromagnetic conducting composites, conducting polymers, development of multiferroic composites for magneto electric coupling, conjugated polymers/ 2D materials for supercapacitors and electromagnetic absorption.

Kuldeep Singh Kakran

Dr. Kuldeep Singh is currently working as a senior scientist at CSIR - Central Electrochemical Research Institute, Chennai, India. His research interests are rooted in the design and development of advanced nanostructured two-dimensional (2D) materials for EMI shielding and energy storage devices especially lithium-ion batteries, solid state lithium ion batteries, lithium sulfur batteries and supercapacitors for Electric vehicles (EVS).

Pradeep Sambyal

Dr. Pradeep Sambyal received his Ph.D. in Chemical Sciences from CSIR-National Physical Laboratory, India. Currently, he is working as a postdoctoral researcher at KAIST, South Korea. Dr. Sambyal has published 21 research articles (citations >1183; h index = 13), 1 review article, 2 books and filed 3 patents. Presently, he is working on MXene based hybrid materials for EMI shielding applications.

www.ingramcontent.com/pod-product-compliance
Lightning Source LLC
Chambersburg PA
CBHW050519240326
41598CB00086B/38